ENVIRONMENTAL GEOLOGY

DONALD R. COATES

State University of New York at Binghamton

JOHN WILEY & SONS
New York   Chichester   Brisbane   Toronto

This book is dedicated to
Jeanne, Cheryl,
Eric, and Lark
So they will know my love for the natural environment

Library of Congress Cataloging in Publication Data:

Coates, Donald Robert, 1922-
    Environmental geology.

    Includes index.
    1. Geology.   2. Environmental protection.
I. Title.
QE33.C66   550   80-21272
ISBN 0-471-06379-7
Printed in the United States of America
10 9 8 7 6 5 4 3 2 1

1 square inch	=	6.4516 square centimeters
1 square foot	=	0.0929 square meter
1 square yard	=	+ 0.836 square meter
1 acre	=	0.4047 hectare
1 square mile	=	2.590 square kilometers
1 cubic inch	=	16.39 cubic centimeters
1 cubic foot	=	0.0283 cubic meter
1 cubic yard	=	0.7646 cubic meter
1 acre-foot	=	1,233.46 cubic meters
	=	325,851 gal
1 cubic mile	=	4.168 cubic kilometers
1 gallon	=	3.784 liters
1 ounce	=	28.33 grams
1 pound	=	0.4536 kilograms

Metric-English Conversions

1 millimeter	=	0.0394 inch
1 meter	=	3.281 feet
1 meter	=	1.094 yards
1 kilometer	=	0.6214 mile
1 sq centimeter	=	0.155 sq inch
1 sq meter	=	10.764 sq feet
1 sq meter	=	1.196 sq yards
1 hectare	=	2.471 acres
1 sq kilometer	=	0.386 sq mile

1 cu centimeter	=	0.061 cu inch
1 cu meter	=	35.3 cu feet
1 cu meter	=	1.308 cu yards
1 liter	=	1.057 quarts
1 cu meter	=	264.2 gallons (U.S.)
1 cu kilometer	=	0.240 cu miles
1 gram	=	0.0353 ounce
1 kilogram	=	2.205 pounds

Temperature

To change from Fahrenheit (F) to Celsius (C)

$$°C = \frac{(°F - 32°)}{1.8}$$

To change from Celsius (C) to Fahrenheit (F)

$$°F = (°C \times 1.8) + 32°$$

[1]From R. F. Flint and B. J. Skinner, 1977, *Physical geology*, 2nd ed.: New York, John Wiley & Sons.

Geology is the science of the earth. Environmental geology is that subject area which relates this science to human activities. In this book, I hope to convey the importance of geology in environmental affairs and to show the close relationship between it and other science disciplines. These themes have been the focus of my teaching and research during the past 30 years, but my interest in these matters is of an even longer duration. Seeing the topsoil of friends' Nebraskan farmlands being blown away during the Dust Bowl years of the 1930s made an indelible impression and inspired in me a conservation-oriented philosophy. This culminated with my decision to move into geology and environmental activism.

The 1970s were called "the environmental decade," and now the 1980s are being hailed as "the energy decade." Clearly, environment and energy are intricately interwoven, and this relationship is a fundamental theme in this book. There are other themes, too, that we will examine. For example, we should realize that environmental concerns act as a two-way street—human activities greatly alter the land-water ecosystem, and natural processes, in turn, produce grievous losses to society. The use of proven geological principles and prudent engineering practices can reduce environmental deterioration and unnecessary costs. Throughout the text, nearly 700 illustrations and tables amplify and document this delicate balance between human society and the environment.

The post-1970 era has witnessed a renewed public and government awakening and a new perception of environmental matters, which have been manifested in many ways. Numerous environmental laws have been enacted to safeguard people and their investments from pollution and other forms of air, water, and land degradation. There have also been massive environmental changes in the twentieth century. These have been produced by a combination of urbanization, population growth, and new machines and fuels.

An important objective of environmental work and study should be to determine which types of construction and resource extraction methods minimize damages to the land-water ecosystem. This objective must be linked to attempts to understand and find solutions to the environmental problems that exist, whether natural or the result of human activity.

There are at least two differences in this book. One is the occasional use of advocacy. I believe it is important for readers to understand that scientists do have definite opinions and even prejudices on controversial topics. Of course, it is necessary that such beliefs be in accordance with known facts, but it is unrealistic to sit in the middle or on the fence on all issues. Another difference in this book is the use of numerous case histories aimed at providing an in-depth approach to the unusual breadth of topics that are fundamental to a comprehensive knowledge of environmental geology.

The book is divided into six parts. The five principal parts have been grouped to demonstrate the content and scope of the subject. Such organizational coherence should also aid the reader to integrate the material.

The first four chapters set the stage for the study of geology and provide a perspective on the topics of environmental geology. A conceptual base for the study is established and is then followed by historical insights and a thorough explanation of those physical systems that are necessary for a full understanding of the material.

Civilization is entirely dependent on the earth's resources and also the energy derived from the land and water. The next four chapters are devoted to showing this vital relationship and the importance of understanding the geologic components that provide the basic ingredients for industry, commerce, and engineering structures.

The four principal geologic hazards—volcanic activity, earthquakes, landslides, and floods—are discussed in the four chapters of Part 3.

The next five chapters demonstrate both the type and the magnitude of our environmental modifications. Such changes result as by-products, sometimes because of lack of knowledge, and sometimes because of an uncaring attitude about human endeavors; often human impacts that are deliberate also have unforeseen consequences.

The chapters in Part 5 discuss the different kinds of plans, policies, and decisions that constitute the fabric of environmental management. Emphasis is placed on positive methods to prevent or remedy problems, and urban affairs and waste disposal problems provide extensive histories and examples. The significance of environmental law is discussed in chapter 21.

Part 6 shows where we have been, where we are now, and where the future may take us. This part provides an evalua-tion of the environmental themes stressed in the book and an assessment of mankind's stewardship of the earth. Medical and military geology are briefly mentioned as two other topics in the larger field of environmental geology.

I am indebted to the following agencies and corporations that have generously allowed me to reproduce work that I developed for them as a consultant: U.S. National Park Service, U.S. Army Corps of Engineers, U.S. Department of Commerce, U.S. Geological Survey, New York State (NYS) Attorney General, NYS Department of Transportation, NYS Energy Research and Development Authority, Consolidated Edison of New York, Niagara Mohawk Power Corporation, and many private mining and law firms. Credits are provided at appropriate places throughout the book, and I extend my thanks to all of these persons for their help and generosity which enriched this book. Many scientists contributed their writing skills as reviewers for separate chapters, but Garry McKenzie, Ron Tank, Sam Upchurch, and Charles Babcock struggled through the entire text and offered numerous suggestions for improvement. For this I am deeply appreciative. Special acknowledgment is due John Conners. The John Wiley staff was exceptionally cooperative throughout the production of the book, and Donald Deneck is especially singled out for his infectious enthusiasm and insights. Thus this book is clearly the result of many minds, but its completion is owed to the moral and spiritual support of my wife, Jeanne.

D.R. Coates

CONTENTS

Volcanic peaks in the Cascade Range. Mt. St. Helens (foreground) is seen emitting gases as a precursor to the explosive eruptions that occurred in May, 1980. Another Cascade volcanic peak is in the far background. (Tom Zimberhoff/Sygma.)

Fundamentals

In order to profit from an exposure to environmental geology, the reader must first become aware and informed of the background rudiments that constitute the fabric of this science, which is a subdiscipline of geology. Therefore, the four chapters that comprise Part One have been designed to provide the necessary overview of the field and to indicate those building blocks that will enhance the understanding of the entire book.

Environmental geology is that area of specialization that takes mankind as the focal point of investigation. Thus it is an applied and practical science. The primary objective of "environmental" work and study should be to determine which types of construction and resource extraction methods minimize distortions of the land-water ecosystem. However, the subdisciplines of economic and engineering geology have now been joined by a consortium of other subdisciplines, and the purists in the various subfields may find themselves in conflict on some of the problems. For example, the economic geologist might point to the necessity for mining a particular resource at a specific locality, whereas the environmental geomorphologist might argue that such an action will cause irreparable damage to flow regimes of the watershed streams. There are no easy solutions for the complex problems of today, but they can be more appropriately addressed when evaluated by teams of interdisciplinary experts . . . not only in science and engineering, but also in the social and cultural fields.

Chapter 1 introduces the topic of environmental geology and shows how its fabric is interwoven into the entire character of nature and ecology. This setting along with the evolution of the subject matter provide a basis that delineates the timeliness of the topic.

Chapter 2 provides a listing of most of the salient themes that comprise the heart of the book. These basic concepts serve as benchmarks that occur repeatedly throughout the highly diverse field of environmental geology. Such coordinating principles show that, in spite of diversity, there is a smaller group of unifying ideas that serve to focus the material into an integrated pattern.

Chapter 3 discusses in panoramic fashion the environmental mood that has persisted throughout history and the feats that have been accomplished. This historical setting is amplified throughout the book to reinstate the old cliché, as paraphrased, "Whoever does not know or understand history is destined to repeat its mistakes." The degradation of the environment was not invented by modern man, because throughout all history mankind has routinely destroyed many aspects of his habitat. Fortunately many segments of society now realize the importance of inhibiting such molestation wherever possible, and the various conservation and environmental crusades bear witness to this mood.

Chapter 4 contains some information on physical geology so that a nonspecialist can read the remainder of the book with greater insight. The emphasis throughout this book is on the physical aspects of the environment; therefore, a basic understanding of the physical materials, and systems that affect them, are vital in the interpretation of environmental matters. Thus the principal purpose of Part One is to give the reader an increased awareness of the role of environmental geology in modern-day life.

Chapter One
Introduction

ACCIDENT AT THREE MILE ISLAND NUCLEAR PLANT JEOPARDIZES SAFETY OF LARGE REGION

The March 28 malfunction of a water pump and valve along with instrumental errors and human confusion at this atomic power facility set in motion a chain of events that has threatened 950,000 lives residing in the four-county area near the plant. (Harrisburg, Pennsylvania, *Times*)

SEVERE EARTHQUAKE DEVASTATES COASTAL RESORT AREA

The April 15 earthquake killed more than 235 people along a 100 km scenic coastline of the Adriatic Sea between Yugoslavia and Albania. (Belgrade, Yugoslavia, *People's Daily*)

DAMAGES FROM PEARL RIVER FLOODING IN EXCESS OF $600 MILLION

April 12 rains that dumped 50 cm of water in this Jackson, Mississippi, area caused the most extensive flooding during the century. The Pearl River crested at 13.2 m, which is 7.5 m above flood stage, and was 1.8 m higher than other floods of the century. Thousands had to be evacuated throughout the entire region and property losses were the greatest in the history of this area. (Jackson, Mississippi, *Ledger*)

RUNAWAY OIL WELL THREATENS ECOLOGY OF GULF COAST REGION

Oil leakage from an uncapped Mexican well in the Bay of Campecho started disgorging on June 3 more than 1 million gallons a day into the beautiful waters of the Gulf of Mexico. Parts of the Mexican and Texas coasts received the brunt of this spillage, which became the largest in history, exceeding the 54 million gallons from the wrecked tanker *Amoco Cadiz* of March 1978 along the French Coast. (Corpus Christi, Texas, *Bugle*)

ACID RAIN CONTINUES TO RUIN LAKES IN EASTERN NORTH AMERICA

New studies by Canadian and American scientists during 1979 have indicated that fallout from acid rain is ruining countless lakes in these two nations. Swedish scientists first called attention to this type of problem in the 1950s when thousands of their lakes had become so sterile that fish were decimated. The cause of the problem is polluted particulates and emissions from coal-burning power plants and industry. (*The Montreal Telegram*)

WESTERN INDIAN DAM DISASTER WORST IN HISTORY

Heavy rains on August 13 deluged the Machu River basin and caused the collapse of the dam and the surge of a 6 m wall of water which overwhelmed this city [Morvi] of 60,000. Total deaths will exceed the previous catastrophe at Vaiont, Italy, where 2200 were lost in 1963. (Morvi, India, *News*)

LARGEST HURRICANE OF THE CENTURY MAULS CARIBBEAN

The week of September 10th saw the spawning and full fury of the killer Hurricane David. Winds in excess of 240 km per hour, heavy rains, and storm surge have created havoc throughout the region and drowned more than 1100 people near Santo Domingo alone. The storm's fury also produced flooding along the American southeast coast. (*The Miami Herald*)

FREDERIC FLAYS GULF COAST

Only a massive evacuation of 400,000 people during the week of September 15th saved most lives in this ruined area. Mobile bore the brunt of the most ruinous storm in the history of the region. Hurricane Frederic with 210 km per hour winds along with waves and rain caused $1.7 billion in property damages throughout the Alabama-Florida coastal area, with $1 billion of this in Mobile. Because of sufficient warning, evacuation, and preparation, the loss of lives was reduced to nine people. (Mobile, Alabama, *Sentinel*)

FREAK TSUNAMI HITS FRENCH RIVIERA

An unsuspected powerful wave raced across the Mediterranean and washed ashore along this vacationer's haven, causing more than $10 million damages and drowning 13 who were not quick enough to race for safety and high ground. (Nice, France, *Journal*)

SOUTHERN CALIFORNIA ROCKED
BY STRONG EARTHQUAKE

California has once again lived up to its reputation of producing large seismic activity. A 6.4 M earthquake, the largest since the San Fernando quake of 1971 in the United States, was felt over a several thousand square kilometer area in southern California. Much of the energy was concentrated in the verdant Imperial Valley during this October 16 event, and damages will run in the millions of dollars (El Centro, California, *Herald*)

These 10 events describe some of the types of environmental events that we witnessed during 1979. The headlines and lead paragraphs are factually accurate statements as they might have been described by the news media for those cities near the areas involved in the action. However, such hazards and disasters represent only one part of the study of environmental geology. Yet these are often the events that attract public interest and even the financial support from government sources, which provides for studies and engineering schemes aimed at protecting society from nature's power—or even nature from our own tampering of the earth's processes.

WHAT IS ENVIRONMENTAL GEOLOGY?

Is the terminology **environmental geology** redundant and unnecessary? Many geologists who view their science as *environmental*—related to the earth and changes on it, whether from natural causes or human activities—would say it is redundant. However, some fields of geology relate to mankind more than others. For example, **economic geology** is concerned with obtaining the natural resources of the earth in order to sustain the endeavors of mankind. **Engineering geology** is involved with the evaluation of earth materials and their stability during and after use in construction, to ensure the safety and welfare of those who use it. Since we live, work, and play on the surface of the earth, nearly all our human activities change or distort the land-water ecosystem. **Geomorphol-** **ogy,** the study of landforms and the processes that transform them is also used by the environmental geologist because it can determine the type and rate of change expected from mankind's alteration of the earth's surface (Fig. 1-1). In similar ways, many other subdisciplines in geology have important environmental roles to play. The **geophysicist** can provide valuable information about earthquakes; the **volcanologist** can assist in giving important information about volcanic hazards; the **geochemist** can give significant data on pollution and waste products; and the **geohydrologist** can offer helpful insights in water resource analysis and management. Thus environmental geology is a collage of many geological subdisciplines. Furthermore, it considers mankind as a force that changes nature. Finally, it is the practical application of the geological sciences in the service of society.

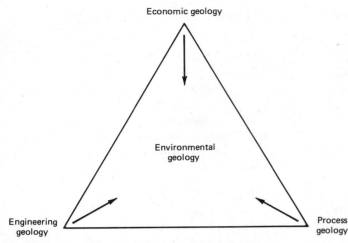

Figure 1-1 The principal fields of environmental geology.

Environment, Ecology, and Ecosystems

Environment describes the entire composition of our human surroundings and all our works. It also includes those conditions and materials that influence the character of the natural setting—such as the weather, water, soils, rocks, flora, and fauna. **Ecology** is the science that deals with the analysis and interpretation of life forms and their relation to the environment. The word *ecology* is derived from the Greek *oikos* meaning "place to live." The emphasis in ecology is placed on the organic, whereas in environmental geology it is placed on the inorganic. Thus, the ecologist is an environmental biologist who is primarily interested in the relationships among organisms, populations, and their community.

A **biotic community** is an assemblage of plants and animals that inhabit a common area and affect one other. The interactions of the biotic community with its physical environment constitute an **ecosystem.** Man as an organism cannot be divorced from these relationships, because he has an integral part in creating changes in both the organic and inorganic components that constitute the systems. Although there are many different types of ecosystems, they have common elements; ecosystems involve the transference of matter and energy into new forms that represent stability for the system. A forest ecosystem can become unbalanced when certain tree species are removed or killed by outside invaders. A drainage basin constitutes one variety of a land-water ecosystem in which the stream characteristics and its channel are adapted to the physical variables inherent in the basin . . . the soils, rocks, topographic properties, precipitation, and so forth. When we alter these components, physical changes will occur in the hydrologic character of the streams, which then affect the channel morphology.

The human environment consists of the earth's natural resources and the cultural, human modifications of them. It includes "the built environment," that is, the structures—buildings, roads, bridges, dams, tunnels, canals—that we *build*, often in a disrupting way. In the broadest sense, **natural resources** are those materials, organisms, localities, and earth processes that are useful or of value to society. If such a definition is adopted, nearly everything on the earth qualifies as a resource—the air, the water, the soil, many minerals and rocks, organisms, and even solar energy. The visual amenities that comprise a beautiful recreational resort are resources for that locality. Similarly some of the earth processes such as waterfalls, geysers, ocean waves, and tides also constitute a resource.

To sustain the human environment, two

different classes of resources are necessary: renewable and nonrenewable resources. Renewable resources are living organisms and those resources that are continuous and enduring. Nonrenewable resources are metals and fossil fuels. When they are mined out, there will be no second crop. These resources are produced at such slow rates that new sources cannot be anticipated within a normal human lifetime. We must remember that the human environment is a **biosphere,** or that part of the planet in which life exists and forms an interacting process. For example, solar energy is a resource that is available and may activate chemicals in water, soil, and rock, which then become the building blocks for living things.

Some Fields of Interest

The scope of environmental geology is so broad that it encompasses not only the subdisciplines of geology but also areas that are of interest to other physical sciences, as well as the biological and social sciences. Later in this book we will look at some of the special aspects of engineering, geography, landscape architecture, and soil science. Thus environmental geology is a multidiscipline with an interdisciplinary character.

The subject matter of environmental geology includes the human-related aspects of earth materials, earth processes, landforms, and certain rate and time considerations. It comprises a wide spectrum of such topics as the location and mining of natural resources; evaluation of the physical changes in the land-water ecosystem when human action rearranges earth materials or interferes with natural processes; assessment of earth forces and hazards that affect human health and safety; and determination of energy and sediment systems for use, storage, and elimination of waste products. Implicit in the knowledge of these topics is the inclusion of such themes as the conservation of materials and processes, and the reclamation of damaged terrain. Thus, environmental geologists become involved in policy matters. They become part

of the decision-making process when it involves planning and management of an arena where they have expertise. This is even true when time is a component and when predictions of the future are necessary. For example, the planning and construction of some developments are linked to certain benefit-cost ratios, which in turn determine the type and size of the structures. Nuclear power plants in the United States are built for a lifespan of 40 years, so the engineering of the buildings is linked to soil and rock stability during that period. Many flood-control projects are designed for the 100-year flood event, and storm water sewers in many cities are designed for the 25-year flood event. The data that the environmental geologist supplies in such enterprises are crucial to the design of the project. A primary objective of environmental geology is to aid in those studies and decisions in order to minimize the human impact on the environment. In accomplishing these goals it is vital that geologists interact with all interested parties and authorities. Only through the establishment of strong communication systems and cooperation can environmental justice be achieved.

OUR RELATIONSHIP TO NATURE

We *are* part of nature. What we are and what we produce are both **natural** and **cultural.** As part of the natural and animal kingdom we have the same needs and drives as other members of this realm, such as protection, nourishment, shelter, and reproduction. Our cultural characteristics are seen in our activities and products. In this book, however, the word **nature** will be used to mean only those parts of the environment that are not our creations. We will talk about the natural forces of erosion. Waves erode beaches and transport sediments, redepositing them elsewhere, just as streams erode their channels and banks. Therefore, change is an integral part of nature. The mountains and hills will eventually be eroded and lowered, regardless of our presence, and such denudation is referred to as **normal, natural,** or **geologic erosion.** However, when man enters the scene, which he may change by design or by careless neglect, setting up slope configurations, disturbing the soil, or modifying processes, a new set of stresses are superimposed on the norm. This additional force accelerates natural process changes and creates what is termed **man-induced** or **accelerated** erosion.

It is important for environmental geologists to know what is normal and what is in harmony with natural systems, for without such a yardstick they would be unable to evaluate abnormal human influences. The prediction of the type, direction, and magnitude of manmade changes in natural systems is therefore dependent on recognition and precise measurements of normal equilibria. Natural forces are constantly at work, changing the landscape, weathering the rocks, producing soils, and supplying products to streams for removal. Such land-water ecosystems achieve certain levels of balance and growth. For example, soil scientists use the terms **tolerable erosion** to define those processes acting on soils in which the production and delivery systems are stabilized. This may amount to as much erosion as 5 tons a year per acre (1835 kilograms per year per hectare) in humid regions soils. Such losses are canceled and not harmful because new soil is being created at similar rates of growth.

Our influence on natural systems has been alarmingly accelerated during the twentieth century. As we have become more populous and our tools of destruction more efficient there are fewer and fewer places that remain untrammeled and unmolested. The exponential growth of the human species when coupled with unmitigated, wanton, and relentless desecration of the earth's air, land, and waters has led to what some observers call **the environmental crises.** Those that champion this view call attention to the deteriorating character of soils, increasing levels of pollution, expanding populations with alarming numbers of starving and malnourished peoples, droughts, floods, energy problems, and so on.

Thus, the two aspects of the relationship that exists between man and nature constitute a **law of reciprocity.** On the one side, man creates multitudinous changes in nature's materials and processes. These impacts invariably cause imbalances, which can lead to problems and hardships. The other side of this relationship consists of those influences and impacts that nature forces on man. These range from the dramatic effects produced by earthquakes and hurricanes to the presence of mountain ranges that restrict man's mobility and habitations. The study of the physical influences of nature on man is known as **anthropogeography.**

Although the term *anthropogeography* is rarely used now, specialization in the subject was in vogue as late as the 1930s. The discipline was initiated in 1817 with the works of Karl Ritter in Germany and amplified in the writings of Ratzel and by Ellen Semple in the United States. The purpose of the discipline is to show the importance of the geographic (environmental) setting for human life, especially in settlement patterns and cultural designs and activities. When extended to the extreme, the environmental setting has been said by some to even affect the human level of civilization and the physical appearance of humans. Hippocrates, writing in the fourth century B.C., attributed the small stature of the Scythians to the severe climate and barrenness of their lands, whereas the Phasians' gross body habit and yellow complexion stemmed from their marshy environment. Even Aristotle used geographic explanations in propounding the assumed superiority of the Greeks over the barbarians, whereas Roman writers explained their ability to reach advanced political and cultural levels by referring to an environmental doctrine. More recently, Ellsworth Huntington throughout his publications had argued that climate is one of the controlling factors of history. He suggested that the development of human races, their attitudes, and their achievement levels is dependent on whether their climate is vigorous and seasonal, or uniform and hot. He believed that the vigorous climates were responsible for the ascendancy of the European peoples, and that tropical climates caused slower development in inhabitants.

There are, however, other and more likely ways in which the environmental setting has influenced human behavior and residence. The physical descriptions of lands have often been directly responsible for colonization and settlement patterns. The early biblical accounts of Canaan as a land flowing with milk and honey was sufficient inducement for the Israelites to invade and possess those lands. The favorable accounts of eastern United States by foreign travelers after the Revolutionary War was responsible for increasing the number of immigrants. In a similar manner the descriptions of the American West, especially after the Civil War, led to the accelerated development and growth of that region. Although the roots of World War II were complexly interwoven with such factors as the harshness of the Versailles Treaty and ideological conflicts, environmental considerations, termed "geopolitical" at the time, were very significant. Germany spoke of the need for "lebensraume" (more living space for their population) and the requirement of more mineral wealth. The war was frequently called the war of the "haves" (the Allies) versus the "havenots" (the Axis nations).

Even the roots of civilization, according to some writers, can be correlated with the human necessity to cope with the environment. For example, Wittfogel developed the thesis that the great early civilizations had one important common element—the need to manage water resources. It was the need to harness and use this resource that provided the unifying force in society. These "hydraulic civilizations" required enormous water importation schemes to irrigate the semiarid lands. Skilled scientists and engineers were needed to evaluate water and food resources and design the canals and water distribution systems. A labor force was necessary to dig the ditches, terrace the lands, and farm the fields. Managerial staff had to deal with the apportionment of the waters, and laws had to be formulated to equalize and set limits to wages, costs, and policies. Thus all the important elements of society were focused and influenced by the "water-based" economy, and the administration of the system revolved around water-based resources.

Environmental settings can be either a positive or negative factor in settlement patterns. Of the 13 largest cities in the United States, only one (Denver) is not located on a major water body. The early population of the United States was concentrated along the Atlantic Ocean or the rivers of the Fall Line (the intersection of the Appalachians and the Piedmont Provinces) where hydraulic energy was available because of steep river gradients. Here the 3 million population was concentrated, with less than 5 percent of the people living more than 10 mi (16 km) from the Fall Line or the coast. Whereas water has attracted settlements, mountains and deserts have generally been barriers or unfavorable for habitation. Thus in mountain terrain transportation is arduous, soils are poor or infertile, temperatures and storms may discomfort inhabitants and produce hazards, and insularity causes poor communications and an unsatisfactory exchange of goods and services. Under such conditions inhabitants become provincial and their rate of "progress" is inhibited.

Finally, what may be called the "human spirit" may also be a reflection of environmental conditions as manifested in literature, art, and music. Sir Archibald Geikie once wrote:

The landscapes of a country, the form, height and trend of mountain-ranges, the position and extent of its plains and valleys, the size and direction of its rivers, the varying nature of its soils and climate, the presence or absence of useful minerals, nearness to or distance from the sea, the shape of the coastline whether rocky or precipitous, or indented with creeks and harbors - all these and other aspects of the scenery of the land have contributed their share to the molding of natural history and character [1905, p. 1-2].

Geikie also pointed out how the "placid scenery" of England influenced the literary works of such writers as Chaucer, Shakespeare, and Milton. The character of the English-Scotch border with its rivers, glens, and dales was clearly the source of inspiration for many of Robert Burns' poems wherein he captures that mood and essence.

INFORMATION SOURCES

Although the term *environmental geology* was first used in the modern sense in 1962 (Betz), those aspects that now comprise the discipline have been used and practiced for millennia. For example, since early times we have mined materials for our use. However, the first book to sound the alarm was written by George Perkins Marsh in 1864, *Man and Nature*. In it Marsh stated that man had become so powerful and destructive that he was placed in jeopardy by his own ability to cope and survive in the deteriorating environment. Similar themes were echoed in such books as *Vanishing Lands* (Jacks and Whyte, 1939), *Our Plundered Planet* (Os-

born, 1948), and *Man's Role in Changing the Face of the Earth* (Thomas, 1956). Dasmann was the first to employ the term *environmental* in the modern context in his book, *Environmental Conservation* (1959). In the 1960s, a new wave of highly popularized paperback books was published. By the 1970s, the movement was in full bloom, and this flowering has led some to call this period **the environmental decade.**

Each of the geology disciplines has made contributions to environmental geology in their more specialized books and journals, and a host of social science books has dramatized the wide range of problems. However, no country, discipline, or set of books or journals has a monopoly on the literature dealing with environmental geology. Indeed, the publications have become so vast and staggering that they are contributing to a printed-page explosion, making it impossible to keep up with all the sources. There are countless journals, magazine symposia proceedings, government reports, agency reports, and a huge "gray literature" in the form of open-file communications, as well as environmental impact statements written by scientists in government, industry, consulting firms, and academic institutions. This bewildering array of documentation on environmental affairs is continuing at a frantic and accelerating pace. To those who specialize in its analysis, I wish them luck and good reading.

Perspectives —WHY STUDY ENVIRONMENTAL GEOLOGY?

Now that we have some background information on the topic of environmental geology, why should we teach or study it? One of the answers should be "because it is here." Only an ostrich or a slumbering Rip Van Winkle can be oblivious to what is steadily happening on the environmental scene. It is a rare week that passes without headlines blaring of droughts, floods, earthquakes, energy crises, nuclear site

disasters, pollution, strip mining, starvation and malnutrition, and many other related topics. For example, the September 13, 1979 front-page headlines of the Binghamton *Evening Press* carried four environmental stories: "Hurricane Frederic rips 4 states," "Indonesia quake dumps town into the sea," "Eruption kills 9 on Mt. Etna," and "Acid rain forecast getting worse."

The status of the environment has also

become a concern with legislative bodies. Each year, the Congress must act on hundreds of environmental bills. State and local governments also must review a great variety of similar legislation. It should be obvious that those citizens that are better informed will be part of a more responsive and representative government. For this to occur, the voters should possess some awareness and perception of the problems. It is also important to obtain certain perspectives, and hopefully, balanced opinions on the major issues. Therefore, people *do* have a stake and an important role to play in those environmental matters that affect their health, welfare, and quality of life.

A second reason for involvement in environmental geology is that the discipline represents a chance to interact with professionals from many diverse fields. As C. P. Snow has pointed out, the world has been moving far too long in the direction of two subcultures . . . that of the scientist and that of the nonscientist. The communications gap has been widening, but environmental geology offers a chance to strengthen communications with many disciplines. Although specialists are needed in all areas of scholarly endeavors, there/is also a significant place for the generalist with the capacity to bridge gaps and understand complex problems of many different facets.

Finally there is the job market. Increasing numbers of positions are opening up in environmental geology and allied fields. Whereas some of the earlier conservation-environmental movements of the twentieth century lost their steam, there is no evidence that the current environmental mood is waning. On the contrary, the strong pace that started in the 1960s is even stronger today. For example, the Environmental Policy Act of 1970 created tens of thousands of new jobs. In turn many State governments have passed their own versions of NEPA, which are creating numerous new positions. Typical of this new wave of employment are jobs in pollution analysis and control, resource evaluation, energy assessment, preparation of environmental impact statements, conservation programs, and reclamation planning and management. In New York State environmental positions have opened because of several new laws and regulations that: require geology reports of all quarries that sell aggregate for use on State roads; mandate mining plans and reclamation plans for all quarries; require environmental impact statements for most public construction and sizable housing developments; and provide incentives to each county that organizes an environmental management council. Thus study and knowledge of environmental geology is challenging and can provide new horizons, but it may also enhance job opportunities and provide marketable skills.

READINGS

Betz, F., Jr., ed., 1975, Environmental geology: Stroudsburg, Dowden, Hutchinson & Ross, Inc., 390 p.

Coates, D. R., ed., 1971, Environmental geomorphology: Publ. in Geomorphology, State Univ. of New York at Binghamton, 262 p.

————, 1972, Environmental geomorphology and landscape conservation, Vol. 1: prior to 1900: Stroudsburg, Dowden,

Hutchinson & Ross, Inc., 485 p.

————, 1973, Environmental geomorphology and landscape conservation, Vol. 3: nonurban regions: Stroudsburg, Dowden, Hutchinson & Ross, Inc., 483 p.

————, 1974, Environmental geomorphology and landscape conservation, Vol. 2: Stroudsburg, Dowden, Hutchinson & Ross, Inc., 454 p.

Dasmann, R. F., 1959, Environmental

conservation: New York, John Wiley & Sons, 375 p.

Day, J. A., Fost, F. F., and Rose, P., eds., 1971, Dimensions of the environmental crises: New York, John Wiley & Sons, 212 p.

De Bell, G., ed., 1970, The environmental handbook: New York, Ballantine, 367 p.

Flawn, P. T., 1970, Environmental geology: New York, Harper & Row, 313 p.

Geikie, Sir Archibald, 1905, Landscape in history: London, Macmillan Co., 352 p.

Helfrich, H. W., Jr., ed., 1970, The environmental crises: New Haven, Yale Univ. Press, 187 p.

Jacks, G. V., and Whyte, R. O., 1939, Vanishing lands: New York, Doubleday, 332 p.

Landau, N. J., and Rheingold, P. D., 1971, The environmental law handbook: New York, Ballantine Books, 496 p.

Laycock, G., 1970, The diligent destroyers: New York, Doubleday, 223 p.

Marine, G. 1969, America the raped: New York, Discuss Books, 331 p.

Marsh, G. P., 1864, Man and nature; or, Physical geography as modified by human action: New York, Charles Scribners and Sons, 560 p.

Marx, W, 1967, The frail ocean: New York, Ballantine Books, 274 p.

McKenzie, G. D., and Utgard, R. O., 1972, Man and his physical environment: Minneapolis, Burgess, 338 p.

Mitchell, J. G., and Stallings, C. L., eds., 1970, Ecotactics: the Sierra Club handbook for environment activists: New York, Pocket Books, 288 p.

Nicholson, M., 1970, The environmental revolution: New York, McGraw-Hill, 366 p.

National Research Council, 1972, The earth and human affairs: San Francisco, Canfield Press,

Osborn, F., 1948, Our plundered planet, Boston, Little, Brown, 217 p.

Ramparts Editors, 1970, Eco-catastrophe: New York, Canfield Press, 158 p.

Rienow, R., and Rienow, L. T., 1967, Moment in the sun: New York, Ballantine Books, 305 p.

Semple, E. C., 1911, Influences of geographic environment: New York, Henry Holt & Co., 683 p.

Stamp, L. D., 1960, Applied geography: Baltimore, Penguin Books, 218 p.

Tank, R. W., ed., 1973, Focus on environmental geology: New York, Oxford Univ. Press, 474 p.

Taylor, G. R., 1970, The doomsday book: Greenwich, Conn., Fawcett, 320 p.

Udall, S. L., 1963, The quiet crises: New York, Holt, Rinehart and Winston, 209 p.

Chapter Two
Basic Concepts

The human legacy. (Photo by Ernst Haas, courtesy of U.S. Dept. of Interior.)

Now that we have been introduced to the subject matter of environmental geology, the next step is to provide some guideposts to serve as focal points for the various topics. Therefore, the purpose of this chapter is to organize the principal themes that will be repeatedly amplified and discussed throughout the book. The 10 concepts that are described do show the biases of the author, but they are intended to indicate the unifying threads that are interwoven into the fabric of this science. These are my favorite guidelines; the list is entirely subjective, and other specialists might offer different personal benchmarks.

Fundamental in a program of environmental geology is the realization that it is a mission-oriented and problem-solving discipline. It is the immediate utilization of current knowledge to deal with specific issues and questions involving society and the environment. Thus there is a specificity to environmental geology studies whereby the work accomplished should have direct and practical applications. This differs from other research because in other fields the scientist has greater freedom to select the problem. In environmental geology, however, *the problem already* exists, or its occurrence can be predicted. So in a manner of speaking, the problem selects the environmental researcher.

In the following concepts, the first five concern physical systems and geological fundamentals. Concepts 6, 7, and 8 deal with the causes and status of problems, whereas Concepts 9 and 10 refer to the resolution of problems and environmental ethics.

Concept One

Complexity is the norm in physical systems.

This concept has many manifestations. One theme has been called by some the **law of variables,** a reference to the many different factors that can influence a given action. Instead of a simple problem solution, as implied by the philosophical rules of the **law of parsimony,** the complexity concept suggests that the correct answer may rest with intricate and compound relationships. Landslides offer a good example of this concept. They can be caused by several mechanisms that can act together when suitably motivated. Landslides may be triggered by excessive rainfall, earthquakes, or human intervention, but these are only the immediate cause. For landsliding to occur, there must also be some underlying and innate weakness in earth materials and structures. Thus one physical system has been superimposed on another physical system. Their combined effect is necessary to cause the final rupture and to create a catastrophic change on the earth's surface.

Another dimension of the complexity concept has been termed the **law of equifinality**— the principle that similar appearing features can be caused by entirely different processes. Alluvial fans and pediments, although not identical, have many similar features; yet these landforms are caused by different forces. Strath surfaces (limited planar features) may also appear in the same shapes as river terraces, but their development may stem from a different set of conditions. In such cases, wrong conclusions about the landscape are possible unless the observer is aware of such complexities.

A third type of error can occur if a blind faith in the **law of uniformitarianism** is adopted. Numerous mistakes are made by those who believe present-day measurements can be directly extrapolated to determine landform evolution. Problems arise for those landforms whose shape is an inheritance from past climates and processes, which may be very different from those at the present. Thus, instead of the present being the key to the past, this simplistic view has not made proper allowance for the relict complexities in the system. Such an observer is measuring an overprint and is studying an effect and not a cause and therefore has become a victim of a **non sequitur fallacy.**

Finally, the very roots of the complexity concept means that environmental geology must be interdisciplinary—the uniting of the geological disciplines with allied fields. Such a realization that nearly everything in the environment is in some manner related to every other thing has been called the **concept of environmental unity.** This idea provides a foundation for all fundamental principles of the study. Such a multidisciplinary and interdisciplinary approach contains an intricate array of complex relationships, but this is the price that must be paid to achieve the knowledge base that is necessary for environmentally sound administrative decisions.

Concept Two

Human-induced changes of the land and water invariably produce environmental feedback systems.

This concept is a special outgrowth of the complexity concept, in which the relatedness of environmental systems is emphasized. **Feedback** is the environmental manifestation of Newton's second law of motion, that is, for every action there is an equal and opposite reaction. Its specific meaning here is that nature has developed a closely tuned symphony of earth forces that generally operate harmoniously. When we enter the system and change one or more of the components, our interference upsets the balance of the orchestration, and noise, not music, is the result. The entire idea of ecology, homeostasis, and the food chain of life illustrates this concept on the biological side of the spectrum. However, there are also many examples from the physical sciences that reveal the importance of this concept in environmental geology.

Feedback operates in different ways, in different scales, and in different time frames. Thus the reaction from some man-induced changes may take years to occur, whereas in other changes the effect may be much more rapid. The location and scale of the reaction may occur at some distance from the source of the dislocation. A river flowing below a dam erodes its channel because of increased energy once the sediment load is impounded behind the dam. The lowered baselevel also affects tributaries, which incise their channels because of the steepened gradients, obeying Playfair's law of stream junctions. Feedback changes can also occur upstream from the dam where the reduced gradients cause increased rates of sedimentation. Along shorelines groins operate as partial dams whereby sediment is trapped on the upcurrent side, allowing the sediment-deprived waters on the downcurrent side to erode. These cases illustrate the importance of regional planning, because a local problem may have been solved but the feedback system can generate a new problem at another location.

Groundwater mining can produce an almost endless chain reaction that shows this domino effect from feedback mechanisms. Excessive pumping and groundwater withdrawal means that recharge of an aquifer cannot keep pace with discharge, and the water table lowers. Unconsolidated sediments can collapse when the buoyant force of water is lost, resulting in subsidence of the land surface. Wells must be drilled deeper and at greater cost, to penetrate new aquifers, leading to higher pumping costs and increased use of energy resources. Furthermore, deeper waters commonly have more dissolved solids because of the longer residence time in contact with mineral matter in the sediments. The salts of such water, when applied to irrigation fields, cause salinization of the soils, which in turn lose their fertility. Thus an entire series of impacts can occur from overpumping, affecting terrain morphology, costs, energy, soil properties, and increased consumer prices of irrigated products.

Concept Three

Geologic thresholds are an ever-present danger in coping with the environment.

Natural systems are interrelated and possess feedback mechanisms that contain built-in checks and balances that act to stabilize the system when subjected to normal stresses. A **critical point,** or **threshold,** is reached, however, when boundary equilibrium conditions are exceeded, thereby initiating rapid and extreme changes. Extraordinary thresholds are reached when catastrophic geologic events occur: earthquakes, volcanic activity, floods, landslides, and so on (Fig. 2-1). However, the word threshold has other dimensions and other applications, in geology and in other disciplines as well. Physicists talk in terms of a change in the system when it has reached a **critical mass;** engineers describe stability of structures in terms of **bearing load;** and geographers evaluate a locale by relating the population to the resource base with regard to the **carrying capacity.** All have the common denominator that the system is in

Figure 2-1 Ancient landslide terrain, Manti-La Sal National Forest, Utah. The landslide diverted the river in the foreground, and the earth materials were displaced from the mountain rock re-entrant (upper left of photograph). More than 100 landslides have occurred in the Manti Canyon, and 460 landslides have been mapped on the west side of the south end of the Wasatch Plateau in the region. Landslides provide a premier example of geologic thresholds. (Courtesy Earl Olson.)

is dependent on only slight changes in gradient and sediment supply. The Hjulström sediment entrainment curves also represent a series of thresholds for different particle size erosion-transport systems. In hydraulics the Froude and the Reynolds numbers are by definition thresholds because they represent conditions whereby bedform and stream power have been radically transformed under supercritical forces. Such thresholds can be described as **extrinsic thresholds,** because they are changes created by an external stress.

Another type of threshold can be referred to as an **intrinsic threshold** (Schumm), whereby the critical limit for rapid change has been exceeded but the threshold was reached not because of external components but because it was initiated from within the system. For example, the usual reasons given for the formation of river terraces are climatic changes or differential uplift, creating new baselevel conditions. Recent studies have shown, however, that terraces can be created by the normal progressive changes that occur within the drainage basin during its evolutionary development. Thus outside factors need not always be resorted to for explaining such landforms. Yet they are threshold forms, because a master stream that was formerly aggrading and building a floodplain has rather suddenly been transformed into one that degrades, and the former floodplain has now been trenched.

When we enter the scene we have the ability to alter drastically the usual balance and, within a landscape, we can even initiate a new erosion cycle. Different terranes may react in a dissimilar fashion to soil and landform displacements. Some climates, such as those in temperate regions, have more self-generating properties and can heal more rapidly than others. Terranes that occur in regions of climate extremes, such as deserts, the tropics, and the arctic, are much more fragile. When disturbed, such climates may contribute to the irreversibility of the changes.

balance with mass, process, and events until conditions transform the system to such an extent that significant changes occur in rate, direction, and magnitude. For example, a fault occurs when the yield strength (a threshold) of the rocks has been exceeded, causing a rapid displacement of earth materials.

Although numerous geological examples of thresholds could be cited, those that occur in surface water hydrology are among the best known. The change in river shape from a straight to a meandering or to a braided channel

Concept Four

The geologic environment contains both open and closed systems.

Whether a system is to be considered open or closed depends on the viewpoints and time frames that are selected. In the context of this book, a system is **open** if there is a free flow of mass and energy into and out of the system. A system is **closed** when no external components enter the system and nothing leaves it. Schumm and Lichty have shown the importance of time when theories of landscape development are compared. When long time periods are involved, the evolutionary scheme of William Morris Davis, using youth, maturity, and old-age cycles of landscapes, becomes a reasonable conceptual model. However, if a shorter time span is considered, the dynamic equilibrium ideas of John Hack, and other steady-state types of theories, seem to fit earth sculpture more adequately. Thus it is possible that such variables as climate, vegetation, and available relief (all factors that aid to determine landscape properties) can be either dependent or independent variables in the system.

Each day the earth receives a finite amount of solar energy, which is transformed into other energy forms. The results of this work, when coupled with gravity forces, constitute those processes known as **exogenic** (acting **on** earth's surface), such as running water, groundwater oceans, glaciers, wind, and mass movement. These forces are arrayed against the **endogenic** processes, which operate **within** the earth and which take the form of diastrophism (movement of solid rock) and vulcanism (movement of molten rock). When viewed on a global scale the hydrologic cycle and the rock cycle seem to operate as open systems. These cycles are possible because the redistribution of energy is constantly being renewed. In the human myopic scale such cycles seem endless, yet they must obey the second law of thermodynamics—that is, the amount of energy available to do work diminishes with time. Ultimately this would lead to stagnation in the system, a time when energy gradients are so reduced that work is not accomplished. Energy input into such a system as the hydrologic cycle can alter its character and effectively stop certain energy transformations and lead to significant climate changes. When we tamper with such systems, as with cloud seeding or the creation of massive land and vegetation changes, we cause perturbations in the system. In extreme cases, deserts may even be formed.

Many systems on the earth may operate within their own realms as open systems, but when two or more become juxtaposed, an aberration results, resulting in absorption or extinction. One aspect of such modification is known as the **law of aphasy,** whereby the climatic environment changes faster than organic adaption to meet the change. If mobility and migration become impossible, this causes either extinction or internal changes within the organic colonies. Thus, when the freedom to operate in an open system is denied the organisms, massive changes become necessary for survival. We live on a finite earth with constraints on space and resources. Although many components within these realms may act as open systems, when unduly influenced they may be so severely altered as to become closed systems, operating only on internal alignments and readjustments. It should be obvious that the recognition of systems and their behavior is a vital ingredient in our successful management of environmental affairs. Our manipulation of systems should be undertaken at only those levels where they interfere least with the natural system, thus reducing chances for irreversible changes.

Concept Five

Resources can be categorized as to whether they are renewable or nonrenewable.

Failure to realize the finite character of many resources has been a significant factor in creating some of the current world problems. Most resources have been used and abused without regard to their exhaustibility. **Resources** are those products that enable human life to be sustained (Fig. 2-2). On the comprehensive scale nearly everything is a resource—solar energy, soil, terrain, hydrographic features (rivers, lakes, oceans, wetlands), forests, grasses, weather, and so forth. In the restrictive sense, resources are those nonliving and living materials that are used or consumed by humans for their health, welfare, and survival. The manner in which these materials are utilized and managed provides an important measure of the quality of life. The types of resources that are used are relative and culturally sensitive. A gold nugget has no use or value in Bushman society and becomes a throwaway, whereas in advanced civilizations the entire culture may worship and be governed by it. Resources may also be time-dependent. What appears to be an infinite resource for one generation may become exhausted by another generation.

The administration of the resource will be determined by certain rules of supply and de-mand, the level of technology for extraction, and economies linked with benefit-cost ratios. Waste debris is the antithesis of a resource, but it is also the result of technology and time-dependent. As some have noted, "Today's waste is tomorrow's resource." In the early mining of iron ore in the Adirondack Mountains the rocks that contained a high-ilmenite content (iron titanium oxide) were discarded onto tailings piles. After World War II new uses for titanium, as in paints, were discovered, so that the old tailings were reworked for the ilmenite. Other waste products such as garbage and materials used in landfills can also be converted into energy and other resources.

Therefore it is important that environmental managers have a full understanding of the finite character of the nonrenewable resources as compared to the regeneration of the living resources. The legal, financial, and sociocultural mix are different when managing the two different resource types. How quickly should natural materials such as iron, petroleum, copper, and others, be mined and with what level of regard for future generations? Should stock-piles be allowed to accumulate as a hedge for future use? The recovery and financial arrangements of nonrenewable resources operate with entirely different guidelines than those for renewable resources. For example, the amortiza-

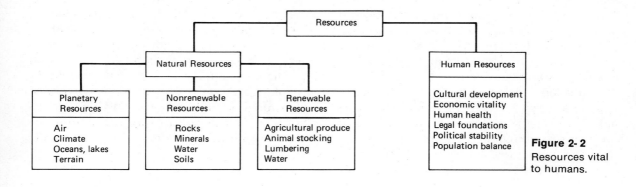

Figure 2-2
Resources vital to humans.

tion and investment schedules obey different monetary rules because of the finite character of the resource and the knowledge of its restriction in time and space. The location of the deposits are fixed, and recovery expenses generally increase with time as deeper and less accessible materials are obtained. Furthermore, there is often a diminution in grade of resource. Such limits to the rate of exploitation of resources thus pose severe problems for management of resources and the economy of a country.

Concept Six

Environmental problems are universal.

Environmental damage is worldwide. All places and all peoples are not afflicted with the same set of problems, but immunity is the exception and not the rule. The problems differ in both type and magnitude, and can be either man-induced or natural. The degree to which we have fouled our own abode has reached such high proportions that it has become fashionable with some authors to term these times "the environmental decade" and to discuss matters as the approach of an environmental Armageddon draws nearer. Individual problem areas have various names: "the energy crisis," "the water crisis," "the pollution crisis." Water provides an example of the variability problem. It rarely is present exactly as man wants . . . there is either too much, too little, or it is too poor in quality. Droughts may be afflicting one part of a nation, while another part is beset by floods, or by salinization.

Because we live on the earth's surface, it is inevitable that our activities cause changes in the land-water ecosystem. These changes range from the creation of microclimates in urban areas to the desertification of huge regions. Industrialized nations have polluted the water and the land in countless ways, whereas agrarian societies have defaced the landscape, as in the use of slash-and-burn clearing techniques, which cause excessive erosion of the soil and siltation in the rivers. We are rarely safe, either from ourselves or from natural processes. Our health can be affected by a variety of toxic wastes in the air and in the water—by our own doing. Geologic hazards are an ever-present threat, and disasters occur at alarming frequencies. Many cities are located adjacent to rivers or shorelines and are periodically subjected to flooding. Those developments that are placed on hillslopes have a high-risk potential because of possible landslides. Earthquakes respect few environments—they occur in arctic as well as tropic terrane, and in the mountains as well as the plains.

The newest combination of threats to mankind and his quality of life is found in the close relationship of population growth, food, and energy. This trinity of environmental problems will severely test human ingenuity in the coming decades. Geology will be an important ingredient in the solution equation because the common base of the problems rests with earth resources and the technology to make them available. To feed expanding populations an adequate food source is needed. This in turn depends on husbandry practices used with land and soil. To grow high-yield crops, fertilizers and pesticides are needed, and these materials are generated from geologic resources. The production of most energy is derived from geologic resources: fossil fuels, radioactive minerals, and water. Thus, the discovery and mining skills of geologists and their ability to construct safe storage and delivery systems with engineering partners are crucial in the resolution of the growing list of environmental problems.

Concept Seven

There is an exponential rate of increase in environmental deterioration.

Prior to the twentieth century, population growth rates were nearly linear, and impact on the environment was conformable to this growth. However, in the twentieth century, an entirely new set of conditions was unleashed and these have led not only to cumulative degradation, but also to an expanding rate of environmental change (Fig. 2-3). One element of this change is that the "population explosion" has required accelerated production of materials and energy. The development of the internal combustion engine coupled with petroleum (the wonder fuel of the twentieth century) provided greater mobility for cars and other machinery. This permitted intrusion into new environments, and the more highly powered and efficient earth-moving machinery made possible removal and displacement of earth materials at an accelerated pace. Thus man's ability to change land-water ecosystems has rapidly increased to alarming proportions.

The exponential growth rate of developments and use of the environment, when superimposed on a finite natural system, causes severe constraints in attempting to maintain equilibrium of earth materials and processes. The utilization of mineral resources and energy is increasing at faster rates than the population, especially in the Western world. The growing of living resources is also being greatly influenced because lands that formerly were allowed to lie fallow for a year or two to regain nutrients are now continuously pushed into production year after year with no chance for healing and regeneration. Thus, all segments of resources and energy are being called on to yield higher and higher outputs to keep pace with the rather gluttonous practices of modern civilization; and supporting this trend is the attitude that has been dubbed by some the "growth philosophy." Society is geared to continuing the increased rates of growth, sometimes measured as the **gross national product,** and economic practices have been traditionally tuned to this frequency.

Concept Eight

Environmental decisions invariably involve and produce internal conflicts.

"You can't please all the people all the time." This axiom is true not only in environmental affairs, but perhaps in all human actions. Even the word *environmentalist* conjures ill thoughts in some people. Those who deal with environmental questions soon discover that, regardless of the decision, "You're damned if you do, and damned if you don't." Conflicts arise largely through the problem of how to decide which system of priorities to adopt. They reflect the lack of homogeneity in people. Although no

single group will act as a monolithic body, typical conflicts arise among the young and the old, the rich and the poor, the urban and the nonurban dwellers, the educated and the uneducated, and so forth. Because of the many diversities, there are few absolutes in environmental management. Decisions and policy become a matter of compromise during which time certain trade-offs are made. By taking such actions it is the hope of environmental administrators that some sort of cushion, or buffer, is created between the implacable and resolute foes.

Garrett Hardin's **principle of the tragedy**

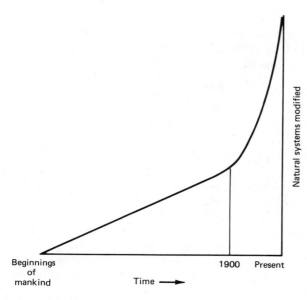

Beginnings of mankind

Time ⟶

1900   Present

Natural systems modified

Figure 2-3 Rate of change of natural systems. Since about 1900, the modification of the land-water ecosystem has been accelerated by human growth and the influence of human enterprises.

of the commons illustrates the conflict-of-interest syndrome. This involves the rights of persons to have perfect freedom without restrictions. When all people are allowed equal access to a common ground, the usage becomes so heavy that the site in question is destroyed and becomes useless for anyone. Thus the management of resources is one of the primary arenas where the conflict battles have been fought. Although this idea will be a recurrent theme throughout the book, we can briefly identify, and oversimplify, this conflict as the utilitarian market view versus the environmental ecology view. Because some of the issues can become clouded, I will try to describe three typical "ethics."

1   The utilitarian ethic   The extreme view of this group holds that man as the supreme ruler of the earth has all the rights and that nature's purpose is to serve and obey the human race. As a justification of this view, some would cite the biblical dictum, "Be fruitful and multiply. Fill the earth and subdue it, and have dominion over the fish of the sea, over the birds of the air, and over every living thing that moves upon the earth." In this view, good environmental management means mastery over nature.

2   The conservation ethic   Those that hold this view cover a wide spectrum of interests and beliefs. Historically, as used by Gifford Pinchot, this view favors the maximum use of resources through time. Obvious difficulties arise as to what time scale should be adopted.

3   The preservation ethic   John Muir was an early leader and champion of this outlook. In the strict sense, this group argues that man should not make alterations of natural areas. In the anthropomorphic sense, nature has rights and the normal processes should not be changed. It is further argued that nature has many qualities that are essential to man's well-being, in terms of aesthetic, recreation, scientific, and therapeutic values. Indeed, if this be true, then a basic dichotomy occurs, because the preservation ethic becomes the most utilitarian alternative to environmental management. The establishment of "wilderness areas" is proof that this group wins some of the battles.

Internal conflicts arise between and within these groups. The classic confrontation occurred between Pinchot and Muir. However, even among subset groups within a given ethic serious differences arise. One typical problem concerns the phreatophytes of the American southwest. Here 17 million acres of phreatophytes annually consume 25 million acre-feet of water. (An **acre-foot** is the volume that would cover one acre to a depth of one foot.) Water conservationists want to eradicate the vegetation and thereby save precious water resources, whereas wildlife conservationists want the vegetation preserved because it provides a necessary environment for living organisms.

The search for a common denominator among the various conflicts and constraints that operate in environmental affairs requires a modern analog of Solomon's wisdom. The need for objectivity must be stressed for those that manage the environment. This means an open system of communications among adversary groups and the avoidance of "tunnel vision."

Policies should be forged in the caldron of dialogue among planners, politicians, and scientists with participation of other citizens in forums that are open and understandable.

Concept Nine

The majority of environmental decisions should be based on benefit-cost analysis.

Decisions in environmental affairs, as in nearly all human activities, are based on some form of benefit-cost analysis. It should be obvious that a given action will not be undertaken if the predicted costs exceed the predicted benefits. The real problems arise when attempts are made to define terms, determine what components should be included and excluded, predict the reliability of future impacts, and decide what time frame to adopt. There is a wide range of possibilities concerning what constitutes benefit-cost analysis. The principle underlying this concept is that objective and measured properties will be evaluated as an economic calculus to derive numbers that will stand up under societal scrutiny. There are many imperfections in such an analysis, but the alternative is to decide matters using subjective and arbitrary techniques.

Benefit-cost analysis and the benefit-cost ratios that are derived do not provide an environmental panacea, because the resulting numbers that emerge are only as accurate as the input data that are supplied for the various equations. Benefit-cost ratios have been applied to many federal construction projects for years. Prior to construction or funding of any dams by the Bureau of Reclamation or the Army Corps of Engineers, studies have to be made to determine what the costs and the benefits of the project will be. This yardstick has also been applied to construction work performed by the Soil Conservation Service (U.S. Department of Agriculture) as stipulated under the Watershed Protection and Flood Prevention Act of 1954 (Public Law 566). There are many who object to the manner in which large benefit-cost ratios are derived (to qualify for financial support the ratio must exceed 1.0). Haveman (1965) and others point to abuses such as (1) the employment of unrealistic discount and financing rates, (2) the overestimation of primary and secondary benefits, and (3) the underestimation of inflation and construction costs. For example, cost overruns are nearly universal in government projects and not restricted to those dealing with the environment. However, a Bureau of Reclamation study of 128 projects initiated between 1935 and 1960 showed more than 75 percent with cost overruns, and the entire group of projects exceeded cost estimates by 72 percent.

A new approach to decision making was ushered in with passage of the National Environmental Policy Act of 1970. This law required more precise data for what constitutes environmental impact in government construction work. It mandated that not only were the predicted impacts to be assessed, but also rationale given as to why other alternatives were not selected. So although a strict formula and benefit-cost ratio may not be involved, the intended construction has had to be evaluated in terms that assess the types of benefits versus the types of environmental costs that will ensue from the action.

Perhaps the most flagrant abuse in benefit-cost analysis has been the failure to internalize the costs. These are costs that are external to the project, which appear elsewhere in the economic system, and whose expenses are born by others. When an industry pollutes a river, the cleansing of the water downstream is a cost to consumers using the water treatment system. The industry has not paid the cost of its own act. However, if the industry is fined, or is forced to discharge only pure water, then it has internalized its costs. Exceptional and controversial policies are now being debated in the United States concerning the relative benefits and costs of using nuclear plants or fossil fuels for energy generation. The decisions that are reached should be made on the basis of consistent evaluations. The true costs involved with nuclear plants should include all phases of the operation—the mining, the health effects of radon on miners, the plant costs and effluent impacts, and the disposal of wastes and other health and risk factors. In similar fashion, all impacts of alternative fuels should be counted—the fatalities in coal mining, the rehabilitation costs of mining operations, the air pollution of plants, the oil spills in drilling and in tanker transportation, and so forth.

The importance of a rigorous adherence to benefit-cost analysis is that it provides a base for decision making, standardizes a method, and makes possible a wider, more understandable concept for people in the political process.

Concept Ten

Environmental stewardship is prerequisite to long-range compatibility of man and nature.

The issue of environmental management is too crucial to be decided by only one spectrum of society. It is the thesis of this concept that representation on policy matters needs the involvement of many disciplines, including scientists. Environmental stewardship implies a constructive liaison between man and nature. It implies that we have a larger responsibility than to ourselves alone, which includes empathy for natural things and processes and consideration for society and future generations. The antithesis of the stewardship concept is the "population explosion," wherein people for purposes of self-gratification are excessively propagating their race without regard for repercusions or the future. Environmental geologists must wear more than one hat. Not only do they need to determine and to interpret facts and data on which decisions may be based, but they should also articulate their findings and be willing to share their judgments in the public forum.

The concept of a *participatory* environmental geology entails difficulties. The traditional role of the scientist has been to report findings and demur from a posture of advocacy. This severe constraint was indulged for fear that any activism would stain a scientific reputation and damage credibility. Such complete divorcement from the "real world" should not be encouraged because, as stated in other concepts, the environment has become so complex, so interrelated with universal problems and threats of new thresholds, that it demands the involvement of the scientific community.

There are many obstacles placed before the path of stewardship, and the greatest of these is human avarice—largely in the form of money and power. While living in Arizona I was greatly disturbed by a lack of ethics on the part of the absentee landlords who instructed their tenant farmers to work the land as fully as possible. Prodigious amounts of water were used to obtain instant profits, with no thought of the future or what it was doing to the land or the water reservoir. The present generation is now paying the price of this dwindling resource. Such an irresponsible approach to nature can be compared with the Amish-Mennonite type of land ethic. These people realize their farms will be handed down from generation to generation. Therefore, the soils are exceptionally well managed with great husbandry skill so that the nutrients will be sustained and capable of production for future times.

The political sector is rampant with programs that involve trade-offs, compromise, and annulment of environmental programs. The history of the Carter administration to deny funding to 17 water projects in the United States is illustrative of the political maneuverings that occurred when a favorite "pork barrel"

project was scheduled for elimination from a congressman's district. Also on several occasions North Dakota threatened to halt all coal shipments to Minnesota unless that state would join them in voting for the Garrison diversion scheme for water allocation.

Environmental stewardship can only be successful if geologists are willing to aid in educating the general public on the issues, to place their data and ideas before the planners and managers, and to become an integral part of the decision-making administration.

Perspectives

Humans are egotistical animals, intoxicated by our own technology. Most of us believe we are infinitely adaptable and can solve any problem if given sufficient money and time. The concepts presented in this chapter suggest there are limitations to what should be exploited and demanded of nature. The complexities, interrelationships, feedback mechanisms, finite resources, and exponential growth rates of the environment are paramount considerations in the assessment of human behavior toward nature.

The standards that we use to make environmental decisions should be based on many factors. Because of so many competing ethics and the presence of internal conflicts, these judgments should take some form whereby benefit-cost analyses are utilized. This should not reach the point, however, where environmental stewardship is neglected. Indeed, the question might be asked whether economic systems that rely on maximum profit incentives are compatible and consistent with the maintenance of a

high environmental quality? Time is a parameter that weaves through all concepts of environmental geology. What is the period over which a given management procedure should endure? Different factors become involved when a development is designed to last 20 years and not 200 years. What type of considerations should be made for the future? The environmentalist would argue that future generations are also a responsibility of the present, and that future environmental needs may be even more critical than they are today. A technocrat might respond by saying the resources that are treasured today may not be needed tomorrow, and that future generations will have increased skills to cope more effectively with the environment; in other words, tomorrow will take care of itself. Perhaps the best answer lies in the middle ground between the arrogance of technology and the idolatry of nature. This position is conveyed in the apt semantics of Ian McHarg, who describes the goal of environmental management as a search for a "design with nature."

READINGS

Abelson, P. H. and Hammond, A. L., 1976, Materials: renewable and nonrenewable resources: Amer. Assoc. Adv. Science, Washington, D.C., 196 p.

Coates, D. R., and Vitek, J. D., eds. 1980, Thresholds in geomorphology: London, George Allen & Unwin, 498 p.

Glacken, C. J., 1970, Man against nature: an outmoded concept: in Helfrich, H. W., ed., The environmental crisis: Yale Univ. Press, p. 127–142.

Haveman, R. H., 1965, Water resource investment and the public interest: Nashville, Vanderbilt Univ. Press., 199 p.

McHarg, I. L., 1971, Design with nature: Garden City, New York, Natural History Press, 197 p.

Prest, A. R., and Turvey, R., 1965, Cost-benefit analysis: a survey: The Economic Jour., v. 75, p. 683–735.

Schumm, S. A., 1973, Geomorphic thresholds and complex response of drainage systems: in Morisawa, M., ed., Fluvial geomorphology, Publ. in Geomorphology, State Univ. of New York at Binghamton, p. 299–310.

Schumm, S. A., and Lichty, R. W., 1965, Time, space and causality in geomorphology: Amer. Jour. Sci., v. 263, p. 110–119.

Smith, H. M., and Stamey, W. L., 1965, Determining the range of tolerable erosion: Soil Sci., v. 100, p. 414–424.

Chapter Three
Historical Perspectives

Excavated remains of the Acropolis, Sardis, Turkey. The structure was built in the third century B.C. and overwhelmed by landslides soon afterwards. (Courtesy Gerald E. Olson.)

INTRODUCTION

Human history is tightly interwoven into the fabric of an environmental setting. What we grow and cultivate depends on soil and climate conditions. We construct homes and buildings out of the nearby available natural materials. Items that are not locally abundant and yet necessary for survival are traded, bartered, or purchased by revenue obtained from materials and labor that are local or imported and turned into saleable items. We select our homesite a' places of convenience, necessity, or safety, depending on our perception of the environment and any risks that may exist. In many instances, convenience and other comfort attitudes supersede the possible dangers that may occur. Thus we build in the shadow of volcanoes because the soil is rich, or on floodplains because the soil is rich, the ground is level, building costs are minimal, and waste disposal in the river is easy. The hazard aspects are therefore disregarded.

In recognition of the importance of earth materials as a factor in human evolution, archeologists have catalogued the "ages of man" according to the ability to use and manufacture products from different natural resource ingredients. This hierarchy reveals a series of evolutionary stages whereby humans developed greater sophistication in the understanding and manipulation of earth materials. The usual succession runs from the Stone Age (Paleolithic and Neolithic cultures), Copper Age, Bronze Age, and Iron Age. Additional advancements have been made as a result of the Coal Age, Petroleum Age, and Nuclear Age. Each of these ages have produced its own special impact on the environment. An alternate way of looking at the human panorama would be the classification of events that led to particular life-styles and organizational patterns of society. Such historical changes would be represented by the Agricultural Revolution, Industrial Revolution, Electrical Revolution, Automotive Revolution, Urban Revolution, and Communications Revolution. There is a close similarity in these two approaches in several instances. For example, the Industrial Revolution was dependent on coal, and the Automotive Revolution needed petroleum products. Clearly mankind and nature are completely intertwined.

THE EVOLVING RELATIONSHIP BETWEEN MAN AND NATURE

Early man was more a part of nature and the environment, for he was just another predator or food obtainer, who existed by the gathering of wild foods and by hunting and fishing. During Paleolithic times man was a vagabond and nomad, moving with the food supply and not staying in one place sufficiently long enough to irreversibly damage nature's ability to heal and regenerate places he had inhabited. However, the first major impact came with the discovery that man could control and create fire. This gave him a new weapon to round up, trap, or drive game by deliberately setting fires. This was not only the start of man-made pollution but also set the stage for the first large increase in erosion rates. Such practices when coupled with the slightly more advanced techniques of slash and burn, which were used in forests to develop crops or to tap forest products, led to startling new rates of erosion and concomitant sedimentation. Dramatic changes occurred in the sedimentary record of many European localities, and this man-induced deposit has been described as **human** or **anthropogene sedimentation** in the European literature. In some instances, man was unusually fortunate because certain climates and soils were sufficiently resilient so that the burning of the forests led to replacement by grasslands, which probably included many of the North American prairies. Such environments were able to support large herds of grazing animals, which in turn helped to inhibit redevelopment of the forests.

The next major impact by man occurred when the nomadic life was changed to the agrarian life. This Agricultural Revolution required a sedentary life-style with concentrations of people. Thus the longer residence time at a single locality allowed for repeated alterations of the soils and water bodies. Man developed tools to artificially disturb the soil, which ushered in the first deliberate and extensive changes in the soil cycle, causing accelerated soil erosion. The Agricultural Revolution was largely synchronous with the domestication of animals for use as food supply. With the larger concentrations of cattle, sheep, and goats, which were forced to subsist in the same area year after year, overgrazing became commonplace. The cutting of forests for use as timber in buildings and for fuel also led to increased erosion of hillslopes. As civilizations grew and flourished, and cities increased in size, the need for animal and plant foods and timber resources led to changes in land use practices. With the need for more foodstuffs, additional water had to be imported in dry regions. In such areas as the Tigris-Euphrates region some of the imported water contained dissolved minerals and, when the water was placed on the soil, high evaporation left behind salt residues that destroyed soil fertility. This inability of the ancients to flush out the salts has been cited by some historians as the prime reason for the decline of several early civilizations. The magnitude of the increased sedimentation is shown by the city of Ur, which thousands of years ago was a coastal city on the Persian Gulf; its ruins are now 320 km inland. Tarsus, now in Turkey, was a major seaport 2000 years ago (the home of St. Paul and the place where Antony and Cleopatra's navies joined and were supplied); it is now 16 km inland due to sedimentation from the barren hillslopes which once were thickly forested but are now severely denuded.

For humans to live in the metal ages, copper, tin, and iron resources had to be located and recovered. Some of these resources were not in great abundance in or near countries that required them, and their importation from other lands led to new discoveries and colonizations. For example, with the decline of the tin mines in the Mediterranean region, the Phoenicians traveled as far as England and developed tin mines at Cornwall. Thus such ages were directly linked with the exploitation of mineral resources, which in turn greatly influenced the strength and wealth of nations.

The Industrial Revolution created massive dislocations of peoples and resources. It caused the influx of miners into coal fields, which then produced the energy and provided the furnaces with the ingredients to mass-produce iron and other metallic products. The Industrial Revolution caused the rapid growth of cities around major plants. It was also the first time that air pollution was experienced on such a massive and concentrated scale, where fumes and debris killed all vegetation for kilometers around the factories. This also led to accelerated soil erosion, as in the Swansea Valley, Wales; Sudbury, Ontario; Ducktown, Tennessee; and Coeur d'Alene, Idaho. The Industrial Revolution also was instrumental in causing the redistribution of wealth, goods, and services and in expediting increased trade and commerce. All of these goods and services required a concentration of management and labor. Such a worker pool became focused in urban areas. So through a series of technological advances, hastened by social and commercial innovations, the seeds were sown for the Urban Revolution.

Each revolution has provided its own special mark on the environment. The Urban Revolution ushered in a new system of management needs. Law and order must be tightly maintained when people live together in congested sites. An entire new bureaucratic structure is necessary to plan and manage this concentration. Not only must the immediate environs be managed but also the outlying areas where road networks are required to maintain a two-way flow of goods and services. Water delivery systems are necessary if water is unavailable in sufficient amounts within the community. Thus construction of immense reservoirs, sometimes hundreds of kilometers from the city, as in New York City and Los Angeles, become vital to the city's maintenance. Plumbing and sanitation systems reduced disease, which in turn increased population. This created an anthropogene series of landforms—roads, buildings, streets—which all combined into a totally new landscape—the **cityscape.**

A new set of problems arose in cities because man still had a need for a certain aesthetic fulfillment. To satisfy this feeling, pressure for open space in the urban environment and parks at distant sites caused a new regard for nature. This respect led to the development of areas where the city dweller could escape the artificial world and obtain relaxation and enjoyment in more pristine environments. The national park movement in the United States offers the best example where such pressures have allowed the development and preservation of a huge park and monument program used by tens of millions annually.

The historical roots of civilization by their very definition commonly made man an antagonist of nature, because in order to survive with expanding populations we had to change nature to suit our growing needs. In our eyes nature became increasingly a provider and servant, to do our bidding on a demand schedule. A manageable servant to be effective must be predictable, so it was incumbent on us to have an advanced understanding of nature's ways so they could best be used to our advantage. Therefore we analyzed nature to learn what laws and principles could be used to predict and control the natural systems. Thus, the growing seasons had to be calculated, the habits of animals along with their cycles and migration patterns had to be learned, and the occurrence of minerals and rocks of practical use had to be discovered.

Throughout history there has been a dichotomy between man and nature. If we use religious writings as a source, they can reveal many contradictions. For example, in the Bible (Genesis 1:26) it is written, "Let us make man in our image, after our likeness; and let them have dominion over the fish of the sea, and over the fowl of the air, and over the cattle, and over every creeping thing that creepeth upon the earth." Such writings have been used to show that man is the supreme ruler and is destined to subjugate all nature. However, the other side of the ledger is found in such passages as, "Woe unto those who lay field next to field and house next to house till there be no place on Earth where they may stand alone" (Isiah 5:8). Such statements can be interpreted as very supportive of the preservation ethic. Other contradictions occur when the habits and ideas of the American Indian are studied. The Indian rapport with nature has been cited by many authors and their sayings such as, "The land is our mother" and "Our fathers received the land from the Great Spirit," have been used to show their reverence to the environment. However, these same Indians drove entire animal herds over cliffs, set fires to drive the buffalo, and when game was plentiful killed many unnecessary animals, eating only the choice parts and wasting the remainder. Sites such as the Mesa Verde cliff dwellings became large garbage dumps. Are these the practices of a consistent environmental attitude? Are they so different from modern-day practices?

THE ANCIENT WORLD

Pre-history

We have gradually become more sophisticated in our survival methods by learning to use nature as a benefactor. If Europe and the Middle East are used as models for the rise of man, the dates of 17,000 to 11,000 ybp (years before present) represent the time that man was a hunter-food collector exploiting wild species of plants and animals. However, with changing climatic conditions and expanding populations competing for the same food sources, man gradually developed more efficient and energy-saving methods for survival. The earliest period of animal domestication dates from 11,000 to 8000 ybp, and the farming period from 8000 to 7000 ybp. Animal husbandry and agriculture were not sudden discoveries, but became necessary when man was expending more energy in hunting, fishing, and food gathering than he was receiving in rewards. In terms of human energy he had reached an impasse, where the benefit-cost ratio had placed him in a precarious position. Thus animals, when domesticated, became labor-saving devices, sources of food in times of need, and valuable as pets or in religious rites. Dung could be used as fertilizer and fuel, so animals became a more predictable resource than wild game. In similar fashion the growing of crops can be more predictable and more efficient by producing more food per area than wild grains and reducing the foraging time necessary to obtain them. Good land with rich soil constitutes the basic ingredient for crops. In the broad valleys of relatively dry areas, as in the Tigris-Euphrates region, crops were abundant, and water supply became a dominant necessity for agriculture. In more forested regions a slash-and-burn technique was developed to tame the "wilderness." Fires deliberately set aided to destroy trees and undergrowth. The residue also produced phosphorus and other materials, which for a time provided soil-enriching products. The scorched earth also temporarily freed the ground from insects and weeds. As pointed out in later chapters, however, each of these environments can suffer deterioration in time when improperly farmed or managed.

Early History

With the development of agriculture, man discovered a more sedentary existence. This gave him more time and led to the development of more arts, crafts, and writing. Furthermore, the rise of larger concentrations of people required the formulation of more social rules and regulations and intensified the need for civil order (civilization). The need for more durable tools and equipment for the construction of permanent dwellings and other necessities for more complex living became mandatory. The successful search and use of minerals that could be utilized in this fashion led to a series of stages, or ages. As early as 8000 ybp copper was being used in an area stretching from western and central Anatolia to the flanks of the Taurus and Zagros Mountains and to the edge of the central desert in Iran. Bronze (copper and tin) came into use about 5000 ybp, and iron about 1500 years later. Although some of the most ancient canals had been constructed east of the Tigris

River at Choga Maminear as early as 7500 ybp, the first major cities with populations in excess of 20,000 did not occur until about 5500 ybp in Sumer. Large cities had also been built in the Indus River valley by 4500 ybp, in the Yellow River valley of China by 3500 ybp, and by 2500 ybp in Central America and Peru.

Environment as a Cause of Civilization Decline

Although the rise and fall of a particular civilization generally has several interrelated causes, environmental factors have contributed to the decline of several. Many authors, including Arnold Toynbee, cite the loss in soil fertility due to salinization (see pages 462–463) as a principal cause of the weakening and ultimate demise of several empires in the Middle East, such as Sumeria and Chaldea. The Mayan empire may have faltered because an expanding population did not permit the forests a sufficiently long fallow time for regeneration. The slash-and-burn agriculture caused soil deterioration, excess erosion, and siltation of lakes and water sources. Two other cases, in Egypt and in Rome, can be examined as perhaps illustrative of what can happen when man is not in tune with the environment.

The geographer Ellsworth Huntington has written extensively of the role that climate has played in the vicissitudes of civilization, but more recently Barbara Bell (1971) has placed new emphasis on drought as the principal cause for two dark ages in ancient history. She especially attributes droughts and the resulting change in the Nile's flow regime as producing the 2220–2000 B.C. dark age in Egypt. The worst part of this period was the 60 years from 2190–2130 B.C., a time when there were more than 31 kings. Sources from this period show all forms of civil disorder were present: strife between districts, looting, killing; crime was prevalent everywhere, as well as social anarchy and famine. The eyewitness reports tell of death by starvation and describe extensive sandstorms, the advance of sand dunes onto fertile lands, and the exceptionally low levels of the Nile. Many children were killed and vandals were common. The people despaired because they could not control the famine, the drought, and the low water levels. And many governments collapsed because the people could not deal effectively with the problems and were powerless to change the agrarian irrigation economy that was so dependent on the climate and hydrology.

The decline of Rome is quite a different matter. Historians have heatedly argued the causes for centuries and there is no universal agreement on the order of importance. Various theories include: (1) laziness of rulers and lack of involvement in the affairs of state; (2) need to use foreign mercenaries, which increased the tax burden; (3) lack of a stable ruling class with inferior emperors; (4) debilitation by disease, malaria, and lead poisoning from eating utensils and pottery; (5) ineffective military power and leadership; and (6) increasing pressure from invading armies from the north. Most of these causes are linked to sociopolitical-economic factors and relate to the decay of the moral fiber of Rome. Corruption and vice were rampant; there was also runaway inflation, gross social inequities with burdensome taxation, and an increasing deficit spending with higher import than export trade. The power base had become reduced at home, and the political borders had become greatly overextended. Against this backdrop of a disillusioned populace, a case can be made that environmental affairs may have proved crucial and provided the stimulus for the ultimate demise of Rome.

Three different theories have been offered wherein the environment may have been significant. They are briefly summarized here.

1 Climate deterioration The climate in northern Europe and western Asia became drier. This caused fewer game and led to massive migration and population pressure by barbarian hordes. Foodstuffs became harder to grow both in the north and in Italy. With dwindling food supplies, competition increased, and warfare was its outgrowth.

2 Mineral resource depletion The military power of Rome and the expansion of the empire were financed by the silver mines in Iberia (now Spain). Silver production had peaked between 50 B.C. and A.D. 100 when 30,000 tons were mined. The Roman legions were furnishing 30,000 fresh slaves each year to maintain the ranks of the miners at 150,000. However, after A.D. 100 silver production continuously declined. By the third century A.D. coins were bronze and only silver coated, and by the fourth century Rome was operating on a barter economy. During these 300 years silver that was in use declined at a 2 percent rate per year because of abrasion, handling, corrosion, reworking, losses at sea, and burial in graves. Thus, the same events that happened in Greece were being repeated. The glory of Greece coincided with high productivity of the vast Laurion silver mines; but when the silver gave out, so did the Athenian power.

3 Soil degradation One of the reasons for early colonization of southern Italy by the Greeks was the high fertility of the soils and the development of extensive wheat agriculture. However, by 200 B.C. the growing of grain in most of Italy had become unprofitable in many areas. The impoverishment of soils and loss in crop productivity can be attributed to several factors. One reason was the widespread belief that soils go through an aging process which is not reversible. By A.D. 60 the Roman agronomist Columella was showing in his writings that careful agricultural methods could prevent loss in soil fertility, but the damage had already been done. Another reason for lower crop yields was associated with the indebtedness of the agrarian class. Heavy tax burdens, on farmers and also on males, had the effect of reducing family size, and an exodus of males from farms to urban areas took place. The small farmer no longer had the manpower or number of sons to manage the lands with the care and time needed. Manuring decreased and soil erosion increased.

With dwindling grain supplies Rome was forced to import heavily from the North African colonies. This in turn caused tax increases and led to spiraling costs of foodstuffs and other commodities. Many farmers were forced from their land because of the rising costs, so that the small farms became incorporated into larger holdings, thousands of hectares in size. Such large farms were not properly managed, and the impoverishment of soils continued unabated. Thus Rome became a grain importer and lost her agricultural market base. This loss in land value and food, along with the unstable and dwindling labor supply and rampant taxation, led to a snowballing effect of the national economy that weakened the entire sociopolitical structure of the society. Such conditions made Rome ripe for revolution and a helpless victim.

The Modern World

Since the emphasis in most of this book is on the modern world, there is little need to provide details at this time. However, several man-made innovations, which have drastically changed our living patterns and technological base, should be briefly mentioned at this point. Not only have such phenomena as the population explosion and urbanization vastly changed the landscape, but other events such as the nuclear race and the energy crisis have caused many problems and confrontation on both national and international levels. Other, more subtle forms of man's inventiveness in communications (radio and television) and the development of computer technology have led to a total new spectrum of social ventures and recreation, commercial, and scientific breakthroughs. The exploration of space, the landing on the moon, and earth satellite imagery would all have been impossible without the computer technology which allowed calculations to be made in a fraction of the time required by other methods. Thus, our progress in space was speeded up by several decades. Indeed, the second half of the twentieth century has been variously referred to as the Nuclear Age, the Space Age, and the Computer Age.

GEOLOGIC CONSERVATION

Geologic conservation can take a variety of forms. In this section, the terms of our discussion embrace the totality of the human effort to deal with the effective management of the earth's surface and subsurface resources, such as soil, water, minerals, and fuel, in addition to the companion aspects of reclamation and rehabilitation of damaged environments. Soil conservation consists of those methods that are used to maintain balance in the soil profile and prevent its excessive erosion or deterioration. Conservation of natural resources involves all phases of the planning, extraction, marketing, and use of mineral, rocks, and energy-creating materials and features for the long-range utilization by man. Preservation, on the other hand, is that ethic which urges minimum consumption of materials and maximum continuation of natural conditions without change by man. Reclamation and rehabilitation are two different aspects of conservation that deal with man's deliberate attempts to change or restore environments. Originally the term reclamation referred to the changing of previously unused lands for some perceived benefit, such as the reclaiming of wetlands or the reclamation efforts to bring production into arid lands. Also rehabilitation meant the renovation of lands already damaged by man (called "derelict lands" in England). However, this renewal is now also referred to as "reclamation" in the literature and by government agencies.

Conservation Impediments

Throughout history society was very slow to develop conservation attitudes and methods for the environment. The following philosophies and priorities have always acted as deterrents for the conservation ethic.

1   Man is the superior being, and the rest of nature is on the planet for his benefit, use, and enjoyment. Thus, instead of a posture of stewardship, man adopts one as master, and even tyrant.

2   Nature is self-healing. This was an attitude of many early cultures that saw constant renewal and regeneration in the seasons and extrapolated this observation to be true of all nature. Thus, even though man might change or destroy things temporarily, nature would have a rebirth and renewal in time. This was a common attitude adopted by most slash-and-burn societies.

3   Nature is cyclic. This is an extension of the second point above, and it adds the observation that all things progress through a series of stages; but in this context man is powerless to change the cycle. The Greeks and early Romans, as mentioned earlier, unfortunately adopted the conclusion that man need not attempt to husband the soil carefully, because it would age and lose fertility regardless of his treatment. The inevitability of natural trends hastened the depletion of many of the ancient world's environments.

4   Man lives in the present. This is another facet of the argument that man has often adopted. Most cultures throughout history have adopted a "now generation" approach to life. People want all the benefits at the moment and during their lifetime. Some societies did not even have a word in their language which meant "tomorrow." Thus all earth materials and features were extravagantly used and abused because no thought need be given to future generations.

5   Land and resources are plentiful and there are always new frontiers and sources. Such a philosophy was widely held by early American settlers. It was the common practice in the tobacco-farmed lands in colonial Virginia, Maryland, and the Carolinas. When soils were exhausted after several years of crops, the farmer simply moved farther west to plant new crops in virgin territory. Such attitudes have carried forth into the twentieth century in a slightly different form. A common viewpoint considers man so intelligent and inventive that, regardless of the depths to which the earth is despoiled and her resources obliterated, man will devise new materials and develop new strategies for survival that are equal or better than those which have been destroyed.

Early Conservation Practices

Various types of reclamation measures were the most usual type of conservation methods that ancient and early historic man practiced. For example, the first Dutch farms and the manipulation of wetlands date as early as 4000 B.C. (see also page 494). Hillside terracing was another technique used for thousands of years in China and was also a method in use in South America, the Philippines, and other lands hundreds of years ago. A different type of conservation and land reclamation occurred in the Negev desert (now Israel) about 3000 years ago. In this method the ancient farmers used very ingenious systems for growing crops in a desolate and arid area. The soils are largely loess derivatives with a veneer of rocks. In many areas these rocks can be seen piled into mounds and into rows at right angles to the hill contours, with intervening man-made furrows. At the bottom of hills, along the wadis (river courses), there are dams still seen. The purpose of this construction, however, remains a controversial topic. Evenari believes these structures were built for the purpose of collecting and conserving water behind the dams. The small reservoirs fed into canals which irrigated the fields. The contrary view is held by Kedar, who argues that soil, not water, was the limiting constraint to agriculture. He theorizes that the land surface was deliberately changed to increase erosion rates. The resulting silt which was washed from the land moved downslope behind the check dams where it formed alluvial flats that could now be used as level land increments for new arable fields. He cites the ancient city of Ovdat in the central Negev with an agricultural district of about 130 km² that contains about 17,000 dams. Formerly the only agricultural lands were the fluvial terraces and alluvial fans which accounted for 0.4 percent of the area, about 50 ha (hectares, or 125 acres). Such little arable land could not support the population of 6000. After the dam-construction period, 700 ha, formerly devoid of soil, became level and arable land, increasing the amount of farmed lands to 5.8 percent of the area. Thus reclamation brought about a 1400 percent change in tillable soils. The rock mounds in the Ovdat district average 70 cm in height with a diameter of 1.6 m and occupy 2700 ha—more than three times as much area as devoted to cultivated land.

The manner of tilling the soils and the methods to prevent their erosion did not appreciably change until about the eighteenth century when soil conservation techniques started to become more prevalent (see also Chapter 16). In similar fashion little progress occurred in the conservation or management of other resources, including timber, and in mineral and rock industries until the latter part of the nineteenth century. Although the development of parks and other sites for recreational purposes occasionally was considered and instituted, programs, sponsorship, and more universal management, along with such thoughts as wilderness areas, became more popular in the twentieth century.

Some Constrasting Ethics

An **ethic** is a perceived or imposed standard of conduct. Ecoethics are those behavioral attitudes that relate man to environmental mandates, and, like everything that man does and thinks, these attitudes are subject to wide controversy and problems. Since the nineteenth century three competing ethics have dominated the environmental scene: the utilitarian ethic, the preservation ethic, and the conservation ethic. The **utilitarian ethic** is associated with a work and development ethic that identifies man's labor and construction enterprises as the primary attributes of society. To subdue and conquer nature is necessary and the crowning achievement of civilization. Change in the natural world is equated with progress, and glory abounds when new dams are completed, highways routed, rivers diverted, and minerals and oil extracted. At the opposite pole is the **preservation ethic,** which contends that modification of natural areas is an abomination, even sinful. The natural world has scientific, therapeutic, recreational, aesthetic, and ecologic values that should be unmolested, both for today and for enjoyment by future generations. The scientific preservationists want to maintain pristine environments so that ecosystems and the ecology remain in equilibrium. They urge that unique geologic features be saved, and that endangered species be protected by law. The therapeutic preservationists hold that nature is physically, emotionally, and spiritually necessary for the health and welfare of mankind. The recreationists want areas set aside for their hiking, hunting, fishing, picnicking, and camping. Loss of such localities becomes an infringement on their rights to enjoy that part of nature which is to their liking. The aesthetic preservationist holds some common beliefs of those whose priority is in nature therapy. However, they stress the beauty and harmony of nature in sights, sounds, and smells. Desecration of these values constitutes a rape of their senses, and such indignities should not go unpunished.

The preservation ethic received much of its stimulus from the early urban monstrosities that resulted from the Industrial Revolution, although the Roman scholar Varro more than 2000 years ago had described cities as unnatural and corrupting. In the United States revulsion to urban areas insensed such men as Thomas Jefferson and Henry Thoreau. Jefferson wrote,

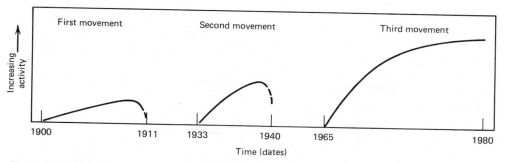

Figure 3-1 Chronology of the conservation-environmental movements in the United States.

"Those who labor in the earth are the chosen people of God . . . The mobs of great cities add just so much to the support of pure government, as sores do to the strength of the human body." Thoreau viewed life as an eternal conflict between industrialism and exploitation of nature on one hand and the serenity, simplicity, and harmony of the natural world on the other.

The **conservation ethic** has undergone an evolution in purpose since its early introduction into the United States by Gifford Pinchot. Although its early flavor was one of self-interest and patriotism, it is now broadly interpreted to mean the careful management of resources for the longest possible time. Thus it embraces not only the appropriate planning strategies for mining, harvesting, and use of raw and living resources in today's world, but considerations for their continual use and supply for the future. The conservationist recognizes there are limits to the earth's resources and urges universal programs of recycling, multiple use, stabilization of population, and a reasoned approach to—and even the abandonment of—a growth-oriented society and "planned obsolescence." We will now investigate the American conservation movements.

CONSERVATION MOVEMENTS IN THE UNITED STATES

The juxtaposition of a variety of events aided the progress of conservation-minded activists (Fig. 3-1). The westward migration of tens and then hundreds of thousands left in its wake in the eastern United States an eroded landscape, denuded hills, and streams polluted by textile mills, lumbering operations, and iron works. Many cities were overcrowded and unsanitary, housing polyglots of peoples and fostering unbridled development. The rapid demise of the American frontier by the late nineteenth century caused many thinking people to start taking stock of what was to become of "the American dream." A growing awareness of the finiteness of land and mineral resources became a special concern of many scientists and some government agencies. Although there were sporadic victories for conservation and land management, such as the establishment of Yellowstone National Park in 1872 and the beginnings of the Adirondack State Park in 1885, the real seeds for the first conservation movement were planted by the passage of the Forest Reserve Act of 1891. Such governmental involvement seemed like a fulfillment of the creeds espoused by such naturalists as John Muir. At this time Muir and Gifford Pinchot, who was to become a driving force behind the early development of the U.S. Forest Service, were good friends. However, passage of the Forest Management Act of 1897 changed the emphasis on forest land management and left no doubt that, instead of being left as wilderness, which was Muir's goal, the forests were destined to be used as tangible and harvestable resources. Thereafter the two men went their separate ways,

Muir trying to awake the country to a preservation ethic and Pinchot championing a utilitarian reasonable-use doctrine.

Many other scientists and citizens would play a part in land management policies, and when placed into a general perspective it can be said that the American conservation movement grew out of firsthand experiences of federal administrative and political leaders who dealt with the problems of the future, the West's economic growth, and the water resource potentials. Grazing lands and forest resources were also important in leading to the first conservation movement, but mineral resources and soil conservation were to become important only in the later movements.

Pinchot in his position as chief of the Division of Forestry in 1898, which became the Forest Service of the U.S. Department of Agriculture in 1905, was very influential with President Theodore Roosevelt. The 1900 inaugural address by Roosevelt contained several of Pinchot's ideas on resource management, and Pinchot was influential in organizing the famous Governor's Conference of 1908. This meeting marked the zenith of the first conservation movement. The massive gathering in Washington assembled all governors or their representatives, members of the Congress and Supreme Court, scores of representatives from private organizations, and outstanding citizens and scientists. Pinchot described the conference as "the first national meeting in any country to set forth the idea that the protection, preservation, and wise use of the natural resources is not a series of separate and independent tasks, but one single problem." Although the conference fell far short of its avowed and hoped for goals, it did aid to foster legislation, such as the Weeks Act of 1911, and to alert the country and some of the rest of the world that conservation of land and water was worthy of governmental consideration and involvement. The conference also vocalized "the gospel of efficient planning" and championed the idea of undertaking an inventory of natural resources that was to be performed by the National Conservation Commission. Within a matter of months the Commission compiled a three-volume report that contained the most extensive cataloging of American resources ever made. The Commission also outlined the specific measures that should be taken by the federal government in leasing public coal and grazing lands, in developing eastern national forests, in containing flood waters, in preventing erosion, in classifying existing public lands in accordance with optimum use, and in repealing outdated land laws.

In spite of an enthusiastic beginning and the apparent national commitment, the conservation movement was soon headed toward oblivion. Although the Governor's Conference spawned such organizations as the National Conservation Congress at their first meeting in August 1909, social scientists dominated the agenda and the conservation theme was greatly expanded, fractured, and dispersed into such topics as the conservation of the morals of youth, the conservation of childrens' lives through the elimination of child labor, the conservation of civic beauty, the conservation of manhood, the conservation of the Anglo-Saxon race, and the elimination of waste in education and war. Furthermore, the National Conservation Congress voted that its purpose was to be "a broad . . . clearing house for all allied social forces of our time, to seek to overcome waste in natural, human, or moral forces."

The political scene did change with the new Taft administration in 1908, but the federal Congress voted few funds to continue the conservation momentum. Pinchot tried to revive the spirit by founding such groups as the Conservation League of America and the National Conservation Association, but such organizations proved to be weak and ineffective and the fervor subsided. Pinchot also tried to start a new journal, *American Conservation,* in February 1911, but it was discontinued after seven months because, as he wrote, "widespread popular demand for a magazine devoted to Conservation does not yet exist." It should also be mentioned that the rising tides and commitments that were a result of World War I and its aftermath riveted public and governmental attention on matters that seemed more relevant and of higher priority. Thus, the first conservation movement had been all but abandoned within a 10-year period. However, another event in 1911, the publication of a pioneer soil erosion study in South Carolina by H. H. Bennett, was to be a harbinger of the second conservation movement.

The stage for the second conservation movement of the 1930s was dramatically set by the 1928 publication of the booklet *Soil Erosion: a National Menace* by H. H. Bennett and W. R. Chapline. A slumbering Congress was finally awakened and took action with an amendment to the Agricultural Appropriations Bill for the 1930 fiscal year. This legislative breakthrough stipulated:

to enable the Secretary of Agriculture to make investigations not otherwise provided for, of the causes of soil erosion and the possibility of increasing the absorption of rainfall by the soil of the United States, and to devise means to be employed in the preservation of soil, the prevention or control of destructive erosion and the conservation of rainfall by terracing or other means, independently or in cooperation with other branches of the Government, State agencies, counties, farm organizations, associations of business men, or individuals, $160,000 of which $40,000 shall be immediately available.

The late 1920s and early 1930s, however, were very difficult and unsettled years. The stock market crash had greatly weakened the economy and created a depression. With the onslaught of severe drought, dust bowl conditions, and ravishing floods, new measures had to be instituted in the husbanding and management of soil and water resources. It was after 1932, when a new Roosevelt became president (Franklin D. Roosevelt), that the new conserva-

tion movement finally took form and substance. Because there was widespread turmoil the government saw the chance to unite many elements into programs that would combine work, stimulate the economy, and still perform the useful service of saving the lands. Thus, totally new directions were given to the protection and repair of lands, and the government for the first time became deeply involved with conservation. These new programs provided the manpower, the Civilian Conservation Corps (CCC) and the Works Project Administration (WPA), along with new emphasis and leadership. This administration saw the beginning of the Soil Conservation Service (SCS) and the Tennessee Valley Authority (TVA).

On September 19, 1933, the Soil Erosion Service was established under the U.S. Department of Interior with a budget of $5 million under the authority of the National Industrial Recovery Act. The agency was invisioned as a temporary expedient, but by 1935 a greatly expanded role was assigned to the operation by passage of the Soil Conservation Act, which established the Soil Conservation Service under the U.S. Department of Agriculture and committed the government to a long-range policy of national involvement in soil conservation. The act states:

It is hereby recognized that the wastage of soil and moisture resources on farm, grazing, and forest lands of the Nation, resulting from soil erosion, is a menace to the national welfare and it is hereby declared to be the policy of the Congress to provide permanently for the control and prevention of soil erosion and thereby to preserve the natural resources, control floods, prevent impairment of reservoirs, and maintain the navigability of rivers and harbors, protect public health, public lands, and relieve unemployment.

The following year Congress was to pass still another historic act of legislation that was a conservation measure of a different sort. Devastating floods occurred in the Mississippi River basin in 1935 and 1936 which provided the stimulus for a vastly expanding role by the U.S. Army Corps of Engineers in the management of major rivers. This Flood Control Act of 1936 established the important concept that flood control must consist of implementation of structures and methods for harnessing surface waters both in the watershed and the river channel. This act stipulates:

It is hereby recognized that destructive floods upon the rivers of the United States, upsetting orderly processes and causing loss of life and property, including the erosion of lands, and impairing and obstructing navigation, highways, railroads, and other channels of commerce between the States, constitute a menace to national welfare. . . . the Federal Government should improve or participate in the improvement of navigable waters or their tributaries, including watersheds thereof, for flood control purposes if the benefits to whomsoever they may accrue are in excess of the estimated costs, and if the lives and social security of people are otherwise adversely affected.

It should be pointed out that this was the strongest statement yet made by the Congress for the mandatory adoption of a methodology that would yield considerations that involved benefit-cost ratios—a strategy that has been adopted in much subsequent environmental legislation. The TVA, another important conservation model (see later discussion, page 243), was another powerful example of government involvement in conservation and other natural resource fields. The conservation movement in the 1930s was primarily in the form of a federal stimulus that initiated work and programs to stabilize soils and water and to manage engineering structures.

But history was again repeated. With the arrival of World War II, the second conservation movement lost its momentum and fervor after 1937. Although there were exceptions, a nationwide appreciation for conservation was not rekindled until the mid-1960s.

The Present Conservation Mood

Several important statutes were passed by the Congress that continued to affirm governmental interest in the types of conservation that had previously been addressed during the intervening years from World War II up to the third conservation movement of the 1960s. Typical of these were the Watershed Protection and Flood Prevention Act of 1954 and the Multiple-use Sustained Yield Act of 1960. (Additional descriptions of these and other acts can be found in Chapters 16, 18, and 21.) However, by the 1960s an entirely different mood was sweeping the United States—a grass roots uprising that put pressures on lawmakers to institute an entirely new spectrum of legislation. This resurgence was not merely concerned with the same traditional problems of soil and water management, but it involved a totally new dimension with involvement in an all-encompassing environmental crusade. For the first time the pollution of the land, water, and air became a major issue. Waste and recycling of resources became rallying cries for many. Another focal point was the legal aspects of fixing responsibility for environmental degradation and instituting fines and injunctions for noncompliance of the new laws on clean air, clean water, and other environmental impact legislation. Thus, the third environmental movement was rooted in the entire environment, exhibiting a deep concern for the essence and quality of life that seemed to be deteriorating everywhere. Although resource management was still crucial, the new arsenal of needs included the health, safety, and welfare of all society.

The Wilderness Act of 1964 was typical of the new awareness within the United States, attesting to the fact that the frontiers had all but vanished, and that the human spirit was something that needed cultivation. Although this act was an extension of some earlier ideas and of the national park system, it sounded a keynote—the new perception that the country was running out of time, space, and resources.

It reinforced the policy that significant parts of our national heritage were in severe jeopardy and would soon be lost unless they were firmly preserved from exploitation and development. We will now briefly explore some of the events and societal changes that caused the third conservation movement to have such a different complexion when compared with the earlier ones.

One event which had profound repercussions was the launching of *Sputnik* by the Soviet Union in 1957. This ushered in what might be called the "space age" or the "space race." It also greatly stimulated and increased industrialization, mechanization, and computerization. Vast sums of money were being spent on what was not a readily tangible and identifiable earthly objective. Of course, there have been many important spin-offs and by-products of these expenditures, but another venture, where even more money was spent, did not have a healthy product—the Vietnam War. This caused great disillusionment and distress with a large part of the population and was especially disheartening to young people. The war effort also caused additional mobilization of resources and their rapid utilization. Few safeguards were mandated, and the push was to produce as much as possible as fast as possible. Bureaucracy swiftly grew unabatedly, and the entire superstructure took on the appearance of a mindless, headless, and senseless dinosaur. Many people felt bewildered and betrayed by the new monstrosities that were emerging through technology, urbanization, and the proliferation of a mechanized society. It was in this jungle of tangled nationalism and utilitarianism that the new ecoethics was born. In a somewhat similar vein, as when Rousseau had stated that the 1755 Lisbon earthquake victims deserved their fate for not living in fields and forests where they would have been saved physically and morally, the new idealists were convinced that the industrial establishment was evil and that a growth-based economy was counterproductive to the sanctity of mankind. Because all aspects of the environ-

ment were under attack, all parts of the environment needed salvation, and the ecological soldiers that marched into this fray were as diverse as the world they were trying to save. The common denominator that united this rainbow of concerns was the goal to make the government more responsive to an environmental era, wherein the measurement of values must consider factors other than the almighty dollar.

Although there were several special-interest national organizations with a long record of environmental concern, such as the Sierra Club, a whole host of new ones sprang up in the 1960s to champion the new causes and to lead the battle against the perceived ills of government and big business. These many groups such as Friends of the Earth, Defenders of the Environment, The Conservation Foundation, The Wilderness Society, along with the older organizations (Nature Conservancy, The National Wildlife Federation, and The National Parks Association) kept a constant pressure on legislative bodies and the general public. A reborn strategy of militancy and confrontation became widespread, and environmental causes found new benefactors in the courts and in the judicial process. More than any other single factor, this activism has been the cornerstone of the third conservation movement and serves to set it apart from the others. It led to countless court battles and the passage of numerous legislative acts, on all levels of government. By the late 1960s the steamroller had become so powerful that Congress passed the National Environmental Policy Act of 1970, and the 1970s were heralded as the "environmental decade." Although complete euphoria does not reign, and the millennium has not arrived, the conservation movement that started in the 1960s still continues and has been highly successful in nearly all theaters of operation. It shows no signs of abating. Indeed, nothing in human history has ever happened to rival this new human attempt to accommodate and develop a rapport with nature.

Perspectives

Since earliest times, human civilization has left its mark on the landscape. The ground was cleared around campsites; the soil was pitted for hearths, sleeping places, and refuse disposal; primitive enclosures were constructed of grass and lumber; and fire, used in clearing fields and driving animals, decimated many forests. Later on ancient civilizations changed flow regimes in rivers, causing increased erosion, siltation, and salinization of soils. Modern man continues to degrade the environment and has added some new twists, both in the magnitude of the changes he causes in the land-water ecosystem and in the type of changes—many of these in the form of chemical pollution.

Just as the type and scope of man's destruction of the environment have changed through history, so have his attitudes toward the natural world. These have shifted from being entirely a predator of wild things to a cultivator of domesticated species. In this process there has been a softening of the dictum of fighting and conquering nature, to one of accommodation and cooperation. Along with this posture, there has been an increased moral ethic that calls for the stewardship of land, water, and resources. Sometimes this ecoethic has gone so far as to admit that it is in man's best interest to live in harmony with natural things, and that they too—both the living and the nonliving world—have some rights of existence. The reaching of this point has not been easy and is not universally adopted or in fashion. It is a conclusion, however, that was learned through pragmatism and the lessons of the preservation ethic.

Human history and progress are intimately linked with the use, and abuse, of the air, water, land, and resources—both geologic and living—of the planet. The level of sophistication with which they are marshaled and employed provides recognition of the heritage and legacy we will leave. Many events shape the destiny of empires and civilization, including resource availability, vitality of the people, and the capacity to deal with nature in the form of climate and hazard conditions. The stages of human development may be described in terms of the resource that is used, such as the Stone Age, Copper Age, Bronze Age, Iron Age, and so on, or in the broader spectrum of the total technology and technocracy of the times, such as the Agricultural Revolution, Industrial Revolution, and Urban Revolution. There are still other large-scale effects that special events and circumstances bring into the cauldron of environmental affairs such as the population explosion, the energy crises, and the conservation movements. Thus it is not surprising that the environment can act as a total focusing point for the evaluation of man and his tenure on the earth.

The Mr. Hyde side of man, wherein he degrades his own living space, has its Dr. Jekyll counterpart that seeks to adjust to the natural world in some form of conservation ethic. In its simplest manifestation, conservation is the beneficial use of earth resources for the greatest long-range period. It takes a variety of forms that range from complete preservation on one end of the spectrum to the deliberate temporary damaging of the environment, but with ongoing plans to rehabilitate the deteriorated areas. Man has periodically assessed his proneness to destroy the environment, and occasionally actions have been taken to mitigate and minimize such tendencies. In the United States when proper recognition was given to the vanishing frontiers, the finiteness of resources, and the fragility of ecosystems, several conservation movements were mounted in the twentieth century. Each one had its own personality and emphasis; the early 1900s was largely aimed at water and forest resources, the 1930s was interested in soil erosion, and the current movement starting in the 1960s deals with the full gamut of ecosystems as they effect the quality of life.

Some optimists are calling our modern times "the age of environmental enlightenment." Only future historians will record whether such a title is apt.

This historical appraisal would be incomplete without mention of the new awareness that "the environment" and its ensuing problems transcend national borders. The mobility of the air masses and ocean currents transport possible pollutants throughout the world. So not only can nuclear fallout affect countries thousands of miles away, but acid rain produced by industrialization can pollute areas far from the original source and contamination site. Societal perceptions and knowledge of physical systems is thereby required for a different type of responsibility, to enhance environmental protection and conservation on the national level, and also to cooperate on projects and programs that are of international scope. Only through such total efforts will human history be able to continue through millenniums of future time.

READINGS

Bell, B., 1971, The Dark Ages in ancient history: Amer Jour. Arch., v. 77, p. 1–26

Bennett, H. H., and Chapline, W. R., 1928, Soil erosion a national menace: U.S. Department of Agriculture Circ. 33, 83 p.

Bryson, R. A., and Baerreis, D. A., 1967, Possibilities of major climatic modifications and their implications: northwest India, a case for study: Amer. Met. Soc. Bull., v. 48, n. 3, p. 136–142.

Chamberlin, T. C., 1908, Soil wastage: in Conf. of Governors in the White House Proc., Blanchard, N. C., ed., p. 75–83.

Coates, D. R., ed., 1972, Environmental geomorphology and landscape conservation, Vol. I: prior to 1900: Stroudsburg, Dowden, Hutchinson & Ross, Inc. 485 p.

Craven, A. O., 1926, Soil exhaustion as a factor in the agricultural history of Virginia and Maryland, 1606–1860: Univ. Illinois Press, 179 p.

Evenari M., et al., 1961, Ancient agriculture in the Negev: Science, v. 133, p. 979–996.

Flawn, P. T., 1966, Geology and the new conservation movement: Science, v. 151, p. 409–412.

Hays, S. P., 1969, Conservation and the gospel of efficiency: New York, Atheneum, 297 p.

Jacobsen, T., and Adams, R. M., 1958, Salt and silt in ancient Mesopotamian agriculture: Science, v. 128, p. 1251–1258.

Kedar, Y., 1957, Water and soil from the deserts: some ancient agricultural achievements in the central Negev: Geog. Jour., v. 123, p. 179–187.

Lowdermilk, W. C., 1943, Lessons from the old world to the Americas in land use: Smithsonian Inst. Ann. Report, p. 413–427.

Roosevelt, N., 1970, Conservation: now or never: New York, Dodd, Mead, 238 p.

Simkhovitch, V. G., 1916, Rome's fall reconsidered: Pol. Sci. Quart. Jour., v. 31, p. 201–243.

Wittfogel, K. A., 1956, The hydraulic civilizations: in Thomas F., ed., Man's role in changing the face of the earth, Chicago Univ. Press, p. 152–164.

Chapter Four
Physical Systems

Yosemite Valley, California. A remarkably glaciated terrane in Yosemite National Park. (Courtesy Robert E. Wallace, U.S. Geol. Survey.)

INTRODUCTION

Geology (literally science of the earth) is becoming increasingly concerned with the study of those processes that relate to the origin of minerals and rocks and how they are changed through time. In spite of the original derivation—earth science—geology is now largely restricted to analysis of that part of the planet known as the **lithosphere**. The broader discipline of earth science includes study of the atmosphere and the hydrosphere in such fields as meteorology, climatology, and oceanography. In its early history, geology was largely descriptive and qualitative. Within the past few decades, however, geology has been transformed into a much more quantitative and mathematical discipline, with extensive use of statistical analysis in all of the subfields.

The Subfields of Geology

All of the separate subdisciplines within geology can play important roles in environmental geology, but some disciplines are more easily and more commonly applicable than others. The different subfields of geology can be divided into those that: (1) deal with materials of the earth: mineralogy, petrology, hydrology, and paleontology; (2) crosscut other subjects: geochemistry, geomorphology, and geophysics; (3)

are specifically related to applications: economic geology and engineering geology; (4) are interested in time: geochronology and stratigraphy; and (5) are involved with dynamics of rock motion: geotectonics, rock mechanics, and structural geology.

Tenor of Geology

Geology is such a large and all-encompassing topic that generalizations about its framework are difficult, because there are always exceptions. As a starting point, however, geology might be described as the science that uses all possible information from other academic disciplines, as well as practical knowledge, to gain a greater understanding of the earth. Such fields as chemistry, mathematics, physics, engineering, and biology contribute basic liaison ideas toward this objective. Specifically, environmental geology emphasizes the linkage of many disparate fields for the common goal of obtaining maximum benefits from the natural system with the minimum disturbance of the land-water ecosystems.

Geology has a kinship with astronomy because it is largely an observational science. Nature is the experimenter and has already, or is currently, performing the experiment. It is up to man to decipher the events, provide a diagnosis, and interpret the clues accurately to discover the scenario of how, why, and when the feature occurred. This is a different philosophy than that of such sciences as physics and chemistry whereby man is generally the experimenter and is mixing various ingredients in order to determine what types of reactions will occur within a controlled set of conditions. Of course, there are many aspects of geology that greatly profit from the experimental approach, where under contrived situations attempts are made to duplicate, or simulate, in the laboratory what nature has accomplished in the "real" world.

Geology has been called the "science of noblemen" because of the character, heritage, and life-style of the early founders of the science. During the formative growth stages of geology 200 years ago, a mix of several necessary factors were required so that geology could prosper in the narrow doctrinaire setting of the times. These factors include the following:

1 Time for observation Most salaried workers labored from dawn to dark to earn a living and they did not have sufficient time, or daylight hours, to make careful field observations. Thus it was largely the wealthy and privileged classes that had ample time to observe carefully the natural world.

2 Time and money for travel Geology is different than other physical sciences, because each geological environment is different. To be a competent geologist one needs to observe many different areas and to compare their similarities and dissimilarities. It is important to discover what conditions are normal, as contrasted to those that are abnormal, and this can only be determined from insights gained from viewing a wide spectrum of localities. Such an approach differs from mathematics or chemistry, because 2 plus 2 equals 4 regardless of where the calculation is made, and the mixing of sodium and chlorine will produce salt whether the operation is conducted in a fancy laboratory or in a phone booth. Thus the early geologists had to have the opportunity, time, and abundant financing to be able to visit a great variety of geologic environments. Walter Bucher was fond of saying, "The best geologist is the one who has seen the most rocks." The early giants of geology were able to see many different rocks.

3 Independence of thought Only those people that did not have to worry about bringing home a paycheck could afford the luxury of disagreement with current dogma. For example, few dared oppose biblical authoritarian ideas such as the Noachian Deluge and the 6000-year time scale for the birth of the earth. It was heretical to differ from such dogma, and a person could be ostracized from society. Thus it took academic mavericks who were fearless and independent of normal society and who could stand up and speak against establishment doctrines. Only those of high position dared such adventurism.

As in all the sciences, geology has the central purpose of cataloging, classifying, understanding, and predicting those phenomena, processes, and features that produce materials and changes on and within the earth. Although some books were written and treatises published of specific aspects of geology prior to the nineteenth century, the first comprehensive book was not published until 1830, written by Sir Charles Lyell. Thus geology is a very young science, the youngest of all the sciences. Another distinguishing characteristic of geology is the time scale and importance of events and materials with which it deals. Like astronomy and other earth sciences, geology must deal with materials that range from the microscopic to those measured in thousands of kilometers, and events that occur in microseconds as well as those that encompass millions of years. The chronology of features, both in relative and absolute dating terms, is fundamental to many subdisciplines and can even be crucial in making policy decisions when we are involved with environmental problems. For example, the siting of nuclear plants is based on the dating of possible seismic events, and the management of floodplains is determined by the statistics of frequency of flooding based on time considerations.

The Habitat of Geologists

The entire earth is the geologist's theater of operations. He can be found on the highest mountains, plumbing the ocean's depths, in the desolate reaches of Antarctica and the Sahara Desert, and even in the heart of major cities. Thus geologists work in vastly different terrains, and they can also count an office or a laboratory their home. Many combine field, office, and laboratory work as necessary for solution of the problems they face. There is also a large variety of types of employment. Geologists may be self-employed and own their business or work as a salaried or commissioned scientist for industry, geotechnical consulting firms, private foundations, and governmental surveys and agencies, or work as teachers and professors in educational institutions. More than half of all geologists are connected with work in the petroleum industry. The U.S. Geological Survey is the largest American employer of geologists, with more than 6000 on their payroll.

Much geologic work continues to be done in traditional ways, but there are several distinct trends that are becoming increasingly important as the science continues to advance. These include: (1) the expanded use of mathematics and the quantification of parameters; (2) the development of models of all types: analog, simulation, digital, conceptual, and physical in attempts to duplicate natural systems; (3) innovation of new types of experimental approaches; (4) increased use of remote-sensing instrumentation to detect changes within the earth as well as orbiting vehicles that produce an array of electromagnetic spectra of features on and near the earth's surface, such as LANDSAT systems, photographic imagery and infra-red, side-looking radar; (5) the analysis of systems from an equilibrium-disequilibrium viewpoint. Boundary conditions, constraints, and limiting stress can all be viewed as geologic thresholds. The threshold concept is becoming increasingly important in geomorphology. The evaluation of when a stimulus will likely exceed the stability index of a system is of vital concern to those who must manage and evaluate environmental systems.

TIME

The earth is calculated to be about 4.5 billion years old, and the oldest rocks found in Greenland and Minnesota dated by radiometric methods are about 3.9 billion years old. Although the rock record is not 100 percent complete, geologists have a good idea of earth history for several billion years. Geological time is divided into different classes (see inside front cover) for the purpose of correlating groups of rocks that occur throughout the world. Although the Precambrian rocks represent 80 percent of all geologic time, they are less well-known than more recent rock histories. The eras are those time divisions based on the complexity of organisms. Thus rocks deposited during the Paleozoic Era contain ancient life forms; sedimentary rocks of the Mesozoic Era have middle-type organisms; and the Cenozoic Era contains modern life forms. The geological periods were named from rocks that were discovered at a specific locality, and the acronym represents some important historical facet, province, or feature of the rocks. Smaller time divisions, such as epochs, were also devised to indicate even finer points of distinction within the period.

Most absolute dating of rocks is accomplished through radiometric procedures. Such a method calculates the ratio of a particular radioactive elemental isotope in relation to its original abundance, or to some ratio with a stable isotope of the same element. For old rocks the potassium-argon method can be used. Potassium-40 transmits to either calcium-40 or argon-40, with a half-life of 1.3 billion years. For rocks that contain organic carbon, the carbon-14 method can be used if the rocks are not older than about 70,000 years. Only this short time interval can be measured because the half-life of carbon-14 is only about 5700 years. Thus any carbon-14 entombed with the sediments has largely disappeared after about 70,000 years. Absolute dating is possible under certain other circumstances—for example, if the materials contain wood and have tree rings, if the rocks have coral limestone that grew diurnally, and if the sediments contain seasonal varve laminations.

The relative age of rocks can be established

Figure 4-1 Intrusion of pegmatite veins into Precambrian crystalline rocks of the Owl Creek Mountains, Wyoming. Photograph taken in Wind River Canyon.

by the use of certain stratigraphic principles. The first, the concept of **superposition**, states that sedimentary rocks are younger than the strata on which they rest. The second, the concept of **cross-cutting relations**, states that an igneous rock is younger than the material it intrudes (Fig. 4-1). The third, the concept of **structural complexity**, states that at a specific locality rocks that are most highly deformed are older than those with minor disturbances.

There are many situations where it is important for the geologist to determine the age of events and the rate of change of processes. Erosion and sedimentation rates must be known for the planning of dams; flood recurrence times are necessary for floodplain management and zoning; and rates of groundwater flow and recharge are important for planning safe annual yield and water use.

MINERALS

Although minerals consist of smaller chemophysical particles, geologists are mostly interested in minerals and the aggregates they form—rocks. **Minerals** are naturally occurring inorganic elements or compounds with definite physical and chemical properties in crystalline form. There are more than 2000 different minerals, but those in six families are the primary rock-forming minerals and comprise most of the mass in the earth's crust; they are: quartz, feldspar, mica, amphibole, pyroxene, and olivine. Since only eight elements compose 98 percent by weight of the rocks—oxygen, silicon, aluminum, iron, calcium, sodium, potassium, and magnesium—it is not surprising that minerals containing these elements are the most abundant.

Minerals can originate in a variety of ways. When molten material solidifies, minerals form by the crystallization of atoms and elements from the ingredients in the melt. Minerals can also form by replacement processes, as in metamorphic rocks. In such environments an older mineral might be unstable and it can change into a new and different mineral that is more in equilibrium with the new set of conditions. In some sedimentary rocks, such as rock salt, minerals form by evaporation of a saline solution and precipitation of the dissolved salts. Replacement processes also occur with sedimentary rocks, wherein some limestones can reform into dolomites (dolostones), and some organic materials such as wood can become "petrified" by replacement of the cellulose with silicon and oxygen, to form agate and other quartz-family minerals.

The geologist's principal interests in minerals are: (1) the way they combine to form rocks, (2) whether they have economic importance, as in the ore minerals, (3) their stability in terms of rates of decay in weathering and soil formation, and (4) their strength and engineering properties for construction purposes.

There are several different schemes for the classification of minerals but the one devised by James Dana, which orders minerals into chemical groupings, is the most usable and widely adopted. This system is largely based on anion or anion-grouping because the negative ions are the dominant particles that determine mineral characteristics and occurrence (page 84). Therefore, the common classes are the sulfides, oxides, halides, carbonates, sulfates, phosphates, silicates, and the native elements. Other less

common classes are the chromates, arsenates, vanadates, tungstates, and molybdates (see also Chapter 5).

ROCKS

With the exception of water, sediment, soil, and organic materials, all of the earth that we can see above and below ground consists of rocks. Deep mines, oil well drilling, and data from earthquake waves which travel through the upper part of the earth further attest that rocks occur to depths tens of kilometers below the earth's surface. Although the ground commonly has a thin veneer of soil, and other unconsolidated sediment and debris, solid rock (bedrock) underlies all valleys, major hills, and mountains. The depth to bedrock is highly variable, depending on the environmental setting. In the desert basins of the American Southwest the thickness of sediments may be several kilometers. Soil as weathered rock (the combined unit is called "regolith") can be more than 100 m, as in Washington, D.C., and certain parts of Georgia, but in hilly or mountainous terrain, rock can be at or near the surface in most places. With few exceptions, such as coal, rocks are formed by aggregates of one or more minerals. Geologists classify rocks on the basis of their origin into three principal groups; they are as follows.

1 Igneous rocks (Appendix B) These rocks form from the crystallization of molten matter (Fig. 4-2). Intrusive (plutonic) igneous rocks originate many kilometers below land surface and are only seen at the earth's surface after extensive erosion has removed the rocks that formerly served as a cover (Fig. 4-3). Granite is the most common intrusive rock. Extrusive (volcanic) igneous rocks form when the molten rock (lava) oozes to the surface, or an eruption causes the molten matter and enclosing rock to be ejected into the air (Fig. 4-4). This airborne debris which later settles to the ground is

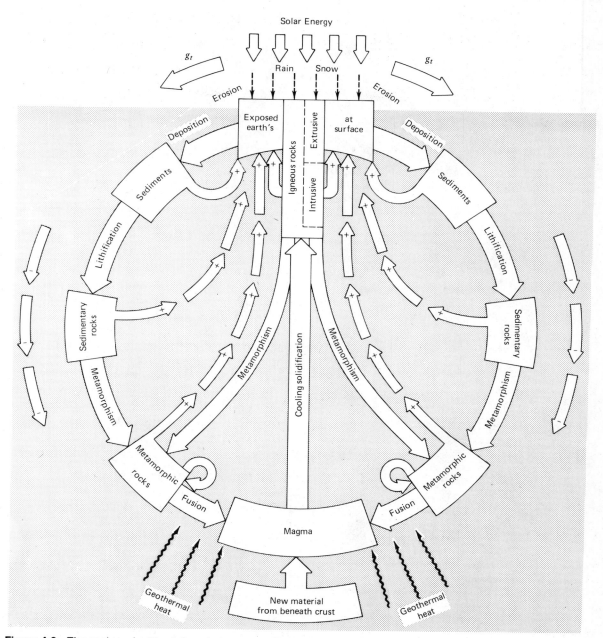

Figure 4-2 The rock cycle. The full cycle is repeated to indicate
earth's spheroidal shape. (After Longwell, Flint, and Sanders, 1969.)

termed "pyroclastic" or "tephra"; volcanic ash is formed in such a manner. The most common extrusive rock is basalt, especially the olivine basalts that occur in the ocean basins.

Figure 4-3 Intrusive igneous dike, Shiprock, New Mexico. The sedimentary rocks into which the igneous rocks were forced have been removed by erosion.

Figure 4-5 A boulder conglomerate in the Wind River Mountains, Wyoming.

Figure 4-4 Devil's Tower, Wyoming. This 250 m high monolith is the remains of an igneous conduit, a modified volcanic neck in which the less resistant rock has been removed by erosion. Note the columnar joints that have formed at right angles to the cooling surface. (Grant Haist, National Audubon Society Collection/ Photo Researchers.)

2 Sedimentary rocks (Appendix C) These rocks form from redeposited broken debris and matter from other rocks that accumulate on the earth's surface (Fig. 4-5). Sedimentary rocks may be either mechanically formed, such as shale and sandstone, from the deposition of solid fragments or may be chemical precipitates, such as rock salt and some limestones. The process of lithification turns the sediments and matter into hard rock by compacting the material more densely and cementing the particles together by the cohesive glue of films of other minerals. Sedimentary rocks are usually stratified (Fig. 4-6) and underlie about 75 percent of the land surface. On the continents they average 2300 m in thickness. Organic rocks such as coal and coral form a special type of organic sedimentary deposit.

3 Metamorphic rocks (Appendix D) This class of rocks originates from the application of heat, pressure, and chemically active fluids on preexisting rocks. Under these changed environmental conditions within the earth the older rocks are transformed into different rocks with new properties. Metamorphism can be caused by the heat and fluids from igneous intrusions and by the large amount of geothermal heat that occurs at great depths. The rocks that result from this type of transformation generally have nonfoliated structures. A second type of metamorphism occurs during mountain building and the introduction of high stress conditions. Such processes produce foliated rocks where there is mechanical alteration of

Figure 4-6 Dipping strata of the Tuscarora sandstone, Lower Silurian age. These hard, resistant beds form the backbone of the Delaware Water Gap, Pennsylvania.

Figure 4-7 Devonian-age metamorphic rocks at Bailey's Island, Maine. These metamorphosed rocks, which are now vertical, were originally in horizontal layers.

constituents with accompanying rock flowage, granulation, recrystallization, and recombination of elements into new mineral assemblages (Fig. 4-7). Metamorphism can be a very intricate process and there are many gradations with igneous rocks. The term "crystalline" is often used to denote some combination of igneous and metamorphic rocks. For example, the shield areas of most continents are composed of crystallines. Most sedimentary rocks have their compositional metamorphic equivalents—for example, shale changes to slate and limestone to marble. Gneiss and schist (Fig. 4-8) are common varieties of foliated metamorphic rocks, and marble and quartzite are usually nonfoliated.

Obviously there is a kinship among all rock types, and they form a constantly renewable system that can be described as the **rock cycle** (Fig. 4-2).

ROCK STRUCTURES

Rock masses are not homogeneous, but instead contain many types of structures and features that interrupt their continuity. The terms **texture** and **fabric** are used to denote the small-scale size, shape, and arrangement of constituent particles and crystals, whereas the term **structure** is used to describe larger scale features created by deformation processes that contorted the rock mass. When the sedimentary rock contains original features such as ripple marks and raindrop impressions, they are referred to as **primary structures**. Diagenesis consists of those processes that involve chemical

Figure 4-8 Manhattan Schist of Ordovician age, Central Park, New York City. The 0.4 m long hammer marks the axis of a syncline. Note the thickening of the individual rock units where the folding is greatest.

Folds and faults are the chief structures produced by deformation in the earth's crust. Folds are bends in the rock and can be classified on the basis of the geometry of the forms (Fig. 4-9) and the relation of the rocks on either side of an axis. An **anticline** is a fold where the strata turn down from a high point, and a **syncline** is a fold with sides that turn skyward from a low point (Fig. 4-10). Other common folds include **monoclines** in which the strata are inclined in the same direction after a change in the rocks (Fig. 4-11), and a **homocline** means all strata are inclined in similar fashion without intervening disruption (Fig. 4-12). For mapping purposes and description of structures, the geologist talks in terms of the **attitude** of the features, and these are defined by the strike and dip (Fig. 4-13). The term **strike** is used to define the bearing or azimuth of a line in a plane on a horizontal surface. Such directions are generally referred to by compass positions, such as a strike N 50° E. When the structure is not horizontal, the angle of inclination from a level surface is called dip, and the amount of dip is always the maximum inclination of that surface (Fig. 4-14). It is expressed in degrees and direction, such as 45° SE. Note that the dip is always at right angles to the strike.

The other common type of structure is a **fracture**, which is a rupture in rock continuity; there are two basic types. A **joint** is a break along which there is little or no differential displacement in the plane of the rupture (Fig. 4-15). A **fault** is a fracture where the two sides

and physical changes before lithification of the sediments. **Secondary structures** are those that result from postconsolidation stress, and this movement of solid rock is termed **diastrophism**.

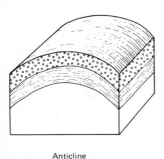

Anticline

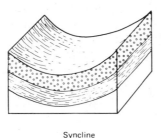

Syncline

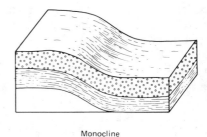

Monocline

Figure 4-9 Terminology for different types of folds.

Figure 4-10 A syncline and anticline in the Folded Appalachians near University Park, Pennsylvania.

Figure 4-11 Monocline in Paleozoic strata of the west flank of the Bighorn Mountains, Wyoming.

have been offset (Fig. 4-16) and the movement occurred parallel to the break. The common types of faults are shown in Fig. 4-17.

Structures come in all sizes, from those that are fractions of a centimeter to those measured in kilometers. Some strike-slip faults such as the San Andreas fault in California have moved rocks that were contiguous millions of years ago a distance hundreds of kilometers (see page 305). Faults also displace rocks several kilometers in the vertical. For example, the eastern front of the Sierra Nevada Mountains has been uplifted by faulting of more than 8 kilometers. On a world wide scale, the displacement of rocks takes on enormous dimensions, wherein entire continents and ocean basins have shifted thou-

Figure 4-12 Homocline (similar dipping rocks) in sedimentary strata of the southern part of the Bighorn Mountains, Wyoming. The rocks all dip east. The ridges are cuestas, with gentle slopes on the dipping beds and steep slopes on the reverse side of eroded rocks.

sands of kilometers. Such grandiose movements are referred to now as **global tectonics**; the earlier term for this phenomenon was **continental drift**, which was first used more than 65

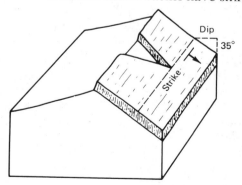

Block diagram

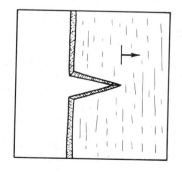

Map view

Figure 4-13 The terms "strike" and "dip" describe the attitude of strata.

Figure 4-14 Measuring the strike and dip of Ordovician strata by use of the Brunton compass, near Richland, Virginia.

Figure 4-15 Jointed Ordovician strata, Virginia. The rocks are dipping toward the road and contain two different joint sets that are at right angles, producing a blocky pattern.

Figure 4-16 This 3 m section of broken rocks shows the complex pattern of faulting. For example, here the center of the triangular rock wedge has moved down, relative to the rocks on either side.

years ago (Fig. 4-18). The structural deformation that is part of the collision of different earth plates creates an entire family of related features, including vulcanism, earthquakes, mountain building, and island arc-trench systems. The understanding of such systems is important in recreating the evolution of the earth's crust as well as the practical consideration for the location of mineral resources.

ROCK MECHANICS

There are many variables that influence the behavior of rocks when they are acted on by forces. The character of rock response is important in the design and construction of all engineering projects and in land management when such hazards as earthquakes and landslides must be considered. Rock mechanics is the discipline that studies the stress-strain (cause-effect) relationships of rock changes, whereas structural geology is mostly involved with the origin, types, and correlation of rock deformation patterns on all scales, from the microscopic to the regional.

The terminology that is generally employed in the description of rock change and failure is shown in Figure 4-19. A behavior is **elastic** if there is complete recovery in size and shape of the body after there is removal of the force. The **elastic limit** is reached if on removal of the force the body is unable to return completely to its original condition. Beyond this a **yield point** occurs wherein permanent deformation occurs in the rock. If the rock ruptures quickly after the yield point, such behavior is described as **brittle**. If on continued application of force the rock mass remains cohesive, although changing in shape, it is **plastic**. Thus a brittle rock ruptures before plastic deformation occurs, and a **ductile** substance is one with a large interval between its elastic limit and the final rupture or failure point. This latter break defines the **strength** of the rock. Many factors influence the strength of earth materials; five of the most important factors are discussed here.

1 Type and amount of force Rocks are stronger under compressive stress than under tension or shear stresses. For example, a typical granite can withstand about 2000 kg/cm² of compressive stress, but will fail with only 40 kg/cm² of tensile stress and will break with 220 kg/cm² of shear stress.

2 Rock type Rock behavior can often be a reflection of the mineralogy, size, and arrangement of crystals or fragments and the type of bonding between the materials. Thus rocks such as slate with preferred orientation of the fabric will respond differently to stress than a granite whose constituents are more homogeneous.

3 Confining pressure The lithostatic pressure or stress caused by rocks being confined by overburden increases with depth. Strength increases under higher confining pressures, so at great depths rocks have a tendency to become ductile. Tests on the Solenhofen limestone showed that under atmospheric pressure of 1 kg/cm² the rock behaved elastically to a compressive stress of 2800 kg/cm² when it failed by rupture. However, when the limestone was placed into a system with a confining pressure of 1000 kg/cm², the rock behaved elastically to a compressive stress of 4000 kg/cm² when it deformed plastically.

4 Temperature Temperature increases with depth and such energy change affects rock behavior, in that rock strength is reduced. Thus rocks at or near the earth's surface deform more readily by brittle behavior whereas rocks at depth deform by plastic response. Experiments on the Yule marble showed that to produce a strain of 10 percent at room temperature a stress of 4500 kg/cm² was required, but when the temperature was elevated to 150°C a stress of only 3000 kg/cm² was required.

5 Time The rapidity with which stress is applied and its duration can be significant factors in the behavior of rock. Deformation can be time-dependent with such substances as asphalt and "silly putty." These materials are brittle when subjected to a short and sudden force, but can flow like a viscous fluid under low pressure when given sufficient time. Repetition can also be a factor in determining **ultimate strength** of materials. A paper clip can be bent back and forth a few times, but repeated bending will cause failure of the metal; thus materials have an endurance or fatigue limit. This ultimate strength is generally about 50 percent the strength for a single stress operation. Another time-dependent rock attribute is **creep**—the almost imperceptible movement of material over long time periods. Creep is a vital consideration in many phases of geology, for example, in the behavior and movement of materials on slopes. When a rock is subjected to a force below its short-time yield point, instantaneous elastic strain occurs (at A, in Fig. 4-20). With continued force a condition known as transient creep occurs with a logarithmically decelerating rate at B. After this a condition of secondary or steady-state creep occurs at a constant rate to point C, when tertiary or accelerating creep is initiated and continues to the ultimate rupture at point D. There is an interesting analog to this behavior pattern in the strain rebound-time curves from earthquake studies. Creep is also a component in such matters as isostatic recovery of rocks after being depressed by glacial ice and in movement of rock into quarries and other rock

Block diagram	Name of fault	Definition
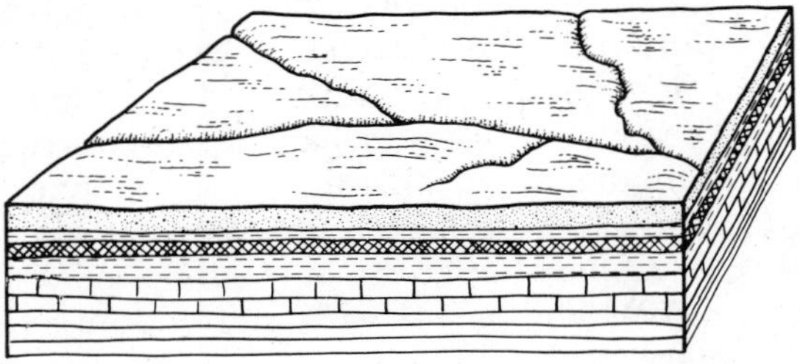		Reference block before faulting. Drainage is from left to right.
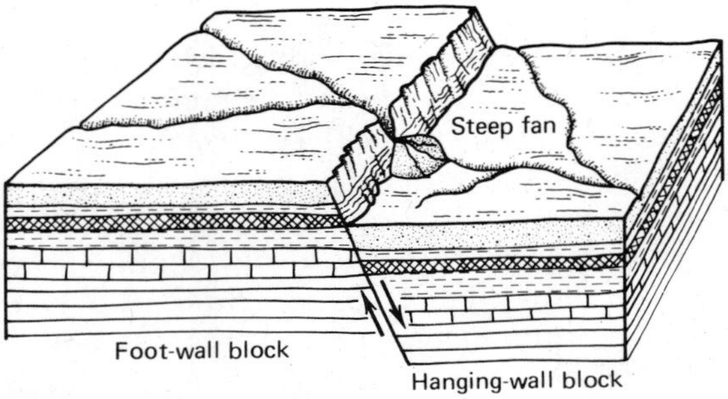	Normal fault	A fault, generally steeply inclined, along which the hanging-wall block has moved relatively downward.
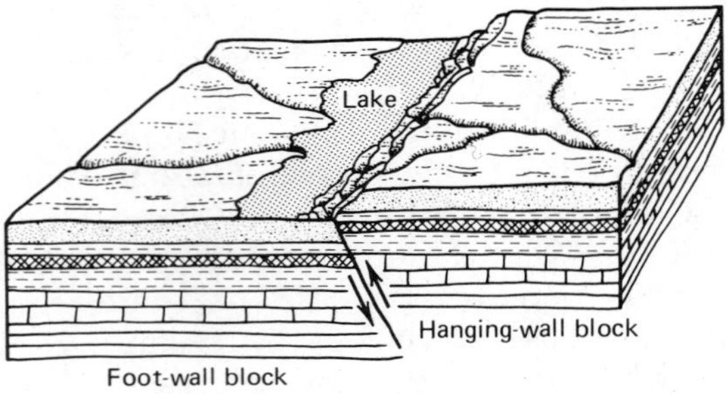	Reverse fault	A fault, generally steeply inclined, along which the hanging-wall block has moved relatively upward. A reverse fault on which the only component of movement lies in a vertical plane normal to the strike of the fault surface is a dip-slip fault.

Figure 4-17 Names and definitions of the principal types of faults. (After Longwell, Flint, and Sanders, 1969.)

excavations. Its consideration is vital in all engineering construction of tunnels and walls of buildings in deep cuts. If the amount of creep is neglected, failure of such projects may occur.

WEATHERING

It was formerly believed that minerals and rocks were vestiges of living matter and lower on the life scale than plants and animals. Such an idea stemmed from observations that the same rock type could be strong and resistant at one locality, but weak and deteriorated at another place. Thus the decayed rock "was obviously old and dead," whereas the fresh rock "was youthful and vigorous." We now know that given the proper circumstances all rocks are destroyed by atmospheric and surface forces. Such changes are referred to as **weathering**—the process whereby solid rocks are changed by physical disintegration and chemical decomposition at or near the earth's surface. Weathering is a

Block diagram	Name of fault	Definition
	Strike-slip fault	A fault on which displacement has been horizontal. Movement of a strike-slip fault is described by looking directly across the fault and by noting which way the block on the opposite side has moved. The example shown is a left-lateral fault because the opposite block has moved to the left. If the ppposite block has moved to the right it is a right-lateral fault. Notice that horizontal strata show no vertical displacement.
	Oblique-slip fault	A fault on which movement includes both horizontal and vertical components.
	Hinge fault	A fault on which displacement dies out (perceptibly) along strike and ends at a definite point.

poor term, because the name weather means short-range air changes; yet it is the year in and year out changes that cause rock breakdown, measured in tens of hundreds of years. **Climating** might be a more appropriate term to describe the concept. Weathering is also often confused with erosion, but differences exist in the type of force and particle motion. Thus weathering should be restricted to passive forces that produce breakdown of rocks in place, whereas erosion is a dynamic force that involves the removal of material from the site and its transportation to a new locale. Weathering should be reserved for the in situ destruction of rocks from the massive to the fragmental state. Weathering performs two important functions: it prepares rocks for easy removal by erosional processes, and it is a necessary ingredient in the formation of soil. The type and rate of weathering is a function of many factors; here are four important ones.

1 Rock type Highly siliceous rocks, such as quartzite, weather much more slowly than rocks

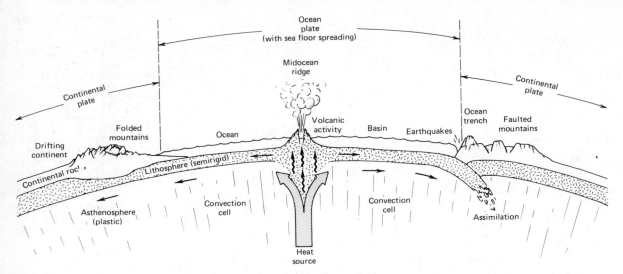

Figure 4-18 Schematic drawing showing the generalized theory of global tectonics. Convection cells driven by interior heat from within the earth drive subcrustal rocks. As new crustal material is formed where the convection cells diverge, the adjacent rocks move away from the ridge. The passive continent is also sent "drifting," whereas the rigid continent forces subduction and the movement of oceanic rocks underneath. Faulting and volcanic activity occur where the lithospheric plates collide.

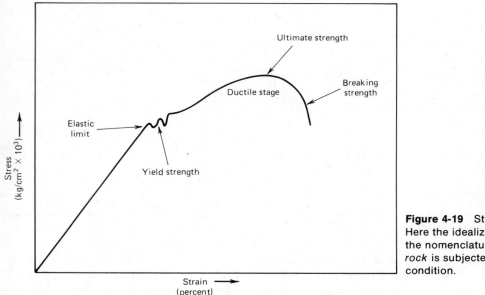

Figure 4-19 Stress-strain curves. Here the idealized relations describe the nomenclature and changes when *rock* is subjected to a stress condition.

with high iron and magnesium content, such as basalt and gabbro. Rocks that are more soluble in water, such as limestone, weather quickly in the presence of moisture.

2 Climate Climate influences weathering because it determines the temperature and availability of moisture, which in turn affect the type of flora and fauna. Climate is so important that

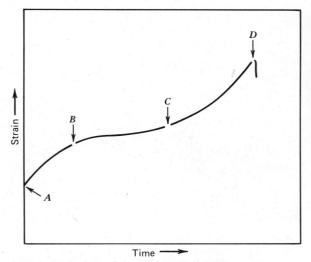

Figure 4-20 Creep (strain-time) curves for rock under a compressive stress condition. *A-B* is instantaneous elastic strain. *B-C* is primary (transient) creep at a logarithmic decelerating rate. *C-D* is tertiary (accelerating) creep that ends in failure at *D*.

Figure 4-21 Sheeting structure in Precambrian rocks, Adirondack Mountains, New York. The joints parallel to the ground surface have formed as a result of the unloading process that relieved the stress when denudational processes removed rocks that formerly covered the area.

its effect on rock destruction can lead to topographic reversals in rock settings. For example, in humid climates limestone commonly occurs in valleys, whereas in arid regions limestone can create a terrain where it forms the cliffs. Climate also determines if frost heaving and ice wedging can be part of the weathering process.

3 Topographic aspect Slope steepness can influence the rate of weathering and soil production, because there is less renewal of fresh rock on flat surfaces than on steep slopes where the combined action of gravity movement and running water constantly cause displacement of debris. Slope orientation is also important since it determines the amount of solar energy that reaches the ground. This is a microclimate effect, and it is common for north-facing slopes to be steeper than their south-facing counterparts in the Northern Hemisphere because of unequal amounts of solar insolation.

4 Flora and fauna The amount and type of living things are reflections of climate, altitude, and latitude. Where there are abundant organisms the acids produced in the living process,

and the movement of animals and microorganisms on and through earth materials, are destructive to the rocks. The growth of plants and trees also provide stress that can accelerate breakdown of materials.

The primary forces involved with weathering can be classed as physical or chemical. The physical or mechanical disintegration of rocks is accomplished by such stresses associated with: (1) unloading—pressure release in rocks by removal of overburden, causing sheeting and massive exfoliation (Fig. 4-21); (2) crystal growth—as with mineral salts; (3) plant growth—the pressure caused by expanding roots; (4) diurnal termperature changes—these can be as much as 50°C, causing alternate expansion and contraction of rock materials; and (5) ice wedging—the crystallization of water in cracks can exert forces up to 4600 kg/cm² in confined areas, which is a stress that can break any rock. The chemical decomposition of rocks (Fig. 4-22), including the biochemical action of organisms, destroys the rocks by producing expansion of the rock mass, decrease in rock and mineral density, and increase in the solubility by such reactions as associated with: (1) oxidation, (2) hydration, (3) hydrolysis, and (4) carbonation. Figure 4-23 shows the typical results of what occurs with alternate wetting and drying and the mineral decay that occurs in the spheroidal weathering process. Figures 4-24, 4-25,

Figure 4-22 Tafoni (honeycomb weathering) in volcanic tuff, Fishlake National Forest, Utah. The weathering of salts has rotted the interior of this outcrop and produced a type of "case hardening" on the exterior surface. (Courtesy Earl Olson.)

Figure 4-24 Ayers Rock, central Australia. This massive monolith shows the effects of differential weathering and erosion. Tafoni is developed along with flared slopes near the ground. This inselberg is 350 m high. (Courtesy Charles Finkl, Jr.)

Figure 4-23 Spheroidal weathering by exfoliation of a basalt boulder, Muddy Gap, Wyoming. The height of the rock is 0.5 m.

4-26, 4-27, and 4-28 depict differential weathering and erosion on a massive scale.

SOIL

The term **soil** means different things to different people. To the pedologist (soil scientist) the term is restricted to weathered rock material capable of supporting rooted plants. To engineers, however, soil is that surface material that can easily be removed by bulldozers without requiring blasting. Such usage by engineers includes, under soil, all unconsolidated or weakly cemented sediments, regardless of whether they have undergone weathering. The geologist regards soil as the upper part of the regolith, which is the total thickness of weathered material resting on bedrock. Most of the land surface of the earth is covered with soil, but exceptions occur on the 10 percent now covered with glaciers, on the desert wastelands that comprise 15 percent of the terrain, and on the bedrock exposures of other surfaces mostly in hilly and mountain regions, which contain about 7 percent of the land. Although the topsoil is generally less than a meter in thickness, the subsoil and weathered zone (regolith) may be more than 100 m in semitropical-tropical environments.

There are many schemes for soil classification. The most comprehensive and the one accepted by the United States and many other nations was formulated in 1960 by the U.S. Department of Agriculture (page 85). This system originally known as the **Seventh Approximation** shows the complexity of soil types. Within the system there are 10 orders, 40 suborders, 120 great groups, 400 subgroups, 1500 families,

Figure 4-25 Bryce Canyon National Monument, Utah. Differential weathering and erosion have occurred along vertical joint directions of the weakly cemented Eocene Wasatch Formation. (Ira Kirschenbaum/Stock, Boston.)

Figure 4-26 The massive limestone that forms the cliff of Sheep Mountain, in the Bighorn Mountains, Wyoming, is slowly retreating because of weathering and erosion of the cliff face. Rockfall of the limestone has created slopes that are now undergoing rock and soil creep.

Figure 4-27 Talus sheet in the Alps, Switzerland.

and 7000 series. The system is based on such factors as soil thickness, maturity, composition, texture, and color.

A well-developed soil will have a series of distinct horizons and this layered system is called the **soil profile** (Fig. 4-29). The top horizon is called A-horizon and contains humus and organic debris that has been thoroughly

Figure 4-28 Talus cones in the Dolomite Mountains, Italy.

leached of many constituents. Below is the B-horizon, which is a zone of accumulation. It contains only the most weathering-resistant minerals. All other materials have been changed, decomposed, and contain particles derived from the other layers. The bottom or C-horizon rests on bedrock and contains the freshest and least-changed constituents. It grades into the unweathered part of the bedrock; some bedrock may project into the horizon, and some rock fragments may be present that are relatively free from weathering.

Soil characteristics and types are of special concern to man because they determine the degree of fertility, the types of crops that can be grown, and the land use practices that should be adopted. Soil properties also influence the amount of erosion and whether water can easily penetrate to replenish the groundwater reservoir. For example, tight and compact soils increase the amount of runoff and erosion, and the low permeability prevents percolation and recharge of underground waters. Soil characteristics also dictate the ease with which the material can be excavated and, when removed, whether the material is suitable for other purposes, such as emplacement into earth-filled dams and construction of artificial levees. Some soils, such as the expandable soils that contain clay minerals (mostly montmorillonite), can cause great damage (see Chapter 13) when improperly used or built on.

The factors that govern soil type, its formation, and its rate of development include: (1) original rock type, (2) climate, (3) topographic aspect, (4) organic process, and (5) time—the same conditions that were influential in determining the weathering process (see also Chapter 16).

GEOMORPHOLOGY AND SURFACE PROCESSES

Geomorphology is the science that analyzes landforms and those processes that produced them. The discipline is primarily interested in the surface processes that include the work of gravity movements, running water, groundwater, coastal forces, glaciers, and wind. These are known as **exogenic processes;** they all sculpture the land surface by **erosion, transportation,** and **deposition** of debris. However, the geomorphologist must also consider the **endogenic processes** of vulcanism and diastrophism, because these forces create new lands which then become subjected to the erosional forces (Chapters 9 and 10). The composite landscape and the landforms that comprise the earth's surface are a reflection of the battle that ensues between these two opposing sets of forces—the exogenic and the endogenic—because one group of processes attempts to tear down the lands (Chapters 11, 12, 15, and 16), whereas the other group of processes tries to build them up. Over long geologic periods these two combatants are closely matched, and the lands are in a state of balance. If this were not true, all land would have been reduced to below sea level conditions hundreds of millions of years ago.

Water

Water is such a special commodity that it deserves to be mentioned in an unusual context. Whoever named our planet "earth" did it an injustice. It is a misnomer because it is the abundance of water on our planet that makes

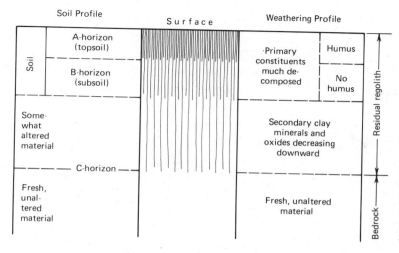

Figure 4-29 Idealized section illustrating the relation between a soil profile and a weathering profile. Because the section is idealized, no vertical scale is given. Soils are generally thin, whereas the thickness of weathered regolith may amount to tens of meters. (After Longwell, Flint, and Sanders, 1969.)

it unique among other members of the solar system. Water covers 71 percent of the earth's surface, and if all the land masses were dunked into the ocean there would still be enough water to cover them to a depth of 4 km! Furthermore, water is the basic ingredient for life, and 75 percent of the human body consists of this fluid. Thus our planet might more appropriately have been named *water, aqua, hydrosphere* or some similar derivative. Water is also the catalyst that aids in keeping the rock cycle churning; it is vital in the character of the geologic cycle (Fig. 4-2) and the entire theme of the hydrologic cycle (Fig. 4-30). Movement of water keeps the atmosphere, lakes, rivers, groundwater, and oceans in balance. An understanding of these relationships is a prerequisite to a thorough knowledge of geomorphology.

Until the seventeenth century human knowledge of the movement and balance of water regimes was sparse, and mostly incorrect. Then in 1670 Pierre Perrault showed that there was sufficient rainfall in France to account for the amount of streamflow in the Seine River, and in 1693 Sir Edward Halley documented, on the basis of studies in the Mediterranean area, that evaporation from marine waters produced clouds that carried the water vapor necessary in the formation of rainfall. Thus were born the important ingredients that would harmonize

into the hydrologic cycle. The movement and behavior of water throughout this system is necessary to understand not only the local level but also the global level pattern. This is especially true as man contemplates the deliberate action of attempting to change weather patterns, and man's inadvertant molestation of climate wherein he accelerates the desertification process (see also Chapters 8, 14, and 17).

GRAVITY MOVEMENTS

The downslope movement of earth materials caused by the gravitational force is universal and the most constant and pervasive of all geomorphic processes. Although gravitational attraction is directed toward the earth's center, all sloping surfaces have downhill components of this stress. Gravity movements come in a wide range of types and scales. The small local movements that affect single slopes are commonly referred to as **mass movement** or **mass wasting.** When the scope includes large land areas, or even entire mountain ranges, it is called **gravitational sliding.** The force of gravity also provides the energy to cause water to flow in streams, ice to move in glaciers, groundwater to develop hydrostatic pressure and even-

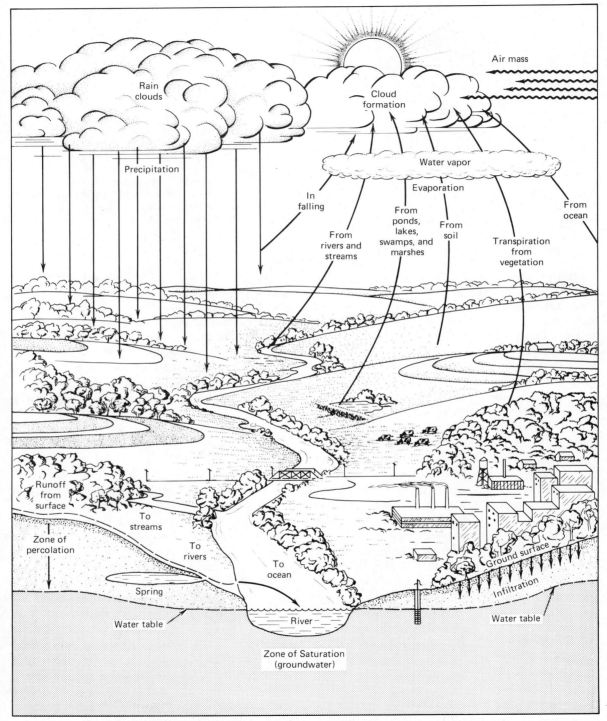

Figure 4-30 The hydrologic cycle. (After U.S. Department of Agriculture.)

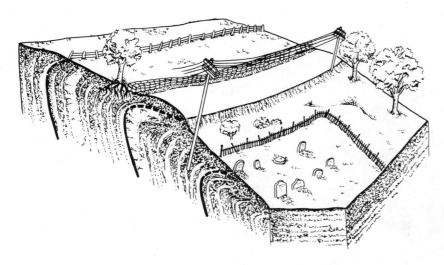

Figure 4-31 Effects of creep. (Modified from C. F. S. Sharpe, 1938, and Longwell, Flint, and Sanders, 1969.)

tually move to lower elevations. In slightly different ways gravity is also the force that produces tides and movements in large water bodies, and it also affects air masses and movements within the troposphere.

Each type of earth material has its own angle of stability, or angle of repose, at which the material is in equilibrium for long periods of time. Resistant bedrock can maintain vertical angles for thousands of years, whereas friable unconsolidated material such as sand and gravel have stable repose angles of about 35°. Such slopes are maintained because the friction and binding force among the grains exceeds the force of gravity for short time periods. For longer time spans, however, there is a downslope movement of all materials, and such motion can be described in various terms that depend on the type and rapidity of motion and the type of materials involved. There are three basic classes of gravity movement: creep, landslides, and subsidence.

Creep

Creep is the nearly ubiquitous, imperceptible, slow gravity movement of surface material (Fig. 4-31). It can involve different types of material that comprise the regolith, and each material

and environmental setting has its own nomenclature. It is called **earth** or **soil creep** when the weathered horizons of the soil profile are involved; **rock creep** when fields of rock are in motion; and **talus creep** when rock fragments at the foot of a cliff are involved (Figs. 4-27 and 4-28). In mountain areas the talus areas may feed extensive rocky masses that spread down from the base and take on the character of flowage motion. Such features are called **rock glaciers.** Another creep variety is **solifluction,** which is most common in areas of permafrost. In such cold regions only the top of the soil thaws during summer, creating a lubricated base on the frozen subsoil which allows for flowage of the surface materials (Fig. 4-32). A general term used to describe gravity-displaced regolith is **colluvium,** in distinction from **alluvium,** which is stream-deposited sediment.

There are several surficial processes that facilitate downhill movement of material: (1) frost heaving—ground frost uplifts soil particles during ice crystal formation at right angles to the cooling surface and, on melting the fragment, moves in the downhill direction; (2) wetting and drying—this produces expansion and contraction of materials and is especially important in clay minerals; (3) heating and cooling—this also produces volume changes with a down-

Figure 4-32 Solifluction lobe in Brooks Range, Alaska. (Courtesy William McMullen.)

slope vector; (4) rock disintegration—when particles weather from a parent rock their fall is downhill; (5) decay of organics—when organisms rot, voids are left that become filled with material from upslope; (6) growth and movement of organisms—microorganisms, worms, burrowing and trampling animals, all displace particles downslope; and (7) solution of material—forms voids that are filled from above.

Landslides

A much more spectacular display of gravity movement occurs when the resistance to movement of materials builds up for a long time and is then suddenly overcome; failure thus occurs, resulting in rapid dislodgement of surface materials. The creation of this landsliding phenomenon may be either natural or man-induced. Landslides involve displacement of material, either regolith or bedrock, and the type of motion may be falling, sliding, or flowing, or any combination (Fig. 4-33). Landslides com-

prise a large array of movements, sizes, and forms and are more thoroughly discussed in Chapter 11.

Subsidence

Subsidence is the general term applied to gravity movement when there is no free face and the displaced material moves essentially vertically. When the rupture occurs, suddenly producing rapid motion, the term **collapse** is used; when the downward moving material is slow and related to man-made structures, the term **settlement** is commonly employed. Subsidence can be natural or man-induced, be slow or catastrophic, in a large range of sizes and localities, and be triggered by a variety of causes. Further discussion of subsidence occurs under the appropriate headings connected to its development, such as sinkhole formation, underground mining, withdrawal of subsurface fluids, and surcharge of loads on the ground (see Chapter 14). Subsidence is becoming an increasingly important factor in land use planning. Its occurrence is becoming more common, and damages in the United States each year are in excess of $2 billion.

RUNNING WATER

The term **running water** is used to denote water in fluvial systems as distinct from the standing water of ponds, lakes, wetlands, and oceans and the groundwater below land surface. The study of running water includes sheetwash, streams, and the drainage basins in which they move. Knowledge of water budgets and stream regimes is a vital component in watershed management. Thus the factors involved in the hydrologic cycle—such as evaporation, transpiration, infiltration, precipitation, and runoff—are necessary adjuncts to the study of running water. For example, the amount of water available for surface runoff is about equal to the amount of

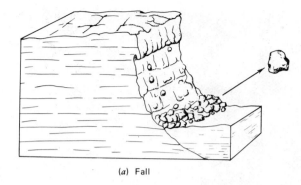

(a) Fall

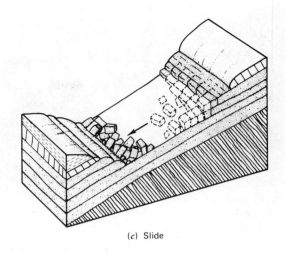

(c) Slide

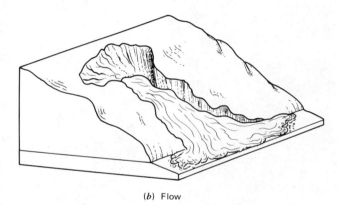

(b) Flow

Figure 4-33   Different types of gravity movement. (After drawings by Mary Ryan.)

(d) Slump

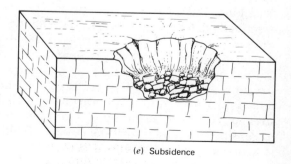

(e) Subsidence

precipitation less the amounts lost by evaporation, transpiration, and infiltration. The average precipitation of the United States is about 76 cm/yr. Nearly two-thirds is lost by evapotranspiration processes, and the remaining third is available for runoff and groundwater increments.

Running water is the primary sculpturer of landscapes. The term **stream** is the general word used for all sizes of channelized flow such as brooks, washes, creeks, and even major rivers. Furthermore, a stream has a definite drainage basin, whereas smaller scale downslope water paths are referred to as rills and gullies (Fig. 4-34). They rarely have a well-defined

Figure 4-34 Shoestring gulleys and rills in the Palouse soils of Oregon. (Courtesy Charles Finkl, Jr.)

Figure 4-35 Grand Canyon of the Yellowstone River, Yellowstone National Park, Wyoming. Yellowstone Falls can be seen in the background. The river has cut its valley in volcanic rocks that are largely pyroclastics (tephra). Such V-shaped canyons can be described as "youthful."

watershed divide, but contain steep sidewalls, usually occur in unconsolidated materials, and are above the water table. Streams in arid terrain are also generally above the water table, have intermittent flow, and are known as **influent** (because their water recharges the groundwater). Streams with continuous flow that intersect the water table are **effluent,** because groundwater discharges into the stream.

Streams are significant in landscape development and carve the valleys in which they reside (Fig. 4-35). Movement of water and debris into the stream comes from the valley walls by sheetwash and gravity movements. At any given location such transportation exceeds that of the stream (Fig. 4-36), but of course the stream is a long-range transportation system and provides the vehicle for the exodus of all material in the basin (except for possible wind deflation). The planimetric form of streams is described in terms of whether the channel is straight, braided, or meandering (Figs. 4-37 and 4-38). Such factors as gradient rock type, water amount, and geologic history determine stream characteristics. Chapters 12, 13, 19, and others provide additional information about streams. However, at this time, it is appropriate to consider the environmental setting in which streams occur—the drainage basin. For management purposes, the study of running water

and the receptical surrounding it are essential to policies that deal with fluvial systems.

Drainage Basins

Drainage basins, or watersheds, reveal important details about streams and their geologic and geomorphic habitat. The form of the basin is often a reflection of the underlying geologic structure of the rocks. Thus, the drainage pattern can reveal the conditions that have prompted streams to behave in a particular manner. A

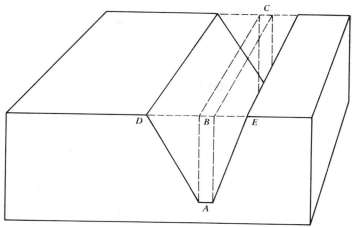

Figure 4-36 Comparison of valley erosion processes. The vertical column *ABC* represents direct erosion by the stream in the valley. The triangular sections *ABD* and *ABE* contained material that was eroded and moved downslope by gravity and sheetwash processes.

Figure 4-37 Stream meanders, Yellowstone National Park, Wyoming.

system of elongated parallel streams may denote uniformly sloping land, whereas angular stream patterns generally develop where there is structural control by folded strata (trellis drainage) or by joint systems (rectangular drainage; Fig. 4-39). When the underlying materials are either homogeneous, as in a granitic terrane, or are horizontally bedded, a dendritic pattern will result. There are many other drainage patterns that can be developed under still different circumstances: radial patterns occur on conical hills and mountains such as isolated volcanos; centripetal patterns occur in structural and other depressions; annular, or ringlike, patterns can be found in such areas as Black Hills, South Dakota, where the sedimentary rocks have been domed by intrusive uplift of the mountain core. Therefore, evaluation of drainage patterns can

Figure 4-38 Entrenched meanders of the San Juan River, Utah.

be extremely useful in diagnosing the geologic character of the fluvial system.

In 1945 a classic paper by Robert E. Horton

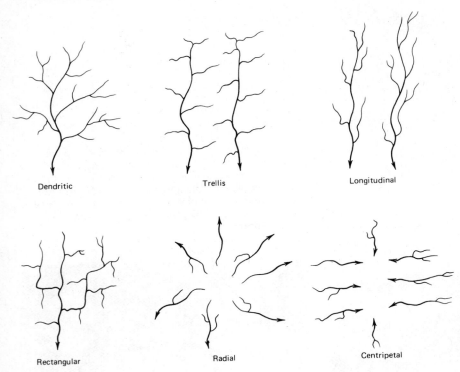

Dendritic

Trellis

Longitudinal

Rectangular

Radial

Centripetal

Figure 4-39 Drainage maps showing the geometry of different stream patterns.

ushered in a new era, and the subdiscipline it fostered became known as **quantitative geomorphology.** Its emphasis was on the quantification of streams, hillslopes, and drainage basins. Prior to its inception, relative and qualitative judgments were made about fluvial systems, and they were often described in judgmental jargon, such as being youthful, mature, or in old age. The new framework, however, started with an ordering system for streams (Fig. 4-40), whereby the fingertip tributaries are called first-order streams (or channels). When two, first-order streams join a second-order stream is formed; and in similar manner when two, second-order streams join a third-order stream is formed. Such a system provides a framework for comparison and statistical analysis of many features of drainage basins. Thus the numbers, lengths, and gradients of streams can be evaluated on a systematic basis. For example, third-order basins in dissimilar terrains can be compared, and

the amount of difference calculated and statistically treated.

Once the pure numbers have been obtained a series of ratios and equations can be used that provide additional meaning and insight into basin properties. Drainage density, the ratio of total stream length to basin area, provides a useful comparison measure. For example, it is possible that two different third-order basins have vastly different areas. Although there can be several reasons for such a discrepancy, in general coarse-grained rocks and soils contain fewer streams per unit area than fine-grained rocks and tight soils. Thus, a sandstone third-order basin would be larger than a shale basin of similar order. Such factors as amount and type of vegetation and climate also influence drainage density. Thick vegetation produces basins with low-drainage density, whereas dry regions generally contain basins with high-drainage density.

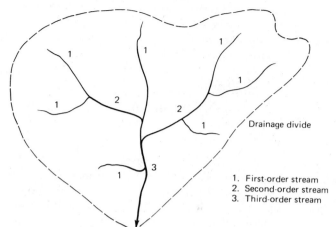

1. First-order stream
2. Second-order stream
3. Third-order stream

Figure 4-40 Stream number and ordering system for a third-order drainage basin. (After Strahler.)

Many other indices are available to the quantitative geomorphologist. The distribution of basin relief and elevations can be important in many ways, as in hydrologic budget studies. One way of determining this relationship is by a calculation of the hypsometric integral, a method that relates the distribution of relief to the areal extent of different elevations. A similar technique is to determine the basin relief ratio *E*, where

$$E = \frac{\text{mean elevation} - \text{minimum elevation}}{\text{maximum elevation} - \text{minimum elevation}}$$

Data obtained from these analyses are also helpful in relating the amount of denudation that has already been accomplished to the residual land mass that is currently undergoing dissection.

Stream relief and basic ruggedness are other parameters that are calculated by quantitative methods. Such numbers determine the amount of energy in the system and provide a measure for degree of topographic variations.

$$S_m = \frac{\text{basin slope} \times 1000}{2 \times \text{drainage density}}$$

where S_m is mean stream relief in meters, basin slope is average slope in percent, and drainage density is the ratio of total stream length in kilometers to the basin size in square kilometers. Amount of sediment yield is related to these factors. The shape of the drainage basin is an important factor that determines the timing and character of sheetflow and tributary flow into a principal stream. For example, the flow regime in a dendritic basin is much different than in a trellis basin. This form determination can be expressed as a circularity ratio (ratio of basin area to the area of a circle with the same perimeter as the basin): $R_c = A_b/A_c$, or as an elongation ratio (ratio of the diameter of a circle equal in area to the basin to the maximum basin length).

COASTAL SYSTEMS

Of all physical environments, the coastal zone is probably the most complex. The reason for this is that processes involved along the shoreline are highly variable and of different types. This land-water interface is subjected to forces that operate on the land as well as those generated in coastal waters and even distances thousands of kilometers away. Thus, streams and gravity movements can influence coastal systems as well as the lake or marine waters themselves.

None of the geomorphic processes contain so many hybrid forms or forces as do coastal features. The shoreline is a conglomeration of landforms that at the present time show: a rising sea level; water and sediment from fluvial systems; gravity slope movements; tidal fluctuations and currents; waves; tsunamis; longshore currents; and oceanic currents. This array of processes produce complex interactions and result in vastly different beach conditions.

The beach and tidal zones of coastal areas differ from features associated with most other geomorphic processes, such as gravity movement and fluvial systems. This is because in the coastal environment the forces may change direction and can cause sediments to be moved in all directions, whereas materials in rivers and on slopes always move downhill. For example, in the foreshore area (page 477) water can move up and down the slopes as well as sideways.

Processes that act in the coastal area produce landforms that change more constantly and regularly than in other environments. This is especially true for beaches, since these features can change twice daily due to differences in tidal heights. In addition, many beaches exhibit large changes in width, slope, type of material, and sand budgets when the winter season is compared with the summer. In most instances, the winter beach is characterized by erosional processes, whereas in the summer the beach generally undergoes a period of buildup and deposition.

When all of the above factors are considered it is not surprising that there is great variety in coastal features, and this dissimilarity has produced many classification schemes that attempt to organize shoreline qualities into a coherent system. The classification proposals are based on the following types of considerations.

1 *Kind of material.* If unconsolidated, beaches can be contrasted on basis of texture, such as muds, sands, and shingle. If consolidated, beaches may be absent, and the land-water interface consists of bedrock. Such rocky shorelines may be combined with those composed of organics, as in coral reefs.

2 *Shoreline configuration.* The shape of the coastline is recognized when such terms as deeply embayed, straight, fiord, ria, barrier island, spit, bar, and so forth, are used to describe the coastal area.

3 *Dynamics or relative movement of sea level.* The apparent rise or fall of the land surface can be described in terms of an emerging coastline for a rising landmass or a submerging coastline with a relative lowering of the land.

4 *Processes that control the coastal environment.* Land-base influences may include rivers that form deltas, as well as volcanic activity and faulting. Organic processes may also control the properties of some beaches (coral reefs). The majority of beaches, however, will respond to forces from the seas.

5 *Combinations of influences.* Many shoreline classifications provide for compound and complex types. Thus they are blends of two or more variables, interacting in the coastal environment.

Important considerations required for appropriate management of the coastal corridor would produce still a different type of classification. This would be based on factors such as: (1) the geologic-geomorphic history of the area, (2) the erosion-deposition flux, (3) subaqueous profile, and (4) characteristics of the seas at and in the vicinity of the shore (see Chapter 15 for further information on coasts).

GROUNDWATER

Groundwater is more omnipresent than surface water, as it occurs under most of the earth's surface. It is located in that part of the lithos-

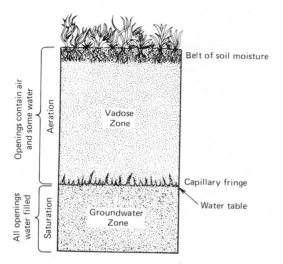

Figure 4-41 Cross section of earth materials showing underground water characteristics in the two prominent zones: the vadose or aeration zone and the groundwater (phreatic) or saturation zone. Except for phreatophytes, vegetation depends on soil moisture and other water above the water table. Capillary fringe height depends on soil type and climate and may range from a few centimeters to 3 m.

phere below the water table where all cracks and materials are saturated (Fig. 4-41). The groundwater zone is under the vadose zone (zone of aeration), where openings contain air and occasional water. The depth to the water table is highly variable and depends mostly on climate. In the humid east of the United States, the depth is rarely more than 15 m and averages about 6 m. In the arid and semiarid parts of the country, however, depths to groundwater are often more than 30 m and may range to 80 m or more. This does not mean that just because there is rock saturation the groundwater can be obtained or used readily. The type of rock and the tightness of fractures may prevent economic pumping from the groundwater zone. Darcy's law is a useful equation that relates the velocity of groundwater flow and the permeability of the host materials. It can be stated in the form $V = P\dfrac{h}{l}$, where V is velocity, h is the difference in elevation of the water pressure head, and l is the length of flow; P is a coefficient of permeability. Thus h/l expresses the hydraulic gradient. In most aquifers, groundwater movement is very slow when compared with stream velocity. Flows of 1 to 15 m/yr are common. The human use and management of groundwater are discussed in several chapters (see Chapters 14, 19, and 20).

Groundwater is a fundamental component in the hydrologic cycle, but its origin was not understood in the ancient world. The Greeks believed the mountains were hollow and that water moved through the openings by some type of poorly conceived spongelike action from the depths or from under the oceans. Because groundwater cannot be seen under normal circumstances, its movement and occurrence have led to a number of problems. Groundwater and surface water relations are shown in Fig. 4-42. As indicated in Chapter 21, groundwater law is very antiquated because it is based on oudated ideas.

In a way groundwater processes are similar to those of the wind—one does not see it directly, only the effects it produces. Groundwater is important to the geologist because it: (1) performs work in the erosion, transportation, and deposition of material; (2) is an essential part of the hydrologic cycle; (3) supplies plants and animals (including humans) with a significant part of their water requirements; (4) influences growth of vegetation and is thus an important factor in determining the amount and type of erosion by surface processes (such as gravity movement, streams, and wind).

Groundwater is both a cementing medium that can bind sediments together into a rock and a destroyer of rocks, especially the carbonates and sulfates. The erosional work of groundwater, with few exceptions, is primarily through the solution of rocks. Such a process enlarges rock openings into caves and caverns (Fig. 4-43) and also results in creating a pitted surface with sinkholes and other depression-type topography (Figs. 4-44, 4-45, and 4-46). In

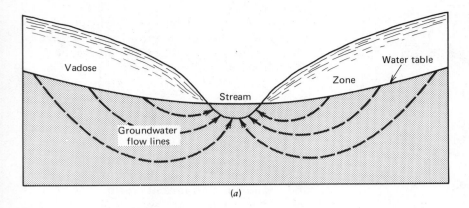

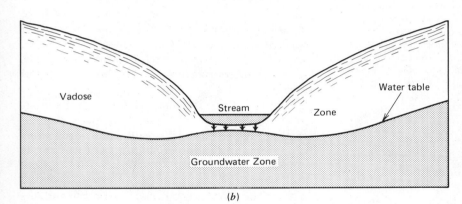

Figure 4-42 Stream-groundwater conditions in two different climate environments. (*a*) Effluent system with groundwater recharging a stream. Such conditions produce permanent streams in humid climates. (*b*) Influent system with streamflow recharging the groundwater zone by percolation in an arid region. Note the water table is below the channel, but that inflow from the stream produces a "groundwater high."

some areas, as in the karst region of Yugoslavia, groundwater processes dominate the landscape and produce a bizarre range of landforms; the term **karst** is used for all terrains where groundwater erosion has created subsidence or collapse features. Most areas are composed of limestone and, less frequently, dolomite.

Another unusual groundwater action occurs in areas with high geothermal gradients. Hot springs and terraces may form at such localities (Fig. 4-47), and under even more rare conditions geysers may form, as in Yellowstone National Park and in New Zealand near the city of Rotorua on North Island.

Groundwater systems can be divided into two main groups. The most common is an unconfined system in which groundwater movement bears a similarity to electric flow through a homogeneous medium. The flow paths are arcs that are a response to a gravity pressure head (Fig. 4-48). Under special conditions groundwater occurs and moves in a confined, or artesian, system. Here the groundwater motion largely occurs in an aquifer with high permeability, such as a sandstone. This stratum is part of a sandwich with upper and lower beds that are impervious and form an aquitard, such as shale.

Figure 4-43 Active stream passage at base level. The tubular cross section is typical of passages that have formed at or below the top of the phreatic zone. During periods of large discharge, the water completely fills the passage. Mammoth Cave, Kentucky, in Mississippian-age limestone. (Courtesy Arthur Palmer.)

Figure 4-45 Sinkhole in Mississippian-age limestone, West Virginia. (Courtesy Arthur Palmer.)

Figure 4-44 Karren topography on Cretaceous limestone, Alps, Switzerland. Such karren, or solutional elongated features, have developed along the joints. (Courtesy Arthur Palmer.)

Figure 4-46 Collapse sinkhole in Mississippian-age limestone, southern Indiana. This opening leads into a cave. (Courtesy Arthur Palmer.)

Figure 4-47 World's largest hot water spring, Thermopolis, Wyoming. This system of terraces is formed of travertine ($CaCO_3$) deposited by the surface discharge of hot groundwater.

GLACIERS

Periods of glaciation are nearly unique, since there have been only six or seven ice ages during the almost 4 billion years of the earth's history. Each period lasts a few million years, so extensive glaciation can be viewed as a catastrophic event. Although the work of glaciers has not been as constant as other geomorphic processes, it is a vital consideration in today's world because 10 percent of the lands are still ice-covered and as little as 18,000 years ago ice covered 30 percent of the continents. Glaciers not only affect lands over which they move, but

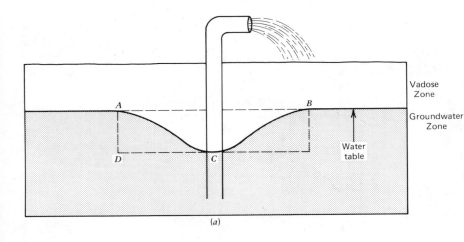

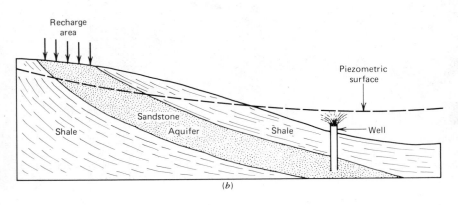

Figure 4-48 Conditions for a well in two different groundwater systems. (*a*) The pumping of a well in unconfined groundwater. When the well is pumped, a cone of depression, *ABC*, develops around the well as the water table is lowered and water drains into the well bore. *AD* represents **drawdown,** or the amount the water table is lowered. (*b*) The artesian or confined aquifer conditions. This includes a permeable aquifer with impermeable (aquiclude) rocks enclosing it. The recharge area acts as a giant plunger, forcing water under great pressure into the aquifer. Water in the well rises to the elevation of the piezometric (pressure) surface.

also influence contiguous areas and in several instances produce global changes. The waxing and waning of ice sheets cause great sea level fluctuations. During the last glacial maximum, there was a lowering of marine waters by 130 m, and if all ice were to melt instantaneously today, sea level would rise about 75 m. Such changes greatly alter sedimentation and erosion rates in coastal environments.

No other single geomorphic process has as wide a range of topographic effects as those produced by processes associated with glaciers (Figs. 4-49 to 4-59). Not only does erosion and deposition by the ice create many different types and scales of landforms, but the side effects of meltwater streams and lakes and wind action also produce another family of sedimentation and terrain features. Glaciers in mountainous regions (Figs. 4-49 and 4-50) cause erosion that can be measured in hundreds of meters and up to more than 1000 m (Yosemite Valley was overdeepened more than 1000 m by ice erosion). They alter the landscape to one that is much more rugged and angular than under a fluvial regime. Continental scope glaciation of ice sheets also erodes, but the dominant theme is to leave the land surface with less relief than before (Fig. 4-54). The large erosional features such as the Great Lakes and the Finger Lakes are unusual, but the combination of erosion and deposition does create thousands of lakes as in parts of Canada, Minnesota, and Finland. The magnitude of erosion by ice sheets can also be enormous. The Finno-Scandinavian glaciers eroded and deposited so much material that if the deposits now resting in northern Europe were reallocated they would fill in the Baltic Sea and all of the thousands of lakes, and there would still be 25 m of material left over.

Average ice thickness in Antarctica and Greenland exceeds 2100 m and reaches a maximum of more than 4000 m in Antarctica. The bulk of the world's ice, 89 percent, occurs in Antarctica; 10 percent is in Greenland and the balance in mountain glaciers, ice fields, and small ice caps (such as on Iceland) throughout the world. The sheer size and amount of ice on Antarctica and its polar climate affect atmospheric and oceanic temperatures, wind, and water patterns throughout the globe. Although changes in ice budgets on Antarctica and Greenland produce sea level fluctuations, the current rise of sea level is not necessarily entirely due to glacier melting. Several ice budget studies of Antarctica have calculated that it is probable that the ice sheet may be accumulating more ice on an annual basis than is lost by wastage.

Glaciers are one of the few geomorphic agents that we cannot directly change. However, we may be inadvertently altering their long-range balance by polluting the atmosphere, which can modify the earth's heat radiation system. So although the impact of humans on glaciers has not yet been demonstrated, the glaciation influence on society has been tremendous—everything from producing the earth's most spectacular scenery, to supplying major water resources (as in the Great Lakes and many aquifers systems), and the deposition of enormous sand and gravel resources (Fig. 4-58). Isolated glacial boulders (Fig. 4-56) are called "erratics." When deposits are not stratified (Fig. 4-60), they are called "till" and result from direct ice deposition (Fig. 4-57). Stratified sediments form from water-associated processes (Fig. 4-58). Even areas near glaciers but never covered by ice may undergo unusual topograhic changes—such as those formed by periglacial processes (Fig. 4-59).

WIND

Wind is the movement of air in the troposphere. The indirect effects of wind are more important than the direct geomorphic changes it produces. Wind is a necessary ingredient in the formation of ocean waves, and waves create more coastal changes than all other processes that act on shorelines. Wind distributes precipitation and is the determining factor that influences storm

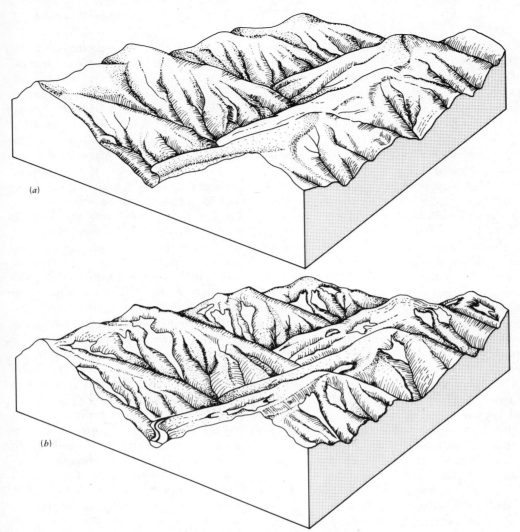

Figure 4-49 Erosion of mountains by valley glaciers. (*a*) Preglacial erosions by fluvial processes. Note smooth rounded forms. (*b*) Snowfields develop with initiation of cirques. (*c*) Glaciers merge to form valley glaciers, and frost wedging acts to sharpen mountain and ridge summits. (*d*) The glaciated terrane now reveals angular topography with valley deepening, widening, and straightening. Hanging valleys, empty cirques, and knife-edged divides with pyramid-shaped peaks typify the region. (After Longwell, Flint, and Sanders, 1969.)

severity. Wind also distributes atmospheric heat and thus becomes an invaluable ally of processes that cause deterioration of rocks by weathering.

Unlike other processes such as running water, wind does not respect topographic divides and can blow up and over obstacles, even mountain ranges. Furthermore, winds have no constant directionality and are free to blow in all directions. Thus, the effects of wind are widespread and are not restrictive.

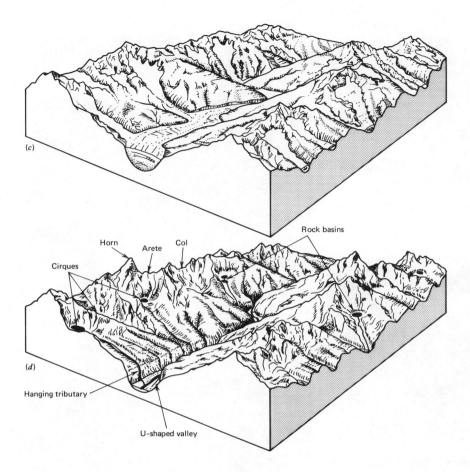

(c)

Horn   Arete   Col   Rock basins

Cirques

(d)

Hanging tributary

U-shaped valley

Figure 4-50 Rhone glacier, Switzerland. Note large crevasses. People give scale on right side.

Figure 4-51 Malaspina piedmont glacier, Alaska. The dark convolutions consist of rock debris that will be deposited when the ice melts. (Courtesy Miles Hayes.)

Figure 4-52 Valley glacier, Chugach Mountains, Alaska. An outwash plain is forming in front of the ice margin. (Courtesy Miles Hayes.)

Figure 4-54 Glaciated bedrock pavement with striatations and grooves. Lake Champlain, New York. (Courtesy Chris Neuzil.)

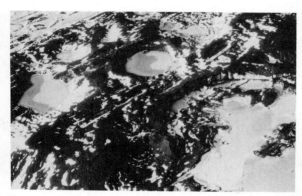

Figure 4-53 Pseudokarst terrane, Alaska. A stagnant mass of ice underlies the surface which contains glacial debris. When the ice melts, it will leave a topography of kames and kettles. (Courtesy Miles Hayes.)

Figure 4-55 Braided meltwater stream, Alaska. Water is derived from the melting ice in the background. Such glaciofluvial deposits will constitute a "valley train" during a nonglacial time interval. (Courtesy Miles Hayes.)

Although the visible work of wind is most prominent in desert and arid terrane, its geologic effects are universal. For example, loess is nearly a worldwide feature, and eolian deposits occur on more than 30 percent of the land surface of the United States. Huge wind deposits also occur in Europe, the Soviet Union, Argentina, and China (where loess attains a thickness of hundreds of meters).

In deserts, sand dunes (Fig. 4-60) are the most prominent wind-derived feature, but one must not think that all deserts are completely covered by dunes. Indeed in the Sahara, the greatest of all deserts, only about 13 percent is dune-covered; the Arabian Desert, with the highest sand dune ratio, is only one-third dunes. Nebraska has the largest dune area (now largely stabilized by vegetation) in the Western Hemisphere—50,000 km². Sand dunes are not restricted to desert regions but are also common

Figure 4-56 The largest New England glacial erratic, in New Hampshire.

Figure 4-58 Glacial stratified materials, Chemung, New York. These deposits occur as glacial outwash in a valley train (see Fig. 4-55). The ice margin was at a sufficient distance away so that the sediments could be well stratified and sorted.

Figure 4-57 Glacial till, Great Bend, Pennsylvania. Note the lack of stratification and the large size range in material from clay to boulders. Such material forms when deposited directly by a glacier without being reworked by other processes.

Figure 4-59 Boulder field at Hickory Run, Pennsylvania. Surface of these boulders slopes less than 2°. This landform resulted from periglacial processes when nearby bedrock hills were attacked by frost weathering and the resulting boulders gradually moved, by gravity, from their original site.

features along many of the world's beaches. Figure 4-61 depicts the variety of forms that develop when sand is shaped by the wind.

The direct work accomplished by wind is performed in three ways: (1) abrasion, (2) deflation, the blowing away of materials, and (3) eolian stress, the force of the wind. Since air has only 1/800 the density of water, normal wind currents generally produce insufficient pressure to cause rock and landform change. However, winds with unusual force, as in tornadoes and tropical storms, are capable of moving some rocks and toppling trees. The uprooting of trees

Figure 4-60 Sand dunes west of St. Anthony, Idaho. (Courtesy Earl Olson.)

Some kinds of dunes based on form

Kind	Definition and Occurrence	Illustration (Arrows indicate wind directions)
Beach dunes	Hummocks of various sizes bordering beaches. Inland part is generally covered with vegetation.	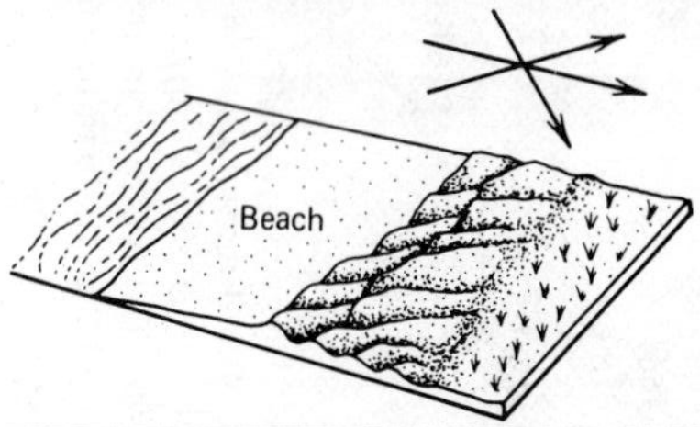
Barchan dune	A crescent-shaped dune with horns pointing downwind. Occurs on hard, flat floors in deserts; constant wind, limited sand supply. Height 3 feet to > 100 feet.	
Transverse dune	A dune forming a wavelike ridge transverse to wind direction. Occurs in areas with abundant sand and little vegetation. In places grades into barchans.	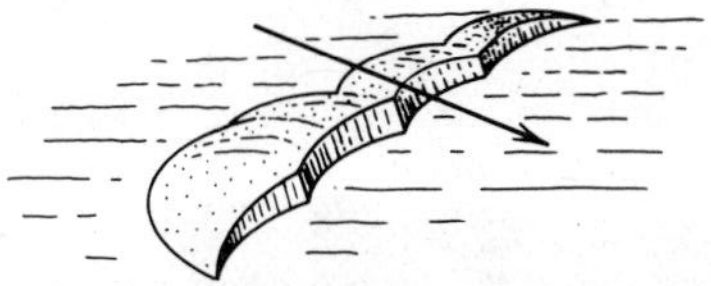
U-shaped dune	A dune of U-shape with the open end of the U facing upwind. Some form by piling of sand along leeward and lateral margins of a growing blowout in older dunes.	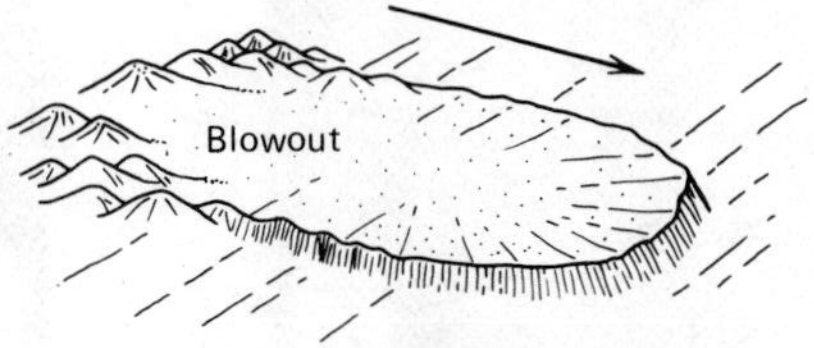
Longitudinal dune	A long, straight, ridge-shaped dune parallel with wind direction. Occurs in deserts with scanty sand supply and strong winds varying within one general direction. Slip faces vary as wind shifts direction.	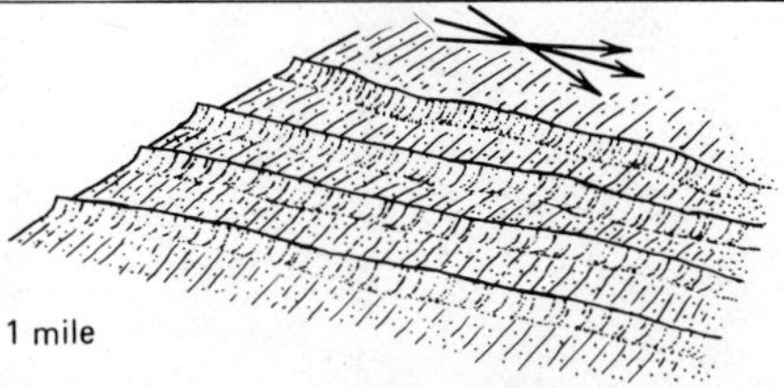

Figure 4-61 Sand dune forms. (After Longwell, Flint, and Sanders, 1969.)

Figure 4-62 Pediment surface, Papago country, Arizona. The bedrock mountains and their piedmont slope have been greatly denuded. The light-colored area in the central part of the photograph is all bedrock or only thinly masked by shallow sediment.

can lead to important landform changes when the disturbed and cratered ground becomes subjected to the other geomorphic processes. In arid regions with mountains it is common to have large expanses of gradually sloping terrain that extend downward from the mountain front. When such surfaces are only lightly veneered with debris they are called **pediments**

(Fig. 4-62). Alluvial fans have many similar features but they are composed of much thicker blankets of sediment, generally have steeper slopes, and a surface that is more regular.

Because wind is invisible it is in the results of the wind, as in the transportation of materials that can be seen, and in the deposits that are formed that we are provided with our best clues about its effects. It is now believed that at least 25 percent of all erosion in the United States is caused by wind. It is the only process capable of transporting material around the world. During volcanic eruptions wind-moved particles may move around the earth several times and stay in the atmosphere for weeks. Such particles may produce weather changes. For example, it is probably not a coincidence that the only five times in recorded history that Cayuga Lake (one of the two largest Finger Lakes) was completely frozen over occurred during years when there had been massive volcanic eruptions with unusually large amounts of pyroclastic materials spewed into the atmosphere.

READINGS

Billings, M. P., 1972, Structural geology, 3rd ed.: Englewood Cliffs, N.J., Prentice-Hall, 606 p.

Birkeland, P. W., 1974, Pedology, weathering, and geomorphological research: New York, Oxford Univ. Press, 285 p.

Bloom, A. L., 1978, Geomorphology: Englewood Cliffs, N.J., Prentice-Hall, 510 p.

Coates, D. R., ed., 1973, Coastal geomorphology: Publ. in Geomorphology, State Univ. of New York at Binghamton, 404 p.

———1974, Glacial geomorphology: Publ. in Geomorphology, State Univ. of New York at Binghamton, 398 p.

Davis, S. N., and DeWiest, R. J. M., 1966, Hydrogeology: New York, John Wiley & Sons, 463 p.

Dennis, J. G., 1972, Structural geology: New York, Ronald Press, 532 p.

Dewey, J. F., 1972, Plate tectonics: Sc. Amer. v. 22, p. 56–68.

Embleton, C., and King, C. A. M., 1975, Glacial geomorphology, Vol. 1, 2nd ed.: London, Edward Arnold, 573 p.

———1975, Periglacial geomorphology, Vol. 2, 2nd ed.: London, Edward Arnold, 203 p.

Flint, R. F., and Skinner, B. J., 1977, Physical geology, 2nd ed.: New York, John Wiley & Sons, 594 p.

Freeze, R. A., and Cherry, J. A., 1979, Groundwater: Englewood Cliffs, N.J., Prentice-Hall, 603 p.

Horton, R. E., 1945, Erosional development of streams and their drainage basins: hydrophysical approach to quantitative morphology: Geol. Soc. Amer. Bull., 56, p. 275–370.

Leopold, L. B., Wolman, M. G., and Miller J. P., 1964, Fluvial processes in geomorphology: San Francisco, W. H. Freeman and Co., 522 p.

Longwell, C. R., Flint, F. R., and Sanders, J. E., 1969, Physical Geology: New York, John Wiley & Sons., 685 p.

Morisawa, M., ed., 1972, Quantitative geomorphology: Publ. in Geomorphology, State Univ. of New York at Binghamton, 315 p.

Ollier, C. D., 1969, Weathering: Edinburgh, Oliver and Boyd, 304 p.

Ritter, D. F., 1979, Process geomorphology: Dubuque, Iowa, Wm. C. Brown, 603 p.

Schumm, S. A., 1977, The fluvial system: New York, John Wiley & Sons, 338 p.

Sugden, D. E., and John, B. S., 1976, Glaciers and landscape: New York, John Wiley & Sons, 376 p.

Yatsu, E., Ward, A. J., and Adams, F., eds., 1975, Mass wasting: East Anglia Norwich, England, Geo Abstracts Ltd., 202 p.

PART TWO

Oil has been a vital resource in the twentieth century. Portable drilling rig in 41 m low water in the Gulf of Mexico. (Courtesy Cities Service.)

Geologic Resources and Energy

The next four chapters present some details on the human use of earth materials for energy and mineral supplies, on which modern civilization is totally dependent. Earth resource problems are enormously complex. It is not possible to achieve a well-informed, balanced view by studying only geology, distribution, and mining aspects of raw materials. Political, economic, social, and ecological consequences and relationships must also be considered.

A basic rule of the economics of man's recent past was that, as the supply of any commodity increases, demand decreases; and as the demand increases, supply rises to meet those demands. As long as the substances in demand are readily and economically available, this tends to be true. But, with the unprecedented growth in human numbers and expectations in the twentieth century, demands for many essential materials have begun to exceed their availability or the rate at which those materials can be supplied to the consuming sector. Thus we are faced with simultaneously increasing demands and declining resources.

With the accelerated depletion of many mineral resources it is becoming abundantly clear that alternative paths must be taken to supply those items necessary to sustain human life. A combination of new science, new technology, and new locations for resource development and procurement will be necessary for even a moderate quality of life standard. A brief listing of such possibilities would contain such ideas as: use of other substances as mineral substitutes; development of nonconsuming and renewable energy sources; programs of mineral and energy conservation; recycling and reprocessing of used and waste products; mining of the ocean floor; and efficient technologies that utilize common source materials.

Chapter 5 presents many of the ramifications associated with the mining and production of minerals. No one can sensibly argue that society can exist without minerals. However, many environmental problems arise from the method of mining and the land disturbance that necessarily accompanies its development. Being forewarned of such impacts, the mining management is in a better position to take precautions that can minimize water and soil deterioration.

Chapter 6 is mostly devoted to the petroleum and coal industries. Their status both nationally and internationally is traced, along with impacts they create on the environment. The emerging technologies associated with such controversial issues as the mining of oil shale and tar sands are described, but whether these possible energy sources will ever be fully utilized is still a highly debatable question.

Chapter 7 contains a spectrum of alternative nonfossil fuel energy sources. Some are already being used, some are on the drawing boards and theoretical, and others are highly controversial. It is obvious, however, as the OPEC nations continue their petroleum strangulation of the economies of many Western countries, that drastic measures must be taken to assure a continuing energy resource that is internally controlled. Because most studies show a clear relationship between energy use and the gross national product, the argument is commonly made that a great abundance of energy will always be necessary to maintain high living standards.

Chapter 8 is purposefully included in Part Two as a reminder that water ranks at the top of all geologic resources as vital to the success of man-made developments and activities. Water is all too often taken for granted because of its common and rather mundane character. Yet this precious commodity is shown to be the determinant in the production of material goods, in agricultural crops, and in the health and welfare of all human societies.

READINGS

Abelson, P. H., and Hammond, A. L., 1976, Materials: renewable and nonrenewable resources: Amer. Assoc. Adv. Sci. Washington, D.C., 196 p.

Amer. Assoc Advancement Science, 1974, Energy issue: Science, v. 184, p. 245–386.

Amer. Assoc. Advancement Science, 1978, Energy issue: Science, v. 199, p. 607–663.

Amer. Assoc. Advancement Science, 1978, Energy issue: Science, v. 200, p. 135–187.

Belknap, R. K., and Furtado, J. G., 1967, Three approaches to environmental resource analysis: Washington, D.C., Conservation Foundation, 102 p.

Cheney, E. S., 1974, Limits to power growth: Geology, v. 2, n. 6., p. 261–265.

Cloud, P. E., ed., 1969, Resources and man: San Francisco, W. H. Freeman, 259 p.

Exxon Company, 1978, Energy outlook 1978–1990: Exxon Company, U.S.A., Houston, Texas, 18 p.

Hammond, A. L., Metz, W. D., and Maugh, T. H. II., 1973, Energy and the future: Amer. Assoc. Adv. Sci., 184 p.

Hayes, D., 1976, Energy: the case for conservation: Worldwatch Paper 4, Washington, D.C., Worldwatch Institute, 77 p.

Hubbert, M. K., 1969, Energy resources: in Resources and man, Nat. Ac. of Sci., Nat Research Council, San Francisco, W. H. Freeman, p. 157–241.

Landsberg, H. H., 1964, Natural resources for U.S. growth: Baltimore, Johns Hopkins Press, 260 p.

———1970, The U.S. resource outlook: quantity and quality: in Revelle, R. and Landsberg, H. H., eds., America's changing environment, Boston, Houghton Mifflin, p. 107–130.

Martinez, J. D., 1976, Environmental influences on and responses to changing patterns in energy supply: Canadian Mining and Metallurgical Bull., Dec., p. 1–10.

Park, C. F., Jr., 1968, Affluence in jeopardy: San Francisco, Freeman, Cooper & Co., 368 p.

Smith, G. H., ed., 1971, Conservation of natural resources, 4th ed.: New York, John Wiley & Sons, 685 p.

Steinhart, John and Carol, 1974, Energy: sources, use and role in human affairs: North Scituate, Mass., Duxbury Press, 362 p.

Turk, J., et al., 1975, Ecosystems, energy, population: Philadelphia, W. B. Saunders Co., 296 p.

U.S. Department of Interior, 1967, Surface mining and our environment: Washington, D.C., U.S. Govt. Printing Office, 124 p.

U.S. Department of Agriculture, 1968, Restoring surface-mined land: U.S. Dept. Agri. Misc. Publ. 1082, 17 p.

Chapter Five
Mineral Resources

Open pit copper mine, Butte, Montana. (Peter Menzel/Stock, Boston.)

INTRODUCTION

In recent years, much attention has been focused on energy shortages. It is important to recognize that energy is but one facet of the overall resource situation (Fig. 5-1). Basic raw materials such as metals and building materials are essential to every aspect of modern civilization, including energy development (Fig. 5-2). Material needs for the U.S. energy-producing industries alone over the next 10 to 15 years include about 187 million tons of iron, 335 million tons of concrete, 26 million tons of barite, 15 million tons of aluminum, 4 million tons of copper, and 2 million tons of manganese (Table 5-1). A major impediment to increased fossil fuel exploration and extraction is the inability to supply necessary drilling and mining equipment as rapidly as it is required. For a similar example on a much smaller scale, consider platinum. Platinum is used in the most efficient fuel cells we have, but it would require more platinum than the world now produces to supply half the U.S. electricity needs using these fuel cells.

The extraction and processing of raw materials accounts for about 25 percent of all U.S. energy consumption and two-thirds of all industrial energy use (Table 5-2). It is a somewhat vicious circle, because supplying the materials needed for increased energy production will

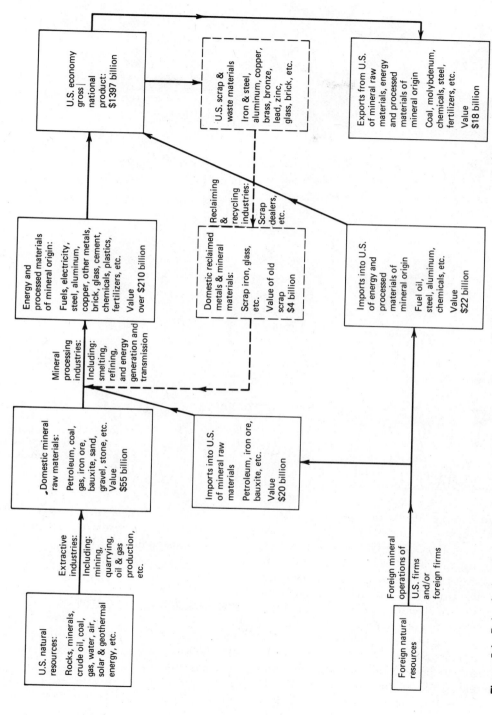

Figure 5-1 Role of minerals in U.S. economy. Estimated values for 1974. (Data from U.S. Bureau of Mines and U.S. Dept. of Commerce.)

U.S. natural resources:

Rocks, minerals, crude oil, coal, gas, water, air, solar & geothermal energy, etc.

Extractive industries:

Including: mining, quarrying, oil & gas production, etc.

Domestic mineral raw materials:

Petroleum, coal, gas, iron ore, bauxite, sand, gravel, stone, etc. Value $55 billion

Mineral processing industries:

Including: smelting, refining, and energy generation and transmission

Energy and processed materials of mineral origin:

Fuels, electricity, steel, aluminum, copper, other metals, brick, glass, cement, chemicals, plastics, fertilizers, etc.

Value over $210 billion

U.S. economy gross national product: $1397 billion

U.S. scrap & waste materials

Iron & steel, aluminum, copper, brass, bronze, lead, zinc, glass, brick, etc.

Reclaiming & recycling industries:

Scrap dealers, etc.

Domestic reclaimed metals & mineral materials:

Scrap iron, glass, etc.

Value of old scrap $4 billion

Imports into U.S. of energy and processed materials of mineral origin

Fuel oil, steel, aluminum, chemicals, etc.

Value $22 billion

Imports into U.S. of mineral raw materials

Petroleum, iron ore, bauxite, etc.

Value $20 billion

Exports from U.S. of mineral raw materials, energy and processed materials of mineral origin

Coal, molybdenum, chemicals, steel, fertilizers, etc.

Value $18 billion

Foreign natural resources

Foreign mineral operations of U.S. firms and/or foreign firms

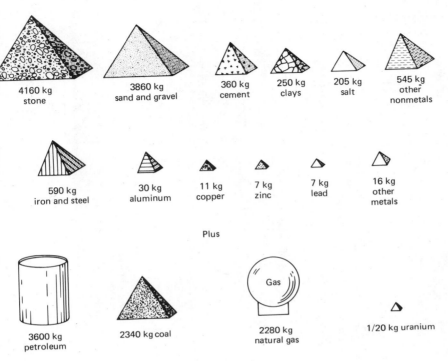

Plus

To generate:
Energy equivalent to 300 persons working around the clock for each U.S. citizen

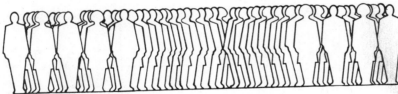

Figure 5-2 U.S. per capita requirements for new materials each year amount to 18,800 kg per person—a total of 4 billion tons. (Data from U.S. Bureau of Mines, 1974.)

itself require huge inputs of energy. The U.S. Geological Survey estimates that energy equivalent of at least 2.5 billion barrels of oil will be needed to provide just 20 selected mineral commodities required by the energy industries from 1975 to 1990. Domestic demand for these same 20 minerals is far greater, requiring some 18.5 billion barrels worth of energy—more than half the known U.S. recoverable petroleum reserves. And this represents only a fraction of the energy required to provide all the mineral commodity demand for the nation for the next 15 years. Thus, minerals comprise a vital link in the lifeline of industrialized society, and adequate supplies are prerequisite for a nation's strength and vitality.

MINERALS AND SOCIETY

Minerals and their derived products permeate all aspects of human civilization—its institutions, its businesses, its welfare, and the very quality of life. Wars have been won and lost over rights to minerals. Exploration for their discovery has

Table 5-1   Minerals needed to achieve U.S. energy goals

Commodity	Short tons	Commodity	Short tons
Aluminum	15,300,000	Iron	187,000,000
Antimony	1,180	Lead	26,800
Asbestos	139,000	Magnesia	532,000
Barite	25,700,000	Manganese	1,930,000
Bentonite	11,200,000	Mica	4,920
Boron	45	Molybdenum	87,200
Cadmium	89	Nickel	289,000
Chromium	279,000	Niobium	44
Cobalt	6,770	Silicon	3,860
Concrete	335,000,000	Silver	418
Copper	3,760,000	Tin	1,090
Fiberglass/Plastic	2,000,000	Titanium	5,220
Glass	7,960,000	Tungsten	38,100
		Vanadium	173
		Zinc	3,530

Source: U.S. Geological Survey, news release, December 8, 1975.

Table 5-2   Energy expenditures in mining and concentrating selected metallic ores

Commodity	Grade (%)	Tons ore per ton metal	Energy per ton product (10^6 BTU) Mining	Energy per ton product (10^6 BTU) Concentration	Percentage of total product energy
Iron	30	4	0.61[a]	2.0	11[b]
Zinc	10[c]	12	5.3	5.2	7
Lead	10[c]	12	4.3	4.5	30
Copper	0.7	157	21.6[a,b]	42.3	57
Uranium	0.2	530	332.4[a]	246.6	58

Source: E. T. Hayes, 1976, In P. H. Abelson and A. L. Hammond, *Materials: renewable and nonrenewable resources,* Washington, American Association for the Advancement of Science, p. 35. Copyright © 1976 American Association for the Advancement of Science.
[a]Surface mining.
[b]Pellets.
[c]Estimated.

led to colonization of new lands. Indeed, the level of society is often measured by the types of usage a nation makes of its own, or imported, mineral resources. The economic health and sustenance of a government often hinges on its ability to distribute effectively the goods that have been manufactured from the mineral industries. Minerals and society are so intertwined as to be inseparable. Thus, the geologist has a significant role in assuring a predictable and plentiful supply of minerals. Unfortunately, the United States must import many minerals (Table 5-3).

Throughout all history minerals have played a vital role in the destiny of man and governments. Flint, chert, quartz, and other

Table 5-3 Net U.S. imports of selected metals and minerals as a percent of apparent consumption[a]

Mineral	1950	1955	1960	1965	1970	1971	1972	1973	1974	1975	Major foreign sources (1972–75)
Columbium	100	100	100	100	100	100	100	100	100	100	Brazil, Thailand, Nigeria
Mica (sheet)	98	95	94	94	100	100	100	100	100	100	India, Brazil, Malagasy
Strontium	100	98	100	100	100	100	100	100	100	100	Mexico, U.K., Spain
Cobalt	90	68	66	92	98	96	98	98	99	98	Zaire, Belgium-Lux., Finland, Norway
Manganese	77	79	89	94	95	97	98	98	98	98	Brazil, Gabon, Australia, South Africa
Titanium (rutile)	W	68	75	97	100	99	100	96	98	95	Australia, India
Chromium	95	83	85	92	89	84	91	91	90	90	U.S.S.R., South Africa, Philippines, Turkey
Aluminum[b]	58	68	68	79	83	85	87	89	88	84	Jamaica, Australia, Surinam, Canada
Asbestos	94	94	94	85	83	83	84	82	87	84	Canada, South Africa
Tin	82	80	82	80	81	81	83	84	84	84	Malaysia, Thailand, Bolivia
Platinum Group	74	91	82	87	78	75	82	87	87	83	U.K., U.S.S.R., South Africa
Fluorine	33	55	48	77	80	79	77	79	81	81	Mexico, Spain, Italy
Tantalum	99	100	94	95	96	96	96	87	87	81	Thailand, Canada, Australia, Brazil
Nickel	90	84	72	73	71	60	65	69	72	72	Canada, Norway, New Caledonia
Mercury	87	20	25	49	41	48	61	78	86	69	Canada, Algeria, Mexico, Spain
Selenium	53	18	25	44	11	27	24	57	59	66	Canada, Japan, Mexico
Zinc	41	51	46	53	54	59	61	64	59	61	Canada, Mexico, Australia, Peru, Honduras
Tungsten	80	NA	32	57	50	46	43	66	68	55	Canada, Bolivia, Peru, Thailand
Gold	25	34	56	72	59	66	71	48	63	52	Canada, Switzerland, U.S.S.R.
Potassium	9	0	E	7	42	46	45	53	58	51	Canada
Tellurium	39	16	15	E	42	20	38	33	41	50	Peru, Canada
Antimony	33	32	43	36	40	36	50	50	44	49	South Africa, P.R. China, Bolivia, Mexico
Cadmium	17	20	13	20	7	31	38	41	46	41	Canada, Mexico, Australia, Belgium-Lux.
Gypsum	28	27	35	37	39	37	38	35	37	40	Canada, Mexico, Jamaica
Vanadium	4	E	E	15	21	21	27	43	36	38	South Africa, Chile, U.S.S.R.
Petroleum[c]	8	10	16	19	21	24	29[d]	35[d]	37[d]	35	Canada, Venezuela, Nigeria, Saudi Arabia[e]
Barium	8	25	45	46	45	36	39	37	38	35	Ireland, Peru, Mexico
Silver	66	58	43	16	26	45	50	66	55	35	Canada, Mexico, Peru
Iron[f]	4	6	15	24	21	32	25[d]	20[d]	24[d]	29	Canada, Venezuela, Japan, Common Market (EEC)
Titanium (ilmenite)	33	40	22	9	24	33	28	28	33	24	Canada, Australia
Lead	40	39	33	31	22	24	24	29	19	11	Canada, Peru, Australia, Mexico

											Foreign sources
Cement	E	1	0	3	3	5	5	7	4	5	Canada, Bahamas, Norway, U.K.
Pumice	3	2	3	5	11	11	14	8	7	4	Greece, Italy
Salt	E	E	2	5	6	7	6	6	7	4	Mexico, Canada, Bahamas, Chile
Magnesium (nonmetal)	4	20	2	1	2	5	5	6	6	4	Greece, Ireland, Japan
Natural Gas	E	E	1	3	4	4	4	4	4		Canada
Copper	31	17	E	15	E	9	11	8	20	E	Canada, Peru, Chile, South Africa

Source: Mining and minerals policy, 1977, Annual report of the Secretary of the Interior.
Note: E = exports; W = withheld; NA = data not available.

a Apparent consumption = U.S. primary + secondary production + net imports. Figures are based on net imports (net imports = imports – exports ± government stockpile and industry stock changes) of metals, minerals, ores, and concentrates. Foreign sources are listed in descending order of amount supplied.
b Largely bauxite and alumina.
c Includes natural gas liquids.
d Revised.
e Major sources in 1975 were Canada, Nigeria, Saudi Arabia, Venezuela, and Virgin Islands.
f Largely iron ore.

hard rock were among the first earth materials used by man for tools, weapons, and utensils. Other softer materials such as soapstone and limestone were used in carvings and construction. Even Paleolithic man was aware of variations in minerals and used more than 13 varieties. Some of these were used for paintings and decoration. In Egypt, stone was used for building material and in the arts because it was locally available. Mining was on a grand scale in ancient Egypt, and the limestone blocks that comprise the Pyramid of Gizeh numbered 2.3 million and average 2½ tons each. However, the Mesopotamian cultures had to use clay and other fine-grained sediments which they fashioned into bricks and pottery, because of the absence of bedrock in nearby alluvial valleys.

In addition to utilitarian use—in tools, utensils, and weapons—some minerals took on special value as gemstones and as currency. Intrinsic mineral worth and the development of metallurgy so that metals could be fashioned into useful forms then necessitated the art of discovery and the study of economic geology. Perhaps the first economic geologist was Captain Haroeris of Egypt who in about 2000 B.C. spent several months prospecting for turquoise in the Sinai. Starting in 1925 B.C., Egyptians dug hundreds of shafts along the Red Sea in search for emeralds, some as deep as 250 m and large enough to permit 400 miners.

As pointed out in Chapter 3, copper, tin, and iron were important in early cultures and responsible for dating the different rise levels of history. Some of the best records that testify to the importance of metals in early history again point to their use in Egypt. The Sinai turquoise and copper deposits, along with gold from Nubia and the Nile cataracts area, and the metals from Cyprus obtained by trade, provided the wealth for Egypt in 2000–1800 B.C. Amenhotep III who ruled Egypt in about 1400 B.C. understood the power of minerals. His armies sacked the countries east of Egypt and returned with 100,000 pounds of copper, one-half ton of gold, and captives for use in the gold mines of

Nubia. His rapid accumulation of mineral wealth coincided with the time of Egypt's greatest power.

The success of Egypt through use of economic mineral power rubbed off on other ancient peoples. Cretan and Phoenician interests turned to seafaring and international commerce and exploration. They were the first to extract mineral wealth from Spain, Sardinia, and Elba that produced gold, silver, and copper. Finally, tin was obtained from Cornwall, England. Such deposits made the Phoenicians an important force in the Mediterranean until the rise of Greece, and later Rome. The defeat of the Persian ruler Xerxes by the Greek Themistocles in 480 B.C. was made possible by the mineral wealth obtained from the silver-lead mines of Laurium near Athens. The silver mines also financed the Athenian treasury for the Peloponnesian wars, and such economic superiority was instrumental in providing the proper political environment for the Golden Age of Greece.

Although the initial rise and beginning of Roman power was not directly related to mineral wealth, it did become a significant factor in sustaining the early Roman Empire. For example, Spanish mines alone yielded hundreds of tons of silver annually, along with iron, copper, gold, and mercury. With the exhaustion of the mines, the purchasing power of Rome declined, trade decreased, and the military machine did not have adequate funding to maintain superiority throughout the empire.

TERMINOLOGY

A variety of terms are used to describe earth resources that are vital to human life. The primary division we use in this book is to include in our discussion the **geologic,** or nonliving resources, and leave the **biologic** or living resources for other specialists. When the geologic and biologic resources are combined, they constitute **natural resources,** as separate from **human resources.** Geologic resources can be further divided into many separate categories. These include such distinctions as **metallic** and **nonmetallic** resources. In its narrowest sense, these materials are considered as mineral resources and their study constitutes the field of economic geology. Sometimes the term **ore deposits** will be used in this connection. To be considered an "ore," the minerals must be able to be mined at a profit. However, this meaning has been somewhat clouded because many materials are mined sometimes regardless of the profit motive, as being necessary for health or political, military, or social needs. Fossil carbon deposits in the form of the **fossil fuels,** petroleum and coal, must also be considered geologic resources. And rounding out the assemblage of geologic materials that are used as resources by society are **water** and **soil.** These ingredients are vital for the sustenance of human life, as well as for the many products derived from their utilization.

CLASSIFICATION

Geologic resources can be classified in a variety of ways. In the broad sense, I have separated them into the following categories which reflect the organization patterns of Part 2 of this book: mineral resources, water resources, soil resources, and fossil fuels (Fig. 5-3). In turn, mineral resources can be further subdivided, depending on such factors as their type of use or geologic origin and occurrence. For example, one classification system groups minerals into metals, industrial minerals, and construction materials. Appendix E indicates one type of mineral deposit classification based on the character of their geologic environment as indicated by the rock association and physical processes that have aided their concentration. Thus, mineral deposits are simply minerals and rocks that are used by man. In order to be economically mined, some of the elements must be in natural

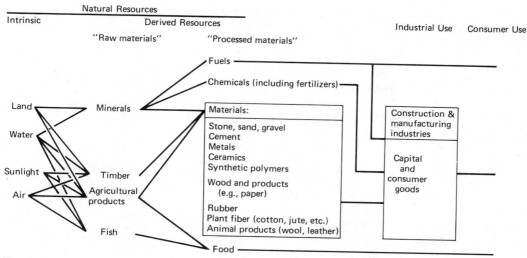

Figure 5-3 Flow of resources through the production system. (S.V. Radcliff, 1976, *in* Renewable and nonrenewable resources, p. 26. Copyright © 1976 Amer. Assoc. Adv. Sci.)

settings where unusual earth processes have greatly concentrated them from their more usual state as part of the earth's crust (Tables 5-4 and 5-5). For example, to be profitably mined, iron must form about 30 to 40 percent of the rock body being extracted, whereas its usual abundance in the earth is only 8 percent. It is important to note that there is nothing mysterious about mineral deposits. They occur in the three different types of rock: igneous, metamorphic, and sedimentary. Furthermore, under special circumstances, they may form from geomorphic processes, such as those associated with the movement of running water and groundwater (and other erosion processes), as well as weathering. Other minerals, and rocks, are simply mined in bulk and the entire deposit is used, as rock salt, sand and gravel, and so on.

Resources and Reserves

The manner of classification of mineral deposits that can be obtained with present knowledge and anticipated production is important in the planning and management of the mineral industries. The system adopted by the U.S. Geological Survey and U.S. Bureau of Mines is predicated on two key criteria: the extent of geologic knowledge about the mineral and the economic feasibility of recovery (Fig. 5-4). Using these guidelines, they classify ore deposits in the following way.

Resources. Concentrations of a mineral in such forms that economic extraction is currently or may become feasible.

Identified resources. Specific bodies of minerals whose location, rank, quality, and quantity are known from geologic evidence supported by engineering measurements.

Undiscovered resources. Unspecified bodies of mineral surmised to exist on the basis of broad geologic knowledge and theory.

Reserve base. That portion of the identified mineral resource from which reserves are calculated.

Reserve. That portion of the identified mineral resource that can be economically mined at the time of determination. The reserve is derived by applying a *recovery factor* to that component of the identified mineral resource designated as the *reserve base.*

Table 5-4 Comparison of minimum-grade ore and common rock that must be mined to yield 1 ton of selected metals

Metal	(1) Crustal abundance (parts per million)[a]	(2) Tons of average continental rock required	(3) Tons of minimum-grade ore required[b]	(4) Concentration above background required (2 ÷ 3)
Aluminum	83,000	12	4	3
Iron	48,000	20	3	6
Titanium	5,300	190	100	2
Manganese	1,000	1,000	3	330
Vanadium	120	8,300	100	83
Zinc	81	12,000	40	300
Chromium	77	13,000	3	4,300
Nickel	61	16,000	100	160
Copper	50	20,000	200	100
Lead	13	77,000	30	2,570
Uranium	2.2	455,000	670	680
Tin	1.6	625,000	6,000	104
Tungsten	1.2	830,000	200	4,150
Molybdenum	1.1	910,000	400	2,275
Mercury	0.08	12,500,000	200	62,500
Silver	0.065	15,000,000	10,000	1,500
Platinum	0.028	36,000,000	330,000	110
Gold	0.0035	285,000,000	125,000	2,280

Source: Mining and minerals policy, 1977, Annual Report of the Secretary of the Interior.

[a]Lee, Tan, and Yao, Chi-lung. "Abundance of chemical elements in the Earth's crust and its major tectonic units." *International Geology Review,* 12(7):778–786, 1970.
[b]Based on current economics and technology.

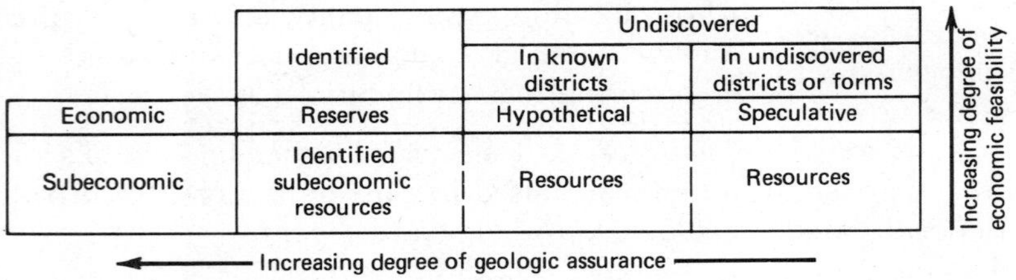

	Identified	Undiscovered	
		In known districts	In undiscovered districts or forms
Economic	Reserves	Hypothetical	Speculative
Subeconomic	Identified subeconomic resources	Resources	Resources

← Increasing degree of geologic assurance →

Increasing degree of economic feasibility ↑

Reserves: Identified resources from which a usable mineral or energy commodity can be economically and legally extracted at the time of determination.

Identified-subeconomic resources: Materials that are not reserves but that may become reserves as a result of changes in economic and legal conditions.

Hypothetical resources: Undiscovered materials that may reasonably be expected to exist in a known mining district under known geologic conditions.

Speculative resources: Undiscovered materials that may occur either in known types of deposits in a favorable geologic setting where no discoveries have been made, or in as-yet-unknown types of deposits that remain to be recognized.

Figure 5-4 Classification system of mineral reserves and resources in use by the U.S. Geological Survey and U.S. Bureau of Mines. (U.S. Geol. Survey Circ. 698, 1974.)

Recovery factor.   The percentage of total tons of mineral estimated to be recoverable from a given area in relation to the total tonnage estimated to be in the reserve base in the ground.

Keeping these definitions in mind, there are still many different approaches as to how one perceives the long-range mineral supply. Because the entire earth is composed of minerals, the mineral resource base is actually the total earth (Table 5-6). However, since ore is only that mineral that can be mined at a profit under prevailing economic, sociopolitical, and technological conditions, there are many variables that determine the character of the reserve for minerals not being immediately mined. Reserves are simply those materials deemed mineable under current or predictable conditions. Most companies are interested in proved reserves for only a 10- or 15-year period of anticipated demand; so calculations for ore in terms of 20 or 30 years are much less accurate than shorter period determinations.

The reader should keep in mind that for the most part the world is not running out of minerals. A single kilometer cube of average rock possesses 2×10^8 tons (metric) of aluminum, 1×10^8 tons of iron, 800,000 tons of zinc, and 200,000 tons of copper. Furthermore, potassium and phosphorus (critical ingredients for fertilizers) are also present in staggering amounts. The argument, therefore, is not one of running out of minerals, but instead what price are we willing to pay and what effect will more costly mining have on other segments of the economy. Although there is an abundance of minerals, there is also a paradox involved if we look closer. It is more costly to mine lower yielding ores but, as quality is lowered, the quantity becomes higher. For example, there are more abundant supplies of iron ore with 35 percent iron than deposits that contain 40 percent iron.

There are several unknowns in the equation for determining future mineral needs and supply. There can be new geologic discoveries or extensions of old deposits that had previously not been considered. Technological breakthroughs can develop that will permit new mining methods or processing techniques that can increase mineral production. For example, the pellet method, for use in refining taconite iron ores, permitted the economic recovery of millions of tons of iron ore, previously thought to be noneconomic. World market prices also determine profitability of mines. The rise in gold prices from $50 an ounce in 1972 to more than $800 in January 1980 permitted the opening of many new mines that formerly had been unprofitable. Silver also escalated from $5 an ounce as late as August 1978 to more than $50 an ounce by January 1980. However by March 1980 gold prices plummeted to $500 an ounce and silver to $12 an ounce.

The degree to which mineral substitutes will be used is another unknown entity in future calculations. Plastics are being increasingly used to replace metal parts that had formerly exclusively used minerals. The case of nickel provides another illustration. Until 10 years ago, most nickel came from sulfide mines that were difficult and expensive to extract. New metallurgy processes, however, have now paved the way for mining nickel in laterite deposits, which are easier to mine and more widespread throughout the world. Other mineral additives can result from more effective recycling of waste products and increased conservation practices.

Nonrenewable Resources

Resources that are replaceable over a short time span, such as food crops, forests, grasslands, and water, are **renewable.** In this chapter, we are concerned with **nonrenewable** resources. Most mineral resources are nonrenewable because they were formed by geologic processes over time spans of millions of years. In terms of human needs, once a copper deposit or an oil field is exhausted, the depletion is permanent.

The term "mineral resources" as used here includes all nonliving naturally occurring ma-

Table 5-5 Abundance, mass, reserves, and resources of some metals in the earth's crust and in the U.S. crust [Abundance in grams per metric ton (g/mt); mass and reserves in metric tons (mt). Calculations = mass (metric tons) × abundance (decimalized) = total content of element]

Element	Total earth's crust				Oceanic crust	
	Goldschmidt[f] (g/mt)	Vinogradov[g] (g/mt)	Lee and Yao[h] (g/mt)	mt × 10^{12}	g/mt[h]	mt × 10^{12}
Antimony	1	0.5	0.62	14.9	0.91	8.1
Beryllium	6	3.8	1.3	31.2	0.83	7.4
Bismuth	0.2	0.009	0.0043	0.1	0.0066	0.059
Cobalt	40	18	25.	600	37	330
Copper	70	47	63	1,510	85	760
Gold	0.001	0.0043	0.0035	0.084	0.0035	0.032
Lead	16	16	12	290	10	90
Lithium	65	32	21	500	20	180
Mercury	0.5	0.083	0.089	2.1	0.11	0.9
Molybdenum	2.3	1.1	1.3	31.2	1.5	14.6
Nickel	100	58	89	2.130	140	1,200
Niobium	20	21	19	460	18	160
Platinum	0.005	—	0.046	1.1	0.075	0.67
Selenium	0.09	0.05	0.075	1.8	0.1	0.89
Silver	0.02	0.07	0.075	1.8	0.091	0.82
Tantalum	2.1	2.5	1.6	38.4	0.43	3.8
Tellurium	0.0018	0.001	0.00055	0.013	0.00088	0.0078
Thorium	11.5	13	5.8	140	4.2	37
Tin	40	2.5	1.7	40.8	1.9	16.8
Tungsten	1	1.3	1.1	26.4	0.94	8.3
Uranium	4	2.5	1.7	40.8	1	7.8
Zinc	80	83	94	2,250	120	1.030
Element	g/mt	g/mt	g/mt	mt × 10^{15}	g/mt	mt × 10^{15}
Aluminum	81,300	80,500	83,000	1,990	84,000	747
Barium	430	650	390	9.4	370	3.3
Chromium	200	83	110	2.6	160	1.4
Fluorine	800	660	450	10.8	420	3.74
Iron	50,000	46,500	58,000	1,392	75,000	667
Manganese	1,000	1,100	1,300	31.2	1,800	16
Phosphorus	1,200	930	1,200	28.8	1,400	12.5
Titanium	4,400	4,500	6,400	153.6	8,100	72.1
Vanadium	150	91	140	3.36	170	1.51

Source: R. L. Erickson, 1973 Crustal abundance of elements, and mineral reserves and resources: U.S. Geological A. Brobst and W. D. Pratt, *United States Mineral Resources.*

[a] U.S. Bureau Mines (1970); 1 short ton = 0.91 mt.
[b] Recoverable resource potential = 2.45 A × 10^6 (abundance A expressed in g/mt).
[c] U.S. Bureau Mines (1970); 1 short ton = 0.91 mt; does not include U.S. reserve.
[d] Recoverable resource potential = 2.45 A × 17.3 × 10^6 (abundance A expressed in g/mt; land area of world is 17.3 times U.S. land area).
[e] U.S. Bureau Mines (1970); data on world basis.
[f] Goldschmidt (1954, p. 74–75).
[g] Vinogradov (1962, p. 649–650).
[h] Lee and Yao (1970, p. 778–786). All calculations are based on this work.
[i] Very high.

Table 5-5 (*Continued*)

Continental crust		Continental crust segments						
		Shield areas		Folded belts		U.S. crust		U.S. crust to 1km depth
g/mt^h	mt $\times$ 10^{12}	g/mt^h	mt $\times$ 10^{12}	g/mt^h	mt $\times$ 10^{12}	g/mt^h	mt $\times$ 10^{12}	mt $\times$ 10^9
0.45	6.8	0.46	4.9	0.43	1.9	0.45	0.41	11.2
1.5	23.8	1.5	16.7	1.6	7.1	1.5	1.4	38
0.0029	0.041	0.003	0.029	0.0025	0.012	0.0029	0.0025	0.07
18	270	19	190	16	80	18	16	440
50	760	52	550	46	210	50	45	1,230
0.0035	0.052	0.0034	0.035	0.0038	0.017	0.0035	0.003	0.085
13	200	13	0.140	13	60	13	12	330
22	320	21	220	23	100	22	20	550
0.08	1.2	0.078	0.81	0.086	0.39	0.08	0.072	2.0
1.1	16.6	1.1	11.6	1	5	1.1	1	27
61	920	64	0.680	53	0.240	61	55	1,500
20	300	20	210	19	90	20	20	550
0.028	0.43	0.031	0.30	0.022	0.13	0.028	0.026	0.71
0.059	0.91	0.054	0.64	0.071	0.27	0.059	0.055	1.5
0.065	0.98	0.067	0.70	0.062	0.28	0.065	0.059	1.6
2.3	34.7	2.3	24.3	2.4	10.4	23	2.1	57.5
0.00036	0.005	0.00038	0.0036	0.00031	0.0016	0.00036	0.00031	0.0085
6.8	0.100	6.6	68	7.1	32	6.8	6	0.160
1.6	24	1.5	16.3	1.7	7.7	1.6	1.4	38
1.2	18.1	1.2	12.7	1.2	5.4	1.2	1.1	30
2.2	33	2.1	22.6	2.3	10.4	2.2	2	55
81	1,220	83	870	77	350	81	73	2,000
g/mt	mt $\times$ 10^{15}	g/mt	mt $\times$ 10^{15}	g/mt	mt $\times$ 10^{15}	g/mt	mt $\times$ 10^{15}	mt $\times$ 10^{12}
83,000	1,242	84,000	869	82,000	373	83,000	74.5	2,000
400	6.1	400	4.3	390	1.8	400	0.37	10
77	1.2	81	0.84	68	0.36	77	0.070	1.92
470	7.1	470	5	480	2.1	470	4.30	11.8
48,000	725	49,000	508	4,000	217	48,000	43.5	1,200
1,000	15.2	1,100	10.6	930	4.6	1,000	0.9	24.9
1,200	16.3	1,200	11.4	1,100	4.9	1,200	0.98	26.8
5,300	81.5	5,500	57.1	5,000	24.4	5,300	4.9	1.30
120	1.85	120	1.3	110	0.55	120	0.11	3

Table 5-5 (Continued)

Element	United States Reserve[a] (mt × 10⁶)	United States Recoverable resource potential[b] (mt × 10⁶)	United States Ratio of potential to reserve	World Reserve[c] (mt × 10⁶)	World Recoverable resource potential[d] (mt × 10⁶)	World Ratio of potential to reserve	Grade[e]
Antimony	0.10	1.1	11	3.6	19	5	Unknown
Beryllium	0.073	3.7	50	0.016	64	4,000	
Bismuth	0.013	0.007	0.5	0.081	0.12	1.5	
Cobalt	0.025	44	1,760	2.14	763	360	
Copper	77.8	122	1.6	200	2,120	10	0.86 percent
Gold	0.002	0.0086	4.1	0.011	0.15	14	
Lead	31.8	31.8	1	0.54	550	1,000	3 percent
Lithium	4.7	54	12	0.78	933	1,200	
Mercury	0.013–0.028	0.20	15–6.8	0.11	3.4	30	
Molybdenum	2.83	2.7	1	2	46.6	23	Unknown
Nickel	0.18	149	830	68	2,590	38	1.5 percent
Niobium	Unknown	49	Unknown	Unknown	848	Unknown	
Platinum	0.00012	0.07	560	0.009	1.2	133	
Selenium	0.025	0.14	6	0.695	2.5	36	
Silver	0.05	0.16	3.2	0.16	2.75	18	
Tantalum	0.0015	5.6	4,000	0.274	97	354	
Tellurium	0.0077	0.0009	0.11	0.054	0.015	0.3	
Thorium	0.54	16.7	31	1	288	288	Unknown
Tin	—	3.9	?[i]	5.8	68	12	0.6 percent
Tungsten	0.079	2.9	37	1.2	51	42	
Uranium	0.27	5.4	20	0.83	93	112	
Zinc	31.6	198	6.3	81	3,400	42	4 percent

	g/mt	mt × 10⁶	mt × 10⁶	Ratio of potential to reserve	mt × 10⁶	mt × 10⁹	Ratio of potential to reserve
Aluminum	8.1	203,000	24,000	1,160	3,519	3,000	
Barium	30.6	980	32	76.4	17	223	
Chromium	1.8	189	387	696	3.26	47	
Fluorine	4.9	1,151	235	35	20	600	
Iron	1,800	118,000	65	87,000	2,035	23	
Manganese	1	2,450	2,450	630	42	67	
Phosphorus	931	2,940	3	15,000	51	34	
Titanium	25	13,000	516	117	225	2,000	
Vanadium	0.115	294	2,560	10	5.1	500	

Calculation of mass of crustal segments

Total earth's crust	24 × 10¹⁸ mt
Oceanic crust	8.9 × 10¹⁸ mt (37 percent of total crust)
Continental crust	15.1 × 10¹⁸ mt (63 percent of total crust)
Shield areas	10.6 × 10¹⁸ mt (30 percent of continental crust or 43.8 percent of total crust)
Folded belts	4.54 × 10¹⁸ mt (30 percent of continental crust or 19.1 percent of total crust)
U.S. crust	0.90 × 10¹⁸ mt (based on United States as 1/17 of land area of world's continental crust)
U.S. crust to 1km depth	24.6 × 10¹⁵ mt (based on average thickness continental crust = 36.5 km; therefore, 1 km is 2.74 percent of U.S. crust)

Table 5-6 Potential world recoverable resources

Mineral	Recoverable resource potential (millions of metric tons)	Recoverable resource potential exceeds known reserves by
Antimony	19	5 times
Aluminum	3,519,000	3,000
Barium	17,000	223
Beryllium	64	4,000
Bismuth	0.12	1.5
Chromium	3,260	47
Cobalt	763	360
Copper	2,120	10
Fluorine	20,000	600
Gold	0.15	14
Iron	2,035,000	23
Lead	550,000,000	1,000
Lithium	933	1,200
Manganese	42,000	67
Mercury	3.4	30
Molybdenum	46.6	23
Nickel	2,590	38
Niobium	848	Unknown
Phosphorus	51,000	34
Platinum	1.2	133
Selenium	2.5	36
Silver	2.75	18
Tantalum	97	354
Thorium	288	288
Tin	68	12
Titanium	225,000	2,000
Tungsten	51	42
Uranium	93	112
Vanadium	5,100	500
Zinc	3,400	42

Source: R. L. Erickson, 1973, Crustal abundance of elements, and mineral reserves and resources: U.S. Geological Survey, Professional Paper 820.

terials that are derived from the lithosphere and are of value to man. Such factors as changing prices, technological advances, and new discoveries often cause the "working stock" of reserves to vary considerably. Mineral resources may be abundant or scarce in the crust of the earth. They may be widely dispersed in the crust or concentrated in only a few unique deposits (see Tables 5-4, 5-5, and 5-6). Only 9 of the 88 natural occurring elements in the earth's continental crust make up 99 percent of its mass.

Metals Metals constitute over three-fourths of the known elements. Their malleability, fusibility, ductility, ability to conduct electricity, and other characteristics make them indispensable to modern man. They may occur in the native (elemental) state, as in gold (Fig. 5-5), silver, and copper, or more commonly as compounds. Iron, aluminum, manganese, magnesium, chromium, and titanium are called the **abundant metals** since each represents more than 0.01 percent of the earth's crust.

Iron is the most valuable single metal and

atoms. Studies in Washington State, where many aluminum plants are located because of abundant hydroelectric power, show that aluminum refining uses 1827 megawatt-hours of electricity per year to produce one ton, compared to 146 for steel.

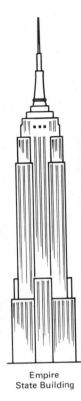

Empire
State Building

Gold

Figure 5-5 Volume comparison of total gold mined throughout history. This production of 87,000 metric tons is more than double the 41,000 metric tons that are mineable throughout the world today. World production had been declining since the 1960s, and two countries dominate— South Africa with 49 percent and Russia with 30 percent. Canada is third with 3¼ percent, and the U.S. is fourth with 2¼ percent.

accounts for about 95 percent of all the metals consumed. Many other metals are valuable primarily as additives to iron. Sedimentary deposits formed by chemical precipitation from solution constitute most of the world's major iron reserves. **Taconite** is a low-grade ore of iron which is the source of most current U.S. iron production.

Aluminum is the most abundant metal in the earth's crust. Its strength, light weight, and durability make it highly desirable for many uses, especially in construction and transportation. In recent years, aluminum has been increasingly substituted for steel, wood, and other materials. Unfortunately, aluminum is a very active element and never occurs in the native state. Thus, to obtain pure aluminum, it must be recovered from compounds that require large inputs of energy to break apart the strong chemical bonds between aluminum and other

UNIQUENESS OF MINERAL RESOURCES

There are several factors that set mineral resources apart from other types of resources or business activities of man. It is important to recognize these differences because they constitute vital components in the planning, managing, financing, and marketing of mineral supplies.

1 **Minerals are nonrenewable.** Unlike agricultural and lumbering resources, there is no second crop. Once mined out, they are gone forever. This finite character of minerals requires entirely different management schemes and amortization schedules of investment than other business enterprises.

2 **Minerals must be mined in place.** Minerals do not always occur where mining is convenient (Fig. 5-6). In fact, often they reside in rather inaccessible terrain in remote regions. Thus, mining costs are high and transportation to major milling and manufacturing sites cause additional problems. They cannot be grown where convenient such as farming produce.

3 **Minerals have high discovery costs.** The location of minerals is becoming increasingly more expensive and difficult. There are also many uncertainties even once a deposit has been located in terms of predictions for optimal mining conditions.

4 **Mining costs generally increase during production.** Unlike mass production of automobiles wherein the unit cost becomes less when mass produced, the cost of mining

"Mining makes a mess of the countryside."

"We have to mine minerals where we find them."

Different views. Polarized. Each seems true, yet in conflict.

Yes, mining disturbs the land. Surface mines overturn vegetation, destroy wild life shelter, disrupt farming, sometimes divert and contaminate streams. Deep mining leaves other scars: waste piles, abandoned shafts. Concerned people fear that mineral mining could destroy the beauty of our land.

Other thoughtful people point to the essentiality of minerals. We must have them for transportation, communications, for food production, houses. And we can only mine these minerals where they are found. We could, it's true, import more of our minerals. But this just moves the problem into someone else's back yard. And ignores the economics of developing our own natural wealth.

What can we do? We must have minerals. So we must have mines. But we must also protect the environment.

We can't mine everywhere. But we should, as a people, support land use decisions that realistically balance economic, social and environmental needs. Decisions that seek greater U.S. mineral self-sufficiency by opening all our lands to exploration and possible development. And we must be willing to pay the price of restoring them to other useful purposes when mining is finished.

With intelligent planning we can have our countryside *and* our minerals.

Raw materials are vital to the production of machines and vice versa. Machines mine the land and help reclaim it. Both mining and reclamation are essential.
essential.

Figure 5-6 Minerals must be mined in place. (Courtesy Caterpillar Tractor Co.)

becomes greater with more extraction. The deeper the mining extends, the more difficult it is to extract minerals. Frequently the grade of ore also diminishes with depth, causing additional mining costs.

Because of the differences in the mining business as contrasted to other business ventures, the federal government recognizes the principle of depletion allowances—a type of tax writeoff that is aimed at permitting mining companies the same type of business advantage as other industries which can claim tax reductions based on obsolescence and depreciation in equipment.

Another disadvantage that can occur with some mineral deposits is a type of "cultural nullification." During urbanization, there can be the loss of important mineral deposits because of paving over by urban communities (see page 616). Because mining operations can produce noise, traffic problems, dust and pollution, zoning restrictions may disallow their development in populated areas. There can also be the withdrawal of mining lands in certain public lands, as in wilderness areas. Thus, any legal imposed restriction on mining can be considered as a social impediment to extraction of important resources, which increases the price of other minerals that must be obtained elsewhere. For example, there is such a variety of laws on Long Island that it is virtually impossible to obtain local sand and gravel supplies. Such commodities must now be imported from outside the region at great cost to users.

MINERAL EXPLORATION

The search for mineral resources embraces the full spectrum of the geological sciences. Because minerals may be igneous, metamorphic, or sedimentary in origin specialists in those disciplines are needed for the discovery and mining of ore deposits. Thus, the origin of mineral deposits is related to the geologic history of the rock body in which they reside. The detection of the ore, its location and abundance, becomes the problem of the economic geologist.

Early discoveries of minerals were usually made on the basis of visual observation of the deposit, tell-tale signs of reactions it may have produced with contiguous materials, or shallow digging so as to expose a likely material. Most prospecters were shrewd geologic observers and developed a "feel for the terrain." For example, the occurrence of "gossan," a type of weathered product, could be indicative that it covered important deposits of some important metals, such as those minerals that contained sulfur. Now that most of the world has been traversed by prospecters, the chance of finding new and unknown large-surface ore deposits is not great. Thus, other techniques must be employed for the discovery of new deposits. These consist of several types. (1) The thorough geologic mapping projects which by careful inference might predict suitable subsurface deposits from the projection of surface indicators; (2) the deliberate incitement of rocks by the application of some type of force, such as in geophysical seismic prospecting; and (3) mapping that employs remote sensing-type instrumentation.

Field Geology Methods

Field geologists use their sense of lithology, structure, and environmental habitat to establish the probability of mineral occurrence (Fig. 5-7). For example, certain minerals are compatible with only certain types of rocks or origins. A search for such minerals as arsenic, bismuth, and antimony in the Canadian shield would be a waste of time because the favorite habitat of such minerals is in young terranes where ore-bearing solution temperatures are low, as in epithermal veins or volcanic areas. To discover placer deposits of the heavy minerals such as gold, platinum, tin, and diamonds, a knowledge of the geomorphology of rivers is necessary. Such deposits would occur, for example, in

Figure 5-7 Old versus new techniques in mapping of earth resources. (a) Members of one of the territorial surveys established in the late 1860s to conduct scientific exploration of the western United States. (b) The processor unit of the EROS (Earth Resources Observation Systems) Data Center, Sioux Falls, South Dakota. The center is managed by the U.S. Geological Survey and contains a central computer complex which controls the data of more than 6 million images and photographs of the earth's surface for use in natural resource and environmental investigations. (Courtesy U.S. Geol. Survey.)

point bars and other positions of slack water in the flow regimes of rivers. It has been known for several decades that some minerals can be assigned to origins in specific metallogenic provinces. However, the new developments in plate

tectonics theory are giving us new geologic insights into their occurrence and possible location. For example, the low-grade porphyry coppers (which produce most of the world's copper) are invariably associated with converging plate boundaries. The extensive deposits in the Andes occur where the westward-moving South American plate overrides the eastward-plunging oceanic crust of the Pacific.

In geophysical prospecting, the aim is to discover an anomaly in the subsurface rocks in the hope the anomalous material will provide an ore deposit. The artifical detonation of explosives produces seismic waves. Recording seismographs can then depict differentials in wave velocity for different rock types. Salt domes would produce slower velocity, whereas metallic rocks would produce higher velocity. Electrical resistivity surveys will reveal differences in rocks because of varying abilities to transmit an electric current.

Remote sensing can involve many different types of instruments, but they have in common the attempt to detect soil and rock differences of measurements by instruments that do not themselves distort rock properties. Air photography, and its interpretation, is one type of remote sensing. With the advent of the space age and satellites, an entirely new field of satellite imagery has developed. One such method used in the United States is termed ERTS (Earth Resources Technology Satellite). These orbiting observatories were specifically designed to aid the location of natural resources needed by man. The instruments are equipped to map with different spectra of electromagnetic energy, visible as well as infrared. LANDSAT-1, which was launched July 23, 1972, had almost immediate success. This multispectral line scanner simultaneously records four images: two visible and two infrared, each covering a 34,000 km² area. Soon this system had located five copper-bearing deposits in a remote part of Pakistan, and studies are now underway to determine whether the deposits are valuable. Other types of remote-sensing instruments include: (1) gravimeters which detect gravity differences, (2) magnetometers which determine magnetic characteristics, and (3) radiometric counters which calculate the amount and intensity of radioactivity.

Therefore, exploration includes the search for new mineral deposits and the detailed examination of potential reserves. The search (propecting) process itself has been responsible for considerable land damage, partly because mining laws state that evidence of active exploration must be apparent for federal U.S. leases to be maintained. Modern mineral exploration can be a highly technical procedure, incorporating geophysical investigation, satellite imagery and other remote-sensing techniques, chemical analyses, and a wide variety of geological studies to determine the shape, extent, and economic feasibility of a deposit. However, digging and drilling are ultimately required to test, determine, and mine the ore.

EXTRACTION METHODS

The method of extraction of mineral resources depends on the location and nature of the deposit. Some of the most severe environmental impacts by man on or near the land surface stem from mineral extraction (mining) procedures.

Undergound Mining

A system of subsurface workings for the removal of mineral matter is termed an **underground mine.** The mining operation consists of removing the ore, delivering it to the surface, and such ancillary operations as providing the workers, supplies, power, ventilation, and water needed to maintain the operation.

Longwall mining This is an extraction method especially applicable to relatively flat-lying, tabular veins or layers of rock. The entire ore zone is removed by advancing along the cutting

Figure 5-8 Subsidence along old slickenside fault planes associated with the mining activities at Magma mine near San Manuel, Arizona. (Courtesy Allen Hatheway.)

face and leaving nothing behind to support the overlying rocks. This usually leads to collapse of the ceiling and a corresponding subsidence of the ground surface.

An alternate method, widely used in U.S. coal and salt mines, is the **room and pillar** method in which ore is removed in a cellular pattern of rooms and passageways with pillars of ore left behind to support the roof. A significant amount of mineable material is left behind in the process. Eventual subsidence (Fig. 5-8) of excavated rooms can produce a land surface pockmarked with collapse depressions (see Fig. 6-13).

Figure 5-9 Open pit iron mine at Mt. Tom Price, Western Australia. Most of the ore is exported to Japan. (Courtesy Australian Information Service.)

Surface Mining

Open pit mining Open pits are an economical method of mineral extraction where large tonnages of reserves are involved, where high rates of production are desirable, and where the **overburden** (waste rock material overlying the deposit) is thin enough to be removed without making the operation uneconomical (Fig. 5-9). The common quarry from which limestone, marble, or granite may be removed is an open pit mine; so are gravel pits. The configuration of many smaller open pit mines depends on the shape of the deposit.

Many of the largest open pit mines, such as those used to exploit major copper deposits, are developed in the shape of an inverted cone with terraced benches along the steep (often about 45°) slopes of the pit. The benches provide haul roads and working platforms from which ore rock may be excavated. Large pits may be many hundreds of meters deep. The major mining operations consist of drilling, blasting, loading, and hauling of rock out of the pit. Over 3 billion tons of iron ore have been removed from open pit mines in the Mesabi Range of Minnesota, transforming parts of a mountain range into an enormous trench.

Area strip mining Strip mining is employed where the mineral deposit, usually coal, occurs relatively close to the surface. The overburden is removed and the uncovered mineral deposit extracted, loaded, and hauled from the mine. If the mineral seam lies at a shallow depth under a relatively level area, the entire area may be strip mined. Figures 6-9 and 6-11 illustrate typical area strip mining procedures. An initial trench or box cut is made, the seam exposed and removed; then a second cut parallel to the first is made, and the overburden is dumped in the adjacent cut left by the first trench. After the area has been mined, a large-scale washboard terrain composed of parallel spoil pile ridges or windows remains. Present laws require recontouring the land, replacing soil materials, and revegetation of the mined area.

The thickness of overburden that can be removed depends on its removal costs relative to the economic value of the mineral deposit to be mined. Some strip mines are hundreds of meters deep. Coal and Florida phosphate account for most U.S. strip mining, although clay, brown iron ore, uranium, and other materials are also strip mined.

Contour strip mining In high-relief terrain, such as the Appalachian Plateau and western mountains (Fig. 6-12), generally horizontal coal seams are too far below the plateau top for the entire area to be stripped. Instead, mining occurs only in the deep valleys where the coal is exposed. The result of this contour stripping is a highwall, bench, and spoil pile sequence that may follow the valleyside contours for many kilometers. The extent and environmental impacts of strip mining are further detailed in Chapters 7 and 18.

Auger mining This method is most commonly used in conjunction with contour stripping of coal. After the highwall has been advanced into the hillslope as far as economically feasible, additional coal may be obtained by boring laterally into the coal seam with a large auger, several meters in diameter. As the corkscrewlike bit rotates, ore is drawn out like shavings from a wood drill.

One problem with strip mining is that the process may weaken rocks, making retrieval of valuable deposits by more thorough underground methods such as longwall very dangerous or impossible.

Placer mining The working and removal of unconsolidated deposits, most commonly alluvium, which contain concentrations of metals or other valuable minerals, is placer mining. Gold, tin, platinum, and gemstones are among the many materials often mined from placer deposits. Basic procedures include preparing the ground (removing vegetation, drilling, blasting, thawing, etc., where necessary), excavating the desired ore material ("pay dirt"), separating and collecting the valuable products, and stacking or redistributing the tailings (ore waste).

Hydraulicking This method, also called hydraulic mining, uses high-pressure jets of water to break down and transport placer sediments to the recovery plant, usually a sluice box, where heavy metals such as gold settle out as the rest of the material is washed away. Hydraulicking is a low-cost method where water is plentiful, but severe environmental impacts have led to it being banned in most developed countries.

Dredging In this procedure, large volumes of unconsolidated material are removed with such machines as the bucket-ladder dredge—which uses a revolving bucketline on a boom that swings slowly from side to side, digging out sediments to depths of over 50 m in some cases; the dragline dredge—which is basically a mobile washing and recovery plant fed by conveyers; and offshore suction dredges—which are used in a wide variety of shallow-water mining operations. As with hydraulicking, dredging is usually undertaken along stream valleys and in or near water where vast amounts of sediment are stirred up and released, creating damage to natural ecosystems.

Other placer mining techniques may utilize such equipment as bulldozers, pipelines, sluiceways, hydraulic elevators, and thawing devices for frozen deposits.

In situ mining In situ mining refers to the utilization or processing of a mineral resource in its natural place of occurrence. It is most frequently used in conjunction with the "in place" burning of coal or the extraction of oil from oil shale with no removal of the actual coal or shale. Only 10 percent of the U.S. coal within 2000 m of the surface is considered mineable. Many otherwise unusable coal seams could be drilled into and their energy exploited by initiating controlled in situ burning. Indians and campers have long made use of the heat escaping through cracks from naturally burning coal seams in the northern Great Plains and elsewhere. In situ combustion may utilize the heat produced directly or, more commonly, be used to generate natural gas, steam, or hot water which is then brought to the surface through pipes or conduits. Precautions need to be taken to control toxic gases that are often produced and especially to control, if possible, the extent of burning.

In situ mining, usually by leaching out the desired minerals, is also important in the extraction of numerous other resources, including halite, potash, uranium, sulfur, and copper. It is usually far less damaging to the environment and, where perfected, tends to be more economical and less energy-intensive than conventional mining operations.

Solution mining Some minerals as salts are mined by solution methods wherein the mineral is dissolved by hot water in wells. Other minerals such as those yielding potash are obtained by solution-precipitation methods (Fig. 5-10).

Ocean mining As resource demand spirals upward and conventional sources are depleted, we find ourselves looking increasingly to the oceans that cover 70 percent of the globe for needed raw materials. Mineral resources may come from the minerals dissolved in seawater, unconsolidated marine deposits, or the bedrock underlying the seas. Seawater is a major commercial source of sodium, magnesium, and bromine salts. Recovery of gold, silver, copper, and other metals from hot metalliferous brines such

Figure 5-10 Evaporation ponds for the production of potash by Texas Gulf, Inc., near Moab, Utah. (Courtesy Burke Industries.)

as those in the Red Sea may someday be feasible. Mining of bedrock deposits beneath the sea is restricted largely to shallow waters at present—for example, Japan and Britain mine undersea coal, and sulfur is mined off the Louisiana coast. For large ore bodies beneath deeper waters, nuclear detonations in subsea drillholes followed by deep-water dredging or chemical leaching to retrieve the ore have been suggested.

Shallow-water dredging for unconsolidated deposits was noted above. Important commodities obtained from shallow-water dredging include sand and gravel (60 percent of all unconsolidated marine production), lime, tin, and about 70 percent of the world's heavy mineral sand production.

The major interest in large-scale deep-ocean mining centers around manganese nodules. These cobble-sized spherical concretions cover a quarter of the abyssal ocean basin floors in places. They are believed to form as colloidal particles of manganese and iron, which grow, perhaps with the aid of bacterial catalysts, by scavenging minerals from the seawater. High-grade nodules contain 27 to 35 percent manganese, 1.1 to 1.4 percent nickel, 1.0 to 1.3 percent copper, and 0.2 to 0.4 percent cobalt, along with numerous other elements. About 70

percent of the anticipated revenues from their exploitation would come from the nickel. Manganese, copper, cobalt, vanadium, zinc, and other metals would also be valuable by-products.

The total quantity of metals in the nodules is staggering—far exceeding all known land deposits of such metals as nickel, cobalt, and copper. (The copper percentage, for example, may not seem great, but it far surpasses the copper content of most land ores now being mined). A 2 million km² area in the Pacific Ocean west of Mexico contains billions of tons of nodules, including 30 times the known terrestrial manganese deposits, 13 times the nickel, and 1.3 times the copper. The estimated value of the Pacific nodule province exceeds $25 million/km².

Other factors are making mining of the nodules look increasingly appealing, including the growing costs and uncertainties over foreign supplies of vital minerals. In 1974, the United States imported 18 percent of the copper used, 90 percent of the nickel, 95 percent of the manganese, and 98 percent of the cobalt at a total cost that exceeded half the U.S. trade deficit!

However, mining the nodules poses horrendous legal, financial, and technical problems. Most nodules lie beneath 4000 to 5000 m of water. Hydraulic siphoning through a 5 km long pipe and a continuous bucketline dredge of equal length are among the recovery methods being considered. Both the risks and potential rewards are high, and estimates are that U.S. industry will invest at least $6 billion in deep-sea floor mining by 1990.

METALLIC DEPOSITS

Although most metals were originally deposited as a result of igneous activity, in some instances the metals have been subsequently reconcentrated by other processes, as in weathering of the porphyry coppers or by fluvial action in placer deposits (see Appendix H). Figure 5-11

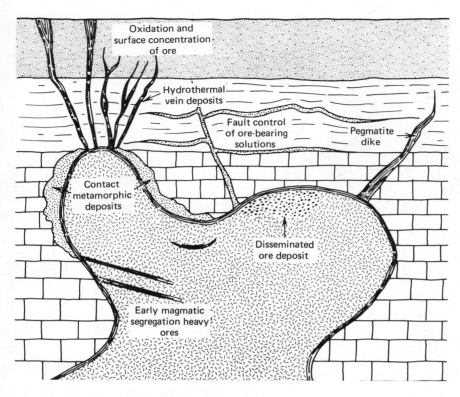

Figure 5-11 Diagrammatic cross section of igneous intrusion showing typical types of ore deposits.

provides a very generalized diagram, showing representative mineral occurrence and rock types. Metallic deposits occur in a great diversity of geographical areas, environmental settings, and rock types. It is now thought that many mining districts originated at plate boundaries where heat flow was at a maximum. Other deposits, however, such as the Mississippi Valley lead-zinc district, are in the more quite, interior zones of continents. A vital part of the exploration for metalliferous deposits is the location of those rock types and structures that act as necessary hosts for the entrapment and storage of the mineral being sought. Thus, an understanding of mineralogy, petrology, and structural geology are important in mineral prospecting.

Igneous Activity

Many different processes associated with magma generation and its allied effects are connected with mineral deposits. Magmatic ore deposits may form either within the main body of the pluton or as an offshoot, such as a vein or dike, that has intruded adjacent rocks. Within the magma the metal-bearing mineral may become concentrated by certain types of differentation, such as by filter pressing or gravitation accumulation. The iron in the Adirondacks and the chromite in the Bushveld of Africa formed in this manner. Sometimes there has been no appreciable mineral concentration, but the ore occurs as isolated fragments within the igneous rock, such as the diamond pipes of the Kimberly district in South Africa.

Perhaps the greatest range of metals that result from ingneous activity is that associated with the dikes and veins that may occur at some distance from the main pluton. Pegmatite dikes are coarsely crystalline, acidic igneous rocks that may contain important amounts of tin, tungsten, uranium, lithium, and beryllium. Hydrothermal veins may occur kilometers from the parent igneous mass. They form from the hot fluids, both liquids and gases, that escaped from the pluton during its crystallization. The hydrothermal deposits may be further classed as hypothermal, mesothermal, and epithermal, depending on the temperatures of their formation. The higher temperature or hypothermal veins are more likely hosts for such elements as gold, silver, tin, and tungsten, whereas lower temperature elements such as lead, zinc, sulfur, and arsenic form in epithermal environments.

Metamorphic Minerals

The most common origin for metals in metamorphic rocks is in the contact zones of metamorphism adjacent to plutonic bodies. Here the heat of intrusion and the various emanations from the magma modify the enclosing rocks by the process known as **metasomatism** (contact metamorphism; Fig. 5-12). The mineralogy of such deposits has a wide range, and metals produced from contact metasomatism processes include iron in the Urals, copper in Mexico, molybdenum in Morocco, and manganese in Sweden.

Sedimentary Rocks and Processes

Surface and near-surface conditions also produce many important metalliferous deposits. Here I include those ores that were formed under low temperature and pressure conditions at or near the earth's surface. The concentration of the ore occurred by waters that redeposited the mineral in an environment of different composition. Such waters could have moved in rivers or through soil and rock by groundwater action. Placer deposits are those where the winnowing action of the fluid (river, ocean, or wind) leave a concentrate of minerals of higher specific gravity. By far the most important placers are those occurring in streams (Fig. 5-13).

Another form of enrichment occurs when groundwater may concentrate the metals, as in the iron ores of the Mesabi District or the bog

Figure 5-12 Metasomatism and hydrothermal alteration at Big Rock Candy Mountains, Marysville, California. (Courtesy Earl Olson.)

ores of Russia, Sweden, and Germany. In the oceans, manganese nodules may form due to the precipitation of the element from seawater and its subsequent precipitation around nuclei on the ocean floor. A somewhat different mode of formation is needed to explain the origin of the porphyry copper ores, the aluminum bauxite ores, and the nickel ores. In these cases, the concentration process is largely vertical. A combination of erosion of igneous rocks, along with downward leaching of the metal ions, causes a gradual enrichment of the metals into concentrated zones that contain metals of sufficiently high grade for extraction.

INDUSTRIAL MINERALS

The grouping of geological resources called "industrial minerals" is a category that relies on the usage of the mineral rather than its chemistry. Such minerals have in common, however, their utilization as a total product, in which the mineral has not been refined for specific use. Instead, the entire mineral or rock mass is used without further alteration, or the elements are used for nonmetallic purposes.

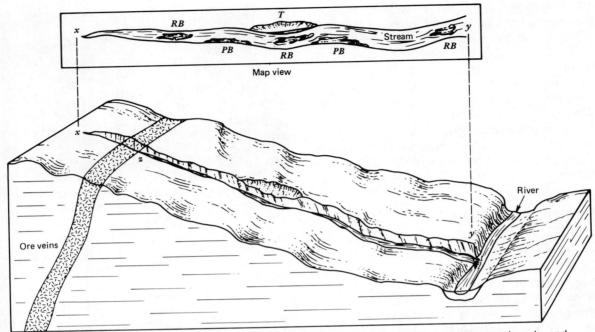

Figure 5-13 Placer deposition. Stream *xy* erodes the ore vein or "mother lode," transports ore minerals, and deposits heavy ore in parts of the stream where slack waters occur. *RB* = river bars; *PB* = point bars; *T* = high-level terrace.

Fertilizers

Mineral fertilizers have become the dominant ingredients to enrich soils and aid crop yields. The most common elements added to soils for their enhancement are calcium, nitrogen, phosphorus, potassium, and sulfur. With the exception of nitrogen, which is also produced by synthetic processes, these ingredients are largely derived from minerals. Calcium is obtained mostly from limestone and is one of the more abundant and least costly of the additives. Nitrogen, in addition to being produced from atmospheric sources, is also obtained from nitrate-bearing nonmarine evaporite deposits. The United States uses about one-third the total world production.

The principal phosphorus-producing mineral is apatite, chemically written $Ca_5(F,Cl,OH)(PO_4)_3$. The United States is the leading consumer and ranks second to Morocco in reserves. Although apatite occurs in many rock types, U.S. production largely comes from the "pebble" phosphate deposits in Florida that annually produce more than 40 million tons.

Potassium is a common ingredient in many rocks, but commercial production is largely restricted to those potassium salts that formed in marine evaporite beds (Fig. 5-14). The extensive reserves in such deposits in North America comprise more than 25 percent of the world's reserves and are ample to use for several centuries. The United States is the greatest producer and consumer of these minerals. Sulfur is also an abundant element in many rocks. The principal U.S. supplies come from calcium sulfate minerals that occur at the tops of many Gulf Coast salt domes; these deposits account for 25 percent of world production. Again, the United States has reserves to last into the twenty-first century.

Chemical Minerals

This group of minerals is related in that they are important for their chemical properties and

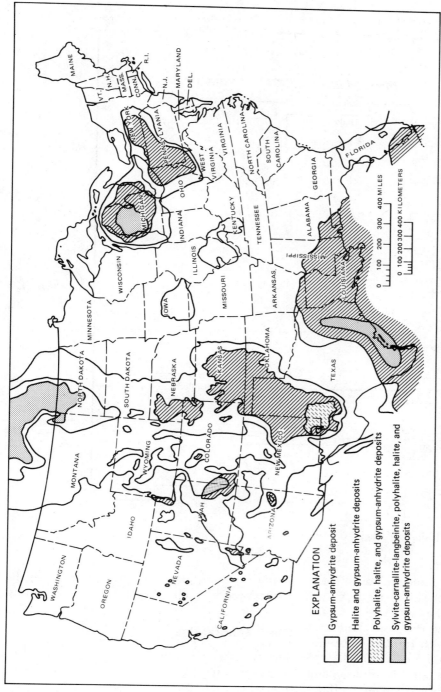

Figure 5-14 Marine evaporite deposits in the United States. (U.S. Geol. Survey Prof. Paper, 820, 1973.)

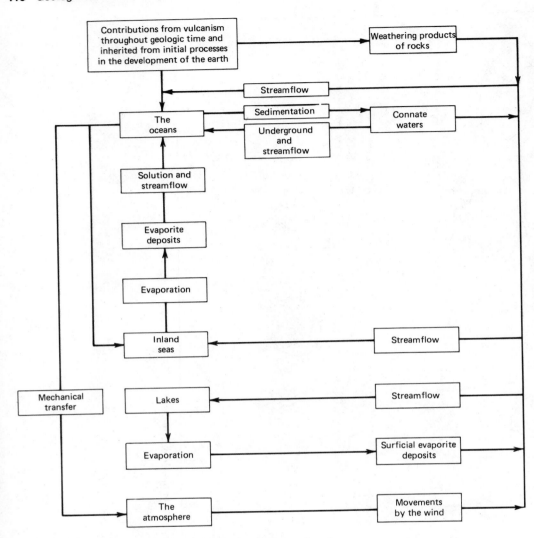

Figure 5-15 The saline cycle. (Courtesy Joseph D. Martinez.)

their applications for inorganic purposes. However, their uses are so broad as to defy easy classification; they cover the "waterfront" in the commercial world. Among the important ones are salt (food preservatives, human consumption, thawing of snow and ice; see Figs. 5-15 and 5-16), clay (ceramics, drilling mud, paint binder, insulators), borates, carbonates, sulfates (glass, dyes, soap, paint, explosives), and fluor-

spar (iron flux, glass, enamel). In general, the United States is self-sufficient in such minerals.

Other Mineral Uses

Society also uses minerals in a variety of other ways. Extra hard minerals are used as abrasives; quartz, garnet, corundum, and diamonds are minerals that are used for cutting purposes

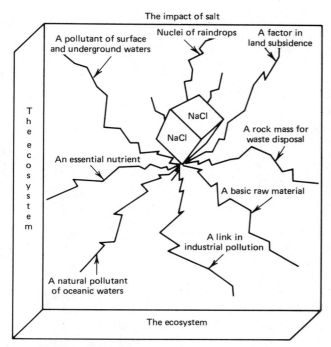

The impact of salt

A pollutant of surface and underground waters

Nuclei of raindrops

A factor in land subsidence

NaCl

NaCl

A rock mass for waste disposal

An essential nutrient

A basic raw material

A link in industrial pollution

A natural pollutant of oceanic waters

The ecosystem

The ecosystem

Figure 5-16   The impact of salt on the human environment. (Courtesy Joseph D. Martinez.)

Table 5-7   Resources, in short tons, of abrasive materials in the United States

Commodity	Identified resources[a] (including reserves)[b]	Hypothetical resources[c]
Diamond	0	0
Corundum	125,000	>2,000,000
Emery	3,000,000	>4,000,000
Garnet	14,000,000	>100,000,000
Triopli (and related materials)	4,500,000,000	>10,000,000,000
Silica sand	Large	Large

Source: R. E. Thaden, 1973, Abrasives: U.S. Geological Survey, Professional Paper 820.

[a]Identified resources: specific, identified mineral deposits that may or may not be evaluated as to extent and grade, and whose contained minerals may or may not be profitably recoverable with existing technology and economic conditions.
[b]Reserves: identified deposits from which minerals can be extracted profitably with existing technology and under present economic conditions.
[c]Hypothetical resources: undiscovered mineral deposits, whether of recoverable or subeconomic grade, that are geologically predictable as existing in known districts.

(Table 5-7). Precious and rare minerals are used as gemstones for jewelry and decorative purposes. The United States must import the great majority of these minerals; they include diamonds, rubies, sapphires, and emeralds. Even semi-precious minerals are in short supply; these include aquamarine, topaz, onyx, opal, jade, and turquoise.

BUILDING MATERIALS

These industrial minerals and rocks comprise the largest tonnage of earth materials used by man and are second only to the fossil fuels in total commodity value. They are in large part bedrock, sand, and clay (Fig. 5-17). They may be used directly as mined, fashioned into specific shapes, or undergo other types of preparation. For convenience, the industrial materials used for building purposes can be divided into two main groups: those materials that are used with alteration only by physical means, and those that are combined with other ingredients and treated by firing, pressure, and then molded into new shapes (Table 5-8). The United States has ample stone industries and sources for decades.

Building Stone

Materials classified as building stone consist of those rocks that because of their physical characteristics can be used without additives or chemical changes in the building and construction industries. All rock types have been used for such purposes. Because their bulk transportation costs are so high, local quarries are used whenever possible. The **zero demand distance** is that distance from the quarry site at which the cost of transportation equals the material

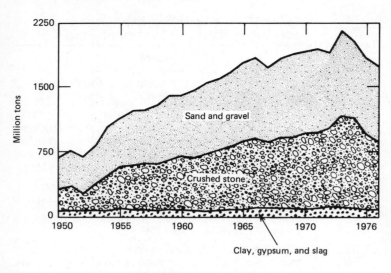

Figure 5-17 U.S. sources of major nonmetallic raw materials for use in construction, 1950–1976. (Data from U.S. Bureau of Mines and National Slag Association.)

Table 5-8 Production of stone, United States (All figures are in millions)

		Dimension stone	Crushed stone	Total (rounded)
1929	Tonnage	4.7	93	98
	Value	$70	$94	$164
1939	Tonnage	2.3	145	147
	Value	$25	$133	$158
1949	Tonnage	1.8	222	224
	Value	$52	$289	$341
1959	Tonnage	2.3	145	147
	Value	$25	$133	$158
1969	Tonnage	1.9	861	863
	Value	$99	$1,326	$1,425
1970	Tonnage	1.6	873	875
	Value	$95	$1,380	$1,475

Source: R. A. Lawrence, 1973, Construction stone: U.S. Geological Survey, Professional Paper 820.

cost at the mined site. For example, in upstate New York, the usual cost of sand and gravel was $2 per cubic yard at the mine, and the transportation cost was 20 cents per mile, so the zero demand distance was 10 mi in 1975. Costs have more than doubled during the 1975–1980 period.

Dimension stone is bedrock that is quarried and cut into measured shapes and sizes and used in the building industry (Fig. 5-18). The famous and much used dimension stone in the United States includes the Barre granite (Vermont), Vermont marble, Salem limestone (Indiana; Fig. 5-19), and Catskill bluestone (New York; Fig. 5-20). Slate from Vermont and Pennsylvania has also been a common dimension stone. The qualities needed for the stone to be valuable include its quarrying ease, strength, color, hardness, texture, porosity, and durability.

Crushed stone can consist of either bedrock that has been broken into suitable sizes by crushing operations (Fig. 5-21 and 5-22) or sand and gravel that has been crushed, sieved, and sorted from unconsolidated sedimentary deposits. When used for roads, as the majority is, it is also called "aggregate" and "road metal." The United States annually uses about 2 billion tons of aggregate, and in some states, such as New York, the economic value of such materials exceeds that of any other mineral product. Limestone and dolostone are the most used crushed bedrock because they are easy to mine and crush, yet are durable in usage. Most sand and gravel deposits in glaciated terranes were formed as kames, eskers, and outwash deposits. In nonglaciated areas, the sand and gravel occurrences are in stream beds, beach deposits, or even ocean sediments. For example, more than

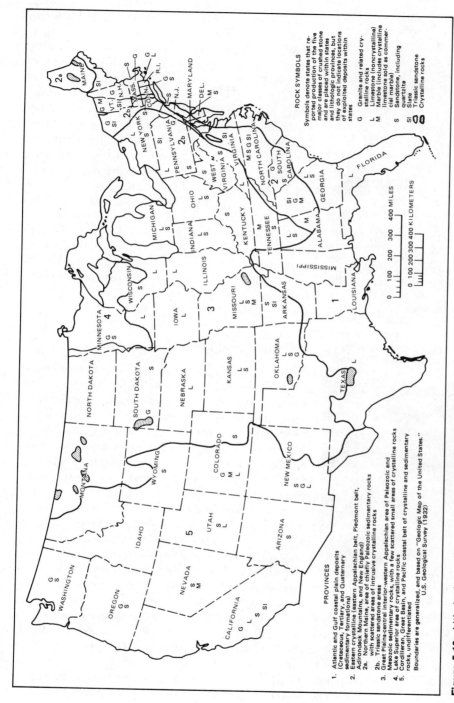

Figure 5-18 Lithologic provinces of the United States as related to occurrence and production in 1969 of principal kinds of dimension stone. (U.S. Geol. Survey Prof. Paper 820, 1973.)

Figure 5-19 Sawed face of a dimension stone quarry (Mississippian limestone) in Salem, Indiana, showing solutionally enlarged joints. (Courtesy Arthur Palmer.)

Figure 5-20 Dimension stone quarry in Catskill ("bluestone") beds, near Sidney, New York. Note the way jointing has controlled mining activity.

95 million yd³ (72 million m³) of sand were mined from the Lower Bay of New York Harbor during the 1950–1973 period. The materials were used in construction of Battery Park City, Port Newark and Port Elizabeth, the New Jersey Turnpike, and Newark Airport. When aggregate is used on state highways, as in New York, there are strict quality controls for the type of material used. One common testing procedure to determine if the materials will endure under hard usage on highways is to use the **magnesium sulfate soundness test.** The materials are placed in a bath of $MgSO_4$ for a prescribed time (such as a 24-hour period), then removed from the bath and placed into a shaking machine and vigorously shaken. This same process is then repeated for several additional cycles; if used for concrete, 10 cycles, and 5 cycles for blacktop. New York specifications call for rejection of those aggregates that have more than a 45 percent loss for use in blacktop and reject materials with 18 percent loss for use in concrete. Domestic sand and gravel production is more than 1 billion tons annually.

Manufactured Mineral Products

This group of mineral uses is almost endless, so we will restrict discussion to the five most important usages and those minerals that form an integral part of the building trade. The common property uniting the use of these products—cement, plaster, clay, glass, and asbestos—is that they must be treated in some particular fashion before usage. Thus, they have been mixed with other ingredients.

Hydraulic cements are those materials that when added to water make a rocklike product. Even the Romans knew that when quicklime was added to volcanic ash and mixed with water that it produced a cement. It was known as **Pozzuolan cement** and widely used in aqueducts, baths, and other structures. Portland cement is made by burning a mixture of lime and clay with some gypsum. When water is added, a gel of hydrous compounds is produced that crystallize and interlock, producing concrete. Limestone and clay are the most common source materials.

Plaster is usually made from gypsum

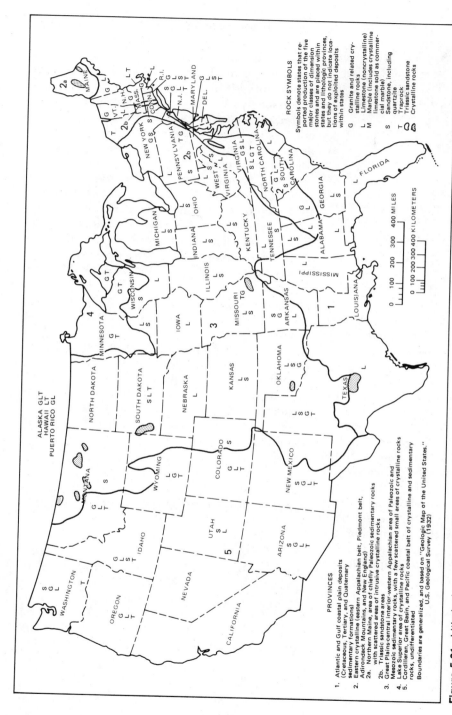

Figure 5-21 Lithologic provinces of the United States as related to occurrence and production in 1969 of principal kinds of crushed stone. (U.S. Geol. Survey Prof. Paper 820, 1973.)

Figure 5-22 Crushed stone quarry in basaltic rocks of the Palisades, New York. These igneous rocks were intruded as a tabular body into sedimentary strata. The vertical fractures represent joints that formed at right angles to the cooling surface.

$(CaSO_4 \cdot 2H_2O)$ or from the processing of anhydrite $(CaSO_4)$. These minerals are rather common and the United States has ample supplies that will last centuries. The extensive occurrence of these minerals in the Paris Basin and the early use in that area established the name "plaster of Paris." Plaster is used as a coating for interior walls, for tiles, and for wallboard.

Clays can be used in a large variety of ways—in paints, drilling muds, cement, refractory materials, bricks, tile, pottery and ceramics, among other uses. Clays may be either specific minerals such as kaolinite, illite, and montmorillonite, or those sediments whose grains fall within the clay range, that is, less than 1/256 mm. Clays are formed by the weathering and breakdown of preexisting rocks and the recrystallization and deposition or precipitation in aqueous media. Most countries have ample clay deposits, and in the United States there are sufficient reserves well into the twenty-first century (Table 5-9).

Glass has become one of the preferred building materials in the twentieth century. It is produced by melting rocks and minerals with sudden cooling in an aqueous medium so that crystals do not have sufficient time to grow. Quartz is the most common raw material, ob-tained from pure sandstones, such as the St. Peter Sandstone, which is found in northcentral United States. Glass is used in a wide variety of ways, for example, in windows, insulators, and conductors.

Asbestos has been an important building material because of its heat-resistant qualities. It consists of threads from such fibrous minerals as chrysotile that are woven together or bound together in a matrix to provide important wear-resistant, heat-resistant, and corrosive-resistant qualities to building materials. Thus, it is used in some portland cements, and for pipe insulation, fireproof shingles, sheet roofing, tiles, wallboards, as well as in asbestos cloth for fireproof clothing, curtains, and warship mats. Within the past several years, however, asbestos fiber inhalation by humans has been linked to cancer, so its use in public exposed areas is being severely limited. Many public places, such as schools where asbestos wallboards were used, are being required to replace the materials with asbestos-free materials.

Many other manufactured mineral products are used in business and commerce—in the pigments of paints, in manufacturing paper, in water purification systems, in sugar refining, in tanning, and so on.

MANAGEMENT OF MINERAL RESOURCES

Grade of Ore

Table 5-4 shows the concentration of metal necessary to become economically mineable and the percent of the metal that can form an ore body. Under special conditions, however, even lower grade metals may be mined when the situation demands it. Such as when:

1 All higher grade ores have been depleted.

2 National or worldwide conditions make production necessary.

Table 5-9 Summary of clay resources of the United States

Type of clay	Reserves[a] (tons)	Potential resources[b] (tons)	Remarks
Kaolin	>300 million	—	Chiefly in Georgia, Alabama, and Mississippi.
	—	5 billion	Lower grade than reserves.
Ball clay	4 million	—	Chiefly in Kentucky. Tennessee and Mississippi.
	—	50 million	Lower grade than reserves.
Fire clay	1 billion	—	Refractories (low to medium duty).
	500 million	—	Refractories (high duty); flint and nodular flint clays.
	—	7 billion	
Bentonite	1 billion	>1 billion	All grades. Limited reserves of high-grade colloidal type. Chiefly in Wyoming, South Dakota, Mississippi, and Alabama.
Fuller's earth	—	2 billion	Chiefly in Georgia, Florida, Alabama, Kentucky, Tennessee, Mississippi, Missouri, Illinois, Texas, and California. Total probably greater.
Miscellaneous clays	Enormous	Virtually inexhaustible	

Source: J. W. Hosterman, 1973, Clays: U.S. Geological Survey, Professional Paper 820.

[a]Reserves: Identified deposits from which minerals can be extracted profitably with existing technology and under present economic conditions.
[b]Potential resources: Identified mineral deposits not profitably recoverable with existing technology and economic conditions and undiscovered mineral deposits whether of recoverable or subeconomic grade.

3 Government subsidy or other financial inducements aid mining costs.

Such factors as political stability, government policies, and social values all play a part in mineral use and production. Also, the profit motive and available financing policies can determine the character of mining methods. **Mineral high-grading** is the mining practice whereby the best ore is always mined first. This results in much higher costs, if and when the lower grade minerals and the minerals more difficult to recover are mined. Reasons for high-grading include:

1 An unrealistically low depletion allowance.

2 Pressure by stockholders to receive quick and high profits.

3 Unstable political conditions.

4 Unpredictable economic conditions.

5 Short-term tax write-offs for first few years of mining.

Such short-sighted practices are very harmful to the industry as a whole, wasteful of resources, and burden the consumer in the long run.

In order to establish a more prudent mineral policy, the following strategies are necessary.

1 Maintenance of adequate stockpiles. In today's marketplace, prices rarely go down, so that continuous production, even during slack times, is generally a benefit.

2 Mineral education programs for the general public that inform the public on the importance of minerals and on the necessity for conservation and recycling.

3 Implementation of methods that will discourage high-grading.

4 Use of more lands for mineral production when the situation requires.

5 Enactment and enforcement of those measures that will permit mining but that will still safeguard the environment.

6 Multiple and sequential use of lands once they have been mined out.

7 Reducing the financial burden in mineral exploration.

Mining and mineral use produce a mixed bag of societal impacts. It is obvious that the use of minerals in today's technological age is a necessity. Civilization could not endure without mineral production. However, some of the other benefits have often been overlooked by environmentalists in their rush to belabor the abuses of the mineral industry. World food supplies would fall far short of feeding the rapidly expanding world population without the use of fertilizers, pesticides, and irrigation— all which require mineral products and its close associate, energy. Furthermore, the use of minerals on farmlands increases production, allowing additional lands to revert to forests and reducing the amount of erosion and siltation. Those countries with sound mineral policies are generally the most stable, economically and politically. However, mining produces many different types of environmental problems that range from despoliation of the countryside to ugly scars on the terrain (Figs. 5-23 to 5-28), which ruin the aesthetic appeal of the landscape.

ENVIRONMENTAL PROBLEMS

Active mine sites are often unavoidably noisy, dusty, unsightly places, and the work involves a higher risk than that of any other occupation (Table 5-10). Blasting, steep pits, vertical shafts, heavy equipment, precarious haulways and, in underground mines, the added problems of ventilation, mine dust explosions, subterranean flooding, possible roof collapse, and other hazards have inevitably taken their toll through the years. Nevertheless, studies show that responsible, safety-conscious companies can reduce the safety and environmental hazards to very low levels.

In addition, many old abandoned mining sites continue to present dangers because of unsealed shafts, sharp cliffs, unstable spoil piles, polluted water drainage, collapse of underground workings, and other hazards (see Figs. 6-12 and 6-13). In 1978, the government identified 22 inactive, potentially hazardous uranium tailings sites in the United States, mostly in the Colorado Plateau area. Cost of decontamination is estimated at $130 million. The link between cancer and workers in mines and mills handling such carcinogens as radioactive materials and asbestos has been well established, and recent studies show higher rates of cancer in the general population of those areas near waste heaps containing carcinogens. In Grand Junction, Colorado, Salt Lake City, Utah, and other places, radioactive tailings have been used as foundation material for homes and public buildings. Millions of dollars have already been spent to remove the hazardous materials. Many additional sites exist where dangerous substances, such as radioactive wastes from spoil and tailings piles, are being distributed by wind, streams, or groundwater, and endangering the long-term health of local residents. Because of the insidious nature of the potential danger, many local residents are uninformed or unconcerned about such hazards.

Although the land directly involved in mining may seem small in comparison to total land area, it is far from insignificant and the impacts often extend well beyond the actual site of operation (Fig. 5-23). In the United States prior to 1965, 1.3 million ha (hectares) of land (an area the size of Connecticut) had been disturbed by surface mining, 56 percent from strip mining, and 35 percent from pit mining. Coal mining accounted for 41 percent of the total, sand and gravel for 26 percent, and stone, clay, gold, phosphate, and iron combined for another 28 percent. The roads serving mining operations also cover hundreds of thousands of hectares.

Tailings from mining and dredging operations can be highly disruptive to the surface environment (see Figs. 5-23 to 5-26). Dredging leaves behind extensive spoil piles and ponds, which can be a continuing source of excess sediment load for rivers unless proper rehabilitation procedures are followed. Tailings left from dredging for gold placer deposits occupy some 80,000 ha of land, mostly in California and Alaska. Ecosystems at the dredging site are totally obliterated, and the sediment carried from the site by water can smother stream channels or marine ecosystems many kilometers away. Excess suspended sediment can impair drinking water supplies, alter water temperature and transparency, shorten the life of lakes and reservoirs, and reduce stream channel water-carrying capacity, thus increasing flood potential.

Hydraulicking has similar and often even more severe impacts. Hydraulic mining for gold in California by the end of the nineteenth century had choked river channels with sediment causing floods, smothered much of the San Francisco Bay with silt, and created extensive shoals, causing navigation hazards.

Ore refining/processing plants, smelters, and other installations closely associated with mining are another source of environmental deterioration, especially air and water pollution. ARCO's Anaconda aluminum plant near Columbia Falls, Montana, provides a good example. The plant was charged with emitting over 1800 kg of toxic fluoride to the air each day, well above the state limit of 392 kg or the 90 kg/day limit which studies have shown is needed to protect the local environment. On November 6, 1978, a federal suit was filed against the plant, stating that the fluoride emissions have "irreparably injured the forest, natural resources and wildlife of the national park [Glacier] and national forest areas." Smelter air emissions have caused extensive damage to veg-

(a)

(b)

(c)

(d)

(e)

(f)

etation, resulting in barren often deeply eroded landscapes in many areas, including Sudbury, Canada; Smelterville, Idaho; Ducktown, Tennessee; and Swansea, Wales (Fig. 5-27).

Many unknowns haunt the spectre of large-scale mining of the ocean floor nodules. The benthic ecosystem, which is not as barren of life as once thought, would be locally destroyed and large quantities of sediment would be stirred up, reducing light penetration and polluting the waters. Many nutrients from the bottom would also mix with overlying waters. Results

(g)

Figure 5-23 *Case study:* Bued River, Luzon Island, Philippines. Two large mines occur in this region. The Thanksgiving Mine is a contact replacement of limestone with grades of 8 percent zinc, with copper, and one ounce of gold per ton. About 150 tons per day are produced by 800 employees. One kilometer away the Black Mountain Mine is a disseminated copper deposit, with a milling capacity of 4000 tons per day by 1200 employees. The mining operation requires much lumber, and the mining families also use forest products as their fuel source. Such high demands on the timber have caused deforestation of much of the region. In addition, families practice slash-and-burn cultivation, and hillsides are burned during the dry season, further reducing forest resources. Such destabilization of the steep hillsides produces vast erosion by both sheetwash and landslidings. It is estimated that 2 million m³ per year are eroded from the watershed to be carried by streams. Furthermore, the Bued must also transport the extra burden caused by the mining. The mines must dispose of 1.5 million tons of tailings yearly, and by law they must be impounded.

However, such waste is merely stacked in the valley in flimsy ponds until a convenient flood occurs during the typhoon season and the valley is flushed out by an "act of God," which is beyond the law. While new ponds are being created, the tailings are being dumped directly into the river. The tailings are largely rock flour ground to a particle size of about 100 microns. Many are eventually swept into the river and out to the Lingayen Gulf. However, other deposits accumulate along the river channel, on adjacent fields, and in the downstream floodplain. But mining is necessary to the economy of the area since more than 50,000 people depend on its revenue. The photographs demonstrate the array of environmental impacts that occurs because of mining and associated human activities. (Data and photographs supplied by the courtesy of John Wolfe.)
(a) Black Mountain Mine.
(b) Bulldozer repairing breached tailings dike until another flood removes sediment at Black Mountain Mine.
(c) Thanksgiving Mine and Mill.
(d) "Rock glacier" caused by deforestation near Black Mountain Mine.
(e) Rockslide area caused by deforestation and road construction. Note mining truck.
(f) View looking north into mouth of Bued Canyon. This is the "hinge" where the active river is 20 to 40 m wide at the degradation site, while changing in the foreground to the aggradational plain 500 to 1500 m wide.
(g) Bued River on the aggradational plain downstream from (f). The heavily burdened river has changed slope by a factor of 20 and dropped its traction load. The white line is an embankment wall designed to protect rice fields from the distributory channels.

of this are largely unknown and could be either good or bad—some concern exists that hazardous disease organisms (viruses, bacteria, etc.) could be introduced to the upper parts of the ocean. Coastal or near-shore port and terminal facilities for processing of nodules would probably constitute a greater environmental threat than the actual mining. Up to 90 percent of the material recovered would have to be discarded as waste.

SOME CRITICAL RESOURCE PROBLEMS

The most critical U.S. mineral resource problems—unequal distribution and impending shortages—are revealed by the figures in Tables 5-1 and 5-3. Shortages of many vital resources are foreseen by many in the near future. Of course, improved technologies will enable us to recover lower grade ores and new discoveries will be made, increasing our reserves.

Unfortunately, as technology and increasing prices make lower grades of ore economically mineable, the total environmental impacts and energy/materials input also tend to increase. In 1900, the lowest grade of copper that could be profitably mined was 3 percent; today it is about 0.3 percent. To produce a ton of copper from 0.35 percent ore requires the removal, transport, and milling of about 300 tons of ore, removal of another 300 tons of waste rock, and about 26,000 kwh (kilowatt-hours) of energy (equal to 4 metric tons of western U.S. coal).

Figure 5-24   Placer mining operations on the ocean front of eastern Australia. This $33 million industry mines the mineral rutile, a basic ore of tin, and exports most of the product to the United States. (Courtesy Australian Conservation Council.)

Figure 5-26   Massive mine tailings, Kalgoorlie gold fields, Western Australia. A permanent reminder and legacy of an anthropogene landscape. (Courtesy Karl H. Wyrwoll.)

Figure 5-25   Mine working area, Kalgoorlie, Western Australia. The gold mines, originally opened in the late nineteenth century, have been largely mined out. (Courtesy Karl H. Wyrwoll.)

(a)

(b)

An analogous situation exists with the new discoveries of fuels and raw materials that are increasingly more difficult to obtain because of deeper deposits, hazardous Arctic or sub-arctic climates, offshore or deep-ocean sites, and other inhospitable conditions. The result is that ever-increasing supplies of raw materials and energy are necessary to produce a given amount of a mineral resource. It is a classic case of diminishing returns, complicated by the increasing costs and scarcity of energy, environmental con-

Figure 5-27   Copper mining section between Ducktown and Copper Hill, Tennessee. Fumes from the smelting of copper and sulfuric acid have destroyed all vegetation and ushered in a human-induced erosion cycle. [(a) Courtesy Library of Congress. (b) Courtesy Dept. of Conservation, State of Tennessee.]

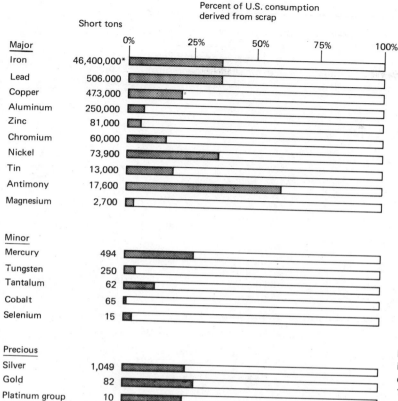

Short tons

Percent of U.S. consumption
derived from scrap

Major
Iron 46,400,000*
Lead 506.000
Copper 473,000
Aluminum 250,000
Zinc 81,000
Chromium 60,000
Nickel 73,900
Tin 13,000
Antimony 17,600
Magnesium 2,700

Minor
Mercury 494
Tungsten 250
Tantalum 62
Cobalt 65
Selenium 15

Precious
Silver 1,049
Gold 82
Platinum group 10

*Includes Exports

Figure 5-28 Scrap metals recycled in the United States in 1972, compared with total use. (Blum, 1976, Tapping resources in municipal solid waste: *Science*, v. 191, p. 671. Copyright © 1976 Amer. Assoc. Adv. Sci.)

straints, and a vast array of international political barriers.

The scramble among countries for finite resources and the inequitable distribution of vital resources are growing sources of contention in the world today. The enormous political leverage and economic benefits reaped by OPEC (Organization of Petroleum Exporting Countries) as a result of their 1973 oil embargo have encouraged other countries that control vital resources to emulate OPEC. In late 1973, Morocco, the leading exporter of phosphate rock, a crucial source of fertilizer, quadrupled the phosphate price from $14 to $64 a ton. This price increase was followed almost immediately by similar increases in price by other exporters. Jamaica, which supplies 60 percent of the alu-

minum ore (bauxite) to the United States, increased its levy on bauxite from $2 to $3 a ton. Exporters, especially the less affluent, of iron, chromium, rubber, mercury, lead, wool, and many other materials have at least attempted to set up cartels similar to OPEC in an attempt to reap financial and political rewards.

Price increases and embargoes inflict serious hardships on other nations, leading to internal instability and inflation, and growing distrust and antagonism toward other countries. All this reminds one of the "wars of redistribution"—of wealth pitting have versus have-not nations against each other, which Heilbroner warns us of in his sobering book, *An Inquiry into the Human Prospect* (1974).

Table 5-10   Mineral exploration costs and results in North America

	Value of metals discovered		Average per discovery (in millions of dollars)	
	Annual average (in billions of dollars)	In dollars per each dollar of exploration expenditure[a] (total industry)	Exploration expenditures	Value of metals discovered
A. Canada				
Nonferrous metals (and asbestos) 1971 dollar value— 1971 metal prices				
Period				
1951–55	4.4	160	2	320
1956–60	5.4	110	6	670
1961–65	5.2	95	6	570
1966–70	5.4	55	14	770
B. Western U.S.A.				
Metals (excluding uranium) 1970 dollar value—1970 metal prices				
Period				
1955–59	2.8	80	NA[b]	NA
1960–64	3.2	59		
1965–69	4.1	45		

Source: Adapted from *Mineral resources and the environment:* National Academy of Sciences, 1975.

[a]Equivalent to efficiency of exploration dollar. The expenditures include expenditures for all exploration programs whether successful or not.
[b]Data not available.

CONSERVATION AND ALTERNATIVES

A "gold mine" of recoverable mineral resources exists in American trash piles (Fig. 5-28). We are adding over 11 million tons of iron and steel, 800,000 tons of aluminum, 400,000 tons of other metals, 13 million tons of glass, 60 million tons of paper, 17 billion cans, 38 billion bottles and jars, 7.6 million discarded television sets, and 7 million junked autos and trucks annually to the U.S. trash heap. Recovery of valuable materials from this solid waste converts a costly liability into an asset. The scrap in U.S. municipal solid waste in 1975 was valued at $1.5 billion. By 1990, these wastes could be providing 6 percent of U.S. production of iron and 1.5 times the domestic production of aluminum. Collecting 70 percent of the glass in municipal trash would supply 30 percent of the industrial need for glass.

Even sludge from wastewater treatment often contains profitably extractable minerals such as gold, silver, phosphorus, tin, copper, and zinc, depending on locality. In 1973, about 7 percent (9.4 million tons) of municipal solid waste was recycled in the United States—the rest was discarded. Only limited progress toward utilizing this valuable resource has been made since then.

Recycling of mineral resources provides great energy savings. The "remelt energy" of magnesium is only 1.5 percent of that needed to obtain the metal from ore. The comparable figure for aluminum is 3 to 4 percent, and for titanium 30 percent. Most metallic mineral resources occur in a stable, oxidized state in nature and to reduce them to the desired elemental form requires great energy input. As the refined metals are used, discarded, scattered over the land, and allowed to deteriorate, the energy that will be required to locate, collect, separate, and re-refine them increases according to the laws of entropy. Rapid collection and recycling prior to "disposal" is the most efficient procedure.

Many of the barriers to efficient reuse of "waste" are institutional, legal, and political. Outmoded tax incentives, laws, subsidies, and attitudes still encourage planned obsolescence and the exploitation and consumption of virgin materials rather than recycling and conservation. To cite just one example, freight rates for scrap iron are set at three times the rates for virgin iron ore—a holdover from the days when the government was trying to encourage development of the young nation's resources.

Perspectives

The critical question is how long can we continue to fulfill our needs for vital resources in a finite world in view of such increasing demands? Some experts hold the optimistic view that, as resources become scarce, substitutes will be found, recycling will increase, and economic and social systems will adjust and persevere. They point out that many early warnings of impending mineral depletions failed to take into account increasing availability of resources made possible by new discoveries and improved technologies, and note that total exhaustion of any mineral resource will never occur.

A less optimistic view states that resource retrieval cannot keep pace with increasing demands, and the collapse of society may be unavoidable without drastic curtailment and conservation measures. This position views technological advances as more of a postponement of the inevitable than a solution, especially since they have tended to accelerate demand and exploitation.

We have been able to keep pace with exponentially increasing demands largely because of the massive inputs of inexpensive energy made possible by the rapid consumption of nonrenewable petroleum, a condition that will never be repeated. In addition, rates of exploitation are constrained by certain unavoidable thermodynamic, economic, environmental, and material limits. Mineral resources of all kinds are limited in absolute quantity, by the rate at which we can extract or recycle them, and by nature's ability to absorb the impacts of accelerating exploitation and consumption. The situation calls for an overhaul of governmental policies and priorities and the awareness that additional important minerals may all too quickly fall into the realm of "endangered species."

READINGS

Albers, J.P., Rooney, L. F., and Shaffer, G. L., 1976, Demand and supply of nonfuel minerals and materials for the United States energy industry, 1975–90—A preliminary report: U.S. Geol. Survey Prof. Paper. 1006-A, 18 p.

Bauer, A. M., 1965, Simultaneous excavation and rehabilitation of sand and gravel sites: Silver Spring, Md., Natl. Sand and Gravel Producers Assoc. 59 p.

Brobst, D. A., and Pratt, W. P., eds., United States mineral resources: U.S. Geol. Survey Prof. Paper 820, 722 p.

Brooks, D. B., and Andrews, P. W., 1976, Mineral resources, economic growth, and world population: in Abelson, P. H., and Hammond, A. L., ed., Materials: renewable and nonrenewable resources, Washington, D.C., Amer. Assoc Adv. Sci., p. 41–47.

Colwell, R. N., 1968, Remote sensing of natural resources: Sci. Amer., v. 218, p. 54–69.

Ernst, W. G., 1969, Earth materials: Englewood Cliffs, N.J., Prentice-Hall, 149 p.

Gilbert, G. K., 1917, Hydraulic-mining debris in the Sierra Nevada: U.S. Geol. Survey Prof. Paper 105, 154 p.

Goudarzi, G. H., Rooney, L. F., and Shaffer, G. L., 1976, Supply of nonfuel minerals and materials for the United States energy industry, 1975–90: U.S. Geol. Survey Prof. Paper 1006-B. 37 p.

Hibbard, W. R., Jr., 1968, Mineral resources: challenge or threat?: Science, v. 160, p. 143–159.

Lovering, T. S., 1969, Mineral resources from the land: in Resources and man, Nat. Ac. Sci., Nat. Research Council, San Francisco, W. H. Freeman, p. 109–134.

Menard, H. W., 1974, Geology, resources, and society: San Francisco, W. H. Freeman, 621 p.

National Academy of Sciences, 1975, Mineral resources and the environment: Comm. on Mineral Resources and the Environment, Washington, D.C., 348 p.

Schellie, K. L., et al., 1963, Site utilization and rehabilitation practices for sand and gravel operations: Silver Spring, Md., Natl. Sand and Gravel Producers Assoc., 80 p.

Secretary of the Interior, 1977, Mining and minerals policy: Annual Report, Washington, D.C., U.S. Govt. Printing Office, 154 p.

Skinner, B. J., 1969, Earth resources: Englewood Cliffs, N.J., Prentice-Hall, 149 p.

U.S. Bureau of Mines, 1975, Mining and minerals policy, 1975: U.S. Govt. Printing Office.

U.S. Geol. Survey, 1975, Mineral resource perspectives 1975: U.S. Geol. Survey Prof. Paper 940, 24 p.

Woolridge, S. W., and Beaver, S. H., 1950, The working of sand and gravel in Britain: a problem in land use: Geog. Jour., v. 115, p. 42–57.

Chapter Six
Energy and Fossil Fuels

A "Texas tower" (oil-drilling platform installation) in the Gulf of Mexico. (Courtesy Bethlehem Steel Co.)

INTRODUCTION

Energy is the ability to do work. **Work** is accomplished whenever matter is moved. The rate at which work is performed is **power.** The energy involved in moving two identical objects from place A to place B is identical but, if one of the objects is moved more rapidly, more power is generated during its movement.

Energy occurs in many different forms, among them mechanical, electrical, radiant, chemical, potential, and nuclear (Fig. 6-1). Energy is readily converted from one form to another, as in the conversion of the sun's radiant energy to heat when it strikes earth, or a green plant's ability to construct complex organic molecules using the same solar energy. Energy is used throughout all activities in civilization (Figs. 6-2 and 6-3). This energy is stored in the plant matter as chemical energy until decay or combustion causes the molecules to break down, thus releasing the stored chemical energy as heat.

The most intense, concentrated manifestation of energy is nuclear—the energy that binds together atomic nuclei. In nuclear reactions, a tiny amount of matter is converted into energy. The energy thus liberated exceeds that of chemical reactions by factors of several millions. For example, 1 kg of nuclear (uranium) fuel releases

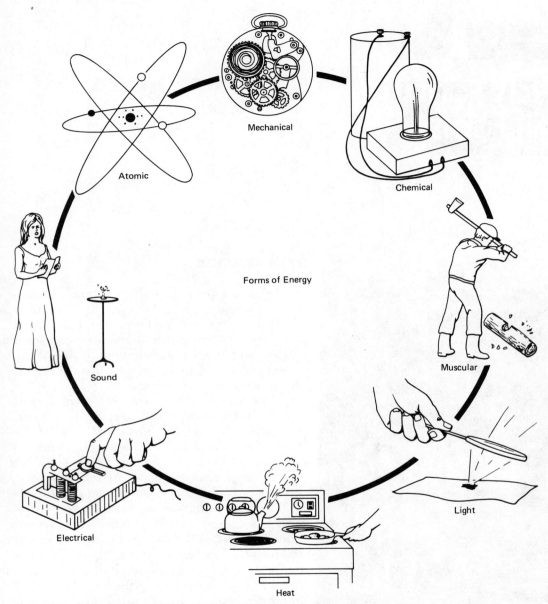

Mechanical

Chemical

Atomic

Muscular

Forms of Energy

Sound

Light

Electrical

Heat

Figure 6-1 Types of energy. (New York State Electric and Gas Corp.)

energy equal to the combustion of 3 million kg of coal.

The **first law of thermodynamics** states that the amount of matter and energy in the universe remains constant—that is, matter and energy can neither be created nor destroyed, although many conversions are possible. Human use of energy is merely a matter of converting energy present in one form into another form, usually into heat or electricity, which is suitable for our purposes.

The **second law of thermodynamics** is

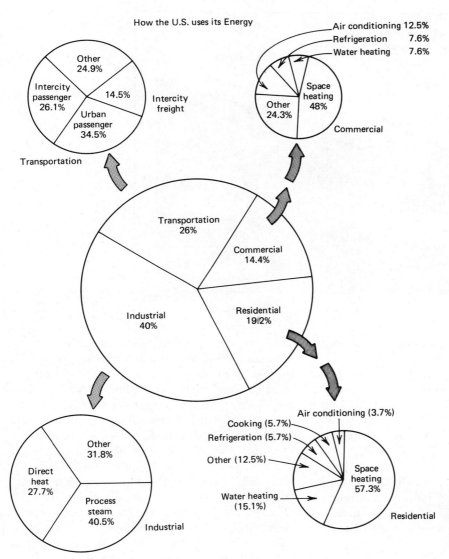

How the U.S. uses its Energy

Figure 6-2 How the United States uses its energy. (Citizens Workshops of the U.S. Atomic Energy Commission.)

more difficult to grasp, but also more important to the study of energy and the environment. In simple terms, it states that entropy increases during all actions involving matter and energy. **Entropy** is the amount of disorder present. Hence, this says that the universe is "running down," being reduced to lower levels of organization. This can be illustrated by thinking of a house built of dominoes which topples at a slight disturbance. The order and the energy level of the toppled dominoes is less than it was to begin with, and only by input of energy from an outside source can the house be rebuilt. In similar fashion, your health would gradually deteriorate and your body die without input of energy in the form of food. And imagine what would happen to earth if the input of solar energy were to cease! Entropy is increased in all energy conversions; that is, useful energy is lost and thus becomes unavailable for use.

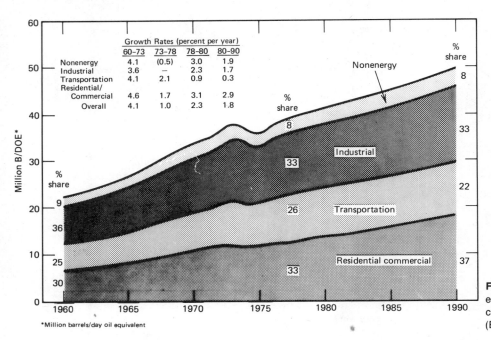

Figure 6-3 U.S. energy demand by consuming sector. (EXXON Corp.)

Therefore, to use the vernacular, not only can you not win, you can't even break even.

When we burn any fuel, we dissipate its energy, converting its energy to a lower state of energy, most commonly heat. Only by an input of energy far exceeding that which is released on burning could the elements be recombined into the original molecules of fuel. Hence, when a pound of coal is burned, that coal is, in a very true sense, gone forever, much like Humpty Dumpty. Yet energy is a basic ingredient of civilized nations, is related to the gross national product, and produces a wide range of benefits as well as impacts.

THE FOSSIL FUEL AGE

Nothing in history has had such far-reaching material effects on human society and the environment as our utilization of fossil fuels. In the developed countries, the escalating use of and misuse of raw materials, agricultural devel-

opments, eating habits, social attitudes, recreational activities, and much more are, to a large extent, by-products of the availability and low price of fossil fuels, especially petroleum. Modern civilization is almost totally dependent on an enormous and continuous supply of inexpensive energy. It is essential for the extraction, refining, and distribution of the mineral resources described in Chapter 5. Our highly complex, and often absolutely necessary, technologies and institutions require vast quantities of energy to function. Yet they cannot respond rapidly to changing conditions. For example, it will require huge sums of money and several decades to redesign our transportation systems so that we are no longer heavily dependent on petroleum-fueled vehicles to transport people and goods. Consider what would happen if liquid petroleum supplies were drastically reduced while no adequate alternative systems exist. The answer suggests that modern civilization is, in many ways, remarkably vulnerable.

Fossil fuels provide about 93 percent of the energy that the United States runs on (Fig. 6-4). And now we are told that the end of the age

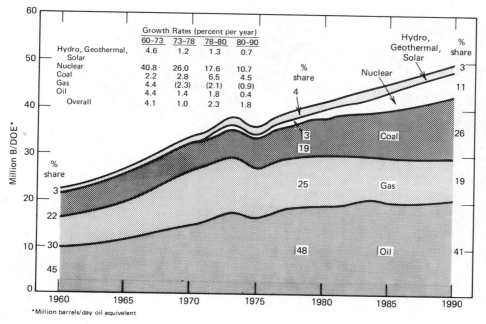

Figure 6-4 U.S. energy supply. (EXXON Corp.)

of abundant, reasonably-priced fossil fuels is in sight—not in a century or so, but within a generation. Some analysts predict a "severe energy crunch" as early as 1985. Already there have been grim reminders of the modern world's vulnerability. Early in 1974, the wheat harvest in India was reduced by 1 million tons (enough to feed 6 million Indians for a year) because they could not purchase needed gasoline for irrigation pumps. In early 1977, an estimated 1.8 million workers were laid off in the United States due to a natural gas shortage in the Midwest and Northeast.

The fossil fuel age was launched in the eighteenth century in England when depletion of firewood forced the substitution of coal. By 1800, Britain was producing about 10 million tons of coal annually and, as a result of coal and the development of the steam engine, was well on its way to becoming the first industrial giant.

In 1850, 90 percent of the commercial energy used in the United States came from wood, wind, and water but, by 1900, fossil fuel was dominant with 75 percent of the energy

from coal. In the twentieth century, oil and natural gas became the dominant feedstock for the world's ever-growing energy appetite (Fig. 6-5). Unlike the earlier switch to coal, the transition to oil and gas was a matter of choice, not necessity. Oil and natural gas were more convenient fuels, cleaner to burn, and easier to transport once an efficient pipeline distribution system was constructed. In addition, massive advertising and a variety of federal economic incentives to producers greatly encouraged the sales and consumption of oil and natural gas at unrealistically low prices, ultimately leading to the present-day dependency on the inefficient use of these precious fuels. In 1950, coal still provided 60 percent of the world's commercial fuel needs but, by 1975, oil and gas were supplying two-thirds of the world's energy budget. In the United States in 1977, oil and gas represented 75 percent of the fuel consumed.

There have always been observers who decried the incredible waste that accompanied the early days of fossil fuel exploitation and warned of early depletion of resources (e.g., Stuart

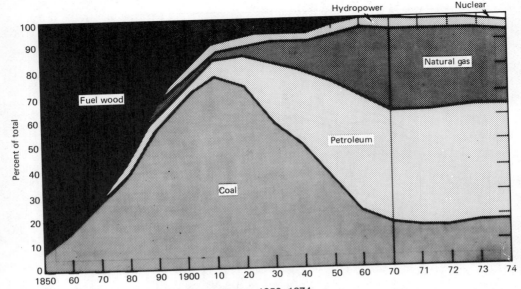

Figure 6-5   U.S. energy consumption patterns, 1850–1974. (Historical Statistics of the United States, Bureau of Census, U.S. Bureau of Mines.)

Chase in *The Challenge of Waste*, 1911), but only recently have such warnings been taken seriously. It might be said that the present period of awareness and alarm began in the 1950s when various respected scientists and economists recognized that the "Petroleum Age" would not last forever—and the decline could come much faster than most expected. By then, about 50 billion barrels of the estimated 200-billion-barrel U.S. oil reserve had been used, and the rest could very likely be gone before the end of the century.

Among the major, and often much maligned, "prophets" was M. King Hubbert. In 1956, Hubbert horrified his fellow employees at Shell Oil with his carefully considered graphs, showing a rapidly approaching peak of U.S. oil production followed by a steady decline back to near-zero production. His predictions also greatly annoyed his superiors.

As it turned out, U.S. oil production did peak, as he predicted, in 1970 (3.6 billion barrels in 1970; 2.8 billion in 1977). Estimates of the U.S. share of the world's petroleum reserves had shrunk from 40 percent in 1937 to 6 percent in 1970 (Fig. 6-6). Since 1970, the United States has become increasingly dependent on imported oil, in spite of the OPEC oil embargo of 1973 (Fig. 6-7). In 1977, the United States was consuming about 18 million barrels per day and importing 45 percent more oil than in 1975. Foreign supplies accounted for 8.6 of the 18 million barrels and cost the U.S. economy $45 billion, the major factor in a trade deficit of $27 billion.

World oil production is expected to peak in about 1990 before beginning its decline. Large new discoveries will do little to change the general picture, except to delay the inevitable by a very few years. It is therefore imperative that the world drastically reduce its consumption of petroleum and launch an all-out search for alternative energy sources (Chapter 7).

Current U.S. government plans hope to encourage conservation by raising fuel costs and decrease dependency on foreign oil to about 6 million barrels a day by 1985, largely by substituting coal and synfuels for petroleum. **Synfuels,** a poor word and misnomer, is the term

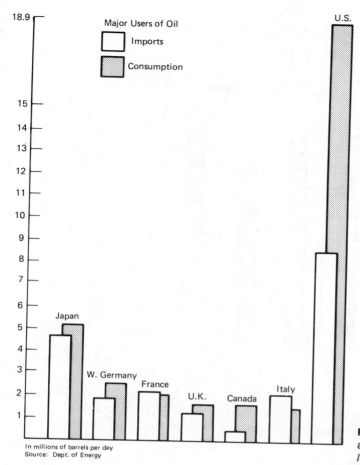

Major Users of Oil
☐ Imports
▨ Consumption

In millions of barrels per day
Source: Dept. of Energy

Figure 6-6 Worldwide petroleum production and reserves. (*Oil and Gas Journal* and *The International Petroleum Encyclopedia*, 1976.)

now used for the gas conversion of coal, oil shale, and tar sands. To meet all U.S. energy industry goals for 1985 will require a capital investment estimated at $580 billion. After this date, energy consumption is expected to continue, increasing at about 2.2 percent per year.

Many individuals are beginning to recognize the impossibility of maintaining even slowly increasing growth and consumption, and urge immediate plans to be made for a transition to a sustainable, steady-state economy. Numerous studies, including a massive document from the federal Government Accounting Office, conclude that the U.S. energy goals are unrealistic when the economic, legal, and especially environmental constraints are considered. Some feel our only way out of the looming energy dilemma is nuclear energy; others favor a "soft" energy path emphasizing conservation and renewable energy sources, such as solar energy. Still others feel that new discoveries and improved technologies will enable fossil fuels, especially coal, to meet most of our needs for several more decades, providing needed time for a smooth transition to the post-petroleum era. We will examine these options in Chapter 7.

Whatever choices are made, whether by plan or default, energy will never again be cheap, and heated controversies between con-

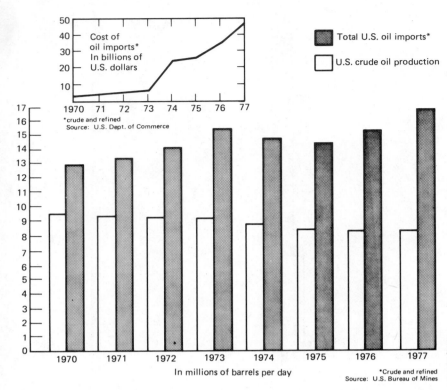

Figure 6-7 U.S. oil imports and production. (U.S. Dept. Commerce and U.S. Bureau of Mines.)

flicting interests are unavoidable. At least, it is now generally recognized that major changes are inevitable and that the age of abundant, cheap fossil fuel is indeed, as Hubbert noted nearly a quarter century ago, a brief, anomalous phase in human history.

COAL

History

Coal is by far our most abundant fossil fuel. Most estimates indicate that from 65 to 85 percent of the world's retrievable fossil fuel energy occurs as coal. Because of depletion of traditional sources of oil and natural gas and political uncertainties concerning Middle Eastern oil, attention is increasingly directed toward coal.

Coal has a long, and occasionally grim, history. It was burned 4000 years ago in Europe, during the Bronze Age, 2000 years ago in China, and 1000 years ago by American Indians. But coal was hard to dig, awkward to handle, and dirty to burn. Wood continued to be the major fuel until it began to run out. London began turning to coal around A.D. 1300 because of an "energy crisis" caused by the limited availability of wood. The soot and foul air that resulted moved Edward I to issue a ban on coal burning in London in 1306. However, there just wasn't sufficient wood to support energy needs in the British Isles and, by 1750, England was the world's first major fossil fuel-powered nation, a fact which allowed it to become the leader of the Industrial Revolution.

In 1850, coal supplied only 7 percent of the energy needs of the United States. As forests and firewood supplies were depleted, coal became more important until, in 1910, it represented 70 percent of what was burned. The early days of coal mining, especially in the Appalachians, comprise one of the most scandalous chapters in American history. Thousands of poor, often illiterate, people sold mineral rights to land speculators for 10 to 50 cents an acre. As a result, years later, people saw their homes and property totally destroyed when bulldozers were allowed to move in under clauses of the notorious "broad form deeds" they, or former owners, had signed. Working conditions in the mines were atrocious—over 100,000 men were killed and over a million permanently disabled in mining accidents in the first 75 years of this century. Even more have suffered from black lung disease. And the land and waters were commonly left in ruin.

After 1910, oil and gas began cutting into coal's percentage of the market and, by 1978, coal accounted for slightly less than 20 percent of the fuel burned in the United States and about 30 percent of world energy production. Although the proportion of coal the world uses has declined, the actual quantity involved has increased from 700 million tons per year in 1900 to 2.7 billion tons in 1977.

Origin and Geology of Coal

About 2 percent of the radiant energy of the sun which reaches the earth is absorbed by plants. They utilize this energy to combine simple, inert hydrocarbons into the complex, organic molecules of living matter. The solar energy stored in this organic matter is normally dissipated by oxidation and decay shortly after death. However, if dead plant matter should fall into an oxygen-poor (euxinic) environment, such as standing water, a small proportion (about 2 percent) of it may be buried by sedimentation and isolated from the degrading surface environment. If the area is subsiding so that more and more sediment can accumulate, the pressure slowly forces water, oxygen, nitrogen, and other impurities out of the organic material until carbon-rich coal is formed. Slowly

subsiding, densely vegetated, marshy coastal and alluvial plains often provided ideal conditions for such burial of plant debris during periods of crustal stability in the geologic past.

The great coal beds of Pennsylvania began about 300 million years ago as large subtropical forests in which grew such "trees" as calamites, related to modern horsetails, and cordates, 30 m high ancestors of the conifers, along with *Lepidodendron* and *Sigillaria*, both ancestors of today's club mosses. There were abundant ferns and shrubs, but no grasses or flowering plants of any kind. This dense, junglelike vegetation grew on an extensive swampy plain along the eastern coast of a large shallow sea. The climate was warm and wet and almost unchanging. It probably required about 30,000 years of continuous plant growth in these ancient forests (perhaps the most luxuriant, fastest-growing forests in the history of the earth) to ultimately form 1 m of coal.

The oldest coals are of Devonian age (about 350 million years). Some important properties of coal are summarized in Table 6-1.

Peat (not a coal) is an abundant, low-grade fuel and an energy source of local importance (see Chapter 7). **Lignite,** or brown coal (it may be black as well as brown in color), is the first-formed, lowest quality variety of coal. Vast quantities occur in North Dakota, which has the largest tonnage of coal of any state. Coal which has been subject to greater heat and pressure and/or residence time during its subsurface interment evolves into **bituminous** or "soft" coal. **Subbituminous** coal represents an intermediate stage between lignite and bituminous and is common in the western United States, especially in the Powder River Basin of Wyoming and Montana (Figs. 6-8 and 6-9). **Anthracite** or "hard" coal formed under more intense heat and pressure represents a low-grade of metamorphism. Anthracite is comparatively rare, pure, clean-burning coal, although some bituminous coals actually have a higher heat value. Heat contents are given in Table 6-1 for comparison. Some woods (e.g., hickory) have a heat

Table 6-1 Coal properties

Coal types	Fixed carbon %	Moisture %	Ash %	Volatiles %	Content BTU/lb per 1,000
Peat[a]					4.5
Lignite	31.7	32.5	5.0	30.8	6–8
Sub-bituminous	44.7	20.7	3.9	30.7	9.5
Bituminous	58.3	3.0	2.4	36.3	12–14.5
Anthracite	90.3	4.2	2.4	3.1	13.5–14

[a]Not a coal, but listed for comparison purposes.

value of 6500 to 7000 BTU/lb, equal to many lignites; a pound of fuel oil yields 19,000 BTU and a cubic foot of natural gas, 1100 BTU.

The **ash** remaining after combustion of coal is composed of inorganic solids, mostly clays which were washed into the swamps during accumulation. Coal also contains numerous chemical impurities, often including water (note the moisture content of lignite in Table 6-1) and over two dozen elements such as sulfur, nitrogen, uranium, and other radioisotopes, and 14 toxic heavy metals (lead, cadmium, mercury, arsenic, etc.). Many of these elements go up the stack and into the air when coal is burned and contribute to coal's Achilles' heel—air pollution.

The sulfur content is especially important to general air quality. Coals vary from 0.5 percent or less (by weight) to over 10 percent sulfur and may be classified as low (less than 1%) medium (1–3%), and high (over 3%) sulfur coals.

The Geography of Coal

Figure 6-10 shows the distribution of coal in the United States. Anthracite is limited to areas near orogenic belts where heat and pressure were greater, as in eastern Pennsylvania. Overall, western coals are younger than eastern coals, were never as deeply buried, and are correspondingly of lower grade. The subbituminous coals and lignites of the Fort Union group of rock formations in Wyoming, Montana, and the Dakotas constitute the most extensive coal deposits on earth. One coal seam, the Wyodak in the Gillette, Wyoming area, is up to 30 m thick.

The world "estimated total resource" of coal is about 16.5 trillion tons according to the U.S. Geological Survey (Tables 6-2 and 6-3). Approximately 6 percent of this amount has been mined. As with most resources, distribu-

Figure 6-9 Aerial view looking northwest at the Western Energy Co. strip coal mine, Colstrip, Montana. Overburden is being removed from the coal seam, and the spoil piles are being reclaimed by blending them into the terrain. (Photo by U.S. Bureau of Reclamation.)

Figure 6-8 Big Sky coal mine of Peabody Coal Co., Montana. (Photo by U.S. Bureau of Reclamation.)

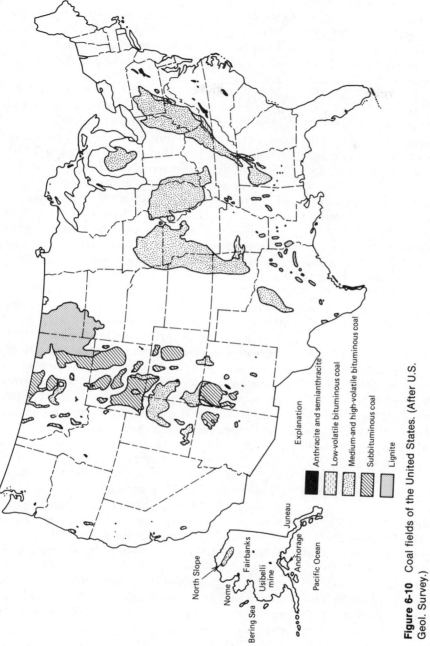

North Slope

Nome

Bering Sea

Fairbanks

Usibelli mine

Juneau

Anchorage

Pacific Ocean

Explanation

Anthracite and semianthracite

Low-volatile bituminous coal

Medium- and high-volatile bituminous coal

Subbituminous coal

Lignite

Figure 6-10 Coal fields of the United States. (After U.S. Geol. Survey.)

tion is uneven, with 11 trillion tons in Asia and 4 trillion in North America. Much of this coal is not economically retrievable. The estimate for "identified resources," which occur in seams 35 cm or more in thickness and within 1200 m of the surface, drop to 6.5 trillion tons for the world and 2 trillion for the United States (Table 6-3). The "reserve base" estimate, referring to coal in seams at least 71 cm (28 in.) thick and less than 300 m underground, is 433 billion tons for the United States (Table 6-4). About 40 billion tons of coal have already been consumed in the United States. With projected increases in the use of coal, U.S. economically recoverable reserves could conceivably be exhausted within 100 years. Of course, new technologies and changing economies will allow less desirable deposits to be mined, but these thinner, deeper deposits will require greater energy input and higher cost.

About 75 percent of the coal consumed in the United States is used to generate electricity, 15 percent for metallurgical processes (mostly steel), and 10 percent for domestic and industrial heat. The oil companies also have a great interest in coal, as shown in Table 6-5.

Current Perspectives on Coal in the United States

In view of declining domestic production of oil and gas, little doubt exists that the United States will continue to increase consumption of its enormous reserves of coal. Government goals in the 1970s consistently called for nearly doubling coal use to about 1.2 billion tons annually by 1985. A 1978 National Coal Association survey indicated 241 new coal-fired plants already in the planning stage—by 1985 these plants would consume 400 million tons of coal annually.

The implications of such unprecedented "development" are staggering. To meet such a goal and to maintain a production of 1.2 billion tons a year, it will be necessary to open 400 to 800 new mines, according to the General Accounting Office in 1977. Over 150,000 new workers will have to be hired and trained by 1985. During the next two decades, $26 to $45 billion of new capital will be needed. Transportation of the coal will require $10 billion in improvements to railroads; probably even more to upgrade and relocate roads over which trucks must haul coal.

Dozens of huge new power plants (some exceeding 10,000 megawatts), envisioned for relatively undeveloped parts of the western United States, would totally alter the economy and life-styles of those farming and ranching regions. This ongoing exploitation of near-surface coal in the West is a source of bitter conflict, with the farmers and ranchers often coming out on the losing side. It is difficult to resist mining coal valued at $300,000 when the acre of wheat-growing land it lies beneath is valued at $300.

Many reasons have been put forth for this great interest in western coal, as opposed to other energy sources such as increased mining of eastern coal. Among them, the arguments that western coal is less expensive; it is low in sulfur and won't violate air quality standards; and strip mining of western coal is safer and faster than that of eastern coal.

However, there is a single basic underlying reason—it has been more profitable for coal producers to mine western coal. There may be 70 billion tons of strippable coal in the West, whereas most eastern coal must be deep-mined. It requires only a third as many workers to produce the same amount of coal from a western strip mine as from an eastern underground mine. However, it is not safer to strip mine coal because deaths per man-hour of work are about equal for surface and underground mining. It takes considerably longer to open a western strip mine (4–15 years) than an eastern underground mine (2½–5 years), largely due to critical shortages of specialized strip mining equipment. Most western miners do not belong to unions and labor disputes are not as common. Western land contains 60 percent of prime U.S. coal reserves and, until 1971, much of it was owned by the federal government or railroad companies. It is far cheaper and easier to lease or buy such land than in the heavily populated East. In 1971, a moratorium was imposed on leasing federal land, but 16 billion tons of coal lay under the land already leased to industry. Only 15 companies controlled 70 percent of the leased land, and they had produced coal from only 7 percent of that by 1977.

Western coal tends to be lower in sulfur

Table 6-2 Estimated total original coal resources of the world, by continents[1]
[In billions of short tons]

Continent	Identified resources[3]	Hypothetical resources[2]	Estimated total resources
Asia[4]	[5]7,000	4,000	[6]11,000
North America	1,720	2,880	4,600
Europe	620	210	830
Africa	80	160	240
Oceania	60	70	130
South and Central America	20	10	30
Total	[5]9,500	7,330	[6]16,830

[1]Original resources in the ground in beds 12 in. or more thick and generally less than 4,000 ft below the surface, but includes small amounts between 4,000 and 6,000 ft.
[2]Identified resources: Specific, identified mineral deposits that may or may not be evaluated as to extent and grade, and whose contained minerals may or may not be profitably recoverable with existing technology and economic conditions.
[3]Hypothetical resources: Undiscovered mineral deposits, whether of recoverable or subeconomic grade, that are geologically predictable as existing in known districts.
[4]Includes European U.S.S.R.
[5]Includes about 6,500 billion short tons in the U.S.S.R.
[6]Includes about 9,500 billion short tons in the U.S.S.R. (Hodgkins, 1961, p. 6).

Table 6-3 World resources of bituminous coal and lignite

Country or Countries	Resources (million short tons)
United States	3,968,264*
Australia	218,879
Belgium	279
Canada	119,906
France	1,511
Germany, West	316,418
India	91,466
Japan	9,511
South Africa, Rep. of	48,875
United Kingdom	179,472
Other Market Economy Countries	109,899
Central Economy Countries	7,534,189
WORLD TOTAL	12,598,669

*The estimated coal resources in the United States total 3,968 billion short tons, of which 1,731 billion tons have been identified and 2,237 billion tons are hypothetical, according to an estimate made by the U.S. Geological Survey in 1974.

than eastern coal, but this is more than offset by its overall lower heat content. By volume, 70 percent of our coal is west of the Mississippi; in terms of more meaningful heat content, 45 percent is west of the Mississippi. It takes more than 2 kg of lignite to produce the same energy as 1 kg of bituminous coal. Thus, bituminous can have twice the sulfur content as the lignite and still produce less sulfur dioxide, fewer solid wastes, and other pollutants. New legislation will require all coal used in power plants to be scrubbed to remove sulfur. This will add to its cost and further reduce any advantages western coal may enjoy by virtue of its low-sulfur content.

The distance from the large energy-consuming markets is also a disadvantage of western coal. Whether by rail, truck, barge, or slurry pipeline, it requires energy to transport coal. Shipping now represents an average of 25 percent of the cost of delivered coal. For eastern Montana coal used by midwestern utilities, shipping may constitute 75 percent of the delivered cost. Even when western coal is used in mine-mouth power plants, large amounts of energy are required to transport the electricity to distant markets.

Unfortunately, there have been many cases of eastern and midwestern utilities burning western coal, often at more than double the price of locally available, low-sulfur coal. This is due to the higher costs that are passed on to the consumers. Utilities are allowed to automatically add 10 percent to such costs for their profit margin; the higher the costs, the higher their profits. In addition, many utilities own their own western coal mines and transportation facilities—in effect, they sell to themselves and earn additional profits at each step in the transaction. Some states have passed laws requiring utilities to use local coal when available.

If western coal is used in mine-mouth power plants, large amounts of energy are still required to transmit the electricity to distant markets.

In spite of the huge strippable reserves in the West, an all-out effort to utilize these de-

Table 6-4 Total estimated remaining coal resources of the United States, January 1, 1972 (In millions of short tons. Figures are for resources in the ground, about half of which may be considered recoverable. Includes beds of bituminous coal and anthracite 14 in. or more thick and beds of subbituminous coal and lignite 2½ ft or more thick.)

State	Identified resources[a]				
	Overburden 0–3000 ft				
	Estimated identified resources remaining in the ground, Jan. 1, 1972				
	Bitumi-nous coal	Subbitumi-nous coal	Lignite	Anthra-cite and semianthra-cite	Total
Alabama	13,342	0	2,000	0	15,342
Alaska	19,413	110,668	d	e	130,081
Arizona	21,246[f]	f	0	0	21,246
Arkansas	1,638	0	350	430	2,418
Colorado	62,339	18,242	0	78	80,659
Georgia	24	0	0	0	24
Illinois	139,124	0	0	0	139,124
Indiana	34,573	0	0	0	34,573
Iowa	6,509	0	0	0	6,509
Kansas	18,674	0	h	0	18,674
Kentucky	64,842	0	0	0	64,842
Maryland	1,158	0	0	0	1,158
Michigan	205	0	0	0	205
Missouri	31,014	0	0	0	31,014
Montana	2,299	131,855	87,521	0	221,675
New Mexico	10,752	50,671	0	4	61,427
North Carolina	110	0	0	0	110
North Dakota	0	0	350,630	0	350,630
Ohio	41,358	0	0	0	41,358
Oklahoma	3,281	0	h	0	3,281
Oregon	50	284	0	0	334
Pennsylvania	56,759	0	0	20,510	77,269
Rhode Island	0	0	0	j	—
South Dakota	0	0	2,031	0	2,031

Tennessee	2,572	0	0	0	2,572
Texas	6,048	0	6,824 0	0	12,872
Utah	23,541[k]	180[k]		0	23,721[k]
Virginia	9,352	0	0	335	9,687
Washington	1,867	4,190	117	5	6,179
West Virginia	100,628	0	0	0	100,628
Wyoming	12,705	107,951	b	0	120,656
Other States	610[m]	32[n]	46[o]	0	688
Total	686,033	424,073	449,519	21,362	1,580,987

Source: P. Averitt, 1973, Coal: U.S. Geological Survey, Professional Paper 820. In D. A. Brobst and W. D. Pratt, *United States Mineral Resources.*

[a] Identified resources: Specific, identified mineral deposits that may or may not be evaluated as to extent and grade, and whose contained minerals may or may not be profitably recoverable with existing technology and economic conditions.

[b] Hypothetical resources: Undiscovered mineral deposits, whether of recoverable or subeconomic grade, that are geologically predictable as existing in known district.

[c] Estimates by H. M. Beikman (Washington), H. L. Berryhill, Jr. (Wyoming), R. A. Brant (Ohio and North Dakota), W. C. Culbertson (Alabama), H. H. Doelling (Utah), K. J. Englund (Kentucky and Virginia), B. R. Haley (Arkansas), E. R. Landis (Colorado and Iowa), E. T. Luther (Tennessee), R. S. Mason (Oregon), C. E. Robinson (Missouri), J. A. Simon (Illinois), J. V. A. Trumbull (Oklahoma), C. E. Wier (Indiana), and the author for the remaining States.

[d] Small resources of lignite included under subbituminous coal.

[e] Small resources of anthracite in the Bering River field believed to be too badly crushed and faulted to be economically recoverable (Barnes, 1951).

[f] Includes coal in the Dakota Formation of the Black Mesa field, some of which may be of subbituminous rank. Does not include small resources of thin and impure coal in the Deer Creek and Pinedale fields.

[g] See other summary reports on coal resources in individual States as follows: Arizona (Averitt and O'Sullivan, 1969); Georgia (Butts and Gildersleeve, 1948; Sullivan, 1942); Illinois (Cady, 1952); Kansas (Abernathy and others, 1947); Missouri (Hinds, 1913; Searight, 1967); eastern Montana (Averitt, 1965); Ohio (Struble and others, 1971); Oregon (Mason and Erwin, 1955; Mason, 1969); Pennsylvania anthracite (Ashley, 1945; Ashmead, 1926; Rothrock, 1950); and Utah (Averitt, 1964).

[h] Small resources of lignite in beds generally less than 30 in. thick.

[i] From Ashley (1944).

[j] Small resources of meta-anthracite in the Narragansett basin believed to be too graphitic and too badly crushed and faulted to be economically recoverable as fuel.

[k] Excludes coal in beds less than 4 ft thick.

[l] Includes coal in beds 14 in. or more thick, of which 14,000 million tons is in beds 4 ft or more thick.

[m] California, Idaho, Nebraska, and Nevada.

[n] California and Idaho.

[o] California, Idaho, Louisiana, and Mississippi.

Table 6-4 *Continued*

	Hypothetical resources[b]			Total resources
	Overburden 0–3000 ft	Overburden 3000–6000 ft	Overburden 0–6000 ft	Overburden 0–6000 ft
Source of estimate	Estimated hypothetical resources in unmapped and unexplored areas reasonably near the surface[c]	Estimated hypothetical resources in deeper structural basins[c]	Total estimated hypothetical resources	Total estimated identified and hypothetical resources remaining in the ground Jan. 1, 1972
Culbertson (1964); T. A. Simpson (written commun., 1972)	20,000	6,000	26,000	41,342
Barnes (1951; 1967)	130,000	5,000	135,000	265,081
Peirce and others (1970)[g]	0	0	0	21,246
Haley (1960)	4,000	0	4,000	6,418
Landis (1959)	146,000	145,000	291,000	371,659
Johnson (1946)[g]	60	0	60	84
Simon (1965)[g]	100,000	0	100,000	239,124
Spencer (1953)	22,000	0	22,000	56,573
Landis (1965)	14,000	0	14,000	20,509
Schoewe (1952; 1958)[g]	4,000	0	4,000	22,674
Huddle and others (1963)	52,000	0	52,000	116,842
Averitt (1969)	400	0	400	1,558
Cohee and others (1950)	500	0	500	705
Robertson (1971)[g]	18,200	0	18,200	49,214
Combo and others (1949; 1950)[g]	157,000	0	157,000	378,675
Read and others (1950)	27,000	21,000	48,000	109,427
Reinemund (1949; 1955)	20	5	25	135
Brant (1953)	180,000	0	180,000	530,630
Brant and DeLong (1960)	2,000	0	2,000	43,358
Trumbull (1957)	20,000	10,000	30,000	33,281
R. S. Mason (written commun., 1965)[g]	100	0	100	434

Reese and Sisler (1928); Arndt and others (1968)[g]	10,000[i]	0	10,000	87,269
Toenges and others (1948)	0	0	0	—
D. M. Brown (1952)	1,000	0	1,000	3,031
Luther (1959; written commun., 1965)	2,000	0	2,000	4,572
Mapel (1967); Perkins and Lonsdale (1955)	14,000	0	14,000	26,872
Doelling 1970, 1971a, b, c, d, e, f, Doelling and Graham, (1970; 1971); H. H. Doelling (written commun., 1971)	21,000[i]	35,000	56,000	79,721
Brown and others (1952)	5,000	100	5,100	14,787
Beikman and others (1961)	30,000	15,000	45,000	51,179
Headlee and Nolting (1940)	0	0	0	100,628
Berryhill and others (1950; 1951)	325,000	100,000	425,000	545,656
—	1,000	0	1,000	1,688
	1,306,280	337,105	1,643,385	3,224,372

posits could largely exhaust them before the end of this century. Far more coal can be retrieved by underground mining than by stripping, and we will eventually have to rely more on our subsurface reserves. There exists a danger of overcommitment to surface mining, resulting in a loss of qualified underground miners and neglect of improvements in underground mining technology. It has also been suggested that large amounts of readily strippable reserves should be conserved for use in national emergencies only.

There is determined opposition to continued expansion of the western coal boom. Many questions concerning socioeconomic impacts, water supply, rehabilitation potential, pollution, and the overall net economy and energy budget of these operations have yet to be answered. Some of these topics are addressed elsewhere in this book. In addition, local and federal governments are imposing stricter environmental regulations on coal mining in an effort to avoid the abuses of the past. After 37 years of trying, a national Surface Mining Control and Reclamation Act was passed in 1977, and many states have even far stricter controls; for example, Montana has imposed a 20 to 30 percent tax on coal mined in that state—and it is being paid as mining continues.

Impacts of Coal

Mining and safety The coal mining industry in the United States formerly employed about 700,000 persons. As a result of automation and the switch to less labor-intensive strip mining, this figure dropped to only 126,000 by 1972. The revival of coal as a major energy source is expected to increase coal mining employment to as many as 243,500 by 1985 and 390,600 by

Table 6-5 Coal production by U.S. oil firms in the United States (millions of tons)[a]

Coal Company	Parent Co.	1964	1965	1966	1967	1968	1969	1970	1971	1972	1973	1974	1975
Consolidation	Continental	45.4	48.6	51.4[b]	56.5	59.9	60.9	64.1	54.8	64.9	60.5	51.8	54.9
Island Creek	Occidental	21.2	20.6	23.7	25.9	25.9[b]	30.3	29.7	22.9	22.6	22.9	20.8	19.4
Old Ben	Sohio	5.1	6.3	9.9	10.3	9.9[b]	12.0	11.7	20.5	11.2	10.8	9.5	9.3
Pittsburg & Midway	Gulf	7.1[b]	8.2	8.8	9.0	9.2	7.6	7.8	7.1	7.7	8.1	7.5	7.3
Arch Minerals	Ashland	2.3	5.1	6.8	7.5	7.0	6.8	6.3	7.2[b]	11.2	12.5	13.9	13.5
Monterey Coal	Exxon	—	—	—	—	—	—	0.3	1.2	2.0	2.7	2.5	2.9
AMAX	SOCAL	8.2	8.3	8.5	8.6	9.3	11.3	14.4	13.3	16.4	16.7	19.9	21.8[b]
Valley Camp[c]	Quaker State	4.3	4.3	4.8	5.5	5.2	5.3	5.5	4.2	4.8	4.1	3.3	3.4
Eight company total		93.6	101.4	113.9	123.3	126.4	134.2	139.8	121.2	140.8	138.3	129.2	132.5
Industry total		487.0	512.1	533.9	552.6	545.2	560.5	602.9	552.2	595.4	591.7	601.0	646.0
Eight company output as percent of total		19.2	19.8	21.3	22.3	23.2	23.9	23.2	21.9	23.6	23.4	21.5	20.5

Source: Exxon Corp.

[a]Includes only those companies with more than 1 million tons of coal production in 1975.
[b]Year acquired by oil company.
[c]Acquired in March 1976.

the year 2000. Productivity of American coal miners peaked at 20 tons per miner per day in 1969. By 1976, this had dropped to around 8.5 tons, due to aging equipment, labor problems, new safety and health requirements, and loss of well-trained miners.

Roughly 52 percent of the coal mined in the United States today comes from surface mining and 48 percent from underground mining. Over two-thirds of the material handled in underground mining in the United States is coal, and about 40 percent of the land disturbance from all surface mining is from coal production.

Mining is the most hazardous of all occupations (100,000 deaths and a million permanent disabilities in U.S. coal mines in the twentieth century). Even with current new safety regulations and improved mining techniques, 4700 miners will die and 350,000 will be disabled in our efforts to meet projected increases in coal production for the rest of this century. The notorious black lung disease, caused by inhalation of coal dust, is costing the U.S. government more than $1 billion a year in compensation payments to victims. In 1975, mining inspectors found that 30 percent of the mines still had conditions conducive to black lung disease.

Land disturbance Some 4 million acres (1.6 million ha) have been excavated by surface mining in the United States (40 percent from coal production) and this rate of terrain change has been steadily accelerating. Strip mining for coal presently denudes about 70,000 acres (28,000 ha) a year in this country. Perhaps the most impressive single strip mine is near Bergheim, Germany, where a 10 km², 300 m deep lignite mine is being worked. There are 500,000 acres (200,000 ha) of "orphaned" lands in the United States (81 percent in Appalachia) that remain in blighted condition from past coal mining (such lands are termed "derelict" in Britain). Costs of rehabilitation (Chapter 18) are often $3000 to $8000 an acre.

Area strip mining can completely alter the topography (Fig. 6-11). Current laws require regrading the land to its approximate original contours (obviously impossible) and replacing soil materials. In the past, many relatively level areas were left with steep ridges (windrows), trenches, and even vertical highwalls, rendering the terrain useless to agriculture and dangerous for grazing. In many regions such as the northern Great Plains, the subsoil or underlying bedrock is rich in sodium or other materials deleterious to plant growth, with the result that nothing of value will grow on unreclaimed spoil banks.

In high relief terrain, such as the Appalachian highlands, **contour strip mining** is a common practice (Fig. 6-12). Haul roads, a major land disturbance in themselves, lead to working benches carved from steep slopes, bounded on the uphill side by 10 to 30 m highwalls and by massive spoil banks on the slopes below. There are more than 40,000 km of contour benches in the United States. A 1000 megawatt power plant annually consumes the coal from a 1 m seam on a bench 70 m wide and 32 km long. Erosion from these steep, debris-laden slopes commonly raises the sediment yields from 10 t/km² (tonnes per square kilometer) for forested sites to 11,000 t/km². More rapid runoff from devegetated mined areas combined with reduced channel capacity causes siltation and increases flood potential along local streams and rivers. Landsliding of loose slope rubble also constitutes a serious hazard to homes and other structures below.

In steep-sided narrow valleys, dams are sometimes constructed of the spoil material. There are many such dams scattered through Appalachia. They are often quite frail and pose a danger to people downstream (see the Buffalo Creek dam disaster, page 243). Reclamation of steep strip mined areas exceeding 14° in slope is rarely successful, and stripping of such terrain should probably be banned.

In many semiarid coal mining areas of the western United States, relatively moist alluvial valley floors are the prime agricultural, wildlife, and water supply areas. Mining in or near these alluvial valleys can disrupt the sensitive hydrology of the valleys with severe results for farms, ranches, small towns, and the natural ecosystem. Unfortunately, easily accessible coal often lies at shallow depth beneath these valley floors—at least 10 mining operations in such valleys are currently underway in Montana. These valleys and water rights in general are a major battleground between pro- and anti-coal development forces in the West. Rehabilitation of disturbed land in such arid and semiarid lands is difficult and can succeed only with good management and large applications of water and fertilizer. Whether long-range ecological stability can be achieved is still highly debatable.

Spoil banks from coal mining cover many areas in the United States. They may be sources of dust, surface and groundwater pollution (as acid mine drainage waters), excess erosion, and are often susceptible to combustion.

In addition to wastes at the mining site, solid wastes generated at coal-fired power plants must also be disposed of. One year's operation of a 1000 megawatt plant using low-sulfur coal yields 71,000 tons of ash in the boiler, 281,000 tons of fly ash collected from the stacks, and 212,000 tons of sulfur solids. Over their predicted 30-year lifetime, the coal plants on line in 1980 will produce enough waste to cover 26,000 ha to a depth of 10 m.

Underground coal mining, which accounts for two-thirds of all subsurface mining, has little initial impact on the land surface. Of course, the usual access roads and spoil piles near the mine mouth cause terrain changes. Long-term effects of subsidence are very serious, however. Sudden collapse into mined-out areas (especially when the room and pillar method was used) has destroyed houses, roads, buildings, and farm areas (Fig. 6-13).

Either rapid or slow subsidence may cause open cracks, small-scale faulting, and surface troughs, depressions, and bulges to form. The tension cracks opened up may encourage spon-

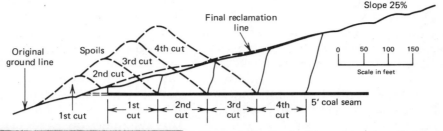

Modern way to strip mine coal. The overburden from the second cut or strip is placed where the coal has just been removed from the first cut.

Figure 6-11 Coal mining and reclamation methods in an Ohio strip mine. Reclamation is started while coal is still being mined. The overburden from the second cut or strip is placed where the coal has just been removed from the first cut. (Courtesy National Coal Assoc.)

taneous combustion of underlying coal beds. More than 800,000 ha of land in the United States has already subsided due to underground mining of coal, and additional thousands are threatened by a similar fate.

Hydrological disruptions Mining, especially of coal, exposes sulfur-containing minerals to the atmosphere and to bacterial action, causing chemical reactions that release sulfuric acids into local waters. Acid mine drainage is extremely lethal to fish and other aqueous life. The red to yellow color of iron precipitates often associated with such drainage gives it the popular name "yellow boy."

Acid waters from mines have contaminated about 10,000 km of streams (90 percent in Appalachia), as well as a large number of lakes,

Figure 6-12 Terrain effects of contour coal mining in the Appalachians. (U.S. Dept. of Interior, 1967.)

reservoirs, and marshes. Abatement methods such as pumping, sealing mines, and lime treatment may need to be continued for many years, adding to the total cost of coal. To clean up streams already contaminated with acid runoff would require in excess of $7 billion.

During area strip mining, natural surface drainage is often diverted through artificial channels to avoid flooding of mine sites. This may cause increased erosion, flooding, and/or siltation along local streams. Similar effects of spoil bank erosion were already noted.

Coal mining, especially in water-poor areas of the West, can seriously deplete groundwater supplies. The water table around Decker, Montana, has been lowered as much as 15 m because of stripping operations. Over large areas of the northern Great Plains, the coal beds themselves are the aquifers that many farmers, ranchers, and small towns rely on. A mining operation can "behead" these very gently dipping aquifers, causing the failure of wells.

The scarcity of needed water for the many large power plants planned in these arid regions may be the limiting factor for development. Dozens of schemes to obtain water are afoot,

Figure 6-13 Subsidence pits dot this landscape above the underground Monarch coal mine near Sheridan, Wyoming, more than 60 years after the mine was abandoned. The pits resulted from the collapse of tunnels and other voids in the mine workings which were as much as 30 m below the surface. Underground mining may produce more terrain damage than surface mining that has exercised appropriate reclamation methods. (Courtesy U.S. Geol. Survey.)

among them major diversions from the Yellowstone, Green, and other rivers, as well as damming numerous streams in nearby mountain ranges. Such operations will have wide-ranging environmental consequences and are facing strong opposition.

Burning coal seams and spoil piles   Spoils containing combustible coal from surface and underground mining, and subsurface coal beds exposed to the atmosphere, often catch fire by spontaneous combustion, lightning, or human activities. Once started, they can be nearly impossible to extinguish. Abandoned coal mines are very susceptible to such fires. There were 500 smouldering spoil piles and over 200 underground coal fires in the United States in 1964.

All such fires pollute the air with particulates and toxic gases, and coal seam fires consume coal supplies, pollute water supplies, and cause eventual subsidence and cracking of overlying rocks, sometimes threatening entire communities. Scranton, Pennsylvania, has spent tens of thousands of dollars in unsuccessful attempts to control such fires. The burning seams cause warm water supplies, toxic gas seeps into basements, health problems for local residents, uneven settling of buildings, and other very costly disruptions. The only sure way to stop underground coal seam fires is to dig a trench deep enough to cut through the burning seam completely around the burning area, isolating it until it burns itself out.

Air pollution   Sulfur dioxide, particulates, and other harmful emissions from coal-fired power plants are a major source of air pollution (Fig. 6-14). Especially significant is the phenomenon that creates "acid rain." In addition, there is growing concern that carbon dioxide buildup in the atmosphere from fossil fuel combustion can effect the earth's climate. These airborne emissions may prove to be the major factor in limiting the amount of coal that can be consumed. For further discussion, see Chapters 17 and 20.

Figure 6-14 Four Corners Power Plant, Farmington, New Mexico. Coal supply in background, with artificial lakes and canals which supply the water source. (Michael Collier/Stock, Boston.)

Coal mining also contributes to significant local air quality deterioration. Coal dust, the same material responsible for black lung disease, is blown from active mining sites, coal storage piles, trucks, and trains. Most of the dust associated with mining is **fugitive dust**—soil and rock particulate matter stirred up by blasting, bulldozing, and dragline operations, trucks rumbling along dirt roads, wind blowing over soil piles and devegetated areas, construction of new railroads and other facilities, and similar activities. The fugitive dust contributes to health and visibility problems and, because about 15 percent of the particles are 10 microns or smaller, effects can be widespread, especially in the West where many surface mining areas are dry and windy. People living near dust sources report increased respiratory ailments in their families and higher rates of cattle disease.

Coal dust is regulated by the federal government but fugitive dust is not. Watering down roads, prompt revegetation, and other dust suppression techniques can help, but it is not possible to reduce dust to zero in large mining operations.

Transportation The expenses and impacts of upgrading roads and railroads to meet demands for more coal have already been noted.

More controversial have been plans to ship coal through pipelines as a slurry (Fig. 6-15). The ability to move oil and gas by pipeline has always been a major advantage of those fuels. To move coal in such a fashion requires large quantities of water—already a scarce commodity in most of the coal mining areas of the West, where most slurry lines have been proposed. It would require about 195 million gal (737,000 m^3) of water to move a million tons of coal annually from the arid plains of Wyoming, South Dakota, or Colorado, where the water is precious, to the water-rich Mississippi midlands. Using saline groundwater has been suggested, but this could still have disrupting effects on water levels and flow systems in the plains. Disposal of the salt water at the end of the line would also be a problem. Less than 1 percent of U.S. coal is now moved by slurry and it is unlikely that this percentage will be exceeded in the foreseeable future.

Socioeconomic impacts Social and economic impacts resulting from the sudden influx of population to new coal mining areas are often severe. In many cases, needs for housing, roads, schools, water supplies and treatment plants, sewage facilities, and health care personnel and facilities far outstrip the local government's ability to supply them. Once-peaceful communities like Rock Springs and Gillette, Wyoming, and Colstrip, Montana, have become modern boom towns with all the mixed blessings that phrase implies. Along with newfound wealth and prosperity for some have come noise, pollution, traffic jams, escalating prices, drastic shortages in housing and public facilities, and skyrocketing increases in alcoholism, suicides, delinquency, divorce, prostitution, and mental disorders. For example, Sweetwater County, Wyoming, which contains Rock Springs, increased from 18,391 in 1970 to 45,000 by 1975 and continues to grow at an annual rate of about 19 percent. Studies indicate that 5 percent is the maximum rate of growth the county can absorb without serious socioeconomic troubles.

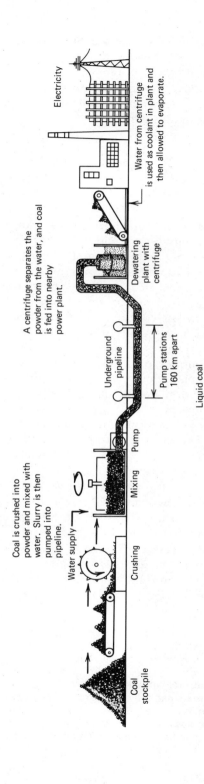

Figure 6-15 Schematic drawing of coal production and slurry transportation method.

(figure labels: Electricity; Water from centrifuge is used as coolant in plant and then allowed to evaporate.; A centrifuge separates the powder from the water, and coal is fed into nearby power plant.; Dewatering plant with centrifuge; Underground pipeline; Pump stations 160 km apart; Liquid coal; Coal is crushed into powder and mixed with water. Slurry is then pumped into pipeline.; Water supply; Crushing; Mixing; Pump; Coal stockpile)

Gasification and Liquidization of Coal

Producing liquid and gaseous fuels from coal is not new. The first commercial coal-gas plant was built in England in 1807, and the Germans made gasoline from coal near the end of World War II when they were cut off from oil supplies. In the late 1800s, some 400 U.S. companies were producing coal gas, mostly for illumination. With the coming of cheaper and more convenient natural gas, coal gas production ceased.

Gas from coal utilizes only 65 percent of the heat value of the coal and is twice as expensive as natural gas. By 1985, the National Petroleum Council projects U.S. demand for natural gas will be 20 trillion ft^3, more than natural domestic supplies can provide. Since we have an abundance of coal, coal gasification might satisfy this excess demand but, at a total cost of about $40 billion, the economics of such operations is in question.

A standard coal gasification plant yielding 250 million ft^3 (7 million/m^3) of coal gas per day would require a minimum of 3 billion gal of water annually—perhaps 15 billion if water is available for cooling. Because most potential sites are in the semiarid West, water will be a limiting factor. As with standard electricity generating plants, there are also pollution, ash, and sulfur removal problems to contend with.

In situ combustion of coal to yield gas is also being investigated and may prove more promising than standard gasification plants (Fig. 6-16).

To produce liquid fuels from coal would also be very costly and thermodynamically inefficient. A ton of bituminous coal will usually yield about 3 barrels of crude oil. By the time this crude is refined to gasoline, it will have taken about 7 kg of coal to provide 40 km of travel in an average car. This "synthetic gasoline" or "synfuel" would sell for approximately twice as much as normal gas. Coal liquidization facilities capable of turning out 3.5 million barrels a day (15 percent of anticipated demand) by 1990 would cost about $100 billion (Table 6-6).

Fluidized-bed combustion Fluidized-bed combustion burns pulverized coal and limestone that is suspended by air forced up through the boiler. There are numerous approaches to this technology and it remains to be seen which will prove most efficient. Basically fluid-bed boilers allow combustion to occur at lower temperatures and hence do not yield the large volume of nitrogen oxides and melted ash that conventional boilers generate. Another advantage is that about 90 percent of the sulfur in coal combines with the limestone to produce calcium sulfate which can then be disposed of as solid waste. Enforcement of strict SO_2 emissions standards should aid fluidized-bed combustion. An Exxon study estimates that the potential fluid-bed market would decline by a third if SO_2 standards were relaxed as large utilities are urging—a good example of how strong environmental regulations can encourage techno-logical innovation. The remaining SO_2, small particulates, CO_2, and various trace elements will still create pollution problems; however, fluid-bed boilers do represent an improvement over conventional systems and should be especially suitable for small-scale heating and electricity generation needs.

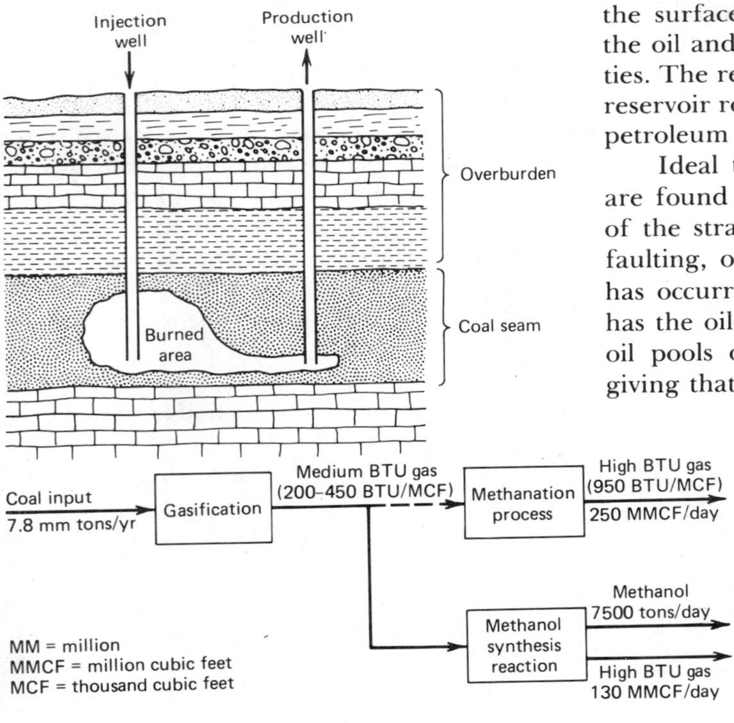

PETROLEUM

Geology

The term "petroleum" encompasses a wide range of gaseous, liquid, and solid hydrocarbon compounds occurring naturally in the earth. The source of most petroleum was aqueous organisms (mostly microscopic plants and animals) that settled to the bottom of shallow nutrient-rich seas on continental platforms. These organic rich sediments were eventually buried, and heat and pressure induced the breakdown of complex organic molecules into various components of petroleum. The lighter gaseous and liquid material migrated toward the surface, where natural "traps" may cause the oil and gas to concentrate in certain localities. The resulting oil or gas-saturated rock (the reservoir rock) is a "pool." A variety of common petroleum traps is illustrated in Fig. 6-17.

Ideal trap areas for petroleum (oil fields) are found where some structural deformation of the stratified rocks, such as gentle folding, faulting, or doming because of salt intrusion, has occurred. The Persian Gulf area not only has the oil-bearing rock and the traps, but the oil pools occur in unusually accessible areas, giving that region a virtual monopoly on inex-

Figure 6-16 Coal gasification process. (a) Cross section of reverse combustion in underground coal gasification. The connection between the two wells has occurred, and the gasification of the coal seam has started. (b) Methanol conversion technology for production of methanol and synthetic natural gas, for a quantity of 7.8 million tons of coal. (U.S. Geol. Survey.)

Table 6-6 Cost comparison of conventional fuels and fuels available from present and future conversion technology

Summary cost comparison	Units	Million (BTU/unit)	Price ($/unit)	Price ($/million BTU)
Conventional fuels[a]				
Crude oil	barrels	5.80	7.00	1.21
Gasoline	gallons	0.12	0.22	1.83
Distillate	gallons	0.13	0.20	1.54
Western coal	tons	16.00	4.00[c]	0.25
Eastern coal	tons	24.00	12.00	0.50
Present technology[b]				
SNG (synthetic gas) from coal	million cu ft	0.95	1.85	1.95
Methanol	barrels	2.80	4.90	1.75
Tar sands	barrels	5.80	8.70	1.50
Future technology				
Coal liquids	barrels	5.80	8.00–10.00	1.40–1.70
Oil shale	barrels	5.80	8.00–10.00	1.40–1.70

Source: Coal and the Energy Shortage, a presentation by Continental Oil Co. to Security Analysts, December 1973.

[a]Typical prices of conventional fuels today.
[b]Typical prices for synthetic fuels escalated to estimate 1978 prices.
[c]Does not include transportation cost (Montana to Chicago, 1200 miles, $10.68/ton).

pensively produced oil during the 1970s (cost about 15 cents a barrel to produce).

Thus there are several requirements that must be fulfilled before it is possible for an oil deposit to occur that is of economic importance. These include:

1 An original environmental setting that favored the rapid accumulation of organic-rich deposits. Such a locale is provided by marine lagoonal sites that gradually subside and act as basins of deposition. Here one-celled plants and animals decayed and the droplet of oil that formed their living material became entombed in the sediments.

2 An appropriate source bed that retains the oil during the sedimentation cycle. The original sediments are generally fine-grained materials. These deposits protect the oil from rapid dissipation or migration and diffusion.

3 A proper reservoir rock. Contiguous or near the source rock should be other strata that can act as a host for the oil when it does move from the source materials. It is important that the reservoir strata should be relatively permeable and porous. Thus, they will release easily the petroleum encased within them when pumped by man.

4 A geologic structure that can serve as a trap or dam to prevent the complete disappearance of the petroleum from the reservoir rock. As depicted in Fig. 6-17 common traps include salt domes (Gulf Coast area), anticlines (Middle East region), faults (California), and stratigraphic traps (Pennsylvania and New York).

5 Crustal stability after emplacement of oil into the geologic trap. If post-oil movement and trapping is followed by additional earth deformation with folding and faulting, or even vulcanism, the oil will be driven away from the site and into other rocks where it will be too disseminated to mine profitably. For example, there are numerous oil seeps throughout the world, places where oil has

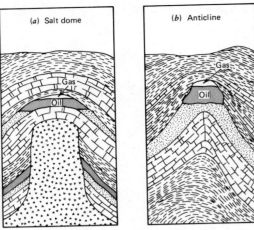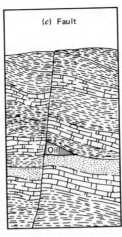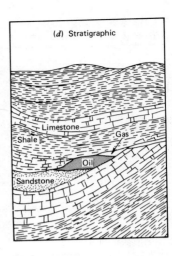

Figure 6-17 Typical geologic and structural traps for oil and gas. Porous rocks such as sandstone can act as reservoirs for fluids if adjacent rocks act as impermeable seals. (a) Salt dome. (b) Anticline. (c) Fault. (d) Stratigraphic trap.

been dispersed, and in which the underlying original petroliferous rocks have been decimated by such earth movements.

Recovery of Petroleum

Location of oil, or gas (one-third of our natural gas comes from oil fields), fields commonly involves detailed studies of stratigraphy and structure, knowledge of the area's geological history, and geophysical investigations (Fig. 6-18). The geophysical investigations include use of gravity meters, airborne magnetometers, and reflection and refraction seismographs.

Despite such sophisticated techniques, only 1 percent of wildcat wells drilled in the United States today strike a field of a million or more barrels—enough to sustain U.S. consumption for 90 minutes. The United States is the most extensively probed of all countries, with seven times the drilling density of the rest of the world.

The first U.S. oil well was drilled by Colonel E. L. Drake near Titusville, Pennsylvania, in 1859. He was thought very odd when he set up a horse-powered rig for the express purpose of finding oil. Find it he did, and the Civil War provided a market. In the 1930s, 92 barrels of oil were found per meter of hole drilled. Following the exhaustion of more accessible oil, discovery dropped drastically to about 8 barrels per meter in 1975.

Oil is obtained by drilling a well in an area where studies have suggested recoverable oil might exist and hoping the educated guess proves correct (Fig. 6-19). Drilling is a very costly and time-consuming procedure. Today, it takes 2 to 10 years to bring a new oil field into production. The petroleum in the pool is commonly under sufficient natural pressure to force about 10 percent of the oil through the well to the surface. In early days of oil drilling, so many holes were drilled into oil pools that subsurface pressure was reduced, rendering the oil more difficult or even impossible to obtain.

Secondary and tertiary techniques are now used in some fields to increase oil production

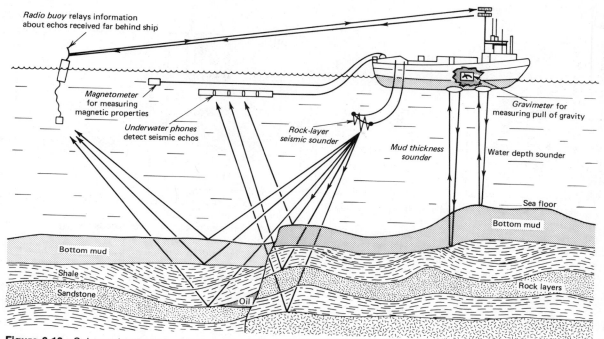

Radio buoy relays information
about echos received far behind ship

Magnetometer
for measuring
magnetic properties

Underwater phones
detect seismic echos

*Rock-layer
seismic sounder*

*Mud thickness
sounder*

Gravimeter for
measuring pull of gravity

Water depth sounder

Sea floor

Bottom mud

Bottom mud

Shale

Sandstone

Oil

Rock layers

Figure 6-18   Schematic diagram showing marine geophysical prospecting methods.

(Fig. 6-20). Underground blasting or hydraulic fracturing aids to release oil from "tight" rock formations. Various chemicals further help to release more oil from the interstices of the rock. Forcing brine, steam, or other fluids into a reservoir may increase pressure and bring more oil to the surface. The use of such methods may increase oil production 30 percent or more.

Figure 6-19   Oil well in Powder River Basin, near Buffalo, Wyoming. Such wells yield only a few hundred barrels per day.

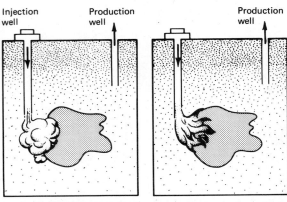
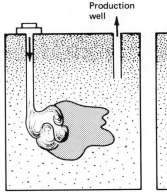
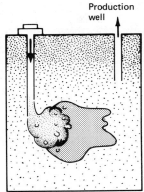

Figure 6-20 Secondary recovery techniques for extra production of petroleum.
(a) Steam drive. Steam is injected at the injection well. As the oil temperature rises, it becomes less viscous and moves more freely to the production well.
(b) Combustion. Hot air is blown into reservoir and produces a fire. The combustion and increased pressure force the oil toward the production pumps.
(c) Miscible injection. A fluid, usually water, that may also contain dissolved CO_2 is injected. The CO_2 mixes with the oil, forming a lighter mixture, and the hydrostatic pressure of the water forces the lightened fluid through the reservoir rocks.
(d) Chemical flooding. Chemical solutions are pumped into the rocks and act as laundry detergents. They wash the oil from the reservoir materials by increasing the mobility and permitting flowage toward the pumps.

The Geography of Petroleum

In 1973 OPEC, which controlled approximately two-thirds of the world's proven crude oil reserves at that time, decided to cut off the oil supply to other countries. Suddenly the often repeated and often ignored predictions of an energy crisis became reality. More importantly, the modern world's vulnerability and dependency on petroleum (two-thirds of the world's commercial energy budget) became shockingly manifest (see Fig. 6.6).

At the close of 1976, proven world petroleum reserves (Table 6-7) were just under 600 billion barrels (55 percent in the Middle East). By 1979, increased estimates for Mexican oil and other new discoveries raised the figure to about 800 billion barrels. Crude oil production in 1976 was 21 billion barrels. At a modest 3 percent annual growth rate, far below the 6.5 percent rate that prevailed from 1940 to 1976, oil production in A.D. 2000 would need to double the 1976 rate, and the total consumed for the 24-year period (1977–2000) would be 750 billion barrels. Recent studies predict that the total ultimately recoverable crude oil reserves will be around 2 trillion barrels (including the 300 billion consumed through 1976). Simple arithmetic reveals that even if these estimates are very conservative, today's most vital fuel will not last long.

The United States, with only about 5 percent of the world's oil reserves and half of that already consumed, uses about 18 million barrels of oil a day (6 billion barrels a year), 8.6 million of it imported. The current oil imports of 48% of total consumption is more than twice the percent (22 percent) it was in 1968. From 1971 to 2000, the United States is expected to consume 276 billion barrels. However, with the continued escalation of world market prices, this figure will undoubtedly be less. OPEC became an overpowering entity in 1971 when they demanded, and received, a 100 percent boost in the $1.80 per barrel price of oil (Table 6-8). They flexed their muscles again with the oil embargo of 1973 which also doubled the price, rising to $11 per barrel by the end of 1975. Prices have continued to rise so that by the end of 1979 they exceeded $25 a barrel. Of course,

Table 6-7 World petroleum reserves and production, 1976–77

Summary	Reserves 1/1/77	
	Crude oil (millions of barrels)	Gas (billion cu ft)
United States	30,942	220,000
Canada	6,200	56,000
Latin America	29,609	90,325
Western Europe	24,539	141,905
Africa	60,570	209,077
Middle East	367,680	513,460
Far East	19,391	120,010
Total free world	538,931	1,350,777
Communist nations	101,100	953,000
Total worldwide	640,031	2,303,777

	Production 1976		
	Crude oil (million barrels a day)	Gas (billion cu ft/yr)	1976 Refining Capacity (million barrels a day)
United States	8,105	19,993	15,043
Canada	1,300	3,157	2,076
Latin America	4,396	2,044	7,690
Western Europe	900	6,239	20,238
Africa	5,598	1,628	1,320
Middle East	21,881	2,003	3,288
Far East	2,672	982	9,778
Total free world	44,851	36,046	59,433
Communist nations	12,360	14,361	12,406
Total worldwide	57,211	50,408	71,839

Source: Adapted from Oil and Gas Journal (12/27/76, 2/28/77, 4/18/77); The International Petroleum Encyclopedia, 1976; and Exxon Corp.

Table 6-8 U.S. reliance on OPEC crude oil, August 1979

	Percentage of crude oil imports	Selling price
Nigeria	15.1	$26.26
Saudi Arabia	14.3	18.00
Libya	8.6	26.05
Algeria	8.5	26.27
Venezuela	7.8	19.31
Iran	6.7	23.50
Indonesia	4.7	23.50
United Arab Emirates	3.6	21.56
Iraq	0.9	21.96
Other OPEC countries	1.7	20.00–27.00
Non-OPEC countries	15.4	25.00

Sources: Petroleum Intelligence Weekly; American Petroleum Institute.

Table 6-9 Comparison of crude oil resource estimates (in billions of barrels) for the entire United States onshore and offshore to 200 meter water depth

Estimator	Original estimate	Undiscovered economic and subeconomic crude oil resources (60 percent recovery)	Undiscovered economic crude oil resources (using present technology)	Ultimate crude oil production from future discoveries	Original crude oil resources (60 percent recovery of discovered and undiscovered crude oil in place)
Hendricks (1965)	400[a]	285[b]	ca 152[c]	—	600[e,g]
Theobald and others (1972)	459[d]	ca 585[e,f]	ca 312[c]	—	ca 900[e,g]
U.S. Geological Survey (1974)	200–400[d]	ca 255–510[e,f]	ca 136–272[c]	—	570–825
National Petroleum Council (1970)	107	199	107[c]	—	514
Resource Appraisal Group (Miller and others, 1975)	50–127 (82 mean)	≥94 to ≤238[h]	50–127 (82 mean)	—	408–553 (468 mean)
Oil Co. A (Weeks, 1960)	—	—	ca 62[i]	—	—
Oil Co. D (National Academy of Sciences, 1975)	89	—	76[e]	—	—
Oil Co. E (National Academy of Sciences, 1975)	90	—	77[e]	—	—
Shell Oil Co. (R. H. Nanz, oral commun., 1975)[j]	65–155	—	65–155	—	—
M. K. Hubbert (1974)	—	—	—	55[k]	—
C. L. Moore (written commun., 1975)	—	—	—	ca 156[k,l]	452

Source: U.S. Geological Survey, Annual Report, Fiscal Year 1975.

Note: If an original estimate was revised for this paper in order to express it in comparable dimensions and if the change involved implicit assumptions, the revised estimate is preceded by ca.

[a] Includes past production and identified resources.

[b] Corrected for past production and all categories of reserves at 60 percent recovery. Hendricks gives 1000 billion barrels of discoverable in-place crude oil.

[c] 32 percent recovery.

[d] Includes NGL.

[e] The original estimate includes NGL and is here reduced by 15 percent in order to subtract NGL.

[f] Estimate was classified as an undiscovered economic resource at the time of release and is reclassified here as an undiscovered economic resource.

[g]Estimates of crude oil on continental slope are subtracted from original estimate.

[h]The 95 and 5 percent probability levels of estimates of economic and subeconomic resources are not arithmetically additive. The proper aggregated 95 and 5 percent probability estimates would be greater than or equal to the sum of the 95 percent estimates and less than or equal to the sum of the 5 percent estimates. Estimates at 60 percent recovery.

[i]Weeks (1960) reported a figure of 270 billion barrels of ultimate resources of crude oil and NGL for the United States. This figure is adjusted by assuming crude oil is 85 percent of this and subtracting cost production and oil categories of reserves as of the end of 1974.

[j]Nanz, R. H., 1975, The offshore imperative—the need for a potential offshore exploration: talk presented at Colloquium on Conventional Energy Sources and the Environment: Univ. of Delaware.

[k]Assumes continuation of pre-1974 socioeconomic conditions.

[l]Calculated by subtracting from C. L. Moore's 471 billion barrels ultimate recovery the 168 billion barrels of past production and present reserves plus an additional 147 billion barrels from identified resources at a 60-percent recovery.

such action has further stimulated investigation for new supplies throughout the world and the reoccupation of some old fields that have now become profitable.

U.S. oil production will not rise significantly in the future but instead will continue to be more costly, since many of the most promising sites for new discoveries lie in deep offshore water and in northern Alaska. When the Alaskan pipeline was proposed in 1969, cost was estimated at $900 million. When completed in 1977, costs exceeded $8 billion.

Continental shelves are submerged parts of continents. Because their subaerial coastal plain counterparts have provided many excellent oil fields, it is reasonable to believe that significant petroleum supplies may occur beneath the shelves. Considerable offshore production already comes from the Gulf Coast, and numerous areas off the Atlantic and Pacific coasts and in Alaska are now under investigation (Table 6-9). Estimates of ultimately recoverable reserves from outer-continental shelf discoveries in the United States vary widely from less than 50 to 150 or more billion barrels. Offshore exploration has been hindered by opposition from people fearing adverse environmental and economic impacts, shortages of equipment and experienced personnel, and the long lead time necessary for such large-scale endeavors in largely untested and hazardous areas.

Oil Pollution

Oil and living things do not mix well (Fig. 6-21). Oil is toxic to most life forms, and major spills at sea tend to turn that local area into a biological desert. The Mexican oil well (p. 5) that took until 1980 to be capped bears testimony to just such dangers. However, good recoveries of life forms have been observed at most sites within 3 to 4 years after spills. In addition to direct toxic effects, oil forms a veneer over water that inhibits oxygen replenishment of waters and reduces evaporation. The

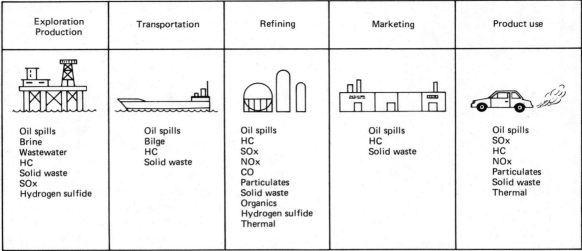

Exploration Production	Transportation	Refining	Marketing	Product use
Oil spills Brine Wastewater HC Solid waste SOx Hydrogen sulfide	Oil spills Bilge HC Solid waste	Oil spills HC SOx NOx CO Particulates Solid waste Organics Hydrogen sulfide Thermal	Oil spills HC Solid waste	Oil spills SOx HC NOx Particulates Solid waste Thermal

HC = Hydrocarbons
CO = Carbon monoxide
SOx = Sulfur oxides
NOx = Nitrogen oxides

Figure 6-21 Major pollutants of U.S. petroleum industries. (Oil and the Environment: The Prospect, Shell Oil Co., 1973.)

ultimate, long-term effects of the enormous quantities of oil being assimilated by the oceans are unknown.

About 650,000 tons of oil enter the oceans each year from natural seeps. With the increase in ocean shipping and petroleum use, the oil entering the oceans has increased dramatically to more than 6.5 million tons. Over one-third of this comes from routine tanker operations such as discharging ballast, cleaning tanks, spills during loading, and transfer operations. Two-thirds of all oil now produced is shipped by sea, and this traffic will probably expand sixfold before peaking.

Tanker accidents spill less than 5 percent of all marine oil but tend to concentrate heavy damage in a small area. Huge new tankers, nearly as long as the height of the Empire State Building and carrying over 3 million barrels of oil are so unwieldly they require 20 minutes and 5 kilometers to stop. About 35 billion bar-rels of oil are afloat on the seas each day in some 7000 tankers.

Offshore drilling contributes only 4 or 5 percent of the oil pollution in the seas but, because of increased drilling in deeper, more hazardous waters, the actual amount of oil from this source is bound to increase.

If the oil from a large offshore spill drifts ashore, as often happens, damages can greatly increase. Coastal waters, estuaries, and saltwater marshes are perhaps the most delicate, productive, and important ecosystems on the planet, and they can be badly decimated by a major oil spill. Economic damages—especially to the shellfish, sport fishing, and recreational industries—along coasts can be devastating. About one-quarter of the oil in coastal waters comes from pipeline leakage. Land-based sources also contribute large quantities of oil to the oceans.

Three-quarters of the oil drained from autos and industrial machinery (around 260 million gal/yr) is discarded in sewers, landfills or on the ground where much of it finds its way into water supplies and ultimately to oceans. From 1960 to 1975, the number of American oil re-refiners dropped from 150 to 45. Recent price leaps are encouraging more re-refining, and we can only hope that at least 60 percent of the 1.4 billion gal of waste oil, which will be generated annually by 1985, will be re-refined. Properly re-refined oil is indistinguishable from new oil.

Leaks from oil and gas storage tanks are frequently to blame for polluting ground and surface water supplies as well as for creating fire hazards. About one-quarter of the oil in coastal waters comes from pipeline leakage. Land-based sources also contribute large quantities.

Tremendous quantities of gasoline and other petroleum products evaporate into the atmosphere with possible harmful effects. Petroleum is also indirectly responsible for large amounts of air pollution from the burning of waste oil, fuel oil, gasoline in internal combustion engines, and similar activities. Oil refineries are the largest industrial air polluters, releasing over 12 billion kg of carbon monoxide, 3.5 billion kg of sulfur oxides, 11 billion kg of hydrocarbons, and 1 billion kg of particulates each year.

Effects of oil spills in cold regions are of growing concern. Oil degrades to less harmful materials very slowly in cold temperatures, and the impact of a major spill in the Alaskan wilderness or other cold areas could be disastrous to the local ecosystem. The Arctic Ocean is believed to contain large petroleum deposits and several countries, including the United States and the Soviet Union, are interested in exploiting them. Drilling in such icy water, with shifting ice packs and icebergs, would be extremely risky, and the probability of major spills very high. A large offshore spill in the Arctic waters would cause a vast layer of dark crude oil to spread out beneath the ice pack. This sea ice accumulates mostly by the freezing of ocean water at the base of the pack. In several years,

the oil could work its way to the surface, increasing absorption of solar energy, and accelerate melting of the Arctic Ocean ice cover. This could have drastic worldwide climatic effects.

Drilling Operations

Compared to many forms of resource extraction, drilling for oil has produced minimal environmental deterioration. Care must be taken to prevent brine and oil seepage at the well head. Blowouts and fires caused by escaping natural gas are infrequent but very dangerous events. Pipeline and road construction cause some environmental degradation, and the odors from producing oil fields are offensive to some people.

The hazards associated with new developments in deep offshore sites are far greater (Table 6-10). Some oil spillage from well blowouts, pipeline, or tanker leaks seem to be inevitable at new offshore sites. There is always the possibility of an oil spill coming ashore in the Southeast Georgia Embayment area along the South Atlantic coast, one of three Atlantic sites with high oil-producing potential (2 to 4 billion barrels of estimated production for most such sites). Here and in other beaching areas for production sites to the north (e.g., Baltimore Canyon, although not highly promising to date) the effects of a major oil spill could cause extensive damage to coastal ecosystems and heavily used recreational areas (including Long Island and New Jersey).

Offshore drilling rigs can be destroyed by hurricanes, blowouts, and the collapse of the drilling platform due to loss of support. Fluids under pressure escaping along the drill hole have been known to produce instability in the ocean floor sediments, causing collapse and even complete engulfment of entire drilling platforms. Development of offshore sites also entails increased shipping, construction of storage areas, and refineries, which often cause more environmental disruption than the actual recovery operations.

Offshore oil development in the Gulf of

Table 6-10 Accident and spillage statistics for the outer continental shelf, 1970–75

Calendar year	Total accidents[a]			Total of all spills		Oil and condensate production (barrels, in millions)	Gas production (cubic feet, in billions)	New well starts	Active leases	Fixed structures (end of year)	Estimated mileage of pipeline supervised
	Number	Fatalities	Injuries	Number	Barrels of oil spilled						
1970	24	33	58	Unknown	Unknown	361	2419	900	1017	1800	4000
1971	35	11	16	1256	2,778	419	2777	841	1083	1891	5000
1972	41	10	9	1161	1,182	412	3038	847	1023	1935	6000
1973	44	9	15	1175	23,096[b]	395	3212	820	1266	2001	6450
1974	39	9	22	1137	23,388[c]	361	3515	816	1590	2054	6700
1975	46	17[d]	14	1128	977	330	3459	882	1792	2079	7150

Source: U.S. Geological Survey, Annual Report, Fiscal Year 1976.
[a]Includes blowouts, fires, explosions, falls, vessel collisions or sinkings, drownings, electrocutions, asphyxiations, blows, etc.
[b]Includes (1) 9935 barrels—structure supporting oil storage tank bent, ruptured tank.
(2) 7000 barrels—barge developed a leak in heavy seas and partially sank, releasing oil.
(3) 5000 barrels—internal corrosion caused numerous pipeline leaks.
[c]Includes (1) 19,833 barrels—dragging anchor snagged and ruptured pipeline.
(2) 2213 barrels— pipeline break apparently caused by hurricane.
[d]Includes six men who died in a blaze resulting from a tanker collision with a platform (escaping oil from tanker ignited and tanker caught fire).

Alaska (estimated recovery 7 to 15 billion barrels) will have to contend with icebergs, vicious storms, and occasional tsunamis, among other problems. The probability of a spill coming ashore in the Gulf of Alaska is greater than 95 percent in winter and 40 percent in summer. Onshore facilities in Alaska will have high earthquake and mass movement risks to contend with, and the shipping crews must be alert for icebergs from coastal glaciers. If the huge Columbia Glacier, not far from the Alaska pipeline terminus at Valdez, retreats from the morainal shoal its snout rests on, it could probably undergo a catastrophic breakup, sending forth a multitude of great icebergs into the seas, seriously threatening shipping and drilling operations over a wide area.

Subsidence

The extraction of petroleum in its various forms, especially oil and gas, has led to severe problems associated with subsidence of the ground surface and earthquakes. Withdrawal of fluid hydrocarbons is different than groundwater mining because the reservoir is invariably in rocks and not sediments and the area of influence is smaller.

Although there was subsidence from groundwater withdrawal in the Wilmington-Long Beach, California, area as long ago as 1928, significant land lowering did not start until major oil production in 1938. The first important subsidence of 0.4 m was measured in 1940, and by 1945 had grown to 1.4 m. By 1951, the annual rate of subsidence had reached 0.6 m per year and was causing enormous damage to buildings, pipelines, railroads, and roads. Extensive diking, filling, and other engineering methods were used to counteract the land lowering that ultimately reached about 10 m. Water injection into the strata, after supportive California legislation had been passed, helped to stabilize the area. In addition to subsidence, earthquakes and faulting have also occurred in the Wilmington oil field. Eight separate periods of seismicity have been associated with the oil production, and the faults have severely damaged hundreds of producing wells with slippage as much as 22.8 cm during a seismic event. Total damages from all subsurface influences is much in excess of $100 million.

The earliest recognized subsidence-fluid withdrawal relationship was made in 1925 of the Goose Creek Oil Field, Texas, when leveling showed subsidence had affected an area 6.5 km long and 3.9 km wide with a maximum depression of more than 1 m. The subsidence area closely corresponds with the extraction area. In addition, faulting has occurred and earthquakes have been recorded. Some ruptures are 700 m long and many form steeply dipping displacements that are localized along the margin of the subsidence bowl. Subsidence in the Lake Maracaibo, Venezuela, oil fields was first discovered in 1933 and was more than 3.3 m in some areas by 1954. The clearest example of extraction-induced seismicity by gas outside North America is the Po Delta, Italy, where production of methane gas in 1951 caused a series of earthquakes. Subsidence in an area 40 km long and 20 km wide has created extensive flooding damage, necessitating construction of higher levees and drainage from the flooded lands. Subsidence of significant proportions also occurs from methane gas wells at Niigata, Japan.

The general model used by investigators to explain hydrocarbon extraction effects is strata compaction by loss of fluid support. In rigid units, low angle thrust faults form in the central area with normal faulting along the periphery.

OIL SHALE

Oil shales contain sufficient hydrocarbon matter to yield a liquid oil when heated. Scotland and Estonia have produced oil from shale since the mid-nineteenth century. The Soviet Union and China crush small amounts of oil shale and burn it directly under boilers.

High-grade oil shale can yield 100 gal or more of oil per ton. There may be about 5 trillion barrels of oil in world oil shales. The 1.8 trillion barrels in the states of Colorado, Utah, and Wyoming are 15 times the U.S. potential for conventional oil reserves (Fig. 6-22). Such figures are meaningless because they say nothing about how much oil can be extracted economically without requiring more energy in the extraction/refining/marketing process than is contained in the final product. The possible development of such "synfuels" is a hotly debated subject.

The richest of the extensive Green River oil shales in Colorado can yield about ½ to 1½ barrels per ton. Estimates of total recoverable oil for the U.S. deposits vary widely but are probably less than 200 billion barrels, although the production cost is approaching world market prices for crude oil (see Table 6-6).

The petroleum in oil shale is mostly **kerogen,** a waxy solid that requires heating to 500°C before liquefying. The shale would have to be mined, crushed, heated, and the resulting shale oil taken to special refineries (Fig. 6-23). Because it has a different chemistry from normal crude oil, refining methods must be altered to handle shale oil. During the recovery operations, the volume of the once-compact shale is increased 20 to 30 percent, which presents a huge disposal problem. Another serious problem is that large quantities of water in a water-poor region would be required. Most of the water would have to come from the Colorado River system which is already overexploited. An oil shale operation yielding 100,000 barrels a day of shale oil would require 12,000 to 18,000 acre-feet of water daily—much more if compaction and revegetation of waste piles is included. Such a plant would have to dispose of 565 million m³ of excess spent shale over a 20-year period—enough to fill a trench 600 m wide, 60 m deep, and 8 km long. In addition, there would be significant degradation of land from mining operations, air pollution from the re-

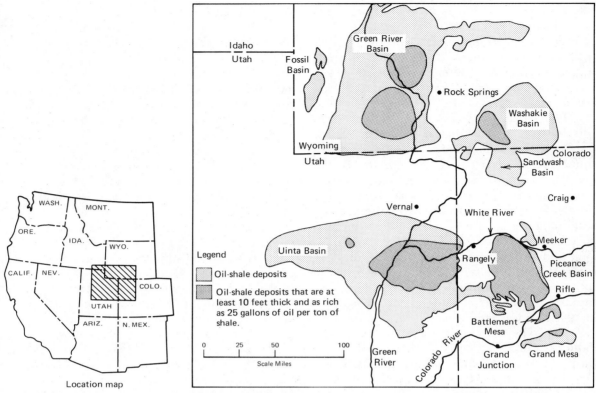

Figure 6-22 Oil shale areas in Colorado, Utah, and Wyoming. (Courtesy Geological Society of America.)

torting process, increased salinity of the Colorado River from water runoff, and associated effects, including those of increased population.

Overall, despite some very optimistic predictions, oil shale operations in the past have been plagued by abnormally high cost accelerations, severe environmental impacts, poor net energy efficiency, and questionable economic viability. *Business Week* magazine on April 25, 1977, concluded that oil shale is "a researcher's dream and an economist's nightmare." However, because of the continuing escalation in crude oil prices on the world market, there is the possibility that alternate technologies such as in situ liquidizing may now improve its prospects. Perhaps the age of synfuels is around the corner.

TAR SANDS

In tar sands (or bituminous sands), another synfuel, heavy asphaltic hydrocarbons act as the matrix in sandstone. The oil tar apparently is the residue left after lighter, more mobile hydrocarbons have escaped. There is an estimated 2 trillion barrels of petroleum in the world's tar sands, of which perhaps 300 billion barrels is recoverable.

The largest known deposits are the Athabasca tar sands in Alberta, Canada, with 250 billion barrels contained in deposits that range up to 60 m in thickness over an area of 77,000 km² (Fig. 6-24). These sands, which have been mined since 1966, can produce about 45,000 barrels of oil from 100,000 tons of sand. As

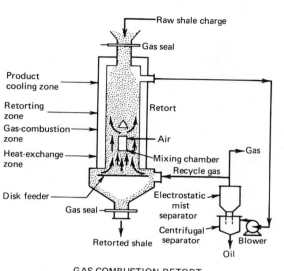

GAS-COMBUSTION RETORT
Recycle gas is mixed with air and burned within the retort. Gases flow upward and shale moves downward.

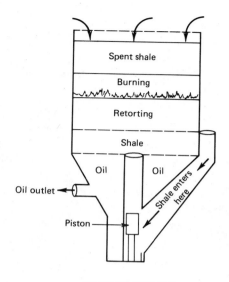

UNION OIL RETORT
Shale is introduced near bottom of retort and forced upward. Air enters at the top and flows downward.

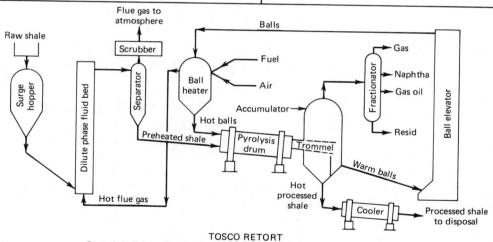

TOSCO RETORT
Ceramic balls transfer heat to the shale. No combustion takes place in retort.

Figure 6-23 Schematic representations of three oil shale retorting processes. (Courtesy Geological Society of America.)

with oil shale, oil from tar sands used to be expensive (about $27 per barrel compared to U.S. crude at $8.52 and OPEC crude at $13.50 per barrel in 1977) and produces similar environmental effects. The price now, however, is nearly competitive with world crude oil prices of more than $25 per barrel. Figure 6-25 presents one method of tar sand processing.

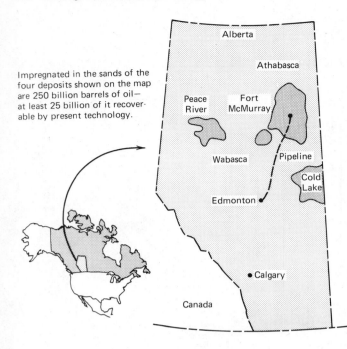

Impregnated in the sands of the four deposits shown on the map are 250 billion barrels of oil— at least 25 billion of it recoverable by present technology.

Figure 6-24 Map of Athabasca tar sand region of Canada.

NATURAL GAS

The gaseous fraction of petroleum is natural gas. The chief component of natural gas is methane (CH_4), the simplest and lightest of the hydrocarbons. Most traditional sources are found in association with oil pools where the gas forms a bubble overlying the oil (Fig. 6-17). The gas results from the chemical breakdown (cracking) of larger organic molecules. This same process, aided by bacteria, forms swamp gas and related forms of biogas (see Chapter 7).

Natural gas is the cleanest burning, highest quality fossil fuel. It provides heat for half of all U.S. homes and industries, represents about 30 percent of U.S. energy consumption, and is a vital raw material in the production of many manufactured products, most importantly fertilizers.

Unfortunately it is also the scarcest of the major fossil fuels. World consumption in the mid-1970s averaged about 15 trillion ft³/yr. Estimates of total world resources vary widely, but tend to cluster in the area of 12 quadrillion ft³. Less is known about natural gas reserves than about coal or oil reserves. Figures 6-26, 6-27, 6-28, and 6-29 provide some comparisons for gas resources and for oil and gas supplies in the United States.

Offshore drilling seems likely to enlarge the known reserve base considerably. Also, vast quantities undoubtedly exist that are dispersed in sedimentary rocks and dissolved in groundwater, but the technical and economic effects of developing these supplies are largely unknown. Projects to release the gas locked in such widely occurring formations as sandstones or black shales by underground explosives, crushing, heating, or chemical reactions have produced mixed results thus far. A highly controversial Atomic Energy Commission program to stimulate gas production in Colorado by underground nuclear detonations met with disappointing results and strong opposition and was eventually abandoned.

The many arguments that were heard in

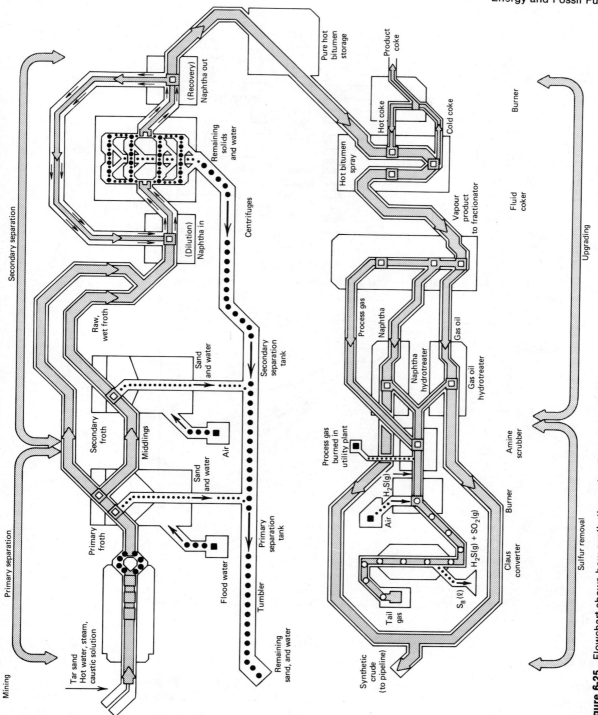

Figure 6-25 Flowchart shows how synthetic crude oil can be produced from tar sands mined in the Athabasca region of Alberta, Canada. (Syncrude Ltd.)

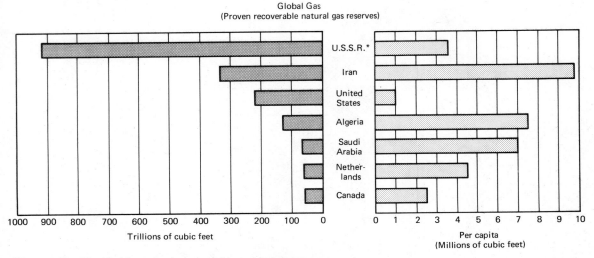

Figure 6-26 Worldwide proved recoverable natural gas reserves.
(*Oil and Gas Journal*.)

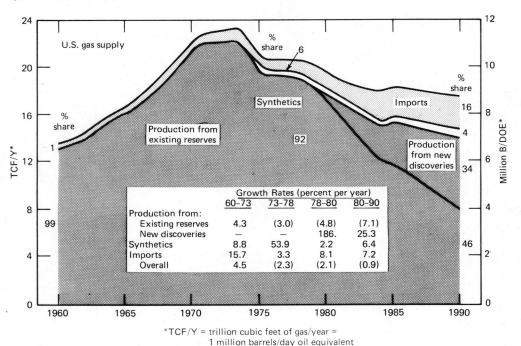

*TCF/Y = trillion cubic feet of gas/year =
1 million barrels/day oil equivalent

Figure 6-27 U.S. gas supply. (EXXON Corp.)

the 1970s regarding shortages versus surpluses of natural gas appear more a product of economic, political, and distribution conditions than the result of a genuine shortage. However, it is likely that declining production from traditional sources and the large costs and lead times required to develop new sources will combine to reduce the domestic supply of this vital

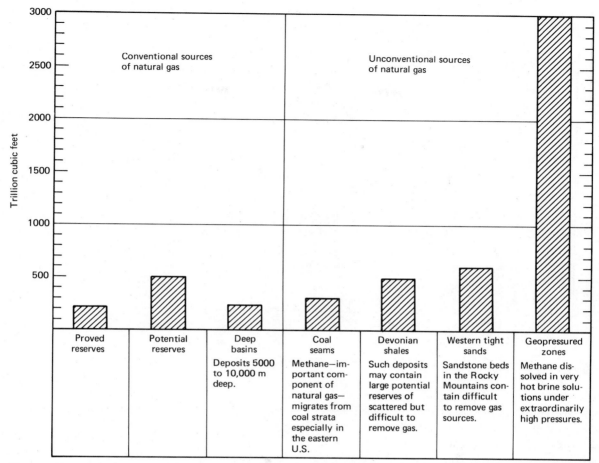

Figure 6-28 Comparison of conventional and speculative sources of new natural gases in the United States. Higher prices and additional technological improvements may make possible utilization of the speculative sources.

fuel in the last two decades of the twentieth century. Some studies state that 80 percent of America's natural gas resources will be depleted by the year 2000.

During the first 50 years of petroleum exploitation, natural gas was regarded as a nuisance and vast quantities were wasted by dissipation to the atmosphere or by burning (flaring off). With the gradual recognition of its value as a convenient, premium quality fuel and the development of a high-pressure pipeline distribution system in the United States, its use grew rapidly during the mid-1900s. Encouraging this growth was a cost to consumers well below that of other fossil fuels, a situation which will change due to the signing of an energy bill in November of 1978, which will gradually phase out government regulation of natural gas prices.

In many other countries, local demand for natural gas is minimal, and large amounts of this precious fuel continue to be wasted. If this gas is to be salvaged, the problem of how to economically transport it across large bodies of water must be solved.

Natural gas can be readily transported overland by pipeline. There are nearly a million miles of underground pipeline in the $50 billion

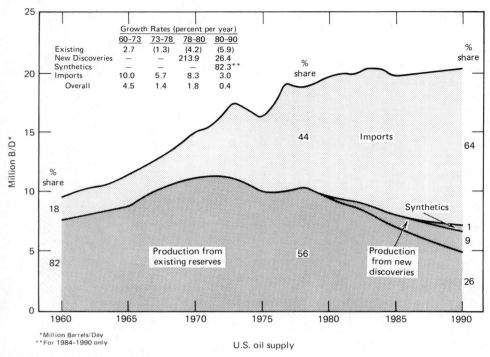

Figure 6-29 U.S. Oil supply. (EXXON Corp.)

gas distribution network in the United States. Oversea shipments from source areas to markets is a far more difficult task, the proposed solution to which is the liquefied natural gas (LNG) tanker or the liquefied energy gas (LEG) tanker. This LNG transportation system may grow extraordinarily fast.

Critics have described LNG tankers as floating bombs. A large tanker with a capacity of 125,000 m³ has the explosive energy of a million tons of TNT—the world's first atomic bomb equaled 20,000 tons of TNT. Should a tanker accident puncture an LNG compartment, the extremely volatile liquid would immediately eject and vaporize, forming a huge methane cloud that could be ignited by any flame or spark, producing a devastating fire storm. In a densely populated area, a major catastrophe claiming tens of thousands of lives could result. Understandably, many busy seaports have made

it clear they do not want LNG tankers entering their harbors.

This has led to a search for relatively remote coastal areas as potential sites for large LNG terminals. The first such site selected for study in California is at Cape Conception and, predictably, many objections and problems have arisen. For example, (1) it is a seismically active area with four known faults passing through or near the site. (2) Local ranchers and real estate developers are concerned about negative impacts on their businesses. (3) American Indians consider the site one of the nation's most important religious places. (4) Since this is one of the few remaining unspoiled sections of the coast, with notable scenic and wildlife values, environmental groups are opposed. (5) Extensive studies on land stability, wind and wave dangers, protection from sabotage, and the like, will have to be undertaken to assure integrity of the site. Similar difficulties can be anticipated for other proposed facilities, no matter where they are located. Valid concerns about the impacts of major projects of all types should be handled fairly and objectively. This usually means many years of delay before such facilities will come into use—a major reason why predictions of rapid solutions to our energy problems cannot be relied on.

Very often, alternatives to a project are not adequately considered. For example, California's LNG Terminal Act of 1977 fails to require investigation of offshore terminals. Improved safety regulations for LNG tanker design and construction as well as navigational procedures would also help calm fears about them.

Perspectives

Since the start of the Industrial Revolution most industrialized societies have not been thwarted in their ability to increase the national growth product by energy shortages. However, this spectre has come to haunt many countries. With the rapid depletion of domestic petroleum supplies, with which the United States had developed a commercial and car-truck economy, the nation is slowly realizing that the days of inexpensive fuel and energy are numbered. Our dependency on foreign imports, and the newly obtained "geopower" of OPEC nations, has sent shock waves throughout America and many other nations. The 1973 oil embargo and continually escalating petroleum prices have caused a new matrix of problems and alternatives for their solution. Energy became the dominant theme of the resource battleground during the 1970s. Petropower, petrodollars, and the geopolitics of oil will continue to be the force that guides many governmental policies during the last decades of the twentieth century.

The twenty-first century will witness an entirely different array of energy resources. The fossil fuels will be largely a memory, but may still be locally important for the first part of the century. Future historians will view the Fossil Fuel Age (1780–2000) as a brief interlude in the affairs of nations.

One problem that was not faced by governments was the finite character of these nonrenewable resources, as well as their lack of abundance. Prior to 1955 it was common practice for oil companies to always keep about 5 to 10 years of reserves "on hand." That is, the discovery rate always kept pace, or ahead, of consumption. Furthermore, geologists were viewed as marginal employees; when they had successfully discovered the amount of necessary reserves, personnel policies of many companies dictated that their employment be terminated. Then, when more reserves needed discovery at some future time, a new group was employed. Although this led to a stop-and-go employment situation, it seemed to work advantageously for profits. Furthermore, when it became apparent that new oil fields were more difficult and expensive to locate in the United States, whereas they could be found more readily in other parts of the world, such as the Middle East, much attention was directed to overseas exploration. Foreign oil also had the advantage that it could be more cheaply developed and processed. This state of luxury existed for about two decades, when finally the non-American countries took matters into their own hands by: (1) nationalizing oil products and (2) marketing and pricing the oil themselves. The stage was set for our present "energy crisis."

Thus the United States and many other industrialized nations are at the mercy and whim of OPEC nations until alternatives to petroleum become economically viable. The environmental impacts of coal will continue to plague society, but its use—along with nuclear energy—will be required until the renewable energy resources have become an engineering, political, and economic reality. Of course, the argument can be made that high oil costs now permit the development of previously marginal petroleum fields and make attractive the use of synfuels and other energy alternatives.

READINGS

Averitt, P., 1975, Coal resources of the United States—January 1, 1974: U.S. Geol. Survey Bull. 1412, 131 p.

Barnes, H. L., and Ronberger, 1968, Chemical aspects of acid mine drainage: Water Pollution Control Federation Journ., v. 40, n. 3, p. 371–384.

Brooks, D. B., 1966, Strip mine reclamation and economic analysis: Natural Resources Jour., v. 6, n. 1, p. 13–44.

Chase, S., 1911, The challenge of waste: League for Industrial Democracy, 32 p.

Collier, C. R., et al., 1964, Influences of strip mining on the hydrologic environment of parts of Beaver Creek Basin, Kentucky, 1955–1959: U.S. Geol. Survey Prof. Paper 427-B, 85 p.

Deasy, G. F., 1960, Terrain damages resulting from bituminous stripping in Pennsylvania: Penn. Ac. Sci., v. 134, p. 124–130.

Fischer, A. G., and Judson, S., eds., 1975, Petroleum and Global Tectonics: Princeton, Princeton Univ. Press, 322 p.

Grey, J., Sutton, G. W., and Zlotnick, 1978, Fuel conservation and applied research: Science, v. 200, p. 135–142.

Mayuga, M. N., and Allen, D. R., 1969, Subsidence in the Wilmington oil field, Long Beach, California, U.S.A.: Pub. Inst. Sci. Hyd., v. 88., p. 66–79.

Pearse, C. R., 1968, Athabasca tar sands: Canadian Geog. Jour. v. 76, n. 1, p. 2–9.

School of Forest Research, 1965, A symposium on coal mine spoil reclamation—scientific planning for regional beauty and prosperity: Penn. State Univ., 141 p.

Squires, A. M., 1974, Clean fuels from coal gasification: Science, v. 184, p. 340–346.

U.S. Geological Survey, 1978, Estimates of undiscovered petroleum resources—a perspective: Annual Report, Fiscal Year 1975, Reprint, Washington, U.S. Govt. Printing Office, 23 p.

Chapter Seven
Energy : Alternative Sources

Aerial view of Grand Coulee Dam, Washington, with Columbia Plateau basalt flows in background. (Courtesy U.S. Bureau of Reclamation.)

INTRODUCTION

In this chapter we investigate non-fossil fuel energy sources. It is impossible to cover adequately the vast amount of information and ideas which abound in this complex area in a single chapter. We will, however, review the more widely considered alternate energy paths with an emphasis on the promises and the problems associated with them.

With the exception of fission, meager budgets have constrained the development of many of these alternatives in the past. Because of this and government and industry prejudices, estimates of the contribution these energy sources can make in the total energy picture are extremely variable. Figures used herein are illustrative of the potential for the various energy alternatives. Such potentials will be realized only with extensive funding and effort.

In addition, most, if not all, statistics used in resource estimates tend to be tenuous. This is well illustrated by the sudden 1974 revision in U.S. Geological Survey estimates of undiscovered recoverable oil and gas resources in the United States. Estimates for oil dropped from the 200 billion to 400 billion barrels figure to 50 to 130 billion barrels, and gas from 1000 trillion to 2000 trillion ft³ to 320 trillion to 655 trillion ft³. In both revised estimates, the *lower* figure is given a 95 percent probability of being accurate and the higher figure a 5 percent probability.

Most published estimates are derived from a bewildering array of government, industry, and miscellaneous sources. Highly divergent results can emerge by using different methods and assumptions, and most investigators use the approach that favors their biased viewpoint.

Everyone has prejudices, but a scientist has a special obligation to control them and to be as open, fair, and objective as possible. Objectivity in controversial matters is sometimes defined as a noncommittal, middle-of-the-road stance. This is a mistake. However, it is necessary to mention both the problems and the need for nuclear power, as should be expected of an environmentally oriented text. Similarly, attention to the promises of many other energy alternatives may seem unduly emphasized. This is partly because, unlike nuclear projects, large-scale development of most alternatives has not occurred and all we have to date is the promise. Furthermore, the anticipated perils of nonnuclear alternates may be less disruptive than the known hazards of radioactivity. Past policies have stressed the nuclear option, whereas many viable alternatives have been unduly and perhaps unwisely neglected.

NUCLEAR FISSION

Following the awesome devastation of Hiroshima and Nagasaki by atomic bombs at the close of World War II, a determination developed to turn this massive energy source to peaceful purposes. The vision was that nuclear energy would provide us with clean, safe, and virtually limitless power in the near future. In the United States, nuclear fission energy became the one major alternate energy source that consumed the overwhelming bulk of research and developmental funds. By the mid-1950s, nuclear reactors were being used to generate electricity in the United States and also in Britain, France,

Figure 7-1 Aerial view looking south to the Niagara Mohawk Power Corp. nuclear power plant facilities at Nine Mile Point, New York. Lake Ontario is in the foreground. The No. 1 plant on the right was completed in 1969 and No. 2 plant is currently under construction. (Courtesy Niagara Mohawk Power Corp.)

and the Soviet Union. In late 1978, with 72 nuclear power plants operating in the United States, and 94 more under construction, nuclear energy was producing slightly more than 12 percent of our electricity (Figs. 7–1, 7–2, and 7–3). There are about 200 commercial fission reactors in the world, with nearly 500 more planned. By the year 2000, 50 or more countries could be producing some 2 million megawatts of electricity from nuclear plants.

But the promise of nuclear energy "too cheap to meter" never materialized, and fears that nuclear power is a "Faustian bargain" which mankind may deeply regret have been growing. Nuclear fission power has become one of the most controversial, vehemently divisive issues in the world today. We will briefly review nuclear energy and some of the issues that have made it the focus of hundreds of protests, some of them violent.

Fission is the splitting of heavy atomic nuclei. When split, a tiny amount of matter in the nucleus is converted into a large amount of energy according to Einstein's famous formula, $E = mc^2$, where E is energy, m is mass, and c is the velocity of light. Very few atoms are fissionable. The only naturally occurring fissionable

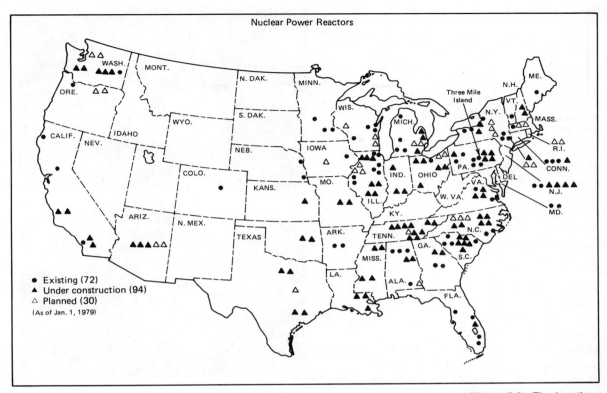

Nuclear Power Reactors

● Existing (72)
▲ Under construction (94)
△ Planned (30)
(As of Jan. 1, 1979)

Figure 7-2 The locations of nuclear power reactors in the United States. (U.S. Dept. of Energy.)

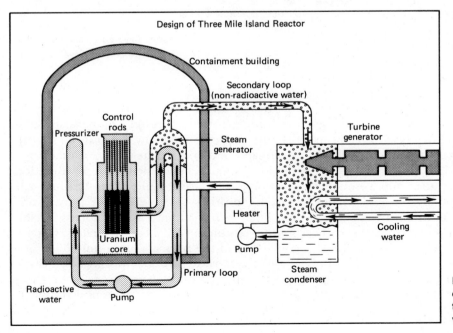

Design of Three Mile Island Reactor

Figure 7-3 Schematic drawing showing how a typical nuclear reactor works.

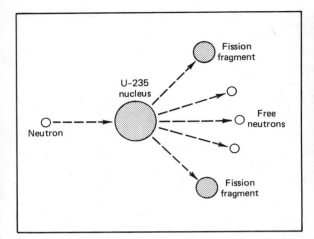

Figure 7-4 Typical atomic fission reaction.

Figure 7-5 Uranium mining. (a) Open pit mine in New Mexico, operated by Kerr McGee. (Georg Gerster/Photo Researchers.) (b) Underground mining in the New Quirke Mine, Ontario, Canada. The holes being drilled will be filled with explosives to loosen the ore. (Courtesy Rio Algom Ltd.)

"fuel" usable in current nuclear plants is uranium-235 (235 being its atomic weight), which constitutes only 0.7 percent of uranium ore. When a neutron is absorbed by a uranium-235 nucleus, the nucleus immediately undergoes fission. One of several possible reactions is: uranium-235 + 1 neutron → barium-142 + krypton-91 + 3 neutrons + energy. The 3 neutrons emitted by this reaction can strike other nearby uranium-235 nuclei, causing them to split and release more energy and neutrons, and so on. This is a **chain reaction.** An atomic bomb utilizes an uncontrolled chain reaction; a **nuclear reactor** uses neutron-absorbing materials and coolants to control the reaction rate (Fig. 7–4). The energy liberated is largely heat, and this is used to generate steam as in a fossil fuel power plant. A pound of uranium-235 can release as much energy as 1500 tons (3 million pounds) of coal.

However, high-grade uranium resources could be nearly exhausted by the end of this century, and the price of uranium has skyrocketed (from $8 to $40 a pound from the early 1970s to the mid-1970s; Fig. 7–5). In 1975, Westinghouse was forced to renege on its uranium supply contracts to electricity-producing power plants because of shortages in production of uranium (Fig. 7–6).

It was hoped that the answer to the uranium shortage problem would be nuclear breeder reactors that could actually generate more fuel (fissionable material) than they consume, thus providing an unlimited energy supply. Breeders can convert relatively plentiful uranium-238 (which comprises 99.3 percent of pure uranium) or thorium-232 into a synthetic fissionable isotope, plutonium-239. Unfortunately, this potential fuel of the future, plutonium-239, is also one of the most hazardous substances on earth. In spite of a continual emphasis on nuclear power programs, the Carter administration in 1978 halted the only major U.S. breeder project

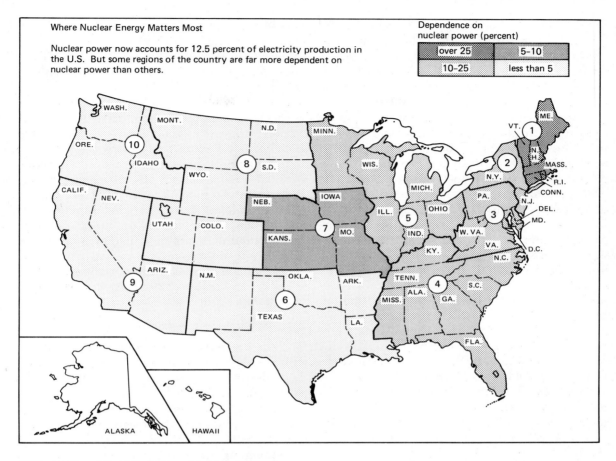

Where Nuclear Energy Matters Most

Nuclear power now accounts for 12.5 percent of electricity production in the U.S. But some regions of the country are far more dependent on nuclear power than others.

Dependence on
nuclear power (percent)

| over 25 | 5–10 |
| 10–25 | less than 5 |

1978 generation of electricity by type of fuel (percent)*

Region	Coal	Oil	Natural gas	Nuclear	Hydro
1	2.5	56.9	.2	35.2	5.2
2	13.5	47.7	.1	20.8	17.9
3	62.6	18.5	.1	17.2	1.6
4	57.3	15.3	4.3	16.0	7.1
5	74.5	6.3	1.2	17.0	1.0

*Preliminary

Region	Coal	Oil	Natural gas	Nuclear	Hydro
6	16.4	8.5	71.8	1.7	1.6
7	70.7	4.6	12.9	8.7	3.1
8	67.0	.8	3.3	.7	28.1
9	11.8	38.4	19.8	4.1	24.3
10	4.5	.3	1.2	3.9	90.0
Total for nation	**44.3**	**16.5**	**13.8**	**12.5**	**12.7**

Figure 7-6 In the United States, nuclear power accounts for 12.5 percent of the total electricity production. However, as shown on the map and table, nuclear dependence varies throughout the country. (U.S. Dept. of Energy.)

at Clinch River, Tennessee, primarily because of the risks involved in large-scale plutonium production, especially its capacity to be fashioned into nuclear bombs. The Soviet Union is presently constructing two large breeder reactors, and a third is planned.

The great difference between nuclear and other power sources is the large quantity of radioactive material produced by nuclear fission. This lies at the heart of the nuclear controversy and it is worth taking some time to understand just why the addition of radioactive materials to the environment is such a crucial issue.

Atoms whose nuclei are unstable may be radioactive, meaning they emit subatomic particles and energy. The three major emissions are heavy **alpha** particles (helium nuclei), **beta** particles (electrons or positrons), and high-energy **gamma** waves. The degree of nuclear instability is highly variable. Some radioactive forms of elements (**radioisotopes** or radionuclides) revert to more stable forms instantaneously; others may require great lengths of time. The time required for half the atoms in a radioactive substance to change into a more stable form is the half-life of that radioisotope. Lead-210 (which is suspected to be a primary cause of bronchial cancer in smokers) has a half-life of 22 years. If we started with 8 grams of lead-210 today, in 22 years 4 grams would remain; after 44 years, 2 grams, after 66 years, 1 gram; and so on. The apparent source of all naturally occurring radioisotopes in the earth today is uranium. Uranium-238 has a half-life of 4.5 billion years; thus, it will outlive shorter lived isotopes.

Not only are the direct products of the fission reaction highly radioactive, but all matter surrounding the reaction area is bombarded with subatomic particles that can enter or disrupt stable nuclei, producing more radioactive isotopes. A large nuclear reactor typically contains about 100 tons of uranium oxide fuel and 2 tons of the radioisotopes of plutonium, iodine, cesium, strontium, and numerous other elements. A thousand times more long-lived radioactive material is present in such a reactor than was produced by the Hiroshima atomic bomb.

Radioisotopes are dangerous to living tissue because their emissions have the ability to **ionize** (strip electrons from) the atoms of the matter they pass through. Because electrons control chemical behavior, ionization of just one molecule in a million will destroy delicate biochemical balances and kill a living cell. Radiation is especially hazardous if taken into the body, as during breathing and eating. The most dangerous form of radioactivity in the body is the heavy particles. Although they cannot penetrate a sheet of paper, alpha particles emitted by radioisotopes such as plutonium lodged within body tissue are extremely hazardous.

Radiation exposure is measured in **rads,** a measure of the energy absorbed per gram of body tissue, or **rems,** a measure of the relative damage to living tissue. Rem values tend to be considerably higher than rad values because a given amount of alpha radiation is about 16 times more damaging to exposed tissue than the same amount of gamma radiation; high-speed heavy nuclei are 20 times more damaging than gamma radiation.

$$\text{Rems} = \text{rads} \times \text{biological damage factor}$$

Average background radiation is about 100 millirems per year. A dosage of about 3000 rems destroys nervous tissue and will produce loss of physical and mental control followed by death within 48 hours. At about 600 rems, actively dividing cells are killed and death usually results from infection, which the body can no longer defend against, or internal hemorrhaging, often preceded by such symptoms as vomiting, diarrhea, and loss of skin and hair. Lower doses can cause leukemia in about 5 years, cancer in 12 to 40 years, and genetic diseases or abnormalities in future generations. In cases of cancer and genetic abnormalities, the cells (specifically the genes which control cellular activity) are damaged by radiation but survive to reproduce other abnormal cells—at a very high rate in many cancers.

Just one damaged cell can initiate cancer. Medical experts generally agree that there is no "safe level" of radiation exposure. This opinion is in disagreement with government and industrial approaches that have based their "maximum permissible concentration" levels for workers and the public on the assumption that there *is* a threshold below which there is no danger. For example, under current (1978) regulations, civilian nuclear workers may be exposed to 5 rads per year. However, radiation exposure is additive. At 5 rads per year, the doubling dose (dose at which the incidence of a disease is doubled) for leukemia will be reached in 1 year, and in 7 to 7½ years for cancer. Recent studies recommend reducing radiation exposure levels by a factor of at least 10, which could spell economic disaster for the nuclear power industry. New studies are revealing that humans exposed to low levels of radiation in the past, such as former uranium miners, nuclear plant workers, and people who live in the vicinity of uranium mining wastes, have significantly higher cancer rates than the general population. Because of the long latency period before many cancer symptoms become apparent, and because we have only had nuclear power for about 30 years, many fear the effects are just now beginning to be manifested.

In view of such evidence, it is possible that low-level "routine" emissions are a much more serious problem than previously believed. Some evidence also suggests that low-level radioactivity is causally related to heart disease.

From the above information, it should be clear that radioactive pollutants are the critical problem for nuclear power. Many nuclear wastes will remain dangerous for thousands of years and must be kept in permanent isolation. The problem of their ultimate disposition has yet to be settled. Various plans for the disposal of radioactive wastes are discussed in Chapter 20.

Even if the waste disposal problem could be solved, heavy reliance on fission energy will require that large quantities of hazardous radioactive material be moved about. Uranium must be mined, refined, processed, enriched, and fabricated into fuel pellets and then placed in reactors. Spent fuel and other reactor wastes must be periodically removed from the reactor and taken to a reprocessing plant where usable radioisotopes are removed and sent on their way while the waste is taken to a containment site. High-level wastes must be stored separately for several years to allow them to "cool down" to a level at which they can be handled and prepared for containment. Every step involves handling and thousands of miles of transport of dangerous nuclear materials. In 1974, there were 1532 shipments involving about 22,500 kg of enriched uranium and 372 shipments of some 720 kg of plutonium in the United States. With nuclear expansions being planned, even a police state cannot guarantee that thefts, terrorism, and accidents involving these materials will not occur.

The risk of major reactor accidents has attracted much attention. The safety requirements surrounding nuclear power development are among the most stringent in industrial history; to date there have been no catastrophic reactor accidents. However, some very serious "close calls" have occurred, such as at Three Mile Island in 1979, and thousands of assorted "incidents" have plagued reactors, often causing shutdowns and greatly reducing the anticipated power output. Up until now, however, nuclear plants have had a remarkable safety record, and the projected deaths when compared with other energy sources is favorable (Table 7–1).

Many reactor operation problems stem from the need to keep the core (where the fission reaction occurs) from overheating. Seven percent of a reactor's thermal output is generated by the decay of radioactive material in its core. Unlike the fission reaction, this heat flow cannot be controlled and, without cooling, the reactor core would become red hot and melt its way downward into the earth beneath the reactor. This so-called "China Syndrome" (the movie by that name has dramatized these fears) would breach containment vessels and release enormous amounts of radioactive materials into the environment. Steam explosions would probably accompany such a meltdown, further spreading the contamination.

A 1964 Brookhaven National Laboratory report for the Atomic Energy Commission es-

Table 7-1 Comparison of risks from hydroelectric, coal, and nuclear electric power technologies

Hazard type	Hydroelectric	Coal	Nuclear
Routine occupational hazard	Construction accidents are significant, but the risks are not as large as for coal mining	Coal mining accidents and black lung disease constitute a uniquely high risk	Risks from sources not involving radioactivity dominate; aggregate risks from all stages of the fuel cycle are less than for coal
Deaths	0.1 to 1.0[a]	2.7[b]	0.3 to 0.6[c]
Routine population hazard	Thought to be benign, although specific cases (for example, the Aswan dam) have produced new health hazards	Air pollution produces relatively high, although uncertain, risk of respiratory injury; significant transportation risks	Low-level radioactive emissions are more benign than corresponding risks from coal; significant transportation risks remain incompletely evaluated
Deaths		1.2 to 50[d]	>0.03[e]
General environmental degradation	Permanent loss of free-running streams, agricultural lands, wilderness	Strip mining and acid runoff; acid rainfall with possible effect on nitrogen cycle, atmospheric ozone; eventual need for strip mining on a large scale	Long-term contamination with radioactivity; eventual need for strip mining on a large scale
Catastrophic hazards (excluding occupational)	Major dam failures have occurred, but rarely in modern structures	Acute air pollution episodes with hundreds of deaths are not uncommon; long-term climatic change induced by CO_2 is conceivable	Risks of reactor accidents are small compared to other quantified catastrophic risks; the problem lies in as yet unquantified risks for the reactors and the remainder of the fuel cycle
Deaths	<1[f]	≃0.5[g]	>0.04[h]

Source: C. Hohenemser et al., 1977, The distrust of nuclear power: *Science,* V. 196, p. 6. Copyright © 1977, American Association for the Advancement of Science.

Note: Deaths are the number expected per year for a 1000 megawatt electric power plant. Man-days lost are converted to deaths by 6000 MDL/death.

[a]This estimate is based on: (1) 10,000 man-years to construct a 1000 megawatt hydroelectric dam and generating station; (2) a heavy construction occupational hazard of 0.34 fatality and 1.34 permanently disabling injuries per 1000 man-years, or about one fatality equivalent per 1000 man-years; (3) distribution of construction fatalities over an assumed 100-year useful life of the project; and (4) hydroelectric generation availability of 10 to 100 percent.

[b]Of the 2.7 deaths, 1.1 are due to mining accidents, including major mine disasters, and 1.6 are due to black lung disease and other injuries.

[c]The range indicated is the generally acknowledged difference.

[d]The lower figure represents transportation accidents only. The higher figure includes an interpretation of uncertain air pollution epidemiology.

[e]The figure is consistent with an average annual exposure of 0.035 millirem per individual reactor, using a cancer risk of 2×10^6 cancers per man-rem. The average exposure applies to reactors only, so is a lower bound for the fuel cycle risk.

[f]The figure is an estimate for dam failure based on all historical incidents. The number is an upper bound because many dam failures will not be connected with hydroelectric generation.

[g]Based on the occurrence of one 500-death air pollution episode per year, with one-fifth of the pollution attributable to coal power plants.

[h]Based on predicted possibilities, but must be regarded as lower bound.

timated that maximum damages from such an accident could kill 27,000 people, cause $17 billion in material damage, and contaminate an area the size of Pennsylvania. A 1972 $3 million Atomic Energy Commission sponsored report (the Rasmussen report) downplayed the dangers and concluded that the chances for the "worst" accident in any one reactor in any one year would be one in a billion. Since then the Rasmussen report has been severely criticized for "arbitrary calculational techniques," highly subjective interpretations, and misleading statements. For example, it assumed that vital components and safety systems would work perfectly (the vital emergency core cooling system had never even been successfully tested at that time), ignored sabotage and terrorism threats, and dismissed accidents that had already occurred.

A voluminous amount of data, both pro and con, has been published regarding reactor safety, but the bottom line remains unsettling: it is impossible to prevent all accidents.

On March 27, 1975, two 1100 megawatt reactors at Browns Ferry, Alabama, were put out of commission and came dangerously close to a core meltdown, because of a fire set by a workman looking for air leaks with lighted candles! In 1978, a workman's galosh fell into the Browns Ferry reactor (one of several such incidents) and precipitated an unsuccessful $2.8 million search. The 1979 accident at Three Mile Island, Pennsylvania, provided the greatest hazard and resulted in a new wave of antinuclear sentiment (Figs. 7–7 and 7–8).

A less exotic, more persistent problem involves leaks in the vital reactor cooling systems. It is very difficult and hazardous to monitor the pipes carrying coolant, especially in the radioactive, high-stress area near the reactor core. This safety problem has been recognized for some time but little has been done to correct it (reactor shutdowns are very costly). On June 17, 1978, an electrical malfunction caused a reactor shutdown at the Duane Arnold plant in Iowa. A routine inspection coincidentally revealed a 10 in. (25.4 cm) diameter primary recirculation cooling pipe with water spurting from a 4 in. (10 cm) long crack. Tests showed that the crack actually extended 270° around the pipe, and ultrasonic tests showed that the seven other recirculation pipes were also developing cracks. There is fear that similar cracks may occur in other reactors. A complete break in these pipes would result in the "worst possible loss-of-coolant accident" and could trigger a core meltdown.

Utilities ordered many reactors when enthusiasm waxed high over the promises of the peaceful atom. Instead of the promised 80 percent operating capacity, actual operations at nuclear power plants have often averaged less than 50 percent of capacity—for example, 42 percent in 1973, 48 percent in 1974, and 45 percent in 1975. Because of the many problems associated with nuclear power, half the reactors in the country have at times been shut down completely. As a result, timetables for nuclear power production have steadily declined. In 1974, U.S. government predictions for nuclear output in the year 2000 were 1.25 million megawatts; by 1976, the predictions had been lowered to 380,000 to 620,000 megawatts (Fig. 7–9).

The costs of nuclear power plants have risen at abnormal rates. In 1978, plant construction costs were comparable for coal power plants; however, fuel consumption costs were less (Table 7–2). From 1969 to 1975, the per kilowatt cost of nuclear plants increased 2½ times as much as did the cost of coal-fired plants. If government nuclear programs grow as predicted, the investment would equal more than a quarter of the nation's entire net capital investments. And these costs do not include such additional expenses as the disposal of nuclear wastes, regulations, and the decommissioning of highly radioactive worn-out nuclear facilities. Due in large part to escalating costs and low overall operating capacities, cancellations and deferrals of planned nuclear plants outnumbered new reactor orders by 25 to 1 in 1975.

The extreme toxicity of radioactive wastes

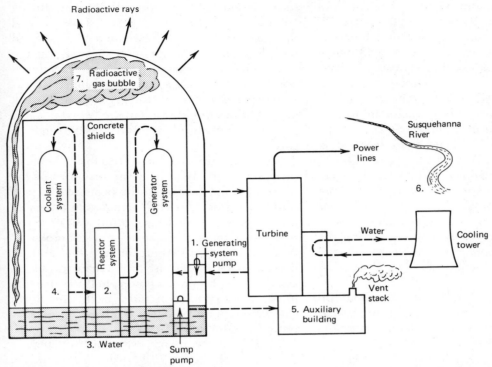

Figure 7-7 Schematic presentation of near-disaster events that occurred at the Three Mile Island nuclear plant, Pennsylvania, in April 1979. 1. Generating system pump stops and cuts off water to steam generator. 2. Reactor system continues to produce heat, raising temperature and pressure and causing reactor to shut down automatically. 3. Valve on pressurizer opens as planned but fails to close. Radioactive water flows into tank and floods floor of containment structure. 4. Pressure and water level in the reactor system drop, triggering emergency cooling system. Operator shuts it off, but some fuel rods overheat. 5. Sump pump transfers radioactive water to auxiliary building which floods, and radioactive steam is vented. 6. Some radioactive water is released into river. 7. Radioactive gas bubble forms at the top of the reactor, raising the danger of a possible explosion or a fuel-rod meltdown.

and the fact that fissionable materials can be fashioned into crude nuclear weapons with relative ease give rise to fears of nuclear terrorism and blackmail. Only 4 kg of plutonium-239, uranium-235, or uranium-233 can be made into a nuclear bomb and only 2 kg of plutonium-239 will suffice as the "trigger quantity" in an implosion bomb. Substantial amounts of fissionable materials have already been diverted illegally, and the reactors the United States plans to sell to other countries in the late 1970s and early 1980s will generate sufficient plutonium to make 3000 small atomic bombs each year. The growing availability of nuclear facilities and materials means that increasing numbers of countries will be joining the nuclear powers. By 1976, at least 15 countries, not including the six that had already exploded nuclear "devices," had the fissionable material and the technical know-how to develop nuclear weapons.

Nuclear plants can become the focus of terrorist attacks. To illustrate the ease of such actions, Werner Twardzik, a West German parliamentary representative, once toured the huge 1200 megawatt Bilbis-A reactor with an undetected 2 ft bazooka under his jacket.

There is also the problem of choosing the proper site for a nuclear plant. It is crucial that

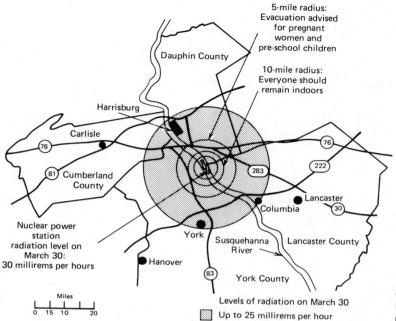

5-mile radius:
Evacuation advised
for pregnant
women and
pre-school children

10-mile radius:
Everyone should
remain indoors

Dauphin County

Harrisburg

Carlisle

76

81 Cumberland
County

283

222

76

Lancaster

Columbia

30

Nuclear power
station
radiation level on
March 30:
30 millirems per hours

York

Susquehanna
River

Lancaster County

Hanover

83

York County

Miles

0   15   10   20

Levels of radiation on March 30

Up to 25 millirems per hour

Less than 2 millirems per hour

Figure 7-8   The danger zone that
developed around the Three Mile
Island nuclear site, and the
precautions that were taken in April
1979.

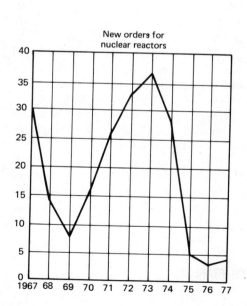

New orders for
nuclear reactors

1967 68   69   70   71   72   73   74   75   76   77

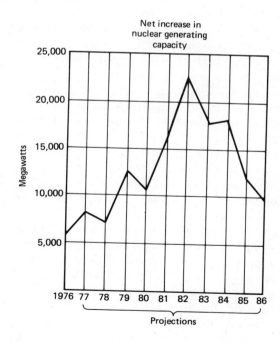

Net increase in
nuclear generating
capacity

Megawatts

1976 77   78   79   80   81   82   83   84   85   86

Projections

Figure 7-9   Decline in nuclear plant construction in the United
States. (U.S. Dept. of Energy.)

Table 7-2 Fuel costs for nuclear and fossil fuel electric power technologies (1977 dollars)

Assumptions	Cost (mills/kWh)
Nuclear fuel[a]	
Yellow cake, $40 per pound	3.5
Uranium for conversion to UF6, $2.75 per pound	0.1
Enrichment (0.20 percent tails assay), $75 per SWU[b]	1.8
Fabrication, $110 per kilogram of uranium	0.7
Net salvage[c]	1.0
	7.1
Fossil fuels[d]	
High-sulfur coal, $1.20 per million BTU, with scrubber	13
Low-sulfur coal, $1.40 per million BTU, with scrubber	16
Number 6 oil, $2.50 per million BTU, without scrubber	26

Source: A. D. Rossin and T. A. Rieck, 1978, Economics of nuclear power: *Science,* v. 201, p. 587. Copyright © 1978 American Association for the Advancement of Science.

[a]Burnup is assumed to be 33,000 megawatt-days per ton for pressurized water reactors and 29,000 megawatt-days per ton for boiling-water reactors.
[b]SWU, separative work unit, as defined by the Department of Energy.
[c]We assumed a net salvage cost (cost associated with the ultimate disposition of discharged or spent nuclear fuel) of 1 mill/kWh.
[d]Cost delivered in the Chicago area, including estimated carrying charges for maintaining a 90-day fuel inventory.

nuclear plants, and the waste disposal areas, be located in tectonically stable areas (see page 665). An earthquake could greatly damage the facility and, if cooling pipes were destroyed and the core coolant backup systems failed, a catastrophe could result. In March 1979, the NRC (Nuclear Regulatory Commission) closed down five nuclear plants in the eastern United States because a computer mistake had led to an improper safety design for cooling pipes. A reanalysis of the construction data revealed that the pipes could not withstand even a moderate earthquake shock.

We have outlined the major problems plaguing the nuclear power program: the low-level radiation threat, waste disposal, accident risks, nuclear weapons proliferation, terrorism, fuel supply, earthquake hazards, and costs. There are also bright spots, of course, such as the excellent operating record of some medium-sized reactors and Canada's reliable CANDU reactor. Nuclear power still has many strong advocates in the United States and around the world. They point out that all energy use involves trade-offs, and nuclear power is the only short-term alternate of sufficient magnitude that can at present replace dwindling fossil fuels. In the United States a severe electricity demand and supply gap would occur if nuclear plants were closed. Such a consequence would drastically affect the national economy and lead to grave socioeconomic problems. Although we have looked at adverse nuclear energy features, positive short-range energy production alternatives have not been developed. Depending on foreign oil is vexing and causes more economic distress than continued development of nuclear power.

NUCLEAR FUSION

Fusion is the joining of nuclei of very lightweight atoms (Fig. 7–10). As with the fission of

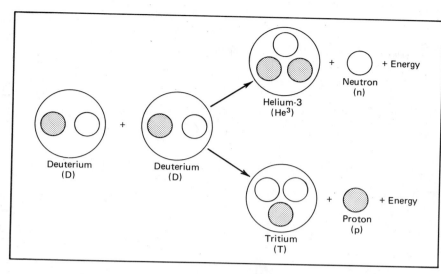

Figure 7-10 Atomic reactions in deuterium-deuterium fusion. The open circles represent neutrons, and the dark circles represent protons. The motion of the neutrons and protons creates energy.

heavy nuclei, a minute amount of matter is converted into a large quantity of energy during the reaction. Fusion generates the sun's energy and is responsible for the incredible destructive force of the hydrogen bomb. Most fusion reactions occur between two of the three forms of hydrogen: **protium** or ordinary hydrogen, **deuterium** or "heavy hydrogen", and **tritium** or hydrogen 3—an artificial radioisotope with a half-life of 12 years.

A temperature of 40 million degrees centigrade must be attained to initiate fusion between deuterium and tritium, the lowest temperature reaction and the one the harnessing of fusion energy will probably rely on. In the hydrogen bomb, the reaction requires an atomic bomb to trigger it. At such high temperatures ordinary matter cannot exist, and atoms break down into a plasma state of free electrons and atomic nuclei. Because no substance can withstand the temperatures involved, fusion reactions must be contained and maintained within a strong energy (probably magnetic) field (Fig. 7–11). Although recent experiments have initiated minute fusion reactions in the laboratory, the production of power from sustained and controlled fusion is still far in the future.

Once the extremely complex problems of controlled fusion power production are solved, it could provide an essentially limitless energy source—the deuterium occurs in seawater and tritium can be produced from lithium which is also available from seawater. The dangers of fusion would be far less formidable than those associated with fission. Tritium and radioisotopes produced by radiation bombardment during the reaction would need to be controlled and disposed, though the quantities would be minor compared to fission. And, unlike the natural heat input of the sun (see following section), fusion would contribute large quantities of additional heat (thermal pollution) to the environment.

DIRECT SOLAR ENERGY

Hydroelectric power, wind, ocean thermal and salinity gradients, organic energy sources, even fossil fuels—all are forms of solar energy. Here we consider only the direct use of sunlight. This use may be **passive,** in which solar energy is used or stored in place, as where the sun strikes a building, or **active,** in which fans or pumps move fluids from a collecting area to storage areas.

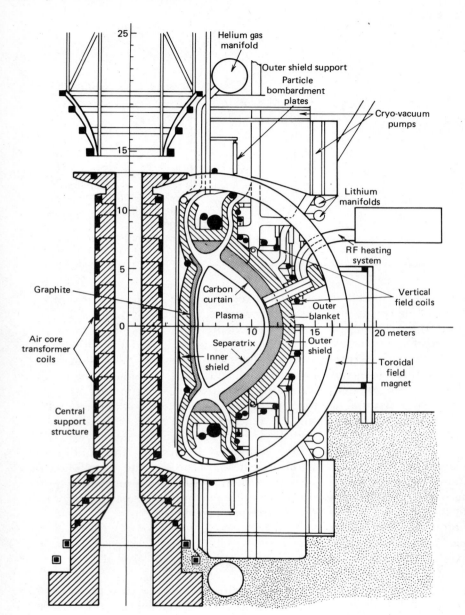

Figure 7-11 Schematic cross section of UWMAK II reactor torus, a possible design for the engineering of fusion power plants. (W. E. Parkins, 1978, Engineering limitations of fusion power plants: *Science*, v. 199, p. 1405. Copyright © 1978 American Association for the Advancement of Science.)

It is clear from their structures that many ancient peoples, such as the Incas and Pueblos, understood well the passive use of solar energy. Focusing the sun's rays to kindle fires has been in use for at least 2500 years. It is believed that Archimedes used large concave metal mirrors in this fashion to set fire to the sails of the Roman fleet that attacked Syracuse in 212 B.C. The fleet was defeated. Among the many applications of solar energy known from the past was a solar oven in 1837, a metal smelting solar furnace in 1868, a solar still in 1872, a 50

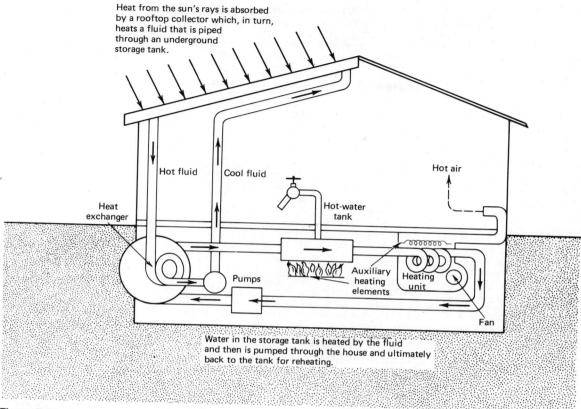

Heat from the sun's rays is absorbed by a rooftop collector which, in turn, heats a fluid that is piped through an underground storage tank.

Hot fluid

Cool fluid

Hot air

Heat exchanger

Hot-water tank

Pumps

Auxiliary heating elements

Heating unit

Fan

Water in the storage tank is heated by the fluid and then is pumped through the house and ultimately back to the tank for reheating.

Figure 7-12 A solar home heating system.

horsepower solar power plant in Egypt in 1912, and a solar irrigation pump along the Nile River in 1913.

In the past few decades, the apparent abundance of fossil fuels and the promise of nuclear power caused development of solar energy to be neglected by industry and government. From 1954 to 1971, while the U.S. government was investing some $3 billion in civilian nuclear projects only $1 million was spent for solar-powered systems! In the ensuing years, government energy agencies have continued to overlook solar energy, so much so that Congress has had to authorize 80 percent more funding than the energy agencies had requested. Solar spending is currently on the upswing, with a 1978 solar-related budget of $400 million. Programs

such as tax incentives for home investments in solar and other energy-saving improvements should guarantee the continued expansion of solar energy in future years.

Some 40,000 U.S. residences now have solar devices, and the number may increase to 2.5 million by 1985 (Fig. 7–12). A 1976 government study concluded that solar heating in new, well-insulated homes was already economically competitive with electric heating. The young and booming business had already marketed products totaling about $200 million in 1978 (Table 7–3).

The new enthusiasm for solar energy is worldwide, with even the oil-rich Mid-Eastern nations showing acute interest. The world's largest solar-heated building is a 325,000 ft²

Table 7-3 Estimates of the heat, electric power, and fuels to be supplied by solar energy in the United States as projected by the Energy Research and Development Administration

Solar technology	1985	2000	2020
Direct thermal applications (in units of 10^{15} BTU = 1 Q per year)			
Heating and cooling	0.15 Q	2.0 Q	15 Q
Agricultural applications	0.03	0.6	3
Industrial applications	0.02	0.4	2
Total	0.2 Q	3 Q	20 Q
Solar electric capacity (in units of 10^9 watts = 1 Gwe)			
Wind	1.0 Gwe	20 Gwe	60 Gwe
Photovoltaic	0.1	30	80
Solar thermal	0.05	20	70
Ocean thermal	0.1	10	40
Total	1.3 Gwe	80 Gwe	250 Gwe
Equivalent fuel energy	0.07 Q	5 Q	15 Q
Fuels from biomass	0.5 Q	3 Q	10 Q
Total solar energy	~1 Q	~10 Q	~45 Q
Projected U.S. energy demand	100 Q	150 Q	180 Q

Source: A. L. Hammond, 1975, Solar energy reconsidered: *Science*, v. 189, p. 539. Copyright © 1975 American Association for the Advancement of Science.

athletic field house in Tabuk, Saudi Arabia. In 1978, the Saudis joined the United States in a $100 million, 5-year joint solar energy research program. And in Japan, as of early 1977, 2 million solar water heaters had been sold.

The potential of solar energy is enormous. The sunlight falling on U.S. roads in 1 year is equal to twice the energy content of all the fossil fuel used by the world during that same time period. Even in the cloudiest parts of the conterminous forty-eight states, the average solar energy received by a 29 × 29 square foot area equals the total energy demands of an average U.S. family.

The question is how to best collect part of that energy and put it to use. There are two major methods of utilizing sunlight directly. The most popular and widespread techniques are **photothermal**—converting solar energy into heat that can then be used in a wide variety of ways, including cooking, space heating, crop drying, air conditioning (using heat pumps), and, where facilities are built to concentrate the

sun's rays thus attaining high temperatures, steam-operated electric generators can be installed.

A wide variety of small-scale thermal solar collectors are available. One common type consists of plates containing an absorbent black surface covered with clear plastic or glass. The black surface is heated when struck by sunlight and the clear cover helps limit heat losses. Behind the black surface, a system of copper pipes contains water that is heated, then circulated or stored for use. Simple solar home and water heaters of this type will pay for themselves in less than 15 years in any part of the United States.

The second major solar energy conversion process utilizes **photovoltaic** (photoelectric) cells. These are made of materials that generate an electric current when sunlight strikes them, a phenomenon first observed by the French physicist A. C. Becquerel in 1839. Photovoltaic cells provide the power for earth satellites, but their costs have been too high for significant earthly contributions. Photovoltaic cells cost $200,000 per peak kilowatt capacity in 1959, but this had dropped to $9000 by 1978. Present photovoltaic cells use silicon crystals that must be carefully grown and hand cut. With mechanization of production and less costly substitutes for the silicon crystals, costs should decline drastically. Some anticipate costs of only $2 per kilowatt in the 1980s. A new amorphous (non-crystalline) silicon-based alloy has been developed by Energy Conversion Devices, Inc., which, with quantity production, should be able to produce electricity at $500 per kilowatt by 1981. Such a cost would give photovoltaic cells a strong economic advantage over fossil fuel and nuclear-generated electricity. Net energy analyses show that photovoltaic generation will pay back its energy debt (i.e., will produce energy equal to the total energy that went into its production) in less than 2 years. Also photoelectric cells have strong advantages: free fuel, long life, no pollution, and minimal maintenance.

Large-scale solar energy schemes include "solar farms" of photothermal or photoelectric collectors, "power towers," and energy-collecting satellites. A centralized 1000 megawatt solar farm plant using present inefficient photovoltaic cells would cover about 10 km² of ground in the southwestern United States. The large amount of land required for such installations has often been the source of criticism, but the area is about the same as that required for a modern nuclear plant (largely for safety reasons) and less than the land that would have to be strip mined to provide fuel to a coal-fired plant of the same capacity for its expected lifetime.

The "power tower" is a large central boiler that would produce electricity from steam generated by the heat of an array of heliostats (movable mirrors) that focus the sun's energy on the boiler (Fig. 7–13). A 10 megawatt pilot plant using 3000 mirrors is being built by the U.S. government near Barstow, California, at a cost of about $100 million, nearly 20 times the cost of coal-fired plants of similar capacity.

Still another centralized solar scheme would use earth satellites to direct the sun's energy to collection areas on earth (Fig. 7–14). However, the future is still uncertain. As with the nuclear program, the solar satellite schemes clearly have military implications that would further encourage their pursuit.

The majority of U.S. government solar research and development funding has been spent on large-scale, costly, centralized, long-term solar installations such as these. Many feel that a much wider range of technologies, especially for small and intermediate-scale devices, should play a much larger role in the government's program. The emphasis on large centralized facilities may be influenced by the huge investments special industries have in centralized systems. Small-scale solar development means energy independence for many people and a loss of customers to electric utilities and energy conglomerates.

Since solar energy is inherently diffuse, development programs utilizing many small collection units in individual homes, apartments, and businesses, rather than large centralized collection sites, have numerous advantages: (1) The environmental disruptions, costs, and energy waste involved in the transportation and distribution of energy are minimized. (2) Diffuse collection technology is known, relatively simple, and can be put into operation almost immediately. (3) A variety of designs allow wide adaptations, which can be efficiently matched to local needs. (4) Many simple but effective solar devices can be constructed at reasonable cost using local materials and labor. (5) It provides people with increased self-sufficiency and independence and helps to avoid widespread blackouts and economic disruptions. These factors are especially important in impoverished countries where the skills, materials, and funds for sophisticated technologies do not exist. Simple, workable technologies could improve living

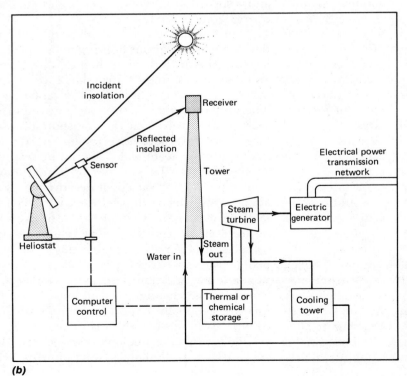

A field of thousands of mirrors
concentrates sunlight on a boiler
mounted atop a 300 foot tower to
produce steam, which is used
to generate electricity.

Electricity

Steam

Steam
turbine

Electric
generator

Timing devices keep the mirrors in alignment
with the sun, and transparent plastic bubbles
enclose them to protect them from the wind.

(a)

Incident
insolation

Receiver

Reflected
insolation

Sensor

Tower

Electrical power
transmission
network

Steam
turbine

Electric
generator

Heliostat

Water in

Steam
out

Computer
control

Thermal or
chemical
storage

Cooling
tower

(b)

Figure 7-13 Use of
heliostats for electric
power generation. (*a*)
About 2000 heliostats
would be necessary to
produce a 10-megawatt
plant. (*b*) Schematic
layout of components in
a heliostat system.

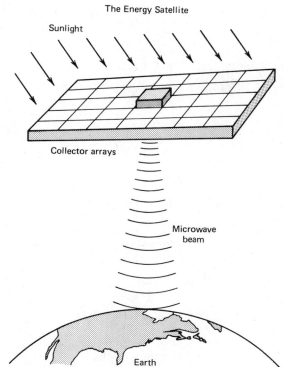

The Energy Satellite

Sunlight

Collector arrays

Microwave beam

Earth

Figure 7-14 Imaginative view of an electrical power system designed for a space satellite, to collect and transmit energy.

Figure 7-15 Windmill used to pump groundwater near Sells, Arizona. The facility also contains a Jensen jack to pump water using fossil fuel when the wind does not blow. The pumping level is 205 m.

standards, help people feed and provide for themselves, and contribute to the stabilization of many potentially dangerous political situations.

WIND POWER

Wind is air set in motion by the uneven solar heating of the rotating earth. There is 20 times more available energy in wind than in hydropower. Man's use of wind goes back thousands of years to the wind-driven seafaring vessels of ancient civilizations. And wind-powered ships could still provide a significant portion of the world's ocean transportation.

Windmills were used in Persia 2000 years ago. By 1900, Denmark had 100,000 windmills and, by 1916, over 1300 wind generators were in operation in that country. In the nineteenth century, over 6 million small windmills dotted rural America where they pumped water, ground grain, and generated electricity. After 1850, they were gradually replaced by steam engines and later by cheap fossil fuels. Today, about 150,000 rural windmills are still used, mostly for pumping groundwater (Fig. 7–15).

The first large wind generator (100 kilowatts) was built near Yalta, Russia, in 1931. On August 29, 1941, a 1250 kilowatt (kW) windmill began operating in Vermont. It was built by a small industrial firm in 2 years and operated smoothly for 16 months. Numerous large wind-powered generators are now operating or are being constructed in the United States and other countries.

Windmills are one of the most efficient ways of utilizing nonpolluting energy sources, with an average energy efficiency of 35 percent compared to 5 to 15 percent for solar units. The force of wind on a propeller blade equals the third power of wind velocity. Thus, a doubling of wind speed equals an eightfold power increase. The World Meteorological Organization has estimated that, by utilizing the best commercially feasible land sites around the world, 20 million megawatts of wind-generated electricity could be produced. By comparison, the total world electric generating capacity in the mid-1970s was about 1.5 million megawatts. The Solar Energy Panel of a National Science Foundation/National Aeronautical and Space Administration (NSF and NASA) study force concluded that wind power could provide 20 percent of U.S. electricity needs by the year 2000.

Whereas some schemes, such as those envisioning thousands of 300 m high windmill towers densely populating the Great Plains, may be exaggerated, wind power on a more modest scale can be an efficient and viable alternative energy source. A windmill "farm" of 20 towers (37 m high with 60 m long propeller blades) could supply the electricity needs for a small town. A machine with 19 m blades in Clayton, New Mexico, operates 90 percent of the time (whenever wind speed exceeds 13 km/h and produces enough electricity to supply 60 average homes.

On a smaller scale, a wide variety of wind generators for individual homes (or small groups of homes or businesses) are available. In 1977, costs of these units were mostly between $4000 and $10,000. With increasing production, these costs should decrease as traditional fuel prices escalate. Two-way grid linkages enable these wind generators to feed excess electricity into the regional grid and credit this to the owner's account. Small windmills, though not inherently as efficient as large models, will be a better choice in some situations because they can be located near the users, can be constructed in less time at relatively less cost, are aesthetically unobtrusive, durable, and can operate even in gentle winds.

Offshore winds tend to be stronger and steadier than winds on land. In New England, wind power averages 150 W/m² on land but ranges from 400 to 700 W/m² on the continental shelf about 100 km off the coast. In 1972, Professor William E. Heronemus of the University of Massachusetts proposed large numbers of floating offshore wind generators, each con-

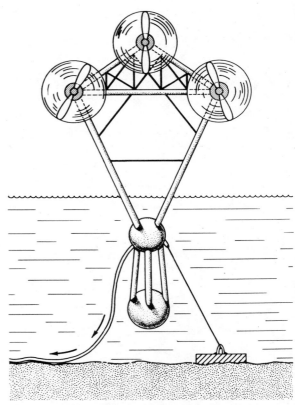

Figure 7-16 Offshore windmills as proposed for the New England region. Wind power produces electricity which makes hydrogen fuel from seawater.

sisting of three 2 MW (megawatt) wind turbines (Fig. 7–16). Excess electricity could be used to generate hydrogen from seawater. The hydrogen could serve as an auxiliary fuel when the wind isn't blowing and may be piped onshore for combustion or storage in fuel cells. An NSF-NASA study in the early 1970s estimated a potential of over 300 billion kWh of electricity per year from offshore New England winds and an annual total for the entire U.S. exceeding 1.5 trillion kWh (equal to total U.S. electricity use in 1970).

Many different windmill designs are available. Most standard designs use a simple two-blade propeller, since this provides the most

Table 7-4 Cost of supplying electricity by means of windmill generators

Base data	
Useful life of windmill generator	15 years
Lifetime of batteries	6 years
Efficiency of batteries	85 percent
Equipment costs	
Windmill generator (10 kW)	170,000s
Wiring, controls	42,000s
Installation	42,000s
Battery storage (100 kWh at 85 percent efficiency)	
118 kWh at 800s/kWh	94,000s
Total equipment costs (approximate)	350,000s
Financing costs[a]	
Windmill generator, wiring, installation,	
254,000s at 10 percent for 15 years	33,000s/year
Batteries, 94,000s at 10 percent for 6 years	22,000s/year
Total cost of financing (approximate)	55,000s/year
Cost of electricity	
Total energy generated	36,000 kWh
Unit cost	1.5s/kWh

Source: N. L. Brown and J. W. Howe, 1978, Solar energy for village development: *Science,* v. 199, p. 655. Copyright © American Association for the Advancement of Science.

[a]Based on amortization of loan in equal yearly installments.

energy per unit cost and is more resistant to damage than more complex types. In the Darrieus wind generator, which resembles an upside-down eggbeater, blades rotate around a vertical rather than a standard horizontal shaft. The Darrieus tends to be more efficient, less susceptible to damage, and less expensive than the standard designs. The confined vortex generator captures and funnels wind to create a tornadolike effect in which strong pressure differences can be exploited to generate electricity.

There are dozens of promising designs that have yet to be thoroughly tested, making cost and efficiency estimates highly speculative. In particular, engineering designs related to metal fatigue in propellers and towers need research under actual operating conditions, as do offshore units. In the Grand Banks area off the New England coast, waves can reach 30 m in height, and practical experience with floatation, mooring, icing, vibrations, and other potential problems is lacking. As with any new technology

trying to expand, large sums of money are needed. When one compares the billions invested in U.S. nuclear and fossil fuel research and development, the $21 million (out of a solar research budget of $290 million) alloted to wind power in 1977 was very skimpy indeed, especially in view of its potential.

The U.S. Department of Energy (DOE) has tended to regard wind power as barely competitive. Based largely on the unreasonably high sums spent on certain government wind projects, which were not well conceived or engineered, DOE has used $10,000 per kW of generating capacity as typical of wind power. Meanwhile, several small independent companies were contracting to build wind generators at $350 to $1800 per kW, and some contend the smaller figure can be cut in half once moderate production is underway (Table 7–4). Congressional appropriations for wind power have exceeded DOE fund requests. In addition, most government grants tend to go to giant corporations whose main interests and expertise lie elsewhere. For example, the only offshore wind energy study contract to date went to Westinghouse whose main energy interests are nuclear.

The environmental effects of wind power are minimal. Windmills need to be located away from the migration paths of birds and where any negative visual impacts are minimized. Offshore sites would cause some disruptions during construction and would need to be avoided by ships. Large windmills can cause television interference within a 30 m to 1.6 km radius, depending on the site. This interference can be cut in half by using durable fiberglass blades (which, once in production, should also cost less than metal). The effects, if any, of large numbers of windmills on weather and climate should be minimal.

HYDROPOWER

Putting running water to work to operate waterwheels goes back many centuries. The first hydroelectric facility began operating in 1882 in Appleton, Wisconsin—it generated 125 kW of electric power. By 1925, hydropower supplied 40 percent of the world's electricity. Today, 15 times more hydropower is produced than in 1925, but it represents only about 23 percent of the world's electricity generation. Some countries (Egypt, Ghana, Paraguay, etc.) obtain almost all their electricity from water. Other countries have enormous potential that is virtually untapped, as in the Himalayan Mountain region.

Approximately 340,000 MW out of a world hydropower potential of 3 million MW has been exploited. Industrialized regions produce 80 percent of all hydropower, but contain only 30 percent of the total potential. About 4 percent of U.S. power is hydroelectric.

By far the most common procedure for exploiting water power from streams is to con-struct a dam at an appropriate location across a valley (see Fig. 14–2). As water from behind the dam rushes through passageways in the dam, it turns turbines which can generate large amounts of pollution-free electricity. One great advantage of hydroelectricity is that, unlike fossil fuel or nuclear power plants, the power can be rapidly turned on and off just by opening or closing control gages. It therefore is valuable in meeting peak periods of demand.

If a nearby upland area is suitable as a reservoir site, electricity can be used during periods of low demand to pump water into the other reservoir; this "pumped storage" can then be released when demand is high, providing additional generation capacity.

Although the generation of hydroelectricity is clean and very efficient, it is not correct to say it is cheap, for the environmental costs of large dams and reservoirs are enormous. Thousands of square kilometers of valuable cropland, forests, wildlife habitat, and scenic treasures have been lost forever beneath reservoirs. In addition (also see Chapter 14), many harmful side effects can result including: disruption of local water tables; severe disturbance of stream regimen, especially below the dam; loss of valuable nutrients to estuaries and seas at the mouth of the dammed river; the triggering of earthquakes; local climate changes; changes in water chemistry harmful to fish and other organisms (e.g., loss of oxygen in reservoir water and increased nitrogen below the dam); and shoreline erosion and stability problems.

Areas highly dependent on hydropower, such as the U.S. Pacific Northwest, are subject to an "energy crisis" whenever a drought reduces streamflow. Accumulation of silt and organic matter limits the useful life span of all reservoirs, and many dams will eventually deteriorate and become potential flood hazards.

In view of such problems, many feel the United States is already badly "overdammed," and even proponents have to admit that choice locations for large dams are scarce. This does not hold for many less-developed nations where hydropower can still be a good choice.

Until recently a very inexpensive, environmentally sound energy source has been largely ignored in the United States. Perhaps because of an overemphasis on "bigness," the contribution of numerous small electricity-generating facilities along streams has been neglected. China is said to have tens of thousands of such generators for local use. A U.S. Army Corps of Engineers study reported 48,000 small dams in the United States are untapped for hydropower. If these small dams were equipped with power-generating capacity, the electricity produced would equal that generated by all nuclear power plants in the United States in 1977. In addition, many large dams in the United States can develop far more electricity than they are currently producing. Because these are already existing dams, the additional costs would be minimal and the environmental impacts of exploiting this energy potential would be small.

OCEAN ENERGY SOURCES

Nearly 71 percent of the earth's surface is covered by oceans, and a large share of the world's population and industry lies within 200 km of an ocean. It is therefore logical that a variety of schemes, both old and new, for exploiting the energy-producing potential of the oceans should be explored.

Wave Power

Offshore wind energy was discussed above. A small fraction of the wind's energy produces some 390 million km² of ocean waves. It is the vertical (up and down) motion of waves that appears most promising for energy development.

A wave power machine was patented by two Frenchmen in 1799. Today, over 100 different mechanical and hydraulic devices have been proposed to extract energy from ocean

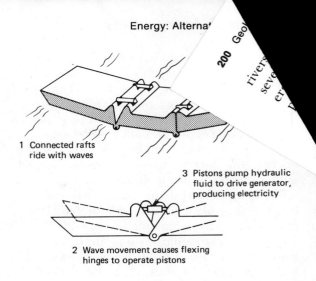

1 Connected rafts ride with waves

3 Pistons pump hydraulic fluid to drive generator, producing electricity

2 Wave movement causes flexing hinges to operate pistons

Energy from the Sea

Figure 7-17 Production of electrical energy using the wave power of the ocean.

waves. Most merely utilize the energy in wave motion to generate an electric current—some by rocking ducklike floats; others by a series of hanging vanes or panels; still others use vertically bobbing rafts (Fig. 7–17). An experimental station in Japan produces 125 kW using large floats that bob up and down in the waves. An attached piston then squeezes air through a turbine.

Japan spent $5 million on wave power research in 1978, and England is investing about $1 million a year in the hope that waves may one day provide a permanent, clean, and safe energy source for the entire United Kingdom. Dr. Steven Salter of Edinburgh University has suggested that 50 MW wave generators (each about the size of a supertanker) placed every 160 km along the Atlantic coastline could provide electricity for all of Europe.

Ocean Currents

The great volume of water in ocean currents possesses enormous kinetic energy. The water moving in the Gulf Stream off Florida exceeds 50 times the freshwater flow for all the world's

. The slow movement of seawater currents rely restricts the efficiency with which energy can be derived from them. Perhaps 4 percent of the energy of the Florida Current segment of the Gulf Stream could be exploited using large, specially designed, slow-speed turbines. This would produce about 2000 MW of energy at a cost that many feel would be presently competitive in the marketplace.

Taking energy from ocean currents on a large scale could produce undesirable climatic effects (e.g., any weakening of the Gulf Stream would produce some cooling in Britain and Europe). Threats to (or from) large ocean life forms (whales, dolphins, etc.) also need to be considered.

Tidal Power

Tides were used in ancient Egypt to turn waterwheels which pumped irrigation water in the Nile delta. In A.D. 1100 the English were using tidal power to grind grain in mills at Bromley-by-Bow, and a similar facility at Woodbridge was operated successfully for 800 years. The world's first major electric generating tidal facility (240,000 kW) was built in the La Rance estuary in France in 1966. Today a number of selected sites are under investigation in several countries. The most promising North American site is in the Bay of Fundy in eastern Canada where tidal ranges often exceed 12 m (Fig. 7–18). Even here, there are serious questions about the economic viability of a large facility.

Good sites are relatively rare, which severely limits the overall contribution tidal power can make to the world's generating capacity. Ideal sites are partly enclosed coastal embayments where a tidal range of at least 5 m is attained. Incoming and outgoing tidal currents can be used directly by paddle wheels and similar devices. More complete exploitation is possible by construction of a damlike barrier across the bay enabling high tide waters to be trapped and then released through turbines during low tides.

Where tidal barriers are constructed, ecol-ogic disturbance of the bay or estuary can be severe. In addition, tidal barriers would block shipping, and periods of high and low tide cannot be controlled to coincide with periods of maximum power demand. Even so, negative impacts may be much less than for most conventional energy sources and the technology is well known, making tidal power an attractive alternative for a small number of appropriate locations.

Ocean Thermal Gradients

In tropical oceans, the sun heats the surface water, creating a layer of water 20 to 25°C warmer than the cold waters below. A heat engine can exploit this temperature gradient by circulating through a closed system a fluid (e.g., propane or ammonia) whose vaporization point lies within the temperature range present. In the warm water, the expanding vapor would drive turbines and then be returned to the area of cold outside water where it would condense back to a liquid and begin the cycle again. The process can be used in cooler climates where warm ocean currents such as the Gulf Stream create strong local contrasts in water temperature. The theoretically harnessable thermal energy potential of the Gulf Stream alone is well over 100 times greater than the total electricity demand of the United States.

Ocean thermal-electric conversion (OTEC) plants would probably be only 3 or 4 percent efficient, and up to one-third of the power produced may be needed to pump water to maintain the cycle. However, since the energy source (solar) is free, these are not critical factors (Fig. 7–19). The French built several OTEC plants for their tropical colonies after World War II but, when their overseas empire collapsed, so did the OTEC facilities. Today many ocean thermal projects are still on the drawing boards. In fiscal 1978, $35 million (20 percent of the solar electric budget) went to various OTEC projects in the United States.

The success of these programs is a matter

"These tides could generate power for 700,000 homes!"

Some see in the tide's ebb and flow a limitless power supply: undeveloped! Others say tides promise too little power, too far from anywhere, too late. Who's right?

Tidal power specialists point to the Bay of Fundy whose tides display majestic power; 100 billion tons of water rising as much as 50 feet, then draining, twice daily—a 200 million horsepower potential each day. They talk of one Fundy site where a tidal generating plant could net 7½ billion KWH. Enough to power 700,000 homes. And Fundy is one of the world's 50 prime sites! True, construction costs are high. But tidal power is pollution-free, environmentally tolerant. Enthusiasts say, "develop it!"

Others doubt tides will ever make a significant contribution. They point to the problems: remoteness of most potential sites, limited application to date. Only two tidal plants exist in the world. None in North America. Tidal power critics call for energy sources promising more power, closer to needs, in less development time.

Where to turn? Petroleum provides 70% of today's energy. Supplies are limited. They'll run low in the forseeable future. We must prepare with an energy policy that encourages development of non-petroleum power sources. Of course, tides should be investigated. Wind and sun too. But let's concentrate on coal for electricity, petroleum substitutes; and on perfecting nuclear power. We urgently need to set energy priorities in terms of a national policy blueprinting responsibilities of consumers, producers, and government.

Figure 7-18 The possibility for energy production by use of ocean tides, such as in the Bay of Fundy. (Courtesy Caterpillar Tractor Co.)

"We must provide energy for 70,000,000 homes."

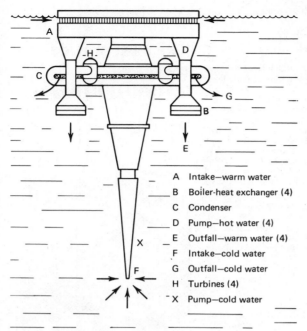

A   Intake—warm water
B   Boiler-heat exchanger (4)
C   Condenser
D   Pump—hot water (4)
E   Outfall—warm water (4)
F   Intake—cold water
G   Outfall—cold water
H   Turbines (4)
X   Pump—cold water

Figure 7-19 Ocean thermal energy conversion (OTEC) utilizes the difference in temperature between warm surface water and colder deep water.

Salinity Gradients

Different concentrations of salt in water produce an osmotic pressure differential that has large energy-producing potential. This pressure difference equals about 24 atmospheres where freshwater enters the oceans. In more practical terms, every cubic meter per second of freshwater entering seawater generates 2 MW of power as the freshwater mingles with the saltwater in an attempt to dilute it. The greater the difference in salinity, the greater the energy involved in dissipation. Where streams enter an extremely saline water body, such as the Great Salt Lake, the power potential is 30 MW or more for each cubic meter of water per second of streamflow. Batteries, semipermeable membranes, vapor pressure chambers, and other devices for harnessing the energy in salinity gradients have not been perfected, but the potential is enormous. Using groundwater brines, or producing brines from abundant salt deposits, can greatly expand the energy potential from this source.

of highly divergent speculation. Cost estimates vary enormously from less than $100 to $4000 per installed kilowatt of capacity. Environmental effects are also disputed. For example, it has been suggested that the mixing of deep, cold waters (about 8°C) with warm surface waters (about 28°C) in tropical seas could: add beneficial nutrients from the cold waters to surface waters, which might then be used to develop fisheries; add significant amounts of carbon dioxide to the atmosphere since the cold waters are rich in carbon; cause either an increase or a decrease in surface water temperature, resulting in climatic changes of unknown extent.

In conclusion, the durability and reliability of OTEC plants is largely unknown, and only firsthand experience will tell whether ocean thermal gradients can provide major or minor amounts of energy in the near future.

ORGANIC ENERGY SOURCES

All fossil fuels were once living organic matter (biomass), and present-day organisms are also storing significant amounts of exploitable energy. Half the trees cut down in the world are burned to heat houses or cook food, and wood-burning stoves are currently enjoying renewed popularity in the United States. Dry cellulose has an average energy content of about 4 kilocalories per gram (60 percent that of bituminous coal).

In addition to wood and dried animal droppings (recall the "buffalo chips" used in the treeless U.S. Great Plains), **peat** has long been a familiar energy source in many parts of the world. The estimated 120 billion tons of peat in the United States contains energy equal to 240

Table 7-5   Major sources of potentially usable biomass residues

Item	Weight (10⁶ dry tons)	Energy[a] Total (Q)	Energy[a] Per acre (million BTU)
Collected			
Urban and municipal solid wastes	160	2.1	
Large poultry and hog operations and cattle feedlots	26	0.3	
Large canneries, mills, slaughter houses, and dairies	23	0.3	
Wood manufacturing	15 to 27	0.4	
Total	~230	~3	
Uncollected			
Cereal straw	161	2.1	22
Cornstalk	142	1.8	28
Logging residues	50 to 75	1.1	130[b]

Source: C. C. Burwell, 1978, Solar biomass energy: an overview of U.S. potential: *Science*, v. 199, p. 1043. Copyright © 1978 American Association for the Advancement of Science.
[a]Residues evaluated at 13 million BTU per dry ton except for wood residues at 17 million BTU/per dry ton.
[b]This assumes that the large branches, stump, and unmerchantable residues are collected and that the total average above-ground material is 9.1 dry ton/ per acre.

billion barrels of oil. Peat deposits up to 10 m thick occupy some 14 percent of Minnesota— enough to supply that state with heat and electricity for 50 years. As with coal, peat and other organic matter can be burned directly or converted into "synthetic fuels." It is estimated that 400 million tons of eastern North Carolina peat could fuel four 400 MG power plants for 40 years or one 80 million ft³/day gasification plant for nearly 50 years. Although it is feasible to utilize only a small proportion of the peat, this still represents a sizable energy source for many regions.

For the entire earth, the biomass produced annually is thought to contain 15 to 20 times the energy we are currently obtaining from all commercial energy sources. The small percentage of this biomass which is exploitable for energy use represents a significant, safe, and relatively clean and inexpensive energy source.

There are two main sources of energy-producing biomass: (1) the organic matter in wastes and (2) crops grown specifically for their energy value. The wastes include animal manure, agricultural residues, logging wastes, spoiled grain, garbage, paper, and sewage (Table 7–5). Plants containing an inherently high-energy content include water hyacinths, seaweeds, algae, sugarcane, sunflowers, and many trees. It is believed that a 1200 km² sea farm growing kelp could yield enough methane (natural gas) to meet current U.S. demands.

Organic matter can be burned directly or converted into other fuels (Fig. 7–20). A wide variety of processes are capable of producing solids (charcoal, wood), liquids (alcohols, oils), gases (methane, hydrogen) and numerous useful by-products (fertilizers, plastics, fibers, detergents, chemicals) from biomass. One-third to two-thirds of the energy in biomass is lost dur-

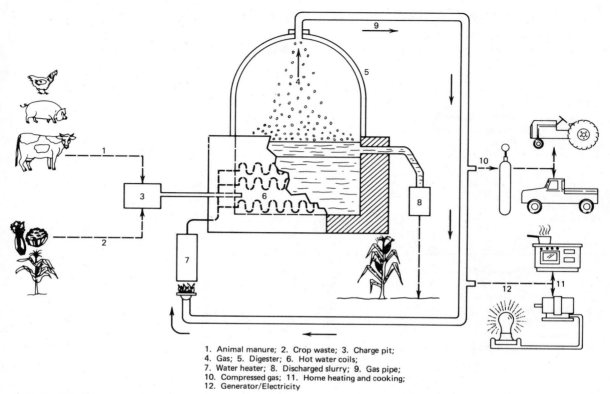

1. Animal manure; 2. Crop waste; 3. Charge pit;
4. Gas; 5. Digester; 6. Hot water coils;
7. Water heater; 8. Discharged slurry; 9. Gas pipe;
10. Compressed gas; 11. Home heating and cooking;
12. Generator/Electricity

Figure 7-20 Electric energy production by conversion and utilization of biomass materials. (After New York State Department of Environmental Conservation.)

ing conversions, but the biofuels resulting are much more efficient and convenient to use, and the by-products may be of even greater value than the fuels.

Anaerobic digestion, a fermentation process carried out by microorganisms in a low-oxygen environment, is the most common and simplest conversion process. Bacteria act on biomass during decomposition, producing "biogas" (mostly methane) and leaving a nutrient-rich residue of great value as fertilizer. Small plants used for biogas, which speed the digestion and collect the methane and fertilizer, are proliferating in poor countries, especially since the 1973 jump in oil prices. China has an estimated 4.3 million small units in operation and India, Korea, Japan, and many other coun-

tries are following suit. India has traditionally burned in open fires most of its annual production of 68 million tons of dry cow dung. This direct burning wastes 90 percent of the heat and all the precious soil nutrients. Anaerobic digesters provide a far superior method of utilizing this material and they can be readily constructed from locally available materials (an old oil drum, a small motor or pump, storage containers, and some tubing).

A wide variety of new integrated or hybrid systems is expanding the potential of organic energy in conjunction with other energy sources. For example, solar energy can provide heat to speed anaerobic fermentation (especially valuable during colder seasons when the process is greatly slowed down) or to yield steam, which can then be used to generate hydrogen gas from the biomass. The residue from a digester can be used to fertilize water in which fish, algae, or other organisms are grown. From the water, a much larger and richer fertilizer, and even food, can then be provided.

It is ironic that much valuable biomass is regarded as waste and often ends up as a serious pollutant and a massive and very costly "disposal" problem. Consider animal wastes, long a major source of water pollution in the United States. Cattle in U.S. feedlots produce some 12 to 14 tons of solid manure each year. The dung of one cow can yield 10 ft³ of biogas per day (approximately equal do the daily cooking needs of an average Indian villager). A small dairy herd could supply all the heating needs for the farm. At least one U.S. farmer heats his house using only the body heat from the cows in his barn. A feedlot of 1000 cattle (some say as few as 100) can profitably be used in a methane generation plant. In May of 1976, an enterprising firm near Oklahoma City with the colorful name Calorific Recovery Anaerobic Process (CRAP), Inc., was authorized by the Federal Power Commission to supply 820 million ft³/yr of methane derived from feedlot wastes to the National Gas Pipeline Company.

Liquid fuels from biomass such as methanol (wood alcohol) and ethanol (grain alcohol) can help replace looming shortages of petroleum-based fuels such as gasoline. The state of Maine is investing in a large-scale program to produce methanol from its vast forests. A ton of wood will yield about 280 gal of methanol for a total yield of 2250 gal/acre from typical Maine forest land. Although methanol and ethanol have less heat value than gasoline, they are also efficient, less polluting, and can be used in standard auto engines that have been adjusted to run on them or in an alcohol-gasoline mixture.

Brazil has embarked on a $500 million program to convert sugarcane and cassava to ethanol. The ethanol will be mixed with gasoline to reduce reliance on imported petroleum. Seven percent of Sweden's total energy budget is supplied by waste from their forest-products industries. And Canadian scientists at the Uni-

versity of Saskatchewan have developed a chemical process for converting biomass into a heavy black oil which they believe can replace "bunker C" heating oil (minus the troublesome sulfur of the regular oil). But despite potential from logging waste and other sources, the United States has tended to downplay biofuels. About 1 billion gal of methanol (and a lesser amount of ethanol) was produced in the United States in 1977 compared to a gasoline consumption of well over 100 billion gal.

Much urban solid waste is combustible; in fact, one-half the waste generated by U.S. cities is paper. Many cities now have plants in which garbage is burned, helping to solve a huge waste disposal problem and generating electricity and heat at the same time. "Garbage power" is in use in, or being planned for dozens of cities, including St. Louis, New York, Chicago, Baltimore, and Milwaukee. In Milwaukee, 30,000 homes receive power from burned garbage, where the incineration operation produces six times as much energy as it uses. A refuse-driven fuel generating plant near Ames, Iowa, generates electricity for 63,000 people of the city, as well as for 12 small communities and the Iowa State University.

Some power plants burn "waste" directly or mixed with coal; others produce biofuels and other by-products. One pyrolysis process, for example, can produce 24,000 ft³ of fuel gas or 1 barrel of oil per ton of refuse. About 7 percent (5 quadrillion BTU) of the U.S. energy budget could be supplied by utilizing urban, agricultural, and forest wastes.

It is possible to reduce the environmental impacts from exploitation of organic energy sources to minimal levels. Biogas often contains such unwanted gases as hydrogen sulfide, carbon dioxide, and water, but these are easily removed by modern techniques. Disease organisms in sewage and garbage can usually be controlled by simple aging. A far more serious threat are toxic chemicals from industrial effluents that are often mixed with other wastes that enter sewage treatment plants. These chemicals can contaminate airborne emissions, pollute soil and water, and be absorbed by living things. They should be treated or disposed of separately, at their source, whenever possible.

However, large-scale use of biomass must be approached with caution. Agricultural residues are often more valuable when left on the fields where they inhibit erosion, enrich the soil and give it tilth, and provide food for animals. Turning large areas of the sea, wetlands, wilderness areas, or other natural ecosystems into monocultural biomass farms could have disastrous environmental effects. Organic energy sources can make a significant contribution to energy supplies, but wise management and foresight are required.

GEOTHERMAL ENERGY

The interior of the earth is hot, probably because of the heat emitted by the decay of radioactive isotopes. For every 40 m of depth, temperature in the solid earth rises an average of 1°C, meaning the boiling point of water would be reached at depths of about 3000 m. Most of the internal heat of the earth is too deep to be exploited for human energy needs. But in many places, as evidenced by volcanic and thermal (hot spring) areas, geologic events and structures have brought hot rock material close to the surface where its heat energy can be tapped and utilized. Young mountain ranges and tectonic or volcanic active areas contain vast geothermal energy potential.

The first use of geothermal heat to generate electricity was in 1904 in an area of natural steam vents at Larderello, Italy. The electricity was used by a chemical plant that recovered boric acid, helium, and other materials from the steam. Commercial electricity production began at Larderello in 1913, and today 400,000 kW of generating capacity has been installed.

The most efficient use of natural hot steam and water (40 to 50 percent more efficient than fuel-heated boiler heat) is in direct heating of

Figure 7-21 Geothermal electric power production at The Geysers, California. These are Units 3 and 4 of the Pacific Gas and Electric Company which went into production in 1967–68. They brought plant capacity to 82,000 kilowatts. In the foreground are steam pipes with expansion loops. The loops allow the pipe to contract when the plant has to be shut down and to expand on startup. The steam condensate rising from the row of five low stacks marks the location of the blowdown valves. When the plant has to be shut down, the steam escapes through these valves. (Courtesy Pacific Gas and Electric Company.)

buildings. Natural geothermal steam or water heats nearly all the buildings in Reykjavik, Iceland, with none of the accompanying air pollution of fossil fuels. Superheated steam also supplies the energy for The Geysers geothermal field in California (Figs. 7–21 and 7–22). Over a hundred wells have been drilled to bring hot steam to the surface where generators convert it to electricity at costs well below those associated with fossil fuel or nuclear power plants. The planned capacity for The Geysers in 1980 is 1180 MW, by far the world's largest geothermal energy producer (Tables 7–6 and 7–7). Today the electrical capacity is about sufficient for a city of 600,000.

In addition to geothermal steam, hot subsurface waters and brines represent a significant energy source (Table 7–8). The geothermal

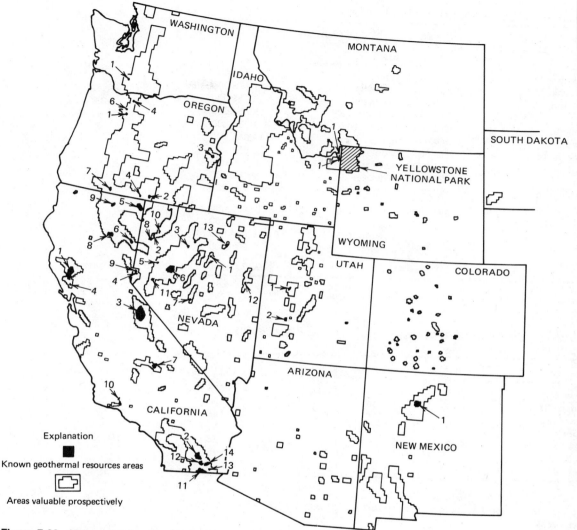

Figure 7-22 Map of the western United States showing lands classified as having potential geothermal resources. (U.S. Geol. Survey Circ. 647, 1971.)

plant at Wairakei in New Zealand runs on hot water, and hot brines (360°C) are abundant in the Salton Sea trough in southern California where considerable geothermal development is planned.

In many places, hot dry rocks occur within drilling range of the surface. The thermal energy released by the cooling of a cubic mile of rock from 350°C to 177°C equals the energy in 300 million barrels of oil! To extract part of this energy, steam could be generated by water injected through holes drilled to the hot rock of a closed water-circulating system (Fig. 7–23). A relatively pollution-free water circulation system has been successfully tested in the Jemez Mountains, New Mexico. It is estimated that

Table 7-6 Geothermal potential in western United States

Known geothermal resources area		Prospective value	
State	Acres	State	Acres
Alaska	88,160	Alaska	11,882,000
California	1,051,533	Arizona	1,473,000
Colorado	20,825	Arkansas	11,000
Idaho	21,844	California	15,754,000
Montana	12,763	Colorado	1,835,000
Nevada	344,027	Idaho	14,845,000
New Mexico	152,863	Montana	3,864,000
Oregon	84,279	Nevada	14,220,000
Utah	37,201	New Mexico	7,482,000
Washington	17,622	Oregon	15,493,000
		South Dakota	436,000
Total	1,831,117	Utah	4,489,000
		Washington	6,082,000
		Wyoming	906,000
		Total	98,772,000

Source: U.S. Geological Survey, news release, October 1, 1974.

Table 7-7 Growth of geothermal generating capacity, by countries, 1900 to 1972

	1960	1965	1970	1972
Italy	300	300	390	395
United States	20	40	195	400
New Zealand	70	175	170	165
Japan	—	20	40	40
Mexico	10	10	12	75
U.S.S.R.	—	10	10	10
Iceland	—	—	9	9

Source: J. P. Muffler, 1973, Geothermal resources: U.S. Geological Survey, Professional Paper 820. In D. A. Brobst and W. P. Pratt, *United States Mineral Resources.*

perfecting hot dry rock systems could supply a nearly unlimited amount of energy for the western United States where several accessible hot rock bubbles (young plutons) have been located (Fig. 7–24). There may also be an extensive body of hot rock not far beneath the Piedmont Province in eastern United States.

Volcanoes themselves are great storehouses of heat. Some have even suggested blasting (perhaps with atomic bombs) holes in volcanoes and injecting water to produce steam as a means for exploiting this heat source!

In areas of heavy, long-lasting sedimentation such as the Gulf Coast, thousands of meters

Table 7-8 Principal utilization (other than for electricity) of geothermal resources

Use	Country	Localities	Quantity
Space heating	Iceland	Reykjavík	1.08×10^{15} cal yr^{-1} in 1969
		Hveragerdi, Selfoss, Saudárkrokur, Olafsfjördur, and Dalvík	8.8×10^{13} cal yr^{-1} in 1969
	Hungary	Various localities	Peak load 3.9×10^{11} cal hr^{-1}; "Optimal useful production capacity" = 1.1×10^{15} cal yr^{-1}
	U.S.S.R.	Caucausus Mountains, Kazakhstan, Kamchatka	Uncertain but large
	New Zealand	Rotorua	~950 wells
	United States	Klamath Falls, Oreg	~3×10^{13} cal yr^{-1}
		Boise, Idaho	Capability of 1–2×10^6 gal yr^{-1} of 75°C water
Air conditioning	New Zealand	Rotorua	5.8×10^8 cal hr^{-1} heat input to LiBr absorption unit
Agricultural heating	Iceland	Hveragerdi	>9.5×10^4 m² of greenhouse in 1960; 5.3×10^{13} cal yr^{-1} in 1969
	U.S.S.R	Various localities	2×10^7 m² of greenhouse in 1969
	Hungary	Various localities	~4×10^5 m² of greenhouse in 1969
	Japan	Various localities	~2×10^4 m² of greenhouse
	Italy	Castelnuovo	3×10^3 m² of greenhouse
	United States	Lakeview, Oregon	2.3×10^3 m² of greenhouse
Product processing			
Paper	New Zealand	Kawerau	~9×10^{14} cal yr^{-1}
Diatomite	Iceland	Námafjall	~2×10^{14} cal yr^{-1}
Salt	Japan	Shikabe, Hokkaido	150 tons yr^{-1} of salt recovered from seawater
By-products			
Dry ice	United States	Imperial Valley, California	>1.84×10^7 m³ of CO$_2$ between 1934 and 1943
Boron	Italy	Larderello	Large production from 1810 to 1966
Calcium chloride	United States	Imperial Valley, California	Uncertain but small

Source: J. P. Muffler, 1973, Geothermal resources: U.S. Geological Survey, Professional Paper 820. In D. A. Brobst and W. P. Pratt, *United States Mineral Resources.*

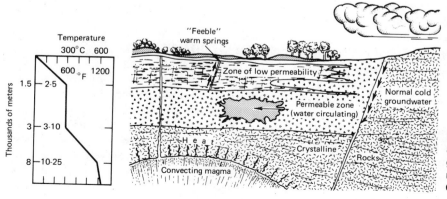

Figure 7-23 Cross section of a natural insulated geothermal reservoir with negligible heat leakage. (U.S. Geol. Survey.)

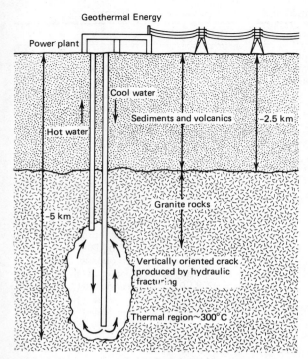

Figure 7-24 Proposed system for extracting energy from a dry geothermal reservoir. (Los Alamos Scientific Laboratory.)

of sedimentary deposits are being compacted and exhibit high temperatures and pressures. Because of the high temperatures, water if forced out of dehydrating clays and into saturated sandy layers can increase the pressure even more. These are called **geopressurized sands.** The top of these geopressurized zones is about 3000 m below the surface in the Gulf area. The temperatures are commonly 150 to 180°C, but may be much higher, and the wellhead pressure is between 4000 and 6000 psi (pounds per square inch). Geopressurized sands contain three potential energy sources: (1) hot water which can be used directly for space heating as is currently being done in Hungary; (2) pressure which can be converted into mechanical energy to operate generators; and (3) natural gas which is frequently dissolved in the hot water while it is under pressure.

Estimates of the energy geothermal sources can realistically supply in the United States by the year 2000 vary from 1 percent to over 20 percent. Over 40 million ha of land in the United States has prospective value for geothermal steam production, and improved technology could eventually raise the potential electric generating capacity from these lands to many tens of thousands of megawatts.

The environmental impacts of geothermal energy vary greatly with the geology of the site and methods employed. Air emissions from geothermal steam often contain sulfur dioxide, hydrogen sulfide, radioisotopes, and other harmful materials that may require special controls. Hot subsurface waters may be extremely saline or contaminated with hazardous chemicals, which, with the thermal pollution, can cause severe surface and groundwater pollution. At many sites, it would be possible to reinject such water back into the ground. Such reinjection can also help prevent subsidence. Geothermal waters often leave behind precipitates that clog pipes and drillholes, requiring new wells to be drilled every 15 to 25 years. However, major geothermal fields should be capable of providing energy for many centuries. There are also the expected visual and drilling operation impacts, but the overall environmental damages should not be severe or uncontrollable.

Other energy potential sources include salt domes (Table 7–9). Such energy could be retrieved because of the osmotic pressure differential.

ENERGY STORAGE

The major difficulty with some of the most promising renewable energy sources, notably direct solar and wind energy, is that they require backup systems for those extended periods when the sun doesn't shine or the wind doesn't blow. This problem can be solved if better ways are found to store energy—a vital field of re-

Table 7-9 Comparison of the energy available from the salt and the oil in selected U.S. salt domes.

Dome	Salt volume (cubic miles)	Oil production (10³ barrels)	Salt energy (MW-years)	Oil energy (MW-years)
High yield				
Thompson (Ft. Bend, Texas)	0.4	259,623	14,000	44,000
Hull (Liberty, Texas)	2.6	156,830	93,000	27,000
Humble (Harris, Texas)	9.8	138,639	350,000	24,000
Medium yield				
Avery Island (Iberia, La.)	4.0	53,054	140,000	9,000
Bayou Blue (Iberville, La.)	4.6	20,806	161,000	3,500
Belle Isle (St. Mary, La.)	1.9	10,316	68,000	1,700
Low yield				
Lake Hermitage (Plaquemines, La.)	0.9	2,475	32,000	420
Bethel (Anderson, Texas)	8.0	1,017	280,000	172
East Tyler (Smith, Texas)	4.3	55	150,000	9

Source: G. L. Wick and J. D. Isaacs, 1978, Salt domes: is there more energy available from their salt than from their oil: *Science,* v. 199, p. 1435. Copyright © 1978 American Association for the Advancement of Science.

search and development generally ignored during, and because of, the era of cheap and abundant fossil fuel.

One of the most promising approaches to the storage problem centers around the production of hydrogen gas. This may be produced by electrolysis from ordinary water or seawater, so the supply is essentially unlimited. When burned, it reacts with oxygen to form water again, avoiding the pollution problems of conventional fuels.

Excess energy from any source such as solar, wind, nuclear, or ocean thermal could be used to produce the hydrogen that can be compressed and stored in compact containers (fuel cells) or in large repositories (including underground) and used as needed. Paul LaCour produced hydrogen using wind power in Askov, Denmark, in the 1890s. The hydrogen was used to light the local high school by directing a hydrogen-oxygen flame at a zirconium element causing it to glow brightly.

Great excitement has been created by prospects of a "hydrogen economy" but, like everything else, it will not be a cure-all. It would require three times the electricity generated in the United States today to produce enough hydrogen to replace the natural gas used in the United States.

Present rechargeable batteries are relatively low-power, short-lived, and expensive, but future improvements may brighten prospects for battery storage.

Common methods of storing heat include hot water tanks or beds of hot rocks. A much more efficient, large-scale variation of this is to inject hot water into underground aquifers, creating an artificial geothermal reservoir. Using waste heat from power plants and industry in this fashion could conserve perhaps 10 percent of the U.S. energy budget. In a related scheme, excess energy, from a solar installation perhaps, could be used to freeze a large block of underground water below a building for air conditioning and other uses.

There are many other possibilities. Energy could be stored as pressurized air, especially in depleted natural gas wells or abandoned mines. Some materials such as eutectic (phase-changing) salts, melted sodium, and various oils have energy-storing potential. A variety of storage facilities using the potential energy in elevated

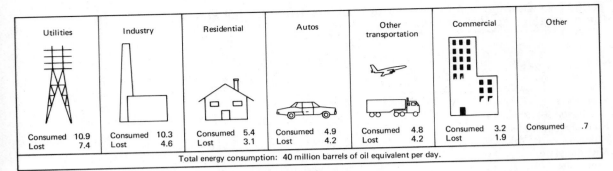

Utilities	Industry	Residential	Autos	Other transportation	Commercial	Other
Consumed 10.9 Lost 7.4	Consumed 10.3 Lost 4.6	Consumed 5.4 Lost 3.1	Consumed 4.9 Lost 4.2	Consumed 4.8 Lost 4.2	Consumed 3.2 Lost 1.9	Consumed .7

Total energy consumption: 40 million barrels of oil equivalent per day.

Figure 7-25 Where waste occurs in the United States, based on the total energy consumption of 40 million barrels of oil equivalent per day. (Brookhaven National Laboratory.)

water are under consideration, including large and small pumped-storage reservoirs that utilize more fully the storage capacity of existing dams. Modern, near-frictionless flywheels can also serve as relatively efficient, compact, and inexpensive storage devices. A flywheel in a wind-powered house could store enough energy for a week of windless days.

ENERGY CONSERVATION

Conservation means wise maximized use. In the case of energy, it means getting the most possible use from whatever energy sources we have. It does not mean living in a cold house or reducing the quality of life; it does mean curtailing energy waste. Conservation is by far the least expensive, least environmentally disruptive, and most rapid method for making useful energy available today. An investment in conservation will yield, on the average, about twice the usable energy compared with an equal investment in new energy-generating facilities.

Some studies have concluded that significant conservation of energy is not feasible without harming the nation's economy and standard of living. This would be true only if all energy producing and using activities were accomplished at maximum efficiency—which is certainly not the case. Energy savings can be realized in virtually all areas of human activity. We will look at a few of these areas.

Increasingly, it is being realized and demonstrated that, in the United States at least, we could cut our consumption of energy in half without significantly changing our life-style (Figs. 7–25 and 7–26). The United States wastes more fuel than the poorest half of humankind uses; yet many predict that we will triple our per capital energy consumption by the year 2000! As with many predictions of increased energy use, this one will (we hope) prove unrealistic. However, future energy-use forecasts have been steadily declining in recent years. For example, a 1972 Federal Power Commission study forecast that the U.S. primary energy demand in A.D. 2000 would be at 160 quads (quadrillion BTU) per year; a 1978 National Academy of Science Committee on Nuclear and Alternative Energy Systems study provided two new forecasts—one projected 96 to 100 quads per year; the other only 67 to 77 quads per year.

If energy growth can be held down in superconsumptive countries such as the United States, several vital overriding questions remain: How will the nations of the world feed and house their people, provide essential services, and power their factories as the production of petroleum and other needed resources on which they have become dependent decreases? And how will these nations, especially the "have-

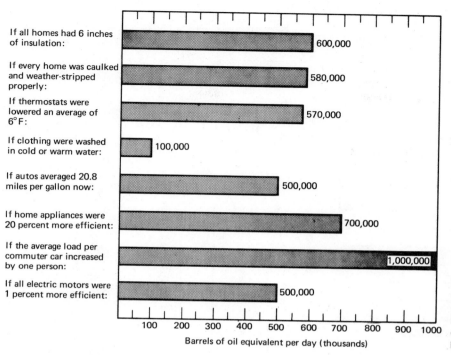

Method	Barrels
If all homes had 6 inches of insulation:	600,000
If every home was caulked and weather-stripped properly:	580,000
If thermostats were lowered an average of 6°F:	570,000
If clothing were washed in cold or warm water:	100,000
If autos averaged 20.8 miles per gallon now:	500,000
If home appliances were 20 percent more efficient:	700,000
If the average load per commuter car increased by one person:	1,000,000
If all electric motors were 1 percent more efficient:	500,000

Barrels of oil equivalent per day (thousands)

Figure 7-26 Methods that could be adopted to conserve energy. Savings shown in barrels of oil equivalent per day. (U.S. Dept. of Energy.)

nots," react to this crisis? Even in the United States, military action against the Arab nations was discussed in secret during the 1973 oil embargo.

Even with full-scale efforts to develop appropriate alternatives, many years of work will be required before these sources can take the place of large amounts of petroleum. The critical factor is time, and every barrel of petroleum conserved buys a little more of the much needed time for a smooth transition to a **post-petroleum era.**

The key to conservation is increased net efficiency in the total energy system. When it takes more energy to find, recover, process, and transport a fuel than can be obtained from it in usable energy, that fuel is essentially depleted and we are only losing ground by such efforts. To this we must also add energy input for building, designing, and maintaining power plants, decommissioning them, transporting the energy, land rehabilitation, and so on.

The United States has rarely considered the entire energy system because such a need was never perceived. The history of the modern energy industry is one of surpluses and how to profitably sell the products in the marketplace, not of shortages and conservation. The notorious U.S. consumer has shown little interest in conservation, as indicated by our continued unparalleled rate of consumption of petroleum products despite the 1973 oil embargo and the "energy crisis" it precipitated. Some cite as justification for large energy consumption its apparent relation to economic indicators. (Fig. 7–27). The abundance of fossil fuels allowed us to do what we wanted with little concern for efficiency. Steam engines and other heat engines (which produce heat, use a small amount of it to do work, and waste the rest) are generally very inefficient, but they remain the major power machines of society. Early steam engines were only 1 percent efficient; modern auto engines are 20 to 25 percent efficient; a well-designed turbine, by comparison, is 85 percent efficient. A 1975 American Physical Society study of U.S.

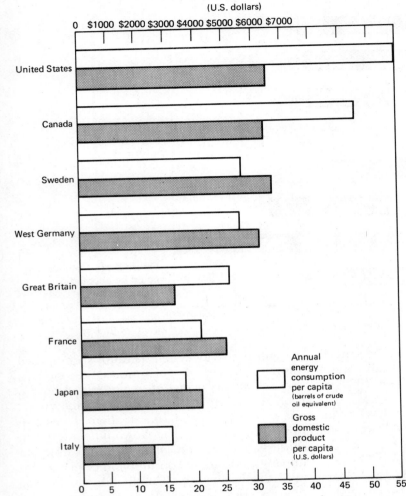

Figure 7-27 Worldwide comparison of energy consumption and gross domestic product. (*United Nations Statistical Yearbook*, 1974.)

energy use in terms of the second law of thermodynamics concluded that our average efficiency is only 10 to 15 percent. Water heating averages only 3 percent efficiency; air conditioning 5 percent; and home heating 6 percent.

Because of the unrealistically low price of fossil fuels in the recent past, many inherently wasteful technologies have emerged and are now so firmly entrenched, with so many powerful interests depending on them for their income, that major changes, however well-conceived, often face strong opposition.

Before concluding this section, we will examine several approaches to conservation and some specific areas in which energy conservation can be achieved.

Agriculture

Modern large-scale agriculture is very efficient because it produces large food yields per person; but this is accomplished only by an enormous expenditure of energy. Traditional Chinese agriculture is some 250 times more energy-efficient (food calorie output compared to total energy input) than U.S. agriculture, which now requires about 9 calories of energy to produce 1 calorie of food.

The electricity use of U.S. farms rose from 15 billion kWh in 1950 to 39 billion kWh in 1975. Irrigation accounts for about 75 percent of the overall energy use of U.S. farms. Some of this water could be supplied by nonconsumptive energy devices such as solar pumps, windmills, small biogas plants, and gravity flow distribution systems.

Less than 20 percent of the energy consumed by the U.S. food system is farm related. The rest is in the processing, distributing, packaging, and preparing of food. The attainable energy savings in nonfarm energy use is enormous: bleached flour, fast foods and convenience foods, overpackaged and overprepared foods, food preparation appliances, even multiple car trips to modern supermarkets (which alone account for 50 percent more energy than does farming!)—all these and more involve large and unnecessary energy waste.

Buildings

The American Institute of Architects estimated that a program to improve efficient energy use in the heating, cooling, and lighting of buildings in the United States could save the equivalent of more than 12.5 million barrels of oil a day by 1990. The program would entail retrofitting existing buildings to reduce energy loss and designing new ones to be 60 percent more efficient than present buildings. Converting existing buildings would be expensive—as much as $1460 billion by 1990; but this investment would save as much as $1900 billion in wasted fuel and unneeded energy-generating facilities.

A Davis, California, building code which requires 6 in. of attic insulation and limits unshaded window area adds only $100 to $400 to the cost of new homes with the result that these homes consume only half the fuel of similar houses without the conservation measures.

Large office buildings are inordinately inefficient. From 1950 to 1970, energy requirements per square foot in new office buildings in New York City more than doubled. Half the space in new office buildings in New York City is unoccupied. One person working late may have to cool or heat several entire floors to regulate temperature in his office. One-half of the air conditioning in an average 10-story Manhattan office building is used to take away heat from overhead lighting; another 35 to 40 percent overcomes heat from overventilation; only 10 to 15 percent fights exterior heat, and much of that comes from neighboring air conditioners!

There are many ways of cutting energy consumption in buildings, for example: develop intelligent budgetary procedures that consider the operating costs over the life time of new buildings; use less glass or use it wisely (20 times more heat is lost through a single pane window as through walls, 10 times more if double-paned); design buildings to maximize passive solar heating (e.g., windows should have overhangs that allow sunlight to enter in winter when the sun is low and shade the interior during hot summers); replace aluminum with steel where practicable (steel production uses only one-fifth the energy that aluminum requires); design efficient ventilation systems and use proper amounts of insulation; whenever possible, develop district heating and total energy systems that recycle waste heat and generate on-site electricity.

Transportation

Transportation, directly and indirectly, accounts for an astonishing 42 percent of all primary energy use in the United States: 25 percent for transportation fuel and 17 percent for road work, vehicle production and repair, and related costs. One of the greatest challenges will be to fit vital transportation needs into a reasonable energy budget. Unfortunately, the overall trends have been toward more wasteful modes of transportation. The two fastest growing types of transport, autos/trucks and airplanes, are also the two least efficient. A railroad can carry a given load 63 miles while consuming a given quantity of fuel; a truck would transport the same load only 11 miles on the same amount of fuel; and an average airplane would transport that load just 1 mile. Pipelines, waterways, and railroads haul over 80 percent of the freight in the United States but when combined consume less fuel than trucks.

World automobile numbers have been rising five times faster than the human population, and the American love affair with the private auto continues unabated despite rising costs. Some 220 million U.S. citizens operate more than 100 million licensed vehicles. The use of mass transit systems in 20 major U.S. cities declined from 24 percent in 1970 to 19 percent in 1976. Over half of all workers drove to work alone in 1976. Over 95 percent of U.S. urban traffic is made up of autos, and autos presently consume 40 percent of our petroleum. If every commuter car carried an average of one more person, 700,000 barrels of oil would be saved each day!

Cars, and especially the proliferating vans, campers, and private trucks, can be made much more efficient. Many of the most promising energy-saving changes involve major redesigns, however, and the giant auto corporations are bound to resist changes that require large-scale revamping of present manufacturing facilities. Economic reality speaks more loudly than efficiency: auto manufacturers are in the business of selling the most profitable vehicles they can, not the most efficient transportation. Major innovations in the auto industry have been minimal in the last 60 years (a 1908 auto race was won by a car averaging 128 miles per hour!). General Motors and Chrysler are presently counting on the diesel engine to keep larger, more profitable cars on the road since it gets up to 35 percent better mileage than standard engines. Unfortunately, the diesel also emits large quantities of hydrocarbons, nitrogen oxides, and 50 to 80 times more particulates than traditional gasoline engines.

Among the auto improvements worth investigating are use of alternate fuels (alcohols, hydrogen, biogas), flywheel propulsion, and improved electric car designs.

Over 50 percent of the land surface in metropolitan areas is used for road vehicles, which are also the major source of urban air pollution. Modern superhighways, shopping centers, and urban sprawl have all continued to worsen the situation and increase dependency on the private auto.

It is obvious that we must redesign our metropolitan areas to minimize the need for autos and make alternate modes of transport (mass transit, bicycles, walking) more pleasant and convenient than they presently are. Some travel can also be reduced at great energy savings by substituting communication via telephone, closed-circuit television, or computer communications for personal contact.

Electricity

The United States and other advanced countries are becoming increasingly dependent on energy delivered as electricity. Electricity use in the United States grew almost sevenfold in the last 25 years. Three-quarters of the coal mined in the United States is used to produce about 45 percent of our electricity. About 30 percent of the energy now consumed in the United States is distributed as electricity, and the U.S. Bureau of Mines estimates that we will use 35 percent of our energy for electric generation by 1985 and more than 45 percent by 2000 (Table 7–10).

Electricity is an important and necessary form of energy in the modern world. It can perform certain tasks more efficiently and better than any other energy form. Unfortunately, much electricity use is highly inefficient and unnecessary. Massive advertising campaigns and wasteful electric devices designed primarily to provide a growing market for electricity were instituted to encourage the continued growth of the electric utility industry. Today, the utility industry is the most capital-intensive of all in-

Table 7-10 Estimates for energy production from different sources by the year 2000

	NSF		NSF/NASA		FEA	
	a	b	a	b	a	b
Solar thermal conversion	1%	—	1%	.43%	1.71%	.73%
Photovoltaic	9.3%	4%	2.97%	1.27%	2.3%	1.1%
Wind energy conversion	25%	6%	1%	.43%	—	—
Solar heating and cooling	12%	—	5.8%	1.69%	—	—
Ocean thermal differences	16.6%	4%	1%	.43%	—	—
Combustion of organic matter	—	—	1%	.43%	—	—
Methane bio-conversion	22%[c]	4%	10%[c]	1.8%	—	—

Source: Energies Magazine, September 1975.
Note: The National Science Foundation (NSF) figures are as cited in March 1975. The NSF/NASA figures reflect that panel's findings in 1972. The FEA figure is based on the "Final Task Force Report on Solar Energy," "Project Independence," November 1974 "(Business As Usual)."

[a]Percentage of estimated total electrical energy.
[b]Percentage of U.S. total electrical energy.
[c]Percentage of estimated total gaseous fuel.

dustries, requiring four times the investment per dollar of revenues as the steel industry. Costs of large new generating facilities have been escalating at a record pace. For example, during the 13 years the Kaiparowits plant in southern Utah was being debated, its proposed size was cut in half but the projected costs rose sevenfold! This was a major factor in the decision not to build it.

About 20 percent of U.S. electricity consumption is for superfluous uses such as commercial overlighting. In addition, electricity is used for many purposes in which other energy forms can be used far more efficiently. Forty percent of U.S. electricity use is for low-temperature heating and cooling. Space heating with a well-designed and adjusted furnace can be 65 percent efficient using oil and 85 percent efficient using natural gas. The maximum efficiency of electricity delivered at a home is about 33 percent. An electric range uses twice the energy of a gas range.

At least 60 percent of the energy in fossil fuel is wasted during the generation of electricity. Part of the energy left must be used to operate the plant (and provide energy for the mining operations if a local coal mine supplies the fuel). Additional energy is lost during transmission of electricity—this loss averages about 10 percent but can be over 50 percent if long distances are involved.

Transmission and distribution represent almost 70 percent of the cost of providing electricity to the average U.S. residence. High-voltage (46 kilovolts or more) transmission lines are proliferating across the American landscape. By 1990, there may be 150,000 km of extra-high-voltage lines (231 to 760 kilovolts) compared to less than 40,000 km in 1970. Higher voltage means greater efficiency in electric transmission. It also means larger, unsightly instrusive towers, stronger electromagnetic fields surrounding the lines, wider right-of-ways, increased "wire noise," television interference, and ozone production. Right-of-way problems and the long-term effects of the electromagnetic field worry many people, farmers in particular, and confrontations between angry rural residents and power line builders are increasing.

Once electricity reaches its destination, its use in the home is responsible for still more waste. As noted above, electric heating and cooling is a very wasteful use of electricity and can be accomplished far more efficiently by other means. Home appliances, on the average, could be made 70 percent more efficient than they presently are, and the ways in which many Americans use them is also extremely wasteful.

About 58 percent of all end-use energy in the United States is needed as heat. Most of this heat requires temperatures less than 100°C; yet it is commonly produced at temperatures of many thousands of degrees in power plants (or millions of degrees in a nuclear reactor), then converted to electricity, distributed, and then turned into low-temperature heat. This represents a very inefficient mismatch between energy source and end use. In the words of Amory Lovins, "we are using premium fuels and electricity for many tasks for which their high energy quality is superfluous, wasteful, and expensive." This is reflected in the fact that electricity supplies only 13 percent of U.S. end-use energy needs yet accounts for 30 percent of U.S. energy consumption.

Improved Technologies and Energy Systems

There is evidence to suggest that the huge (1000 MW or more) power plants being built today are significantly less efficient than smaller plants. Smaller plants tend to be more reliable; the overall energy supply system is less vulnerable to failure and sabotage; the capital costs are less; they can be better suited to local end-use needs; and there is less wasted energy in transmitting electricity to distant markets, since they can be built nearer the market (or a town, industries, and other consumers can locate near the plant).

On the average, over two-thirds of the heat generated at a typical power plant is wasted by transferral to cooling water or air. This waste heat is the cheapest energy source we have. There is considerable potential in such waste heat from both power generation and industrial sources and this energy source can be put to work heating nearby homes; operating solar greenhouses; warming nearby lakes, coastal waters, irrigation water, or even soil in which food can be grown; generating electricity by cogeneration; and in other ways too numerous to list here. Increasingly, ventures of this sort are being attempted and are proving successful.

Studies by the National Coal Policy Project concluded that the electric utility industry could save up to $5 billion per year in capital costs, reduce consumer electric bills by $4 billion per year, and cut oil imports by 1 million barrels a day through cogeneration. Physicist Robert Williams conservatively estimates that cogeneration of processed heat or steam in U.S. factories could produce 208 gigawatts of electricity by the year 2000. The price of cogenerated electricity is roughly equal to 1978 electricity rates, and rising fuel prices will make cogeneration relatively more economical in the future.

Magnetohydrodynamics (MHD) is an electricity-producing design that can increase efficiency from the standard maximum of about 40 percent to about 60 percent. In this process, hot air charged with electrons rushes through a magnetic field, then through a conventional turbine, generating electricity during both steps. The Soviet Union now operates a 250,000 W MHD generator near Moscow.

Pricing

The cost of energy was never cheap despite the low price tag. The costs of preventing or reducing material, structural, and health damages are gradually being internalized and reflected in the consumer price. Even more important in determining costs, however, are various political and socioeconomic factors.

As stated earlier, the big upswing in oil prices resulted from actions taken by OPEC in 1973. However, it is unlikely that either the embargo or the price hikes would have been successful without the cooperation of large multinational oil companies who control most of the world's production, distribution, and marketing facilities. Company profits soared following the embargo and have continued to rise.

Long before the oil embargo, there was sharp criticism about the number of subsidization policies that, directly or indirectly, took public funds and made them available to oil (and other energy) corporations. One example is the tax depletion allowance—a reward to petroleum landholders that is intended to encourage new development of resources to replace those depleted. But studies have shown that, for the annual $1 billion subsidy to oil and gas producers, only $150 million worth of energy reserves were annually added. Other hidden subsidies such as intangible development costs and foreign tax credits have also been strongly criticized as handouts to giant corporations, which only helped keep prices artificially low and encouraged wasteful consumption.

Following the 1973 embargo, rising costs and increased profits also spilled over into other energy-related industries; for example, net profits for coal production was about 20 cents a ton in early 1973; by late 1974, they were $2.80 a ton.

There are many who feel that a handful of giant corporations whose main interest is maximizing profits should not be allowed to dominate such a vital industry. Instead, energy resources should be managed with the long-range needs of people as the foremost considerations. In many countries, the government carefully controls energy and resource activities. The United States is still largely dependent on private corporations for much basic data on resources, making long-range planning and policies difficult.

In spite of a sixfold increase in oil prices in the last decade, fuel costs in the United States are still well below world averages. For example, the average retail price of gasoline in the United States in 1972 was 40 cents a gallon. By 1977, it had risen to 63 cents and to $1.35 in 1980. But the 1977 price in France and Japan was $1.67 a gallon, $2.05 in Italy, and by 1980 had increased almost 80 percent more.

Since the oil embargo, the U.S. government's major method of encouraging conservation has basically been to allow retail petroleum prices to rise, which will hopefully discourage consumption while encouraging energy companies to use the funds for increased exploration and development. Bear in mind that the costs of finding and developing new fuel reserves, especially petroleum, are vastly more expensive now that the readily available U.S. resources are nearly exhausted.

Price rises are warranted, but they tend to place unfair burdens on low- and middle-income people. Many feel a well-organized rationing plan would be a more effective and a fairer means of encouraging conservation.

At long last, efforts are being made to halt reduced per-unit pricing in which the cost of a given amount of electricity, or other commodity, is decreased as more is consumed. In this widely used system, designed solely to encourage consumption, conservation-minded consumers pay more for using less and, in effect, subsidize big users. A better system would be to charge more per unit of energy used above a reasonable base quantity, thus penalizing wasteful consumption.

Miscellaneous Energy-Saving Strategies

Of the multitude of other schemes, methods, and systems that can add to energy savings, only a few additional ones can be noted.

1 Load leveling (spreading out demand) can decrease the need for new power plants, many of which are needed only to meet brief periods of high demand.

2 Some utility companies are granting conservation loans, often interest-free, to individuals for installation of insulation and other energy-saving devices and are finding that it is mutually beneficial.

3 A Ford Foundation Energy Policy Project found that the U.S. paper industry could

cut its fossil fuel consumption 75 percent by using its wood wastes as fuel and adopting presently available energy-efficient technology.

4   A 1 percent increase in the efficiency of electric motors would save a million barrels of oil a day. Also new technologies now available in the steel, paper, and cement industries will use 50 percent less fuel than most present operations.

5   Recycling of metals, paper, bottles, among other materials, at levels now economically practical can save energy equal to more than 3.3 billion gallons of gasoline per year. Recycling steel cans alone would save energy equal to the output of eight 500 MW power plants. If all glass containers were reused six times, the energy saved would equal the output of nine 500 MW plants. This means, in effect, that these nine plants would not need to be built, and the capital costs, mining, reclamation, and pollution problems would be avoided The money saved could go to other uses, such as developing renewable energy sources which, incidentally,

would also create far more jobs than the capital-intensive power plants. Recycling also reduces the extra pollution, waste disposal problems, water needs, and the like, which would be needed to supply new metal, cans, bottles, and so forth.

6   Much could be accomplished by producing well-made, durable goods that are easily repaired and have readily replaceable parts. Planned obsolescence is criminal (inexcusable at least) in a world of dwindling vital resources and staggering imbalances between rich and poor peoples, and rich and poor nations.

Conservation is the most immediate and environmentally sound method of expanding usable energy and prolonging the critical time available to develop safe, dependable alternates to the fossil fuels (Fig. 7–28). Conservation is also the soundest long-term energy investment at the present time. However, for energy conservation to be a large-scale factor in the energy budget, intelligent laws and implementation, as well as public support and cooperation, are needed.

Perspectives

Although most of its manifestations to date have been political, economic, or institutional in origin, the "energy crisis" is real. Most will agree today that the age of profligate use of energy and materials is fast coming to an end. Increasing demands coupled with increasing costs and decreasing availability are causing havoc with the world economy and straining at certain physical and environmental limits.

It is important to recognize that real needs exist that must be met, and this may entail some decline in environmental quality as well as social costs. On the other hand, it is also true that

unrealistic levels of consumption are continuing despite clear warning signs. The complex economy of countries like the United States is tuned to the marketplace syndrome and an expanding pace of consumption. For those who make policy, it can be politically unrewarding to take unpopular stands that encourage people to use less energy and to change their habits and lifestyles. Only when severe peril becomes obvious do most governments spring into action in attempts to alleviate deteriorating energy conditions. In addition, there remain many solidly entrenched and outdated mechanisms—political, economic, and social—that make it very

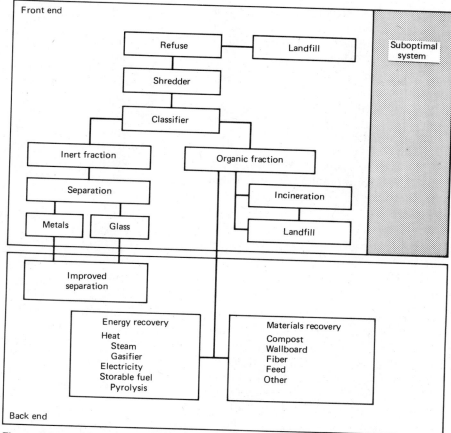

Figure 7-28 A modular approach to energy and materials resource recovery. Front end refers to materials recovery. Back end refers to direct utilization or conversion of the organic part of the waste. (J. G. Albert et al., 1974, The economics of resource recovery from municipal solid waste: *Science*, v. 183, p. 1053. Copyright © 1974 Amer. Assoc. Adv. Sci.)

difficult to adopt different approaches, such as a conservation-oriented, steady-state economy.

Government tax incentives and subsidies strongly favor fossil fuel and nuclear energy development which discourages investment in alternative energy options and makes them appear economically uncompetitive. Politics and economics may not be the topics of this book, but the fact remains that these are the arenas where decisions are made. It is all too often the short-term economic and political considerations that determine resource policy and action,

not intelligent and objective scientific evaluations. But scientists must get into the decision-making matrix.

Each year, about $100 billion of U.S. government subsidies go to support conventional energy systems, that is, those relying on fossil fuels and nuclear fission. This reflects the bias of decision makers in Washington, most of whom are from the traditional, conventional, energy establishment. Certain alternate energy paths, such as those favoring decentralization, are a threat to established interests and their huge investments. Energy paths are to a degree mutually exclusive—only so much money is available and a million dollars spent searching for more petroleum is a million dollars that is not available for solar research or other options.

It should be clear from the foregoing discussion that a wide variety of feasible and often desirable options are available. No single power source or method of production should be relied on to provide the bulk of our energy, especially if it incorporates the use of nonrenewable fuels and entails high environmental risks. There is far greater stability and security in diversification of energy sources.

Large-scale development of all new energy sources will be expensive. But with most, it is worth recalling that, even if initial costs should be higher than for a conventional power facility, these units will not have the high continuing expenses of fuel, waste disposal, pollution controls, mining, and land reclamation.

Conservation has been strongly emphasized in this chapter, but let us remember that conservation should be applied to all resource use. As vital as energy is, such resources as water, productive soils, and a life-sustaining ecosystem are even more important to human welfare.

READINGS

Abert, J. G., Alter, H., and Bernheisel, J. F., 1974, The economics of resource recovery from municipal solid waste: Science, v. 183, p. 1052–1058.

Barnes, J., 1972, Geothermal power: Sc. Amer., v. 226, n. 1, p. 70–77.

Bethe, H. A., 1975, The necessity of fission power: Sc. Amer., v. 234, v. 1, p. 21–31.

Burwell, C. C., 1978, Solar biomass energy: an overview of U. S. potential: Science, v. 199, p. 1041–1048.

Charlier, R. H., 1969, Harnessing the energies of the ocean: Parts I and II, Mar. Tech. Soc. Jour., v. 3, n. 3., p. 13–32; n. 4. p. 59–81.

Calvin, M., 1974, Solar energy by photosynthesis: Science, v. 184, p. 375–381.

Clark, W., 1973, Interest in wind is picking up as fuels dwindle: Smithsonian, v. 4, n. 8, p. 70–78.

Cranston, A., 1974, A bright future for solar energy: Nat. Parks & Conservation Mag., v. 48, n. 10, p. 10–13.

Daniels, F., 1967, Direct use of the sun's energy: Am. Sci., v. 55, n. 1, p. 15–47.

Godwin, L. H., et al., 1971, Classification of public lands valuable for geothermal steam and associated geothermal resources: U.S. Geol. Survey Circ. 647, 18 p.

Gough, W. C., and Eastland, B. J., 1971, The prospects of fusion power: Sc. Amer., v. 224, n. 2, p. 56–64.

Gray, T. J., and O. K. Gashus, eds., 1972, Tidal power: New York, Plenum Press, 630 p.

Grose, L. T., 1972, Geothermal energy: geology, exploration, and developments, Part 2: Mineral Ind. Jour., v. 15, n. 1. p. 1–16.

Hammond, A. L., 1971, Breeder reactors: power for the future: Science, v. 174, p. 807–810.

Hildebrandt, A. F., and Vant-Hull, L. L., 1977, Power with heliostats: Science, v. 197, p. 1139–1146.

Holdren, J. P., 1978, Fusion energy in context: its fitness for the long term: Science, v. 200. p. 168–185.

Hohenemser, C., Kasperson, R., and Kates, R., 1977, The distrust of nuclear power:

Science, v. 196, p. 25–34.

Marvinney, S., 1974, Power from the manure pile, The Conservationist, v. 28, no. 5, p. 7–9.

Maugh, T. H., II, 1972, Hydrogen: synthetic fuel of the future: Science, v. 178, p. 849–852.

Metz, W. D., 1977, Ocean thermal energy: the biggest gamble in solar power: Science, v. 198, p. 178–180.

Rossin, A. D., and Rieck, T. A., 1978, Economics of nuclear power: Science, v. 201, p. 582–589.

Parkins, W. E., 1978, Engineering limitations of fusion power plants: Science, v. 199, p. 1403–1408.

Rex, R. W., 1971, Geothermal energy—the neglected energy option: Atomic Sci. Bull, v. 27, n. 8, p. 52–56.

White, D. F., and William, D. L., 1975, Assessment of geothermal resources of the United States—1975: U.S. Geol. Survey Circ. 726, 155 p.

Wolf, M., 1974, Solar energy utilization by physical methods: Science, v. 184, p. 382–386.

Chapter Eight
Water Resources

Water siphon method of irrigation at the Salt River Project, Arizona.
(Courtesy U.S. Bureau of Reclamation.)

INTRODUCTION

Water is the lifeblood of human existence and at the heart of civilization. It has been instrumental in guiding the destiny of history and determining the socioeconomic behavior of governments. In urban areas the activities and engineering works associated with water delivery and disposal are the most costly items in the budget. Industry and agriculture are overwhelmingly the greatest users of water, and when denied the resource economic and social catastrophe result. All industrial nations use prodigious amounts of water, and in the last 75 years per capita use in the United States has tripled. The worldwide concern has been manifested in numerous meetings among nations; in fact, the 1970s were dubbed the International Hydrologic Decade.

Although the presence of water makes earth a unique planet, water is not equally abundant on the land. Only a little more than half the land can be considered as well watered. Those living in water-deficient areas must depend on imports, either from other areas or by withdrawal from underground sources. The use of water is generally an internal affair of nations and, in spite of its ranking as a basic resource, it is not a commodity that is sold as such on the international level. Instead, the sagacious use of water is manifested in the products derived from its utilization, and such an outfall is reflected by the economic strength of a country.

Water problems can be summed up as threefold: too much, too little, and too polluted. Floods and droughts are much more common than we would like, and contaminated water—whether polluted by humans or from saline sources—is an ever-present danger. Water control and distribution systems modify the water balance and regime of hydrologic systems and may produce unwarranted and undesirable deleterious impacts on the environment. Their management needs the utmost care in the design and implementation of those works that would change the natural hydrology of an area.

Because water is a renewable resource by means of the hydrologic cycle, it is very easy for people, or even nations, to take it for granted. In the United States it is almost akin to a national heritage, so that when a water tap is turned on it is always expected that there will be the flowage of good, clean, clear water in abundance. However, as Lord Byron observed 150 years ago, "Til taught by pain, men know not what good water's worth." It is only when water is not present in its proper amount that true appreciation of this precious commodity is understood. Water is such a vital component in all phases of industrial society that it is being used in ever-increasing amounts. The extravagant consumption of water and the staggering abuses place it in the realm of the proverbial "two-sided coin." On the one side, water is basic to life and civilization, so when used properly it becomes mankind's most needed servant. However, when improperly managed and abused, water can become a despoiler and turns tyrant. It can aid in causing grievous environmental degradation through erosion, land subsidence, loss in soil fertility, siltation, and by damages inflicted during floods and other hazard-producing forces.

WATER PROPERTIES

Not only is water in unique abundance on our planet but its many unusual physical properties allow its use in all fauna and flora and permit a vast number of activities that utilize its attributes. Thus, its properties qualify it as nature's most important resource. The following list suggests some of its many significant qualities:

1   It is the only material that exists in all three physical states at the earth's surface: that is, as liquid, solid, and gas.

2   It is a renewable resource because of the hydrologic cycle.

3   In liquid form pure water is colorless, odorless, and transparent. All these properties are necessary for maximum benefits, such as plant growth. For example, the transparency of water permits light penetration in streams and lakes so vital in sustenance for aqueous animals and plants.

4   Water has very low viscosity. This mobility allows easy transportation, as well as use as an avenue for movement of boats and ships. The buoyancy of the fluid enhances its effectiveness as a medium for transportation.

5 Water is one of the very few compounds that expand when freezing. This fortunate circumstance allows lakes and rivers to freeze at the top down, instead of from the bottom up, as would be the case if frozen water were more dense than the liquid form. Such a condition would mean that in cold climates completely frozen lakes and rivers would have a disastrous effect on fish, wildlife, and flora—the area would become lifeless.

6 Water has the highest heat capacity of all natural substances except ammonia. This ability to absorb a great deal of heat without itself becoming extremely hot enables oceans, and large lakes, to be huge reservoirs of solar warmth. These moderating climate effects are noticeably lacking in deserts where the temperature extreme fluctuations provide a precarious existence for living things.

7 Water has uncommonly high surface tension that not only permits the substance to stick to itself, but also to support objects with higher density and to wet other materials. The combination of tension and adhesion produces a force that operates in capillary fashion in narrow tubes which enhance circulation of water in plants, soils, and bloodstreams of the body.

8 Because water is wet, can adhere to other substances, and has a chemical bonding that is eager to be reactive with materials it contacts, it is considered the universal solvent. Given sufficient time, water can dissolve all substances; the mutual need of the atoms for obtaining electrons and the type of bonding determine the boiling and freezing qualities of water.

Occurrence and Distribution

The world's supply of water moves through the hydrologic cycle (Fig. 4-30) and is present in the atmosphere, in rivers, in lakes and oceans, and underground. Wetland is an environment that contains saturated ground at the land surface. The oceans contain the great bulk of the earth's water, 97 percent (Table 8-1), and since another 2 percent is locked up as ice in glaciers this leaves less than 1 percent as being nominally usable by man without extensive engineering changes. Because groundwaters at depths greater than 1 km are for the most part economically inaccessible or saline, only about 11 percent of water on continents can be considered as available. The yearly renewal rate can be unpredictable from a tenuous atmosphere which only contains water that would be 2.5 cm deep if converted to a liquid.

The 48 conterminous states receive an annual precipitation of about 30 in., or 4400 billion gallons per day (bgd; 1370 mi³; see Figs. 8-1 and 8-2). Of this only 8.5 in., or 1200 bgd, occurs as streamflow (Fig. 8-3), and the remaining 21.5 in., or about 70 percent, is returned to the atmosphere through evaporation and transpiration. This 70 percent includes all water used to grow crops, pastures, forests, and other vegetation (Figs. 8-4 and 8-5). The Great Lakes contain about 22,600 km³ of water, and all freshwater lakes in North America have about 33,100 km³ or nearly 25 percent of all freshwater on the planet.

HISTORICAL BACKGROUND

It could easily be argued that the story of water is the story of mankind. Water is so pervasive throughout history and the developing stages of mankind that an environmental volume could be organized with this as its central theme. Water has been instrumental in decisions of man's settlement patterns. It has guided policies on colonization, and descriptions of water abundance or absence has decided the fate of how terrains would be utilized. The oceans, lakes, and rivers have formed the hydrous highways of trade and commerce. Their existence has

Table 8-1 Distribution of world's estimated water supply

Location	Surface area (square miles)	Water volume (cubic miles)[a]	Percentage of total water
Surface water			
Freshwater lakes	330,000	30,000	0.009
Saline lakes and inland seas	270,000	25,000	0.008
Average in stream channels	—	300	0.0001
Subsurface water			
Vadose water (includes soil moisture)		16,000	0.005
Groundwater within depth of half a mile	50,000,000	1,000,000	0.31
Groundwater—deep lying		1,000,000	0.31
Other water locations			
Icecaps and glaciers	6,900,000	7,000,000	2.15
Atmosphere (at sea level)	197,000,000	3,100	0.001
World ocean	139,500,000	317,000,000	97.2
Totals (rounded)	—	326,000,000	100

[a] One cubic mile of water equals 1.1 trillion gallons

Source: U.S. Geological Survey, news release, August 13, 1972.

also commanded the attention of strategists who have used water as a bulwark against enemy encroachment. The use and abuse of water forms a large chapter in the chronicle of human history, and water management and legislation constitute a major part of environmental rules and laws.

Water was being used for irrigation purposes in Egypt as early as 3400 B.C., and before 3000 B.C. inhabitants of the lower Euphrates River valley near the Persian Gulf were draining and reclaiming wetlands and irrigating crops from canals. By 2000 B.C. Egyptian hydrologists had established gaging stations and were fore-

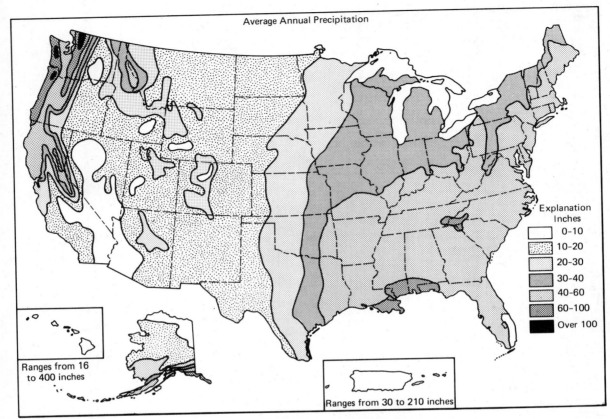

Average Annual Precipitation

Explanation
Inches

	0–10
	10–20
	20–30
	30–40
	40–60
	60–100
	Over 100

Ranges from 16 to 400 inches

Ranges from 30 to 210 inches

Figure 8-1 Average annual precipitation in the United States. (U.S. Geol. Survey.)

casting Nile water levels as aids for water users downstream. Hammurabi's conquest of Mesopotamia in about 1760 B.C. led to a strong centralized control system of water and also established a legal framework for the use of water—some of which was pronounced in the Code of Hammurabi. For example, each person had to keep his part of the ditch system in repair, and failure to do so meant he must compensate those flooded by his inaction. Wittfogel's thesis of the "hydraulic civilizations" is that water was responsible for accelerating the organization of mankind into a structured governmental society. He feels it was no accident that the early great civilizations were centered in semiarid lands where there was a dependence on water, because all elements of society were needed in the proper management and allocation of water resources (see p. 11).

The first dam on record was built by Egyptians about 5000 years ago to store drinking water and for irrigation use. However, the 106 m long dam with a crest height of 12 m was poorly designed and failed soon after construction. It wasn't until 2000 years later that other peoples constructed significant-size dams. Jacob's well was excavated through solid rock 30 m deep more than 3000 years ago and is still in use. About 950 B.C. King Solomon directed construction of large aqueducts in Israel. The Tukiangyien water system, built in China more than 2200 years ago, diverted the Min River whose source was in the high plateau of Tibet. A series of dams and dikes separates the river

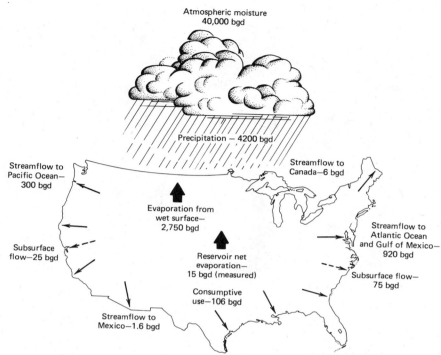

Atmospheric moisture
40,000 bgd

Precipitation — 4200 bgd

Streamflow to
Pacific Ocean—
300 bgd

Streamflow to
Canada—6 bgd

Evaporation from
wet surface—
2,750 bgd

Subsurface
flow—25 bgd

Streamflow to
Atlantic Ocean
and Gulf of Mexico—
920 bgd

Reservoir net
evaporation—
15 bgd (measured)

Subsurface flow—
75 bgd

Consumptive
use—106 bgd

Streamflow to
Mexico—1.6 bgd

Water Budget of the Conterminous United States

Figure 8-2 Water budget of the conterminous United States. (U.S. Geol. Survey.)

flow, and the water was used to irrigate about 400,000 ha.

By the start of the Christian era Rome had the largest water works with eight aqueducts: five were fed by springs and the others diverted streamflow. However, the Romans made no public provision for storage, but the poorest quality waters were used for fountains and irrigation. The first extensive modern long-distance importation of water was to Kalgoorlie, the western Australia mining center. Here the water is derived from a reservoir 1025 km away at the Mundaring dam in the coastal hills. The project was completed in 1890.

New York City first imported water in 1842 when the Croton water system went into operation, ending the use of local wells and ponds for public supply on Manhattan. Construction of the very extensive Catskill water supply system was authorized in 1905 and begun in 1907, and the first water reached New York City in

1915. The Hudson system was completed in 1927 and plans begun on the Delaware system. Construction of dams on the Delaware began in 1937, and the first reservoir and dam were completed in 1951, called the Roundout. At present the New York water supply system depends on eight different watersheds: the Croton, the Bronx River, the Esopus, the Schoharie, the Roundout, the Neversink, and the East and West Branch Delaware rivers (page 656). The drainage area is about 5200 km², half as large as Rhode Island, and water occurs in 28 reservoirs and moves through 640 km of aqueducts and tunnels. To build the system required the obliteration of 20 villages, 60 cemeteries, and removal of 10,000 caskets and 6000 living residents. In order to tap Delaware River water, New York City had to agree to maintain a flow of 1525 cfs (cubic feet per second; 43 m³/sec) at the U.S. Geological Survey's gaging station at Montague, New Jersey. This is the minimum

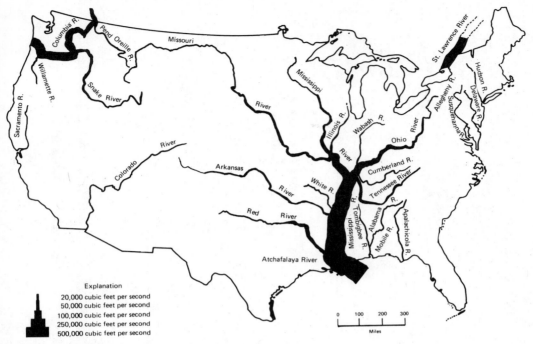

Figure 8-3 Streamflow volume for major rivers in conterminous United States with average flow at the mouth of 17,000 cfs or more. (U.S. Geol. Survey.)

amount of water necessary to provide adequate water for community water supplies, sewage disposal, pollution control, fishing, and maintenance of freshwater to Delaware Bay as a control for salinity and the shellfish industry.

Water quality has also played an important historical role, and as early as 400 B.C. Hippocrates recognized the dangers to health of polluted drinking water. He recommended the filtering and boiling of water for safety. Waterborne epidemics have especially afflicted cities with dysentery, cholera, and typhoid throughout much of history. The easy contamination of surface water sources was one of the prime motivations for drilling wells, which are more independent of pathogenic organisms. Poughkeepsie, New York, constructed the first slowsand filter in 1872, and by the 1880s the drilling of artesian wells at favorable sites was becoming common in the United States. The first deep well in Chicago was drilled in 1864, but by 1890

so many hundreds had been drilled, with withdrawals equaling 10 million gpd (gallons per day), that the artesian head no longer flowed above ground level. By 1915 the water level was 45 m deep, and by 1958 it was 120 m deep with usage of 58 million gpd.

A complete history of water would also need to include such massive structures as the Suez and Panama canals and the reclamation efforts of the Dutch in turning wetlands and coastal waters to productive uses for society.

WATER RESOURCES

It is easy to eulogize water as a resource because it is basic to all living things and the cornerstone of civilization. Because of the hydrologic cycle water is generally considered a renewable resource. This is not true, however, in areas where

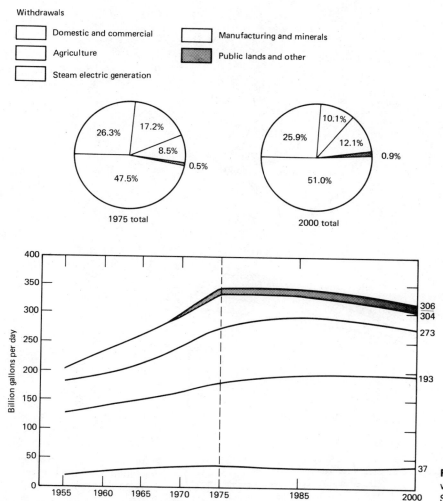

Withdrawals

☐ Domestic and commercial   ☐ Manufacturing and minerals

☐ Agriculture   ▨ Public lands and other

☐ Steam electric generation

1975 total

2000 total

Billion gallons per day

Figure 8-4 Total freshwater withdrawals in the United States. (U.S. Geol. Survey.)

heavy groundwater pumpage depletes the underground storage; hundreds or even thousands of years would be required to restore water levels to their natural state. In such situations water is considered to have been "mined." The use of water as a resource can either be "in-place use" or "displaced use," depending on whether the water is used at the site or is transported to another locale. Water use can also be described as being "returnable" or "consumptive." For example, industrial water can usually be recycled and is reusable water, whereas water used by growing plants is consumptive and irretrievable.

Water use in the United States continues to escalate at alarming proportions (Tables 8-2, 8-3, and 8-4). This rise is reflected by figures for daily per capita use which was 600 gal in 1900, 1300 gal in 1950, and 1900 gal in 1975. The U.S. Geological Survey determined total use in the country to be 435 bgd in 1977 and estimated the use would be 450 bgd by 1980. The four principal off-channel uses are (1) public supply (for domestic, commercial, and industrial uses), (2) rural (domestic and livestock), (3) irrigation, and (4) self-supplied industrial (including thermoelectric power). Withdrawals for these uses in 1975 was 11.5 percent higher than uses in

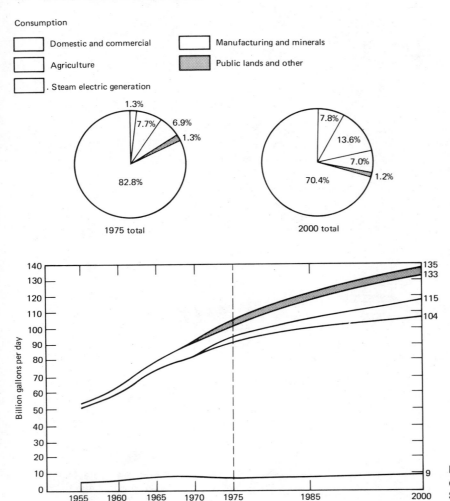

Consumption

☐ Domestic and commercial ☐ Manufacturing and minerals
☐ Agriculture ▨ Public lands and other
☐ . Steam electric generation

1975 total

2000 total

Figure 8-5 Total freshwater consumption in the United States. (U.S. Geol. Survey.)

1970. The fifth principal withdrawal use, hydroelectric power (an in-channel use), showed a 20.7 percent increase from 1970 to 1975. Furthermore, there was a 10 percent increase during the 5-year period in consumption of freshwater. Waterborne intercity commerce in the nation constitutes about 15 percent of all freight transported. Recreation constitutes another in-channel use, and about one-fourth of all outdoor recreation is dependent on water: swimming, boating, fishing, ice skating, and so forth.

Human Use

Water is vital to the life process of living things. The human body contains 70 percent of aqueous solutions. A healthy adult can live for a month or so if deprived of food, but will perish in about a week without water. In temperate zones 5.5 pints of water are needed daily to replace losses from perspiration, exhalation, and excretion; the figure may vary with age, weight, and level of activity. Most life functions depend on water—even the embryo floats in a

Table 8-2 Changes on water withdrawals and water consumed in the United States, in billion gallons per day, 1950–75 (partial figures may not add to totals because of independent rounding)

	1950	1955	1960	1965	1970	1975	Percent increase or decrease 1970–75[a]
Total population (millions)	150.7	164	179.3	193.8	205.9[b]	217.5[c]	5.6
Total withdrawals	200	240	270	310	370	420	11.7
Public supplies	14	17	21	24	27	29	7.9
Rural domestic and livestock	3.6	3.6	3.6	4.0	4.5	4.9	10.3
Irrigation	110[d]	110	110	120	130	140	10.9
Self-supplied thermoelectric power use	40[e]	72	100	130	170	190	18.0
Other self-supplied industrial use	37[e]	39	38	46	47	44	−5.6
Sources from which water was withdrawn							
Fresh groundwater	34	47	50	60	68	82	21.7
Saline groundwater	(f)	0.65	0.38	0.47	1.0	1.0	−6.0
Fresh surface water	160[g]	180	190	210	250	260	5.1
Saline surface water	10[g]	18	31	43	53	66	30.9
Reclaimed sewage	(f)	0.2	0.1	0.7	0.5	0.5	2.2
Water consumed by off-channel uses	(d)	(6)[f]	61	77	87[h]	95[h]	9.9
Water used for hydroelectric power	1100	1500	2000	2300	2800	3300	20.7

Source: U.S. Geological Survey, 1978, The nation's water resources 1975–2000: U.S. Government Printing Office.

[a]Calculated from original unrounded computer printout figures for the two years.
[b]Including Puerto Rico.
[c]Including Puerto Rico and Virgin Islands.
[d]Including an estimated 30 bgd in irrigation conveyance losses.
[e]Estimated distribution of 77 bgd reported by MacKichan (1951).
[f]Data not available.
[g]Distribution of 170 bgd of freshwater and saline water reported by MacKichan (1951).
[h]Freshwater only.

water container. Water is required for breathing, digesting, glandular activities, and for heat dissipation and secretion of waste products. Water acts as a lubricant, helping to protect certain tissues from external injury and by aiding flexibility of muscles, tendons, cartilage, and bones. It is important in metabolic processes, in regulating the body temperature, and in nourishing the tissues.

Although total per capita use of water in

Table 8-3 Total water withdrawals and consumption, by functional use, 1975–2000, million gallons per day

Functional use	Total withdrawals			Total consumption		
	"1975"	1985	2000	"1975"	1985	2000
Freshwater:						
Domestic:						
Central (municipal)	21,164	23,983	27,918	4,976	5,665	6,638
Noncentral (rural)	2,092	2,320	2,400	1,292	1,408	1,436
Commercial	5,530	6,048	6,732	1,109	1,216	1,369
Manufacturing	51,222	23,687	19,669	6,059	8,903	14,699
Agriculture:						
Irrigation	158,743	166,252	153,846	86,391	92,820	92,506
Livestock	1,912	2,233	2,551	1,912	2,233	2,551
Steam electric generation	88,916	94,858	79,492	1,419	4,062	10,541
Minerals industry	7,055	8,832	11,328	2,196	2,777	3,609
Public lands and others[a]	1,866	2,162	2,461	1,236	1,461	1,731
Total freshwater	338,500	330,375	306,397	106,590	120,545	135,080
Saline water,[b] total	59,737	91,236	118,815			
Total withdrawals	398,237	421,611	425,212			

Source: U.S. Geological Survey, 1978, The Nation's water resources 1975–2000: U.S. Government Printing Office.

[a]Includes water for fish hatcheries and miscellaneous uses.
[b]Saline water is used mainly in manufacturing and steam electric generation.

the United States is more than 1900 gpd, the actual domestic house use is about 170 gpd. Typical water home use for different activities include:

Flush a toilet	5 gal
Tub bath	30–40 gal
Shower bath	15–30 gal
Wash dishes	10 gal
Run washing machine	20–30 gal

Industrial Use

In 1950 irrigation constituted the largest single water use, but by 1955 the amount withdrawn by industry equaled irrigation use, and by 1975 industrial use had continued to escalate and was 167 percent more than that for irrigation. Water used by thermoelectric power plants in 1975 was about 200 bgd or about 81 percent of all industrial usage. Of the total water withdrawn

for self-supplied industry, 92.6 percent was used for cooling, and 29.3 percent of the water was saline.

In the manufacturing of products the processing industries use more water than the fabricating industries (Table 8-5), with the largest users being in metal, chemicals, paper, and fossil fuels. Typical amounts of water (in gallons) to produce 1 ton of the product include: steel, 65,000; paper, 120,000; rubber, 85,000; aluminum, 300,000; and nylon, 350,000. Although industry uses prodigious amounts of water, only about 6 percent is consumed, the remainder being returned to disposal sources and rivers for reuse. Of course, some of the water is imperfectly treated (see page 644) and is a major source of pollutants. In 1977 there were only 15 food processers in New York that used wastewater disposal for spray irrigation, and the largest disposal area was 32 ha.

Hydroelectric power production in the

Table 8-4 Groundwater withdrawals and percentage of overdraft, 1975

Water resources region	Total withdrawal (mgd)	Overdraft Total (mgd)	Percent	Subregions Number in region	Number with overdraft	Range in overdraft (percent)
New England	635	0	0	6	0	—
Mid-Atlantic	2,661	32	1.2	6	3	1–9
South Atlantic-Gulf	5,449	339	6.2	9	8	2–13
Great Lakes	1,215	27	2.2	8	1	30
Ohio	1,843	0	0	7	0	—
Tennessee	271	0	0	2	0	—
Upper Mississippi	2,366	0	0	5	0	—
Lower Mississippi	4,838	412	8.5	3	3	7–13
Souris-Red-Rainy	86	0	0	1	0	—
Missouri	10,407	2,557	24.6	11	10	4–36
Arkansas-White-Red	8,846	5,457	61.7	7	7	2–76
Texas-Gulf	7,222	5,578	77.2	5	5	24–95
Rio Grande	2,335	657	28.1	5	4	22–43
Upper Colorado	126	0	0	3	0	—
Lower Colorado	5,008	2,415	48.2	3	3	7–53
Great Basin	1,424	591	41.5	4	4	7–75
Pacific Northwest	7,348	627	8.5	7	6	4–45
California	19,160	2,197	11.5	7	5	7–31
Subtotal	81,240	20,889	25.7	99	59	1–95
Alaska	44	0	0	1	0	—
Hawaii	790	0	0	4	0	—
Caribbean	254	13	5.1	2	1	5
Total	82,328	20,902	25.4	106	60	1–95

Source: U.S. Geological Survey, 1978, The Nation's water resources 1975–2000: U.S. Government Printing Office.

United States increased 22 percent during the 1970–1975 period, but the water required to generate the electricity is almost immediately returned to the hydrologic system so is not included in water use figures. It is interesting to note, however, that the 3300 bgd of water passing through the electric turbines is 275 percent more than the average annual runoff in the conterminous United States.

Municipal Use

The quantity of water withdrawn for public water systems was 29 bgd in 1975, or an average of about 170 gallons per person served. This water includes domestic use and municipal uses such as firefighting, street cleaning, public-building use, park maintenance, and water main leaks (which amount to about 20 percent of the total used). These municipal systems served about 175 million people. Of the nation's 100 largest cities, 66 obtain water from surface sources entirely, 20 from groundwater entirely, and 14 combine surface and groundwater. Another way of separating the 100 cities shows that 37 use reservoirs entirely, 20 use rivers entirely, 10 use the Great Lakes, and 33 use a mixture of sources.

Rural Use

About 42 million people had their own domestic water supply source in 1975, and the quantity

Table 8-5 Quantity of water used by manufacturing industries

Industry group	Number of employees	Annual water intake (billions of gal)	Gallons per employee per day (thousands)	Intake water appearing as wastewater (percent)
Processing Industries: SIC				
20 Food and kindred products	1,589,380	812	1.400	91
24 Lumber and wood products	489,354	161	1.146	82
26 Paper and allied products	583,234	2,078	9.762	94
28 Chemicals and allied products	734,261	3,899	14.584	94
29 Petroleum and coal products	152,470	1,400	25.157	94
30 Rubber and plastic products	406,777	168	1.439	95
32 Stone, clay, and glass products	550,451	264	1.434	88
33 Primary metal industries	1,122,911	4,587	11.196	94
Weighted average	—	—	6.507	—
Fabricating Industries: SIC				
21 Tobacco products	76,989	4	0.168	67
22 Textile mill products	854,543	158	0.644	91
25 Furniture and fixtures	360,882	8	0.079	94
31 Leather and leather products	322,747	20	0.215	94
34 Fabricated metal industries	1,058,954	76	0.249	93
35 Machinery, except electrical	1,424,432	172	0.421	95
36 Electrical machinery	1,502,324	114	0.264	87
37 Transportation equipment	1,593,285	252	0.551	95
38 Instruments and related products	301,650	31	0.363	90
39 Miscellaneous manufacturing	371,858	19	0.175	93
Weighted average	—	—	0.378	—

Source: U.S. Geological Survey Data gathered for manufacturing establishments with six or more employees. Adapted from Reid, 1971, pp. 250–251. From J. H. Feth, 1973, U.S. Geological Survey Circ. 601-I.

used increased 7.4 percent from 1970 to 1975. Human use in 1975 was 2.8 bgd and livestock use was 2.1 bgd. Whereas only 4.7 percent of rural domestic water came from surface sources, 42 percent of water for livestock was surface water.

Irrigation

When the old dictum "the rain follows the plow" was proved false, the early settlers in the American West then realized that to be successful crops would have to be irrigated. Gardens along the Arkansas River in Colorado were irrigated as early as 1832, and by 1848 the Mormons in Utah were employing irrigation methods on a rather large scale. John Wesley Powell was a staunch advocate of reclaiming the West by irrigation from dammed surface waters, and finally in 1902 the Reclamation Act was passed by Congress for the initiation of massive irrigation projects.

Agricultural lands with irrigation farming have several advantages over rainfall farming. Irrigated lands produce higher yields because water intake can be programmed to reach plants at the optimum growth time. This dependability and also the chance to grow more than a single crop a year produce more products per area per year. Of course, irrigation farming is more costly in terms of equipment, water systems, and maintenance and can lead to higher fertility loss through soil salinization. Water is used not only for plant intake, but also as frost protection and for control of high temperatures on specialty crops.

In 1955 there were 34 million acres of irrigated lands in the United States that used an average of 32 inches of water or 91 million acre-feet. Surface water composed 70 percent and groundwater 30 percent of water used. By 1973 the irrigated acreage had expanded to 51.5 million acres with 86 percent, or 45.6 million acres, in the West. Plants need a varying amount of irrigated waters; the following are representative: (in inches) alfalfa, 22–52; corn 19–29; cotton, 24–31; grains, 12–18; oranges, 18–32; pasture, 19–25; rice (in Louisiana and Arkansas), 18–36. Irrigation is largely consumptive, with about 83 percent of water lost in terms of other usage. Irrigation is currently practiced on about 10 percent of American farms. However, according to a General Accounting Office 1976 report inefficient and wasteful uses are responsible for large losses. They cite that 50 percent of the water supplied to farms by the Bureau of Reclamation is wasted, about 11 billion gpd. Such losses occur by absorption from weeds, oversaturation of the land causing drainage problems downstream, degraded water quality due to minerals, fertilizers, sediment, and pesticides. Prior to June 1973 the Bureau of Reclamation had spent $1.3 billion in construction, irrigation projects, and rehabilitation of lands in the West. Table 8-6 shows the water expenditure budget for another government agency for all its investigations—the U.S. Geological Survey.

Methods   In the West about 70 percent of irrigated waters come from surface water impounded in reservoirs, and 30 percent from groundwater. In some states, however, such as Arizona, groundwater use comprises more than 60 percent of total use. The Colorado River is the single largest source with claims of 16.5 million acre-feet of water annually, on the average flow of about 13.5 million acre-feet. Under a compact the upstream states of Colorado, New Mexico, Utah, and Wyoming were allocated 7.5 million acre-feet, and the downstream states of Arizona, California, and Nevada the same total amount, with Mexico receiving 1.5 million acre-feet. The only thing preventing a major controversy over what could be a 2.5 million acre-foot deficit is that all states are not using their full allocated amounts. There are four principal methods in use to distribute the irrigation water on farmland.

1   The oldest method for large-scale irrigation is the diversion of water by canals and ditches into furrows where crops are planted. The American method generally has a large central ditch, preferably lined to prevent infiltration losses, higher than the surrounding land. Water from the ditch is directed into the adjacent fields by siphon hoses (Fig. 8-6). Various devices, weirs, and gates control the water flow, and in some projects it is especially important to have measuring equipment to monitor flow volumes. Of course, other early irrigation was done by human labor, by carrying and distributing water to fields at different levels (Fig. 8-7).

One problem with this irrigation method is the salinization of soil in semiarid environments when the application of water and its disposal is inadequately engineered (see page 462). Each year throughout the world, thousands of hectares are abandoned because salt buildup has destroyed their productivity (Figs. 8-8 and 8-9). To aid in preventing this occurrence, tiles are placed below the ground to prevent waterlogging, and excess water is applied to the fields during the nongrowing season to flush salts from the soil (Fig. 8-10).

2   Irrigation by sprinklers is a more recent method used in more humid or only moderately semiarid climates. Exessively hot periods of low humidity and high winds cause evaporation losses, because a water spray is shot into the air.

Table 8-6 U.S. Geological Survey water appropriation for the fiscal year 1979

	Fiscal Year 1979
Coordination of national water data activities	$ 841,000
Data collection and analysis	15,328,000
National Water Data Exchange	1,196,000
Regional Aquifer System Analyses	11,373,000
Airborne positioning system	1,200,000
Hydrologic research	5,430,000
Hydrologic investigations on public lands	241,000
Improved instrumentation	1,200,000
Training, publications, and other supporting services	3,355,000
Total	$40,164,000

The Critical National Water Problems program of the U.S. Geological Survey (federally funded) will also collect and interpret hydrologic data required for the evaluation and solution of urgent existing and emerging water problems. During FY 79, the USGS will be extensively engaged in obtaining data on water demands for new energy developments. The program components and funding are:

Energy related programs:	
Coal hydrology	$ 8,283,000
Nuclear waste hydrology	5,612,000
Oil shale hydrology	1,163,000
Groundwater recharge	1,278,000
Subsurface waste storage	1,333,000
Flood hazard mapping	396,000
Total	$18,065,000

The appropriation breakdown for the previously mentioned cooperative program:

Collection, analysis, and dissemination of streamflow, water quality, groundwater, and sediment data	$32,027,000
Coal hydrology	2,811,000
Water use studies	2,201,000
Total available for matching state funds:	$37,039,000

Source: U.S. Geological Survey, news release, December 15, 1978.

Figure 8-6 The majority of crops, such as this field of cotton, irrigated by the Salt River Project, Arizona, use the siphon or rill method. (Courtesy U.S. Bureau of Reclamation.)

Originally sprinkler irrigation was accomplished by a central nozzle that rotated in place and shot water spray in an arc around the nozzle (Fig. 8-11). Later an elongated pipe with holes for water dispersion was moved by a tracked vehicle along the parallel rows of crops. The most recent innovation has been the development of the central-pivot irrigation sprinkler system, whereby the elongated pipes move in giant circles around a pivot point (Figs. 8-12 and 8-13). This system has been labeled by some as the most significant mechanization innovation in agriculture since the replacement of draft animals by tractors. This method has many advantages that include: (1) accurate con-

Figure 8-7   Terraced gardens at Hotevilla, Arizona. Similar methods for crop cultivation were used by the ancient Indians throughout this region. (Courtesy Soil Conservation Service.)

Figure 8-9   Abandoned citrus grove, Coachella Valley, California. Thousands of once-fertile hectares in this region had to be abandoned because of salinization of the soil. (Courtesy U.S. Bureau of Reclamation.)

Figure 8-8   Salt damage to the Coachella Valley, California carrot crop. Insufficient rainfall and water use have not allowed through-leaching of the salt, resulting in incrustation and lack of fertility of the soil. Water sources in the All American Canal are from the Colorado River. (U.S. Bureau of Reclamation.)

Figure 8-10   Man-made lakes in the Coachella Valley, California, for the purpose of flushing salts from surface soils by leaching to underground materials. Here the leaching process consists of sinking 1.2 m of irrigation water through the soil to remove accumulation salts into drainage pipes 2.1 m below the surface. (Courtesy U.S. Bureau of Reclamation.)

trol of the application rate and frequency; (2) more uniform distribution of water; (3) can be adapted to rolling terrain and sandy soils; (4) more effective use can be made of herbicides and insecticides; and (5) total costs can be cheaper than other methods. Some new lands, formerly considered nonirrigable, can now yield high-crop production. In other areas the older methods of irrigation are being replaced by sprinkler irrigation; in all, about 250,000 ha of spray irrigation lands are being added each year.

3   Drip irrigation is a relatively new method discovered accidentally in Israel about 40 years ago when it was observed that a large tree near a leaking faucet exhibited more vigorous growth than other trees in the area. Limited use was made of the method in Australia in the 1930s to irrigate peach trees; it was also used for growing tomatoes in England in 1948. The first use in the United States occurred in the early 1960s when the method was used for orchards and some row crops in California.

Drip irrigation is a low-water pressure system for delivering small amounts of water directly to the roots of plants at frequent intervals (Fig. 8-14). The use expanded rapidly when plastic tubes of a small diameter could be used, thus providing greater flexibility to the system.

Figure 8-11 Sprinkler irrigation on the Bear River Project, Utah–Idaho. A pump supplies the pressure for the system at this operation near Garland, Utah. (Courtesy U.S. Bureau of Reclamation.)

Figure 8-13 Central-pivot sprinkler irrigation near Custer, Montana. The system distributes 1000 gal/min (3.8 mpm) and can be set for a 24-hour to a 4-day cycle, giving 1 to 5 cm of moisture. This automatic irrigation method covers a 650 m circle and irrigates 64 ha. (Courtesy U.S. Bureau of Reclamation.)

Figure 8-12 Central-pivot, self-propelled sprinkler system being used to irrigate alfalfa northeast of Dillon, Montana. Source of water is the Clark Canyon Dam on the Beaverhead River, which flows through the East Bench Canal. (Courtesy U.S. Bureau of Reclamation.)

Figure 8-14 Drip or trickle irrigation near Mendotta, California, part of the Central Valley Project. Water is delivered by a buried plastic pipe with risers. The emitter releases 5 liters of water per hour. (Courtesy U.S. Bureau of Reclamation.)

Worldwide 162,000 ha have drip irrigation, with 54,600 ha in the United States and 24,290 ha in California alone. Drip irrigation is especially effective in very arid areas, with high labor costs, and where only poor quality of water is available. It is not feasible for use with closely planted crops.

Drip irrigation has many advantages over other types in the growing of some crops. (1) Ditch irrigation wastes 30 to 60 percent of the water that plants do not receive, and sprinkler irrigation is about 75 percent efficient, whereas drip irrigation water is about 90 percent used. (2) Plants grow without stress because moisture is more constant, and higher yields are pro-

duced. Tomatoes produce 163 percent more with drip than with sprinkler irrigation. (3) Less water is used, labor costs are reduced, and fertilizer effectiveness is enhanced. (4) Inferior terrain and soils can be used. (5) Wind, har-

Figure 8-15 Horizontal irrigation of an almond orchard near Los Banos, California. This Central Valley Project has a variety of different types of irrigation methods that provide water to more than 200,000 ha. (Courtesy U.S. Bureau of Reclamation.)

vesting, and weed control operations do not interfere with the irrigation cycle. (6) An unexpected benefit was the discovery that a poorer quality of water, with relatively high salt amounts, could be used. It is not unusual for drip irrigation crops to effectively use water with a salt content two to three times higher than that used with other methods, because the drip irrigation moisture pushes salt below the root zone.

4 Horizontal irrigation, also called "level basin irrigation," is another new method that is in use under special circumstances. This method is used on level fields, surrounded by dikes, with the application of ponded water to the entire crop area (Fig. 8-15). The method is well adapted to soils with relatively low water-percolation rates and on terrain that is perfectly flat. Once the system is installed less labor is required than in most other systems, because water control can be maintained at a single intake. Soil erosion is minimal, confined to a small area near the inflow, and it can be controlled by an energy-dissipating structure. Uniform application of water is assured, and there is the elimination of a tailwater reuse operation, which must often accompany ditch irrigation.

All nutrients and water are contained onsite, with no degradation of waters to adjacent lands.

WETLANDS

Wetlands have not always been treasured as a resource. Instead, such areas as bogs, swamps, and marshes were always viewed as a place to either drain, dredge out, or fill. The Swamp Land Acts of 1849, 1850, and 1860 gave 64 million acres (26,000 ha) of federal wetlands and overflow flood area to 15 states for their disposal and sale. Of the total original 127 million acres in the 48 states, less than 70 million acres (28 million ha) remain, and these are being consumed at a rate of about 1 percent per year. In the southeastern states about 4 million acres of bottomland forest floodplains have been destroyed since 1950 by federal projects. Wisconsin had lost 54 percent of the wetlands in the southeastern counties by 1958, and an additional 7 percent was lost between 1959 and 1968. Nebraska has lost 15 percent of sandhill wetlands and 65 percent of rainwater wetlands. Similar losses have occurred in many coastal wetlands. Greatest losses were in California and Florida, but in New York, New Jersey, New Hampshire, and Connecticut 15 percent have been lost, and in the 1950–1969 period 650,000 acres were lost.

There are both saltwater and freshwater wetlands, but they are equally important as resources and in the functions they perform in both the physical and biological aspects of the environment. Along the Atlantic and Gulf coasts, two-thirds the cash value of fish and shellfish harvested are dependent on the spawning grounds of wetlands and estuarine environments. The shrimp catch was valued at $226 million in 1975. The 1975 fish harvest was $48.5 million in New England coastal waters, and 98.5 percent of the species are linked to a wetland life cycle. And in Louisiana recreational fishing contributes $160 million to the state each year.

Wetlands perform critical ecological functions in the breeding, feeding, resting, molting, and wintering of all migratory waterfowl. They are also essential to nonavian wildlife from big game such as moose, elk, and bear to smaller fur-bearing animals such as beaver, mink, and otter.

In addition to their importance as part of the food chain in biological systems, wetlands perform a variety of other functions.

1   They retain stormwater and reduce the severity of flooding, hurricanes, and ocean storms. They also serve as a hydrologic buffer to reservoirs, by releasing water during dry periods and storing water during wet periods.

2   They aid to mitigate pollution by trapping, retarding, and transforming silt, pesticides, toxic metals, and organic matter. The microorganisms in wetlands break down air pollutants such as sulfates, and water pollutants such as nitrates thus act as a living filter and at the same time generate significant amounts of oxygen.

3   The storage ability of wetlands acts as a groundwater reservoir to winnow out deleterious materials so that downstream water can be cleaner and clearer, thus enhancing those resources. When all the benefits of wetlands are totaled it would cost us $50,000 per acre ($20,000 per ha) to engineer the same advantages that we reap because of wetlands.

DAMS

Reservoirs impounded behind dams have become a preferred storage method for water resources in most industrialized nations during the twentieth century. Such water storage is a predominant supply for all hydroelectric plants, for much of the world's irrigated lands, for

Figure 8-16   Dworshak Dam, Idaho. (Courtesy Corps of Engineers.)

Figure 8-17   Pristine terrain of the scenic river and valley now inundated by the reservoir from Dworshak Dam. (Courtesy John Conners.)

many municipal water works, and also for industry and water sports activities (Figs. 8-16 and 8-17). Another prime function for building dams is to control downstream flooding. In this chapter we will emphasize the resource and safety aspects of the dam-reservoir-conduit system, and in Chapter 13 we will explore the environmental impacts produced by dams.

There are 49,422 large dams in the United States more than 7.5 m in height, and of these 1400 impound major reservoirs. The first dam for a water-powered grinding mill was built in Milton, Massachusetts, in 1634, and by the nineteenth century the "old millstream" had become part of every New England town. Dams are currently being constructed at a rate of five each day, many of them for private real estate ventures. In addition to the many private, municipal, and state dams, the federal government has four different agencies that contstruct dams: the Corps of Engineers, the Bureau of Reclamation, the SCS, and the TVA (see also pages 603–605). However, the construction of dams is invariably steeped in controversy—their safety, their economic value, and their environmental impacts (Fig. 8-18).

Safety

Throughout history there have been numerous dam failures, so nearly any site selected for construction may become a potential hazard. A recent study of 300 dam failures throughout the world revealed that 35 percent were caused by floods in which flood stage exceeded spillway designs; 25 percent were due to foundation problems such as piping, cracking, settling, pore pressure changes, and faults; and 40 percent were attributed to poor design, construction, operation, maintenance, quality of materials, and other miscellaneous reasons. As early as the period of 1864–1876 there were 100 dam failures by the undermining of water-bearing beds below the foundation. Although the United States has had many notable dam disasters (see page 369), and although the world has also witnessed several recent tragic calamities (at Vaiont, in 1963, when 2200 were killed; at Malpassett Dam, France, in December 1959, when 421 were killed; and the Oros Dam, Brazil, March 25, 1960, when more than 1000 were killed), it took the Buffalo Creek and Rapid City catastrophes in 1972 to sufficiently stir the Congress to action. A National Dam Inspection Act was passed August 8, 1972, authorizing the Corps of Engineers to provide an inventory and inspection assessment of United States dams. Unfortunately, no funds were allocated to the program and it took the Toccoa

disaster in 1977 before renewed action was taken in a presidential directive on December 2, 1977. This had the force of promising funds for the 4-year period it will take to examine the 9000 dams that are listed as "high hazards," and the estimated cost is $70 million.

In the first preliminary report of 230 dams that were studied, the Corps of Engineers classified 10 as being so unsafe that the reservoirs were drained. The corps also evaluated their own 500 dams, and said that 61 of them " . . . may not perform as well during extreme floods as intended." The Bureau of Reclamation found that 17 of its 300 dams " . . . require modification to prevent their failure." For example, the Navajo Dam leaked 1.8 million gpd during late 1977, and $3 million were requested to seal off the leakage.

Economic Value

The economics of dams is a hotly debated issue. I believe there are many cases where dams are necessary, but in many others their importance and need has been greatly overrated, and their economic benefits exaggerated. Figure 8-18 illustrates a typical type of controversy that is enjoined when dams are planned or constructed. Dams have become notorious in terms of "pork-barrel legislation" and questionable benefit-cost calculations associated with them. This was especially visible during the congressional-presidential dam controversies in the United States during 1978.

More than 30 new dams and water projects were scheduled for construction if a water bill passed Congress, but it received only a partial veto by President Carter. The following cases typify projects that were dropped as a result of a compromise solution, allowing construction of more than half the projects. The proposed $48 million construction of a dam on Glover Creek in Oklahoma and the creation of Lukfata Lake was rejected. Purpose of the project was reported as flood control and water supply. However, the 6000 people the reservoir would serve

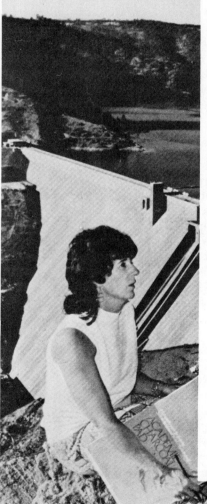

"That dam messed up the valley's wildlife."

Few water projects are built without conflict. Idaho's Dworshak Dam was no exception. One side cited a billion watt electric potential and urgently needed flood protection. Others objected to environmental disruption. Who was right? The decision was difficult.

The dam brought change. Fifty-three miles of river and valley disappeared. It blocked steelhead trout in their upstream run to age-old spawning grounds. It flooded winter grazing lands of one of our few elk herds. Deep reservoir waters brought marinas, campers, tourists. Unaccustomed life style replaced the quiet woods.

But the dam has prevented flood damage to homes and businesses. Power brought new jobs, better schools, roads, recreation, tourist dollars and increased tax revenues. Dworshak's total economic contribution, an estimated $50 million a year.

Dworshak shows we can have both wilderness resources and water power. A $20 million fish hatchery was created to supplant the steelhead run and stock the reservoir with rainbow trout and kokanee. The elk herd lost a grazeland. But, a new preserve was established to replace it. Reservoir waters opened up recreation in formerly inaccessible land.

Without question America should respect its untamed lands. Some wilderness should remain wilderness. But we need power, too. All kinds. Hydroelectric power doesn't pollute air or water, it's reliable, controllable, leaves no waste products. An ideal power source, but there aren't many sites left. We should develop as many of these power resources as prudent, striving always to balance drawbacks with compensating benefits.

Figure 8-18 Descriptive comparison of the benefits and costs of dam construction. (Courtesy Caterpillar Tractor Co.)

"That dam can generate power to light 30,000 homes."

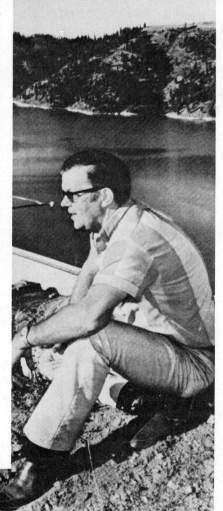

already had sufficient water from another reservoir to supply 90,000 people. The proposed Hillsdale Dam in Kansas, which would cost $55.7 million, was to protect 2800 ha (7000 acres) of farmland from flooding, but the reservoir would flood out 5600 ha (14,000 acres) of already productive farmland above the dam. The Bayou Bodcau Project in Louisiana was designed as flood protection for downstream landowners, but the investment would be $240,000 each. At an interest rate of 6 percent the Fruitland Mesa Irrigation Project showed a benefit-cost ratio of 0.3 (much below the accepted 1.0 figure). It would have benefitted only 69 farmers who would have paid only one-twentieth of the cost for the irrigated waters, and the dam would have cost $127.9 million.

Few dams have been steeped in as much controversy as the Garrison Dam and its accompanying Garrison Diversion Project. Waters from such an enterprise were the dream of North Dakotans even before statehood. As early as 1887 Congress had been requested to construct a canal from the Missouri River in Montana across the length of the North Dakota region. The drought of the 1930s and the Flood Control Act of 1944 converged into a project known as the Pick-Sloan Plan, which would enable the Corps of Engineers to build the dam and the Bureau of Reclamation to be in charge of irrigation and hydroelectric generation. Construction of the Garrison Dam on the Missouri River in western North Dakota was begun in 1947 and finished in 1955 at a cost of $300 million and the loss of 120,000 ha (300,000 acres) of rich bottomlands, hundreds of historic sites, and large areas of valuable wildlife habitat. To compensate for the loss of land, a diversion was authorized in 1965. Its purpose was to irrigate 100,000 ha (250,000 acres) in central and eastern North Dakota. This would necessitate an additional loss of 25,000 ha (63,000 acres) of farmland and construction of 7400 km of canals and laterals with two regulating reservoirs and 141 pumping stations. Costs rose from $250 million in 1968 to $335 million in 1973. The canals, some as deep as 35 m, drained aquifers, lowered the water table, and caused wells to fail. By 1976 the project was 19 percent complete, and the estimated costs for total completion since its origin were $1 billion. To justify the diversion project the Bureau of Reclamation had claimed a benefit-cost ratio of 2.51 to 1, which was based on interest rates of 3.25 percent, the inclusion of many indirect benefits, and an addition of wetland acres. Instead, however, there would be 7057 ha (17,425 acres) of wetlands lost, another 7800 ha (19,000 acres) would be drained, and 2000 ha (5,000 acres) lost from subsurface drainage. If a more realistic figure of 10 percent interest is used, as recommended by the Office of Management and Budget, and indirect benefits excluded from consideration, the benefit-cost ratio is 0.42 to 1.0. Because of the mounting furor, an out-of-court settlement was reached whereby the 118 km long McClusky canal could be completed (it was 90 percent completed in 1977), but further work was halted pending the filing of a comprehensive Environmental Impact Statement. (The deleterious impacts of dams are discussed in more detail in Chapter 14, and further information on environmental management involving dams is contained in Chapter 18.) One trick in the use of water as a resource is the impoundment of water, and this can take the form of either surface or subsurface storage. The following account briefly presents a case history that is a marvel in water planning and engineering.

The Snowy Mountains Scheme

Near the border of New South Wales and Victoria, Australia, the Snowy Mountains rise to heights of 2100 m (Figs. 8-19 and 8-20). They are snow-covered 5 to 6 months a year and are the source of three main rivers: the south-flowing Snowy River and the west-flowing Murray and Murrumbidgee Rivers. Although proposals to supplement flow of the inland rivers date back to 1880, the first large-scale dual-purpose plan for both irrigation and hydroelectric development was not made until 1944. In 1949

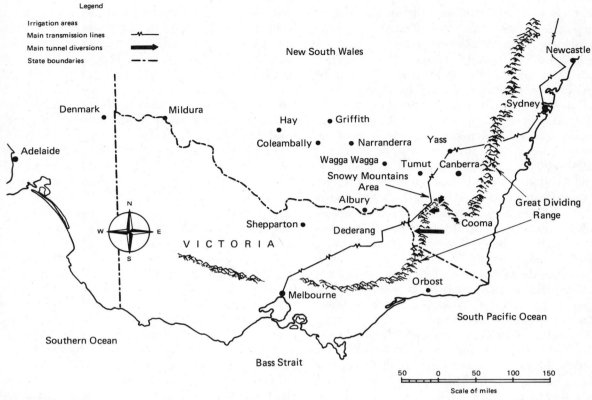

Figure 8-19 Map of southern Australia showing Snowy Mountains. (Courtesy Snowy Mountains Hydro-electric Authority.)

Figure 8-20 Snowy Mountains. (Courtesy Snowy Mountains Hydro-electric Authority.)

the Snowy Mountain Hydro-electric Power Act was passed by the Commonwealth government which established the Snowy Mountain Hydro-electric Authority as the body responsible for the detailed investigation, design, and construction of the Snowy Mountains Scheme.

By 1974 in an area of over 5200 km² there had been built 80 km of aqueducts, 144 km of tunnels, 16 large dams, a pumping station, and 7 power stations (Figs. 8-21, 8-22, and 8-23). Hundreds of transmission lines interconnected the power stations and transmitted the electricity to New South Wales and Victoria.

The Scheme diverted waters of the Snowy River and its tributary, the Eucumbene River, through two transmountain tunnel systems driven westward under the drainage divide. The waters feed the two inland rivers with increased supplies for irrigation expansion in their valleys, and its path moves through shafts (Fig. 8-24), tunnels, and power stations. The total generat-

Figure 8-21 Tutmut Pond Dam. This is one of the 16 major dams in the Snowy Mountains Scheme. It is 85 m high and has a reservoir capacity of 42,800 acre-feet. (Courtesy Snowy Mountains Hydro-electric Authority.)

Figure 8-22 Tutmut Three power station and dam. This station has a capacity of 1.5 million kilowatts. (Courtesy Snowy Mountains Hydro-electric Authority.)

Figure 8-23 Eucumbene Dam. This dam is 115 m high and has the largest reservoir capacity of all dams in the Snowy Mountains Scheme— 3.89 million acre-feet. (Courtesy Snowy Mountains Hydro-electric Authority.)

ing capacity is 3.7 MkW, which produces 5000 million kWh annually. The waters will provide the Murrumbidgee with 1.1 million acre-feet per year and 800,000 to the Murray region. Cost of the scheme has been $800 million.

This project was constructed in a remote and seldom developed area, but numerous steps were taken to maintain it and prevent erosion and degradation. More than 500,000 trees, hundreds of thousands of pegs and cuttings,

Figure 8-24 Tutmut Three power station and pipeline, Snowy Mountains Scheme. This is a closeup view of the station shown in Figure 8-22. (Courtesy Snowy Mountains Hydro-electric Authority.)

places as the Prairie Provinces in Canada, the Great Lakes Basin, and the High Plains region in the United States. The centerpiece would be a 500 mi long reservoir in the Rocky Mountain Trench. The artificial lake would be 3000 ft above sea level, and from it water would flow south through a series of gigantic aqueducts. Costs at that time were estimated at $100 billion, and construction would take 25 years. Similar continental-scope projects have been proposed in the Soviet Union. One project hopes to divert the north-flowing Ob-Yenisei-Irtysh river systems southward to the Caspian Desert area. Such highly controversial and costly projects at this time are still in the realm of dreams, but certainly by the twenty-first century we will have to embark on major water projects that very easily might change and derange the present orientation and configuration of our hydrologic resources.

WATER PROBLEMS

The dual problem of water abundance and water scarcity has afflicted mankind throughout history. Flooding of low-lying areas is an ever-present danger to inhabitants and their property (see Chapter 12). However, periods of water deficiency—droughts—can also produce great economic loss and create personal tragedy and loss of life (see also page 543).

Droughts

Our losses because of drought are steadily increasing throughout the world. The reason is that humans are using and needing more water than ever before. With the increased population and loss of new frontiers, we have less mobility in the twentieth century, and we may be a contributor to the increasing length and severity of drought years by our own alteration of weather conditions (see Chapter 17).

The four-year drought that affected central

and 20,000 shrubs were grown in various Authority nurseries and planted in the field. During early construction years the reclamation work was seriously handicapped by the presence of stock, but animals were prohibited in 1958 and since that time the 200 tons of grass and clover and 1500 tons of fertilizers have aided in the establishment of slopes that are more erosion resistant. However, in recent years large irrigated areas that use the water have begun to experience salinization problems.

Even more grandiose schemes have been proposed to divert vast amounts of water to distant areas, but these have not gotten off the drawing board. The North American Water and Power Alliance (NAWAPA) was proposed in an April 1964 publication of the Ralph M. Parsons Company. The purpose of this project was to take water that was flowing unused into the Arctic Ocean and reverse the flow and redistribute it throughout the continent to such

United States in the 1930s and created the Dust Bowl was the worst natural disaster in the country's history in terms of property loss and human suffering. More than one-half million people emigrated from the area, and millions of acres of rich farmland were eroded beyond a rehabilitation threshold. Again in 1944 a severe drought produced great damage to much of Latin America. Drought also brought great losses to Australia and Europe in 1944–45. During the 1950–54 period the southern and southwestern states in America were hard-hit by drought, and in 1955 the East had a great water shortage. Brazil was particularly devastated by droughts in 1954 and also in 1958 when more than one million refugees fled from the northeast part of the country to such cities as Rio de Janeiro and Sao Paulo. Famine is invariably a result of droughts in third-world nations, and in the Ganga plain (Ganga is the new term for Ganges) of India 25 million died during the 1957 drought.

Water shortages caused by the 1961–66 drought in the eastern United States produced more losses there than any other disaster. More than 500,000 km² in 14 states were subjected to precipitation deficits that averaged about 100 cm for the six-year period. Numerous wells failed and New York City, Philadelphia, and the northern New Jersey metropolitan areas had serious water problems. Streamflow was abnormally low, and pollution levels increased to hazardous proportions. Hay and other crops suffered; the dairy industry was hit with serious losses; and many farmers quit because of irrecoverable losses. Industrial output had to be curtailed and unemployment rose. Camping and hunting activities had to be restricted because of the fire hazard, and swimming and boating were reduced due to low water levels and pollution. Forest fires in Maine and New Jersey approached record levels, reaching a fire intensity of the 200-year forest fire event. More recently the western drought of 1977 vividly showed the fickleness of weather patterns and how dependent the population is on water resources. Although 1976 was also a dry year for the West, by 1977 the situation had become greatly aggravated. Losses caused by water shortages amounted to more than $1 billion in California alone. The U.S. Geological Survey termed the California drought "the worst dry period in (its) history." The drought not only affected crop production, but because of the need to increase groundwater withdrawals the heavy pumping brought an accelerated movement of saline waters into coastal area wells. For example, at Klamath the town well rose from 1700 ppm (parts per million) to 2800 ppm in chloride concentration. The dissolved oxygen content in streams was greatly reduced, causing losses to aquatic life, and pollution and sewage discharge also produced environmental dam-

Water Quality

The maintenance of a sufficiently pure water supply is becoming an increasing problem throughout the world (Table 8-7). Contamination of water ranges from physical incorporation of sediment, to the biochemical pollutants of organic waste and mineral decomposition products. Eighty percent of all world disease is caused by water pollution, and 1.2 billion people do not have safe drinking water. Another 200 million have no water with which to maintain a level of sanitation.

Water quality is affected by geologic, hydrologic, and biological factors. Whereas the preponderance of water impurity is from geologic sources, man-produced pollution has been accelerating at phenomenal rates. This pollution consists of waste discharges (including heat) from domestic, industrial, and municipal sources, as well as sediment and other wastes from agricultural lands, mining, logging, and construction activities.

The pollution in rivers is so severe that the National Stream Quality Accounting Network (NASQAN) was established by the United States, with primary monitoring responsibility assigned to the U.S. Geological Survey. The 345

Table 8-7 Comparison of chemical constituents in the drinking water standards of the World Health Organization and the U.S. Public Health Service

| Chemical constituent | Concentrations in milligrams per liter | | | |
| | WHO International (1971) | | U.S.P.H.S. (1962) | |
	Highest desirable level	Maximum permissible level	Recommended limit	Maximum allowable
Anionic detergents	0.2	1.0	0.5	—
Arsenic	—	0.05	0.01	0.05
Barium	—	—	—	1.0
Cadmium	—	0.01	—	0.01
Calcium	75	200	—	—
Carbon chloroform extract	—	—	0.2	—
Chloride	200	600	250	—
Chromium (hexavalent)	—	—	—	0.05
Copper	0.05	1.5	1.0	—
Cyanide	—	0.05	0.01	0.2
Fluoride	Same as U.S.P.H.S.	—	0.8–1.7[a]	1.6–3.4[a]
Iron	0.1	1.0	0.3	—
Lead	—	0.1	—	0.05
Magnesium	150[b]	150	—	—
Manganese	0.05	0.5	0.05	—
Nitrate (as NO_3)	—	45	45	—
Phenolic compounds (as phenols)	0.001	0.002	0.001	—
Selenium	—	0.01	—	0.01
Silver	—	—	—	0.05
Sulfate	200	400	250	—
Total solids	500	1500	500	—
Zinc	5.0	15	5.0	—

Source: J. H. Feth, 1973, U.S. Geological Survey, Circ. 601-I.

[a]Recommended limits and maximum allowable concentrations vary inversely with mean annual temperature.
[b]If there are 250 mg/l of sulfate present, magnesium should not exceed 30 mg/l.

stations installed by 1975 had increased to 525 in 1976. These monitoring sites measure 46 physical, chemical, and biological water quality characteristics, including temperature, specific conductance, dissolved minerals, trace elements, nutrients, and a variety of organic and bacterial constants.

Salt is a particularly vexing problem in irrigated areas and in coastal wells. A combination of salt encrustation in soils and water logging helped cause the demise of the Hohokam Indian civilization in central Arizona during the A.D. 1100–1300 period. Today there are

more than 26,000 km of tile in the San Joaquin Valley to aid in draining away deleterious salt buildup during irrigation of the land. Initially tiles were placed 60 to 100 m apart, but in recent years with increased salt content tiling is commonly placed 15 to 30 m apart. Excess amounts of water are constantly needed to wash the salt from the soil (see Fig. 8-10). On the 500,000 acres (200,000 ha) of irrigated lands, 3 million acre-feet of water are being applied per year—enough water to cover the land about 1 m deep every 6 months. In the Phoenix region the excess salt buildup in the soil was 140,000 tons in 1944–45 and had increased to 350,000 tons by the 1950–51 period. The decline of the water table averaged 45 m throughout the area.

WATER SALVAGE AND CONSERVATION

The entrapment of water and its containment in storage for later use have become major goals in water resource management. For example, in rivers that receive sewage effluent it is important even during low flow conditions that the pollutants be properly diluted. This may necessitate some form of "low flow augmentation" whereby standby water can be released into the main channel to prevent dangerous levels of toxicity. There are a number of strategies that have been used (and proposed) to enhance water availability or to increase our technical efficiency. The following six points illustrate the variety of techniques being put to use.

1 Recharge of used water into wells to prevent saline intrusion in coastal areas.

2 Scarification of the land surface and excavation to produce surface basins so that surface waters can infiltrate and recharge groundwaters. Such water harvesting can take place in a number of different environments using a range of techniques. On Long Island, New York, more than 2500 recharge pits have been constructed for the purpose of impounding storm sewer water to allow its penetration into the substrate. In California water spreading for subsurface infiltration has been designed for such rivers as the Santa Clara, San Gabriel, San Antonio, Cucamonga, Santa Ana, and Whitewater. Here ditches, furrows, and pits are constructed to divert the surface water into the ground. In western Australia road catchments provide a means for water storage. Catchments are areas of compacted parallel ridges of earth that have cambered surfaces with impervious materials. The drains near the roads are given grades that permit water to reach the storage area with minimum erosion. Such systems are especially suited for undulating terrain with clay soils near the surface. Farming with the use of runoff, as practiced in ancient Negev (page 35), is still another application of this method whereby water from the hills is directed to collection basins and the surface smoothed from rocks to accelerate sheet runoff.

3 The floor of basins is crucial in deciding the purpose of the excavation. When infiltration is desired the floor should be kept pervious and open to penetration; however, when the need is for the surface water, or when its retention for other processes is required, it is then important to assure an impervious floor. This can be accomplished by use of various types of sealants—natural materials such as clay, or manufactured coatings and water repellents including bitumens, oil, silicone, butyl rubber, plastic film, and flexible linings (Fig. 8-25).

4 In arid and semiarid regions, water storage is a problem because of the high rate of evaporation—as much as 3 m/yr in such reservoirs as Lake Mead. Different types of surface coatings and covers have been tried including paraffin wax, polystyrene, foam, and a variety of other artificial films. The technology is still in the initial stages, and such products have been used only on small and restricted scales, with minimal success.

5 Cloud seeding to increase precipitation is another technique for the possible augmentation of water resources.

6 The use of icebergs as a freshwater resource was done on a small scale between 1890 and 1900 in South America. Small icebergs were towed by ship from Laguna San Rafael, Chile, to Valparaiso, and as far north as Callao, Peru, 3850 km away. With the high cost of salinization plants for coastal cities, rising costs of water, and new countries that can afford experimentation the 1970s witnessed renewed interest in water from icebergs. In 1973 a U.S. Geological Survey study estimated that towing an iceberg from Antarctica to such a locale as

Accordion folded Watersaver membrane lining is spread to full panel width for final positioning. Linings fabricated by WATERSAVER COMPANY, INC. P.O. Box 16465 Denver, CO 80216 303-623-4111

(a)

Watersaver membrane lining receiving protective earthen cover. Equipment works from bottom up on the slopes on minimum 12" of the cover. Linings fabricated by WATERSAVER COMPANY, INC. P.O. Box Denver, CO 80216 303-623-4111

(b)

Figure 8-25 Installation of a flexible membrane lining for a water-holding facility in Colorado. The PVC (polyvinyl chloride) cover consists of 193 cm widths that are bonded to form an impervious floor for water and liquid waste storage basins. (a) The accordion-folded membrane liner is being emplaced from the carton on a lowboy trailer and spread to full panel width for final positioning. (b) The liner is being covered by a protective layer of sediment. (Courtesy Watersaver Company, Inc.)

Australia or the Atacama Desert in South America would cost $0.007 per thousand gallons, or one-hundredth the cost of desalinization of seawater. Another study by the Rand Corporation concluded water could be produced from icebergs if towed to California at a cost of $30 per acre-foot, as compared with the $65 cost for aqueduct water and $100 for desalinized seawater. A conference to study the feasibility of iceberg use in 1977 determined the optimum iceberg would be 1 mi long, 1000 ft wide, 900 ft deep and would contain 20 billion gallons of water. It would cost $30 million to make a plastic cover to shield the ice surface and an additional $100 million of research would be necessary to work out all the engineering and design details. The favorable aspect of iceberg use is that such waters are in a sense already wasted, and the environmental impact would be less serious than many other alternate schemes for additional water manufacture.

Perspectives

Investment in water resources in industrialized nations always constitutes a big part of the budget and, as in the United States, all activities associated with water management constitute major costs to cities. However, there are many ways the average person could aid in water conservation, but too often this precious commodity is taken for granted and unnecessarily wasted. For example, leaking faucets in the United States account for hundreds of millions of gallons lost daily.

The use of water is a vital necessity to human life and to civilization. Even as early as 1790, 95 percent of the 4 million Americans lived along the Atlantic seaboard, and now 12 of the 13 major cities are adjacent to major water sources and rivers. Water management involves (1) water supply, (2) flood control, (3) water quality improvement, (4) watershed care, (5) beach, shore, and riverbank protection, (6) land drainage, and (7) conservation. It must always be remembered by the policymakers that no water plan is good unless it is economical, socially valuable, and politically feasible. Thus, evaluation of water resources and accompanying problems cuts across many disciplines and involves people from different backgrounds with a variety of specialties.

Water production, storage, delivery systems, use, and disposal constitute the realm of water resources. Each of these can involve abuse of the natural environment, so the determination of priorities, and which techniques to incorporate, provide the policymakers with problems that need careful consideration. The inescapable prognosis is that water will become an ever-growing problem throughout the world (Fig. 8-26).

READINGS

Feth, J. H., 1973, Water facts and figures for planners and managers: U. S. Geol. Survey. Circ. 601-I, 30 p.

Haveman, R. H., 1965, Water resource investment and the public interest: Nashville, Vanderbilt Univ. Press, 199 p.

Hunt, C. A., and Garrels, R. M., 1972, Water, the web of life: New York, W. W. Norton & Co., 308 p.

James, G. W., 1917, Reclaiming the arid west:

More people crowd our world each year . . .

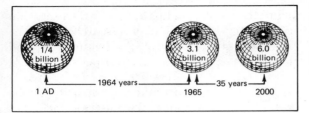

. . . and energy . . .

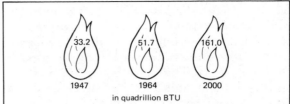

in quadrillion BTU

. . . and our nation too . . .

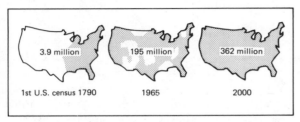

. . . our nation's water will be used at a steadily increasing rate . . .

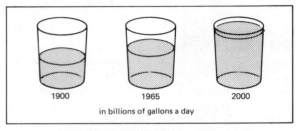

in billions of gallons a day

. . . in the United States this expanding population will . . . earn and spend more money . . .

U.S. Gross National Product—in billions of 1954 dollars

. . . this same population will place greater demands on our National Parks and fish and wildlife resources.

millions of visits per year

. . . on products using minerals . . .

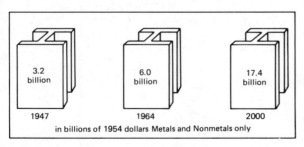

in billions of 1954 dollars Metals and Nonmetals only

| | | Hunting: | 12.0 | 14.1 | 24.2 |
| | | Fishing: | 12.6 | 20.2 | 48.4 |

millions of paid licenses per year

Figure 8-26   Population pressures on natural resources. (U.S. Dept. of Energy.)

New York, Dodd, Mead, 411 p.

Niering, W. A., 1868, The ecology of wetlands in urban areas: Garden Jour., v. 18, n. 6, p. 177–183.

Piper, A. M., 1965, Has the United States enough water?: U. S. Geol. Survey, Water-Supply Paper 1797, 27 p.

Snowy Mountains Hydro-electric Authority,

Power from water, The Snowy Mountains Scheme, 11 p.

U.S. Department of Agriculture, 1955, Water—the yearbook of agriculture 1955: Washington, D.C., U.S. Govt. Printing Office, 751 p.

U.S. Water Resources Council, 1968, The Nation's water resources: Washington, D.C., U.S. Govt. Printing Office, var. pages.

U.S. Water Resources Council, 1978, The Nation's water resources 1975–2000: Second Water Assessment, Washington, D.C., U.S. Govt. Printing Office, 86 p.

Weeks, W. F., and Campbell, W. J. 1973, Icebergs as a freshwater source: an appraisal: U.S. Army Cold Regions Res. Eng. Lab. Res. Report No. 200, 22 p.

Weinberger, L. W., Stephan, D. G., and Middleton, F. M., 1966, Solving our water problems—water renovation and reuse: Ann. New York Ac. Sci., v. 136, p. 131–154.

White, G. F., Chairman, 1968, Water and choice in the Colorado River Basin: Washington, D. C., National Academy of Sciences Publ. 1689, 107 p.

PART THREE

Volcanic activity became a particularly dangerous geologic hazard
during the twentieth century. Eruption of Irazu Volcano, Costa Rica,
in 1963. (Courtesy United Nations.)

Geologic Hazards

It is appropriate to devote a special part of this book to the discussion of geologic hazards because they constitute the most dramatic evidence of nature's dominance over us. A geologic hazard is a phenomenon associated with geologic processes that can produce a disaster when a critical threshold is exceeded and can result in significant loss in life or property. The distinguishing mark of a geologic hazard compared with other natural damages is its short duration. Erosion, sedimentation, expansive soils, all produce enormous losses, but most losses occur over time periods measured in months or years. We must not forget, however, that we are now capable of producing hazards—floods, earthquakes, and even landslides.

It has only been in the last decades that we have started to take serious measures to abate imminent hazards. These measures take the form of investigations that seek to identify hazardous conditions. Mapping programs delineate hazard-prone areas. Books and articles are written to pinpoint dangerous locales. New planning operations are conducted, and management programs become instituted to eliminate or reduce hazard risks; for example, laws may prohibit the occupance of some areas. But these precautions are still not sufficiently widespread, and in many instances we continue to occupy hazardous terranes—near volcanoes, on floodplains, on hillsides, in earthquake-known sites. It is true that if all potentially dangerous areas were considered, there would be few places for settlement. Thus, in a manner of speaking, we have become gamblers—we have learned to assign risks, make priorities, and then take our chances. Nearly all earth areas are subject at some time to either short- or long-range geologic disturbance. In some instances hazards may be prevented or mitigated, and damage potential reduced. In other cases the hazard may be so overwhelming that no amount of construction can prevent loss. The only other course of action is to avoid the problem in the first place, or abandon the site.

Although it is an academic exercise some people argue over which hazard is the most frightening and awesome, but more important it is necessary to evaluate hazards in terms of their suddenness, severity, affected area, potential for losses, degree of warning possible, and the level of control that may be exerted. In such terms, earthquakes are probably the most feared because they invariably strike without warning, transpire in a matter of seconds, allow no time for evacuation, and can involve thousands of kilometers. On a worldwide basis, flooding affects more people than any other hazard and also causes the greatest total losses. Landslides have become more severe with our help, whereas volcanic activity is more rare and affects the fewest people.

Hazard perception has become a favorite topic for social science research and fortunately such research has led to increased awareness by public officials. The basis for decisions concerning what type of human adjustments to take involve such variables as:

1 Magnitude, intensity, and frequence of the hazard.

2 Recency of the last disaster event.

3 The degree of personal involvement and experience with such a hazard.

4 Assessment of the probability for occurrence.

5 Human traits associated with behavioral mores, religion, and life values.

It should also be pointed out that many other hazards to mankind are not discussed in Part Three, such as tornadoes, fire, frost, blizzards, and so forth. Disease and pestilence are other candidates that have been omitted. However, it should be borne in mind that some disasters, in human terms, such as famine and starvation, may be indirectly caused by geologic events; for example, as a result of droughts crops may be ruined. All of these topics are covered by certain federal legislation in the United States such as the Civil Defense Act of 1950 and the Disaster Relief Act of 1970. The latter act authorizes the president to assist states in the development of comprehensive plans and programs for: (1) preparation against major disaster, (2) relief assistance to individuals and businesses, and (3) long-range recovery and reconstruction assistance. Many states, and even some cities, have special disaster and emergency acts that cover damages from geologic hazards.

Chapter 9 deals with volcanic hazards. The volcano, a natural process, although highly spectacular when in full fury, causes the least hardship; yet it is difficult to impossible to control under most circumstances. The volcanic areas of the world are now well-known so there is little surprise as to where such activity will take place.

Chapter 10 discusses another internal earth force—earthquakes. This hazard is uncontrollable under present technology. However, in the future the possibility for more finely tuned earthquake predictions will undoubtedly become a reality in some areas. Furthermore, knowledge of seismic activity is advancing so rapidly that attempted control of some events may be made in the decades that lie ahead.

Chapter 11 which focuses on landslides concludes that landsliding will greatly escalate in future years because we are being forced to occupy hillsides as a result of a growing population. In many cases we have become the instigators of a disaster, by building on unstable ground in a manner that accelerates displacement of the surficial materials.

The last chapter of this part, Chapter 12, discusses how the incidence of flooding is increasing, along with the escalation of losses that occur on nearly a yearly basis. The water regime has been more altered by human activity than any other single geologic entity, so it should be no surprise that flooding continues to be the bane of civilization. Because we have chosen to occupy the natural province of rivers—the floodplain—our presence has changed the normal flow regime.

READINGS

Bolt, B. A., and others, 1977, Geological hazards: New York, Springer-Verlag, 330 p.

Burton, I., and Kates, R. W., 1964, The perception of natural hazards in resource management: Natural Resources Jour., v. 3, p. 412–441.

Leet, L. D., 1948, Causes of catastrophe: New York, McGraw-Hill, 232 p.

Mark, R. K., and Stuart-Alexander, D. R., 1977, Disasters as a necessary part of benefit-cost analysis: Science, v. 197, p. 1160–1162.

Oakeshott, G. B., 1976, Volcanoes and earthquakes—geologic violence: New York, McGraw-Hill, 143 p.

Office of Emergency Preparedness, 1972, Disaster preparedness: U. S. Govt. Printing Office, var. pages.

Olson, R. A., and Wallace, M. M., eds., 1969, Geologic hazards and public problems: Office of Emergency Preparedness, Washington, D.C., U. S. Govt. Printing Office, 335 p.

White, G. L., ed., 1974, Natural hazards: New York, Oxford Univ. Press, 288 p.

Chapter Nine
Volcanic Activity

Eruption of Mt. St. Helens, Washington, on May 18, 1980. The volcano blew its top and caused the greatest number of deaths and the most destruction to property in U.S. history. (United Press International.)

INTRODUCTION

Volcanoes provide one of nature's most intriguing and spectacular displays. Volcanoes and related features were present on earth long before any life had evolved. Since man first arrived, over 4 billion years later, he has been fascinated and awestruck by both the serene beauty and the cataclysmic violence of volcanoes. From earliest times, men regarded towering volcanoes such as Fujiyama in Japan, Mt. Etna in Sicily, and Mt. Mayon in the Philippines

Figure 9-1 Mt. Mayon, Philippines, June 1978. (Courtesy John Wolfe.)

Figure 9-2 Mt. Mayon, Philippines, in background and Cagsawa Church remains in foreground, largely destroyed by the blast and eruption in 1814 killing more than 100 people. The last major eruption occurred in 1968, with tephra that covered 100 km² to a depth of 5 cm and killing 6. (Photograph taken by John Wolfe in June 1978)

(Figs. 9-1 and 9-2) as sacred places, the homes of gods both wrathful and beneficent. In fact, the vast body of folklore surrounding such places is an important source of information on prehistoric volcanic activity. For example, fiery volcanic eruptions gave rise to the belief that hell must be somewhere deep inside the earth. The term "volcanic" is derived from the Latin name Vulcan—the Roman god of fire.

Volcanoes, along with earthquakes, are geologic hazards that originate from endogenic forces within the earth. Whereas earthquakes result from the movement of solid rock, **vulcanism** is the movement of liquid rock (molten magma). There are several differences that we can note if we compare volcanic activity to other geologic hazards. Also, there are several paradoxes that we will discover when we analyze the impacts of volcanoes on man. The volcanic legacy to man contains a greater variety of debits and credits than other hazards. Although volcanoes can erupt with awesome fury and can destroy nearby properties and life, their long-term record of total losses is less than that of other hazards. Furthermore, volcanic activity is the only hazard that has such an extensive array of beneficial impacts to man. Soils that have evolved because of volcanic activity are among the richest on the planet. They can be the source of water supplies and geothermal energy. Volcanoes are also major sources for new production of minerals, and the edifices they construct provide some of the most aesthetically pleasing sights of all terrains.

Volcanic side effects in other regions are more damaging than the immediate destruction of areas close to the volcano. There is increasing evidence that major eruptions entrain huge amounts of aerosols into the atmosphere that can cause abnormal and costly weather conditions. The other far-traveled damaging effect can be from tsunamis that may yield losses to people and property thousands of kilometers from the initiating volcanic site.

Although exceptions occur, impending serious volcanic eruptions usually have given sufficient warning so that the local populace can flee to safety. However, volcanoes, like earthquakes, render us helpless and unable to prevent their action. The prevention of damages and the control of volcanic activity are generally impossible tasks, but in some instances minor lava flows have been altered by man.

OCCURRENCE AND DISTRIBUTION

Volcanoes are hills or mountains constructed by the expulsion of solid or fluid matter from

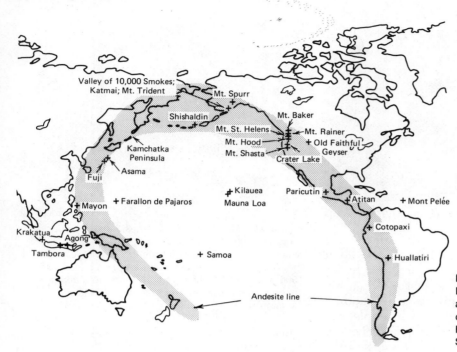

Figure 9-3 "Ring of fire." Most of the earth's volcanic activity occurs in the nearly circular belt surrounding the Pacific Ocean. (Flint and Skinner, 1977.)

the earth's subsurface. They may be composed of **lava** (extruded molten rock) or **tephra** (pyroclastic material thrown into the air by explosive eruptions; also called **ejecta**). There are about 850 major active volcanoes in the world today, over three-fourths of them in the "ring of fire" or Circum-Pacific belt surrounding the Pacific Ocean (Fig. 9-3).

Most volcanoes are located near the boundaries of lithospheric plates. Where divergence of plates is occurring along mid-ocean ridges, basaltic extrusions—reflecting the mafic composition of the oceanic crust—dominate. Where convergence of plates is occurring, as in the vicinity of ocean trenches (subduction zones), intermediate-to-felsic extrusives dominate (Fig. 9-4). Their composition reflects the acidic nature of the weathered and eroded sedimentary

rock materials deposited in the subduction trenches and of the continental crust which is commonly near the subduction zones. Linear volcanic mountain ranges tend to develop landward of the trenches, either as island arcs (e.g., Japan, the Aleutians) or, if on the continent, as volcano-capped ranges a few miles inland (Cascades, Andes).

Volcanic activity is quite common wherever the earth has been deeply fractured by tectonic movements. On continents, many volcanic terranes occur along young rift zones where the crust is (or was) under large-scale tension, as in the Rio Grande valley in New Mexico, Snake River Plain in Idaho, and parts of the Basin and Range province in the southwestern United States. A more pronounced rift zone is the famous African Rift valley.

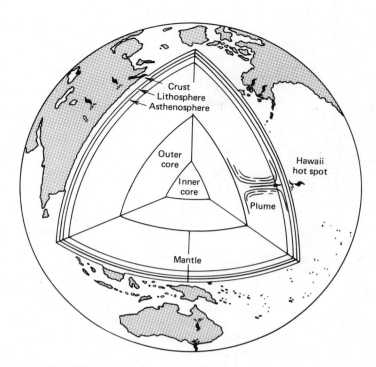

Figure 9-4 Representation of the earth's interior showing "hot spot" and plume of magmatic material rising to the surface.

Isolated areas of vulcanism, such as the Hawaiian Islands, Yellowstone National Park, and the Tibesti Mountains in north-central Africa, are surface expressions of deep-seated **mantle plumes** of hot, low-density material. Where the plume reaches the surface, a localized **hot spot** develops (Figure 9-5). The movement of lithospheric plates over plumes can sometimes be charted by a line of extinct volcanoes that increase in age as they move away from the present active volcanic hot spot, as in the Hawaiian Islands.

CLASSIFICATION

A number of schemes can be used to classify volcanoes. They have generally been classified on the basis of their period of activity, type of activity, and type of landforms created.

1 *Period of activity.* Active if eruption has occurred during historic times, otherwise inactive, dormant, or extinct.

2 *Type of activity.* Largely on the basis of severity of the eruptions and degree of explosivity. The names used are derived from the locality of the volcano that exemplifies the type of activity.

(a) *Pelean*—Mt. Pelée, Martinique; an extreme in explosiveness.

(b) *Vesuvian*—Mt. Vesuvius, Italy; intermediate in explosiveness.

(c) *Strombolian*—Stromboli, Italy; eruption of incandescent fragments accompanied by white clouds.

(d) *Hawaiian*—Hawaiian Islands; essentially lava outpourings from craters (with minimal ejecta of explosiveness).

(e) *Icelandic*—Iceland; essentially lava outpouring from fissures, rather than a central vent.

3 *Topographic character of the edifice.*

The simplest system that classifies accord-

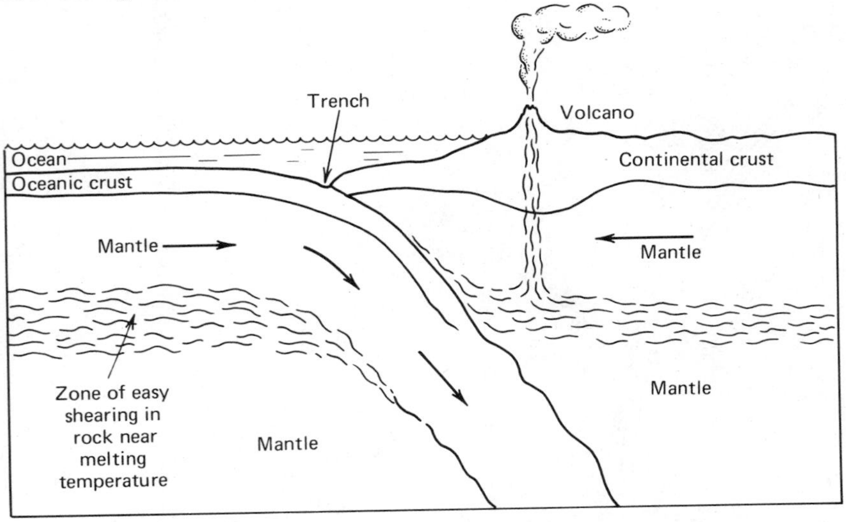

Figure 9-5 Schematic view showing origin of volcanism along margins of subduction zone.

ing to period of activity—as active, dormant, or extinct—is also the least satisfactory. Many a supposedly extinct, or dead, volcano has returned to life with a vengeance, yielding catastrophic results for people living in false security nearby (Table 9-1). Volcanoes can remain dormant (quiescent) for hundreds or thousands of years before another eruptive phase commences. In A.D. 78, a beautiful, smooth-sided conical mountain overlooking Italy's Bay of Naples was believed to be extinct since no historic volcanic activity had ever been associated with it. In A.D. 79, however, the dormancy period ended and Mt. Vesuvius began a series of eruptions that have continued at erratic and unpredictable intervals to this day. Three nearby cities (Herculaneum, Pompeii, and Stabiae) were engulfed by up to 15 m of volcanic rubble along with poisonous fumes that killed more than 2000 people.

CHARACTERISTICS

Topographically, we may separate volcanoes into four basic categories. **Cinder cones** are simple steep-sided piles of pyroclastic materials that collect about a central vent (Fig. 9-6). An inverted cone-shaped crater often overlies the vent at the top of the cinder cone. The pyroclastics range from fine ash and cinders to huge blobs of lava and blocks of bedrock which were thrown into the sky during eruption. Cinder cones may form very rapidly—the cone of Paricutin, which erupted from a Mexican cornfield in 1945, attained a height of 300 m within a month. Thousands of cinder cones, most well under 400 m feet in height, dot volcanic landscapes around the world. Lava flows often originate at the base of the cinder cone following the ejection of pyroclastics.

Shield volcanoes have broad dome-shaped features with gentle side slopes (2 to 10°). They are composed almost entirely of basaltic lava flows and frequently have a large **caldera** (wide, flat-floored craters) at the summit. Shield volcanoes can attain massive size—the largest single mountain on earth is Mauna Loa in Hawaii (Fig. 9-7). The dimensions of this great shield are approximately 96 km long, 48 km wide, and 9000 m high (above the Pacific Ocean floor), with a total volume of some 67,000 km³. The magmas from which the basalts of shield volcanoes are derived tend to be very fluid. They flow easily, release gases (volatiles) readily, and generally produce gentle eruptions of lava, often through fissures along the flanks of the volcano.

Table 9-1 Major volcanic disasters

Name and location of volcanic activity	Remarks
Santorini, island in the Mediterranean, 1500 B.C.	Destroyed cities on the island of Thera and devastated adjacent lands and coastal areas
Vesuvius, Italy, A.D. 79	Killed more than 2000 and buried cities of Pompeii and Herculaneum
Mt. Etna, Sicily, 1669	Killed 20,000 and devastated thirteen towns
Skaptar Jokull, Iceland, 1783	Resulted in death to more than 10,000, many dying of starvation
Tamboro, Indonesia, 1815	Killed 12,000 by direct effects, and more than 70,000 by starvation when crops ruined
Krakatoa, Indonesia, 1883	36,000 died, most as result of tsunami
Mt. Pelée, Martinique, 1902	Killed 30,000 by ash and gases; only two people survived
La Soufriére, St. Vincent, 1902	Killed 2000 and caused Carib Indian population to be decimated
Kelut, Indonesia, 1909	Killed 5500
Taal, Philippines, 1911	Killed 1300
Mt. Lamington, Papua, 1951	Killed 6000
Mt. Agung, Bali, 1963	Killed 1500
Taal, Philippines, 1965	Killed 500
Hekla, Iceland, 1970	Gases killed thousands of sheep
Nyragongo, Zaire, 1977	Killed 70
Mt. St. Helens, Washington, 1980	Killed about 70 and caused more than $2 billion in damages

Stratocones, also called **composite volcanoes** or **stratovolcanoes,** are an entirely different matter. These are typically symmetrical, steep-sided mountains composed of alternating layers of intermediate lavas, such as andesite and pyroclastics. They commonly attain heights of several thousand meters and are among earth's most spectacular scenic attractions (Figs. 9-1 and 9-8). They are also among the earth's most dangerous and unpredictable features. The intermediate-to-felsic magmas feeding stratocones are highly viscous; they tend to retain gases and congeal rapidly, blocking vents and causing interior pressures to build up. The result can be sudden explosive eruptions in which the volcanoes can literally "blow their top" (Fig. 9-9). The 1200 m deep caldera now half-filled by Crater Lake in Oregon is a classic example of violent stratocone activity. Nearly 7000-years ago, a 3600 m high stratocone, Mt.

Mazama, much like Mts. Hood, Shasta, Ranier, St. Helens, and others in the Cascade Range, stood where Crater Lake (elevation 2056 m) now lies. About 6600 ybp, a series of violent

Figure 9-6 Cinder cone near St. George, Utah. Note off-road vehicle trails in center slopes. (Courtesy Earl Olson.)

EXPLANATION

‒ ‒ ‒ ‒ ‒ ‒ ‒
Boundary between volcanoes

‒ ‒ ‒ ‒ ‒ ‒ ‒
Boundary between island districts

Rift zones

Figure 9-7 Island of Hawaii showing major volcano boundaries. (U.S. Geol. Survey.)

eruptions accompanied by subsidence of the entire upper portion of the volcano created the 9.6 km wide caldera of Crater Lake. Ash from the eruption covered the entire northwestern United States and adjacent Canada, where it is now a valuable marker horizon used in dating recent sediments and soils (Fig. 9-10). Most of the approximately 114 km³ of mass lost by Mt. Mazama collapsed within the crater due to release of pressure and subterranean drainage of the magma beneath the mountain. Stratocones are huge pressure cookers that may remain dormant for many centuries between cataclysmic eruptions.

A fourth family of volcanoes might be called **volcanic domes.** These result when extremely viscous rhyolitic magma is forced through a vent to the surface. Because the magma is so viscous, or even partly solidified, it cannot flow for any distance and tends to build up around the vent. In **extrusive** domes, the magma reaches the surface and the volcano grows by a series of thick, massive lava flows; **intrusive** domes grow from within as magma is squeezed slowly from the earth, usually causing the overlying layers of hardened volcanic rock to break into a great jumble of angular blocks as the dome expands upward and outward. One of the most famous dome volcanoes is Mt. Lassen in northern California—in 1915 the site of one of the latest volcanic eruptions in the 48 contiguous states (Fig. 9-8).

Variations on the above types of volcanoes are legion, and not all volcanoes fit readily into

Figure 9-8 The Mt. Lassen, California, eruption, called "the big umbrella." This view of the explosion was taken 80 km away on May 22, 1915. The column of vapor and tephra rose to heights of 8 km and was the last eruption in the United States prior to those of Mt. St. Helens in 1980. (Courtesy U.S. National Park Service.)

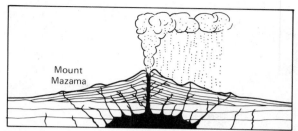

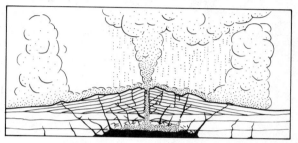

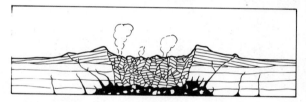

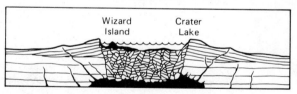

Figure 9-9 Origin of Crater Lake, Oregon. Original Mount Mazama is subjected to increased subsurface volcanic activity. The rising molten rock is temporarily dammed by enclosing material, but eventually builds up sufficient pressure to "blow the top" with a massive explosion, and remaining material subsides into a jungle of broken rocks back into the caldera. Wizard island, a volcanic cone, is the last vestige of the volcanism. (Redrawn from Longwell, Flint, and Sanders, 1969.)

any classification. In addition to volcanoes, there are a plethora of volcanic features, some of which are mentioned later in this chapter. At this point, we will look at only one—lava flows.

The largest of all subaerial volcanic landforms are the great basalt plateaus such as the Deccan Plateau in India and the Columbia Plateau in northwestern United States (Fig. 9-11). In the geologic past, highly fluid **flood basalts** have emanated from great fissures in the crust to cover many thousand of square kilometers. These massive outpourings apparently are related to early phases of deep-seated rifting apart of lithospheric plates. The flood basalts of the Columbia Plateau cover an area of over 518,000 km^2, reach a maximum thickness of over 3000 m, and contain an estimated 400,000 km^3 of basalt.

Individual lava flows vary widely in their characteristics. The more viscous lavas tend to produce stubby, thick flows with rough, jagged surfaces. In the most viscous type, great blocks of hardened, angular lava are often pushed along from behind by an advancing lava stream, **block flows.** Broken fragments of rock are carried along on top of an advancing flow by moderately viscous lavas, producing **aa flows.**

Low-viscosity mafic lavas often produce a smoother, undulating surface; a thin scum of lava may form a ropy, **pahoehoe,** surface. Lava tubes formed by movement of fluid lava through pipes and tunnels and fissures beneath a solidified surface are common in fluid basalt

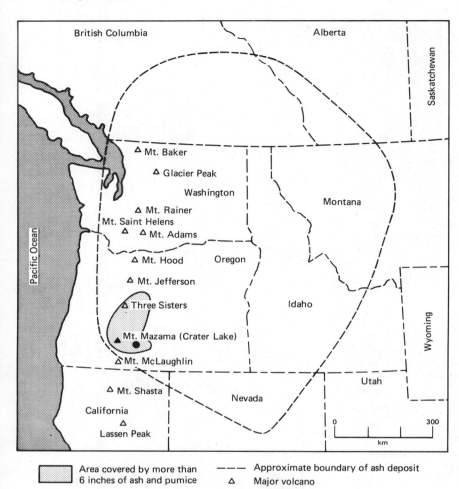

Figure 9-10   Location of major stratocones in the northwest United States, an area covered by a thick ash cover from the eruption of Mt. Mazama. (After Williams and Goles, 1968.)

[legend] Area covered by more than 6 inches of ash and pumice

△ Major volcano

— — — Approximate boundary of ash deposit

flows. As in a limestone karst topography, care must be taken to avoid danger from the collapse of such subterranean chambers when developing these areas. The congealed lavas can provide important water resources. The joints, fissures, tubes, and other openings permit the storage and movements of groundwater.

IMPACTS ON SOCIETY

Approximately 200,000 people have died in the past 500 years as a consequence of volcanism, not including volcano-related tsunami deaths (see Table 9-1). Today, millions of people reside in areas where severe hazards from volcanic activity are an ever-present threat (Figs. 9-12 and 9-13). In 1978, only about 40 of the world's 850 active volcanoes actually produced eruptions, and only one death was attributed to volcanism, reflecting the relative quiescence of most volcanoes in recent years. As with earthquakes, however, long rest periods (often 100 years or more) seem to prevail prior to exceptionally violent outbursts from a given volcano. Hence, the present complacency of most volcanoes should perhaps be viewed with some trepidation, especially after the Mt. St. Helens disaster of 1980.

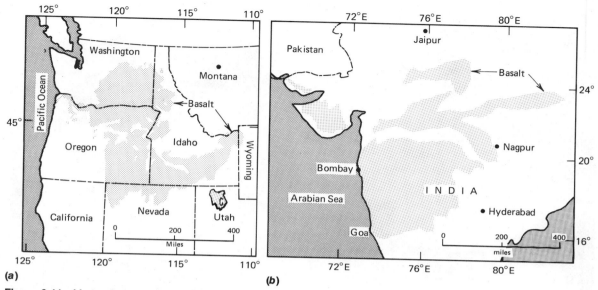

Figure 9-11 Maps of two great basalt plateaus. The light gray stippled areas are the only remnants of the original deposits, reduced in area by erosion and covered by younger sediments. (a) Columbia Plateau, northwestern United States. (b) Deccan Plateau, India. (Redrawn from Longwell, Flint, and Sanders, 1969.)

Figure 9-12 Taal volcano, Philippines, during an eruptive phase in September 1976. (a) Aerial view. (Copyright© by Josephus Daniels/Photo Researchers.) (b) Ground-level photograph showing the 1976 activity, with the *aa* flow of 1968–69 covered with tephra from the 1976 eruption. (Taken by John Wolfe.)

(a)

(b)

Figure 9-13 Mount Pelée, Martinique. *(a)* Eruption June 1902. *(b)* View of the completely demolished city of 30,000 after the eruption in 1902. (Courtesy United Press International.) *(c)* Modern view of St. Pierre from the same position as shown in Fig. 9-13*b*. (Courtesy French Government Tourist Office.)

Benefits

In spite of the violence and obvious dangers of volcanic activity, the benefits it has yielded through time far outweigh the hazards. Gaseous emissions from volcanic activity in the geologic past gave us the atmosphere and hydrosphere. Furthermore, volcanic materials have been a major builder of the land surface.

Volcanic terranes often produce exceptionally fertile soils, especially in warm moist climates where rapid weathering of fine volcanic materials makes nutrients from the new earth readily available to plants. Some of the longest maintained, highest yield agricultural lands in the world are in areas where volcanic eruptions, especially when accompanied by fine ash falls, are frequent, as along the Indonesian island arcs. This is a mixed blessing, however. High volcanic activity contributes to more fertile soils, which in turn encourage high population dens-

(c)

ities, thus increasing the risks from volcanic hazards.

Many volcanoes are scenic treasures, and extensive resorts, ski areas, campgrounds, and related developments have been built around them (Fig. 9-14). Most of the major volcanoes of the Cascade Range are either national parks or wilderness-recreation areas which attract increasing numbers of visitors each year, again

Figure 9-14 Tiwi Hot Springs National Park, Luzon, Philippines. *(a)* Park pavilion area. *(b)* Geothermal well with estimated production of about 300 megawatts of electricity when development is completed. (Courtesy John Wolfe.)

increasing the risks in the event of a sudden eruption.

Volcanic activity also provides sources for many valuable mineral commodities. Early man used obsidian for cutting tools and projectile points. Hardened lava, especially the solid, evenly jointed interior part of thicker flows, is often quarried for building stone or aggregate. Many cinder cones are literally disappearing as the loose cinders are hauled away for fill material in buildings and transportation facilities. Most cinder cones consist of light, frothy rocks (pumice and scoria) which make them highly desirable for some uses, as in lightweight concrete (for building domes, etc.), but poor for

other uses where weight, substance, and durability are important. Unfortunately, many scenic cinder cones are being destroyed merely because they are the cheapest immediate source of material.

A variety of valuable minerals are associated with volcanic processes, ranging from metals in volcanic rocks to the elemental sulfur and other minerals which often accumulate around volcanic vents or in thermal areas. The great pressure deep within the crust along vents (diatremes) serving as conduits for gaseous volcanic discharges have produced most of the world's diamonds. Bentonite is an important clay formed by the alteration of volcanic ash. The list could go on and on.

Volcanoes as a potential source of geothermal energy (see Figs. 7-21 and 9-14) were discussed in Chapter 7, and the influence of volcanoes on weathering was noted in Chapter 4.

Weather Effects

The first scientist to go on record as saying that volcanoes could influence weather was Benjamin Franklin. He believed the smoke from Iceland's volcanic eruptions in 1783 was responsible for the loss of solar heating during that summer and led to the severely cold winter of 1783–84. Later workers linked the 1815 Tambora explosion in the Dutch East Indies to the low Northern Hemisphere summer temperatures in 1816. During that summer in New England, 1800 people froze, snow and frost killed crops, and farmers had to sell homes and livestock at great loss. In Europe, 1816 was called "the year without a summer." Crops failed to ripen and famine hit England, France, and Germany. Other unusual weather phenomena also occurred following the explosions of Krakatoa in 1883 and Mt. Katmai, Alaska, in 1912. Thus it is probably no coincidence that the only five times that the major Finger Lakes of Cayuga and Seneca were completely frozen over followed times of major volcanic eruptions.

After slumbering for 120 years, Mt. Agung, Bali, exploded in 1963 and expelled the greatest amount of volcanic ash into the atmosphere of any volcanic activity in the last 68 years. This event provided a good opportunity to observe and measure volcanic weather effects because the explosion occurred at a time when the tropospheric temperature measurements could be obtained, accurate measurements of the aerosol optical depth were being made at several observatories in both hemispheres, and there was direct sampling of the composition of the stratospheric aerosols. After study of all the data, Hansen et al. (1978) concluded that the eruption caused tropospheric cooling by a few tenths of a degree Celsius for one to two years.

The spectacular sunsets and other unusual effects of visible light produced by volcanic ash prove the energy-scattering effects of the ejecta. Cooling is another major effect because the added aerosols in the atmosphere and on earth's surface contain sulfuric acid and other particulate matter which is highly reflective of solar radiation. These aerosols tend to decrease the amount of energy absorbed by the earth's system. Such a blocking effect produces a type of albedo that leads to cooling of the atmospheric layers.

Cataclysmic Eruptions

In 1883, most people regarded Krakatoa in the strait between Java and Sumatra as an extinct volcano. For over 200 years, no eruptive activity had occurred. In May of that year, the period of dormancy ended. The initial eruptive phase waned, only to be followed by three months of activity which climaxed in catastrophic fury on August 26 through August 28, 1883. It began with a series of stentorian explosions, the loudest of which, at 10 A.M. on August 27, may have been the loudest sound in historic times—it was heard 4800 km away! The disturbance of the sea from this great blast produced a tsunami that reached heights of 36 m in some bays and took the lives of 36,000 people.

Clouds of ash from the eruption rose 80 km into the atmosphere and produced a total, sulfurous darkness that lingered for 57 hours at a distance of 80 km from the volcano. Winds transported dust from Krakatoa around the globe and produced eerie and fantastic sunsets all over the Northern Hemisphere during the ensuing months. Following the eruption, a great crater extending 270 m beneath the sea was found where the land had previously stood at 120 to 420 m above sea level. As at Crater Lake, much of the vanished rock material had collapsed within the crater following the eruption.

Many other examples of cataclysmic eruptions could be described. Deaths directly attributable to the explosions themselves, heavy falling debris, or moving lava are relatively rare. Most fatalities and damages result from a wide variety of associated phenomena that will be described individually below.

Lava Flows

Lava flows quite slowly—about 300 m per hour for a typical Hawaiian basaltic flow and perhaps only a few meters per day for thick, viscous block flows. Velocity depends on the temperature and fluidity (composition) of the lava and the steepness of the slope. Some flows can move faster than others, but usually there is adequate warning and people are rarely caught completely unaware. Great property damage results when lava encroaches on developed areas. It buries farmland, blocks and diverts rivers, and engulfs buildings. In the case of viscous block flows, it is likely to cave in walls and bulldoze its way through structures. The largest extrusion of lava within historical times was in Iceland in 1783 when a single eruption covered 560 km² of land, including several farmsteads. During Miocene time (20 million years ago), individual flows of the Columbia River flood basalts covered as much as 26,000 km²!

Control of lava flows involves diversion or blockage of the lava before it reaches valuable property. One method is to bomb the flow. A well-placed explosion can open the congealed lava along the side of a flow, allowing liquid rock to spread out in a new direction. If a lava pool in a crater is feeding the flow, bombing the crater wall can open up a new avenue of escape onto less valuable land. Bombing can also help congeal a stream of liquid lava, retarding its forward progress.

Another method is to construct artificial barriers at strategic locations to divert the flow. Such barriers must be made of heavy materials and have a wide base with gentle slopes to the crest, to resist the great pressure exerted by the lava. Dams to temporarily impede or impound the flow, accompanied where possible by artificial channels to carry lavas away from populated areas, represent a related approach.

The entire city of Hilo, Hawaii (population 27,000) is highly vulnerable to lava flows. Extensive diversions and plans for bombing flows have been initiated here. The success of all such efforts depends heavily on the local topography and the volume of lava extruded.

Spraying water on an advancing lava flow has also been shown to help slow down the advance. This technique helped halt the flows that advanced toward the town of Vestmannaeyjar in Iceland in 1973.

Ash Flows and Nuées Ardentes

When a volcano erupts, the sudden release of pressure experienced by the molten rock allows gases within it to expand violently. As a seething mass of boiling magma erupts into the atmosphere, it begins to solidify while continuing to release pent up gases. This often results in the disintegration of the congealing magma into suspended fragments. Although the total mass of fragments may be great, the expanding superheated gases give the mass great fluid mobility.

Such a cloud of hot expanding gas and assorted sizes of tephra may roar forth from a volcano and tear down its slopes as a **nuée ardente**—a "glowing avalanche" of incandescent lava blocks, cinders, and ash. An **ash flow** is much the same but has finer fragments—basically an emulsion of hot glassy pumice and ash fragments. The intense heat may cause deposits from hot tephra flows to be welded together in a solid mass, for example, as welded tuffs. As with other phenomena of explosive origin, these nuées ardentes and ash flows are associated primarily with stratocone and dome volcanoes. Ash flows appear to be very rare with only one definite occurrence on record—at Mt. Katmai in Alaska. Nuées ardentes are fairly frequent during explosive eruptions, and a brief description of the most destructive one on record follows. (Fig. 9-13)

Mt. Pelée is a 1335 m high composite volcano overlooking the town of St. Pierre on the island of Martinique in the Lesser Antilles island arc of the Caribbean region. Volcanic activity commenced in April 1902 with sulfurous odors, ash falls, and steam emissions from the summit. Floods of boiling mud discharged into local rivers, and a series of violent explosions added to the activity in early May. People were assured by the government and the local press that there was no cause for alarm. At 7:50 A.M. on May 8, a series of explosions occurred and two great clouds burst forth from the volcano's summit. One cloud emerged almost horizontally through a notch in the crater and roared down the mountain toward St. Pierre at about 160 km-h. In two minutes, the glowing avalanche reached the city and, during the next two or three minutes, 30,000 people died. Only two people are definitely known to have survived the carnage. The force of the blast totally destroyed the city, stripping branches and bark from trees and tearing down 1 m thick masonry walls. The heat, estimated at perhaps 700 to 1000°C, was so great that wooden ships were set afire in the harbor, glass began to melt, and human skulls split open at the sutures as water in brain tissues vaporized and expanded.

Since nuées ardentes move downhill by gravity and follow valleys, at least partial diversion and lessening of impact may be possible by construction of barriers and channelways at critical locations. However, the best hope for saving lives in threatened settlements lies in careful monitoring of active volcanoes and efficient evacuation procedures. Unfortunately, such events do not always give adequate advance warning.

Poisonous Gases

The most common gas emitted during eruptive activity is water vapor, but dangerous gases such as sulfur dioxide, carbon monoxide, hydrogen sulfide, and hydrofluoric acid may also be present and have claimed human lives in several instances. Considerable damage to plants, animals, and property may also be caused by hazardous emissions. Acid rain resulting from the fumes of the 1912 Katmai eruption in Alaska burned peoples' skin in Seward, tarnished brass near Cape Spencer (1100 km away), and were still damaging clothes in Vancouver, British Columbia, a month after the eruption. The suffocating gases also explain many of the deaths from the Vesuvius eruption in A.D. 79 and the Mt. Pelée catastrophe of 1902.

Tehpra Falls

Falls of fine pyroclastic debris from the skies usually accompany volcanic eruptions. In most cases, the debris falls gently and the ash and pumice fragments are no longer hot. People caught in tehpra falls may be asphyxiated by noxious fumes or oxygen depletion. This is apparently what happened to many of the 2000 persons who were trapped in Pompeii during the eruption of Mt. Vesuvius in A.D. 79. The eruption blanketed the town with 4 m of ash and pumice. Fortunately, 18,000 people had taken the warnings and fled to safety before the explosion.

Heavy ash falls may be similar to lava flows in that all vegetation and structures can be obliterated and a new land surface created. The ash from the Irazu volcano, Costa Rica, caused $150 million loss to farms in 1963–65. Light

falls can contaminate water supplies, choke rivers and streams, and harm crops. The glassy ash can damage livestock grazing on ash-covered vegetation, and some ash contains poisons that can be a threat to plants and animals.

Little can be done to guard against tephra falls aside from such obvious steps as sealing homes to keep dust out, wearing dust masks to prevent inhaling dust, shoveling ash off roofs to prevent collapse, and shaking it from valuable trees and plants. The wisest step, if the volcano is nearby, is to evacuate the area until the activity ceases. In the summer of 1978, 20,000 tourists and 7000 residents were forced to evacuate the area around Mt. Usu in Japan because of a heavy ash fall. The damage to the land is usually temporary, and the benefits of ash falls, as noted above, in reviving the soil far outweigh the damages over the long term.

Lahars

Lahars are volcanic mudflows. The abundance of ash and other loose ejected debris on the steep slopes of stratocones provides ideal conditions for the development of mudflows. Lahars may be hot, as when accompanying an eruption, but more commonly they are cold and not related to a contemporaneous eruption. They are known to travel at speeds up to 90km-h and can move 160 km or more from their source area. Lahars have probably done more total damage to property than any other single volcanic process.

Lahars may be initiated by normal rainfall, the explosive ejection of a crater lake, and rapid melting of ice or snow. Lahars from the ejection of a lake on Kelut Volcano in Java took 5000 lives and devastated 200 km² of farmland in 1919. Ice-mantled Mt. Ranier in Washington has a long history of lahar activity—at least 55 in the last 10,000 years.

Floods from Glacier outbursts

Jökulhlaups (literally "glacier bursts") are catastrophic discharges of water from glaciers.

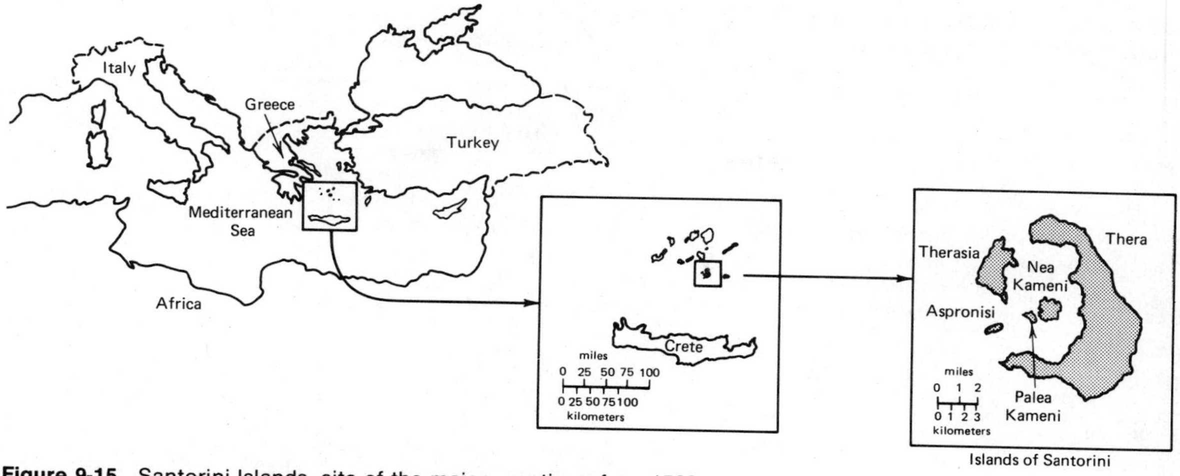

Figure 9-15 Santorini Islands, site of the major eruption of ca. 1500 B.C. that may have hastened the destruction of the Minoan civilization.

Those on Iceland originate as a result of partial glacier melting by volcanic heat. Nearly all known examples of this particular phenomenon occur in Iceland, where the term originated. Where glaciers cover volcanic mountains, heat may melt large quantities of ice inside the glaciers. Eventually, the pressure of the melted water exceeds the strength of the damming ice, and a catastrophic flood bursts forth.

Though short-lived, the discharges can be enormous—one discharge from the ice-buried volcano Katla in Iceland in 1918 was estimated to be far greater than the average discharge of the Amazon River. Fortunately, very few people live in the susceptible parts of Iceland. But what if glacier-draped Mt. Ranier or Mt. Baker in Washington or similar volcanoes were to heat up?

CASE HISTORIES

Santorini (Thera), Greece

Plato's account of the destruction of the empire of Atlantis as the result of volcanic activity has fascinated the general public for generations. Few scientists rallied in defense of such an explanation, however, until the twentieth century. Santorini is a name applied to five Aegean Sea islands, and Thera is the largest (Fig. 9-15). Thera Island is semicircular with a central caldera that averages 8.5 km in diameter and is 390 m deep. Archeologists have long known that the first true Bronze Age civilization in the Mediterranean was the Minoan civilization. This nation controlled a powerful empire centered on Crete with communities throughout the Aegean region. By 1500 B.C., they were at the zenith of their strength. However, something highly unusual must have happened by 1400 B.C. because by that time the Minoans were dominated by Mycenaean Greece. K. Y. Frost, an Irish scholar, was perhaps the first to suggest in 1909 that the demise of the Minoan civilization was catastrophically caused by the eruption of the Santorini volcano and that these islands should be considered as "the lost continent."

Not until the extensive excavations and published works by the Greek archeologist Spyridon Marinatos in 1939 did science scholars decide to take another look into the matter. Then, two breakthroughs occurred. First, geologists Dragoslav Ninkovich and Bruce C. Heezen discovered in 1965 a very extensive tephra horizon in deep-sea cores that were taken throughout the Aegean and Mediterranean re-

Table 9-2 Number of eruptions on the island of Hawaii and their relationship with hazard zones, during recent time to about 4000 years ago

Area	Historic time (since approximately 1800)			Recent prehistoric time (5000-year interval prior to 1800)	
	Number of times vents have erupted within area	Number of times lava flows have covered land within area	Percentage of land covered within area	Number of times vents have erupted within area (estimated)	Number of times lava flows have covered land within area (estimated)
A	0	0	0	0	0
B	0	0	0	0	Less than 5
C	0	0	0	Less than 5	Less than 5
D	0	0	0	0	More than 10[a]
DE	1	2	6	More than 10	More than 10
E	1	35[a]	15	About 10	More than 100[a]
F	80	More than 80	50	About 2000	More than 2000

Source: U.S. Geological Survey, 1975, Natural hazards on the island of Hawaii.

[a]Most lava flows that entered areas D and E erupted from vents in area F.

gion. And second, in 1967, the remains of a large Minoan city were found on Thera, which had been buried much like Pompeii beneath many meters of volcanic ash. Radiocarbon dating established a data of ca. 1410 B.C. for the event. Perhaps the strongest advocate for uniting these ideas into a cataclysmic sequence is the Greek geophysicist George Galanopoulos. He further reasoned that the volcanic explosion and caldera collapse triggered immense tsunamis that aided in wiping out coastal communities throughout the region. Thus, he champions the theory that the eruption on Thera not only was responsible for the near annihilation of most Minoans and their culture, but that the event indeed was the true lost Atlantis of Plato. Less enthusiastic converts admit the geologic gravity of the event, which would have rivaled the explosive force of a Krakotoa (or even exceeded it), but point out there was ample warning for the people and that most of them had escaped to safer ground on the Greek mainland. Being scattered they became easy prey to the Mycenaeans.

Hawaiian Volcanoes

The Hawaiian volcanic chain is a very youthful earth feature. The principal shields have formed within the last 500,000 years (Table 9-2). These edifices are composed mostly of lavas, with less than 0.5 percent consisting of fragmental material (Fig. 9-16). Mauna Loa is the world's most studied volcano and the site of the U.S. government's Volcano Observatory. Here eruptions begin with a series of earthquakes that accompany new fractures along a rift zone. Lava fountains spurt on the floor of the caldera and extend into one of the rift zones (Fig. 9-17). This may be followed by large billows of steam and other gases and the production of lava that flows in thicknesses of about 1 m. Spatter cones and pahoehoe lava are the normal results of this activity. The longest historic flow occurred in 1859; it lasted 10 months and moved to the sea a distance of 53 km. Another flow in 1881 reached the outskirts of the village of Hilo after flowing 46 km. The historic eruptions produce lava at the rate of 1 to 5 million

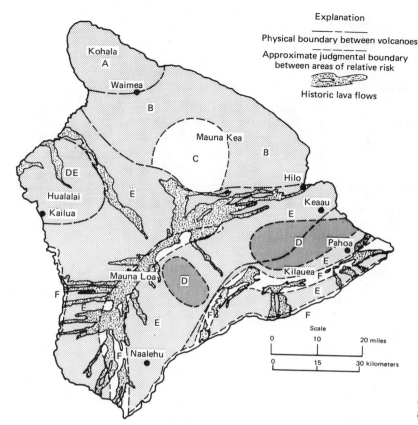

Explanation

— — —
Physical boundary between volcanoes

— — — —
Approximate judgmental boundary
between areas of relative risk

Historic lava flows

Figure 9-16 Hawaii lava flows and risk map. Risk increases from A through F. Lava flows on the map occurred between the years of 1800 and 1974. (U.S. Geol. Survey.)

tons per hour, and since the first recorded eruption in 1832 they have an average occurrence of every 3.6 years.

The location of the city of Hilo has been unfortunate because it is in the path of all-too-common lava flows that have threatened destruction in 1852, 1855, 1881, 1899, 1935, and 1942 (Fig. 9-16). When the 1935 flow jeopardized the community, the Air Force tried bombing the flow in order to divert it to another direction. The bombing was partly successful because it blocked the main feeding tube and lava overflowed at another more remote spot. Six days after the bombing, the flow stopped. Again in 1942, bombing was employed. Since the streamlike lava develops a type of natural lava levee, the strategy was to breach the levee,

thus diverting the flow. This was accomplished by the bombing, and the new lava course spilled through the break and flowed downslope parallel to the old course, rejoining the old course several kilometers farther downslope. However, there seems to have been a reduction in the production of lava which increased its viscosity and thus shortened the distance it was able to travel. None entered Hilo. However, elsewhere in the Islands, a 1960 eruption burned and crushed the village of Kanoho.

Mauna Loa had important eruptions in 1949 and 1950, but then was quiet until the eruption of July 5–6, 1975. The 25-year dormancy was the longest quiescence in Mauna Loa's recorded history. Because of the nature of the activity and the associated thousands of

Figure 9-17 Fiery fountains of lava about 30 m high. Molten lava streams cascaded from this rift of Mauna Loa on the Island of Hawaii during the July 5–6, 1975 eruption. This activity was considered by resident volcanologists a precursor to future major eruptions predicted to occur before July 1978. (Courtesy U.S. Geol. Survey.)

(a)

(b)

Figure 9-18 Volcanism at Puu Kiai–the newest volcanic hill in the United States formed during the September–October 1977 eruption of Kilauea. The eruption began September 13, 1977, with fiery fountains (a) and lava outpourings and lasted until October 12. Flows from the eruption threatened the nearby village of Kalapena, but stopped 700 m short. All vegetation was killed in this area.

earthquakes, this rather small eruption of 30 million m³ of lava was viewed as a precursor to larger eruptions in the near future. The U.S. Geological Survey extended its monitoring stations and by 1976 had installed 19 tiltmeter stations. Although the larger eruption predicted to happen by 1978 did not materialize from Mauna Loa, a significant event did occur at the Kilauea site. After 22 months of slumber, Kilauea began erupting on September 13, 1977 (Fig. 9-18). This event had been forecast during the 1977 summer when swelling was detected around an old spatter cone and additional movement of underground magma was monitored. The initial eruption phase produced 70 m lava fountains which reached a crescendo September 26, with heights of 300 m. By September 28, the 325-m/h flows had turned toward the village of Kalapana, and the village was soon evacuated. The 12 m thick and 300 m wide fiery mass of cascading lava continued flowing toward the city but finally hardened and stopped 700 m from the nearest home. Although Army engineers had attempted to stem the flow by exploding water bombs designed to cool the molten rock and dam its flow, the effort was ineffective. The natives prided themselves in stopping the flow by flying over the conduit and sacrificing three bottles of gin to the angry fire goddess Pelé.

In a number of instances, barriers and dams to divert Hawaiian lava flows have been attempted. During the 1955 eruption of Kilauea, small hurriedly constructed barriers had limited success in modifying the flow. Again in 1960 a series of barriers, largely damlike structures, were emplaced in attempts to provide additional, far more extensive downslope diversion. Although the barriers were eventually overridden, they did have some success in helping to limit the amount of destruction.

Heimaey, Iceland

At 2 A.M., January 23, 1973, a series of volcanic events was initiated on the small island of Heimaey, 20 km off the south coast of Iceland (Fig. 9-19). Ironically less than a year previously, Icelandic scientists had officially pronounced the parent volcano "dead" because it had been inactive more than 600 years. The first event in this volcanic series was the opening of a 1.6 km long fissure that extended the length of the island. Spectacular lava fountains predominated during the first eruption phase. Within 2 days a cinder-spatter cone rose 100 m, and soon tephra was being ejected at rates of 100 m³/sec. Volcanic ash was blown into the town of Vestmannaeyjar, completely burying homes on the

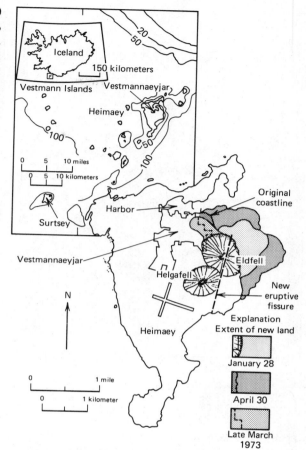

Figure 9-19 Map of Vestmann Islands and Haimaey, Iceland. (U.S. Geol. Survey.)

east side. By the end of February, the cinder cone had grown to a height of 200 m, and the central conduit disgorged massive *aa* flows that moved north and east. By early May, the flow was 100 m thick in places, averaging 40 m, and had essentially stopped. The tephra also diminished after February and by mid-April was being emitted at a rate of 50 m³/sec. The ground thickness of the materials varied from 0.3 m on the northwest part of town to more than 5 m in the southeast and up to 100 m thick on the extreme city outskirts. By July, the eruption had nearly ceased and had produced 220 million m³

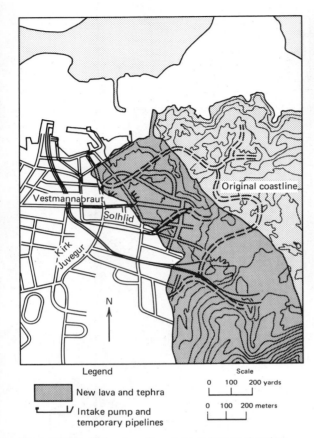

Legend

New lava and tephra

Intake pump and
temporary pipelines

Scale

0   100   200 yards

0   100   200 meters

Figure 9-20 Map of northeastern Heimaey showing
deployment of pipes along northwestern edge of new
lava flows on April 15, 1973. (U.S. Geol. Survey.)

of lava and 20 million m³ of tephra. Lava
temperatures reached 1080°C. Poisonous gases
also accompanied the activity, but only one person died and several were overcome but were
saved from the fumes.

The damage was twofold: (1) the tangible
destruction to homes, public buildings, installations, commercial property, and harbor infill;
and (2) social and economic impacts by disruption of services and fishing. Within six hours
after the eruption started, nearly all the 5300
inhabitants had been safely evacuated to the
mainland. By early May, 300 buildings had been
engulfed by lava flows or gutted by fire, and
another 70 homes have been devastated by

(a)

son died and several were overcome but were
saved from the fumes.

The decision to "fight" the lava flows came
from advice by Icelandic geologists. Using calculations derived from the study of Surtsey (a
volcanic island created off the shores of Iceland
less than 10 years previously), an operation was
set in motion to cool and harden the lava by the
spraying of seawater. This effort became the
single most ambitious program ever attempted
by humans to control volcanic activity and minimize damages. Two different techniques were
used in the attempt to diminish lava encroachment: the spraying operation and the construction of a lava barrier. Within 15 days after the
start of the eruption, the city water supply was
used effectively in slowing the lava advance and
causing the front to thicken and solidify. By
March, large pumps from the United States
were being used to pump seawater at rates of
1.0 m/sec (Figs. 9-20, 9-21, 9-22, and 9-23). This
water cooled about half its volume of lava to the
solidification temperature of basalt. The cooling
of the flow margin was used in conjunction with

(b)

Figure 9-21 Volcanic activity at Heimaey. *(a)* Eruption of January 25, 1973. (Copyright © photograph Associated Press.) *(b)* Steam rising from the cooling lava. The volcanism nearly closed the harbor of Vestmannaeyjar, Iceland's most important fishing port, and the populace had to be evacuated. (U.S. Geol. Survey.)

bulldozed diversion barriers of scoria. The cooled lava tended to pile up against the barrier rather than burrow under because of reduced viscosity.

The water-cooling program had a noticeable effect on the main lava flow. Before watering, the flow was blocky with partly welded scoria and volcanic bombs. After watering, the flow became much more jagged, because the cooling caused the more mobile interior part of the flow to break upward and disrupt surface materials. The spraying operation was halted on July 10, 1973, after having delivered 5.6 million m³ of seawater with 75 men working

Figure 9-22 Steam rises from Heimaey as rescue workers pump millions of cubic meters of water over tongues of advancing lava during the 1973 eruption. (Courtesy U.S. Geol. Survey.)

Figure 9-23 Rescue workers on Heimaey lay water pipe and bulldoze a road across piles of volcanic debris in their remarkably successful efforts to chill and slow the lava flows that endanger Vestmannaeyjar. (U.S. Geol. Survey.)

(a)

Figure 9-24 *(a)* Mt. Baker, a 3284 m high stratocone in northwest Washington. *(b)* On March 10, 1975, unusually large amounts of steam and ash clouds were emitted from Mt. Baker. An extensive monitoring program was initiated by government scientists, both by remote sensing and by ground observation measurements, as seen in this photograph. Emissions ranged from 10 to 100 times the normal amounts and displayed the greatest activity since the volcano last erupted in the nineteenth century. At this site, the Sherman Crater, south of the summit, the steam emission caused rapid breakup of ice and snow in the 500 m diameter crater. (Courtesy U.S. Geol. Survey.)

(b)

The Pacific Northwest

If the 48 conterminous states were to be inflicted with a volcanic disaster, most geologists believed the Pacific Northwest to be the most likely locale for such a scenario. The groups of spectacular stratocones in the Cascade Mountain Range (Fig. 9-10) are the most probable candidates for eruption and form the greatest volcanic hazard region in the country. Prior to 1980 the only major volcanic explosion in historic times occurred near the southern extremity of the range at Mt. Lassen in 1915 (Fig. 9-8).

In March 1975, Mt. Baker began producing spectacular emissions of steam that were 10 to 100 times the normal activity (Fig. 9-24). This early level of activity did not increase, however, and had largely returned to normal by April 1976. The volcanic heat did cause continued melting of the glacier inside the Sherman Crater

day and night for several months. The cooling operation was remarkably successful. The water had converted 4.2 million m³ of molten lava to solid rock. Data from borehole instruments showed that the lava cooled 50 to 100 times more rapidly in the sprayed areas than where lava was self-cooling. Total cost of the operation was $1.5 million. The work had saved the remainder of the city and, by March 1975, the population had grown to 4300, fishing was reestablished, and a major tourist industry began to flourish. Scientists still continue to study all aspects of this remarkable experiment.

(a)

(b)

Figure 9-25 Mt. St. Helens, Washington. *(a)* and *(b)* Photographs taken during early phases when mostly gases were being emitted, April 1980. (Courtesy Tom Zimberhoff.) *(c)* Photograph taken during the catastrophic eruptions on May 18–19, 1980. The plumes

of smoke and tephra rose to heights of 18 km while this photograph was taken. Note the flatter summit of the peak area, when compared with *(a)*, because of blowing off 500 m of the crest. (Copyright © United Press International.)

near the top. The melting back of the ice also uncovered fumarole vents that emitted some of the newly visible steam. These events caused a debris avalanche in August 1977 on Sherman Peak that traveled 1.6 km down the Boulder Glacier. This area has produced other similar avalanches about every 2 to 4 years during the past few decades. At present, the major danger is to mountain climbers.

Within the last 10,000 years, Mt. Ranier has produced at least 55 large mudflows, several hot avalanches of rock debris, at least one period of lava flows, and 12 eruptions of volcanic ash. Within historic time, many small-scale eruptions have occurred between 1820 and 1855, and the last eruption of steam and smoke was in 1894. One spectacular mudflow occurred about 6000 years ago and involved 610 million

m³; it flowed down the White River a distance of 50 km. The Osceola mudflow, about 5000 years ago, contained 2 billion m³, covered an area 325 km², and extended into Puget Sound lowland where 40,000 people currently live. Mt. Shasta has erupted at least once every 800 years for the past 10,000 years, and once every 600 years in the last 4500 years. The last eruption was about 200 years ago.

Mt. St. Helens now takes its place as the most damaging volcano in U.S. history. (Fig. 9-25). Although its record extends back more than 37,000 years, virtually all of the visible volcano has formed since 500 B.C., and most of its upper part within only the last few hundred years. Studies by Crandell and Mullineaux indicate that the volcano has never been dormant more than five centuries during the past 4500 years. The previous known eruptions can be grouped into four periods: 2500 to 1600 B.C., 1200 to 800 B.C., 400 B.C. to A.D. 400, and A.D. 1300 to A.D. 1850. Tephra from Mt. St. Helens rivals amounts from some of the best-known volcanoes. For example, the eruption in 1900 B.C. produced a volume of 3 billion m³, compared with the 1707 eruption of Mt. Fuji which produced 750 million m³.

On March 28, 1980, the long slumbering Mt. St Helens finally came alive, and extensive volcanic activity began for the first time since 1857. Plumes of steam, gases, and ash were blown to heights nearly 5000 m above the summit, accompanied by earth tremors that reached 4.0 M. Soon a new 70 m opening developed in the 600 m wide crater, and these signals were considered sufficiently dangerous so that governmental agencies ordered the evacuation of people within a 24 km radius of the mountain. Continuing eruptions, ejecta of ash and other debris, and earthquakes (up to 4.9 M) occurred during the following weeks. At one stage the mountain mass seemed to have expanded several meters, and the floor of the crater subsided about 300 m. Winds carried the ash to distances as great as 80 km where it coated all ground objects. The worst immediate hazard was that the heat might melt the 5 m snowpack, or the 30 m of glacial ice, which could trigger avalanches and mudslides. However, all of this previous activity was only "small potatoes" when compared to events that occurred on May 18 and 19. During these days the 3000 m mountain exploded with the greatest fury of historic times and blew the northern part of the summit off. Here the mountain lost 500 m of its crest, and the newly formed crater was 1.6 km wide and hundreds of meters deep. In all, more than 1 billion m³ of material was blown to the surrounding countryside, to heights of 18 km. The force of the explosion equalled 500 Hiroshima-type atomic bombs and caused an extraordinary shock and heat wave. Trees were blown down in a 390 km² area, and the loss in the Gifford Pinchot National Forest was $500 million. Humans were killed by the heat to distances of 40 km, and nearer people and installations were either vaporized and disintegrated or buried by the tephra and ensuing mudflows. About 70 people died, including a volcanologist with the U.S. Geological Survey who had been monitoring the mountain from a distance of several kilometers. During the 127 years of relative quiescence, Mt. St. Helens— the subterranean andesitic rock—had been building up pressure. The earthquakes aided to relieve the load of covering materials and prepared the way for the rise of magma and the ultimate demolition of the summit (Fig. 9-25c). At the time of this writing, it is estimated that total damages will probably exceed $2 billion, including losses of $1 billion in Washington to private and state lands and $200 million to the salmon and sport fishing industries.

Scientists who have studied these beautiful Cascade stratocones argue that the general public cannot afford to become complacent or oblivious to the possible threat of renewed activity. Especially when the magnitude of the possible eruption is considered.

EVALUATION AND PREDICTION

The risks to people and property near volcanoes may be better appraised by studying the geology and recorded history of the volcano. Many volcanoes tend to repeat past performances and this can be used in the prediction of future activity. Many also reflecd an apparent long-term evolution of the originating magma toward a more viscous state, leading to progressively more explosive eruptions.

Maps may be prepared to indicate the risk potential in locations near volcanoes in the event of an eruption (Fig. 9-26). We can only hope that these will be used in land planning and will discourage development in highly vulnerable locations. Unfortunately, a broad gap in communication and in goals often exists between safety-minded geologists and development-minded landowners.

In rare instances, as at Mauna Loa, consist-

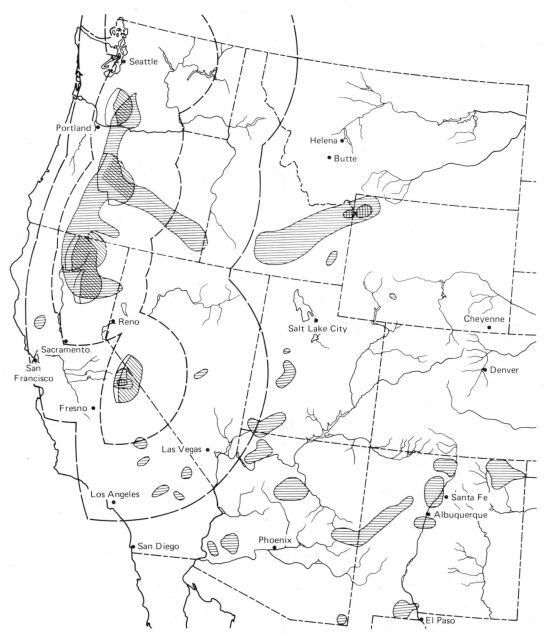

Figure 9-26 Map of western United States showing volcanic hazard zones. Shaded zones with vertical lines are volcanic vent areas that had one or more extremely explosive and voluminous eruptions within the last 2 million years. Zones with horizontal lines are subject to lava flows and small volumes of ash from groups of volcanic vents called "volcanic fields." Zones with diagonal lines would get most of the ashfall from nearby relatively active and explosive volcanoes. The inner dashed line encloses areas subject to 5 cm or more of ash from a large eruption, and the outer dashed line encloses areas subject to 5 cm or more of ash from a very large eruption. (U.S. Geol. Survey.)

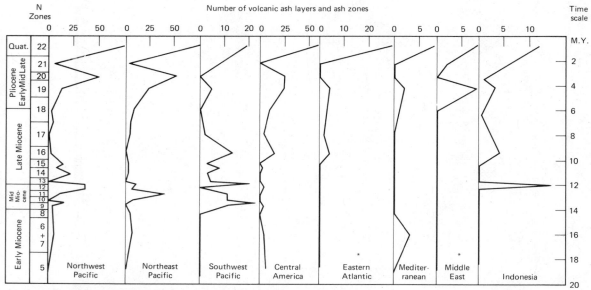

Figure 9-27 Total number of volcanic ash layers throughout the world from early Miocene time. (J. P. Kennett and R. C. Thunnell, 1975, Global increase in Quaternary explosive volcanism: *Science,* v. 187, p. 500. Copyright © 1975 Amer. Assoc. Adv. Sci.)

ent periodic behavior allows frequent predictions of volcanic activity. However, it is not possible to predict when most volcanoes will reawaken. To date, no single set of activities gives completely reliable clues to a volcanoe's future behavior, although there are a variety of events that *may* signal a forthcoming eruption.

Earthquakes and microseisms precede many eruptions. They may begin years in advance or less than an hour before volcanism. They sometimes cease altogether before the eruption. Among the other activities which may herald approaching volcanic activity are:

● Changes in the characteristics and behavior of fumaroles, hot springs, and crater lakes.

● Changes in the strength and orientation of the earth's magnetic field.

● Increased geothermal heat flow from the area—sometimes detectable by infrared photography.

● Alterations in local electric currents within the earth.

● A tumescence or bulging upward of the volcano due to magma buildup—detectable by tiltmeters on the ground.

● Erratic behavior of animals and other signals indicative of possible impending earthquake activity.

Scientists have shown the widespread occurrence of volcanic events since Miocene times (Fig. 9-27). To date, however, very few volcanoes have received detailed studies of their behavior prior to eruption. With such studies, it is probable that many eruptions can be anticipated and procedures taken to prevent loss of life and property. Continuous monitoring of the world's hundreds of potentially hazardous volcanoes would be prohibitively expensive but, perhaps, some of the most threatneing can be closely observed. The problem is complicated by the fact that volcanoes frequently produce ominous signs—venting of steam, seismic activity, assorted rumblings—which turn out to be false alarms. People living around volcanoes should be informed of the dangers posed by volcanoes and be well-versed in what actions to take in the event an eruption becomes imminent. Eight thousand people were evacuated from a high-risk area near the Mayon volcano in the Philippines in 1978, and this foresight saved all lives during the ensuing eruption.

Preventive measures, notably land use and settlement regulations, are the least expensive and most successful approaches to reducing risks from volcanic activity. Although such measures may clash with certain business interests, these would be minor inconveniences compared to the catastrophic consequences of a Pelée or Krakatoa.

Perspectives

As long as we insist on inhabiting the nearby areas of volcanoes, we will forever be faced with potential disastrous eruptions. Human safety and salvation therefore become dependent on the ability of scientists to predict an impending eruption. It is imperative that monitoring networks be established at those sites of known dangerous volcanoes that are near populated areas. Volcanologists now possess a wide range of instruments for detection of renewed magma activity in a threatening volcano. When this technology is coupled with the studied phases that all volcanoes share, the combined predictive scenario will in most cases save the lives of residents.

Even prior to our modern understanding of volcanoes, there were numerous examples of large-scale evacuations that have saved countless lives. The A.D. 79 eruption of Mt. Vesuvius had sufficient precursors so that 18,000 nearby inhabitants were saved; however, 2000 still died. In 1914, Sakurajima volcano in a bay of the Japanese Island of Kyushu heralded the upcoming event with smoke, earthquakes, and increased flow in springs. The local authorities forced the complete evacuation of the area, thus saving lives of the 23,500 inhabitants. When the explosion occurred, ash clouds exploded to heights of 6000 m, and 12 percent of the entire mountain was demolished. Another more recent illustration of foresight was exhibited in the spring of 1979. La Soufriére, a volcano at the northern end of the 27 km long Caribbean Island of St. Vincent, had killed more than 2000 in 1902 and once again was showing threatening signs. Geologists had been monitoring various precursor events and became sufficiently convinced of an impending eruption; they urged Premier Milton Cato to order an evacuation. This resulted in the movement of 20,000 people from the area. Prior to dawn on Good Friday (April 13) the eruption did occur, and with an explosive roar a great cloud of hot gases and debris shot forth. Several ensuing eruptions occurred the following day, creating a mushroom-shaped cloud 12,000 m high and cascading debris as far as the island of Barbados, 145 km away! Fortunately, no one was killed this time because of the early warning and evacuation to safety.

Thus, an effective early warning system is required for those who insist on living in the shadow of known hazardous volcanic sites. To move poeple out prematurely, however, can cause resentment and financial hardships. After all, such areas are generally richly endowed with the heritage of important resources—both the physical manifestations of soil, water, and energy, as well as the breathtaking grandeur of the architectural forms that the fire gods have constructed.

READINGS

Aspinall, W. P., Sigurdsson, H., and Shepherd, J. B., 1973, Eruption of Soufriére Volcano on St. Vincent Island, 1971–1972: Science, v. 181, p. 117–124.

Bullard, F. M., 1962, Volcanoes in history, in theory, in eruption: Austin, Univ. of Texas Press, 441 p.

Crandell, D. R., and Mullineaux, D. R., 1969, Volcanic hazards at Mount Ranier, Washington: U. S. Geol. Survey. Bull. 1238, 26 p.

Crandell, D. R., Mullineaux, D. R., and Rubin, M., 1975, Mount St. Helens volcano: recent and future behavior: Science, v. 187, p. 438–441.

Crandell, D. R., and Waldron, H. H., 1969, Volcanic hazards in the Cascade Range: in Olsen, R., and Wallace, M., eds., Geologic hazards and public problems: Office of Emergency Preparedness, U. S. Govt. Printing Office, p. 5–18.

Francis, P., 1976, Volcanoes: Middlesex, England, Penguin Books, 368 p.

Hansen, J. E., Wang, W., and Lacis, A. A., 1978, Mount Agung eruption provides test of a global climatic perturbation: Science, v. 199, p. 1065–1067.

Kennett, J. P., and Thunell, R. C., 1975, Global increase in Quaternary explosive volcanism: Science, v. 187, p. 497–503.

Luce, J. V., 1969, The end of Atlantis: new light on an old legend: San Francisco, McGraw-Hill, 224 p.

MacDonald, G. A., 1972, Volcanoes: Englewood Cliffs, N.J., Prentice-Hall, 510 p.

———, 1958, Barriers to protect Hilo from lava flows: Pacific Science, v. 12, p. 258–277.

Moore, J. G., Nakamura, K., and Alcarez, A., 1966, The 1965 eruption of Taal Volcano: Science, v. 151, p. 955–960.

Perret, F. A., 1924, The Vesuvius eruption of 1906: Carnegie Inst. of Washington, v. 339, 151 p.

Toon, O. B., and Pollack, J. B., 1977, Volcanoes and the climate: Natural History, v. 86, n. 1, p. 8–26.

U.S. Geological Survey, 1975, Man against volcano: the eruption on Heimaey, Vestmann Islands, Iceland: USGS Inf-75-22, Washington, D. C., U.S. Govt. Printing Office, 19 p.

U.S. Geological Survey, 1975, Natural hazards on the island of Hawaii: USGS Inf-75-18, 15 p.

Vogt, P. R., 1977, Hot spots: Natural History, v. 86, n. 4, p. 37–44.

Chapter Ten
Earthquakes

Destruction of Harvardian School, Cotobata, Mindanao, Philippines, by an earthquake on August 16, 1977 that measured 7.9 M (Richter scale). (Courtesy Robert E. Wallace, U.S. Geol. Survey.)

INTRODUCTION

Earthquakes are the most fearful and diabolic of all nature's weapons that can change the earth's surface and thereby cause great havoc to human civilization. Whereas there generally is some indication of impending doom with other geologic hazards, possible earthquake indicators are mostly too subtle for the populace to take warning. For example, people that live on floodplains, along the shores, or near volcanoes know that periodically danger is lurking close at hand, so that they have come to expect hazardous conditions. The frightening element of earthquakes, however, is their extreme suddenness and the inability to move to safety during the few seconds that the quake takes its toll.

Whoever called our planet earth "terra firma" did it a grave injustice because the earth is anything but stable, with more than a million earthquakes of all sizes occurring annually. However, the term "earthquake" is exceptionally appropriate because the earth literally does shake when this phenomenon happens. This is the surface manifestation of a catastrophic event usually occurring within the earth's crust. Most often earthquakes result from fault movement, and the sudden release of energy takes the form of shock waves and elastic vibrations (seismic waves) within the earth materials.

All throughout human history humans have speculated about the mysterious and puzzling violent movements and sounds of earthquakes and have incorporated the phenomenon into their religions and literature. Most often earthquakes were associated with the gods of the basic support system of the planet and so were beyond a human's ability to predict or control such forces. Although modern man has hoped and tried to learn ways to predict earthquakes, only within the past two decades have new areas of research opened doors that may lead to a more sophisticated understanding of the timing and occurrence of earthquakes. It is imperative that these efforts be increased because their urgency is dictated by the continuing urbanization of the planet and the growth in areas that are earthquake-prone. Additional investigation of all facets of earthquakes is vital because now we are providing the trigger in many cases—for example, in water-flooding methods for secondary recovery of oil and in construction of mammoth reservoirs. Man-in-

duced quakes can have a beneficial side, as those that are deliberately caused by detonation of a charge for the purpose of creating seismic waves so that the rock strata can be mapped and perhaps reveal important economic resources.

There is no abatement in the annual disasters that occur from earthquakes and, as recently as 1976, 655,000 were killed in China alone. Property damages amount to billions of dollars every year, and the accompanying human suffering is immeasurable. Our ability to control earthquakes is highly questionable, so the only accommodation is to perfect prediction and warning systems and to adopt stringent codes of occupance for those areas known to be hazardous.

PHYSICAL PROPERTIES OF EARTHQUAKES

Although earthquakes may occur in the earth to depths of 700 km, those that effect us result from displacements of rock within the earth's crust. Such ruptures release some of their energy in the form of sudden vibrations, or seismic waves, that spread out from the disturbed area. Most energy, however, is locally absorbed during the moving and heating of the rock. The severity of an earthquake depends on the length, breadth, location, and type of slippage along the rupture zone or fault. Other factors involved include the type of rock, amount of displacement, and the length of time that forces have been accumulating prior to passing the threshold for rock stability.

The seismic waves may manifest themselves by producing changes in surface materials, topography, and water bodies, but they can also be detected by instruments called **seismographs** (Figs. 10-1 and 10-2). When an earthquake occurs the measurement and character of the zigzag trace on the **seismogram** provide a fingerprint and blueprint for the size and location of the disturbance.

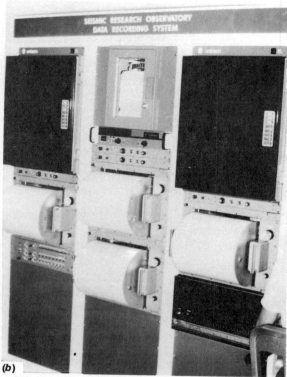

Figure 10-1 Earthquake recording system at the U.S. Geological Survey's seismological laboratory, Albuquerque, New Mexico. The system detects signals from the borehole seismometer *(a)* about 100 m deep, which are transmitted as seismic viibrations to the seismograph-seismogram hookup *(b)*. Such motions are recorded by a stylus or light beam which marks the vibrations on the drums. (Courtesy U.S. Geol. Survey.)

There are two principal types of vibrations: **surface waves,** which are transmitted along the ground, and **body waves,** which travel through deeper earth materials (Figs. 10-3 and 10-4). The strongest motions are produced by the surface waves, which produce most of the damages. There are also two types of body waves: compression and shear. Particle motion of **compressional waves** occurs in the direction of the energy impulse; so these waves travel with greater velocity and arrive at the surface first. Thus, they are also called **primary** or **P waves.** **Shear waves,** or **secondary S** vibrations, displace material at right angles to their travel direction and are also termed **transverse waves.**

The arrival times of the P and S waves, the ratio of their speed, and the amplitude as determined by the seismogram record yield information about earthquake severity, the epicenter, and focal depth. Thus, the first indication of an earthquake will be a sudden thud, which is the arrival of the compression waves, followed in turn by the shear waves, and then the ground roll of the surface waves.

Earthquake Severity

The size or severity of an earthquake is generally given in terms of its magnitude or intensity. Magnitude, as expressed by the **Richter Scale,**

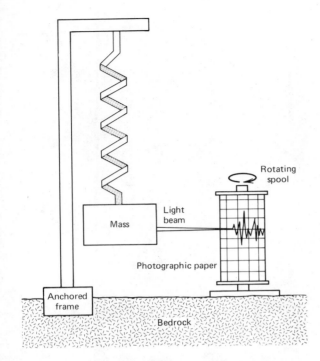

Figure 10-2 Basic seismograph principles. The low inertial momentum of the weighted mass remains essentially stationary while the rotating drum vibrates during the earthquake. A light beam or stylus scriber on the motionless pendulum makes a photograph or a barograph that shows the amount and type of vibration.

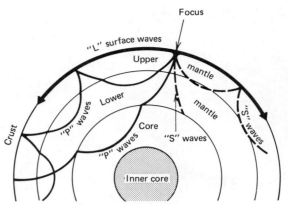

Figure 10-3 Cross section of the earth showing different zones, types, and movements of earthquake waves. (U.S. Geol. Survey.)

is an absolute measure of the amplitude of the seismic waves which are dependent on the amount of energy released. Intensity, usually expressed by the **Modified Mercalli Scale** (Table 10-1), is a relative cataloging of features that describe the degree with which the shock is felt and the damage it produces.

The Richter Scale was developed by Dr. Charles F. Richter in 1935 and provides a measurement of earthquake magnitude at its source. Theoretically the scale consists of Arabic numbers that range up to 10, but the most severe quakes have been 8.9 M and the smallest microseisms are −3 M. This magnitude scale varies logarithmically so that each whole number represents about 31 times more energy than the preceding number and a tenfold increase in the amplitude of the quake (Fig. 10-5). Therefore,

an 8.8 M earthquake is not twice as large as a 4.4 quake, but instead 10,000 times larger, and the energy released is about one million times greater. A 2 M earthquake is generally the smallest felt by humans, and quakes greater than 6 M can be disasters if they occur in highly populated regions.

In 1902 Mercalli, an Italian seismologist, devised a rating scale using Roman numerals I to XII as a measure of earthquake intensity as revealed by its effects at a given locality. This system was updated in 1931 by the American seismologists Harry Wood and Frank Neuman to account for modern structural buildings. The ranking method is now called the *Modified Mercalli Scale*. This method has more meaning for the general public because it is based on actual observations and damages. A drawback is the dependence on firsthand and relative reports; so there is a lag time before accurate intensity maps can be prepared for the earthquake event. Furthermore, the intensity becomes a function of all those features that determine the amount of damage. Thus, earthquake destructiveness depends on a host of factors that include magnitude, focal depth, distance from epicenter, local geologic conditions, topography, the design of man-made structures, and population density. For example, an area underlain by unstable ground (as sand or clay) is likely to

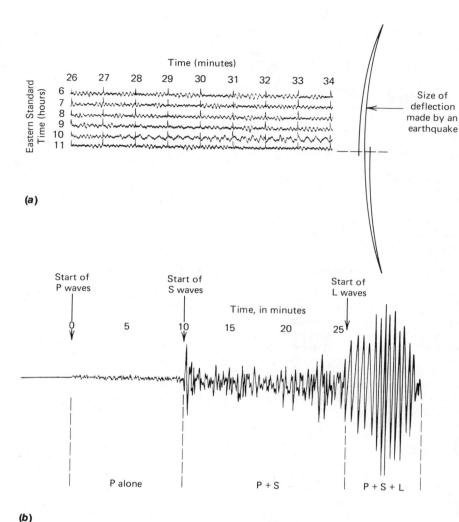

(a)

(b)

Figure 10-4 Examples of seismograms. *(a)* Traces of microseisms recorded at Troy, New York, January 1, 1967, compared with deflection made by an earthquake on the same day. (Rensselaer Polytechnic Institute.) *(b)* Earthquake recorded at Cambridge, Massachusetts, December 26, 1939, with epicenter at Erzincan, Turkey. P = primary waves, S = secondary waves, L = surface waves. The elapsed time of 10 minutes and 45 seconds, between the start of the P waves and that of S waves, registers a distance from epicenter to the station of 88° 30′, or about 9700 km. (After Geophysics at Harvard.)

experience more noticeable effects than an area equally distant from the epicenter but underlain by firm ground such as granite (Fig. 10-6).

Other important indicators of the strength of an earthquake and its ability to cause damage involve the acceleration of ground shaking. An instrument called the **accelerograph** measures this and records the measurement on an **accelerogram.** When equated with the acceleration due to gravity, the higher the acceleration, the greater the damage potential. Accelerations of 1.0 g would produce a catastrophe of major proportions.

Distribution and Occurrence

Earthquakes are neither random nor uniform in their geographical positions throughout the world. Instead, the great majority of quakes are confined to two principal belts (Fig. 10-7) which are the most prominent and dangerous to man. The circum-Pacific belt is commonly referred to as "the ring of fire" because it largely girdles the Pacific by a nearly continuous series of young volcanoes and is also the site for much of the crustal unrest and tectonism in island arc systems and the west coastal areas of the West-

Table 10-1   Modified Mercalli Intensity Scale of 1931

I.   Not felt except by a very few under especially favorable circumstances.

II.   Felt only by a few persons at rest, especially on upper floors of buildings. Delicately suspended objects may swing.

III.   Felt quite noticeably indoors, especially on upper floors of buildings, but many people do not recognize it as an earthquake. Standing automobiles may rock slightly. Vibration like passing of truck. Duration estimated.

IV.   During the day felt indoors by many, outdoors by few. At night some awakened. Dishes, windows, doors disturbed; walls make cracking sound. Sensation like heavy truck striking building. Standing automobiles rocked noticeably.

V.   Felt by nearly everyone, many awakened. Some dishes, windows, etc. broken; a few instances of cracked plaster; unstable objects overturned. Disturbances of trees, poles, and other tall objects sometimes noticed. Pendulum clocks may stop.

VI.   Felt by all, many frightened and run outdoors. Some heavy furniture moved; a few instances of fallen plaster or damaged chimneys. Damage slight.

VII.   Everybody runs outdoors. Damage negligible in buildings of good design and construction; slight to moderate in well-built ordinary structures; considerable in poorly built or badly designed structures; some chimneys broken. Noticed by persons driving automobiles.

VIII.   Damage slight in specially designed structures; considerable in ordinary substantial buildings with partial collapse; great in poorly built structures. Panel walls thrown out of frame structures. Fall of chimneys, factory stacks, columns, monuments, walls. Heavy furniture overturned. Sand and mud ejected in small amounts. Changes in well water. Persons driving automobiles disturbed.

IX.   Damage considerable in specially designed structures; well-designed frame structures thrown out of plumb; great in substantial buildings, with partial collapse. Buildings shifted off foundations. Ground cracked conspicuously. Underground pipes broken.

X.   Some well-built wooden structures destroyed; most masonry and frame structures destroyed with foundations; ground badly cracked. Rails bent. Landslides considerable from river banks and steep slopes. Shifted sand and mud. Water splashed (slopped) over banks.

XI.   Few, if any structures remain standing. Bridges destroyed. Broad fissures in ground. Underground pipelines completely out of service. Earth slumps and land slips in soft ground. Rails bent greatly.

XII.   Damage total. Practically all works of construction are damaged greatly or destroyed. Waves seen on ground surface. Lines of sight and level are distorted. Objects are thrown upward into the air.

Note: The first scale to reflect intensities of earthquakes was developed by de Rossi of Italy and Forel of Switzerland in the 1880s. This scale, with values from I to X, was used for about two decades. A need for a more refined scale increased with the advancement of the science of seismology and, in 1902, the Italian seismologist, Mercalli, devised a new scale on a I to XII range. The Mercalli Scale was modified in 1931 by American seismologists Harry O. Wood and Frank Neumann to take into account modern structural features. (U.S. Geol. Survey.)

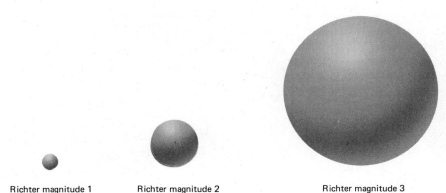

Richter magnitude 1   Richter magnitude 2   Richter magnitude 3

Relationship between Earthquake Magnitude and Energy

Figure 10-5 Relationship between earthquake magnitude and energy. The volumes of the spheres are roughly proportional to the amount of energy released by earthquakes, in the magnitudes given, and illustrate the exponential relationship of magnitude and energy. At this same scale the energy released by the San Francisco earthquake of 1906 (on the Richter scale 8.3 M) would be represented by a sphere with a radius of 33 m. (U.S. Geol. Survey.)

Figure 10-6 Aftermath of an earthquake shaking at Varto, Turkey. Buildings within the area marked as "bench" were not affected, whereas much damage was sustained by buildings within the area of "old channel," which consisted of unconsolidated sediments. Recent high groundwater levels further amplified the difference between the two dissimilar foundation materials. (R. Wallace, U.S. Geol. Survey, Circ. 701, 1974.)

ern Hemisphere. The second earthquake belt forms an arcuate corridor that extends from Spain on the west through the Mediterranean region into the Middle East and the Himalayan Mountains. These earthquake concentrations are not fortuitous, but instead are largely coincident with the lithospheric plate margins as defined by the geography of global tectonics (see page 53 and Fig. 10-8) and known positions of young mountain building. However, few regions can afford to be complacent and assume that they are entirely safe from quakes. In the United States severe earthquakes have occurred in Boston, Chicago, Missouri, and Charleston. North Dakota is the only state that has no record of a major earthquake within the past 300 years.

The **focus** or **hypocenter** is the point within the earth where the earthquake's energy originates. The **focal depth** is that distance from the focus to the surface at a point called the **epicenter.** The epicenter defines the geographic position of the quake and is the point most commonly plotted on maps to show the location of the tremor. Earthquakes with focal depths from the surface to 60 km and classed as shallow are the ones that produce most damage to humans. Quakes with focal depths from 60 to 300 km are intermediate, and those from 300 to 700 km are deep earthquakes. Each year there are more than 100 earthquakes with magnitudes greater than 6 M, which would be very damaging to inhabited areas. Fortunately most occur in the ocean or at land locations that are sparsely populated. Earthquake frequency does not seem to be dependent on the weather, time of day, or time of year.

Earthquake "Belts" of the World

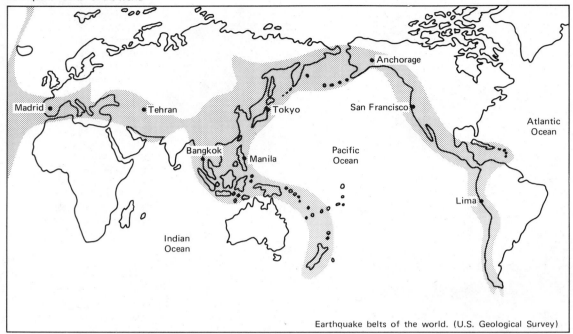

Earthquake belts of the world. (U.S. Geological Survey)

Figure 10-7 The locations of seismic belts and epicenters of major earthquakes. Symbols indicate depth of focus. (After L. Hiersemann, 1956, *Freiberger* *Forschungschefte,* v. 24, based on data in Gutenberg and Richter, 1974.)

CAUSES OF EARTHQUAKES

Early Ideas

Ancient peoples observed that the earth seemed to be immobile whereas the rest of the heavens were revolving around the planet. They reasoned that something was necessary to hold the earth in place and keep it from falling, so they imagined a variety of animals in the role. When the animal grew weary of its load and moved, the earth shook as a result. Thus in Bali, Borneo, and Bulgaria the animal was a buffalo; a tortoise to the Algonquin Indians; a hog in the Celebes; a serpent in the Moluccas; a crab in Persia; and a frog in Mongolia. The Japanese believed the animal was a great spider or catfish, and natives of Siberia's quake-prone Kamchatka Peninsula blamed tremors on a giant dog named Kosei who kept tossing snow off his fur.

In Greece, Pythagoras believed earthquakes were caused by the dead fighting among themselves, but it was Aristotle who first gave a process-oriented theory. He believed the rumblings, and the resulting quakes, were caused by hot air masses trying to escape from hollowed out parts of the earth's interior. His logic was so convincing that it was still in vogue during Shakespeare's time; the immortal playwright wrote in Henry IV that unruly winds within the earth toppled "steeples and moss-grown towers." Only after the eighteenth century were other more convincing theories finally adopted.

Another line of reasoning used by the ancients ascribed earthquakes to the work of a vengeful deity. This idea stems in Jewish and Christendom beliefs from the Bible, where in Psalms 18:7 it is written, "Then the earth shook and trembled; the foundations also of the hills moved and were shaken, because he was wroth."

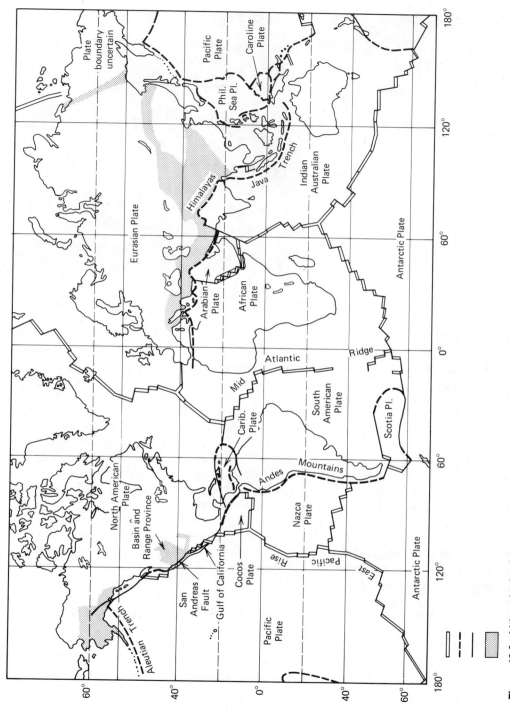

Figure 10-8 Lithospheric plates of the world, showing boundaries that are presently active. Double line: zone of spreading, from which plates are moving apart. Line with barbs: zone of underthrusting (subduction) where one plate is sliding beneath another; barbs on overriding plate. Stippled area: part of a continent, exclusive of that along a plate boundary, which is undergoing active extensional, compressional, or strike-slip faulting. (U.S. Geol. Survey, *Plate tectonics and man,* 1978.)

Figure 10-9 Fault scarp, Pleasant Valley, Nevada. This earthquake fault produced a vertical movement of 5 m and a horizontal trace along 35 km. The earthquake destroyed adobe houses and other structures in the area. (Courtesy Robert E. Wallace, U.S. Geol. Survey.)

As late as 1750, Thomas Sherlock, the Bishop of London, told his parishioners that the two recent earthquakes were directed at sinners and told the people to repent.

Modern Ideas

Seismologists now have delicate instruments and infinitely more data to use than the ancients in developing new ideas for the genesis of earthquakes. The new theories must explain the entire spectrum of observations such as foreshocks, principal shocks, and after shocks. Duration of the quake may be from seconds to minutes, and the shaking can affect millions of square kilometers. The common phenomenon of earthquakes occurring in association with faulting and volcanic activity has led to theories that find volcanoes and faulting as primary causes of most natural earthquakes. In 1908, H. F. Reid formulated the **elastic rebound theory** which provided a descriptive explanation for the release of energy during faulting. In this theory, the immediate cause for the shaking stems from the continual buildup and storage of energy along a plane of weakness in the earth's crust. Finally the stress exceeds the frictional resistance of the rocks which yield and break along a fault (Figs. 10-9 and 10-10). The two sides are snapped loose but retain certain quasi-elastic properties that produce a partial

Figure 10-10 Closeup of upthrown side of the Pleasant Valley, Nevada, fault scarp. (Courtesy Robert E. Wallace, U.S. Geol. Survey.)

springing back. This "rebound" results in the nearly instantaneous release of the deformational energy in the form of shocks or seismic waves. Within the past 20 years new theories provide additional details of the earthquake process and these are based on information obtained from: (1) the movement of surface rocks above the focus, (2) behavior of samples stressed in the laboratory at high pressures and temperatures, (3) radiation patterns of seismic waves, and (4) new insights into precursor events. Study of these phenomena with special emphasis on precursor events has produced a dramatic breakthrough in our knowledge of earthquakes and has led to a new **dilatancy-diffusion model** that provides some hope in the establishment of earthquake predictions.

However, current thinking still associates the majority of earthquakes as the aftermath of faulting. Although the cause of faults can be ascribed to various tectonic events, those that are especially damaging to humans occur primarily along convergent (subducting) and strike-slip plate boundaries within those parts of the continents undergoing intraplate deformation. A typical example of a great subduction-zone earthquake was the Alaskan quake of 1964. In this area the Pacific lithosphere is moving northwest, tips down at the Aleutian Trench, and slides with gentle inclination beneath a wedge, similar to Figure 10-8. This drags the base of the wedge and results in sudden faulting that breaks violently through to the surface.

HUMAN CAUSES OF EARTHQUAKES

Our ability to initiate earthquakes has dramatically accelerated during the twentieth century. We now have the potential to induce earthquakes in a variety of ways—some of which are inadvertant by-products of certain activities and others which are deliberate and planned. Some of the quakes have been very damaging, whereas beneficial results can occur when care is exercised to control the size and location of the quake.

Dams and Reservoirs

The connection between artificial reservoirs and earthquakes was first discovered in Greece in 1931. Since then more than 40 cases have been substantiated where reservoirs have triggered quakes. Only 0.3 percent of the world's 11,000 large dams (higher than 10 m) have produced seismicity. However, for reservoirs deeper than 90 m, 10 percent have had related seismic events, and 21 percent of reservoirs deeper than 140 m have had significant earthquakes. The shocks range from microseisms to those of 6.4 M.

The Marathon Dam in Greece first impounded water in 1929 and by 1931, when the reservoir reached its highest level, earthquakes were noted. From 1931 to 1966 strongest seismicity was always associated with periods of rapid rise in water level.

In the United States, Lake Mead, the Hoover Dam reservoir, started to fill in 1935, and the first tremors in what had previously been an aseismic area occurred in 1936. In 1937 a seismic network was installed, and about 100 earthquakes were recorded. The largest event occurred in May 1939 of 5.0 M, but activity in the years after 1942 decreased. Also, in India the filling of the Koyna Reservoir in 1962 caused tremors in an area previously mapped as aseismic. Five important quakes were felt prior to December 10, 1967, and on that date a 6.3 M event killed about 200 people, injured more than 1500, and left thousands homeless. Bombay, 230 km from the epicenter, was severely shaken, and the shutdown of the hydroelectric plant paralyzed industry. Additional quakes occurred through 1973 and their timing showed strong correlation with water levels in the reservoir. Other reservoirs that have produced significant quakes include: Monteynard, France; Grandvale, France; Mangla, Pakistan; Contra, Switzerland; Kariba, Zambia; Kremasta, Greece; Manic, Canada; Hendrick Verwoerd, South Africa; Nourek, U.S.S.R.; Kurobe, Japan; Kamafusa, Japan; Hsinfengkiang, China; and Camarillas, Spain. Although the principal cause for the Vaiont, Italy, disaster is still being debated, most researchers ascribe to multiple causation and some feel the reservoir filling was largely responsible. The facts speak for themselves. The filling started in 1960 and a seismograph installed at the time recorded 250 tremors during the 1960–63 period, with epicenters 3 to 4 km from the dam. Furthermore, tremor frequency could be closely correlated with water levels. Large bursts of activity especially followed the three greatest rises in water level, with tremors subsiding afterwards.

Water Injection

We have also caused earthquakes in some instances when we have pumped fluids, either water or contaminants, into bedrock. At the Rocky Mountain Arsenal of the U.S. Army, chemical warfare products were originally disposed in reservoir ponds in the Denver region. By 1961, 19 years after the original imponding, groundwaters had become contaminated, so a new system was started in March 1962. This disposal method consisted of pumping the toxic wastes into a 3671 m well that was drilled into weathered schist and fractured granites and gneisses of Precambrian age. In April 1962, Denver, Colorado, felt the first earthquakes in what had formerly been a quiescent area. The injection program was stopped in September 1965 when proof established a relationship between injection and quakes; shocks were still felt as late as 1969 when there were two 3.5 M events and fourteen 2.5 M events. In all, more than 1500 tremors occurred during the 1962–67 period, ranging from 0.7 to 4.3 M with epicenters that were mostly within an 8 km radius of the well.

Water flooding of oil-bearing strata by injection wells has become an important method for the recovery of additional petroleum. At Baldwin Hills, California, a new reservoir was constructed in 1951. However, the site is surrounded by oil wells that constitute part of the Inglewood oil field. A pilot water injection experiment was so successful in recovering more oil in 1954 that a major program was inaugurated in 1957 and, by 1963, there were 22 injection wells. Although the first fault from this operation occurred in May 1957, eight additional ones had been activated by 1963. On December 14, 1963, water burst through the foundation and in a matter of hours had emptied the 8 ha, 20 m deep reservoir. The dam failure deluged the communities below, damaged and destroyed 277 homes, killed 5 persons, and caused the total destruction of more than $15 million in property and utility losses. Seismic tremors were not reported as accompanying the fault movements, but a $25 million lawsuit brought by the Los Angeles Department of Water and Power against the oil company was settled out of court for $3,875,000.

The Chevron Company started successful secondary oil recovery by water-flooding methods in 1957 at the Rangely field, Colorado. In 1962 a newly installed seismological station 65 km away immediately pinpointed small earthquakes in the Rangely area. In the fall of 1967, the U.S. Geological Survey put in an array of seismographs in the region and recorded 40 small earthquakes in a 10-day period. In October 1969 the Survey started an experiment with four wells that were periodically injected with water, shut down, and then pumped and backflowed. During the first year of the project 900 earthquakes occurred, with 367 quakes within 1 km from the well bottoms. By 1973, when the water had again been pumped out of the well area, there was nearly an immediate cessation of seismic activity. One-third of the shocks during the four-year experiment were more than 1 M.

Earthquakes have also been caused by water injection at Matsushiro in central Japan, in the Snipe oil field of Canada, and during hydraulic salt mining in deep wells at Dale, New York.

Other Man-Induced Earthquakes

Inadvertant seismic waves are produced during some mining operations. The excavation of rock within the earth causes a readjustment of the stress field in the vicinity of the new hole. When sudden failure occurs, instead of the more normal rock creep, a phenomenon known as "rock burst" causes fractured materials to explode, and this quick release of energy can produce local earthquakes. There is a long record of these bursts from mines in Canada, South Africa, and even deep coal mines in eastern United States.

The detonation of nuclear devices below the ground can produce quakes of sufficient magnitude to be recorded by seismographs thousands of kilometers away. For example, stations in the United States have usually felt explosions from the U.S.S.R. testing program. Some of these reach a size of 7 M. At the Nevada test site a 1.1 megaton nuclear explosion produced thousands of shocks, with a 5.0 M that extended to 7 km depths.

Figure 10-11 Knik Arm, Cooke Inlet, Alaska. This area near Anchorage was affected by the Alaska Earthquake of March 27, 1964. Note landslide in foreground and on the opposite side of the inlet—all triggered by the earthquake. (Courtesy Wallace Hansen, U.S. Geol. Survey.)

Seismic activity is produced by humans in a variety of other ways. For example, heavy truck traffic on highways produce tremors that can be recorded by nearby seismographs. Most blasting operations, when building roads, tunnels, and other excavations as in mining, also cause local seismic activity. And we have learned to put our technology to work during seismic exploration to uncover mineral and petroleum resources. The deliberate detonation of small dynamite charges along a geophysical array of instruments can provide useful information about the character and thickness of different rock units in the earth's crust. Such data can then be interpreted in light of rock composition and structure, so that subsurface maps can be produced of visually unseen rocks. This type of remote sensing has proved to be the most important method for oil discovery in rocks that may be some depths below ground surface.

EARTHQUAKE EFFECTS

Earthquakes produce numerous changes on the land (Fig. 10-11); cause destruction to property and death and injury to people; and induce many side effects to people, the ecology, and the environment. Many of the results attributed to earthquakes are not directly due to the quake itself—which is the vibration of the earth—but to the after effects. Thus earthquake damage and effects can be classed as primary, when the vibration causes the destruction, or secondary, when the shaking initiates other processes such as landslides and tsunamis. Earthquakes are manifested by the ground motion from the seismic waves. Concomitant with this motion is the lurching and tilting of segments of the crust, uplift or subsidence, offset along faults, and development of new fractures in rock masses. These phenomena can then trigger other processes and events that are invariably more damaging than the initial shocks. It should be made clear, however, that many misconceptions exist about earthquakes that are not true. For example, the cracks associated with the quake do not open and swallow individuals, houses, or entire communities. It *is* true that small opening may occur in some fissures and that during certain types of landslides, such as liquefaction flows, people and houses may be engulfed within the deformed debris.

Unlike other natural hazards such as floods and landslides, earthquakes themselves would seldom be lethal except for their catastrophic side effects and the structures that people build which crumble from the shaking or after effects. Earthquakes result in physical and psychological injury; loss of life; destruction of property; economic disruption and indirect losses; and ecological changes and damage. Most deaths and injuries are caused by collapsing structures and falling debris, such as bricks and glass; inundation of communities by tsunamis; overwhelming of communities by landslides, mudflows, and avalanches; floods from collapsed dams and levees; fires and smoke; and the release of toxic and chemically reactive contaminants. Much of the property loss is due to the rupturing of lifelines such as public utilities.

Several factors determine the actual loss of life and property damage. Such casualties depend on the earthquake magnitude, the time of occurrence, the stability of geologic foundations, the vulnerability of human structures, and the population density. For example, if the Alaska, Good Friday earthquake of March 27, 1964, had occurred during midday on Thursday instead of at 5:36 P.M., the loss of life would have been many times the 114 that were killed. The late hour and holiday time meant that schools were out, offices empty, and many had left the urban area because of the holiday eve. Furthermore, this is an off-season for fishing, so that few boats were in the water and the canneries were closed. The mild weather also assured that many people would be outdoors.

The principal factors that influence the amount and type of damage to man-made structures include: the strength and intensity of the earthquake waves reaching the ground; the duration of the earthquake motion, including the acceleration and aftershocks; the design and type of construction and materials; and the geology of the foundation. The quake damage to construction usually provides more information about the type of structure than it does about the earthquake.

Damage generally occurs because earthquakes exert horizontal stresses against structural elements that most often are built to withstand only vertical loads. Few structures are homogeneous, so differential stress is exerted between the dissimilar materials (Fig. 10-6). The ultrarigid and weak elements are fractured and tear loose from their linkages, and thus the entire structure may collapse because of a few vulnerable parts. Wood and steel frame structures have a good safety record, whereas unreinforced masonry, because it is inflexible, produces much damage. Reinforced concrete is excellent because it combines the compressional strength of concrete with the tensional strength of steel. The character of the geologic foundation is a vital concern to structural integrity because unconsolidated materials such as alluvium, fill, sand, and clay transmit a greater amount of seismic motion than bedrock.

Ecological changes can be as varied as the terrestrial and aquatic environments affected. The New Madrid, Missouri, earthquakes of 1811–12 changed the level of land as much as 6 m over thousands of kilometers. It drained swamps, altered the course of the Mississippi River, and created new lakes, such as Reelfoot Lake, Tennessee. The 1964 Alaska quake raised intertidal areas far above the reach of the highest tides, resulting in nearly immediate destruction of all nearshore aquatic life. Surges of water scoured out beds of clams and deposited mud and debris that suffocated underlying life in other areas. Estuaries were uplifted or submerged, destroying the nesting grounds for ducks and geese. Elevation changes also ruined many salmon-spawning environments, and submarine pressure changes in the water killed numerous fish and other organisms.

Bizarre Phenomena and the Human Psyche

Many sightings have been made of strange lights, now called "earthquake lights," associated with some earthquakes (Fig. 10-12). The first published investigation into the phenomenon was of the Idu Peninsula earthquake in Japan

Figure 10-12   Photographs of earthquake lights taken by T. Kuribayashi during the Matsushiro, Japan, earthquake swarm which lasted from 1965 to 1967. (U.S. Geol. Survey.)

on November 26, 1930. The lights, which appeared similar but longer lasting than sheet lightning, also produced auroral type streams, beams and columns, fireballs, and a ruddy glow that were most conspicuous during the middle shock period, but some of these phenomena also occurred before and after the main shocks. Similar lights have also been observed in connection with the Santa Rosa, California, quake of October 1, 1969; the Hebgen, Montana, earthquake of August 17, 1959; and the great Chinese earthquake of July 28, 1976. A current explanation of the phenomenon attributes the lights to a "piezoelectric effect" of quartz-bearing rocks. The stress changes are believed to produce a high electric potential difference in the rocks that is released by passage beyond a critical value into the atmospheric gases.

Earth sounds are sometimes heard before and during earthquakes. These noises are quite distinct from those associated with structural damages and have been described as similar to low rumbles or roars, thunder, heavy traffic, isolated gunshots, artillery, or tearing of cloth.

Such sounds have been recorded moments, minutes, hours, and even days before the occurrence of the earthquake as well as during the quake. Other effects, some of which are detected only by instruments, are changes in atmospheric pressure, including waves or blasts, and disturbances in the earth's magnetic field.

Earthquakes may injure people in ways apart from the physical damage they inflict. It is not unusual for people to become dizzy and nauseated or to vomit during severe tremors. The psychological impact can be great, even when there is little damage to life and property. Earthquakes are a rather unique geologic hazard because they do not arrive from some place else, as do floods, landslides, and volcanic materials. Instead, they surround the inhabitant and distort the human feeling of solidarity with the earth. Severe psychological distress persisted long after the event in uninjured children and many adults who experienced the San Fernando earthquake in 1971. Since human response is not directly related to the intensity of the earthquake, potential damage to the human mind is not predictable. Of course, earthquakes can also greatly affect animals, and many different species exhibit unusual and uncharacteristic behavior, as described later in this chapter.

EARTHQUAKE DISASTERS

Although earthquake disasters are less frequent than those caused by flooding, they account for more human deaths but less property damage. Earthquakes have probably killed about 100 million people throughout history, and the twentieth-century annual toll is about 15,000. Table 10-2 provides a partial listing of some of the major earthquake catastrophes, and Tables 10-3 and 10-4 indicate damages from some earthquakes in the United States. Included within these listings of deaths and damages are figures for all losses, regardless of whether they were caused by vibratory motions or secondary effects induced because of the disturbance. The following accounts of several earthquakes were selected because of the special circumstances associated with the disaster; several spectacular landslides that were triggered by the shocks are also discussed.

New Madrid, Missouri

The center for a series of shocks that started December 16, 1811, was the village of New Madrid. The earthquakes continued intermittently for two days, and then even larger shocks occurred on January 23, 1812, and again on February 7. New Madrid was completely destroyed and the quake was so severe that it was felt over a larger area, more than 5 million square kilometers, than any other in North American records—from Alabama to Canada, and from the Atlantic coast to the Rocky Mountains. Chimneys were toppled in Cincinnati and Richmond, the Mississippi River course was changed in several places, and new topography was created over thousands of square kilome-

Table 10-2 Major earthquake disasters throughout the world

Year	Locality	Deaths
856	Greece, Corinth	45,000
1038	China, Shansi	23,000
1057	China, Chihli	25,000
1268	Asia Minor, Silicia	60,000
1290	China, Chihli	100,000
1293	Japan, Kamakura	30,000
1531	Portugal, Lisbon	30,000
1556	China, Shensi	830,000
1667	Caucasia, Shemaka	80,000
1693	Italy, Catania	60,000
1737	India, Calcutta	300,000
1755	Northern Persia	40,000
1755	Portugal, Lisbon	60,000
1759	Lebanon, Baalbek	30,000
1783	Italy, Calabria	50,000
1797	Ecuador, Quito	41,000
1811	U.S., New Madrid, Mo.	Several
1819	India, Cutch	1,543
1822	Asia Minor, Aleppo	22,000
1828	Japan, Echigo (Honshu)	30,000
1868	Peru and Ecuador	25,000
1875	Venezuela and Colombia	16,000
1886	U.S., Charleston, S.C.	60
1896	Japan, Sea Wave, Sanriku Coast	22,000
1897	India, Assam	1,542
1905	India, Kangra	20,000
1906	U.S., San Francisco, Calif.	700
1906	Chile, Valparaiso	1,500
1908	Italy, Messina	75,000
1915	Italy, Avezzano	29,970
1920	China, Kansu	180,000
1923	Japan, Tokyo-Yokohama	143,000
1932	China, Kansu	70,000
1935	Pakistan, Quetta	60,000
1939	Chile, Chillan	30,000
1939	Turkey, Erzincan	23,000
1946	Eastern Turkey	1,300
1946	Japan, Honshu	2,000
1948	Japan, Fukui	5,131
1949	Ecuador, Pelileo	6,000
1950	India, Assam	1,500
1953	Northwestern Turkey	1,200
1954	Algeria, Orleansville	1,657
1956	Northern Afghanistan	2,000
1957	Northern Iran	2,500
1957	Outer Mongolia	1,200
1957	Western Iran	2,000
1960	Morocco, Agadir	12,000
1960	Southern Chile	5,700
1962	Northwestern Iran	10,000
1963	Yugoslavia, Skopje	1,100

Table 10-2 *Continued*

Year	Locality	Deaths
1964	Southern Alaska	131
1965	Chile, El Cobre	400
1966	Eastern Turkey	2,529
1967	Venezuela, Caracas	236
1968	Northeastern Iran	11,588
1970	Western Turkey	1,086
1970	Northern Peru	66,794

Source: Office of Emergency Preparedness, 1972, Disaster preparedness: U.S. Government Printing Office.

Table 10-3 Lives lost in major U.S. earthquakes

Year	Locality	Lives Lost
1811	New Madrid, Mo.	Several
1812	New Madrid, Mo.	Several
1812	San Juan Capistrano, Calif.	40
1868	Hayward, Calif.	30
1872	Owens Valley, Calif.	27
1886	Charleston, S.C.	60
1899	San Jacinto, Calif.	6
1906	San Francisco, Calif.	700
1915	Imperial Valley, Calif.	6
1918	Puerto Rico (tsunami from earthquake in Mona Passage)	116
1925	Santa Barbara, Calif.	13
1926	Santa Barbara, Calif.	1
1932	Humboldt County, Calif.	1
1933	Long Beach, Calif.	115
1934	Kosmo, Utah	2
1935	Helena, Mont.	4
1940	Imperial Valley, Calif.	9
1946	Hawaii (tsunami from earthquake in Aleutians)	173
1949	Puget Sound, Wash.	8
1952	Kern County, Calif.	14
1954	Eureka-Arcata, Calif.	1
1955	Oakland, Calif.	1
1958	Khantaak Island and Lituya Bay, Alaska	5
1959	Hebgen Lake, Mont.	28
1960	Hilo, Hawaii (tsunami from earthquake off Chile coast)	61
1964	Prince William Sound, Alaska (tsunami)	131
1965	Puget Sound, Wash.	7
1971	San Fernando, Calif.	65

Source: Office of Emergency Preparedness, 1972, Disaster preparedness: U.S. Government Printing Office.

ters. Along one stretch of the Mississippi River for a distance of 56 km, numerous landslides created depressions 30 m deep and 30 m wide. Only a few lives were lost because of the sparsely populated area.

San Francisco, California

This was the first major earthquake of modern times that made Americans aware of nature's awesome power, and it has been a continual reminder of how fragile the California landscape can be. The surprising thing about the April 18, 1906, quake, which occurred at 5:12 A.M., was that the people of the area had learned little from previous severe quakes that had also greatly damaged the city in 1836, 1838, and 1865. All quakes were associated with the San Andreas Fault system, which can be traced nearly 1000 km in California (Figs. 10-13 and 10-14). The earthquake resulted from largely strike-slip motion along the fault that reached a maximum displacement of 6.5 m. The fault moved the ocean-side block northwestward throughout a 560 km distance. The affected area was about 112 km wide, and the quake was felt in a region of about 1 million km². The principal shocks lasted about 1 minute, with the greatest tremor occurring 40 seconds after the start. Total deaths numbered at least 700, with 315 in San Francisco. About 85 percent of the $400 million damage was caused by the ensuing fires (Fig. 10-15). Inadequate wiring systems were easily broken, causing electrical shorts at numerous sites. The resulting fires could not be controlled because of ground failures and breakage of the main water lines. Much of San Francisco was built on loose sands and clays and poorly compacted fill. Such areas were severely shaken, with displacements of more than 2 m both horizontally and vertically. Liquefaction flows also added to the damage (Figs. 10-16 and 10-17), and sand boils spurted sand geysers to heights of 6 m. Thus, the major oscillatory damage occurred in the soft sediments that

Table 10-4 Property damage in major U.S. earthquakes (in millions of dollars)

Year	Locality	Damage
1865	San Francisco, Calif.	0.5
1868	San Francisco, Calif.	0.4
1872	Owens Valley, Calif.	0.3
1886	Charleston, S.C.	23.0
1892	Vacaville, Calif.	0.2
1898	Mare Island, Calif.	1.4
1906	San Francisco, Calif.	24.0
	Fire loss	500.0
1915	Imperial Valley, Calif.	0.9
1918	Puerto Rico (tsunami damage from earthquake in Mona Passage)	4.0
1918	San Jacinto and Hemet, Calif.	0.2
1925	Santa Barbara, Calif.	8.0
1933	Long Beach, Calif.	40.0
1935	Helena, Mont.	4.0
1940	Imperial Valley, Calif.	6.0
1941	Santa Barbara, Calif.	0.1
1941	Torrance-Gardena, Calif.	1.0
1944	Cornwall, Canada-Massena, N.Y.	2.0
1946	Hawaii (tsunami damage from earthquake in Aleutians)	25.0
1949	Puget Sound, Wash.	25.0
1949	Terminal Island, Calif. (oil wells only)	9.0
1951	Terminal Island, Calif. (oil wells only)	3.0
1952	Kern County, Calif.	60.0
1954	Eureka-Arcata, Calif.	2.1
1954	Wilkes-Barre, Pa.	1.0
1955	Terminal Island, Calif. (oil wells only)	3.0
1955	Oakland-Walnut Creek, Calif.	1.0
1957	Hawaii (tsunami damage from earthquake in Aleutians)	3.0
1957	San Francisco, Calif.	1.0
1959	Hebgen Lake, Mont. (damage to timber and roads)	11.0
1960	Hawaii and U.S. West Coast (tsunami damage from earthquake off Chile)	25.5
1961	Terminal Island, Calif. (oil wells only)	4.5
1964	Alaska and U.S. West Coast (tsunami damage from earthquake near Anchorage–includes earthquake damage in Alaska)	500.0
1965	Puget Sound, Wash.	12.5
1966	Dulce, N. Mex.	0.2
1969	Santa Rosa, Calif.	6.3
1971	San Fernando, Calif.	553.0
	Total	1,862.1

Source: Office of Emergency Preparedness, 1972, Disaster preparedness: U.S. Government Printing Office.

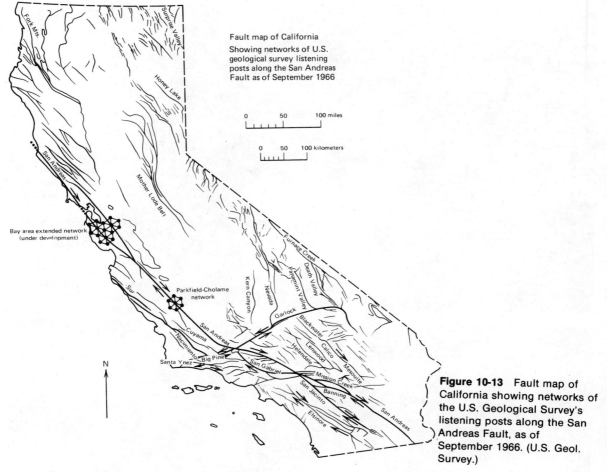

Fault map of California

Showing networks of U.S.
geological survey listening
posts along the San Andreas
Fault as of September 1966

Figure 10-13 Fault map of
California showing networks of
the U.S. Geological Survey's
listening posts along the San
Andreas Fault, as of
September 1966. (U.S. Geol.
Survey.)

were much wetter than usual because the area
had been experiencing an exceptionally rainy
season with more than twice the usual precipitation.

Areas outside of San Francisco and near
the fault corridor experienced much damage to
installations and the offsetting of numerous
structures and even rivers (Fig. 10-14). The
triggering of numerous landslides further
added to the destruction. The largest single
landslide produced a scar 1.6 km wide and 0.9
km long. One of the largest slumps occurred
near San Pablo, with a 450 m width of scarp
that was 15 m high and extended 120 m down-

slope. Many earthflows were set in motion, the
largest being 810 m long, 30 m wide, and 1 m
deep.

Kansu Province, China

December 16, 1920, marks the day of the greatest landslide catastrophe in human history when
in Kansu Province 180,000 people were killed.
Here a 160 by 480 km area was devastated by
loess flows that were dislodged from the hillsides by a massive earthquake. According to
survivors, the silt-size materials cascaded like
water, overwhelming everything in their path.

Figure 10-14   Aerial view of San Andreas Fault in the Tremblor Range, California. Note the offsetting of some stream channels. (U.S. Geol Survey.)

Figure 10-15   Unusual aerial photograph taken of San Francisco after 1906 earthquake. George R. Lawrence suspended a large panoramic camera below a battery of 17 kites, secured to ship in harbor. Camera weighed more than Wright brother's airplane, including pilot, and shot picture at altitude of 650 m. Empty and burned shells of buildings are clearly shown on right side of photograph. (Library of Congress.)

Madison, Montana

Although this earthquake with resulting landslides killed only 28 campers, on August 17, 1959, it is a significant event because it marked the first documented and thoroughly studied rock avalanche attributable to seismic activity. At this locality the rock structure favored inst-

ability, with Paleozoic dolomites (Fig. 10-18) dipping 40° toward the valley on a basement Precambrian complex of gneiss and schist. The weak and weathered rock was suddenly transformed by the earthquake shock into a mass of debris that flowed into the valley with speeds of 180 km/h. The principal slide displaced a volume of $28.3 \times 10^6 \mathrm{m}^3$, moved 1.6 km, and

Figure 10-16 Ground ruptures at Moss Landing, between Monterey Bay and the former Salinas River. This deformation was caused by shaking during the San Francisco earthquake of 1906 which produced liquefaction of the substrate at this locality. (U.S. Geol. Survey photo.)

Figure 10-17 Differential slumping in artificial fill on Union Street between Pierce and Steiner Streets, San Francisco, California. The man-made fill areas were particularly damaged by shaking during the 1906 earthquake. (Courtesy U.S. Geol. Survey.)

Figure 10-18 This Madison dolomite block (3000 tons) was moved by the rock avalanche that resulted from the Madison earthquake of August 17, 1959. The force of the landsliding motion transported the rock to a height of 30 m above the valley on the opposite side of the valley where the slide had originated.

Anchorage, Alaska

The Alaska (Good Friday) earthquake of March 27, 1964, produced the greatest damage in Anchorage (Fig. 10-21), although it did significant damage throughout a 129,000 km² region and was felt in an area 10 times as large. It measured 8.5 M and is the largest recorded earthquake in U.S. history. The early shocks lasted 1.5 to 7 minutes, and for a 69-day period more than 12,000 aftershocks of 3.5 M were recorded. Alaska property damage exceeded

blocked the river, forming Hebgen Lake (Figs. 10-19 and 10-20). In addition to the main slide, other landslides up to 267,000 m³ and rockfalls were common. The quake also initiated an earthflow 120 to 240 m wide and 0.9 km long that moved a distance of 30 m.

Figure 10-19 This landslide scar was caused by the Madison earthquake. Newly formed Hebgen Lake is in the foreground. (Courtesy Robert E. Wallace, U.S. Geol. Survey.)

Figure 10-20 Slump along Hebgen Lake, Montana. This resulted from a seiche in the water which initiated many local landslides at water level. The Madison earthquake and fault downdropped the lake at this locality and destroyed the road. Local damage amounted to $11 million. (U.S. Geol. Survey photo.)

$500 million, and 114 were killed of whom only 9 died from effects of shaking. The greatest single area of damage was in the Turnagain section of Anchorage, where 280 ha with 750 homes were destroyed, or 14 percent of the city. This damage was caused by the earthquake triggering liquefaction in the Bootlegger Cove Clay, and failure occurred by flowage and slab sliding (see also Fig. 11-8). The quake also set off the Sherman rock avalanche, initiated numerous subaqueous slides—some of which may be the largest slides on the planet (see Table 11-2), and led to extensive damages in coastal areas from the sliding and the tsunami that followed.

Figure 10-21 Government Hill School, Anchorage, Alaska. The 1965 Alaska earthquake damaged this grade school, but the building maintained much of its structural integrity. The damage resulted from the downdropped section caused by the landsliding. (Courtesy Wallace Hansen, U.S. Geol. Survey.)

Figure 10-22 Statue of Jesus Christ at Cemetery Hill overlooking Yungay, Peru. The statue together with four palm trees and parts of a burial crypt are all that remained of the village after the May 31, 1970 earthquake and its resultant landsliding and mudflows. (Courtesy U.S. Geol. Survey.)

Tsunami waves traveled as far south as coastal areas in the Pacific Northwest states and Mexico, and one 15.6 m wave destroyed much of the harbor area of Whittier, Alaska.

The source of the quake occurred about 30 km deep under Prince William Sound, and the fault zone extended 800 km, roughly parallel to the Aleutian Trench. The oceanic plate in this region is underthrusting the continental plate that includes Alaska. The earthquake probably changed the terrain as much as nearly any earthquake of recorded time. On the land east of the fault 129,000 km² were elevated to heights of 10 m, whereas on the west side the land was lowered as much as 2 m. An additional 284,000 km² of sea floor was warped and deformed 15 m.

Peru

A major catastrophe occurred throughout the Peruvian Andes on May 31, 1970, from a 7.7 M earthquake with its epicenter in the Pacific Ocean and 85 km distant from the destruction sites. Nearly all the 70,000 deaths were due to the secondary effects of landslides, floods, and building collapse. The greatest concentration of death and destruction from this tragedy was in the towns of Ranrahirca and Yungay, which were totally demolished by the rock avalanches and mudflows that killed the 21,000 inhabitants (Fig. 10-22). The extraordinarily strong shaking dislodged rocks and glacial ice near the top of Mt. Huascaran, a 6600 m elevation, that had a freefall of 600 m and then crashed into morainic materials below. On impact the snow and ice partly melted and, along with debris incorporated from the unconsolidated sediments, the entire mass rapidly flowed 2700 m on a 23° slope. At this point the landslide hit the Llanganuco River and continued its fury at super speeds in the form of a giant rock avalanche-mudflow. Total distance moved was 14.5 km in 3 minutes, an average speed of 400 km/h! No wonder that the two cities in its path were completely obliterated. Part of the movement seems to have been airborne, because in places the rock avalanche destroyed vegetation at

Figure 10-23 Destruction of adobe houses in Huaras, Peru, resulting from ground shaking during the 1970 Peru earthquake. Most deaths were in the older section of the city where the collapse of two- and three-story adobe buildings killed both occupants and people who fled into the narrow streets attempting to find safety. In contrast, the brick buildings with reinforced concrete in the new section of the city were only partly damaged. (Courtesy U.S. Geol. Survey.)

heights of 21 m above the valley, and yet at Yungay a few trees were totally untouched. In addition, the quake produced thousands of other rockslides within a 100 km area of Chimbote, Peru, and extensive damage occurred to buildings because of the common use of inflexible mortar (Fig. 10-23). A total of 200,000 buildings were destroyed and 800,000 people were left homeless.

San Fernando, California

This earthquake provides a very significant case study because it showed what a moderate-size 6.5 M quake can do to a modern, urbanized area. It has become the most studied earthquake in history, and the lessons learned are continuing to provide guidance and legislation regarding construction and settlement in earthquake-prone terrane. Although a destructive earthquake had hit the Long Beach, California, area on March 10, 1933, it had been thought that the locality of San Fernando was more immune to such seismic activity. There were not immediate or recognizable foreshocks, but on February 9, 1971, at 6:01 A.M. the earthquake struck with strong motion that lasted about 12 seconds.

One highly unusual aspect was the severity of the acceleration, which was 0.7 g on the vertical component and 1.25 g horizontally—both figures that far exceeded other recorded events in California's history. The epicenter was in the San Gabriel Mountains, about 8 km north of the San Fernando Valley, which is the northwest section of the Los Angeles area. Movement occurred along a 45° dipping reverse fault and caused surface ruptures throughout the region. Along the trace of the fault all structures were broken, but 99 percent of the damage was the result of heavy shaking and side effects far removed from the fault plane. Fortunately the earliness of the hour of faulting greatly reduced the toll in lives. However, 47 of the 65 deaths were patients of the Veterans Hospital, which was not an earthquake-resistant structure; the patients were unable to move swiftly enough to protect themselves. Some damage occurred to even the most strongly constructed buildings (Fig. 10-24), but most of the damage was to other buildings with construction that did not comply with up-to-date building codes, homes

Figure 10-24 Olive View Hospital, San Fernando, California, was severely damaged by the 1971 earthquake shocks. Although part of the structure was unscathed, other parts were destroyed and some sections completely toppled over. (Courtesy Robert E. Wallace, U.S. Geol. Survey.)

Figure 10-25 Collapsed overpass connecting Foothill Boulevard and Golden State Freeway, caused by the 1971 San Fernando earthquake. (Courtesy U.S. Geol. Survey.)

and furnishings, roads, and various utilities (Fig. 10-25). The original survey to estimate damages officially announced a total cost amounting to $511 million (Pacific Fire Rating Bureau, 1971), with $271 million in the governmental sector and utilities, and $240 in the

Figure 10-26 View of the Lower Van Norman Dam, San Fernando, California, showing the slump on the north side and the concrete collapse into the reservoir in the wake of the 1971 earthquake. The water level had already been lowered in the city at the time of this photograph, to protect people and property in case of renewed shocks. (Courtesy U.S. Geol. Survey.)

private sector. The largest displacement of earth materials was about 2 m. In addition, more than 1000 landslides in terrain above the San Fernando Valley were triggered, and some sand layers that had undergone liquefaction moved down failure surfaces of only 2.5 percent gradient for distances of 1.6 km.

An interesting sidelight is afforded by the near catastrophe associated with the Van Norman Reservoir. The San Fernando Dam complex consists of three earthen dams that impound water and form the Van Norman Reservoir complex. These constitute about 80 percent of the city's water supply. The lowermost dam experienced the greatest damage, when the earthquake tremors liquefied 610,000 m³ of hydraulic fill on the upstream part of the embankment; a seiche also resulted that aided to erode other parts of the dam. Fortunately the reservoir had been operating at levels 4.5 m lower than normal, and the top of the dam was 10 m above water level. However, the earthquake events left only 1.5 m of freeboard, and

the dam would undoubtedly have failed if the tremors had persisted several seconds longer. Facing an imminent disaster, the 80,000 residents in the valley below were evacuated as quickly as possible and only allowed to return after the water had been drained (Fig. 10-26).

Another by-product of the earthquake was discovered by John Sims, a geologist with the U.S. Geological Survey. He reasoned that the San Fernando earthquake was also sufficiently severe to produce deformation structures in sediments. Thus, when the Van Norman Reservoir was drained, he studied the bottom silts that had been accumulating in the sedimentation trap caused by the impondment. The investigation paid off when it was discovered that, in addition to the contorted structures produced by the 1971 quake (Fig. 10-27), two other quakes produced similar features. On the basis of rainfall records that were used to determine the annual sedimentation rates, it could be calculated that known quakes in 1930 and 1952 had also caused the sediment contortions. Thus,

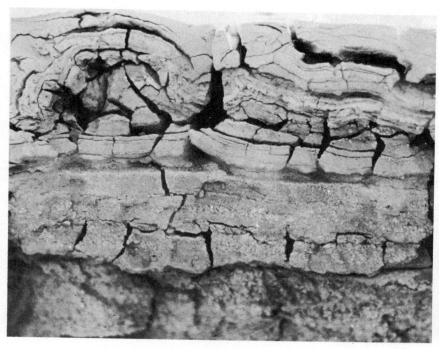

Figure 10-27 Deformation structure in the upper 4 cm of sediment of Lower Van Norman Reservoir, San Fernando, California. This soft-sediment distortion formed as a result of the earthquake shocks which produced liquefaction of a IX Intensity (Mercalli Scale) at this site. (Courtesy U.S. Geol. Survey.)

this technique when carefully applied to othe settling basins can be used to obtain data fo the recurrence intervals of major quakes. Suc information can be a vital part in making lar use decisions.

Other Earthquakes

It is impossible for a single chapter to list or discuss all the important earthquakes that have occurred throughout history or to capture the suffering and pathos they cause. However, no listing would be complete without the mention of the Tangshan, China, earthquake on July 28, 1976, that killed 655,000, injured 779,000, and largely destroyed the industrial city of 1.6 million people. It can be seen in Table 10-2 and Appendix I that earthquake catastrophes are not abating and have caused particularly large losses within the past 10 years. However, throughout human history earthquakes have been an ever-present danger (Tables 10-5 and 10-6).

TSUNAMIS

Although more rare than other ocean-initiated catastrophes, perhaps the most frightening and feared waves of all are those associated with tsunamis (often mistakenly called "tidal waves"). They are not related to tides but instead are mostly caused by submarine earthquakes or occasionally by submarine slides or volcanic activity. There were 181 tsunamis (Japanese for "waves in harbor") recorded in the Pacific Ocean from 1900 to 1970. Of these, 34 caused local damage near the source and nine were destructive both locally and at a distance.

Tsunamis travel fastest in deep water, reaching speeds of 900 km/h at 7200 m depths, with a reduction of speed to 180 km/h in 140 m of water and 50 km/h at 20 m depths. What constitute ocean swells of 1.5 m heights may become 30 m high waves at the shore. In 1896 a tsunami took a toll of 27,000 lives in Japan. The April 1, 1976, tsunami was generated by an earthquake in the Aleutian Trench. It

Table 10-5 Proposed general scheme of description if there is suspected archeoseimic damage

1. Location and size of site
2. Main periods of occupancy
3. Age of damaged structures
4. Nature of excavation works (rescue and salvage operation, preliminary, single-season, continuing, etc.)
5. Mode and mechanism of excavation (equipment employed, amount of overburden removed, program of operations, etc.)
6. Extent of excavated area and number and size of the exposed buildings and structures
7. Type and quality of construction of the damaged buildings and structures (e.g., masonry, stone, adobe, etc.; type of cement, reinforcements, and fundaments)
8. Type of damage (e.g., collapse, oriented collapse, tilting, breakage, subsidence, fractures, and displacement)
9. Extent and distribution of damage across the site (number of damaged elements, changes in amount and intensity of damage, direction of features of damage and of any possible alignment of the fallen components, etc.)
10. Occurrence of similar damage at other contemporary sites
11. Differences between the observed features of damage and those characteristic of man-induced damage
12. Physiographic setting of the site (relief, distance from cliffs and slopes, slope characteristics, distance from watercourses and shores, etc.)
13. Type and composition of the ground (e.g., rock, alluvium, clay; depth to bedrock, etc.)
14. Features of recent ground instability (e.g., slides, creep, rockfalls, desication cracks, erosion gullies and rills, occurrence of karst features)
15. Structural settings of the site (e.g., distance from faults and their orientation, occurrence of joints and their orientation, inclination and structural position of the strata

Source: I. Karcz and U. Kafi, 1978, Evaluation of supposed archaeoseismic damage in Israel: *Journal of Archeological Science.* Copyright © 1978 Academic Press, Inc., London with permission.

plucked the Scotch Cap lighthouse on Unimak Island from its position on a 13.5 m high cliff and thrust it into the sea along with the five inhabitants. In less than five hours the waves reached Hawaii, killing 133 and injuring 163. The May 22, 1960 Concepcion, Chile, earthquake caused a powerful tsunami that rampaged throughout the Pacific, killing more than 1000 people. It traveled the 10,600 km distance to Hawaii at an average speed of 710 km/h and swept many city blocks in Hilo clean of buildings, easily moving 20-ton rocks and killing 61 people. Unusual effects can occur in local areas, such as at Lituya Bay, Alaska, where a landslide from an earthquake caused a 510 m high wave in the immediate vicinity, but was not propagated outside the bay. The tsunami associated with the Alaskan earthquake on Good Friday in 1964 caused $104 million (in 1967 dollars) damages in Alaska and the Pacific Northwest.

Perhaps the most famous historical tsunami was the one generated by an earthquake off the shore of Portugal on November 1, 1755. It devastated much of the city of Lisbon. The 10 m water surge destroyed countless buildings and killed more than 60,000 inhabitants. The tsunami was so powerful and enduring that it caused waves to travel across the Atlantic, hit-

Table 10-6 Damage attributed to earthquakes at archeological sites in Israel

Site	Evidence	Time
Ai	Widespread destruction, collapse of buildings, tiled walls, breakage	27th century B.C.
Avdat	Damaged, subsequently repaired city walls, cracks, damaged buildings	A.D. 5–6th century
Beit Shean	Disturbances and collapse in the northern cemetery	?
Caesarea	Displaced offshore moles and port installations, tiled walls	A.D. 2nd century
Ein Hanaziv	Directed collapse, oriented fallen masonry	A.D. 7th century
Hazor	Tilted columns and walls, collapse of buildings	8th century B.C.
Jericho (Tel el Sultan)	Slides and features of collapse	?
Jericho (Tel Abu Aleik)	Tilted and distorted walls, subsidence, collapse and breakage in buildings and water installations	1st century B.C.
Jericho (Khirbet Mafjar)	Widespread destruction and collapse	A.D. 8th century
Khirbet Qumran	Breakage and displacement in a water reservoir, cracks and damage to buildings	1st century B.C.
Khirbet Maqari	Violent destruction	7–8th century B.C.
Khirbet Shama	Tilted and distorted walls, collapsed buildings	A.D. 4th century
Kyrpos	Features of collapse and sliding, imbricated fallen masonry	?
Massada	Damaged and cracked floors	1st century B.C.
Massada	Tilted walls, oriented fallen masonry, collapse of parts of buildings over the rock cliffs	1st century B.C. and later shocks
Shivta	Collapse and damage, and subsequent repairs in the northern area	A.D. 5–6th century
Susita	Oriented fallen columns, tilted walls and buildings	?
Tel Apheq (Antipatris)	Tilted and distorted walls, collapse, subsided arches	A.D. 5th century
Tel Masos	Collapse and oriented fallen masonry	
Telleilat Ghassul	Breakage and displacement, cracks, tilts	40th century B.C.
Tiberias	Desertion of the southern part of the city	A.D. 11th century

Source: I. Karcz and U. Kafi, 1978, Evaluation of supposed archaeoseismic damage in Israel: *Journal of Archeological Science.* Copyright © 1978 Academic Press, Inc., London with permission.

ting the Azores and the West Indies. Their northward moving components produced water changes in canals and coastal areas of Holland and Sweden. In Scotland, the Firth of Forth rose more than 20 cm in a four-minute period. The earthquake shock was also felt in distant places and even rattled chandeliers in Atlantic coastal cities of the United States.

PREDICTION AND CONTROL OF EARTHQUAKES

Twenty years ago the possibility of predicting earthquake occurrence was only a dream of seismologists, but events within the past 15 years provide much encouragement that under some conditions certain types of earthquakes may be anticipated with moderate accuracy. However, the visionary goal of predicting all earthquakes will continue to be elusive because of the sheer number of possibilities and the anomalous variations in earthquake behavior.

The single event that triggered rejuvenation in the prediction field was the 1969 announcement made in the Soviet Union about the Garm region in Siberia. This locality had been the site of a damaging earthquake in 1946, and thereafter a group of scientists installed seismic instruments in the area to learn more about the seismic history of that desolate region.

After a period of more than two decades of painstaking research, the effort paid off with the observation that a significant change in the velocity ratio of the P and S waves invariably occurred prior to an important earthquake. The usual ratio of about 1.75 would drop as low as 1.6, and after the quake the ratio would again increase to normal. When this news reached American scientists, several reexamined previous seismograms of known earthquakes with encouraging results. For example, the San Fernando earthquake, which was thought to have had no felt foreshocks, did have a seismic history for a change in the P/S ratio of microseisms prior to the major tremor. The theory was put to the test at Blue Mountain Lake, in the Adirondacks of New York. The Lamont-Doherty Observatory geophysical group of Columbia University had installed a network of seismographs in the vicinity in 1971. By 1973 Yash Aggarwal was so confident of the data that on August 1, 1973 he predicted a 2.5 M earthquake "in a couple of days." Sure enough, on August 3, 1973, a 2.5 M event occurred. Since then the system has been used in a number of other prediction attempts. An earthquake near Riverside, California, on January 30, 1974, was predicted to occur in about the general time period it happened, three months in advance. Whereas the prediction called for a 5.5 M event, the actual earthquake measured 4.1 M. A quake at Hollister, California, on November 28, 1974, was also forecast within the time error of prediction, and a 3.2 M earthquake on December 8, 1976, near San Jose, California, was also predicted.

All of these findings have given a renewed impetus to prediction studies and have also provoked new theories to explain the physical behavior of rocks and materials associated with earthquake-producing phenomena. For example, two similar theories are in current vogue: the dilatancy-diffusion theory favored by most American seismologists and the dilatancy-instability theory used by Russian scientists. The basic ingredient to both ideas is the occurrence of dilatancy as a result of rock strain in the earth's crust. During the buildup of stress, tiny fractures develop that produce a volume expansion in the rock mass—**dilatancy,** and this occurs prior to the final shearing along the major fault plane. Rocks that undergo dilatancy create other changes such as those of electrical resistivity, in addition to the velocity ratio differences.

Precursory Events

Because all earthquakes do not exhibit the same precursors, a variety of different types of observation is now being used in an attempt to provide more precise predictions about the timing of a major event. Precursors can be classified under four types: changes in seismicity, physiochemical changes (Fig. 10-28), landform changes, and animal behavior. Furthermore, some precursors are short range whereas others, such as the seismic gap theory, are long-range types of forecasting.

Changes in seismicity

1. Change in velocity ratio of P and S waves. Although this may occur in certain types of earthquakes, it is not a universal phenomenon.

2. Change in the number and frequency of foreshocks or microseisms. Such earthquake swarms do herald upcoming events for some quakes, but again many have an insufficient seismic differential to make firm predictions. Sometimes the earthquake swarms occur at one locality and the major quake and damage occurs some distance away, as happened with the Izo Peninsula, Japan, earthquake of January 15, 1978.

Physiochemical changes

1. Change in electrical resistivity. This factor depends on the amount of water the rock

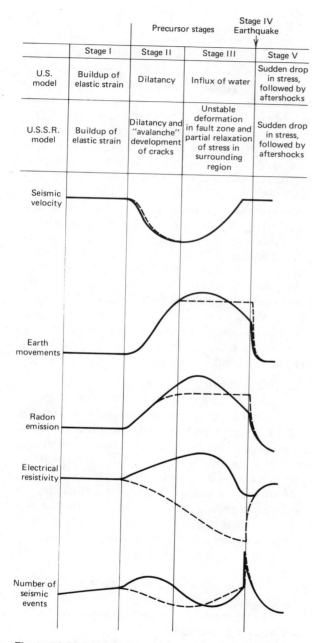

	Stage I	Precursor stages		Stage IV Earthquake	Stage V
		Stage II	Stage III		
U.S. model	Buildup of elastic strain	Dilatancy	Influx of water		Sudden drop in stress, followed by aftershocks
U.S.S.R. model	Buildup of elastic strain	Dilatancy and "avalanche" development of cracks	Unstable deformation in fault zone and partial relaxation of stress in surrounding region		Sudden drop in stress, followed by aftershocks
Seismic velocity					
Earth movements					
Radon emission					
Electrical resistivity					
Number of seismic events					

Figure 10-28 Comparison of U.S. and U.S.S.R. models for precursor stages of earthquake activity. The curves are drawn diagrammatically to show either increasing or decreasing measurement of the phenomena. (After Frank Press, "Earthquake Prediction", 1975. Copyright © 1975 by Scientific American. All rights reserved.)

contains. When the rocks are saturated, there is a reduction in resistivity (or an increase in conductivity) during dilatancy. Because the opening of small fractures allows more water to penetrate the disturbed area, a decrease in electrical resistivity should mark the coming of a seismic event. This was the case with earthquakes occurring in the Garm region.

2. Changes in water levels in wells. It has been known for more than 30 years that earthquakes can cause fluctuations in water wells. An observation well of the U.S. Geological Survey reached amplitudes of 4 m in Milwaukee, Wisconsin, after the August 1950 Assam earthquake. Water levels in several Florida wells fluctuated sharply as a result of the 1964 Alaska earthquake. There is now evidence that some wells respond to water level changes at the site *during* an earthquake. This happened during the Izo-Hanta-Oki, Japan, earthquake of May 1974 and the October 1968 Perth, Australia, earthquake. More recently water level changes occurred prior to the Haicheng, China, earthquake on February 4, 1975, and such observations played an important role in prediction of the quake. In California water changes preceded one quake by a few hours.

3. Emission of radon gas. The release of this inert gas from deep wells or even from fault zones in seismic areas has been noted in several instances, such as at the Tashkent 1966 earthquake in the Soviet Union. Here radon content in a deep well increased to twice its normal value several years prior to the quake; after the event it returned to its normal value. The data are readily explained in terms of dilatancy. The radon method, however, is not widely applicable because of the absence of records for normal amounts, the difficulty in continual monitoring, and the infrequent number of

sufficiently deep wells in many earthquake localities.

Landform changes Probably the most commonly reported precursory features are anomalous terrain changes. The most thoroughly documented and long-range example concerns the 1964 Niigata, Japan, earthquake that measured 7.5 M. Precise measurements for the area started in the late nineteenth century and revealed a slow steady rate of uplift from 1898 to 1955. In 1958 a higher rate of uplift occurred, but thereafter little movement until the earthquake. These various phases fit the model for changes as shown in Figure 10-28. Tiltmeters have been used to record several precursory changes at several sites. At Odaigahara, Japan, where an earthquake 6 M occurred in 1960, tiltmeters indicated uplift up to 100 km from the epicenter which increased in rate six months prior to the quake. One month before the event several stations recorded tilting toward the epicenter. This reversal might be viewed as a short-term precursor due to slight compaction of the dilatant volume. The Danville, California, 4.5 M earthquake of 1971 also had precursory tilting that preceded the quake by one month, with a reversal in tilt toward the epicenter 10 hours before the event.

The most provocative and perplexing uplift in an earthquake-prone terrane was first discovered in 1976 by a U.S. Geological Survey scientist near Palmdale, California (Fig. 10-29). This locality is 64 km from Los Angeles and lies astride the San Andreas Fault zone. Since its discovery, one-third of the total national effort in earthquake prediction has been focused in Southern California, and the Palmdale bulge covers an 84,000 km² area. The regional uplift began commencing about May 1959, and by mid-1961 the area had risen abruptly as much as 25 cm, gradually increasing another 10 cm in some places during the following decade. Between late 1972 and early 1974 the uplift had expanded to the southeast, where a maximum elevation increase of 45 cm occurred near Yucca Valley. Subsequently from 1974 to 1976, much of the uplifted area subsided, with the region just north and east of Los Angeles being 10 cm lower than its 1955 elevation, whereas the area around Palmdale is about 20 cm higher. This unusual behavior poses difficult interpretation problems, because different parts of the area change elevations differentially with erratic movement rates. Study of unconsolidated sediments 24 km from Palmdale showed previous earthquakes occurred during the years of 575, 655, 860, 965, 1190, 1245, 1470, and 1745. However, the question remains, since earth surface changes have presaged several important earthquakes throughout the world and the Palmdale bulge is situated along a known major fault system, will a devastating earthquake occur, and if so when? It should also be noted that between 1897 and 1914 an uplift of the Mojave block along nearly the same trend caused a 30 cm displacement, which then collapsed prior to a releveling survey in 1926. Such changes *did not* result in an earthquake.

Animal behavior It has now become amply documented that many different animals exhibit unusual behavior prior to some earthquakes. The Chinese have especially developed the art of animal watching for telltale signs of erratic movements that might signal an impending earthquake. Thousands of people in all major cities have been trained to monitor the behavior of animals and to report to central headquarters any unusual signs. Part of a six-page 1973 booklet issued by the Seismological Office of Tientsin, China, carries this verse:

Animals are aware of precursors before earthquakes;
Let us summarize their anomalous behavior for prediction.
Cattle, sheep, mules, and horses do not enter corrals,
Rats move their homes and flee.
Hibernating snakes leave their burrows early,

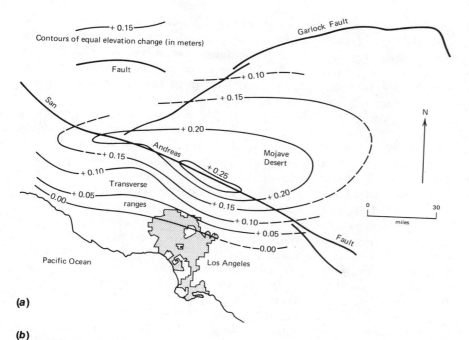

(a)

(b)

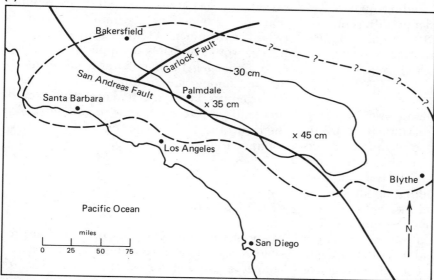

Figure 10-29 Maps of the Palmdale region, California. *(a)* Original map drawn after the "Palmdale bulge" was first discovered in 1976. *(b)* Conditions in the area in 1978. The dashed line shows the approximate position where there was zero elevation change between 1959 and 1974. Solid contours enclose the area, uplifted more than 30 cm, and "x" shows the maximum heights attained. (Courtesy U.S. Geol. Survey.)

Frightened pigeons continuously fly and do not return to nests.

Rabbits raise their ears, jump aimlessly and bump things,

Fish are frightened, jump above the water surface.

Every family and every household joins in observation,

The people's war against earthquakes must be won.

Although animals are not the only precursory items that are monitored, their erratic behavior—along with other signals—provided China an unusual prediction year in 1976, when three earthquakes were successfully forecast. These earthquakes were in the Yunnan province on May 29, in the Szechuan province on August 16, and in the Szechuan-Yunnan border region on November 7. All the quakes were about 7 M. However, the crowning achievement had been reached with the extensively documented prediction of the Haicheng earthquake of February 4, 1975 that measured 7.4 M. The city was totally destroyed but very few people were killed, even though about a million lived near the epicenter, because the population had been warned and was evacuated. Snakes were some of the earliest oddities when in mid-December normally hibernating snakes left their holes to slither to the ground surface where they froze on the ice. However, animals are not infallible indicators of impending quakes, because they didn't act sufficiently different, nor were other precursors used to predict the disastrous Tangshan 1976 earthquake.

Earthquake Control

Our ultimate victory over nature would be the complete control of those earthquake hazards that jeopardize life and property. In limited and small-scale activities we have already exerted some control over earthquakes—we can produce them by water flooding operations or postpone them by dewatering the substrate. However, it is not the small magnitude and intensity quakes that pose problems; it is the *major* ones that reach 6 M and more that provide the catastrophes. One method of earthquake control would be the deliberate initiating of small events in order to prevent the continued accumulation of stress which would ultimately cause a single major event. Another approach is the dewatering by extensive pumping operations those fault zones that are saturated and permeable. By removing the lubricating medium—water—there would be greater frictional resistance of the rocks, thus slippage along the fault plane might be prevented. Others have suggested a series of nuclear detonations along the proved hazardous faults, using the theory that the intense microfracturing would occur to sites that could absorb stress buildup, relieving the zone of a single long-distance displacement. Some combination of these three strategies has been advocated to provide internal crustal barriers for earthquake prevention.

There are grave consequences, both in human terms and in legal uncertainties, that face any operation that attempts control methods. Many earthquakes are still fickle and unpredictable, and there probably is no universal cure that is applicable to all. It is possible that one particular area might be controlled in part, but stress buildup might occur at an unsuspected locality where seismic activity might be enhanced. Thus prevention at one spot might cause a disaster at another. The legal ramifications of such actions are immense, and the typical environmental laws of trespass and negligence would probably be applicable (see page 673). Other legal and political issues would be debated among cities, counties, and states. Until the scientists know much more than they do now, and until technology is further advanced, the control of earthquakes on a large scale will continue to be only a dream.

LAND USE PLANNING

Many of the same planning tools that are feasible in dealing with earthquake hazards are equally applicable for other geologic hazards. The basic ingredients in such programs consist of an educational program to increase public awareness and perception of the potential dangers for occupance of disaster-prone areas; sufficient funding to meet the objectives of any management scheme; as precise data as possible for the reports and maps that describe and delineate dangerous areas; and a strong political and legal framework that not only enlists the cooperation of interested parties, but mandates appropriate measures related to zoning and building codes.

In the United States there has been incredible lethargy in activating programs and laws that deal effectively with earthquake hazards. Even the San Francisco 1906 earthquake seemed to have no lasting effect for the establishment of consistent building and settlement guidelines in California or the country (Fig. 10-30). One exception was passage of the California Field Act in 1933 which set rigid standards for the mitigation of risk in new public schools. Proof of its success was documented by damage to schools from the 1971 San Fernando earthquake. Fifty of the 568 older schools that lacked safety requirements were so badly damaged they had to be demolished, whereas almost all of the 500 new ones that met seismic resistance requirements suffered no structural damage. However, as a result of this earthquake, and also bolstered by the 1964 Alaska earthquake, an entirely new assessment of potentially hazardous areas has been formulated in many states and is most advanced in California—the state that stands to lose the most.

The two most effective measures for safeguarding people and property are education and legislation. The education programs must deal with appropriate methods for the dissemination of information relating to hazard perception and awareness, in addition to prepardness training both during and after a disaster. A firm legal basis is vital to ensure stringent legislation and ordinances, as well as their enforcement. Of course, when possible, new construction should be prohibited from those corridors with a known history, or a probable potential, for large-scale faulting. Older structures should be "earthquake-proofed," with provision for tax relief and other monetary incentives. All new construction must adhere to the latest building codes and must use materials that are most durable and designs that cause

Figure 10-30 Aerial view, southeast along the San Andreas Fault, California, showing how communities have brazenly (or bravely) sprung up adjacent to the fault zone, which extends through the center of the photo, just left of San Andreas Lake. (Courtesy U.S. Geol. Survey.)

maximum stability for the structure. Part of any long-range master plan for land use in earthquake-prone regions must be based on maps that delineate high-risk areas. Such localities can then be appropriately zoned, but still placed in open space, for use as recreation areas (including golf courses, nurseries, horseback and bike trails, ball fields, etc.), freeways (but not interchanges), parking lots, parks, and agricultural lands (including orchards, crop lands, and timber lands when appropriate).

A number of important publications in the United States have laid the groundwork for land use programs. They include *The San Fernando Earthquake of February 9, 1971 and Public Policy, Disaster Preparedness,* and *Earthquake Prediction and Public Policy.* In addition, California has especially been active with legislation that deals with seismic hazards. The Geology Hazard Zones Act of 1972 requires cities and counties to include within their general plans provisions for identification and appraisal of seismic hazards and measures for protecting the public from ensuing fires. In response to this, the San Francisco Department of City Planning developed a Community Safety Plan that became part of the Comprehensive Plan of San Francisco on September 12, 1974. It accepts as inevitable that an earthquake will occur so it seeks to: (1) understand and identify geologic risks, (2) determine the degree to which existing structures are likely to be damaged and endanger human life, (3) develop ways to correct existing structural deficiencies and ensure new buildings can withstand seismic activity, (4) prepare emergency plans to handle crises during and after the earthquake, and (5) prepare plans for reconstruction of the city. Various counties have also enacted legislation such as the Seismic Safety Plan of Santa Clara County, California, and towns have passed ordinances; for example, Portola Valley legislation prohibits human occupance of land within 15 m of San Andreas Fault traces.

As mentioned in the introduction to Part Three, the federal government has sponsored legislation concerning disasters, and more recently in 1977 passed the Earthquake Hazards Reduction Act. This act sets up a $200 million fund, over the period 1978–80, with which the president is to organize a "coordinated earthquake hazards reduction program." Among the objectives are: development of technologically and economically feasible earthquake-resistant designs for new and existing structures; a prediction capability for areas where earthquake risk is high; and the development of emergency services to warn of impending quakes and aid in rehabilitation and reconstruction after such an event. Under this act $14 million was allocated for prediction investigations, whereas in 1976 only $5 million was budgeted for such studies. The act also established a National Advisory Committee on Earthquake Hazard Re-

duction with a $11 million fund for the two-year period. All programs emphasize the necessity for producing national maps (Fig. 10-31) that depict the most hazardous regions, plus detailed local maps that can be vital to individual communities.

Nuclear Power Plants

An especially violent controversy has emerged during the last several years regarding the siting and development of nuclear power plants. There are two main arguments that are used by the opponents of nuclear power generation. One group contends that such plants are not needed and that alternate energy sources can fulfill the nation's needs. They fear radioactive contamination on the local level and point to the long-range dangers, citing that a 100 percent safe disposal method has not yet been achieved. The second group is instrumental in mandating such a large number and variety of technological and engineering safeguards, in siting and plant construction, that their effect greatly prolongs the licensing procedure, which is now about 14 years. Such actions are proving to be increasing impediments in the nation's attempt to narrow the energy gap of petroleum imports and to become self-sufficient.

Important considerations for the siting of a nuclear power station include: a location removed from dense populations; presence of abundant water; and a geologically safe environment. Therefore, the determination of geology, lithology, and structure is necessary, and the presence of faults and earthquake potential is vital in site selection. Strict requirements are mandated by the Federal Nuclear Regulatory Commission (NRC) to assess the area for the possible occurrence of "capable faults." For example, it is inferred that a capable fault under the proper stress condition might produce an earthquake and provoke a disaster at the plant. Thus, the mapping of faults is vital for licensing of a nuclear plant. In the current language of

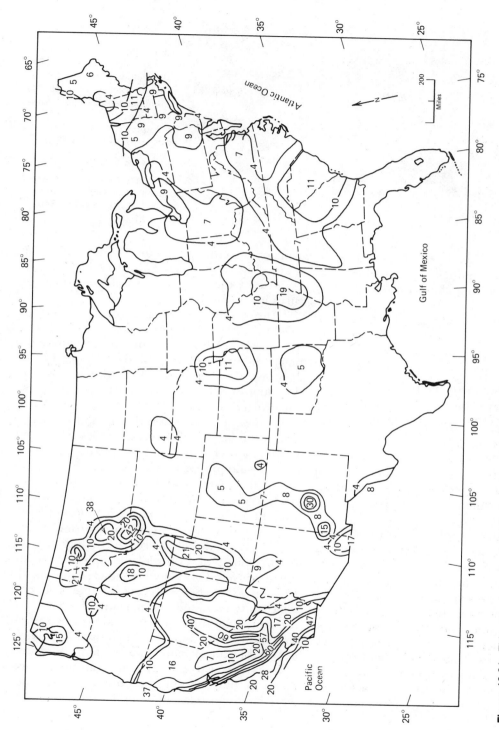

Figure 10-31 This map drawn by the U.S. Geological Survey shows expectable levels of earthquake-shaking hazards. The contour lines depict ground-shaking levels in percentages of the force of gravity for the maximum likely to occur at least once in a 50 year period.

Figure 10-32 This site for a nuclear reactor at Bodega Bay, California, was abandoned after expenditure of more than $25 million and several years of site preparation. Insufficient geological data were available prior to development, but construction proceeded regardless. Subsequently, a nearby fault was located which caused abandonment of the project. (Courtesy U.S. Geol. Survey.)

regulations by the NRC, under Part 100-Reactor Site Criteria (page 10.006), a capable fault is one with the following characteristics:

1 Movement at or near the ground surface at least once within the past 35,000 years or movement of a recurring nature within the past 500,000 years.

2 Macro-seismicity instrumentally determined with records of sufficient precision to demonstrate a direct relationship with the fault.

3 A structural relationship to a capable fault according to characteristics (1) or (2) of this paragraph such that movement on one could be reasonably expected to be accompanied by movement on the other.

Several possible sites have had to be abandoned, and in some cases construction had already been done and costly surveys performed (Fig. 10-32) when potentially dangerous faults were later discovered near the site. The Bodega Bay, California, site had been developed before the new guidelines were adopted, and there had not been a thorough geology investigation. Millions of dollars were spent on preliminary work for a nuclear site in western New York, which had to be discontinued once it was discovered that the Clarendon-Linden Fault was

a major structure in the area. The Sears Point, Maine, site was rejected in 1977 by the NRC because of the presence of deformation structures in nearby earth materials.

During the past several years I have been involved with several power companies to locate a new, safe nuclear plant site; to determine if a newly discovered fault would be a threat to an existing plant; and to assess the earthquake potential of a fault near a nuclear plant being constructed. It should be pointed out that, even if it can be demonstrated that a fault did move within the last 35,000 years, it does not necessarily follow that any accompanying seismic activity would necessarily endanger the 40-year life of the plant. This depends on the type of rock, type of fault, and the character and amount of movement. During a routine petrologic mapping study near the Hudson Highlands, a fault was discovered at the Indian Point reactor plant of Consolidated Edison, 80 km north of New York City on the east side of the Hudson River. Not only was an intensive geologic study mounted to determine the nature of the faulting at this site, but a major investigation was also launched on the much broader subject of whether the Ramapo Fault system constitutes a safety threat to the area. Several million dollars were devoted to these studies and hearings were held which lasted several weeks. The final

(a)

conclusion and the results of the hearings concurred that the faults did not jeopardize the safety of the nuclear plants.

More recently, during the period 1977–80, an intensive study by Dames & Moore was completed of a fault that was discovered in a cooling water trench during construction of Niagara Mohawk Number 2 Nuclear Generating Plant at Nine Mile Point, New York (Fig. 10-33). This site is on the south side of Lake Ontario. The bedrock of nearly horizontal Paleozoic sedimentary rocks had always been previously mapped as devoid of faulting, and the site has no historical record of earthquake activity. Although the $4 million investigation did show there had been some movement along the fault within the last 35,000 years, the data indicated such movements were very minor and incapable of strong vibratory motion. In my report I concurred with the conclusion of the investigation:

(b)

NNE

SSW

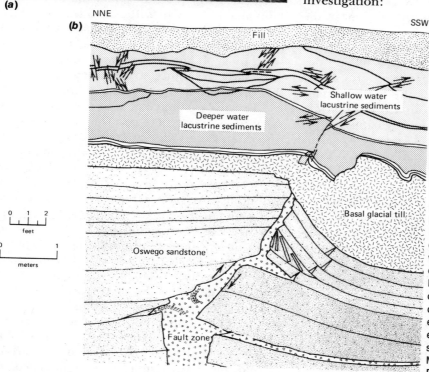

Fill

Shallow water
lacustrine sediments

Deeper water
lacustrine sediments

Basal glacial till

0 1 2
feet

0 1
meters

Oswego sandstone

Fault zone

Figure 10-33 Fault at the Niagara Mohawk Power Corp. Nine Mile Point nuclear reactor site. The fault was discovered while trenching for a cooling water pipeline. A $4 million geologic investigation was undertaken to determine whether the structure would pose a seismogenic hazard to Plant Unit 2 (see Fig. 7-1). The evidence showed that the fault would not produce a risk during the life of the plant. (a) Photo of fault soon after discovery. (b) Diagrammatic cross section of fault at the east wall of the trench dug to expose characteristics of the structure. (Courtesy Dames & Moore and Niagara Mohawk Power Corp.)

all of the evidence identified during this investigation indicates that none of the geologic structures encountered at the site represents a seismic hazard.

Such a finding is highly significant because two other nuclear generating stations are also located at this site, Niagara Mohawk Number 1 and the FitzPatrick Nuclear Plant (Power Authority for the State of New York). The total investment for all plants is about $4 billion, so it is crucial that their safety and integrity be assured.

Perspectives

Although earthquake activity is concentrated primarily in two broad belts on the earth, there are few places entirely safe from seismic activity because severe quakes have occurred in nearly all terrain. One exception is the Precambrian shield areas. Earthquakes provide the least warning and are the most difficult to predict of all geologic hazards. Yet this is one of the most promising research areas in geophysics, and governmental funding for prediction studies has greatly increased in the past few years. Unlike China, which in 1975 predicted a major quake and evacuated more than 1 million people, the United States has not yet faced the problem of what to do, or the legal entanglements that would arise from a prediction of a possible devastating earthquake in a major metropolitan area. If the populace were evacuated and the earthquake did not occur, it would prove highly embarrassing to science and highly costly to the community and the government. There could be accidents from a fleeing and panic-stricken public; the arrival into the area of thrill seekers and looters; a future drop in tourism and property values; the nonrenewal of earthquake insurance policies; and massive litigation dealing with business losses and other claims. So as seismologists develop ever more precise tools for detecting and calculating the time of earthquakes, close liaisons must be maintained with social scientists and government policymakers. Prediction in the United States is more difficult than in areas that are more seismically active. For example, of the 160

earthquakes studied of 7 M or greater, only half had foreshocks more than 4 M. Some of these occurred 2 to 3 months before the main shock, but most came less than a day before it. The use of microseisms as precursors is promising, but their detection requires a very dense network of seismographs. Funding for such instrumentation will always be in competition with other necessary programs, and the benefit-cost ratio analysis may be difficult to justify in light of reduced government spending. As with all hazards, citizens who live in areas hundreds or thousands of kilometers away, and thus feel safe, show resistance to spending on projects that provide no direct benefits to them.

Because earthquakes generate both primary and secondary effects, the damages produced, and the wide variety of chain reactions they induce, increase the difficulties of trying to prevent heavy losses. Several types of land use practices must be adopted to cope with mounting possibilities for major disasters in metropolitan areas. Additional legislation and construction and building surveillance are necessary in earthquake-prone localities to minimize losses. Geologists can play an important role in providing essential data, maps, and reports that delineate the most hazardous areas. They should also aid those forums and agencies that seek the adoption of policies to restrict and control land use practices in such sites. The entire spectrum of public safety needs careful monitoring, and those activities and structures that jeopardize human welfare should be mitigated whenever and wherever possible. Ordinances should prohibit high-intensity use of zones known to be part of the earthquake corridors. Nonconforming buildings that are in violation of safety building codes in earthquake-prone terrane should be strengthened, renovated, or removed. Special attention must also be devoted to public places, such as schools, hospitals, auditoriums, and office buildings. Although such stringent guidelines may be diffi-

cult in the United States, the goals are obtainable. Other parts of the world where earthquake intensity is a grave problem may have greater difficulty in attempting earthquake-proofing measures because of greater antipathy and lack of options or money to undertake such programs. The next several decades will continue to have as many earthquake disasters as occurred in the 1970s. The millennium has not yet arrived.

READINGS

Anderson, D. L., 1971, The San Andreas fault: Sc. Amer., v. 225, n. 5., p. 52–67.

Bolt, B. A., 1978, Earthquakes, a primer: San Francisco, W. H. Freeman and Co., 241 p.

Cook, N. G. W., 1976, Seismicity associated with mining: Eng. Geol. v. 10, p. 99–122.

Eckel, E. B., 1970, The Alaskan earthquake, March 27, 1964; lessons and conclusions: U.S. Geol. Survey. Prof. Paper 546, 57 p.

Evans, D. M., 1966, Man-made earthquakes in Denver: Geotimes v. 10, p. 11–18.

Fuller, M. L., 1912, The New Madrid earthquake: U. S. Geol. Survey Bull. 494, 119 p.

Gates, G. O., ed., 1972, The San Fernando Earthquake of February 9, 1971 and public policy: Joint Committee on Seismic Safety, California Legislature, 127 p.

Hadley, J. B., 1964, Landslides and related phenomena accompanying the Hebgen Lake earthquake of August 17, 1959: U. S. Geol. Survey Prof. Paper 435-K, p. 107–138.

Haicheng Earthquake Study Delegation, 1977, Prediction of the Haicheng earthquake: EOS, Amer. Geophys. Un. Trans., v. 58, n. 5, p. 236–272.

Healy, J. H., Hamilton, R. M., and Rayleigh, C. B., 1970, Earthquakes induced by fluid injection and explosion: Tectonophysics, v. 9., p. 205–214.

Howell, B. F., 1973, Earthquake hazard in the eastern United States: Earth and Mineral Sci. v. 42, p. 41–45.

Iacopi, R., 1973, Earthquake country, 3rd ed.: Menlo Park, Calif., Lane Books, 160 p.

Khattri, K., and Wyss, M., 1978, Precursory variation of seismicity rate in the Assam area, India: Geology, v. 6, n. 11, p. 685–688.

National Academy of Sciences, 1975, Earthquake prediction and public policy: Washington, D.C., National Academy of Science, 142 p.

Pakiser, L. C., Eaton, J. P., Healy, J. H., and Raleigh, C. B., 1969, Earthquake prediction and control: Science, v. 166, p. 1467–1474.

Press, F., 1975, Earthquake prediction: Sc. Amer. v. 232, p. 14–23.

Scholz, C. H., Sykes, L. R., and Aggarwal, Y. P., 1973, Earthquake prediction: a physical basis: Science, v. 181, p. 803–810.

Sims, J. D., 1973, Earthquake-induced structures in sediments of Van Norman Lake, San Fernando, California: Science, v. 182, p. 161–163.

Steinbrugge, K. V., et al., 1971, San Fernando Earthquake February 9, 1971: San Francisco Pacific Fire Rating Bureau, 93 p.

Wallace, R. E., 1974, Goals, strategy, and tasks of the earthquake hazard reduction program: U. S. Geol. Survey Circ. 701, 27 p.

Chapter Eleven
Landslides

Wright Mountain landslide, California—a two-stage landslide. The first landslide was 500 years ago; the second, more recent, was in 1967–69. More than 13 million m³ of rock was involved. (U.S. Geol. Survey.)

INTRODUCTION

Landslides belong to that family of short-lived and suddenly occurring phenomena that comprise catastrophic geology. These processes—earthquakes, volcanic eruptions, floods, and hurricanes—produce extraordinary landscape changes in a short time period and can cause great destruction to property and people. **Landslide** is an unusual name because it is used both for the geomorphic process, which involves rapid gravity movements, and for the resulting landform that is created by the displaced material. Furthermore, the term *landslide* is also a misnomer because the process of landsliding has come to include a broad range of different types of motion whereby earth material is dislodged by falling, sliding, *and* flowing. Even rapid subaqueous gravity movements are generally included under the category of *land*slide events.

Landslides are known in nearly all terrains and can occur in different climates and in most assemblages of earth materials. All that is needed for their formation is a sufficiently strong triggering mechanism to overcome the natural stability of the rock and soil. When this shear resistence threshold is exceeded, landsliding results, and it can take a variety of movements and forms. Landslide types have been named to reflect the kind of earth material, the type

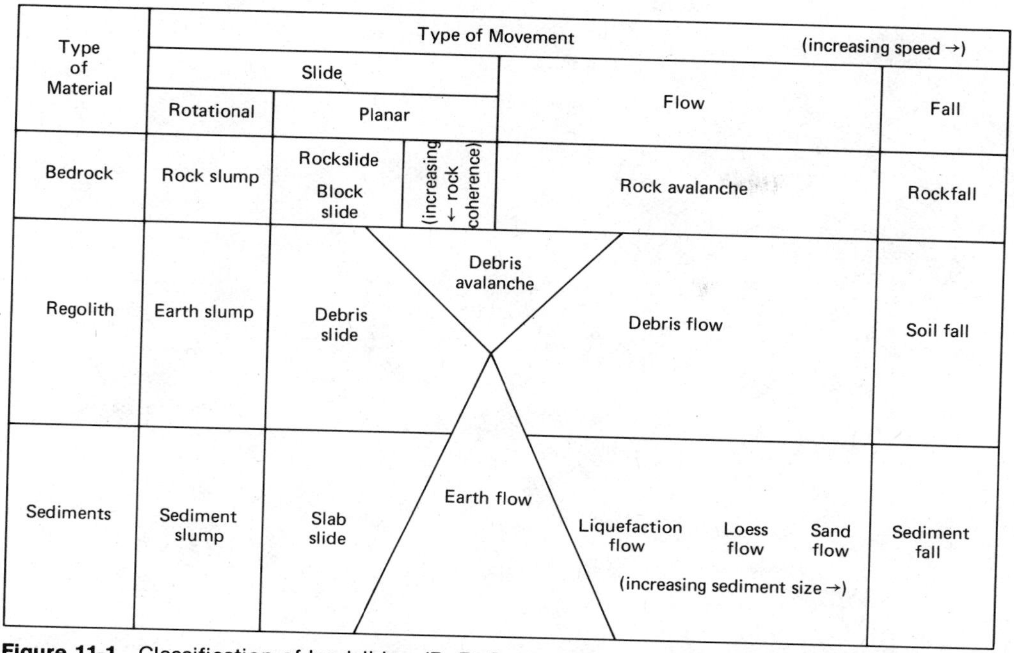

Type of Material	Type of Movement				(increasing speed →)
	Slide			Flow	Fall
	Rotational	Planar	(increasing ↓ rock coherence)		
Bedrock	Rock slump	Rockslide / Block slide		Rock avalanche	Rockfall
Regolith	Earth slump	Debris slide	Debris avalanche	Debris flow	Soil fall
Sediments	Sediment slump	Slab slide	Earth flow	Liquefaction flow / Loess flow / Sand flow (increasing sediment size →)	Sediment fall

Figure 11-1 Classification of landslides. (D. R. Coates, 1977. Copyright © Geological Society of America.)

and rate of movement, the moisture content, and the character of the failed surface (Fig. 11-1). The size of landslides range from those with displacements of only a few cubic meters to rock avalanches that involve hundreds of millions of cubic meters and affect entire mountainsides. Extensive landslides are present on the moon and on Mars.

The phenomenon of landsliding is less studied than other geologic hazards such as earthquakes, volcanic activity, and floods. For example, there are only six books on the subject, and each year there are generally less than 50 publications throughout the world written on the topic. However, this is changing because a new awareness is emerging about the dangers of landslides and the problems encountered as we expand our habitat more and more onto hillsides.

Landslides are those events or forces that are among the most rapid of all mass movements (Fig. 11-2). Other aspects of gravity-produced features are discussed elsewhere in this book (see Chapter 13, for example). Thus creep and solifluction do not qualify as landslides because their movement is too slow, they generally affect the entire hillslope area, and the locus of displacement is usually a transitional zone rather than a sharply defined shear surface. Subsidence, a gravity phenomenon, is too slow an event, but it, along with a collapse, does not have a free exterior surface for movement.

THE IMPORTANCE OF STUDYING LANDSLIDES

Population pressures and increasing urbanization, along with local zoning laws, have been forcing man from flat-lying areas and floodplains onto the adjacent hillslopes. The hillslope as an environmental setting can help us avoid the severe flooding of the low areas, but can be equally hazardous because of potentially unstable earth materials. Therefore, recognition of

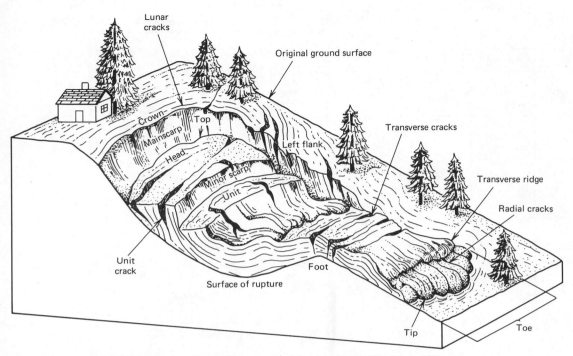

Figure 11-2 Landslide terminology for various parts of the landform. (Sketch by Mary Ryan.)

landslide-prone topography is becoming increasingly important in land use decisions.

Under the proper set of circumstances landslides can affect most hillslopes. The three principal causes for landsliding are excessive rainfall, human activities, and earthquakes. Even nearly flat terrain, when underlain by fine-grained unconsolidated sediments, is vulnerable. For example, prolonged precipitation produced the disaster on the level ground at St. Jean Vianney, Canada (Fig. 11-3), and the terrace in the Turnagain Heights area of Anchorage, Alaska (see Chapter 10) was devastated by movement resulting from the 1964 earthquake. Thus, there are few areas that are completely safe from landslides, although the probability for their occurrence is greatly enhanced by such factors as steep slopes, weak bedrock, unstable regolith, absence of trees, and high moisture content. However, the important factor in landsliding is to determine if the driving forces that are seeking to change hillside equilibrium exceed the resisting forces that operate to maintain slope stability.

Each year landslide damages in the United States amount to more than $1 billion and several times that amount throughout the world. Highway damages alone are more than $100 million yearly; in addition, hundreds of millions of dollars are spent in preventative and avoidance measures to minimize losses. Although the loss in life from landslides in the United States for the 50-year period of 1920–1970 was fewer than 300, about 100 times as many people were killed in other countries (Table 11-1). In California it has been estimated that unless there is stringent management of hillslope development the total damages from landslides in that state will amount to $9.85 billion in the 30-year period from 1970–2000.

Fluvial processes are generally considered to be the dominant denudational force working to lower the landscape, but in many regions the landsliding process may be even more significant. Studies in such localities as the San Francisco Bay area, parts of Brazil, and several

(a)

(b)

Figure 11-3 (a) Aerial view of the landslide scar from the north side of St. Jean-Vianney, Quebec. This liquefaction phenomenon occurred May 6, 1971 and resulted in more than 30 deaths and extensive property damage. (Courtesy Geological Survey of Canada.) (b) Closeup view of destruction at St. Jean-Vianney, Quebec. (Courtesy Geological Survey of Canada.)

regions of Central America indicate landsliding is the principal process causing landmass reduction.

LEGAL AND ADMINISTRATIVE AFFAIRS

Recognition of the importance of landslides was first realized on a legal basis in California. Here grading ordinances were adopted in the Los Angeles area in 1952, and, by 1963, the statutes had been greatly enlarged and strengthened. A series of devastating rainstorms in 1951 and 1952 caused millions of dollars of damages on hillsides and on other urban developments (Fig. 11-4; see also Chapter 19). Although the area was again hit by damaging storms during the winter months of 1968–1969 it is significant that of the total damages of $6.5 million, only $182,400 in losses occurred in developments that adhered to the grading ordinances. The

Table 11-1 Landslide disasters

	Date	People killed	Remarks
Brenno Valley, Switzerland	1512	600	Rockslide dammed valley; dam broke in 2 yr causing destruction
Tour d'Ai, Switzerland	1584	300	Landslide devastated village of Yvorne in Rhone valley
Mount Conto, Switzerland	1618	2,430	Rockslide
Goldau, Switzerland	1806	457	Landslide destroyed village
Mt. Ida, Troy, New York	1843	15	Sediment slump and flow
Elm, Switzerland	1881	115	Rock avalanche also demolished 83 houses
Trondheim, Norway	1893	111	Liquefaction flow in marine clays
Frank, Canada	1903	70	Rock avalanche destroyed most of town
Kansu Province, China	1920	100,000– 200,000	Earthquake caused loess flows
Nordfjord, Norway	1936	73	Rockfall created 74 m wave
Kobe, Japan	1938	461	Rocky mudflows
Kure, Japan	1945	1,154	Rocky mudflows
Yokahama, Japan	1958	61	Rocky mudflows
Madison, Montana	1959	28	Rock avalanche buried campers
Vaiont, Italy	1963	2,600	Rockslide into reservoir created wave that flooded below dam
Anchorage, Alaska	1964	114	Combined toll from landslides and earthquake
Aberfan, Wales	1966	144	Man-made mining spoiled hill; landsliding buried mostly children
Brazil	1966–67	2,700	Combined toll from landslides and floods
Nelson County, Virginia	1969	150	Combined total from debris avalanches and floods
Huascaran area, Peru	1970	21,000	Combined rock avalanche and debris flow buried two cities
St. Jean Vianney, Canada	1971	31	Slab flows buried people and houses

Sources: Blank, 1971; Browning, 1973; Brunsden, 1974; Close and McCormick, 1922; Hansen, 1965; Heim, 1882; Jones, 1973; Kiersch, 1965; Mathews, 1960; Nakona, 1974; Newland, 1916; Williams and Guy, 1971; Zaruba and Mencl, 1969. (From D. R. Coates, 1977. Copyright © Geological Society of America.)

same storms caused damages of $25.4 million in the nine counties of the San Francisco Bay area where grading ordinances were not in effect.

Brazil was the first country to pass national decrees in 1955 that stipulated the requirement of an investigation to determine the stability of sloping land before construction could be undertaken. This study must also include what the effects of such a development would be on land and buildings further downslope. The Brazil Forest Law of 1959 made construction work

Figure 11-4 Damages in an urban area from improper land use planning. This mud and debris slide began during heavy rains on December 19, 1969, and ultimate failure occurred along bedding planes of shales and siltstones. The building in the background is the CNA Building at 6th Street, Los Angeles, California. (Courtesy Allen Hatheway.)

illegal above certain specified slopes in order to preserve the stabilizing influence of vegetation. In 1967, the Brazil Law of License of Construction in Uneven Terrain regulated construction on steep slopes and other unstable localities and demanded that the contractor must obtain proof of slope stability before construction was allowed to begin.

In the United States, the Flood Disaster Protection Act of 1973 (Public Law 92-234) was basically an amendment to the Housing and Urban Development Act of 1969 (Public Law 90-448), and it included for the first time insurance coverage for damages caused by mudslides. Interestingly enough, the terminology used to describe mudslides is equally applicable to the great majority of landslides:

Mudslide means a general and temporary movement down a slope of a mass of rock or soil, artificial fill, or a combination of these materials, caused or precipitated by the accumulation of water on or under the ground.

Furthermore, in the future, we can anticipate more lawsuits, as already seen in California, where court decisions are made to assess the liability when landslide damage occurs.

Another future trend will be the increasing governmental involvement in making surveys and publishing information in the interest of public safety. For example, a vital part of the five-year San Francisco Bay investigation by the U.S. Geological Survey (in cooperation with the Department of Housing and Urban Development) was the mapping of slope instabilities of the entire nine-county area. This has resulted in numerous publications and the delineation of six types of landslide-prone topography in the region. Recently the West Virginia Geological Survey has completed mapping of nonstable terrain in the seven major urban areas of the state, and similar mapping was done in the Pittsburgh region by the U.S. Geological Survey.

The importance of local governments in addressing landslide problems should not be overlooked. In the 1960s in Fairfax County, Virginia, about 30 homes built on slopes were badly damaged by landsliding. Thus, the county passed an ordinance that required developers to retain a consulting soils engineer. Housing plans must then be scrutinized by a peer review board comprised of other consultants. As a result, in this area there are now practically no losses from landsliding.

LANDSLIDE CHARACTERISTICS

Because of their exceptional diversity, it is impossible to provide many generalizations about landslides (Table 11-2). Gravity is the principal force involved, and the movement must be moderately rapid geologically—on the order of about 1 ft/yr (0.3 m/yr; Gary and others, 1971, p. 396). The movement may include falling, sliding, and flowing, with one free face so that the motion is down and out. The plane or zone of movement is not identical with a fault, and the displaced material may include parts of the regolith and/or bedrock. Such materials are invariably broken and reordered so that they destroy their original continuity. Although most

Table 11-2 Landslide terrane properties for large events[a]

Locality	Date	Volume (10⁶ m³)	Area covered (km²)	Vertical displacement or fall (m)	Horizontal movement (km)	Runup distance (m)	Estimated velocity (km/h)	Landslide type
Huascaran, Peru	May 31, 1971	10		4,000	14.5		400	Rock avalanche–debris flow
Little Tahoma Peak, Washington	Dec. 14, 1963	10.7		1,890	6.9	90	152	Rock avalanche
Elm, Switzerland	Sept. 11, 1881	12.7	0.6	610	1.4	103	160	Rock avalanche
St. Albans, Canada	Apr. 27, 1894	19.1						Liquefaction flow
Madison, Montana	Aug. 17, 1959	28.3	0.5	400	1.6	130	180	Rock avalanche
Sherman, Alaska	Mar. 27, 1964	28.3		600	5.0	137	185	Rock avalanche
Frank, Canada	Apr. 29, 1903	36.5	2.5	870	4.0	120	175	Rock avalanche
os Ventre, Wyoming	June 23, 1925	38.2	0.8	640	1.9	106	164	Rock avalanche
Goldau, Switzerland	Sept. 2, 1806	40		550	1.7			Rock avalanche
Apollo 17, Moon	—	200	21	2,000	5.0			Rock avalanche
Silver Reef, California	Prehistoric	226		790	5.3	46	104	Rock avalanche
Vaiont, Italy	Oct. 9, 1963	260		1,220	1.8	240		Rockslide
Blackhawk, California	Prehistoric	283			8.0	64	120	Rock avalanche
Gohna, India	Sept. 1893	290		1,470	1.6			Rock avalanche
Martinez, California	Holocene	382		2,000	7.6			Rock avalanche
Ticino River, Switzerland	During glacial retreat	500						Rock avalanche
Upper Garhwal, India	Sept. 22, 1893	566		1,520	3.2			Rock avalanche
Sawtooth Ridge, Montana	Prehistoric	650		360				Rock avalanche
Klonsee, Alps	Interglacial	770						Rock avalanche
Tin Mountain, California	Prehistoric	1,795						Rock avalanche
D'Onsoi, Pamir Mtns.	Feb. 18, 1911	2,080	10.3	1,160	3.6			Rock avalanche
Flims, Switzerland	Interglacial	12,000	41.4	1,980	16.1			Rock avalanche
Lake Tahoe, California–Nevada	—	10,000						Subaqueous slide
Saidmarreh, Iran	Holocene	20,000	165.7	1,650	14.5	457	338	Rock avalanche
Samar Island, Philippines	2000 y.b.p.	135,000						Block glide
Gulf of Alaska, Alaska	Mar. 27, 1964(?)	590,000						Subaqueous slide

Sources: Browning, 1973; Crandell and Fahnestock, 1965; Harrison and Falcon, 1938; Heim, 1882, 1932; Howard, 1973; Hsü, 1975; Hyne and others 1973; Kent, 1966; Kiersch, 1965; Mudge, 1965; Newland, 1916; Shreve, 1966, 1968a; Zaruba and Mencl, 1969; and data from this volume in Bock; Molnia and others; Wolfe; and personal observations and calculations.
[a](From D. R: Coates, 1977. Copyright © Geological Society of America.)

landslides are local and affect only a part of a hillside, Wolfe mapped a landslide on Samar Island, Philippines, that covered an 18 km by 25 km area with a slip plane of 0.6°. Molnia and others (1977) mapped a submarine slide in the Kayak Trough of the Gulf of Alaska that is 15 km by 18 km with a 1° slope. Other size relations are shown in Table 11-2, including data on representative large slides throughout the world. Therefore, because of so many landslide variations, and because they contain different types of material and fragment sizes, different amounts of moisture, and are subject to movements that may involve falling, sliding, or flowing, it is important to use a classification system that incorporates most of these features (Fig. 11-1). Although many different techniques have been used in the literature to classify landslides, for our purpose we will use a system that recognizes the importance various types of earth materials play in the formation of different landslide types, whether bedrock, regolith, or sediment. Other considerations employed in the classification reflect the type and rate of movement, moisture content, shape of the failed surface, and particle size. Alternate landslide classification schemes have been developed by others that are based on such factors as the geometry of the resulting landforms, mechanics of motion, degree of landslide activity, and the environmental setting.

Landslide movement can consist of earth materials that fall, slide, flow, or contain combinations of these motions. **Falls** are abrupt free-fall movements from cliffs and steep slopes where material is not in continuous contact with the ground but instead can roll, bound, and ricochet downslope. They can be classified as **rockfalls, debris falls,** and **soilfalls. Slides** are those mass movements where the displacement is along a definite **shear surface,** and the ruptured material moves with some semblance of unitary motion. The shear surface may be either planar or concave (Fig. 11-5). A **slump** (Fig. 11-2) is a **rotational slide** in which the materials commonly retain coherence and move along a shear plane that is concave upward. **Planar slides** are grouped on the basis of type of material and degree of coherence during movement (Fig. 11-6). **Block glides** occur when the displaced bedrock maintains its integrity and general orientation. **Rockslides** occur when the detached bedrock becomes fragmented during rapid movement over the failed surface. They can be very large, measured in many millions of cubic meters, and when combined with **rock avalanches** form the most awesome and spectacular of all landslide types (Fig. 11-7).

Figure 11-5 Tread of the Cottonwood landslide, Manti Canyon, Utah. Note grooves which may have similar appearance to grooves created by glacial action or by faulting. (Courtesy Earl Olson.)

Figure 11-6 Planar or translational landslide, Pennsylvania. (Courtesy Jesse Craft.)

The **flow** transport mechanism of landslides involves motion of materials as a viscous mass. Such landslides are further classified in terms of the type of earth materials and moisture content. When the material is unconsolidated, but contains bedded sediments that fail by a lateral spreading process, such landslides are **slab slides.** They did more than $100 million of damage at Anchorage, Alaska, where the Bootlegger Cove Clay was liquefied during the 1964 Alaska earthquake (Fig. 11-8). Other types of flow landslides include **liquefaction, sand,** and **loess flows.** However, the largest controversy over landslide movement mechan-

Figure 11-7 Gros Ventre rock avalanche, Wyoming. This latest landslide occurred June 23, 1925. The valley-dipping shales and sandstones (Cretaceous age) provide a smooth plane for movement. Deep-melting snows and heavy rains had completely saturated the area, and a slight earthquake tremor provided the trigger to dislodge the mountainside. Landsliding speed was in excess of 160 km/h. Two years later, in May 1927, the dam built across the Gros Ventre River to hold landslide debris broke and flooded the town of Kelly 6.5 km downstream. Six lives were lost, ranch homes were inundated, and livestock killed. (Courtesy Wyoming Travel Commission.)

ics concerns the somewhat anomalous transport of materials in rock avalanches.

A **rock avalanche** is an exceptionally large mass of broken bedrock with extraordinary rapid movement that terminates with dry flow (Fig. 11-9). One of the problems is how to explain the unique motion that may be in excess of 160 km/h. It is important to compare the various theories that attempt to explain this motion, because this type of slide has caused many of the greatest catastrophes, such as those that occurred at Elm, Switzerland, Frank, Canada, and Madison, Montana, among others (Fig. 11-10). P. E. Kent believes they move by a fluidization process whereby air is entrapped between the materials. In explaining the Blackhawk and Sherman landslides, however, Shreve postulated the slide moved on a cushion of air that acted as a lubricant for the mass and reduced friction there, allowing the material to obtain high speeds. K. J. Hsu proposed a third idea—that the rapid motion is caused by a type of flow in which the mass operates as a fluid medium with cohesionless grains. It is the dispersion of the fine debris among the larger blocks that

provides an uplifting stress and buoyancy of the materials, which act as an instital fluid that reduces the effective pressure of the entrained grains. The study of lunar and Martian landslides may aid in providing some of the answers. For example, one rock avalanche on the moon moved more than 5 km, and the thickness of the debris was only a few meters at the distal end. As K. Howard observes, "Evidently lunar avalanches are able to flow despite the lack of lubricating or cushioning fluid."

Debris avalanches constitute another type of flow that produces movement of materials down a long and relatively narrow tract (Figs. 11-11 and 11-12). They occur on steep and mountainous terrain in humid to tropical climates. Invariably they are triggered by exceptionally heavy rains such as those that caused the severe devastation in Brazil in 1966–67 and the landslide in Virginia from Hurricane Camille in 1969. **Debris flows** differ from debris avalanches in that the scar head is much wider; a larger amount of water is incorporated into the earth materials so that the lowered viscosity permits much longer travel (Fig. 11-13). **Liq-**

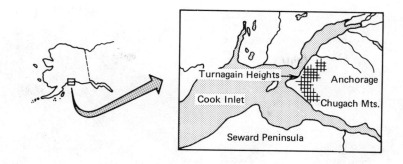

Two minutes after the earthquake begins, cracks appear near the edge of the bluff.

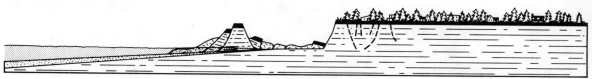

Ten seconds later, a block has slid 250 feet toward the sea.

After 30 seconds, the first block has moved farther and more are following.

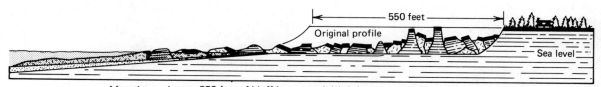

After three minutes, 550 feet of bluff is gone and debris has moved 700 feet into the sea.

Figure 11-8 Evolution of landslide that occurred at Turnagain Heights, Anchorage, Alaska as a result of the 1964 Alaska earthquake. (Redrawn from Smithsonian Institution drawing.)

uefaction flows occur in stratified sediments where the interstitial water has separated from the particles, so that the entire body moves as a flowing viscous mass. These slides are espe- cially prevalent in some of the glaciomarine clays found in Norway and in the Leda Clay of the St. Lawrence region. Both areas have been subjected to serious disasters, and engineering

(a)

(b)

Figure 11-9 Hope rock avalanche, British Columbia, Canada. This landslide occurred on January 9, 1965, contained 47×10^6 m³ of schistose metavolcanic rocks, and buried a highway with a thickness of 75 m of debris. (a) Overview of Hope landslide scar and debris in the Canadian Rockies. (Courtesy British Columbia Dept. of Highways.) (b) Closeup of some of the rock material at the base of the landslide. (Courtesy Geological Survey of Canada.)

Figure 11-10 Frank rock avalanche, Alberta, Canada. On April 29, 1903, about 70 million tons of rock slid down the 400 m high face of Turtle Mountain, and in 100 seconds 66 people died in the valley. For a 3.2 km distance across the valley, the slide was 30 m thick. (Courtesy Geological Survey of Canada.)

Figure 11-11 Debris avalanche scars, Franconia Notch, New Hampshire.

problems occur whenever the clay is encountered during construction. The term **earthflow** is a catchall term that refers to a wide range of unconsolidated materials that have undergone some type of flow. **Mudflows** occur in stream channels and are a hybrid of water and sediment. They are generally caused by abnormal sheetwash on hillslopes where the entrained debris is moved into a more channelized flow regime at the bottom of the slope. **Snow avalanches** constitute still another gravity phenom-

Figure 11-12 Debris flow in the Salmon National Forest, Utah. Bedrock is the Germer Tuff. (Courtesy Earl Olson.)

Figure 11-13 In La Crescenta, California, workmen stand on the roof of a house that was buried in February 1978 by debris flows that contained mud, rock, and cars. A combination of heavy rainfall, geologic conditions vulnerable to water, and earlier destruction of protective vegetation by wildfires produced events that led to this mudslide of 16,000 m³, which flowed out of a 160,000 m² area in the adjacent canyon. Other cars are buried below the visible one. (Courtesy U.S. Geol. Survey.)

enon and are very serious hazards in the Alps, the Rocky Mountains, and other alpine areas.

LANDSLIDE CAUSES

Landslides occur when there is a loss of support under or in front of the shear surface, or when there is a change in the physiochemical constitution of the earth materials. The three factors that influence stability of hillslopes are:

1 *The internal properties of the earth materials.* These include the type of material and its structural character—such as degree of consolidation and cementation, thickness and arrangement of different rock types, strike and dip of zones of discontinuity, and size and distribution of faults, joints, and other fracture and bedding surfaces.

2 *The geomorphic setting and environment.* Hillslope characteristics include the amount of relief, steepness of slopes, shape of the land surface, slope orientation and aspect, nearness to rivers, vegetation, antecedent moisture conditions, and time of year.

3 *Independent external factors.* These are termed

the **triggering mechanisms** because they provide the immediate stress that initiates movement of the mass. Prior to displacement, there is an accommodation equilibrium that exists between the earth materials and the geomorphic and environmental setting. The three most common disturbances that cause the critical threshold of cohesion to be exceeded are excessive precipitation, human activities, and earthquakes.

Excessive Precipitation

Antecedent moisture conditions determine whether large amounts of rainfall will successfully trigger a landslide. When earth materials already contain much water, the severity of precipitation from a new storm can be less and still trigger landsliding. The magnitude, intensity, and duration of the storm all play a role in determining whether a hillslope will fail. Excessive precipitation weakens earth materials by displacing air and increasing the pore water pressure along shear surfaces. Conditions that enhance ultimate failure occur when surface

(a)

(b)

Figure 11-14 *(a)* and *(b)* Two views of a rockfall on Interstate 70 in foothills west of Denver, Colorado. Landslide was triggered by heavy rains on May 5–6, 1973, which saturated rock materials at this oversteepened roadcut in Precambrian gneiss. (Courtesy of Wallace Hansen, U.S. Geol. Survey.)

materials are porous and permeable and are underlain by material of low permeability. These conditions are common in the loose friable regolith of crystalline rocks in humid to tropical climates, where they have led to many disasters (Fig. 11-14).

In Brazil, exceptionally heavy rains in 1966 and 1967 ravaged the midsouthern part of the country, from Rio de Janeiro to a region 50 km west. One account described that "Landslides numbering in the tens of thousands turned the green vegetation-covered hills into wastelands and the valleys into seas of mud." The 1967 storm "laid waste by landslides and fierce erosion a greater land mass than any ever recorded in geological literature" (Jones, 1973). This devastation affected a 25 km long and 7–9 km wide banana-shaped area. The 1967 storm lasted only 3½ hours, but deluged the area with an average rainfall of 240 mm. Total lives lost in these two years exceed 2700 (1000 in 1966 and 1700 in 1967). A large variety of mass movements were created, including earth and mudflows; debris slides, flows, and avalanches; and rockfalls and rock avalanches. One mudflow in Floresta Creek valley killed hundreds of people; 50 debris avalanches and debris flows nearly demolished the power plant that serves Rio de Janeiro; and a single rockslide in the Laranjeiras district killed 132 people. The precipitation so weakened the earth materials that delayed reactions are still affecting the area years after the storm events.

In the western United States, heavy rains in California during the 1968–69 winter season caused landslide damages of $25,400,000 in the San Francisco Bay area and $6,500,000 in the Los Angeles area. In the eastern United States, Hurricane Camille on August 10–11, 1969, produced a 710 mm rainstorm in eight hours that Williams and Guy (1971) described as the worst natural disaster in central Virginia history. Property damage in Nelson County alone was more than $116 million, and the combination of debris avalanches and flooding killed 150 people. Most died from broken bones and blunt force injuries rather than by drowning. A detailed study of 186 of the hundreds of debris avalanches (Williams and Guy) showed that the avalanches:

1 Followed preexisting depressions on hillside slopes steeper than 35°.

2 Produced head scars at the steepest part of the hill where the convex slope changed to a concave or planar slope.

Figure 11-15 (a) and (b) Two views of debris avalanches in Virginia. These occurred as a result of torrential rains on August 19–20, 1969, from Hurricane Camille. (Williams and Guy, in D. R. Coates, 1971, *Environmental Geomorphology*.)

3 Were more numerous on north, northeast, and east-facing slopes.

4 Caused devastating surges of water and sediment in stream channels.

The usual dimensions of typical scars were about 150 m in horizontal length, 15 m wide, and 10.7 m deep. Some were as short as 6 m and others longer than 300 m; some were up to 60 m wide and 6 m deep (Fig. 11-15).

Numerous landslides and associated mass movements and flooding have occurred throughout Asia. In July 1938, heavy rainstorms in Kobe, Japan, produced landslides and mudflows that killed 461 and destroyed 100,000 houses. In September 1945, at Kure, rainfall from the Makurazaki typhoon produced rocky mudflows that resulted in 1154 deaths, and in Tokyo the Kanogawa typhoon of September 1958, produced the heaviest rainfall in the recorded history of Tokyo, 392.5 mm in 24 hours. It created 1029 landslides and land collapses and killed 61 people. In March 1961, excessive moisture in Yui Shizuoka Prefecture in central Japan created a massive landslide that required $56 million in remedial work to keep the railroad route open.

Human Activities

We are affected both by landslides that are naturally occurring and by those that we have triggered through our own carelessness (Figs. 11-16 and 11-17). There has been a sharp increase in the number of man-induced landslides during the twentieth century. Because of expanding population, urbanization, and construction, we have been rapidly modifying the landscape on a massive scale. New engineering techniques, heavy land-moving equipment, and petroleum which increased machinery mobility have permitted the metamorphosis of slopes throughout the world on a scale of extraordinary dimensions. The nearly universal construction of dams in most countries (largely absent from pre-1900 terranes) and the building of immense highway systems have added their toll to the increase in landslides. Most man-made landslides are caused by changes in the slope configuration or by activities that permit additional moisture to enter the substrate material. Thus, any activity that increases hillside gradient, undercuts earth materials, adds weight to the slopes, or produces more water content can lead to instability and set the stage for landsliding (Figs. 11-18 and 11-19).

Figure 11-17 "Slipoff" landslide caused by improper road drainage, Pennsylvania. Seepage undermined the roadbed and produced saturated materials which, in turn, undermined the road. (Courtesy Jesse Craft.)

Figure 11-16 Landslide generated by water flow from a culvert on Highway 50, Utah. (Courtesy Earl Olson.)

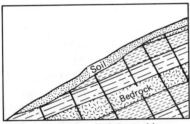

(a) Undisturbed slope with bedrock inclined subparallel to slope and jointing inclined into hillside.

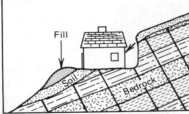

(b) Typical excavation and fill to create level ground for house.

(c) Typical slope failure as result of poor construction practice.

1. Upper part failed at soil bedrock interface, due to removal of slope support at excavation.

2. Lower part failed along fill-soil and soil-bedrock interface due to overloading.

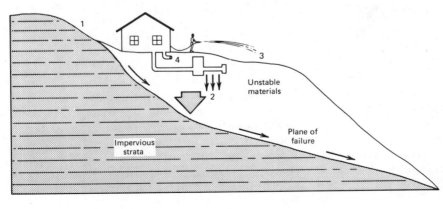

Figure 11-18 Typical ways for causing instability in earth materials. (a), (b), and (c) A sequence of events leading to landsliding by oversteepening the terrain. (d) Shows how household activities may induce landsliding.

Figure 11-19   Landslide in Mancos Shale (Upper Cretaceous age) at Bostwick Park, 10 km northeast of Montrose, Colorado. This complex slump developed overnight in June 1966. Cracks first occurred at the top of the slide a year before, and 10 days before the landsliding the ground dropped 0.3 m at the head of the slide. This slide is 360 m wide and 100 m from head to toe, with a headwall scarp 10 m high. Seepage from irrigation which saturated the unstable earth materials was the principal cause for failure. (Courtesy Wallace Hansen, U.S. Geol. Survey.)

Figure 11-20   Landslide blocking the eastbound lane of Interstate 80 near Echo Junction, Utah, May 12, 1968. This slide necessitated the permanent relocation of the highway. The oversteepened rockcut is in the Echo Canyon Conglomerate (Cretaceous age). The landslide is 200 m wide, 270 m long, and 85 m high. (Courtesy Richard Van Horn, U.S. Geol. Survey.)

Highway construction   Road building is one of the most common ways that we have directly caused landsliding. The majority of landslides occur where a hillside cut is made and the design of slope stabilization structures is inadequate (Fig. 11-20). Thus, when the underground water flow is changed and the shear resistance of earth materials lessened, such artificial excavations become prime candidates for earth movements. Because of the high costs of land acquisition for highways, sometimes the corridor purchase is too narrow to allow for sufficiently low gradient slopes that would increase stability. In the United States, the 68,400 km Interstate system was unduly rushed in some parts with insufficient route design and materials testing. For example, near Rockwood, Tennessee, along Interstate 40, there were 20 major landslides within a 4 mi (6.4 km) section.

Mathewson and Clary (1977) describe multiple landslides that occurred along Interstate 45 in road cuts near Centerville, Texas. The bedrock is composed of overconsolidated, low- to high-plasticity sedimentary rocks. The landsliding was caused by a combination of increases in pore-water pressures and the oversteepening of hillslopes. On November 7, 1972, a rockslide of 13,000 m³ fell on an uncompleted section of Interstate 93 near Woodstock, New Hampshire. It curtailed construction and necessitated a total redesign of the highway section through the rock cut. Extensive remedial and preventative measures were taken to protect the highway from further damage. For example, (1) rock benches were cut, (2) high-strength steel tendons were emplaced along the toe, (3) spot rock bolts were installed at the crest of bench, and (4) an extensive system of rock drains was installed. In the Province of Cosenza, Italy, an investigation of 104 landslides that occurred in 1966–68 revealed that nearly all were initiated by road construction.

Hillside development   Building on hillsides can clearly be risky to structures, especially when they are improperly designed or poorly located (Figs. 11-21, 11-22, and 11-23). In

Figure 11-21 Effects of landslides in the Pittsburgh area, Pennsylvania. Bedrock consists of weakly cemented and fine-grained sedimentary rocks that become highly unstable with excess moisture and human intervention. (*a*) House in center has recently been abandoned because of continuing headward growth of landslide. (*b*) One year later, road remnant has now been lost to landsliding, and the house has been vandalized. (Courtesy Jesse Craft.)

Figure 11-22 Landslide effects near Pittsburgh, Pennsylvania. (*a*) House has just been abandoned because landsliding on the hillside has placed it and occupants in jeopardy. (*b*) One year later, head of landslide has engulfed part of the house and eroded further upslope. (Courtesy Jesse Craft.)

Southern California, it was realized that the large number of hillside structures, many inadequately designed, caused landslides; in response to this growing problem, the ordinances of 1952 and 1963 were passed. Although single buildings rarely cause significant or large-scale landsliding, housing developments such as those on the slopes of the Palos Verde Hills, California, can be instrumental in producing major landsliding. Here in the Portuguese Bend area the eastern part of an old landslide was reactivated in the summer of 1956 and involved about 160 ha. The most likely probable cause for the remobilization was the increase in pore-water pressure from lawn watering and septic

tank effluent. In a court case against the County of Los Angeles, however, the property owners claimed that the landslide was caused by a county road fill at the top of the slope, and they received a judgment for damages of $5,360,000.

Dams and Reservoirs

The building of dams and the development of artificial lakes and reservoirs create another set of conditions that can provide fertile settings for landslides. The primary cause of these new instabilities is the artificially induced changes in contiguous water tables. The construction of Grand Coulee Dam was delayed at various times during the 1934–1952 period by more than 100

(a)

(c)

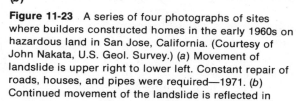

(b)

(d)

Figure 11-23 A series of four photographs of sites where builders constructed homes in the early 1960s on hazardous land in San Jose, California. (Courtesy of John Nakata, U.S. Geol. Survey.) (a) Movement of landslide is upper right to lower left. Constant repair of roads, houses, and pipes were required—1971. (b) Continued movement of the landslide is reflected in major slump structure. Note scarp on the right—1974. (c) Major foundation damage necessitated demolition of the house in the foreground—1977. (d) The toe of the landslide, with the overthrusting of pavement prompted closure of the road. Even major road repairs have now ceased and the area abandoned—later in 1977.

separate landslides that cost $20 million to remedy.

. . . a large part of it could have been saved had presently known geologic facts been available and had engineers applied them in designing and construction work [Jones et al., 1961].

In addition, as the reservoir began to fill, more landsliding occurred. Jones et al., (1961) studied 321 of these landslides and concluded:

The principal cause of the landslides in the area was the weakening of sediments by ground water. The sediments generally have a lower shear strength when saturated and partly saturated than when dry.

Fortunately, none of these landslides caused loss of human life. Unfortunately, the same cannot be said for the inhabitants of the Vaiont, Italy area.

The calamity at Vaiont is one of the largest single man-induced disasters on record. Here a large reservoir was imponded behind a spectacular thin-domed cupola-arched dam in the Vaiont Valley. But on October 9, 1963, a landslide mass of 260 million m³ catastrophically plunged into the reservoir and created giant waves more than 100 m high that overtopped the dam, and in the valley 1 km below the dam the wall of water was still 70 m deep. Although the dam remained intact (but of course is now unusable), more than 2000 people in the valley lost their lives in this debacle. Most investigators found that human influence was the significant factor in the disaster. Man-controlled changes in the reservoir level of many meters had been made at various times in the months prior to October, and these changes altered the bank storage conditions which decreased the shear resistance of the rocks. Other factors that contributed to the landslide were the geologic structures which consisted of sedimentary strata that dipped toward the valley. The rocks also possess abundant arcuate joints and other fractures that roughly parallel ground slopes, thus enhancing the possibility for downhill motion. Heavy rains during August and September 1963, produced increased percolation raising the groundwater level. This recharge added to the bulk density of the rocks. Thus, the extra weight of the water and the increased lubrication of rock discontinuities caused a reduction in the shear strength of the materials and, with probable seismic tremors, led to slope failure.

Mining   Mining operations have also provided the immediate triggering mechanism for some landslides. Slate mining was an important contributing cause for the disastrous Elm landslide in Switzerland that killed 115. The mining of clay at Haverstraw, New York, produced a landslide in the unstable glaciolacustrine sediments that killed 20 people on January 8, 1906. Here the oversteepening of a cut into a terrace produced the failure. During the fall of 1976 two massive debris avalanches were initiated by quarry blasting at two sites on central Mitkof Island, southeast Alaska. Both landslides occurred on slopes that were 35 to 45 percent heavily forested, started below quarries, were triggered by the blasting, and involved 30 to 60 cm of colluvial soils that were displaced along the contact with granodiorite bedrock.

Lumbering   Timber operations, especially clearcutting, can accelerate mass-wasting processes and produce a variety of different types of landslides. In a study of the harvesting methods in Oregon, Swanston and Swanson (1976) determined that clearcut areas had 2 to 4 times the amount of debris-avalanche erosion than other areas (Fig. 11-24). They also found that road-related debris-avalanche erosion was increased by 25 to 340 times the rate of normally forested areas. Landslides can also be produced by multiple events. In the construction of Lookout Point Dam, Oregon, all of the adjacent forest cover was removed, and along one side a trackway was dug high on the slope and the excavated material dumped farther downslope. Heavy rains in October 1950 became the final straw in producing landslides in this area. Thus, the surcharge of additional load coupled with diminution of shear resistance of trees and the increased water content exceeded the strength of the earth materials, and failure was the result.

Other man-made landslides   There are other ways in which human activities create landsliding; for example, when streams are diverted and undercut, a hillside becomes oversteepened and fails. Landslides also occur in the artificial settings that humans have created. The calamity at Aberfan, Wales, was caused by heavy rainfall on a large hill of mining spoil from coal operations. In the Netherlands Province of Zeeland, there were 229 slides between the years 1881 and 1946. These were related to the system of dikes built to protect the lands. The largest slides were 1.1×10^6 m³, and most were associated with the extra water pressure on the fine-grained sediments caused during exceptionally high tides.

We have also interfered with natural processes during the construction of off-shore oil drilling platforms and pipelines. Because many of these environments are near the instability threshold, they are easily triggered into failure by landsliding, as in the Mississippi delta. Since 1974 the U.S. Geological Survey has been map-

Figure 11-24 Landslide, Drift Creek, Oregon. This rotational slump was caused by clearcutting timber operations. (Courtesy Douglas Swanston.)

ping this hazardous area because of its bearing on oil developments. The mass movement phenomena include slumps, mudflows, bottleneck slides, and retrogressive slides. The initially high pore-water pressures associated with rapid deposition of the fine-grained sediments produce highly unstable conditions and, with only slight increments of force—new deposits, storm waves, or pile driving for platform supports— they can cause subaqueous slides on slopes as low as 0.2°. One single mudflow can be more than 30 m thick with a frontal length of 50 km. Such submarine slides did great damage to the port facilities of Valdez, Alaska, during the 1964 earthquake.

Earthquakes

Earthquakes are a major immediate cause for some landsliding and can produce disastrous effects. The relationship between earthquakes and landslides is discussed in Chapter 10.

RECOGNITION AND OCCURRENCE

In the past landslide areas were often either not recognized or were developed regardless of former tragedies. Today, however, landslide investigations are receiving increased support by government agencies as well as the general public. This stems from the growing awareness of their destructive potential. Zoning laws, grading ordinances, and environmental impact statements are becoming more common and reflect the increased level of sophistication in the public's understanding of earth phenomena. Mapping of landslides and landslide-prone topography is being done on both the regional as well as the local scale. For example, an aerial photography analysis of Colorado revealed that more than 25,000 km² of the state contained either landslides or were susceptible to landsliding.

Regional Mapping

Although landslides can occur in nearly all types of terrain, earth materials, and climates, certain environmental settings are more apt to produce landsliding than others. Mollard (1977) describes the four major landslide terranes in Canada as occurring on (1) mountain slopes in the Cordilleran region of western Canada, chiefly steeply dipping bedded and foliated rocks; (2) valley sides in Upper Cretaceous argillaceous bedrock in the western interior; (3) river banks and terrace bluffs of eastern Canada that contain fine-grained glaciomarine deposits (Leda Clay); and (4) valley walls, escarpments, and deep thaw basins in northwestern Canada.

Mountain glaciated terrane is a favorite locale for landslide occurrence because the creation of U-shaped troughs by ice erosion oversteepen the slopes. Such sites are then affected by a wide range of mass movements that include rockfall, slides, and avalanches, as well as talus and soil creep. When ice occupied the valleys, often more than 1000 m thick, an additional stress was created by this weight on the rocks. With deglaciation of the valley, the rocks readjust with the development of unloading joints that dip valleyward. These conditions have led to numerous landslides in Norway fjords, the Alps, the Rocky Mountains, and other glaciated mountainous terrane throughout the world. Such abnormally steepened hillslopes and the fracture planes have combined to provide the vulnerable setting for some of the major landslide tragedies, such as the one at Vaiont, Italy.

Local Mapping

Careful investigation of the stability of hillslopes should always precede human development on or below them. There is no substitute for prior geologic, hydrologic, geomorphic, and soil-rock mechanics studies by qualified experts. Although each type of landslide has its own fingerprint that needs interpretation, there are several diagnostic symptoms that generally indicate hazardous conditions. These surface features include (1) unusual contour changes and breaks in slope profiles (Fig. 11-25); (2) frontal bulges; (3) cracks in the ground and lateral tears; (4) steplike ground features; (5) hummocky ground; (6) anomalous moisture conditions and seeps; and (7) sharply defined vegetation changes. If some development has already occurred, the interested potential homeowner or developer should look for some of the following conditions that might betray the presence of unstable conditions: (1) doors and windows that stick and jam; (2) new cracks in plaster, tile, brickwork, foundations, and roadways; (3) outside walls, walks, or stairs pulling away from the buildings; (4) slowly developing cracks in the ground; (5) leaking of swimming pools; (6) movement, tilting, and out-of-plumb features such as trees, utility poles,

fences, walls, and other structures. When building new structures they should not be placed on steep slopes or close to the edges or in front of such slopes. Construction should also be avoided near or in the pathways of swales or drainages that would be capable of transporting large amounts of water and debris.

We cannot afford to be complacent and assume that landslides cannot occur even where the ground is reasonably level. The environmental setting for landslides that move, as in liquefaction flows or slab slides, is vastly different from most other landslide-prone terrane, because here the movements occur on nearly flat and featureless surfaces that give no hint of the hazard. For an understanding of these landslides that can lead to careful management of the terrain, knowledge of the stratigraphy and sedimentology of the strata is vital. Such movements only occur in fine-grained stratified sediments that contain abundant interstitial waters and that in a sense have been underloaded. Favorable locations where many of these have occurred are in glaciolacustrine and glaciomarine environments and in deltaic regions. Various processes can initiate the slides, but most are triggered by rapid changes in water content, loading, or earthquakes (and of course human intervention, as at Haverstraw). In Canada there have been many liquefaction flows on even level terrain such as at St. Thuribe on May 7, 1898, where nearly 3 million m^3 flowed through a 60 m opening in less than four hours and left a scar 500 m by 900 m and a depression that ranged to 15 m deep. The Nicolet slide in Quebec occurred November 12, 1955, and killed three people and caused $5 million property damage. An even worse disaster occurred on May 4, 1971, at St. Jean Vianney, Canada (Fig. 11-3). Here a new housing development was built on a terrace underlain by a 30 m thickness of the infamous Leda Clay (the glaciomarine material that has caused most landslides in the St. Lawrence region). Although a few cracks appeared on some of the blacktop streets on April 23, the real tragedy did not happen until 11 days later:

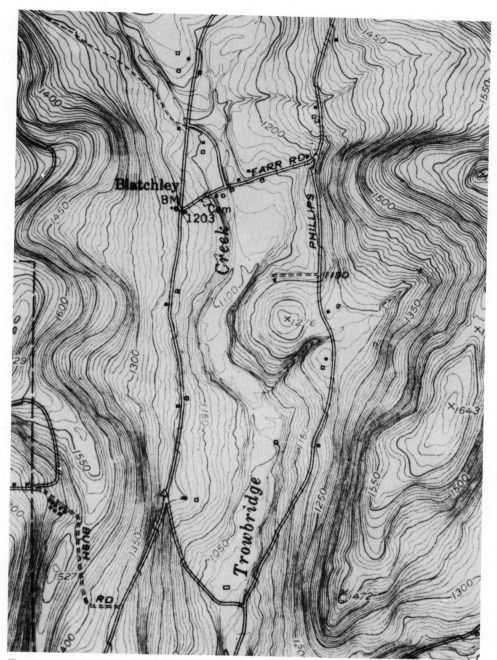

Figure 11-25 Blatchley Hill, New York. The hill in Trowbridge Creek valley occurred as a massive sediment slump, moving west from the reentrant concave slope of the mountain. It formed in immediate postglacial times. Other concavo-convex forms that extend from the reentrants along the north-south trending ridges are also periglacial features and primarily moved as solifluction lobes. (Topographic map of the Windsor Quadrangle, New York, U.S. Geol. Survey 1 : 24,000 series.)

The earth simply dissolved to a depth of nearly 100 feet; in the canyon thus formed, a river of clay—sometimes as deep as 60 feet—flowed at a rate of 16 miles per hour toward the Saguenay River, two miles away. At its widest, the canyon was a half mile across, and it extended for approximately one mile [Blank, 1971].

The few hours of destruction claimed the lives of 31 persons, and 38 homes disappeared into the liquefied material. Exceptionally heavy rains had overloaded the sediments, thus producing the conditions for landsliding.

Animal Behavior

Animals at times appear to feel the precursors of landslides. Before the Vaiont landslide cattle became exceedingly agitated, and at the St. Jean Vianney landslide dogs were the first to indicate unusual behavior patterns.

LANDSLIDE ASSESSMENT, PREVENTION, AND CONTROL

Geological engineering designs and solutions for the remedy of landslide problems need a strong data base to determine which control and prevention methods are most appropriate. Detailed studies are necessary to assess and calculate such factors as the size and shape of the unstable mass, the nature and composition of rock types and their structures, the attitude of joints and bedding features, and the water characteristics of the area. Thus, a combination of geologic, geomorphic, and hydrologic studies with soil and rock mechanics is necessary. These data can then be evaluated in terms of a total benefit-cost ratio for the project to determine the level of safety index—the quantitative analysis of the resistive forces versus the shearing forces. When this ratio is 1.0, the area is stable, but generally a safety factor of more than one is necessary and desirable. Some of the typical features that are looked for and assessed include the following.

Earth materials Cohesive fine-grained soil and sediments that are likely to become saturated. These materials are especially sensitive when disturbed by the removal of vegetation or by excavation.

Unconsolidated materials with low shear strength that can readily change to new angles of repose.

Inclined bedding, foliation, and schistosity that dip toward the valley or proposed development site.

Decomposed crystalline rocks that are susceptible to a large increase in moisture content. Weathering of such rocks may quickly reduce their shear strength to critical limits.

Fracture systems that possess low strength and can act as media for rapid water infiltration and lubrication.

Unfavorable rock sequences wherein the various units may easily uncouple and move differentially.

Water conditions Surface water accumulation in sag ponds; depressions.

Groundwater seeping through the soil or rocks or saturating materials so they are under high pore-water pressure. Also, the blocking and piping through pervious layers may suddenly increase hydrostatic conditions and cause slope failure.

Undercutting of banks on the outside bend of meanders by wave action in lakes, oceans, and reservoirs.

Sudden release of bank storage by rivers after flood stage.

Rapid drawdown of reservoirs, also causing sudden release in pressure of the bank storage.

Surcharge of water by heavy rainstorms that

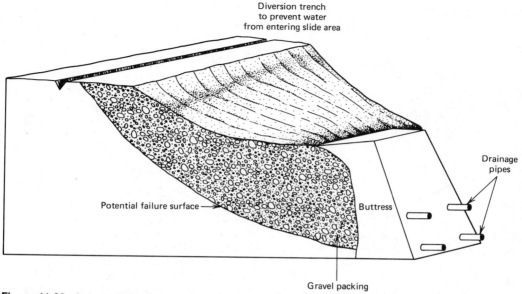

Figure 11-26 Landslide control by use of drainage methods and buttresses.

saturate the earth materials quickly to the threshold of shear failure.

Reduction of evaporation, causing failure in swelling clays by absorption of water.

Lines of springs and seeps may indicate permeable materials that rest on impermeable materials and thus form an unstable condition.

When the assemblage of data is complete, it needs to be evaluated to determine the magnitude and frequency of the potentially hazardous site. After the costs of preventive or remedial measures have been calculated and the legal aspects considered, a decision must be reached whether to avoid the unstable area or to attempt an engineering solution of the problem. The great diversity of landslide types and the numerous variables that are involved provide great challenges to the engineering geologist.

Avoidance Methods

There are many situations where the only viable solution is complete avoidance of the area. This strategy has the advantage of being the safest alternative—especially if other reasonable sites can be used in the area. In highway construction changing the grade in nearby corridors may aid in bypassing a dangerous location. If bridging over the potential hazard is required, great care must be taken so that the construction activities of blasting and the vibrations of pile driving do not trigger landsliding.

Water Control Methods

Water is the dominant ingredient in producing landslides. Thus techniques for its control and removal in the earth materials are the most widely applied measures to prevent landslides and to aid in the stabilization of hillslopes (Fig. 11-26). As in all control systems, the basic fundamentals for landslide remedy and prevention are to (1) decrease stress within the system, and (2) increase the shear resistance of the earth materials.

Surface water   In dealing with surficial water, the strategy is to prevent it from entering the

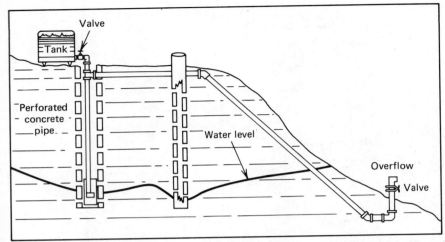

Figure 11-27 Siphon method for control of subsurface water. (Drawn by Mary Ryan.)

landslide area whenever possible and to drain off all water that is on the ground. Surface ditches, circumferential around the crown of the slide, divert the sheetwash that would otherwise enter the slide area and produce a water surcharge on the unstable materials. Ditches and collector systems on the slide may also be used to drain ponds and remove standing water. Some slopes may be regraded to allow more uniform drainage from fragile areas and to prevent its impoundment on the land surface. Cracks and other openings are filled with grouting or sealant to prevent water penetration into the regolith or bedrock. Sealants include such impermeable materials as clay, concrete, or bitumens. In exceptional cases, the entire slide area may be paved to allow rapid runoff and to prevent percolation, as was done at the Ventura Avenue oilfield in California.

Subsurface water For managing subsurface waters, the technique is to dewater the earth materials whenever possible and to lower the water table and pressure head when appropriate. There are a variety of ways for accomplishing this. Galleries and tunnels can be effective if size and cost are not important considerations. Horizontal drains and tile can also be installed near the surface. These are especially effective

in granular, highly permeable materials. In some cases pumping or collection of water by vertical drains, holes, and wells is necessary. These devices act as sumps for water removal. Occasionally a continuous siphoning system can be installed especially in permeable aquifers (Fig. 11-27). In other instances pipes and trenches may be used to divert undergound water away from the hazardous area.

Excavation Methods

These methods are used when an exceptionally dangerous condition has developed or failure has already occurred, and when the slide area is small or of modest size. Thus man-made cuts are deliberately produced for the purpose of changing the slope geometry and reducing shear stress of the materials.

1 Removal of slide This can only be done for small slides or those where the expense can be justified on a benefit-cost ratio. It guarantees safety.

2 Unloading head of the slide This method is only usable where there is easy access to the upper part of the slide and for those landslides that move because of such superimposed pressure, as in various types of slumps.

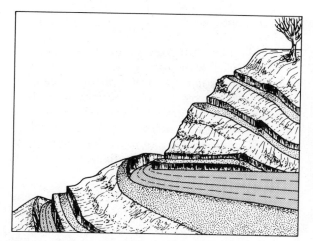

Figure 11-28 Stabilization of highway rockcuts by hillside benching. (Drawn by Mary Ryan.)

3 Regrading and slope reduction Such methods may be helpful when it has been determined that slope irregularities are the principal cause for building up stress differentials that can lead to landsliding. Slope changes also provide better drainage and equalization of pore-water pressure.

4 Hillside benching Construction of manmade terraces and berms can be effective in reducing landslide hazards in some instances (Fig. 11-28). Such methods are used when the slopes are especially long or steep and the toe of the slide has been undercut or altered. Benches prevent the buildup of slope-long stresses, but they should be designed for easy maintenance and coupled with drainageways to divert water from collecting.

Restraining Structures

These structures are used to increase the resistance to slide movements and are generally installed at or near the toe of the unstable area (Fig. 11-29). They are used as part of an overall design for hillside excavations or as a last resort when an area is already undergoing landsliding.

Steel
plate
Bolt

Figure 11-29 Use of rock bolts to stabilize inclined rock strata. (Drawn by Mary Ryan.)

Their advantage is in a relatively low cost, but they can be used only for small slides that are already known to exist. To be effective such structures must be coupled with other methods, especially with drainage techniques, because restraining devices are like barriers or dams and water is invariably imponded behind them. Thus diversion of water is essential to their success. Their stability is also dependent on whether the structure can resist sliding on or below its base, overturning, and failure within the structure. There are five restraining methods now in use; a brief discussion of each follows.

1 Buttresses These structures consist of rock or earth fill material that is placed on or into the toe of the slide to provide additional weight to increase the shear strength of materials in the slide. When properly designed, buttresses can be effective with landslides having rotational movement because the added mass produces a binding force that can feed back into the system and provide extra stability for upslope materials. They can also be effective in the slip-out type of landsliding. Buttresses can fail if they do not project sufficiently deep and where drainage was not adequately designed. After Interstate 70 was under construction in Colorado, the Loveland Basin landslide was discovered. By placing 46,000 m³ of fill at its base, the safety factor of the area was increased to 1.1 at the bottom and to 1.2 at the top of the slide.

2 Shear keys These are prismatic-shaped excavations that are filled with compacted fill. They are installed on the slide, generally near the top or where the increased uniformity will provide stability. They have been used at many California sites and successfully stopped movement of the Vista Verde landslide.

3 Retaining walls There is almost an infinite variety of styles and materials that can be used in attempts to dam the downslope movement of a landslide. Such structures are also called dikes, cribs, bulkheads, and walls, but their purpose is always the same—to impede motion of an unstable area. They can be composed of timber, concrete, stone, gabions (wired networks filled with stone rubble), grout, metal ribbing, and other types of fencing. Retaining walls are restricted to small areas, and their effectiveness is generally limited. To be successful they must be anchored with tie rods to adjacent stable terrain, and the back side must contain adequate drainage to divert water build-up.

4 Rock bolts The use of some type of metal dowels has become popular in certain situations in the past 20 years. They work best in rocks that are jointed or bedded with planes of discontinuity inclined downslope. They are usually composed of steel rods, drilled and inserted at angles to the planes of weakness, and contain a wedge or expansion device to secure them to the rock. In unusual instances they may exceed 12 m in length. They have proved successful throughout the Alps and in many rock cuts in the United States.

5 Piles The installation of piling or some type of vertical plug has occasionally been employed in slides that contain unconsolidated material. They are not in wide use and are often unsuccessful because material can move between the piles. The vibration during their driving and emplacement can also produce accelerated movement of the earth materials. Their use is restricted to very local situations.

Miscellaneous Methods

A variety of other techniques have been used to stabilize hillslopes where regolith materials are involved in landsliding. These methods have as their objective the increase in shear resistance by altering the physiochemical properties of the material; three methods are discussed here.

1 Grouting Under this method new materials are emplaced on or into the regolith in order to cement, harden, and stiffen materials, or release water. Portland cement can be used in sorted and granular materials to bind them together. Chemicals such as sodium silicate have been used to aid the bonding strength of some siliceous minerals, and lime has been used to reduce the plasticity of clayey materials (for example, montmorillonitic clays). Lime applications have been successful in stiffening the clay and eliminating water in landslides in California, Oklahoma, Iowa, Panama, and Brazil.

2 Electro-osmosis This method is rarely used because it is costly and is effective in only a very narrow range of materials. However, it has been used effectively on a small scale to produce accelerated dewatering of silty soils in Norway and Canada. Purpose of the method is to drive out pore water by passing an electrical current through the materials.

3 Temperature changes Other methods for slope stabilization have been used that depend on causing new temperature regimes in the regolith. In order to allow construction on unstable slopes when building the Grand Coulee Dam in Washington, the materials were frozen by the circulation of freezing fluids through pipes in the sediments. In the loess soil of Rumania construction was possible only after the soils were baked and hardened.

Perspectives

The general public is finally starting to realize the hazards of landsliding. This is indicated by the increasing number of laws that pertain to hillside development and the increased funding level of government agencies to investigate and map landslide-prone terrane. Because of the nearly ubiquitous character of landslides, there are few regions that can afford to ignore them in land use planning. Fortunately, excellent mapping models are available such as the San Francisco Bay region study by the U.S. Geological Survey and the study of urban areas by the West Virginia Geological Survey in that state. Laws such as the California grading ordinances and the National Flood Disaster Protection Act of 1973, and of course the National Environmental Policy Act (NEPA), show an increased governmental commitment to public safety and the necessity of dealing with the landslide problem. Such policies will make additional demands on the scientific and engineering professions for accurate prediction of unstable areas, for more precise calculations of benefit-cost ratios, and for the development of safety measures at new construction sites.

Fortunately, the disciplines of soil and rock mechanics, which had very slow growth during their formative years, are now coming of age. The data that such specializations can produce when coupled with interdiscipline investigations—including geology, hydrology, engineering, and geomorphology—can yield scientific conclusions that must be used in the decision-making process for landslide management.

The forces that are causing the topic of landslides to come under increasing scrutiny today will probably still be influential in tomorrow's world. With continued population growth, urban and suburban areas will expand into more marginal and potentially dangerous terrain. Furthermore, there will be increased development in regions whose slope stability is more fragile than many of those in current use. For example, the permafrost areas and tropical regions will become more densely populated. The LDCs (Less Developed Countries) will need much expertise as they develop their economy and resources. Land use planning in such nations will need careful evaluation if tragedies are to be avoided. And the cost of land acquisition, as needed for highway projects and in other developments, is rising sharply. Thus, it becomes vital to know how large a safety corridor to purchase so as to maintain the integrity of the investment. It makes a big difference whether slopes are to be designed with 25 percent or 50 percent grades. Only careful regolith and rock studies can supply the answer.

READINGS

Bailey, R. G., 1971, Landslide hazards related to land use planning in Teton National Forest, northwest Wyoming: U. S. Dept. Agri. Forest Service Intermountain Region, 131 p.

Beland, J., 1956, Nicolet landslide, November 1955: Geol. Assoc. Canada Proc., v. 8, p. 143–156.

Bjerrum, L., 1967, Progressive failure in slopes of overconsolidated plastic clay and clay shales: Jour. Soil Mech. Found. Div. Amer. Soc. Civil Eng., v. 93, p. 1–49.

Blank, J. P., 1971, The town that disappeared: Reader's Digest, v. 99, Dec., p. 86–90.

Cleaves, A. B., 1961, Landslide investigations, a field handbook for use in highway location and design: Washington, D.C., U.S. Dept. Commerce Bureau of Public Roads, 67 p.

Close, U., and McCormick, E., 1922, Where the mountains walked: Nat. Geog. Mag., v. 41, p. 445–464.

Coates, D. R., ed., 1977. Landslides: Geol. Soc. Amer. Reviews in Engineering Geol. Vol. III, 278 p.

Crandell, D. R., and Fahnestock, R. K., 1965, Rockfalls and avalanches from Little Tahoma Peak on Mount Ranier, Washington: U. S. Geol. Survey Bull. 1221-A, 30 p.

Crawford, C. B., 1968, Quick clays of eastern Canada: Eng. Geol-Internat. Jour. v. 2, p. 239–265.

Cruden, D. M., 1976, Major rock slides in the Rockies: Canadian Geotech. Jour., v. 13, p. 8–20.

Eckel, E. B., ed., 1958, Landslides and engineering practice: Natl. Research Council, Highway Research Board Spec. Report. 29, 323 p.

Hsu, K. J., 1975, Catastrophic debris streams (Sturzstroms) generated by rockfalls: Geol. Soc. Amer. Bull. 86, p. 129–140.

Jones, F. O., 1973, Landslides of Rio de Janeiro and the Serra das Araras Escarpment, Brazil: U. S. Geol. Survey Prof. Paper 697, 42 p.

Jones, F. O., Embody, D. R., and Peterson, W. L., 1961, Landslides along the Columbia River Valley, northeastern Washington: U. S. Geol. Survey Prof. Paper 367, 98 p.

Kerr, P. F., Stroud, R. A., and Drew, I. M., 1971, Clay mobility in landslides, Ventura, California: Amer. Assoc. Petroleum Geologists, v. 55, p. 267–291.

Ladd, G. E., 1935, Landslides, subsidences and rock-falls: Am. Ry. Eng. Assoc. Bull., v. 37, 72 p.

Leighton, F. B., 1976, Geomorphology and engineering control of landslides: in Coates, D. R., ed. Geomorphology and engineering, Stroudsburg, Dowden, Hutchinson & Ross, p. 273–287.

Mathewson, C. C., and Clary, J. H., 1977, Engineering geology of multiple landsliding along I-45 road cut near Centerville, Texas: in Coates, D. R., ed., Landslides, Geol. Soc. Amer. Reviews in Engineering Geol. Vol. III, p. 213–223.

Mollard, J. D., 1977, Regional landslide types in Canada: in Coates, D. R., ed., Landslides, Geol. Soc. Amer. Reviews in Engineering Geol. Vol. III, p. 29–56.

Morton, D. M., and Streitz, R., 1967,

Landslides: Calif. Div. Mines. Geol. Mineral Inform. Ser., v. 20, n. 11, p. 135–140.

Nilson, T. H., Taylor, F. A., and Brabb, E. E., 1976, Recent landslides in Alameda County, California (1940–71): an estimate of economic losses and correlations with slope rainfall, and ancient landslide deposits: U. S. Geol. Survey Bull. 1398, 21 p.

Nilson, T. H., Taylor, F. A., and Dean, R. M., 1976, Natural conditions that control landsliding in the San Francisco Bay Region: U. S. Geol. Survey Bull. 1424, 35 p.

Olson, E. P., 1978, Landslide investigation Manti Canyon, in 2 parts: U. S. Dept. Agriculture-Forest Service, Intermountain Region, 51 p.

Perla, R. I., and Martinelli, M., Jr., 1978, Avalanche handbook: U. S. Dept. Agriculture-Forest Service, Agri. Handbook 489, U.S. Govt. Printing Office, 254 p.

Sharpe, C. F. S., 1939, Landslides and related phenomena: New York, Cooper Square Pub. Inc., 137 p.

Shreve, R. L., 1968, The Blackhawk lands Geol. Soc. Amer. Spec. Paper 108, 47

Swanston, D. N., and Swanson, F. J., 1976, Timber harvesting, mass erosion, and steepland forest geomorphology in the Pacific Northwest: in Coates, D. R., Geomorphology and Engineering, Stroudsburg, Dowden, Hutchinson & Ross, Inc., p. 199–221.

Terzaghi, K., 1950, Mechanisms of landslides: Geol. Soc. Amer.: in S. Paige, Chairman, Application of geology to engineering practices, Berkey Volume, p. 83–123.

U. S. Department of Agriculture, 1978, Avalanche handbook: Forest Service-USDA Agri. Handbook 489, 254 p.

Varnes, D. J., 1958, Landslide types and processes: in Eckel, E. B., ed., Landslides and engineering practice: Natl. Research Council, Hy. Research Board Spec. Rept. 29, p. 20–47.

Williams, G. P., and Guy, H. P., 1971, Debris avalanches—a geomorphic hazard: in Coates, D. R., ed., Environmental geomorphology, Pub. in Geomorphology, State Univ. of New York at Binghamton, p. 25–46.

Zaruba, Q., and Mencl, V., 1969, Landslides and their control: Amsterdam, Elsevier, 205 p.

Chapter Twelve
Floods

Flooding results of Hurricane Camille in Virginia.
(Courtesy Garnett Williams, U.S. Geol. Survey.)

INTRODUCTION

Floods are not a newly invented phenomenon sent to afflict only modern man. Their occurrence has been common since the first rivers formed, and they can be an entirely natural event. Ancient man was so impressed with floods that they became an important part of his legends and religion. Furthermore, since the time of Noah and the ark, mankind has attempted in some manner to adjust to flooding and the damages it causes. Not only do floods cause grievous harm to man, his belongings, and his property but, in some cases, man can also be the instigator of floods as well as the cause of their amplification. Thus there can be several causes and types of flooding. They can result from excessive precipitation or from the bursting of a dam. Unlike other geologic hazards, floods affect only lowlands. It is also necessary to make a distinction between river floods and coastal floods, because the only element they have in common is the inundation of the lands. They differ in cause, length of time on the ground, and predictability.

Floods are the most ubiquitous of the geologic hazards and affect more people than all other hazards combined (Table 12-1). In the United States, more than 20 million ha are subject to river flooding, where 10 million people live in the principal hazard area and 25

Table 12-1 Flood losses throughout the world

Year	Locality	Lives lost	Remarks
1969	India, southeast coast	600+	Cyclone
	Taiwan	177	Typhoon
1970	Martinique	50	Tropical storm Dorothy
	Philippines	2,011	Typhoons George, 670; Kate, 107; Sening, 583; Titang, 526; Patsy, 125
	Bangladesh, delta	300,000	Cyclone; second worst disaster of 20th century
	Rumania	200	Floods
1971	Philippines	40	Tropical storms Wanda, 25; Dadang, 15
	South Korea	178	Typhoons Olive, 78; Hester, 100
	Hong Kong	25	Typhoon Rose
1972	Peru	12	Floods
	Japan	115	Floods
	Hong Kong	87	Floods
	Philippines	427	Floods, 2 million homeless
	India, east	100	Floods
	Nepal	105	Floods
	South Korea	296	Floods
1973	Algeria	20	Floods
	Argentina	20	Floods
	Bangladesh	900+	Floods
	Mexico	130	Floods
	Philippines	54	Typhoon
	Pakistan	290	Floods
	Sicily	10	Floods
	South Vietnam	60	Floods
	Spain	190	Floods
	Tunisia	86	Floods
1974	Argentina	60	Floods
	Australia	15	Cyclone, and flooding, greatest natural disaster
	Bangladesh	2,000	Floods, inundated one-half of lands
	Bangladesh	300	Cyclone
	Brazil	1,500	Floods and landslides
	Honduras	8,000	Hurricane Fifi
	India	302	Floods
	Mexico	20	Floods
	South Korea and Japan	88	Typhoon Gilda
	Philippines	149	Tropical storm and floods
	Caribbean Islands	48	Tropical storm Alma
1975	Argentina	20	Floods
	Egypt	15	Floods
	India	17	Floods; more than 50 died from cholera caused from flooding
	Pakistan	63	Floods
	Thailand	131	Floods
	Yemen	70	Floods
1976	Baja California	500	Hurricane Liza
	Colombia	58	Floods
	Hong Kong	500	Floods
	Indonesia	136	Floods

Table 12-1 *(Continued)*

Year	Locality	Lives lost	Remarks
	Japan	93	Floods
	Mexico	50	Floods
	Pakistan	316	Floods
	Philippines	215	Floods
	South Korea	25	Floods
1977	Bolivia	16	Floods
	India	35	Floods
	India	8,400	Cyclone
	Greece	26	Floods
	Italy	15	Floods
	South Africa	12	Floods

million others are on contiguous lands that could also be affected (Table 12-2 and Fig. 12-1). Since 1955, the annual average of lives lost from floods is about 100, and property damages run about $1.5 billion per year. Flood losses are the price mankind must bear for inhabiting hazard-prone terrain. Floodplains are highly prized because of their rich agricultural soils, efficient layout for housing and industry, and easy access for transportation and service facilities. The rationale for such development is generally related to a perception or calculation that the benefits derived from such occupance outweigh the costs that can be incurred in the case of disaster.

One positive element about most floods is that their forecasting and warning can be more predictable than other geologic hazards, such as earthquakes and landslides. Excessive precipitation is the cause for most disastrous floods, and it usually takes a day or more to accumulate sufficient runoff to initiate major calamities. Sudden flash floods and dam failures are exceptions where advance alerts are not possible.

There are many negative elements about floods, and the strategies for dealing with them are subject to more controversies in land management than any other single natural phenomenon. The conflicts can be on the governmental level—for example, deciding which agencies should have jurisdiction of potential problems.

Whether big dams are superior to many small dams is a continuing question of debate. Such arguments are also part of the battle between those who live in upstream versus downstream areas. Thus, there is not agreement on which methods are most effective in flood prevention, nor on the merits of dams when other environmental impacts they produce are considered. As long as we continue to inhabit floodplains and low areas, we will always face the threat of floods. Indeed, floods are the handmaiden of civilization and the hazard phenomenon that challenges our ingenuity in compromising with nature.

PHYSICAL CHARACTERISTICS

The majority of the world rivers flow in channels with walls that are steeper than the land. The adjoining flatter area is commonly called a **floodplain** because it is moderately level terrain that provides the safety valve for movement of high flows when they exceed the capacity of the channel to hold them (Fig. 12-2). When the flow spreads out from the normal channel it is referred to as **floodwater**, and the river at this time is said to be in **flood** or in **flood stage**. It is normal for rivers in humid-temperate climates to overflow their channel once in about every

Table 12-2   Severe flood disasters in the United States

Year	Place	Lives lost	Million dollars lost	Cause
1831	Barataria Isle, Louisiana	150		
1844	Upper Mississippi River			Rainfall-river flood
1856	Isle Derniere, Louisiana	320		Hurricane tidal flood
1875	Indianola, Texas	176		Hurricane tidal flood
1886	Sabine, Texas	150		Hurricane tidal flood
1889	Johnstown, Pennsylvania	2,200	20	Dam failure
1900	Galveston, Texas	6,000	25	Hurricane tidal flood
1903	Passaic & Delaware rivers	100	25	Rainfall and dam failure
1903	Missouri River basin		50	Rainfall-river flood
1903	Heppner, Oregon	247		Rainfall-river flood
1906	Gulf Coast	151		Hurricane tidal flood
1909	Gulf Coast-New Orleans	700		Hurricane tidal flood
1913	Ohio River basin	467	150	Rainfall-river flood
1913	Brazos & Colorado rivers, Texas	177	128	Hurricane rainfall-river flood
1915	Louisiana and Texas	550		Hurricane tidal flood
1919	Louisiana and Texas	284		Hurricane tidal flood
1921	Arkansas River	120	13	Rainfall-river flood
1926	Miami & Clewiston, Florida	350	70	Hurricane tidal and river flood
1926	Illinois River			Rainfall-river flood
1927	New England		50	Rainfall-river flood
1927	Lower Mississippi	100	284	Rainfall-river flood
1927	Vermont	120		Rainfall-river flood
1928	Lake Okeechobee, Florida	2,400		Hurricane tide and waves
1928	Puerto Rico	300	50	Hurricane tide and waves
1932	Puerto Rico	225		Hurricane tide and waves
1935	Susquehanna-Delaware rivers	52	36	Rainfall-river flood
1935	Florida Keys	400		Hurricane tide and waves
1935	Republican River, Kansas, Nebraska	110		Rainfall-river flood
1936	Northeastern U.S.	107	221	Rainfall-river flood
1936	Ohio River basin	137	150	Rainfall snowmelt flood
1937	Ohio River basin		418	Rainfall-river flood
1938	New England streams	200	125	Hurricane tidal and river flood
1938	California streams	79	100	Rainfall-river flood
1939	Licking & Kentucky rivers	78	1.7	Rainfall
1940	Southern Virginia, Carolinas, E. Tennessee	40	12	Rainfall
1942	Mid-Atlantic coastal streams		28	Rainfall-river flood
1943	Central States	60	172	Rainfall-river flood
1944	South Florida		63	Hurricane tidal & river flood
1944	Missouri River basin		52	Rainfall-river flood
1945	Hudson River basin		24	Rainfall-river flood
1945	South Florida		54	Hurricane tidal and river flood
1945	Ohio River basin		34	Rainfall-river flood
1947	South Florida		60	Hurricane tidal & river flood
1947	Missouri River basin	29	178	Rainfall-river flood
1948	Columbia River basin	35	102	Rainfall-river flood
1950	San Joaquin River, California		32	Rainfall-river flood
1950	Central W. Virginia	31	4	Rainfall
1951	Kansas River basin	28	883	Rainfall-river flood

Table 12-2 (*Continued*)

Year	Place	Lives lost	Million dollars lost	Cause
1952	Missouri River basin		180	Snowmelt floods
1952	Upper Mississippi River		198	Rainfall-river flood
1954	New England streams		180	Hurricane tidal floods
1955	Northeastern U.S.	185	684	Hurricane tidal & river flood
1955	California & Oregon streams	61	271	Rainfall-river flood
1957	Louisiana	556		Hurricane tidal floods
1957	Ohio River basin		65	Rainfall-river flood
1957	Texas rivers		144	Rainfall-river flood
1959	Ohio River basin		114	Rainfall-river flood
1960	South Florida		78	Hurricane tidal & river flood
1960	Puerto Rico	107		Hurricane tidal & river flood
1961	Texas Coast		300	Hurricane tidal floods
1963	Ohio River basin	26	97.6	Rainfall
1964	Montana	31	54.3	Rainfall
1964	Florida		325	Hurricane tidal & river floods
1964	Ohio River basin		106	Rainfall-river floods
1964	California streams	40	173	Rainfall-river floods
1964	Columbia River-N. Pacific		289	Rainfall-river floods
1965	South Florida		139	Hurricane tidal and river flood
1965	Upper Mississippi River	15	158	Rainfall snowmelt river flood
1965	Platte River, Colorado-Nebraska		191	Rainfall-river flood
1965	Arkansas River, Colorado-Kansas		61	Rainfall-river flood
1965	New Orleans and vicinity		322	Hurricane tidal flood
1965	Sanderson, Texas	26	2.7	Flash flood
1967	S. California	12		Rainfall-river flood
1969	California	103	399.2	Rainfall
1969	Northern Ohio	30	87.9	Rainfall
1969	James River, Virginia	154	116	Rainfall
1971	Chester, Pennsylvania	10		Rainfall-river flood
1972	Rapid City, S. Dakota	245	200	Rainfall-river flood
1972	E. United States	118	3,500	Hurricane Agnes
1973	Mississippi basin	11	1,500	Rainfall
1974	U.S. Midwest	28		Flash floods
1976	Big Thompson Canyon, Colorado	140	16.5	Flash floods
1977	Johnstown, Pennsylvania	49	117	Flash floods
1977	Texas	31		Rainfall
1977	Kansas City, Missouri	23	50	Rainfall

Source: Office of Emergency Preparedness, 1972, Disaster preparedness: U.S. Government Printing Office; and other accounts.

2.33 years. The term **floodway** is used to define the area of the channel and the immediately contiguous flat ground that provides the avenue for most floods. In large valleys with well-defined floodplains, that part of the valley adjacent to the hillslopes is the **floodfringe**; this area may be inundated by exceptionaly high floodwaters that constitute the major floods which occur on a frequency of about every 100 years (Fig. 12-3).

Engineers consider a river to be "in flood" when its water has risen to an elevation (flood

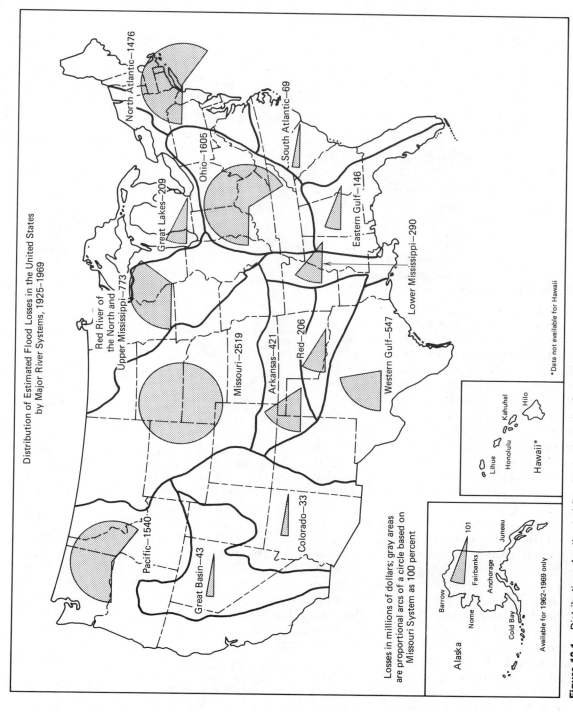

Distribution of Estimated Flood Losses in the United States
by Major River Systems, 1925–1969

North Atlantic—1476

South Atlantic—69

Ohio—1605

Great Lakes—209

Eastern Gulf—146

Red River of
the North and
Upper Mississippi—773

Lower Mississippi—290

Missouri—2519

Arkansas—421

Red—206

Western Gulf—547

Pacific—1540

Colorado—33

Great Basin—43

Losses in millions of dollars; gray areas
are proportional arcs of a circle based on
Missouri System as 100 percent

Lihue
Honolulu
Kahuhal
Hilo
Hawaii*

*Data not available for Hawaii

Alaska
Barrow
Nome
Fairbanks
101
Anchorage
Juneau
Cold Bay

Available for 1962–1969 only

Figure 12-1. Distribution of estimated flood losses in the United States
by major river systems, 1925–1971. (U.S. Dept. of Commerce,
Climatologic Data, National Summary, 1972.)

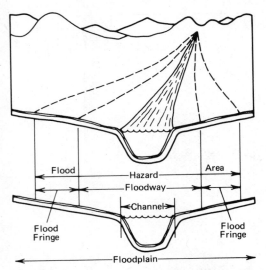

Figure 12-2. Valley cross section showing different components of the floodplain.

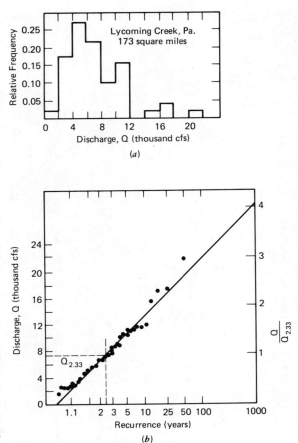

Figure 12-3. Relative frequency (a) and probability of occurrences and recurrence period (b) of floods for Lycoming Creek, Pennsylvania. (After Reich in Coates, 1971, *Environmental Geomorphology*.)

stage) at which damage can occur in the absence of protective works. When planning engineering construction, the term **standard project flood** is used as a bench mark against which other floods can be measured. Its frequency is variable but it is usual to express it as the magnitude that recurs every 100 or 200 years. However, most engineering flood prevention measures are gaged to control the **design flood.** This is the peak discharge value adopted as the basis for the planning of a specific project. It is chosen on the basis of engineering considerations and judgment.

Further distinction on flooding should be made of river floods that result from excess precipitation and/or snowmelt and of coastal floods that inundate low-lying areas with ocean or lake waters. River floods may be of two types (1) **floods** that develop and crest over a period of 12 hours or more, and (2) **flash floods** that develop suddenly and crest within a few hours. For example, many of the local and high watershed damage from Hurricane Camille occurred from flash flooding where streams crested up to 6 m above normal flood stage in 4 to 8 hours. More than 2500 communities in the United States are subject to flash flooding. Finally, flooding can also be caused by our upsetting of the natural hydrologic regime and streamflow, and by bizarre events associated with other geologic processes which will be described in the following section (see also page 406 and page 616).

Origin of Flooding

Although the majority of flooding is created by heavy rainfall (Fig. 12-4), there is a wide variety

Figure 12-4. Newspaper headlines of the September 1970 flood in Arizona, Utah, and Colorado. (U.S. Geol. Survey.)

of both geologic processes and processes induced by human intervention that produce flood conditions.

Geologic causes

1 *Excess precipitation.* This is the principal cause of most floods. Stream channels are often developed during the fluvial cycle of erosion to hold only the runoff caused during heaviest rainfall periods, which occur about every two years. Thus storms of greater magnitude or intensity that recur every three years or more are capable of exceeding the bank storage, thereby overflowing onto the adjacent floodplain or other terrain.

2 *Snowmelt.* In regions of heavy snowfall if the melting time is accelerated by abnormal weather and coupled with either already saturated ground or frozen ground then chances for extreme flood conditions are greatly increased. These circumstances occur in many northern states and have often

led to floods—for example, in early spring in the Red River of North Dakota.

3 *Ice dams.* Flooding is a common occurrence in northern climates where river ice builds up and prevents normal river flow. The backup of such water forms a temporary lake condition that then reaches a critical threshold of impoundment, breaching the ice barrier and resulting in downstream floods. Many rivers in Alberta Province, Canada, display these characteristics.

4 *Landslides.* When landslides dam rivers, the imponded waters may burst the barrier and cause downstream flooding, as in the Kelly slide, Gros Ventre, Wyoming (Fig. 11-7).

5 *Glaciers.* The most spectacular flooding on earth was associated with what were originally termed the Spokane floods. The flooding was produced by the spillage of glacial Lake Missoula through a glacial ice barrier. Modern jökulhlaups occur periodically in

Figure 12-5. Fire or clearcutting may so completely destroy vegetation that contour-trenches may be necessary to obtain immediate control of runoff and aid in revegetation and stabilization of the soil. (Courtesy of U.S. Dept. of Interior.)

Iceland glaciers and in Alaska. For example, in Alaska the Lake George breakout is nearly an annual event. Here in the Chugach Mountains the 48 km long lake often breaks through its dam of ice between mid-June and late August and deluges the Knik River valley with about 4 billion m³ of floodwater.

Human activity causes

1 *Urbanization.* By encroaching onto the floodplain and making the ground impervious to percolation, we have caused more and larger floods (Chapter 19).

2 *Deforestation and poor cropping practice* (Fig. 12-5). Stripping vegetation from previously forested slopes and agricultural and grazing abuses of soil increase the size and frequency of floods.

3 *Channelization.* Confinement of a river by engineering works may aid the immediate area but can produce higher flood peaks in downstream reaches (Chapter 13).

4 *Dams.* Faulty construction, poor geologic foundations, and induced environmental changes have caused many dam failures and tragedies. (Table 12-3, and pages 243, 375).

5 *Weather modification.* Although the claims are controversial, the opinion is held by some that cloud seeding in South Dakota was partly responsible for producing the exceptional rainfall that caused the Rapid City flood of 1973.

6 *Mining.* Surface and strip mining produce excess sediment that clogs channels and increases flood peaks (Fig. 5-23).

FLOOD DISASTERS AND DAMAGE

Each year, floods produce staggering losses throughout the world, killing thousands of people and countless animals and costing billions of dollars in property destruction and commercial dislocations (Appendix I). Such losses range from minor and local inundations to flooding of thousands of kilometers. The Huang Ho

Table 12-3 Flood losses from dam failure

Date	Location	Killed	Millions of dollars lost	Remarks
1874	Connecticut River watershed	143		
1889	Johnstown, Pennsylvania	2200	10	Severe rainstorm
Mar. 12, 1928	St. Francis, California	350	10	Erosion of bedrock foundation
Oct. 9, 1963	Vaiont, Italy	2600		Landslide into reservoir created flood wave that overtopped dam, structure not damaged, but now abandoned
Dec. 14, 1963	Baldwin Hills, California	5	12	Combination of faulting and differential settlement fractured earth dam
Mar. 24, 1968	Lee, Massachusetts	2	10	Piping in earthfill foundation
Feb. 26, 1972	Buffalo Creek, W. Virginia	125	50	9.4 cm rainfall overloaded coal-waste fill dam
June 9–10, 1972	Rapid City, S. Dakota	245	200	25 cm rainstorm in 9 hours; Canyon City Dam breached but this contributed only a part to the total destruction
June 5, 1976	Teton River, Idaho	14	1000	Differential settlement and piping through foundation fractures, undermined earthfill dam
Nov. 1977	Toccoa, Georgia	38	5	Rainfall overloaded 40-yr-old weakened earth dam and devastated college and other structures downstream
Aug. 11, 1979	Morvl, India	5000	Millions	Excessive rainfall caused collapse of dam on the Machu River; dam had been built in 1978

(Yellow River) of China is known as "China's Sorrow" because of the almost annual floods that devastate the region. The 1931 flood of the Yangtze River affected an area the size of the British Isles, killed millions, and ruined the homes and belongings of 60 million. Flood losses are generally assessed as being either primary or secondary. Some losses are easy to calculate in terms of property destroyed, but other losses are more difficult to analyze and might be described as "intangibles," such as life and health. Although most major floods can now be predicted, they still cause ruinous societal damages.

Primary Effects

These are the direct losses attributed to inundation and contact with floodwaters and the strength of the currents. People and animals may be killed or seriously injured by the water and the debris-laden flow. The swift-moving floodwaters cause structural damage to buildings, roads, bridges, and railroads (Figs. 12-6

Figure 12-6. The Mississippi River flood in 1973 inundated many communities. (Photo near Memphis, Tennessee, by Corps of Engineers.)

and 12-7). Power lines, gas mains, sanitary lines, water lines, and telephone lines may be damaged or destroyed. When transportation and communication systems are ruined or interrupted, the rate of recovery to normalcy is impeded, which causes additional losses. Railroad and highway beds are often vulnerable to the erosive forces and can be undermined. Bridge abutments may also wash out (Fig. 12-8), causing collapse of the structure. Debris on roads and airport runways also interfere with

quick restoration of services. Industrial and business inventories may be destroyed, and personal belongings lost. Crops and trees can be removed; farms can lose storehouses of grain, feed, equipment, and buildings; and farmlands may be deeply eroded by new channels, lose topsoil, or be waterlogged for long periods so that cropping for the season is lost. The loss to farmers alone contributed to hundreds of millions of dollars during the 1973 Mississippi River flood (Fig. 12-9).

Secondary Effects

The floodwaters can produce additional damages on services, facilities, and goods even though they do not come in direct contact with the water. Malfunctions may occur when power and gas lines are broken that have significant impact on adjacent areas. Electrical fires can be produced by short circuits, and fire and explosions may be initiated by damaged gas lines. Chemical pollution can contaminate the waters if storage areas and containers become dislodged and subsequently incorporated into the adjacent lands and water.

Figure 12-7. Mississippi River Flood of 1973 interrupted communications and covered rich farmland. (Photo near Memphis, Tennessee, by Corps of Engineers.)

Figure 12-8. Hurricane Agnes floodwaters destroyed Day Hollow Bridge in 1972, Binghamton Metropolitan Area, New York. The upstream reach of the stream had been previously channelized, which contributed to the amount of local flooding.

Both the primary and secondary effects may further cause chain-reaction effects that greatly increase the costs and disruptions from the flooding. The dislocations of normal services, goods, and supplies can lead to hunger, disease, and other human suffering. Families may be separated, people left homeless, and their economic productivity lost. Continuing health hazards may occur from contamination and breakage of water and sanitary systems. Wild animals, snakes, and rats may be driven from their normal habitats, infest the human community, and introduce additional health problems. Local schools and businesses may be closed with losses to education, commerce, and employment. Trade may be delayed or lost when industrial goods cannot be shipped to other communities. Fields and pastures may lose productivity through either erosion of rich topsoil or siltation of deleterious debris. New river channels may form to the detriment of property lines and navigation. Although most wildlife is tuned to a certain amount of flooding, large floods may destroy food sources and breeding grounds for several species.

The amount of flood damage is a function of the magnitude of the flood, the rate of rise and fall of floodwaters, the velocity of the water, the sediment load, and the time of year. For example, a winter flood produces less damage than a summer flood in northern states because the ground is frozen and is not easily eroded and the crops have already been harvested and most grains marketed. The size of the flood is dependent on watershed conditions which include:

1 *Antecedent soil moisture.* If the ground is already saturated, a new storm of even moderate intensity can produce rapid flooding.

Figure 12-9. Extensive damages occurred to farmlands, such as this one in Mississippi, and loss of revenue to farmers was great because inundation lasted more than two months and prevented the planting of new crops. The levee maintained its integrity at this locality but had been damaged upstream. (Corps of Engineers photo of Mississippi River 1973 flood, taken April 30.)

2 *Vegetation types and density.* In general, forested terrain can produce a dampening effect on storms of moderate size, whereas barren or devegetated ground will accelerate conveyance systems and cause higher peak flows.

3 *Topography.* The length, steepness, and shape of hillslopes all influence the character of storm runoff.

4 *Geology.* The composition and texture of the soil and bedrock determine the percolation rate and storage capacity of water. With porous and permeable materials, such as sandy soils, the holding ability of the materials may inhibit rapid delivery of water to stream channels and act as a buffer against excessive and quick streamflow.

CASE HISTORIES

There have been so many floods throughout human history that it is difficult to select as case histories those that are representative. Although flooding is a worldwide phenomenon, Tables 12-2 and 12-3 show the status of flood losses in an industrialized and increasingly urbanized society—the United States. The following discussion of flood case histories contains disasters that have occurred from different causes, but emphasizes the point that flood catastrophes are not diminishing in recent years. Indeed, losses continue to climb and they pose greater threats today than they ever have.

United States

The year 1972 has been hailed as "the year of the floods" because three separate floods, all distinctly different, ravaged three parts of the country.

1   On February 26, 1972, the most destructive flood in West Virginia's history swept through the Buffalo Creek valley in the southwestern corner of the State. A coal-waste dam collapsed on Middle Fork in the watershed and released 511,000 m^3 liters of water. The flood wave, up to 6 m in height, traveled the 24 km long valley at a speed of 9 km/h. The small settlement of Saunders and 16 mining camps were destroyed, 125 lives were lost, 500 houses ruined, 4000 people left homeless, and total property loss exceeded $50 million. Whereas the 9.4 cm rainfall produced the 10-year flood in adjacent watersheds, the discharge in Buffalo Creek was 40 times greater than the 50-year flood. The design of the dam was defective in that construction did not make allowance for the water pressure, it contained no spillway for high headwater, and the mined refuse which comprised the structure had not been properly compacted during construction.

2   The 50,000 population of Rapid City, South Dakota, was augmented by thousands of tourists awaiting the start of "Dakota Days" in early June 1972. On June 9, a stationary group of thunderstorms in the adjacent Black Hills produced more than 25 cm of rainfall in six hours, causing the highest floods ever recorded in South Dakota. Rapid City was in the center of the flood belt which was 64 km long and 32 km wide on the eastern slopes of the Black Hills. Rapid City was severely flooded on June 9 and 10, and on June 9 floodwaters rose 1 m in 15 minutes. The flood peaks traveled 10 km/h. Total discharge through the city was 16 billion liters during the two-day period, and the floodwaters killed 245, destroyed 1335 houses and 5000 automobiles (Fig. 12-10), and caused damages of about $200 million. Mud covered more than 20 percent of the city, and 80 blocks of paving were ripped away by the force of the currents. The arguments concerning the previous cloud seeding west of the storm area and its relationship, if any, to the heavy precipita-

Figure 12-10. Floodwaters neatly stacked these cars during the June 9–10, 1972, flood in Rapid City, South Dakota. A total of 5000 cars were destroyed and many more were damaged by the inundation. (Courtesy *Rapid City Journal*.)

tion and the role played by collapse of the Canyon Dam are still being debated.

3   The weak tropical disturbance first detected on June 14, 1972, over the Yucatan Peninsula gave no hint it was to become the single most devastating storm in the history of the United States. The storm rose to hurricane status as it moved over the Gulf of Mexico but, after hitting the Florida Panhandle on June 19, it weakened and never regained hurricane intensity. Although called Hurricane Agnes, it moved northward toward central New York and as it traveled joined forces with an extratropical storm centered in northeastern United States. This reinforced combination system stagnated over parts of New York and Pennsylvania for about 24 hours and produced rainfalls of 38 cm during this period. The extraordinary precipitation and large storm diameter of 1600 km caused unusual flooding and ravished parts of 12 states, with much of the damage in Pennsylvania. In the Susquehanna Valley from New York to Chesapeake Bay, this was the greatest recorded flood since 1784. Peak flows from Harrisburg downstream exceeded 1 million cfs (cubic feet per second, or 28,300 m³/sec). At Richmond, Virginia, the flow was 313,000 cfs (8850 m³/sec), more than the passage of Hurricane Camille, and the highest flows since 1870. The losses were staggering—206 counties and 27 cities were declared disaster

areas by presidential proclamation. At least 118 people were killed, 116,000 houses were damaged or destroyed, and total destruction amounted to at least $3.5 billion.

Several interesting footnotes can be told of what "might have been." In Pennsylvania the Tioga and Cowanesque dams had been authorized in 1958, but the monies were not appropriated for the $151.3 million they would cost, although preliminary field studies had been completed by 1967. If these dams had been "on line," their reservoir pool could have imponded sufficient water to greatly reduce the $50 million losses suffered by the New York cities of Painted Post, Corning, and Elmira. School damage alone exceeded $11 million. Another *if* concerns the Binghamton metropolitan area. This urbanized community would have suffered $950 million loss, instead of the $1.5 million, if the adjacent watershed had received the 35 cm of rainfall that fell 60 km to the west instead of the 21 cm that did fall.

Mississippi River flood of 1973   The Mississippi River and its tributaries drain 41 percent of the 48 states and have been subjected to flooding throughout human history. Particularly great floods occurred in 1849, 1850, 1882, 1912, 1913, and 1927. Of these the 1927 flood was the most disastrous, inundating 67,300 km². Levees were breached, and cities, towns, and farms laid waste. Crops were destroyed, and industry and transportation paralyzed. Hundreds lost their lives, hundreds of thousands were displaced from homes, and total damages exceeded $2 billion in 1979 dollars. This led to the first Flood Control Act of 1928 which committed the federal government to a definite program of flood control. However, further acts, such as the one of 1936, were necessary to provide additional jurisdiction and funding for more extensive flood control measures. Billions were spent on flood prevention from 1928 to 1972, so the entire country was shocked with the results of the 1973 flood, called by those who had forgotten prior history, "the year of the big flood."

Figure 12-11. View looking north at a junction of the Ohio River on the right and the Mississippi River at Cairo, Illinois, during flood of 1973. (Photo taken April 5 by Corps of Engineers.)

The rainfall pattern responsible for the 1927 flood was duplicated by the precipitation throughout the fall and winter months of 1972–73. The prolonged and incessant precipitation for a period of more than five months had completely saturated the ground, filled many reservoirs, and caused high water conditions on numerous streams. At Cairo, Illinois (Fig. 12-11), the junction of the Ohio and Mississippi rivers, water rose 6.5 m during the first half of March, and flood stage was reached at St. Louis on March 11. To add to the waters in the Lower Mississippi, the northern part of the state of Mississippi received 27.6 cm of rainfall

Figure 12-12 Break in man-made levee on March 18, 1973, during Mississippi River flood. (Courtesy Corps of Engineers.)

during a 30-hour period in mid-March. The waters continued their buildup all throughout April, and another bad storm hit the Mississippi Valley on May 1 (Figs. 12-12, 12-13, and 12-14). Finally, on May 4 the river crested 16.7 m at Cairo. The crest reached Memphis on May 8, was at Vicksburg on May 10, and at New Orleans on May 14. The river's fall was slow and lingering and was not back within its channel at Vicksburg until June 20. In all, more than 68,000 km² was inundated, and damages exceeded $1 billion. Although only 11 lives were lost, 69,000 were made homeless.

A wide variety of damages occurred. Because of the long residence time of floodwaters on agricultural lands, as much as three months, many spring crops could not be planted. This caused hundreds of millions of dollars in lost farm revenue. In Mississippi alone, 1.2 million ha of rich cotton lands were flooded, and the cotton loss exceeded $100 million. In Louisiana the petroleum industry lost $300,000 a day when oil and gas wells were flooded out of production. However, levees built at St. Louis in 1955 at a cost of $80 million prevented losses that would have totaled $340 million. Without construction of the Mississippi River and Tri-

Figure 12-13. Repairing and heightening levees (a) and (b) during Mississippi River 1973 flood. (Courtesy Corps of Engineers.)

Figure 12-14. Polyethylene sheets being used to provide impermeable coating for levee during 1973 Mississippi River flood. (Photo taken by Corps of Engineers on March 25, 1973.)

butaries Project (MR&T), an additional 59,000 km² would have been inundated and total damages would have exceeded $15 billion. (For another side of the story see pages 405–406.) The flood also produced a number of geologic changes. Lateral erosion was 271 percent more than the previous nine years. Average thickness of deposits on Louisiana levees amounted to 53 cm, of which 68 percent was sand. The backswamp area received 1.1 cm of sediment, with 97 percent silt and clay.

Two important floods, one natural and one man-induced, occurred in 1976. The first was on June 5, 1976, when the Teton Dam reservoir was nearly filled and the dam failed and disaster struck. The 302.8 million m³ of imponded water released a flood peak nearly equal to the highest peaks of the Mississippi River and inundated 780 km², of which 400 km² was prime agricultural land. The cities of Rexbury and Sugar City, Idaho, were devastated, leaving 25,000

homeless and killing 14 people and 20,000 livestock. Railroad tracks were ripped up for 51 km, and total damages amounted to about $1 billion. By March 1977 the federal government had already paid $138 million in the more than $400 million direct claims.

Teton Dam was designed and built by the Bureau of Reclamation for the multipurpose of flood control, power generation, recreation, and irrigation water supply. The entire structure did not collapse (Fig. 12-15), but structural undermining occurred on the right side of the dam, where subsequent studies pinpointed the weakness. A combination of geologic conditions and engineering judgments was cited by the investigating panel as the reasons for failure, and they included: (1) numerous open joints in the abutment rocks; (2) the use of substandard silts for the impervious zones; (3) complete dependence for seepage control on a combination of deep key trenches filled with wind-blown silt; (4) selection of a geometric configuration for the key trench that encouraged arching, cracking, and hydraulic fracturing in the brittle erodible backfill; (5) primary reliance on compaction methods of the impervious materials to prevent piping and erosion along and into the open joints; and (6) inadequate provisions for

Figure 12-15. View looking upstream at Teton Dam site, Idaho. This earthfill structure failed on June 5, 1976, causing a loss of lives and $1. billion property damages. During the break the emptying reservoir poured 58,800 m³/sec through the breached dam. The fractured volcanic bedrock, with joints and fissures, formed the abutment, on left side of photo. The earthfill core of dam is to the right. (Courtesy Robert E. Wallace, U.S. Geol. Survey.)

the collection and safe discharge of seepage and leakage through foundation materials. The primary trigger for the collapse was an open passage through the key trench. This had been initiated by piping of the silt through the joints and differential settlement. The Teton Dam failure provided the main stimulus for a renewed government program that investigated the safety of the nearly 50,000 dams throughout the country.

The second important flood of 1976 occurred on the afternoon of July 31, 1976, when 2500 residents and tourists were enjoying the spectacular scenery and climate of Big Thompson Canyon, Colorado. They were unprepared for the deluge that would occur from 6:30 to 11:00 P.M., with peak rainfall intensity reaching 30.5 cm/h. The exceptionally steep mountain slopes and the thin soil mantle possess negligible water-holding qualities, with the result that the majority of rainfall immediately discharges into streams of the 155 km² watershed which feed the Big Thompson River. Whereas the streamflow was 3.9 m³/sec (137 cfs) before the storm, during the high discharge the flow was 883 m³/sec (31,170 cfs) and filled the canyon to 9 m

depths. Although high discharge lasted only a few hours, 140 were killed and property damage in the recreational area was $16.5 million. Along the 40 km of river canyon many parts of Highway 34 were destroyed, numerous vacation homes and other dwellings were ruined or damaged (Fig. 12-16), and 1530 m³ of rock were washed out of the Olympus Dam. Big Thompson Diversion Dam was heavily damaged, and siphon pipe at the mouth of the canyon was completely demolished, and a power plant and tailrace channel were inundated with debris. The force of the water was so great that it moved 3 ton boulders and removed the bark from trees.

Although the loss of life could easily have been much greater, two factors combined to cause losses. (1) the heavy rain began at 6:30 P.M., but the National Weather Service did not issue a flash flood watch until 7:30 P.M. and the flash flood warning was not given until 11:00 P.M.—after most flood damage had already occurred! (2) Many of the deaths were to those people who got in cars and tried to outrun the floodwaters, instead of abandoning vehicles and climbing to high ground.

A somewhat similar but less well-known flood occurred in June 1965, in Sheep Creek Canyon, Ashley National Forest, Utah. Eight vacationers died in the torrential deluge. Debris floods from such rainstorms are a hazard in all mountainous terrain, as well as on the lands adjoining them (Fig. 12-17).

Another locale where people often have a false sense of security is the desert region of the Southwest but, as shown in Fig. 12-4, even here flooding can occur with disastrous results. The September 4–7, 1970, floods killed 25 and damaged millions of dollars of property.

Recent floods As further indication that flooding in the United States is a continuing and ever-present problem the following examples can be cited.

Four memorable floods occurred in 1977. In July Johnstown, Pennsylvania, was hit with a third flood disaster within a century. A 30 cm

Figure 12-16. Four views of the destruction caused by the Big Thompson Canyon flood in Colorado. A "cloudburst" caused flash flooding on July 31, 1976 in this narrow valley, killing 140 persons and causing millions of dollars in property damage. (a) Highway 34 suffered extensive damage. (Courtesy Don Doehring.) (b), (c), and (d) Vacation homes bore the brunt of personal property losses. The saltation of huge boulders caused great havoc, and homes broken by the impact of the boulders were sometimes completely set adrift by the raging floodwaters. (Courtesy Wallace Hansen, U.S. Geol. Survey.)

rainstorm in 6 hours killed 49, left 50,000 homeless, and caused damages of $117 million. Kansas City was also hit with a 30 cm rainfall in 24 hours that killed 23, left 1200 homeless, and caused damages of $50 million. A manmade dam at Toccoa, Georgia, collapsed in November, overwhelmed Toccoa Falls Bible College, and killed 38. The largest flooding occurred in the states of Kentucky, West Virginia, and Virginia in April of 1977, when the Big Sandy, Kentucky, and Cumberland rivers were in flood stage (Fig. 12-18). At least 23 deaths occurred, and 25,000 were made homeless, and officials conservatively estimated the damages at $275 million.

Several major floods occurred in 1978. In August the area between Corpus Christi and Brownesville, Texas, was inundated by severe storms that killed more than 31, ruined crops, destroyed buildings, and eroded and silted terrain that had not been under water for 100 years. The September 12 rainstorm that deluged the Little Rock, Arkansas, area killed 10 and caused damages of many millions of dollars.

Thus, there seems to be no abatement in floods and their damages in the United States. The only encouraging note is that there have been no recent tragedies in loss of life to match those in Johnstown, Pennsylvania, where 2200 were killed in 1889; those in Galveston, Texas,

Figure 12-17. Sedimentation from debris flood, Wasatch Mountains, Utah. (A.C. Crofts, 1967, Rainstorm debris floods: University of Arizona.)

where 6000 were drowned in 1900; and those in Lake Okeechobee, Florida, where 1928 was doomsday for 2000.

Other Countries

The United States has no monopoly on disasters from flooding. Unfortunately even worse calamities are all too frequent in many other nations (Table 12-1). The exceptionally large casualties in some of the countries can be attrib- uted to such factors as: (1) high population density, (2) lack of a warning system, (3) inadequate preparedness or transportation facilities for movement to safe havens, (4) absence of flood prevention engineering, (5) increased human intervention in despoilation of watershed areas, (6) fatalistic attitude toward nature, and (7) occurrence of more tropical storms and monsoon rainfall. From a study of Appendix I one can see that the frequency of typhoons and cyclones is greater throughout the Pacific and

Figure 12-18. Pikeville, Kentucky. Three meters of water cover much of the city after water overtopped the floodwall, April 6, 1977. Torrential rains created floods throughout the Cumberland River valley, causing millions of dollars of damages. (Courtesy *Courier Journal* and *Louisville Times.*)

Indian oceans than comparable hurricanes spawned in the Caribbean and Gulf of Mexico. Some of the great floods of China occur in those rivers where the combined natural and manmade levee systems are 15 m or more higher than adjacent lands. The inland floods in India and Bangladesh are invariably heightened and accentuated by the deforestation that has occurred on a massive scale within the past 40 years. For example, the September 1978 floods of the Ganges and Jamuna rivers originated in a severely eroded and denuded area. The rivers have become so clogged with silt that they no longer can hold moderate storm water, and as a result more than 1000 drowned, 600,000 dwellings were damaged, and crop losses exceeded $100 million. One-half the city of Benares with a 300,000 population was inundated, along with thousands of kilometers of agricultural lands.

HISTORICAL BACKGROUND OF FLOOD CONTROL PROGRAMS

The flow regimes of all rivers are marked by periodic flooding, and for millennia we have paid the price of losses for our occupance of the floodplain lands. In the United States, however, starting in the eighteenth-century we have constantly sought government intervention for our protection instead of bearing the costs of constructing our own means of shelter. Efforts to control floods within what is now the United States began with the French settlements in the alluvial valley of the Mississippi River. The artificial levees were constructed in New Orleans in 1717, and 10 years later the governor boasted of its mile length and the 18 ft height. At first landowners along the river built their own part of the levee with their own resources, but the King of France decreed that inhabitants several miles from the river must also work on the levees under the supervision of the local government. By 1824 the U.S. Corps of Engineers had begun work on navigational improvements of the Mississippi River, and in 1850 Congress authorized additional surveys of the river that would determine the most practical plan for ensuring against floods. However, little was done until 1879, even though Congress had also passed the Swamp Land Acts of 1849 and 1850. Although these acts were largely aimed at land reclamation, they did provide that proceeds from the sale of federal swamps and flooded lands that had been granted to the states could be used for drainage, reclamation, and flood control projects. The acts were not very successful because of the lack of coordination within and between states.

The year 1879 really marks the first major federal step toward the protection of communities from flooding, when the Mississippi River Commission was established as a permanent agency of the War Department and was charged with the planning and implementation of flood control projects on the lower Mississippi River. Such action was prompted by the total failure of uncoordinated levee systems which had led to abandonment of much of the fertile floodplain lands. This program was initially confined to the repair and strengthening of existing systems and was not significantly expanded until 1928. The first Flood Control Act was passed by Congress in 1917 for construction along the Mississippi River and the Sacramento River in California, but little came of the project. It took the disastrous Mississippi Flood of 1927 to awaken Congress to pass the second Flood Control Act of 1928, which authorized the federal government to undertake massive protective works in the Mississippi Valley. To obtain a wider base for support the act also included projects related to river flow regulation, with benefits accruing to conservation, irrigation, water supply, hydroelectric power, navigation, and recreation. The Corps of Engineers could also undertake surveys in other areas when so designated by Congress. By 1933 other parts of the country had shown their need for flow regulation of rivers, and the Tennessee Valley Authority Act was passed by Congress (see page 599) which had profound impact on flooding in that region and has served as a model for other water programs throughout the world.

Other parts of the environment had also received some attention to aid in the alleviation of flooding. The Clarke-McNary Act of 1924 and the McSweeny-McNary Act of 1928 represented federal legislation directed toward flood control by reforestation and soil conservation

within the watersheds. In 1935, the Soil Conservation Act led to the formation of the Soil Conservation Service and the development of a wide variety of measures that could aid soil preservation and flood inhibition. Also the Fulmer Act of 1935 authorized the Secretary of Agriculture to enter into cooperative agreements with the various states to assure better forest land management. The 1937 Bankhead-Jones Farm Act also sought the control of soil erosion and the mitigation of floods by watershed planning.

Disastrous floods in the Ohio River Basin in 1935 and 1936 generated the momentum for federal passage of the third and fourth Flood Control Acts of 1936 and 1938. Whereas the Corps of Engineers were authorized in 1936 to undertake appropriate flood prevention works on all navigable streams on a cost-sharing basis with the states and local governments, the 1938 act also allowed federal funds to be used for relocation of threatened developments to higher ground. By 1940 the Supreme Court had passed on the constitutionality of almost all federal acts in the field of flood control and upheld actions that such works were not limited to navigable rivers. The Flood Control Act of 1944 established that the U.S. Department of Agriculture, through the Soil Conservation Service, would have primary upland watershed responsibility for small flood control measures in 11 designated watersheds throughout the country, but this was greatly expanded by the Watershed Protection and Flood Prevention Act of 1954, which allowed the construction of dams in areas with less than 250,000 acres. This greatly strengthened the power base of the Soil Conservation Service, but the Corps of Engineers had also fared very well with passage of the Flood Control Act of 1946, which authorized the construction of 123 new projects by the Corps of Engineers at a cost of $772 million.

A new stratagem emerged in 1960 when Congress authorized the formation of the Flood Plain Management Service as part of the Corps of Engineers. Their work was to be coordinated with other federal, state, and local government agencies for the purpose of supplying flood information to hazard-prone communities. By 1971, 5200 communities with problems had been identified, and 440 Flood Plain Information Reports had been published, covering 1300 localities. Another innovation was enactment of the Water Resources Planning Act of 1965 which established the National Water Resources Council and a series of interstate commissions to oversee interstate river systems. These commissions consider problems of flood hazards on a regional rather than on a local scale and investigate the entire range of water control and related land use patterns. This network is modeled in part after the TVA. The New England River Basins Commission, established in 1967, has published numerous reports on the water systems of that region.

Executive Order 11296 of 1966 mandates that flood hazards be analyzed for all federal decisions dealing with disposal of lands, awarding of grants or loans, and construction of federal buildings. The National Flood Insurance Acts of 1968 and 1973 (also called the Flood Disaster Protection Act) limit federal subsidies for insurance to those communities that have adopted permanent land use and control measures that account for flooding hazards and regulate development as prescribed by the acts.

The most recent government involvement in flood control, in 1968 and 1973, established the flood prediction and warning system that is now the statutory responsibility of the National Oceanic and Atmospheric Administration (NOAA) of the Department of Commerce. The National Weather Service (NWS) of NOAA prepares the official forecasts and issues public warnings on floods throughout the country, with the exception of the Tennessee River Basin, where the TVA has responsibility. The River and Flood Forecast and Warning System of NWS essentially covers 97 percent of the nation, and in general works very well. Failure did occur, however, during the Hurricane Camille flooding of 1969, when the storm intensified instead of

weakening as predicted, and also during the Big Thompson disaster.

FLOOD PREVENTION AND LOSS REDUCTION

Few other areas in environmental geology contain as many controversies and problems as those associated with strategies to inhibit and control flooding. Table 12-4 and Fig. 12-19 present a digest of the different measures available in the management of floods, along with their advantages and disadvantages. Such methods can be grouped under the following five principal headings.

Hazard Perception Education

In order for the populace to avoid serious losses, appropriate education measures are necessary to instill an awareness and perception of the dangers of flooding. Unless such an information program is effective people will not take the required steps to assure safety and minimum losses. Only when a proper base for action has been developed will people react positively to the onslaught of a potential flood and to communications that provide forecasts, watch, alert, and warning of the flooding.

Legal Actions

Under a variety of mandates governments may enact rules and regulations that pertain to floodplain use and occupance (Fig. 12-20). **Zoning** can be an effective tool in minimizing flood damage. Urban areas can be assessed as to the degree of potential hazard present, for example, as high-, medium-, and low-risk areas. This will govern the type of development permitted in each zone. **Building codes** can be required that set standards for construction to protect the health and safety of inhabitants. **Floodproofing** (Figs. 12-21 and 12-22) can be man-

Table 12-4 An overview of floodplain management techniques[a]

Tool	Purpose	Approach to flooding threat	Incidence of costs	Advantages	Limitations
Land use regulations	1. Foster health and safety 2. Prevent nuisances 3. Prevent fraud 4. Promote wisest use of lands throughout a community	1. Require individual adjustment of uses to the flooding threat	1. Landowner must bear cost of adjustment; community bears cost of adoption and administration of regulations	1. Low costs 2. Promote economic and social well-being 3. Promote most suitable use of lands 4. Can be put into effect immediately 5. May remain effective for long periods if adequately enforced	1. Must not violate state and federal constitutional provision 2. Can't prevent all losses 3. Generally do not apply to governmental uses 4. Limited application to existing uses
Dams, reservoirs, levees	1. Reduce flood losses, protect safety, promote economic well-being 2. Protect existing uses 3. Promote navigation, water recreation 4. Make new sites available for development, increase tax base	1. Adjust flooding threat to land use needs	1. Generally public at large pays for benefits which accrue to landowners, local communities	1. Reduce wide range of flood losses 2. Protect existing uses 3. Promote navigation and recreation 4. Permit regional approach to problems	1. Federal subsidy leads to private gains 2. High costs 3. Construction may take many years 4. May not be consistent with community plans, environmental quality 5. Maintenance required 6. Sedimentation may reduce effectiveness 7. Catastrophic losses may result from failure of dam or levee 8. No site may be available for dam, or levee; geology wrong
Land treatment (to retain precipitation)	1. Prevent future increases in flood heights; reduce existing levels 2. Promote water and soil conservation	1. Reduce existing flood conditions; prevent future increases in flood heights in frequent floods	1. Expense largely public; however, landowners may bear portion of costs	1. Limited cost 2. Attack flood problem where it begins 3. May be consistent with broad community needs	1. Not applicable in many instances 2. Effectiveness limited to relatively frequent, small floods

Public open space acquisition for parks, wildlife areas, floodways	1. Reduce flood losses 2. Achieve broader community recreation and conservation goals	1. Adjust use to threat	1. Public pays but receives multiple benefits	1. Multiple benefits 2. No problem of constitutionality 3. Permanent 4. Active public use of lands possible 5. B.O.R. and other federal grants may be available for open space acquisition 6. Particularly attractive in urban areas	1. Acquisition costly 2. Flood losses to open space uses (e.g., campgrounds) remain 3. Sites not always suitable for recreation wildlife 4. May create shortage of land needed for businesses, industry, etc. 5. Creates public land management requirements
Flood insurance (National Flood Insurance Program)	1. Promote flood regulations 2. Promote long-term cost-bearing by individual occupant	1. Require individual cost-bearing 2. Adjust use to threat	1. Public pays, in part, for subsidized insurance 2. Private landowner pays for unsubsidized insurance	1. Spread cost of flood losses 2. Promote regulation 3. Encourage consideration of flood costs in private decision making	1. Subsidized insurance may promote continued use at primarily public rather than private expense 2. May undercut floodway regulations to abate existing uses
Warning systems	1. Warn property owners of impending threats 2. Permit advance evacuation, installation of temporary flood abatement measures	1. Adjust use to threat	1. Public bears costs (usually)	1. Can permit adjustment to threat 2. Useful in combination with regulations	1. Of no use unless floodplain occupants are willing and able to take necessary protection measures 2. Systems must be adequately operated and maintained

Source: Office of Chief Engineers, 1976, A perspective on floodplain regulations for floodplain management: U.S. Government Printing Office.

[a] These and other floodplain management tools such as permanent evacuation and relocation, flood proofing, and flood emergency and recovery measures are usually used in combinations.

Floodplain enroachments increase flood heights

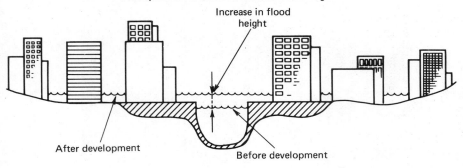

Increase in flood
height

After development

Before development

The regulatory floodway depends on the size of the flood
chosen as a basis for regulation

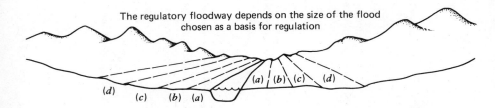

Natural floodway limits for a flood which on the average would occur:

(a)–(a) Once every 25 years.
(b)–(b) Once every 50 years.
(c)–(c) Once every 100 years.
(d)–(d) Once every 500 years.

The width of the regulatory floodway for the same size
flood will vary depending on permissible backwater
effects, that is, on increased flood height acceptable under ordinance

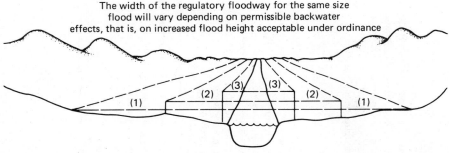

(1)–(1) Natural floodway boundaries for a flood which on the average would occur once every
100 years.
(2)–(2) Regulatory floodway for 1 meter of acceptable increase in flood heights for the same flood.
(3)–(3) Regulatory floodway for 2 meter acceptable increase in flood heights for the same flood.

Figure 12-19. Application of floodway computation factors for land use planning. (Corps of Engineers.)

dated for all public places, such as hotels, office buildings, apartment houses, and stores, as well as those dwellings that occupy flood-prone sites. **Flood insurance** can be required for those who insist on habitation, whether business or dwelling, in hazard areas. The burden thus falls on the individual rather than on society to reimburse the nonprudent owner. **Relocation** in-

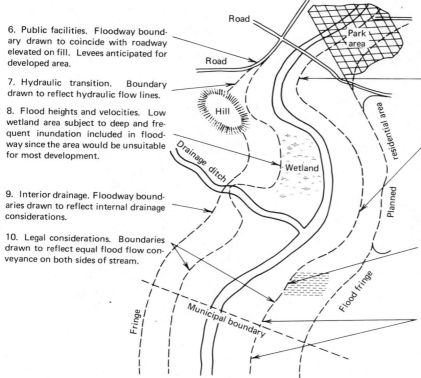

6. Public facilities. Floodway boundary drawn to coincide with roadway elevated on fill. Levees anticipated for developed area.

7. Hydraulic transition. Boundary drawn to reflect hydraulic flow lines.

8. Flood heights and velocities. Low wetland area subject to deep and frequent inundation included in floodway since the area would be unsuitable for most development.

9. Interior drainage. Floodway boundaries drawn to reflect internal drainage considerations.

10. Legal considerations. Boundaries drawn to reflect equal flood flow conveyance on both sides of stream.

1. Discharge of regulatory flood (measured in cfs). Used to define the flood magnitude to be used in establishing the boundaries of the regulatory floodways.

2. Permissible increases in flood heights. Narrow floodway limits reflecting 1 meter of permissible increase where no damage threatened to park area.

Broader floodway limits reflecting no permitted increases in flood heights due to damage posed to planned upstream residential area.

3. Community comprehensive plans. For significance of planned residential area see discussion of "permissible increases" above.

4. Existing development. Floodway boundaries drawn to exclude intensively developed areas to be protected by future levees.

5. Plans of adjacent communities. Broad floodway boundaries with small permissible increases in flood height and velocities to reflect plans of adjacent community to establish residential community.

Figure 12-20. Regulatory aspects of floodway control. (Corps of Engineers.)

volves the permanent removal of buildings and other structures from vulnerable areas. This can serve the dual purpose of assuring that loss of life is lessened and providing less constriction for floodwaters. **Tax adjustment and incentives** can be offered to those who fulfill all required regulations and aid in protecting their property from loss, as well as for actions and construction projects that reduce community losses.

Upstream Watershed Land Management

The condition of the land in a watershed and the use that is made of it influence the rate and amount of runoff that reaches the rivers. The objectives of watershed management schemes are to improve the ability of the land to hold

water, and this can be accomplished in several ways that include: reforestation, reseeding of denuded areas, mechanical treatment of slopes to reduce gradient by terracing, contour ditches, construction of small check dams and reservoirs, contour plowing, strip cropping, rotation farming of crops, planting of cover crops, and emplacement of mulch and other materials to prevent barren surfaces from exposure to rain impact and runoff. Such land use practices can reduce flood damage in at least three ways. (1) The improvement of tilth and soil structure promotes an increase in the volume of water retained in the soil. (2) Tilth and structure enhancement in soil plus vegetation protection increases rates of infiltration into the soil; this percolation delays the delivery of water and so acts as a storage reservoir to dampen and re-

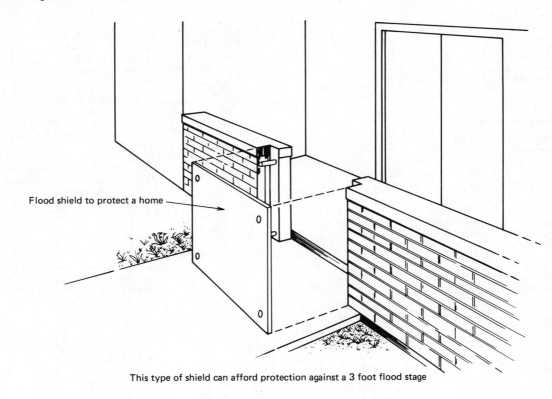

Flood shield to protect a home

This type of shield can afford protection against a 3 foot flood stage

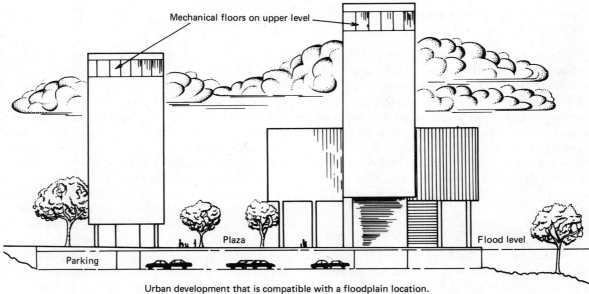

Mechanical floors on upper level

Plaza

Flood level

Parking

Urban development that is compatible with a floodplain location.
In this development, the uses of the building have been adjusted to
avoid uneconomic flood losses.

Figure 12-21. Types of floodproofing. (After Corps of
Engineers.)

Explanation

1. Permanent closure of opening with masonry
2. Thoroseal coating to reduce seepage
3. Valve on sewer line
4. Underpinning
5. Instrument panel raised above expected flood level
6. Machinery protected with polyethylene covering
7. Strips of polyethylene between layers of cartons
8. Underground storage tank properly anchored
9. Cracks sealed with hydraulic cement
10. Rescheduling has emptied the loading dock
11. Steel bulkheads for doorways
12. Sump pump and drain to eject seepage

Figure 12-22. A floodproofed structure showing the many items that must be considered for floodproofing in order to reduce damages. (After Corps of Engineers.)

duce flood peaks. Furthermore, the infiltrated water which eventually reaches the river carries no sediment burden. (3) There is a reduction in erosion losses. The number and size of rills and gullies are less, so less sediment moves through the system and sheet runoff processes are slowed and inhibited. Plant cover can be vital in protecting the land and diminishing flood runoff because (1) energy of falling raindrops is absorbed by litter and leaves, and (2) plants transpire water and thus make less available to pass through the system.

Effective land use management is especially effective for small to moderate storms, but is not a panacea for large storms. This is because great floods are almost invariably associated with long periods of rainfall during which the soil becomes saturated. Infiltration rates and water retention possibilities are also minimized during periods of snowmelt or frozen ground. Watershed management works best on small streams where the time between rainfall and peak runoff is not more than two or three hours. Under such conditions upstream management for a small watershed can reduce peak discharges 20 to 40 percent for 2 to 5 cm rainfall storms and 15 percent for 7 to 12 cm storms. Upstream reservoirs can also subdue a part of flood discharge, as in the case of an 8.6 cm storm over the 186 km^2 watershed of the Middle Colorado River, Texas, October 3, 1953. Five detention reservoirs in the area reduced peak flow discharge from 600 to 420 m³/sec—a lowering of 30 percent.

Downstream Structures

Structural measures are concerned with the building of engineering works and landscaping within the floodway. Their purpose is to restrict waterflow, to prevent inundation of adjacent lands, and to increase water velocity (see page 686). The following techniques are used for these purposes, and each is designed to do a specific job and must be designed for the task at hand.

Floodwalls, revetments, levees, and dikes   These are constructed essentially parallel to the river to confine the flow to the channel. When such structures are overwhelmed, however, the floodwaters cause greater damage than ever because adjacent development may be extensive under the opinion that such lands were safe. It is common to plan such structures to accomodate the 50-year flood (Fig. 12-23).

Channel improvements   This work includes straightening the streambed (channelization), enlarging the channel cross section, riprapping the channel side and bottom, regrading the bed, removing debris and vegetation, and **river-training** devices such as emplacement of dikes and groins. The principal purpose of this work is to allow the channel to transport larger water volumes and to expedite through-flow by increasing the velocity—the rationale being that the sooner floodwaters evacuate an area, the less damage they can do.

River diversions   During high water, there may be basins or auxiliary water corridors where the flood can be diverted and thus provide an alternate path that will not be destructive to the community.

Major dams   There are two types of downstream dams that are in use—those that block drainage in the main river, and auxiliary dams that control imponded waters in secondary di-

(a)

(b)

(c)

Figure 12-23.   The battle to protect Minot, North Dakota, from floodwaters of the Souris River in April 1976. (*a*) Overview of the river and levee being hastily heightened and reinforced. (*b*) The ''borrow'' area where material is being excavated for emplacement on levees. (*c*) Trucks moved fill for a two-week period, 24 hours a day. A total of 90 km of levees were constructed at a cost of $7 million. The project was successful and damage was minimal. (Photos by John Conners.)

version works. Although there is great controversy regarding the merits of big dams versus small dams, most data support the view that major reservoirs reduce downstream flood flows more effectively than multiple upstream reservoirs. The reason is simple: because of the lower river gradient and wider floodplain, large dams can store far greater volumes of floodwater. Furthermore, the reservoirs may serve several functions—for recreation, as a fish and wildlife sanctuary, water supply, and in hydroelectric generation. Of course, some of these functions would be competitive; for example, reservoirs are most effective deterrents when empty and most efficient for electric power when full. Serious social disruptions can also occur because there would be the loss of the most valuable floodplain lands, and large dams also produce a large number of deleterious impacts on the physical environment (see also page 444), such as initiation of earthquakes, changed flow regimes in rivers with siltation in the upstream part, and erosion in downstream areas. Construction of major engineering works is also subject to much "pork-barrel type legislation" in Congress.

Do nothing Another alternative to management of flood hazards is **to do nothing** about potential losses. There are many reasons for adoption of such a position. Some people may be simply unaware or uneducated that such an event could produce harm. Others may decide, or even entire communities, that when all factors are considered the costs of preventive measures are not worth the investment. Such views range from those who are willing to gamble losses will not occur or will be less than predicted—the syndroms that "it won't happen to me"—to those localities that are too poor to solve their own problems. In the United States many people believe it is the government's responsibility to protect them and their property, regardless of location, and that if losses occur they will be reimbursed through some type of relief program or subsidy to indemnify for the losses.

Perhaps in the future serious efforts will be made to alleviate floods by using other large-scale methods—for example, major river diversions to other watersheds or weather modification.

MISSISSIPPI RIVER AND TRIBUTARIES FLOOD CONTROL PROJECT (MR&T)

The largest flood control project in the world is the one designed by the Corps of Engineers to prevent major damages on the lower Mississippi River; it is known as the Mississippi River and Tributaries Flood Control Project (MR&T), (Fig. 12-24). The program was initiated when it was realized that levees alone could not inhibit great flooding, and in an attempt to curb the large number of disastrous floods that have devastated the basin. The MR&T Project is designed to control the **project flood**—a result of combining some of the most severe storms and placing them in a pattern to produce the greatest possible flood having a reasonable probability of occurrence. The project flood would produce an estimated 3 million ft^3/sec (85,000 m^3/sec) at the latitude of Old River, half of which would pass down the leveed Mississippi River Channel to the Gulf of Mexico, with the other half down the Atchafalaya Basin. There are four major elements of the project: levees, floodways, channel improvements, and major tributary improvements.

Levees

The main levee system includes about 3500 km of levees, floodwalls, and revetments that average about 9 m in height. The main-line levees begin just below Cape Girardeau, Missouri, and extend to the Gulf. On the east bank between Cairo, Illinois, and the Gulf, the levees alternate with natural bluffs. Gaps occur where tributaries enter the river. Revetments are only used at the most critical sites because of great cost.

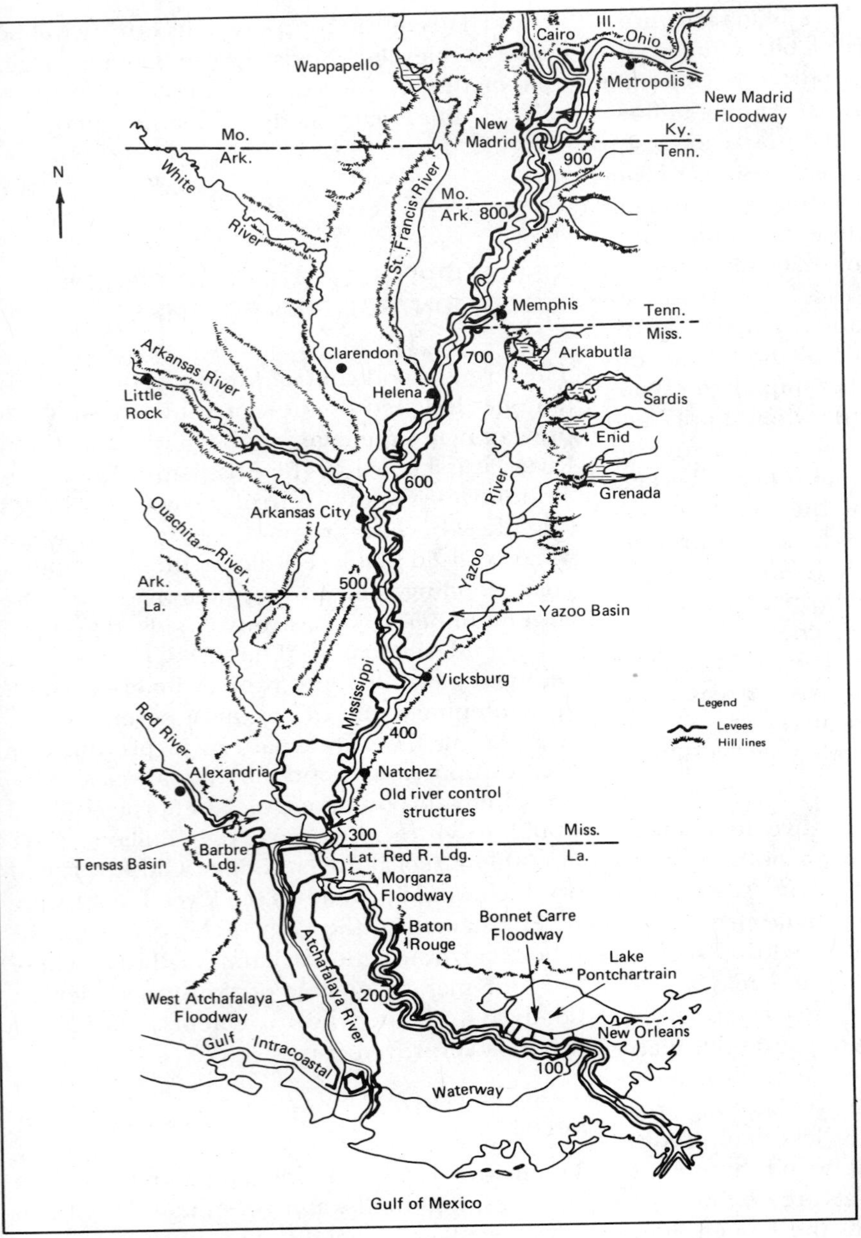

Figure 12-24. Map of Mississippi River & Tributaries Flood Control Project, Corps of Engineers.

Floodways

Four floodways have been built, one in Missouri and three in Louisiana (Fig. 12-25). They are located at strategic positions to divert excess floodwaters, and some were completed as early as the mid-1930s. The New Madrid Floodway was used in 1937, 1945, 1950, 1973, and 1975. The Morganza Floodway (Fig. 12-26), completed in 1954, was opened for the first time in 1973. The West Atchafalaya Floodway has not yet been used.

Figure 12-25. Bonnet Carre Floodway and Spillway of the MR&T Project, Corps of Engineers.

Figure 12-26. Morganza Floodway of the MR&T Project, Corps of Engineers.

Channel Improvements

The MR&T Project includes those channel changes designed to stabilize the alignment of the channel and improve its flood-carrying capacity. By 1942, a total of 16 cutoffs (Fig. 12-27) and two major chutes had been engineered.

One effect of these improvements was to lower the project-flood river stage about 4.8 m at Arkansas City and 3 m at Vicksburg. Other channel works include installation of revetments (Fig. 12-28), dikes, and dredging operations. Dikes are used to direct the river into more desired locations where damages will be less

Figure 12-27. Coushatta Cutoff, Mississippi River. The old meander is on the extreme left of the photo, and the channeled cutoff is on the right. (Courtesy Corps of Engineers.)

Figure 12-28. Revetment sinking unit for Vicksburg District Corps of Engineers. At this locale, a 3.2 km long bankline is being stabilized on the west bank of the Mississippi River upstream of the Algiers Navigation Lock. Cost was $3.2 million. (Courtesy Corps of Engineers.)

severe; this is called **river training** (see Fig. 13-7), and dredging is necessary to maintain river depths for navigation as well as increasing flow velocity.

Tributary Improvements

The principal component of this phase has been the construction of five reservoirs in the hills that border the valley. Those in the Yazoo Basin (Enid, Arkabutla: Sardis, and Grenada) played particularly important roles during the 1973 flood.

The MR&T Project is more than 41 percent complete, and total expenditures have exceeded $9 billion. However, during the years it has been in operation it has reduced flood damages by billions of dollars (see page 374). However, construction of man-made systems has also increased flood stage. There are also many disadvantages for such massive operations.

Perspectives

Floods are the most universal of all geologic hazards, capable of occurring in all river valleys and in low-lying coastal areas. The mechanics of rivers requires their periodic overflow, so not only is the frequency of floods greater than other hazards but, with few exceptions, their buildup is subject to more precise forecasting. However, some floods are so large that we cannot prevent their rise—Like King Canute, we can try, but we also fail. Historically floods have not been all "bad." Egyptian farmers depended for at least 5000 years on annual floods to water the fields and prepare soil for cultivation. Floods now, however, are invariably perceived as dreaded threats to life and property. Instead of damages decreasing, they are increasing because of the continuing development in floodplains, population increases, and growth of urban areas.

People, and governments, are often apathetic about natural events or even man-induced environmental changes. Only when particularly devastating losses occur do they become sufficiently motivated to transcend their lethargy. Much of the "disaster-type legislation" in the United States has been prompted by exceptional tragedies, and this is especially true of floods. The St. Francis Dam calamity in 1928 which killed 500 prompted California to enact regulations that mandated geological engineering studies of dam construction in that state. The many floods of the Mississippi River and tributaries have been the main stimulus for the Flood Control Acts by the federal government and have led to numerous construction projects and flood abatement programs by many different agencies—the Corps of Engineers, TVA, Soil Conservation Service, Bureau of Reclamation, and others. Dams can also be important in some flood control projects (Fig. 12-29). However, the recent U.S. dam failures of the Teton Dam and the one at Toccoa, Georgia, have

Figure 12-29. Shasta Dam, California. The dam is 180 m high, with a 1040 m crest length, and can store 4.5×10^9 m^3 of water. This multipurpose dam produces electricity, regulates floods, and stores surplus water runoff for many uses—irrigation, maintenance of navigation levels, conservation of fish, and prevention of coastal area salinization. (Courtesy U.S. Bureau of Reclamation.)

prompted new laws and provided increased funding to examine the safety of 50,000 dams. Of these, 20,000 can be considered dangerous, and 9000 are in the high-hazard category. Thus we can be our own worst enemy, creating risks where we should be providing the required safety.

Since the year 1837 when Lt. Robert E. Lee built the first confinement dikes at St. Louis, where the harbor area was being ruined by floods, the Mississippi River has been a prized artery where extensive construction has attempted to control the damage from the all-too-frequent floods. The MR&T Project is the largest inland flood control program, just as in the Netherlands the Delta Scheme, and its forerunners, constitutes the most massive work to protect coastal lands from ocean flooding. Our use of low areas, with their many advantages, is obtained only at a price. It is a problem that governments must face. They must decide when that price is too great, and what remedial measures must be taken to alleviate the threat, or they must require the occupation of less hazardous terrain.

READINGS

Baker, V. R., 1976, Hydrogeomorphic methods for the regional evaluation of flood hazards: Env. Geol. v. 1, p. 261–281.

Belt, C. B., Jr., 1975, The 1973 flood and man's constriction of the Mississippi River: Science, v. 189, p. 681–684.

Bue, C. D., 1967, Flood information for flood-plain planning: U. S. Geol. Survey Circ. 539, 10 p.

Davies, W., Bailey, J. F., and Kelly, D. B., 1972, West Virginia's Buffalo Creek Flood: U. S. Geol. Survey Circ. 667, 32 p.

Dougal, M. D., ed., 1969, Flood plain management: Ames, Iowa, Iowa State University Press, 270 p.

Hess, W. N., ed., 1974, Weather and climate modification: New York, John Wiley & Sons, 842 p.

Hoyt, W. G., and Langbein, W. B., 1955, Floods: Princeton, N.J., Princeton Univ. Press, 469 p.

Kates, R. W., 1965, Industrial flood losses: Dept. of Geography Research Paper 78, Univ. of Chicago, 76 p.

Leopold, L. B., and Maddock, T., Jr., 1954, The flood control controversy: New York, Ronald Press, 278 p.

Noble, C. C., 1976. The Mississippi River flood of 1973: in Coates, D. R., ed., Geomorphology and engineering, Stroudsburg, Dowden, Hutchinson & Ross, p. 79–98.

Rahn, P. H., 1975, Lessons learned from the June 9, 1972 flood in Rapid City, South Dakota: Assoc. Eng. Geol. Bull., v. 12, p. 83–97.

Schwartz, F. K., et al., 1975, The Black Hills—Rapid City flood of June 9-10, 1972: U. S. Geol. Survey. Prof. Paper 877, 47 p.

Sheaffer, J. R., 1960, Flood proofing: an element in a flood damage reduction program: Dept. of Geography Research Paper 65, Univ. of Chicago, 160 p.

Sheaffer, J. R., Ellis, D. W., and Spieker, A. M., 1969, Flood-hazard mapping in metropolitan Chicago: U. S. Geol. Survey Circ. 601-C, 14 p.

White, G. F., 1945, Human adjustments to floods: a geographical approach to the flood problem in the United States: Dept. Geog. Res. Paper No. 29, Univ. of Chicago Press, 225 p..

Williams, G. P., and Guy, H. P., 1973, Erosional and depositional aspects of Hurricane Camille in Virginia, 1969: U. S. Geol. Survey Prof. Paper 804, 80 p.

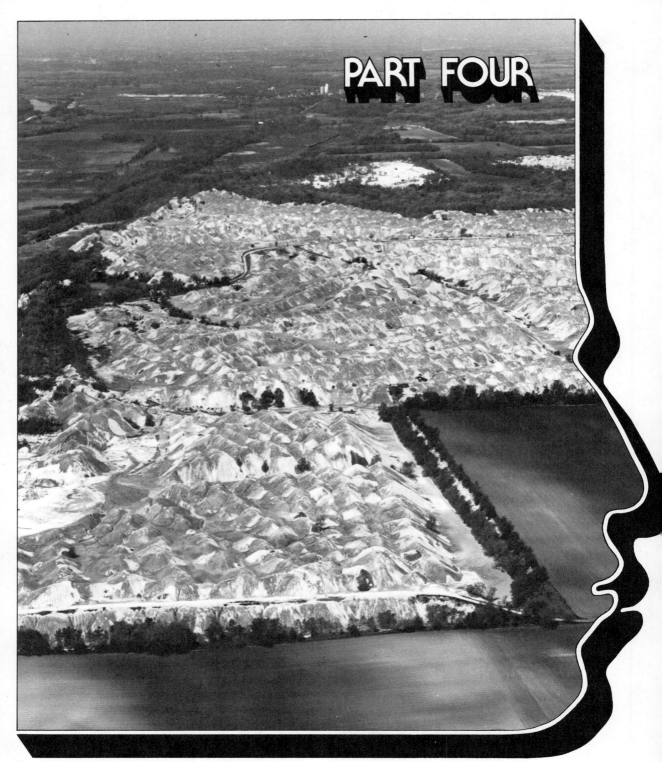

PART FOUR

Humans vastly change rural landscapes, as in
this coal strip mining operation near Utica, Illinois.
(Courtesy U.S.D.A. Soil Conservation Service.)

The Human Modification of Nature

Whereas Part Three dealt primarily with the way nature affects humans, Part Four discusses how we affect nature. It should be remembered, however, that this latter theme runs throughout the entire book. For example, we produce many changes in materials and in the processes of the earth in all our endeavors—when we extract mineral wealth from under and on the ground, when we construct cities, and when we dispose of wastes. The emphasis in Part Four is on both the deliberate and inadvertent alterations that humans cause to the land-water ecosystem whenever they rearrange the topography for the purposes of engineering, development of living resources, or use of the atmosphere. Indeed, such a section could also be titled "How we have changed the land, the water, and the air."

Part Four also has as its theme the manner in which engineering and geotechnology have attempted to remedy harmful environmental situations and to lessen the environmental damages that occur because of human activities on the land. This is truly environmental geology because people are now acting as a physical process and are inducing a metamorphosis of the landscape. In so doing, we not only produce an anthropogenic landform, but we change the manner in which processes operate, altering both the magnitude and frequency.

The encouraging element in this discussion is that we now possess most of the basic tools and skills to be able to predict the character of our disturbance. For example, it now comes as no surprise that when a dam is constructed that a large array of changes, both downstream as well as in the reservoir area, will occur. Or that when great amounts of groundwater are withdrawn that some deleterious impacts can result. The changes that are created by man run the full gamut and include excessive erosion, harmful sedimentation, groundwater aberrance, soil salinization, terrain deformation, damage to water supply systems, pollution, and even desertification.

Chapters 13 and 14 both emphasize the engineering aspect of environmental geology. Whereas Chapter 13 covers several of those endeavors where humans have deliberately altered the topography, Chapter 14 specializes in the environmental impacts that occur when water regimes are modified. The purpose of these chapters is to demonstrate the essential significance of engineering geology. It is important, however, for those who practice this discipline to realize the myriad types of environmental feedback that operate as a result of their action. It is not enough to build a structure that is technically suitable; it is also important to make sure its location is at an environmentally suitable site that is the most advantageous of an entire series of alternative possibilities. Indeed, the $64 question is whether the development was necessary in the first place. Are there other strategies that could have been used to accomplish the same type of mission? Is the mission truly beneficial for the majority of society?

Chapter 15 describes the elements that comprise the coastal corridor and the various techniques that are used to modify the beach-coastal environment. We are becoming increasingly interested in coastal areas, and all too often we have viewed our job as one of subduing the natural processes that operate at the land-water interface. Under the prevailing worldwide geologic conditions that now operate on marine waters, there will be a continuing rise in sea level and a concomitant loss in beach areas. This is a reality of life that we must learn to accept and live with. Thus it becomes a question of short-range priorities as to how much increased investment we should make in beaches, if in the long run such areas are doomed to flooding.

Chapter 16 may at first glance seem out of place in an environmental geology text-

book, but when viewed from a dynamic posture it is totally relevant. Soil is formed by the geological process of weathering, and our use of the soil constitutes a change in the material and further affects rates of erosion, sedimentation, as well as producing changes in streamflow properties. Furthermore, soils are a basic resource, and to protect our investment the agriculturalist very often uses engineering geology and geomorphology techniques to control erosion processes—for example, in using terracing, hillside drainages, check dams, and so on. Also, in the growing of crops many geologic products are required: water, rock fertilizers, among other natural resources.

Chapter 17 includes material that is rarely presented in geology books. However, because the air and its characteristics of pollutants, water vapor, and temperature affect the realm of man and the land and water, their inclusion rounds out the full discussion of the hydrologic cycle. Knowledge of weather and climate will become an ever-pressing matter as we pursue our policies of changing the land, the forests, and the water surfaces of the planet.

READINGS

Bouillenne, R., 1962, Man the destroying biotype: Science, v. 135, p. 706–712.

Brown, E. H., 1970, Man shapes the earth: The Geog. Jour., v. 131, pt. 1, p. 74–85.

Bryan, P. W., 1933, Man's adaptation of nature: New York, Holt, Rinehart and Winston, 386 p.

Coates, D. R., ed., 1973, Environmental geomorphology and landscape conservation, Vol. III, Nonurban regions: Stroudsburg, Dowden, Hutchinson & Ross, Inc., 483 p.

Dansereau, P. 1957, Man's impact on the landscape: in Dansereau, P., ed., Biogeography: An ecological perspective, New York, Ronald Press, p. 258–293.

Detwyler, T. R., ed., 1971, Man's impact on environment: New York, Mc-Graw Hill, 731 p.

Douglas, I., 1971, Dynamic equilibrium in applied geomorphology: two case studies: Earth Sci Jour., v. 5, p. 29–35.

Eckholm, E. P., 1976, Losing ground: New York, W. W. Norton & Co., 223 p.

Garrels, R. M., Mackenzie, F. T., and Hunt,

C., 1975, Chemical cycles and the global environment, assessing human influences: Los Altos, Calif., William Kaufmann, Inc., 206 p.

Nelson, H. J., 1959, The spread of an artificial landscape over southern California: Ann. Assoc. Amer. Geog., v. 49, n. 3, pt. 2, p. 80–99.

Marsh, G. P., 1864, Man and nature: edited in 1965 by D. Lowenthal, The Belknap Press of Harvard University, 472 p.

Sherlock, R. L., 1931, Man's influence on the earth: London, T. Butterworth, 256 p.

Stall, J. B., 1966, Man's role in affecting sedimentation of streams and reservoirs: in 2nd Ann. Amer. Water Resources Conf., p. 79–95.

Thomas, W. L., Jr., ed., 1956, Man's role in changing the face of the earth: Chicago, Univ. of Chicago Press, 1193 p.

Trask, P. D., ed., 1950, Applied sedimentation: New York, John Wiley & Sons, 707 p.

Whyte, W. H., Jr., 1968, The last landscape, Garden City, N.Y., Doubleday, 376 p.

Chapter Thirteen
Surface Changes and Engineering

Eucumbene Tutmut Tunnel, Australia. (Courtesy Snowy Mountains Hydro-electric Authority.)

INTRODUCTION

Nature abounds with marvelous engineering works—bee hives, termite mounds, spider webs, bird nests, beaver dams, coral reefs and prairie dog towns. Thus we are not a unique animal. We simply build larger edifices on land and thereby change natural systems by a much larger measure. Engineering endeavors are undertaken because of a real or perceived need by society. The job of the engineer is to assure us those services that are necessary for our livelihood and sustenance. Because earth materials are used and terrain is involved, geologists rightly view their role as significant in the planning of such disturbances to the geologic environment.

As in all affairs of society, there are right and wrong ways of accomplishing objectives. Engineering activities may be viewed the same way as hunger—many different foods may quiet the hunger pains, but not all foods are equally nutritious for the body. There may be different engineering techniques that can be used to perform a project, but some will produce greater environmental damage than others. Thus, an effective liaison between geology and engineering must be established to assure the minimum of deleterious land-water impacts whenever we cut, tunnel, dredge, fill, load, and alter natural systems. Those engineering specialties that are especially relevant are civil, hydraulic, and rock and soil engineering. Teamwork with geologists is often vital to the success of the project, and the field of geology comprises many subdisciplines, including engineering geologists, geomorphologists, and geotechnology specialists.

Unfortunately, some engineering projects are "site specific" wherein the problems that are addressed and solved have constraints that include time, money, size of investigation and building area, and social and political priority considerations. Thus, the design engineer may not have infinite flexibility for the project, but instead must develop the plans within the framework of priorities dictated by the verities of civilization. When projects are planned and constructed with such limitations and in isolation from contiguous areas that may be affected, undesirable feedback may occur that jeopardizes stability of the system or adjacent lands. We address some of these effects in this chapter but, because engineering activities permeate all spectra of mankind, many other illustrations are cited throughout the book—as in topics dealing with urban areas, landslides, mining activities, soils, and so forth. Engineering is so universal that it *is* part of the human way of life.

GEOLOGY AND ENGINEERING

Engineering has been described as the business of planning, designing, constructing, and managing machinery and facilities useful to society. The geologist enters the scene when engineering activities are undertaken that involve earth materials or processes. Society has charged the engineer with the responsibility for constructing arteries and systems that maintain the free flow of goods and services that are necessary to preserve the health, safety, welfare, and sustenance of mankind. Thus, the engineering profession has clearly defined goals with the mission to solve those problems of civilization that are solvable through construction enterprises. When geologists also address these problems and become involved with providing data necessary for the success of building in or on the earth, they are performing engineering geology work. This type of work may also be called geoengineering or geotechnology. This encompasses such a broad array of topics that a single chapter does not do them justice. In addition to discussing these aspects in this chapter and in Chapter 14, such matters are also described in most chapters throughout this book (for example, see pages 483–494 and 522–523). Our primary concerns in this chapter are with the impacts produced by those construction activities that create the land and water highways and the machines that use them. Such building invariably affects the land-water ecosystem because we have imposed a new, artificial environment onto a natural setting.

Geological investigations that serve engineering purposes must cover a wide spectrum of specialties because no two areas have exactly the same type of rocks, structure, topography, or physical processes that operate within them. Most investigations take at least three basic forms.

1   *Reconnaissance investigation.* This provides an overview and the generalized range of problems to be encountered. Alternate options should also be considered in this phase.

2   *Feasibility investigation.* This provides the detailed basis for the project—the plans, the budget, the working schedule, the source of materials, the geologic constraints.

3   *Construction investigation.* This is an ongoing study that continues for the duration of the project. For example, excavation work may reveal unanticipated problems that need resolvement in a different manner than that planned.

The *preliminary* work that is essential includes such facets as a thorough review of all previous relevant reports; inspection of the geology, hydrology, topography, and acting processes; and verification of features by test holes as well as geophysical and other sensing instruments. During the detailed study more precise geologic information is needed: the location of trouble spots and hazard conditions that must be stabilized so that construction is not placed in jeopardy; a determination of bearing strength of materials, in situ stress, and pore-water pressure; and an accurate assessment of all logs and other structural and material indicators. And finally, there is no substitute for the **continuous monitoring** of all systems to assure conformance with all design specifications. Structural failures can usually be traced to inadequacies in determining the correct order of magnitude of stability of the geologic environment. Even when the design specifications have been met, however, environmental degradation can occur because of inability to control all feedback mechanisms that operate in the complex battery of phenomena that operate in the natural environment.

Erroneous Engineering Appraisal of Geologic Features

Elsewhere in this book poor engineering judgment of the geologic setting is mentioned, but it might be informative to site two case histories at this point that illustrate what can happen when teamwork is lacking in engineering endeavors.

Although geologic studies had indicated the presence of till in the substrate of the St. Lawrence region, contractors who were building the St. Lawrence Seaway failed to recognize this material in their test borings. Instead, they interpreted the sediment as sand and gravel, which are materials that can be easily excavated. Because of the toughness of the till they had to spend an extra $27 million in construction costs and brought claims to recover such monies. Eventually a settlement was reached whereby they were paid nearly $5 million.

In the Boston, Massachusetts area the usual explanation for the rounded hills is that they are **drumlins,** a glacial landform with deposits that often are composed of till. Because till is usually compact, water does not flow through it readily. The contractors for three major buildings—a large apartment, the Saltonstall State Office Building, and the Boston Common Garage—encountered serious problems because of failure to interpret correctly the drilling logs. They had assumed the hills were drumlins with till and little water. Instead, however, the materials were gravelly and contained abundant water. Such sediment did not support the pilings and costly delays resulted when the projects had to be redesigned to accommodate the changed situation. This case illustrates the law of equifinality—the mistaken landform in this case was a terminal moraine and not a drumlin.

Other problems can emerge with engineering projects when the cost estimates are not reliable indicators of actual and final costs. Although such cost overruns may be due to a variety of causes, they occur all too frequently with government contracts. A 1976 study of 178 Corps of Engineers projects showed a total cost overrun of 110 percent—$12.7 billion in estimated costs and $26.7 billion in final costs. Typical large overruns were for the Fire Island Inlet, 325 percent; the Tennessee-Tombigbee Waterway, 321 percent; and the Sacramento River Project, 356 percent.

Engineering Credits

Engineering endeavors have provided civilization with the wherewithal to build new societies with unrivaled opportunities. The most massive infusion of engineering construction ever undertaken occurred in the United States as a result of the depression years in the 1930s. Government-financed programs provided a quantum jump in building that helped alleviate the depression years and led to a new liaison with government and the building and engineering professions. The Public Works Administration (PWA) built 11,000 highway projects; 1800 sewer systems; 2600 water supply systems; 470 flood control, waterpower, and reclamation projects; 375 electric power projects; and 11,800 schools and other public buildings. During the same period the Works Project Administration (WPA), which employed one in five male workers at some time during its tenure, built the following: 965,606 km of roads; 116,000 bridges and viaducts; 600 airports; 110,000 schools, libraries, and auditoriums; 1500 nursery schools; thousands of sewer and water supply systems; 8000 parks; 13,000 playgrounds; and planted millions of trees. The Civilian Conservation Corps (CCC) also performed prodigious feats in providing labor and materials for thousands of projects, and all of these endeavors changed the face of the land throughout the country.

STREAM CHANNELIZATION

Throughout recorded history whenever man decided to occupy floodplains he has busily engaged himself in making modifications of the parent river. Such changes include flow diversion to other sites, impoundment behind an artificial barrier, or stream channelization. We will use the term **stream channelization** to denote deliberate attempts to alter the entire stream, whereas the terms **canal** and **ditch** apply to only parts of the stream that have been diverted away from the main water body. Chan-

Figure 13-1. Man-made sedimentation and channelization of streams along the Interstate 88 construction project, Broome County, New York.

Figure 13-2. Channelization and floodwall of Chenango River, Binghamton, New York.

nelization is an engineering technique that is used to accomplish a variety of objectives that include:

1 *Drainage.* The reclamation of wetlands by lowering the water table.

2 *Flood control.* Increasing the capacity or velocity of the stream.

3 *Navigation.* Providing straighter and deeper channels.

4 *Erosion control.* Emplacement of artificial materials.

5 *Agriculture.* Straightening channels so that farmlands are bigger and more manageable.

6 *Engineering structures.* For bridge, highway, and other alinement purposes (Figs. 13-1 and 13-2).

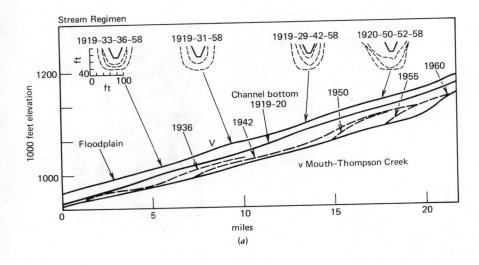

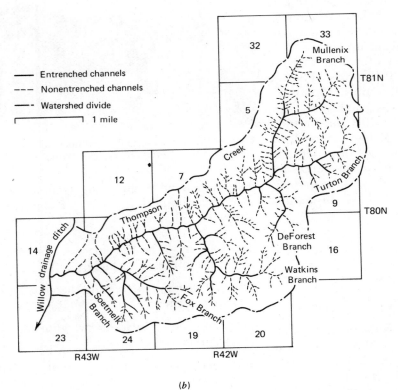

Figure 13-3. Effects of channelization on Willow Creek, Iowa. (a) Longitudinal and transverse profiles of the new Willow Drainage Ditch during specific years. (b) Drainage net of Thompson Creek tributary system, showing extent of major entrenchment. (R. V. Ruhe, 1971.)

Channelization is generally accomplished by some combination of deepening, widening, straightening, clearing brush and debris, and channel lining and armoring (Figs. 13-3 and 13-4). Channelization was being practiced in California as early as 1871 in attempts to prevent flooding. Here streams were enlarged by using horse-drawn scrapers and Chinese labor-

Figure 13-4. Aerial view showing channelization of Kissimmee River, Florida. Light area shows the extent of the destruction to the soil as result of the construction. (Courtesy South Florida Water Management District.)

ers. On a vastly different scale, the Corps of Engineers have been channeling the Mississippi River since the 1870s. Although other river systems were eventually brought under channelization action for flood control purposes, it was not until the mid-1950s that large-scale alteration of waterways for agricultural purposes got seriously underway. The impetus was provided by the Watershed Protection and Flood Prevention Act of 1954. Under this statute the Soil Conservation Service (SCS) helped farmers to channelize more than 13,000 km of streams throughout the country. During the same time the Corps of Engineers straightened another 2400 km of waterways—most of them considerably larger than those modified by SCS. In the meantime, the SCS has congressional approval for channelization of 20,000 km of streamways, but the rate of man-made changes has slowed because of objections by environmentalists.

Damaging Impacts

The deleterious effects of stream channelization occur in both the physical and biological systems of the watershed. The amount of damage depends on the character of the stream regime changes and the floodplain environment (the original type of topography and geology). In a 1973 study the A. D. Little Company analyzed 42 channelization projects that contained 3700 km of channel alteration, in which more than one-half had been completed. They concluded that 11 had produced significant wetland changes; 5 created dramatic loss of hardwoods; water tables and stream recharge capability were adversely affected in 6; downstream flows were changed by 4; and 19 produced unsightly landscapes. In no instance was channelization found to be helpful to the aquatic life in the stream or to the plants and animals of the natural bottomland ecosystems. Indeed, numerous studies show the harmful effects on fishing populations. The following three case histories are illustrative of changes that can occur from channelization projects.

The Mississippi River Of all rivers in the United States none has been more modified than the Mississippi. Starting in 1837 with the first confinement dikes at St. Louis, this river has been repeatedly constrained by human attempts to control its behavior. The magnitude of this modification is discussed on pages 389–393, but in this section we are interested in possible changes that have been produced. For example, during the 1973 flood, the river overflowed its banks from Burlington, Iowa, to Cape Girardeau, Missouri—a distance of 560 km. Here the previous flood stage record occurred in 1884, but the 1973 flood was 0.6 m higher. However, the 1973 flood contained 35 percent less discharge! The higher stage has been attributed by many scientists to levee confinement and the wing dikes and revetments that were constructed for navigation stabilization channels. Such structures cause a reduction in channel width and also can produce additional deposition which further confines cross-sectional area. So although at other sites the river had undergone downcutting, this was insufficient to balance channel constriction. The average cross-sectional area along this reach was reduced nearly one-third by the channelization program. In addition to the changes in this section, the increased downcutting at certain sites produced increased sedimentation further downstream. Such deposits must then be dredged, creating a type of self-defeating feedback loop.

The Willow River The Willow River Drainage Ditch in Harrison County, Iowa, has become a classic example of the degradation associated with the human manipulation of a natural river. During the early 1900s the Willow River, a tributary of the Missouri River, commonly flooded. To alleviate this problem the channel was reconstructed in three phases starting in 1906. Thus, the original stream was artificially made into a drainage ditch essentially by straightening the channel. Prior to construction the 26.3 mi stream flowed in a 20.2 mi valley. Bank width was 60 to 100 ft and channel depth was 10 to 12 ft. In the lower reach the gradient averaged 5.2 ft/mi and was 7.5 ft/mi in the upper reach. The project was completed in 1920, with slightly different designs for the lower, middle, and upper reaches. However, they all called for channel deepening, narrowing, and straightening.

Two major changes have taken place in the channeled Willow River since its alteration. In the lower reach filling occurred so that periodic cleaning and rebuilding have been necessary. In the main reaches the ditch became entrenched (Fig. 13-3). Note in particular that deepening is progressively greater upstream—5 to 10 ft at mile 0, 10 to 15 ft at mile 5, 15 to 20 ft at mile 10, 25 to 30 ft at mile 15, and more than 30 ft near mile 20. The deepening process was accomplished by headward erosion when a series of knickpoints progressively sapped the channel as they migrated upstream. Thus, this provides a classic example of how stream adjustment changes channel geometry when man interferes with the natural stream regime and configuration.

In addition to in-channel changes of the master river, Willow River, changes in tributaries were also produced. Thompson Creek, a major tributary, became entrenched along with some of its tributaries. Between 1919 and 1957, Willow River incised its channel 18 ft at the junction with Thompson Creek. These new conditions produced feedback into the Thompson Creek drainage, à la Playfair's law of stream junction concordance. Knickpoints traveled upstream into Thompson Creek and its tributaries. The mouth of Thompson Creek cut down to be in accordance with the Willow Ditch and entrenched to depths of 40 ft the upper parts of the 6 mi long watershed. Thus, the channelization ushered in a new erosion-gully cycle of denudation. Conservation programs are now in operation attempting to control the gullying and hillside erosion. Other problems and expenses have also occurred. Roads and bridges have had to be changed and repaired. Wells have also had to be deepened due to the lowering of the water table.

The Sacramento River This River is a meandering stream that drains 68,000 km^2 of the central part of California. To protect crops farmers in the 1800s built low levees and straightened parts of the channel. Vegetation was also removed. However, the area continued to have floods, culminating in a 1955 failure that resulted in the loss of 40 lives and $48 million in property damages. In the 1960s the Sacramento River Bank Protection Project was initiated. The project has produced vast lengths of sterile rock-lined channel and the destruction of nearby vegetation.

Many other studies have been made of the deleterious effects of channelization. Emerson cites the case history of the Blackwater River in Missouri. In 1910, a new channel was dredged for 29 km with a gradient of 31.0 m/km, shortening the river 24 km. The old river had been 53.6 km long with a 1.67 m/km gradient. The meanders became entrenched in tributaries, serious erosion problems developed along the banks of the channelized stream, and headward erosion of gullies took place. Bridges had to be widened because of increase in valley width. Coates described the channelization of the Susquehanna River near Binghamton, New York, and the lawsuit brought against the Corps of Engineers because of damage to homeowners initiated by the changed river configuration. New channels must invariably be created during highway construction and these commonly lead to new erosion-siltation regimes of the revised stream.

The case against channelization also reaches into the popular literature with such articles as the "Rape on the Oklawaha" (Miller, 1970) and "Plague on all your rivers" (Madison, 1972). The Oklawaha Project—a combined canal and river diversion in Florida—received so many adversary votes that, even though $50 million in federal funds had been committed, construction was halted by a presidential order. Another Florida channelization project is now receiving great criticism. In October 1956 the Corps of Engineers released a report citing the need for channelization in the Kissimmee River Basin of central Florida. Extensive and costly flooding had particularly occurred in 1948, 1951, and 1953, and the expanding agricultural economy indicated a need for increased protection of farmland. However, soon after completion strong objections were raised over the destruction of a unique meandering river, decline of fish and waterfowl resources, degrading of water quality, and perhaps acceleration of the eutrophication of Lake Okeechobee into which the river drains (Fig. 13-4)

Channelization Benefits and Design

In considering stream channelization it is important to evaluate the entire range of impacts that can result—not only the detrimental aspects but also the beneficial results. After some floods it may be necessary to undertake remedial measures in attempts to minimize future flood trajectories. Thus, channelization and rerouting may be essential to the community. Adverse impacts from channelization may also be a reflection of inadequacies in the engineering design of the project. Those works that do not consider stream mechanics and hydraulic geometries are likely to produce aggravated environmental impacts.

Many rivers throughout the United States, and indeed the world, already reflect human intervention—some deliberate and other inadvertent. So many channelization projects have influenced rivers that at the present time few rivers are in their natural form. The clue to proper geomorphic engineering practices for river modification is to provide those configurations that replicate as closely as possible the equilibrium regimes of water flow for that particular environment. The flow characteristics of a stream, and the channel geometries that have formed, provide a fingerprint that is diagnostic of prevailing conditions. For example, the great majority of streams have pool and riffle sequences that are predictable. Generally, the distance between successful riffles (Fig. 13-5) is about six times the channel width. Thus, whenever man decides to alter this normal condition, it is axiomatic that the stream will seek to reestablish this sequence regardless of the im-

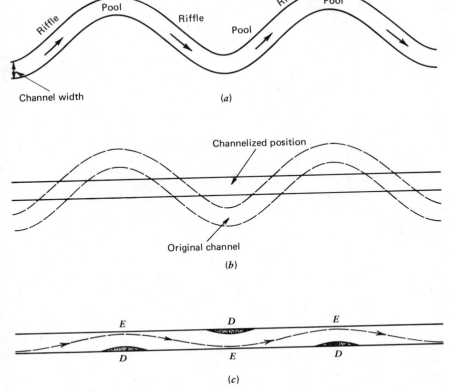

Riffle Pool Riffle Riffle Pool

Pool

Channel width

(a)

Channelized position

Original channel

(b)

E D E

D E D

(c)

Figure 13-5. Sketches showing before and after effects of channelization. (a) Map of a natural meandering stream with normal pool and riffle sequences. (b) Map of new channeled stream and location of original channel. (c) Streamflow in the straightened channel is modifying the bed by developing a modified thalweg and starting the formation of meanders and point bars. E = erosion; D = deposition of alluvial material.

pediments placed by man. Free-flowing rivers with floodplains generally contain meanders, although in drier climates braided streams may be more common. Many researchers have shown that a meandering stream has the greatest tendency toward equilibrium, so if we change the meandering configuration once again the water flow will try to revert to its original shape. Thus, the changing of meander curvatures and amplitude distances will cause disturbances within the river that will ultimately have repercussions in the system.

One other characteristic of streams is that for a given terrain there may be a characteristic hydraulic geometry that signifies balance. When the flow equation for discharge (Q) is considered, there may be a unique manner wherein the width (W), depth (D), and velocity (V) are related. We can write this relationship as $Q = WDV$.

It would be foolhardy to design a new channel that would force a different relationship of values in this equation whereby the artificial channel would have a different width or depth characteristic than the original and natural one. This would be especially damaging in urban areas and in immediately downstream reaches because it is known that urban rivers invariably produce stream widening. This results from the increase in flood magnitude, flood frequency, and flood routing time caused by the artificial drainage in urban environments. Thus, to minimize channelization effects, it is vital that we design our changes with the full knowledge that rivers are open systems, can produce feedback to other parts of the system, and that the safest plan is to duplicate natural conditions whenever possible.

A prime consideration of channelization projects, as in most all other engineering ventures, is a complete analysis of the benefit-cost ratios. Economic and safety justifications should be mandatory for all projects. Channelization of the Chenango and Susquehanna rivers in the southern New York area have produced premiums in both diminution of flood losses and in the increased safety of residents. There are numerous cases of very high agricultural yields where farmlands have been usually saved from damaging floods. Ruhe describes a case history of beneficial changes produced by the Corps of Engineers in their channelization of the Missouri River at the Otoe Bend area in Nebraska and Iowa (Fig. 13-6). Here starting in the mid-1930s the Corps began a series of manmade improvements to stop the perennial flooding of the area and also create additional farmland. A series of permeable pile dikes (Fig. 13-7) were emplaced to check flow velocity and permit bed and bank accretion downstream of the dikes. Such sediment permitted subsequent upstream deposition that ultimately filled and covered the dikes, welding them into a single landmass. Natural vegetation growth has aided to stabilize the features. Since completion of the project, flooding has been virtually eliminated. At the same time adjacent farmlands have been saved from flooding, and new farmlands added to the productive role of the region. Navigation has also been made safer and more predictable for this reach.

On an overall balance river channelization may be absolutely necessary in some cases, but often it has been demonstrated that the societal costs outweigh the benefits. The courts are increasingly recognizing such damaging effects and, in 1978, a federal court halted governmental channelization of the Obion and Forked Deer rivers in western Tennessee.

HIGHWAYS

In the developed and industrialized nations extensive highways have been constructed to link all major cities, for the transportation of people and especially products in the rural regions. As with all situations, whenever we cut or fill the original landscape, upsets in the natural systems occur. Wanton building practices accentuate terrain degradation whereas careful planning can reduce these impacts. Interstate 70 in the Vail Pass portion of this Colorado highway is proof that highways can be built in environmental harmony. This 22.5 km segment was designed to be an integral part of the total landscape rather than an offensive intruder. It was accomplished by an interdisciplinary team of planners, scientists, and engineers that represented the Colorado Division of Highways, the Federal Highway Administration, U.S. Forest Service, Colorado Division of Wildlife, and the International Engineering Company. Criteria that were adhered to in this 3230 m elevation at the Continental Divide included design and construction procedures that (1) created minimum terrain damage during construction; (2) made maximum use of precast and fabricated units; (3) assured maintenance of aesthetic qualities of the route; and (4) prevented all roadway encroachments and embankments near streams.

Automobiles first became commercially available in the United States in 1903 and, by 1910, there were 60 companies producing them. As numbers grew so did the demand for roads. In 1916 Congress created the Bureau of Public Roads which focused on constructing "farm to market" roads. Until the 1950s the great majority of road construction was financed and built by local and state governments. In an attempt to combat the Great Depression, Pennsylvania began construction of the Pennsylvania Turnpike in the 1930s. The section from Harrisburg to Pittsburgh was completed in 1940 and had been built on a long-abandoned railroad right-of-way. To finance the project the old practice of charging tolls was reinstituted. After World War II many other states followed suit and, by 1960, there were 3000 mi (5000 km) of toll roads. In the summer of 1954 President Eisenhower's administration set forth "a grand plan"

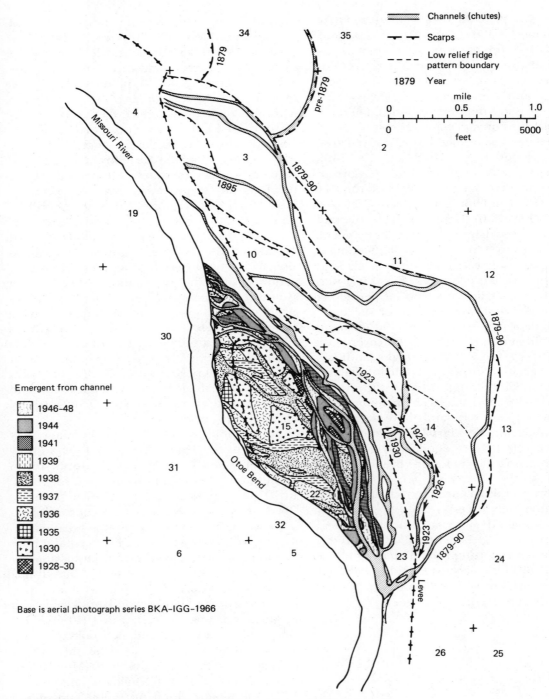

Figure 13-6. Otoe Bend area, Missouri River Valley, in 1930 and 1940. The new configuration of channel and dikes (constructed in the 1930s) is superimposed on the 1930 map for comparison purposes. (R. V. Ruhe, 1971.)

Figure 13-7. Missouri River, Upper Monona-Lower Monona Crossing (vicinity mile 701), looking downstream along a crossing area between two bends. At this site the Missouri has been "trained" using permeable pile dikes and bank stabilization structures. (Courtesy Corps of Engineers.)

to build the most ambitious engineering project in history—The National System for Defense and Highways. This was to receive the name The Interstate Highway System, and the original design was to produce 64,000 km (40,000 mi) of limited-entry freeways that would criss-cross the country uniting nearly every major city in the 48 states. Construction costs were to be met by excise taxes on auto gasoline and tires, with the federal government paying 90 percent. The system is now 99 percent complete with 68,400 km of four lanes or more of highways, though constituting only 1 percent of total traffic. The total cost is $114.3 billion!

Highways produce a vast array of changes— economic, societal, and environmental. But they are vital as arteries of commerce in farming and industry. They also change population distribution, land use, property values, taxation assessments, type and volume of business, safety, and political boundaries and zoning. However, rapid and unregulated growth may be generated with chaotic developments and high gov-

ernmental costs. Severance damages to land parcels and the displacement of people by eminent domain can create personal hardship and grief. The possible range of environmental impacts is staggering, and such influences can be either short-range upsets and transient or long-term and create permanent degradation. The following points illustrate some of the ways that road construction can alter the environment.

Beheading of Aquifers

Shallow aquifer systems where the water table is intercepted by the road excavation can cut off the water supply to downslope wells, springs, and water systems. A new spring line and seepage develops on the cut hillside which must be accommodated by roadside drainage culverts (Fig. 13-8).

Groundwater Sink

When the road cut is deep, the excavation may be considerably below the entire water table of the site, thus producing a groundwater sink (Fig. 13-9). This will produce a lowering of the water table as the flow regime establishes a new gradient. The water level decline will extend in time to adjacent areas and effect other groundwater divides and reduce the volume and flow of groundwaters to nearby water supply systems. Of course, the additional waters that enter the sink must be discharged by an effective drainage system.

Slope Stability

The artificial hillside cut often gives rise to slope failures (Fig. 13-10; see also Figs. 11-14, 11-15, and 11-19). Rockfalls, rockslides, slumps, and other failure modes are commonplace and are especially prevalent when the upslope section contains permeable materials with a high water table. The loss of stability may also be caused by slope wash, gullying, and piping processes. Such processes produce siltation problems that clog drains and water supply systems. With fine-grained sediments, their removal is facilitated

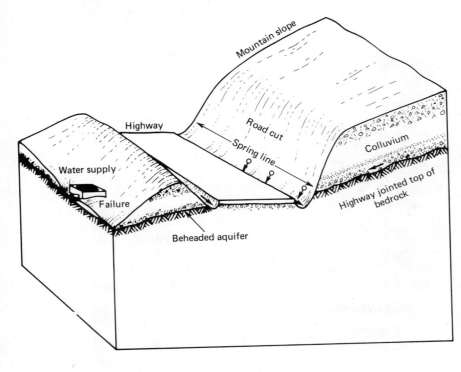

Figure 13-8. Beheaded aquifer caused by the disruption of the groundwater flow in a new roadcut. (R. R. Parizek, 1971.)

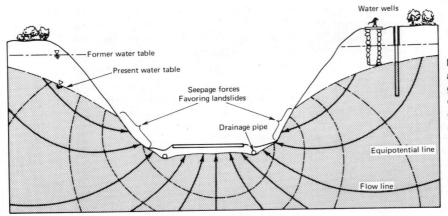

Figure 13-9. New roadcut area acts as a giant groundwater sink, with a disturbance to the groundwater flow, causing a discharge into the man-made terrain depression. Construction in such areas must make allowance for effective drainage to prevent waterlogging. (R. R. Parizek, 1971.)

by grain-to-grain transport, aided by the seepage pressures (piping). Such pressure may also add to the driving force to produce potential surfaces for landsliding. These forces are directed downward and can overcome the shearing resistance of the earth materials.

Siltation and Erosion

Highway construction is notorious for producing accelerated rates of erosion with the concomitant problem of siltation. Maryland was the first state to develop comprehensive laws in

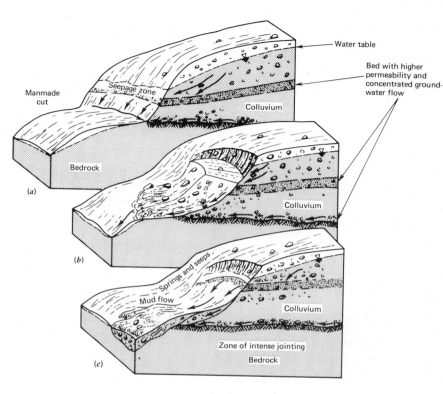

Figure 13-10. Slope failure by piping caused by man-made hillside cut. (*a*) Erosion of fines from a thin aquifer is initiated by groundwater discharge to the new cut. (*b*) Piping continues until the slope fails by slumping. (*c*) With prolonged groundwater discharge and surface runoff, the slumped mass may give rise to a mudflow, and the exposed slope is subjected to a new erosion cycle. (R. R. Parizek, 1971.)

1970 that required strict sediment control measures during all types of construction, in highways, and in the building trades. Divided land highways require the exposure of 10 to 35 acres per mile during construction and produces as much as 3000 tons of sediment per construction mile (Fig. 13-1). Sediment may be derived from fresh road cuts, embankments, and borrow pits. The siltation problem is most severe prior to seeding and emplacement of pavement. The increased discharge and redistributed surface runoff from pavements and embankments help to promote additional erosion on adjacent lands and streams. The resulting sediment may be deposited in streams, causing reduction in streambed infiltration rates. This may cause streams to divert from their natural pathway, increase the flooding hazard, and increase turbidity in water supply systems. Of course, highways are subject to erosion by streams when improperly designed (Fig. 13-11) or subjected to abnormal flood conditions (see Fig. 12-7).

Acid Drainage

Rocks exposed along road cuts that contain such minerals as pyrite and other sulfides are espe-

Figure 13-11. Highway erosion, West Virginia. The stream meander has undercut the road on the outside bend of the meander, and the channel is restricted by the point bar deposits on the inside bend or right side of photo. (Courtesy Library of Congress.)

cially vulnerable to weathering and leaching processes that can yield highly acid waters, just as in surface coal mining. The problems are most acute where the receiving bodies of water afford little if any dilution or neutralization, and where such waters drain into public water supplies (Fig. 13-12).

Man-induced Pollutants

Contamination of water supplies and adjacent terrain and vegetation is produced by exhaust fumes from cars and trucks, spraying of pesticides and herbicides to control insects and weeds, and materials placed on roads to control ice and snow buildup.

Our need for unrestricted travel during winter months in freezing climates prompted the use of deicers, the first of which were such abrasives as cinders and sand. However they pose removal problems because they clog drains, culverts, and catch basins. Salts were first used in very small amounts to combat freezing of stockpiled sand. As its applications to roads grew, its virtues were extolled by highway maintenance operators because the use of salt (1) eliminated the expensive summer clean-up because it dissolves and is washed away naturally, (2) lowers the freezing point of the solution and prevents ice buildup, and (3) reduces accidents from slippery highways. The "bare pavement" policy of many highway crews is causing increased amounts of salt usage on the nation's highways. In general salt (NaCl) is used when the temperature is above $-9.4°C$, calcium chloride ($CaCl_2$) when it is below $-17.8°C$, and a mixture of the two when the temperature is between. Average salt application may range from 200 to 400 lb per lane mile, and during a season it is not unusual to have total applications of 20 tons per lane mile. Seventy percent of deicing salt is used in northeastern states, and salt used for deicing purposes in the country comprises 20 percent of total salt production. However, there are environmental impacts from use of highway salts; here are seven major results of icing.

1 Soil When sodium constitutes more than 15 percent of the ion exchange capacity of the soil, the soil structure begins to deteriorate. The result is reduced flocculation of soil particles, causing a reduction in their permeability and water-holding capacity. This in turn increases surface runoff, decreases infiltration, and diminishes soil aeration for plant growth.

2 Plants An increase in salt concentration in soil water inhibits water intake by plants because the water flow is in the direction of greater salt concentration. This impairs root aeration, causing water deficiency and nutrient imbalance. At Newton, Massachusetts, an average of 500 trees per year were killed between 1965–70 from road salting. The effects included growth depression, leaf burn and shoot dieback, and death.

3 Wildlife Salt excesses in animal habitats when ingested interfere with osmotic balance and disturbs normal body processes.

4 Groundwater Sodium chloride values in a New Hampshire study ranged from 140 ppm (parts per million) in a control area 100 ft from the highway to 1203 ppm in the upper 6 in. of soil 3 ft from the edge of the road. Because of moderate precipitation in the northeast, many salts are leached into the adjacent groundwater. This causes hardness in well supplies. People most affected by sodium-contaminated drinking water are those requiring low sodium diets for control of hypertension, kidney disorders, and obesity. Although water hardness appears to reduce heart failure, excess sodium increases them. The recommended sodium content for drinking water is set at 20 ppm by the American Heart Association, but a study in Massachusetts showed many near-road wells to have levels of 117 ppm from road salting. Of course, heavy metal toxic substances—for example, cadmium, lead, zinc, and nickel—from car exhaust also become entrained in groundwater systems. Such materials were observed in the soils of the Washington-Baltimore Parkway 32 m from the roadbed.

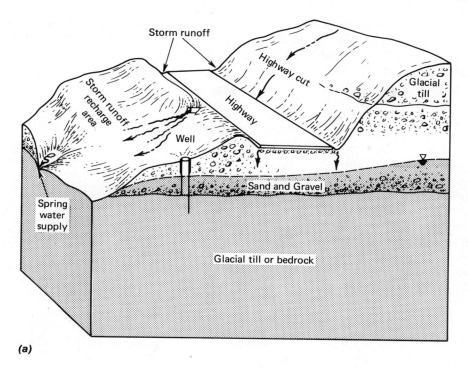

(a)

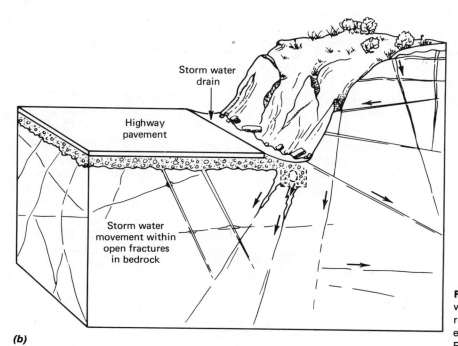

(b)

Figure 13-12. Schematic views (*a*) and (*b*) showing how roadcuts enable pollutants to enter water supply systems. (R. R. Parizek, 1971.)

5   Lakes   A study of First Sister Lake, Michigan, showed that incoming salt-laden waters caused salt buildup in the bottom waters and prevented the spring turnover. This can lead to oxygen depletion in the hypolimnion (the lower levels in a lake which remain at constant temperature during summer) and result in elimination of benthic organisms and fishkill. Lakes may become permanent-stratified because salt introduction is becoming cumulative and exponential, doubling every five years. Lake Minnetonka just west of Minneapolis has increased in chloride fivefold from 1940 to 1970, with 25 ppm at the present time. In similar fashion Irondequoit Bay near Rochester, New York, has also increased five times in the last 20 years. The period of summer stratification has now been prolonged more than 30 days by the salt density gradient.

6   Car corrosion   Salt facilitates rusting of cars. It produces an electrochemical reaction resulting in the oxidation of iron in the presence of salt and water. The mobile chloride ions increase the electrical conductivity of the solution, and costs average $50 per year for those cars that use salted highways.

7   Pavement deterioration   Destruction of paved roads is hastened by crystal growth and the thermal expansion of salt compounds. The salt crystals grow and expand in openings and cracks and are especially damaging with large diurnal temperature changes, which are accentuated on sunny days. Such disintegration also permits penetration of water that also expands on freezing and thus accelerates the pavement breakdown. A February 1978 study of potholes by The Road Information Program showed there were 116.6 million potholes on U.S. highways that averaged 16 in. in diameter and 5 in. deep. It takes 110 lb of asphalt to fill the typical pothole, and the repair program, plus car repairs and lost gas by sudden starts and stops, was calculated at $882 million. All pavement deterioration is not caused by salting, however. Other contributing factors include poor snow-plowing techniques, aging roads, inadequate quality control of highway materials, and normal attrition because of heavy usage (Fig. 13-13).

OFF-ROAD VEHICLES (ORV)

The phenomenal growth in the use of off-road vehicles (ORVs) has largely occurred in the last two decades. Part of this began in 1947 when Soichiro Honda, a 41-year-old mechanic in Japan, strapped a small war surplus generator to a bicycle. The result was so successful that soon he had exhausted the supply of surplus generators and had to modify the ingredients to form a new type of motorcycle. By 1960 Honda was marketing them for $300 in the United States when domestic models cost $1500 or more. By 1971 Honda had captured more than half the U.S. market. In 1976 there were 8.3 million U.S. motorcycles in use, with 5.4 million being ridden in off-road ventures.

German engineers in the 1950s refined the two-stroke internal combustion engine which produced more horsepower per pound of engine than the traditional four-stroke engine. This made the mass production of snowmobiles practical. J. Armand Bomardier in Quebec built the first commercial ones and sold 259 in 1959. By 1969 there were 265,000 snowmobiles in North America, and today there are 5 million.

Other types of off-road vehicles include 250,000 dune buggies in the United States and 3 million four-wheel drive vehicles. In addition, several varieties of two-wheel drives are used for off-road purposes along with front-wheel drive cars. These more than 10 million vehicles are producing numerous changes of the terrain they traverse and are modifying plant and animal life. Such vehicles in a sense create their own roads.

The federal government was caught totally unaware of this new phenomenon and has been snail-slow in reacting to it. The Bureau of Out-

"Most of our roads need immediate work."

One looks at America's freeways and marvels. But another talks secondary roads: many outdated, in disrepair. Which reflects reality?

We have 3.8 million miles of roads. Almost two million miles paved. 37,000 miles are wide, sweeping Interstate Highways joining 42 state capitals, 90% of all towns over 50,000 people, carrying 20% of our traffic. Urban freeways speed traffic into and around cities. Highway fatality rate on our Interstates is about half other roads. Many expressways are beautifully landscaped, such a pleasure to travel, you're tempted to conclude: "America's roads are great. The work is finished!"

Driving secondary roads and city streets is another matter. Most were built generations ago. Many are narrow, .broken-up, patched, re-patched, crossing antiquated bridges, and unguarded tracks. Hosting 13.5 million accidents yearly. Years behind today's needs. 60% in need of modernization or repair. Travel these roads and you say, "Our roads are in bad shape!"

Truth is, we have an excellent road system. But roads wear out with age and use. Some Interstate Highway miles are 20 years old. They need maintenance, repair. Older secondary roads are in worse shape. The total repair and maintenance will take over $200 billion according to the Department of Transportation. Every delay increases the cost in terms of vehicle damage, higher construction cost later on and inefficient transportation. Near 90% of America's intercity passenger travel is by road. We must be willing to fund a national policy that gives priority to maintaining that important national asset.

Figure 13-13. Highway advantages and costs. (Courtesy Caterpillar Tractor Co.)

"We have the world's best highway system."

Figure 13-14. ORV land desecration. Nearly 65 cm of soil mantle has been lost, leaving exposed sterile bedrock. Redwood Road, East Bay Regional Park District, California. (Courtesy Howard Wilshire, U.S. Geol. Survey.)

Figure 13-16. Erosion caused by ORV traffic. In this case, ORVs destroyed junipers on steep slopes by direct vehicle impact, and subsequently root systems were exposed. Ballinger Canyon, Los Padres National Forest, California. (Courtesy Howard Wilshire, U.S. Geol. Survey.)

Figure 13-15. Deflation by ORVs in the California desert terrain. (Courtesy John Nakata, U.S. Geol. Survey.)

Figure 13-17. Extensive stripping of grass cover by ORVs. Remnants of the grass are seen around the large plants and at the margins of the main use zone. Ballinger Canyon, Los Padres National Forest, California. (Courtesy Howard Wilshire, U.S. Geol. Survey.)

door Recreation does not even include ORVs in their 1971 report, and the first nationwide outdoor recreation plan offered by the Department of Interior in 1973 barely mentioned ORVs. Thus land use managers had no common strategy for dealing with such activities; ironically more than half the travel is on federal land, largely governed by the Bureau of Land Management. In 1977, 43.6 million Americans engaged in some form of ORV activity—25 percent of the total population. Indeed, such recreation has now surpassed the majority of other types. It is a phenomenon that must be reckoned with.

ORVs have damaged every kind of ecosystem found in the United States—sand dunes and beaches at Cape Cod and Fire Island (Figs. 13-14 to 13-19); pine and cyprus woodlands in Florida; hardwood forests in Indiana; prairie grasslands in Montana; chaparral and sagebrush in Arizona; alpine meadows in Colorado; conifer forests in Washington; tundra in Alaska; and desert vegetation in California. The problem is that the very type of terrain preferred by ORV users is the terrain which is most fragile

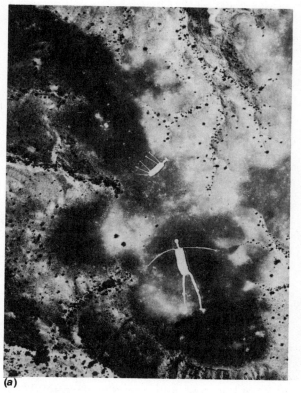

(a)

and vulnerable to long-term degradation. The ready availability of federal lands has profoundly affected usage, and inadvertently it has stimulated ORV growth. Furthermore, such federal policy inhibits private enterprise from development of commercial lands. For example, in the entire Los Angeles region, where ORV use is heavier than any other part of the country, there are only four commercial ORV parks.

Terrain Damage

Since 1968 ORVs have damaged half as much land as was damaged by all U.S. mining operations from the birth of the country. Most of this occurs in the drylands (arid and semiarid regions) where soil is so fragile it takes 250 to 500 years to generate 1 cm. Such scarred lands regenerate so slowly that wagon wheels that rutted the terrain more than 100 years ago in the Mojave Desert are still visible, as are the jeep and tank tracks from General Patton's World War II maneuvers. In the Panoche Hills of the Diablo Range, California, an area that had been open to ORVs for three years was then closed. Four years later the ORV-impacted areas were still losing soil at the rate of 3 t/ha (tons per hectare), while erosion in adjacent

(b)

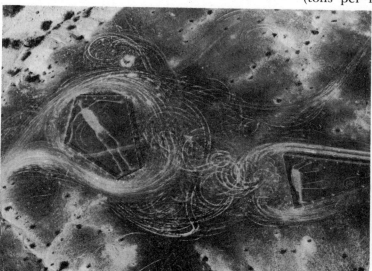

Figure 13-18. ORVs and intaglios (Indian pictographs), Blythe, California. (Courtesy Howard Wilshire, U.S. Geol. Survey.) (a) A 1932 photograph showing typical, unspoiled character of the largest figures, about 50 m long. (b) A photograph taken in 1975 showing the extensive damage to intaglios.

Figure 13-19. Aerial view of constructed and vehicle-produced ORV trails on the west side of the San Andreas Fault, Hollister Hills State Vehicular Recreation Area, California. (Courtesy Howard Wilshire, U.S. Geol. Survey.)

undisturbed land was too small to measure (Fig. 13-14).

ORV wheels churn the topsoil into loose aggregate that flows downhill with the first rain. On dry land the topsoil turns into a type of slurry, and on wetter land there is constant attrition by raindrop impact and removal from the site. With continued use hillsides become carved and washed into a badland topography, and in some California areas the gullies are now more than 3 m deep. The slope wash materials are deposited at the base of hills where they bury and destroy vegetation.

On flat dry areas ORV traffic exposes the soil to another powerful force—the wind (Fig. 13-15). These sites become favorite locales for heavy usage and races. The traffic not only pulverizes the surface layer but also compresses the soil beneath. This diminishes water infiltration into the substrate and accentuates the dryness; at the same time it depletes moisture for plant growth and sustenance. When vegetation is killed the change in albedo increases soil temperature and further degrades the material. Not until September 1972 did the Bureau of Land Management (BLM) begin requiring or-

ganizers of competitive racing events to obtain special land use permits. During the next 12 months there were 151 such events in California with 67,000 participants and 200,000 spectators. Earth Resources Technology Satellite (ERTS) photographs of January 1, 1973, showed several dust plumes that originated from such events, some up to 75 km long. In 1974 the Barstow to Las Vegas motorcycle race yielded more than 610,000 kg of airborne particulates. We are clearly contributing to a desertification process in these areas.

Vegetation Damage

In addition to soil destruction, ORVs also stunt, damage, and kill vegetation. This is done in several ways. Direct contact with plants break branches. ORV use around a plant may be so heavy that the soil erodes and the plant collapses from lack of any supporting medium (Fig. 13-16). Compaction injures the root system and can kill even large perennials. Germinating seeds on or within the ground may be crushed, and other plants may be smothered by wash or deflated debris. Plant life in even moderately used desert sites declines more than 50 percent within a few years and can be nearly completely destroyed in heavily used sites. Recovery is nearly impossible; for example, creosote which under natural conditions may live 600 to 700 years takes 80 years to mature. Restoration is also made difficult because of a reduction in residual moisture and abnormal temperature changes, which are environmentally sterile to most plant growth, for example, there is water depletion of about 100 percent and diurnal temperature differences of more than 12°C when compared to nonimpacted areas (Fig. 13-17).

Wildlife Damage

Silence is a resource. Some animals cannot adjust to noise and commotion. Unnatural noise places undue stress on animals and has been shown to be destructive to such species as the iguana. In the California terrain used by ORVs there formerly were 40 bird species not found elsewhere, but their numbers are decreasing in relation to the use of an area. Terrestial animals decline 60 percent with slight use and more than 90 percent in heavy-use areas.

Archeological Damage

ORVs provide increased mobility into areas that previously were seldom seen or traveled. Vandalism to petroglyphs and intaglio sites has become a severe problem that has no easy answers (Fig. 13-18).

A Rationale for Planning

Policies that govern ORV use need to balance benefits and costs that result from this form of recreation. ORV enthusiasts argue that such vehicles provide much pleasure to participants, aid the economy of communities and services that support their use (gas stations, motels, restaurants), enable an escape from the hustle of urban life, and enhance family ties and togetherness (Fig. 13-19). Several states have weighed such factors and have still prohibited ORVs on state land, as in Indiana, or confined it to designated trails, as in Massachusetts.

It must be realized by decision makers that ORV damage is an external cost and that the general public pays for the benefits enjoyed by ORV personnel. If damage is to be reduced there must be (1) recognition of the magnitude of the enforcement problems and measures taken to address it; (2) a stiffening of procedures that expand federal management roles; (3) a separation of ORV use from other types of use in open space land; and (4) incentives applied for proper ORV use and punishment for vandalous actions. Two executive orders provide the legal framework for accomplishing these objectives. Executive Order 11644 "Use of Off-Road Vehicles on the Public Lands" dated February 8, 1972, sets guidelines for

action and assigns responsibilities, and Executive Order 11989 of May 24, 1977, expands the role of government agencies and requires public land managers to close areas or trails to ORV use whenever it is determined that "The use of off-road vehicles will cause or is causing considerable adverse effects on the soils, vegetation, wildlife, wildlife habitats, or cultural or historical resources . . . ."

PIPELINES

Pipelines have become a preferred method for transporting fluid resources: water, natural gas, oil, and even slurried coal. When improperly constructed, they can produce damages to the environment, but no study of possible engineering impacts would be complete without mention of the biggest private construction job of all time—the Trans-Alaska Pipeline Systems (TAPS). This was built by Alyeska Pipeline Service Company at a cost of more than $8 billion; it extends from Prudhoe Bay in north Alaska on the Arctic Ocean to the port of Valdez (a deep-water harbor) in south Alaska. Although the project was first proposed in 1969 more than a year after the Atlantic Richfield Corp. had struck oil, it was not until spring 1974 that bulldozers moved the first earth. The project was completed by June 1977 when the first month-long journey of oil started its southern trip.

TAPS is 1270 km long and traverses three mountain ranges, more than 800 streams (including the giant Yukon River), and is in permafrost terrane for 950 km (Fig. 13-20). To obtain authorization for the project a mammoth Environmental Impact Statement (EIS) had to be prepared in accordance with NEPA regulations. The EIS was in six volumes and weighed 11 kg. Intervenors held up construction for more than a year, but as a result of all the hearings a more environmentally sensitive project was developed. Here are some of the considerations that were adopted.

Figure 13-20. The Alaska Pipeline. The gravel and insulation workpad underlying the pipeline prevents degradation of the installation from permafrost. This aerial view is along the north slope near Pump Station 3. (Courtesy Alyeska Pipeline Service Co.)

1 More valves per kilometer than any other long-distance pipeline—151 in all. They were always placed upstream or upslope of key rivers, population centers, and ecologically fragile areas. Thus, if a leak develops, the impact will be minor: a maximum of 50,000 barrels is all that would be lost.

2 The pipeline is 680 km above ground (the longest bridge in the world) with 78,000 vertical supports. More than two-thirds of the supports contain built-in heat radiators designed to keep the permafrost from melting. Furthermore, where the pipe is buried in permafrost, refrigerating insulation is emplaced to prevent permafrost disturbance.

3 Pipeline sections are flexible and are capable of withstanding earthquakes up to 8.5 M.

4 Pump stations were moved as much as 16 km so as not to disturb peregrine falcon nesting areas. Although at some locales the cheapest construction would have been to elevate the pipeline, it was instead buried at 23 such sites. This allowed migration freedom to caribou and gave elk and bear greater freedom. A total of 400 underpasses and pathways over buried pipeline provides wildlife mobility. Construction was stopped on many occasions for wildlife, to avoid hibernating bears and nesting swans, for example.

5 More than 14,000 ha of right-of-way land was seeded along the pipeline corridor, and the plants were selected to match the local chemistry of soils and the temperature and precipitation conditions. More than 1.5 million willows were planted in the northern reaches in an attempt to provide a habitat for big game grazing.

Total costs that were made for environmental reasons rather than technical reasons added $1.5 billion to the project. This does not include the most extensive siting program ever undertaken for a long-range pipeline. Complete aerial photograph interpretation was made of a 3.2 km wide strip on a 1:12,000 scale. More than 3500 soil borings were drilled, with 2100 on centerline, and 33,700 samples were analyzed in the laboratory. The final product has been a 122 cm diameter pipeline capable of transporting 1.6 million barrels a day to the delivery stations at Valdez. This is about 7 percent of the total U.S. petroleum demand.

The next project is to construct a natural gas pipeline to deliver the north Alaska gas to Canada and the United States. Three companies investigated this proposal. Arctic Gas Co. spent seven years and $200 million preparing their plan, but lost the contract to a smaller consortium titled the Northwest Pipeline Corporation of Salt Lake City. The Arctic Gas Co. was rejected by the Canadian government which called it "environmentally unacceptable." The other bidder which spent even more planning money, the El Paso Gas Co., also lost because their plan was not as environmentally feasible as Northwest's. Estimates for the project are now at the $14 billion mark.

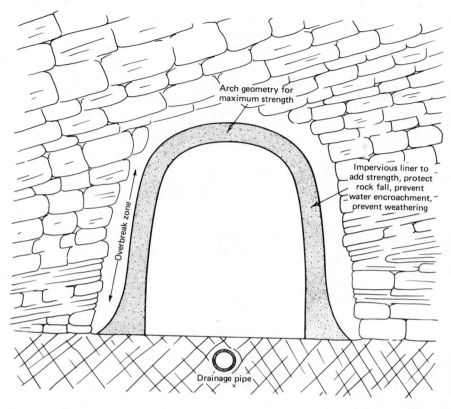

Arch geometry for maximum strength

Impervious liner to add strength, protect rock fall, prevent water encroachment, prevent weathering

Overbreak zone

Drainage pipe

Figure 13-21. Schematic cross section showing characteristics of typical tunnel construction.

TUNNELS

A tunnel is an underground passage. It is a conduit whose entrance is different from its exit. Tunnels provide avenues of commerce for the transportation of people in cars, railroads, and subways or for water resources and communication facilities. Perhaps the first extensive tunnel was constructed more than 2500 years ago under the Euphrates River and was 3.5 m wide and 4.5 m high. The Romans built many tunnels and the most extensive was built for drainage of Lake Fucino. It was cut through limestone a distance of 5.6 km and was 2 m × 3 m in cross section. It required 11 years to build, employing 30,000 laborers.

Of all engineering construction, tunnels require the most geologic information, because they occur entirely in earth materials. Knowl-edge of rock type, structure, and water conditions are vital to success. Extensive boring programs are necessary to evaluate the conditions that will be encountered for planning purposes. Tunnel size and shape depend on their use. If not used for human occupance circular tunnels are made because they are the strongest. When used for human transportation a flat floor is needed with an arched roof for greatest support (Fig. 13-21). An appropriate lining along with drainage systems comprise some of the most important considerations. Lining is necessary to prevent cave-ins and to protect rock from fumes and weathering so it doesn't deteriorate. Water tunnels must be also lined with smooth surfaces to reduce friction losses, and the lining must be impervious to prevent water loss and to prevent water contact with rocks which could be eroded. The cost of tunnel lining is often about 25 percent of the total expenditure. Shattered and decomposed rock, along with groundwater seepage, constitute the principal problems in tunnel construction. Such phenomena are the responsibility of the geologist who must predict their character and magnitude. If erroneous conclusions are made, cost of construction is greater, unnecessary delays occur, and ultimate failure may result. It is also important to determine the rock stress that will be encountered. Pressure occurs not only from the weight of the overburden, but also from in situ stress conditions within the rock that result from their characteristic elasticity and creep properties (see page 55). Any type of rock excavation requires that the host rock will undergo some transformation to adjust for such an indignity, as in **Newton's second law of motion.**

Tunnel lengths vary throughout the world. Those in the Alps range up to 20 km whereas the water aqueduct tunnels in the Catskill Mountains range to 50 km. The idea for construction of a Washington, D.C., subway with a tunnel system is more than 100 years old. In a lengthy article of December 5, 1909, the *Washington Post* asked the question " . . . why isn't a subway for Washington just the thing?" After 10 years of design and construction this was answered on March 29, 1976, when Phase I was completed with 7.2 km of track and five stations placed in operation. The Metro opened 18 more kilometers of track and 23 additional stations in 1977. The geology of the tunnel was very difficult to appraise because of structural and lithologic complexities. The different rock types ranged from Precambrian solid rock, to thick zones of weathered saprolite and weak Coastal Plain sediments. The large tunnel stations were all placed at extra depths to assure location in solid rock for adequate support.

EXPANSIVE SOILS

A variety of terms are used in the literature to describe the shrinking and swelling properties of certain earth materials, but materials with such properties are usually referred to as **expansive** soils, sediments, and clays. Many different rock types and sediments comprise the family of earth materials which when water is either added or withdrawn undergo significant volumetric change. However, the parent materials associated with the phenomenon generally fall into two categories.

1   Basic igneous rocks. In these soils the feldspar and iron-magnesium minerals decompose to form such clay minerals and other secondary products as montmorillonite.

2   Sedimentary rocks. The constituents may already contain montmorillonite or other hydrous-reactive minerals.

The problem of expansive soils was not recognized by soil engineers until the late 1930s. Prior to 1920, the majority of lightly loaded buildings in the United States were of frame construction. Such dwellings could withstand considerable displacement without developing cracks or breaks that threatened the integrity of the structure. If cracking did occur it was usually explained to be the result of poor materials or design, or was blamed on the foundation settlement. The U.S. Bureau of Reclamation was the first government agency to recognize the soil expansion problem in 1938 and, since then, engineers have slowly become aware of it. But it was not until the 1940s that the problem took on new scope. This was caused by the increased use of concrete slab-on-ground construction which prevented "breathing room" for substrate materials. It has been estimated that expansive soil damages in the United States amount to more than $2.255 billion annually. It is, however, a worldwide phenomenon with extensive damages reported in such countries as Argentina, Australia (where it is called "gilgai," an aborigine term), India, Mexico, South Africa, and Spain. Thus, these damages rank high on the listing of earth forces that produce destruction to human property. Extensive damages occur to buildings, highways, pipelines, pilings, retaining walls, and lightly loaded structures. Typical signs of this process include doors that won't close, sticking doors and windows, cracks in walls and ceilings, separation of walls and ceiling intersections, and dwelling alignments that are out of plumb.

How the Process Works

Expansive soils predominantly occur with those earth materials that contain the clay minerals, because they have the property of absorbing certain anions and cations and retaining them in an exchangeable state. Montmorillonite is 10 times more reactive than other clays such as kaolinite. In climates where the soil is saturated, or where moisture content does not significantly vary, the swelling soil problem is minimal. Only in dryland environments, where soil moisture radically changes during the seasons, is there usually extensive damage from expansive soils.

Numerous factors influence the behavior of expansive soils: (1) water table depth, (2) stress conditions in the soil, (3) magnitude of moisture changes, (4) vegetation, (5) topography, (6) drainage conditions, and (7) type and quality of construction. Climate controls such variables as the water table, active moisture zone, vegetation, and soil properties. Thus many of these effects are predictable.

The character of vegetation, both before and after construction, will affect degree of soil expansion, or shrinkage. Large trees use much water, so when an area is cleared for construction the old vegetation will have produced a suction pattern that will be disturbed on removal. New plantings or remnant vegetation near structures can draw moisture from the new site and cause shrinkage (Fig. 13-22). Large trees should be planted at distances away from foundations that are more than one-half their mature crown height. Elm, poplar, and willow are species that pose many problems.

The structure itself has an impact on the soil. Foundation slabs and roads stop all evap-

oration of soil moisture below them. This means the soil attains an equilibrium through time with no moisture variations. This does not assure safety because during the period for establishing equilibrium volumetric changes can occur. After construction the edges of the structure will still be subject to moisture differences, while the center is cut off from the surface environment. The edge-soil moisture regime spreads inward under the structure. In dry areas center lift or doming is more apt to cause slab failure, whereas in more temperate climates end lift is likely to occur (Fig. 13-22). Center lift generally takes longer to be manifest than end lift which may occur within a short time after the foundation is cast. If construction is started before the soil moisture has had time to adjust to the new conditions, greater structural damage will occur than when there is a delay period between site preparation and finished development.

Improper grading and landscaping may cause ponding, with localized swelling at building sites and highways. Many highways are now designed in attempts to avoid expansive soil changes. Special drainage care is taken to reduce moisture changes to a minimum under the highway and shoulders. In Texas highway design now requires at least an 8 percent slope away from the foundation for a distance of 3 m. However, some roads were designed before these requirements were instituted (Fig. 13-23). Even highways where the problem is anticipated have been subject to problems. Mathewson and Clary (1977) discuss multiple landslides on Interstate 45 in Texas which were caused by the buildup of pore pressure in the expandable clay of the road cut. More than $200,000 were spent to remedy the problem. Font (1977) has described failures caused by expansive soil units in the Eagle Ford and Washita Groups of sedimentary strata near Waco, Texas. Here the South Bosque shale and Del Rio Clay members are the most notorious for producing landslides. In a sense these rock units can be considered to be "overconsolidated"—that is, the internal

(a)

(b)

(c)

Figure 13-22 The effects of expansive soils on home construction. (a) A newly constructed home. The foundation at x acts as a barrier, preventing capillary loss of soil moisture to the atmosphere. (b) Trees planted too close to the house create "center lift." Tree evapotranspiration at E causes loss of water to the system when soil moisture at y dehydrates via roots. (c) Downspouts are too close to house, creating excessive swell and producing "end lift" with accompanying cracked foundation and wall fractures. Rain aids to increase soil moisture, R, around house, while there is a moisture deficit under the house because water cannot penetrate below foundation.

stress has not yet accommodated a changed condition in overburden pressure. In such materials additional expansion and uplift pressure result.

Overconsolidation of rock units may occur in many different environmental settings, in fact, any place where the soil or rock undergoes a rather dramatic pressure change when overburden materials are eliminated. This rebound tendency may not be released until high-moisture content differentials allow the release of the mechanical energy that is stored. In Seattle, Washington, the pressure of glacial ice over-loaded the underlying clay strata, which were mostly composed of illite and chlorite with smaller amounts of montmorillonite. When tunnels were constructed through the clay, cave-ins resulted along with surface subsidence. In 1966 a downtown Seattle highway was constructed using only conventional retaining walls. After the first cut a slide occurred, followed by six instabilities in other reaches—all due to expansion of the host sediment from moisture addition.

In some instances time is a factor in the structural design. Foundation damage in many

Figure 13-23 Scene in Waco, Texas, showing the disruption and destruction of a road as a result of expansive soils. The Eagle Ford Shale is overconsolidated and, when modified either by removal of overburden or by increased moisture infiltration, it swells and causes damage to projects. (Courtesy Robert Font.)

cases takes 3 to 5 years to develop. Highway planners are especially concerned with the time element because it may be less expensive to fix the damage than to spend additional monies to reinforce the road, especially if the soil stabilizes during the construction period. In other cases the useful life of the structure may be less than the time it takes to produce the expansive changes.

Remedial Measures

There is no substitute for determination of the soil and rock properties prior to any construction activity. Three limits, under the **Atterberg limit system,** define the plastic properties of soils. The **liquid limit** is the moisture content expressed as percent by weight of the oven-dried soil at which point the soil will begin to flow when slightly jarred. **Plastic limit** is the lowest moisture content expressed as percent by weight of oven-dried soil that can be rolled into threads 1/8 in. in diameter without breaking. Soils that cannot be rolled into threads are considered nonplastic. **Plasticity index** is the difference between the liquid limit and the plastic limit. It is the range of moisture content in which the material is plastic. Although permeability and porosity are not directly related to the soil swelling, they do influence rates of moisture migration and liquid retention. Thus, their condition in soils is also vital, because when a clay is dry it can desiccate and allow water penetration but when wet is impermeable.

Once the soil and rock characteristics are known, design engineering has two options—the structure can be built to be sufficiently massive to resist volume soil changes, or soil and its environment can be modified or stabilized to produce minimum volume changes. Of course, a decision may be reached to live with the problem. The techniques for prevention of expansive soil damage include (Fig. 13-24):

1 Replace soil with nonswelling earth materials.

2 Provide sufficient deadload pressure on all footings and slabs to withstand expansive forces.

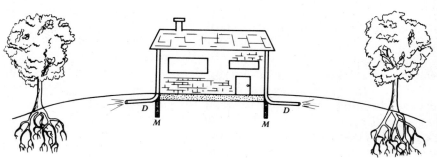

Figure 13-24 Proper landscaping and house design in area with expansive soils. House is graded with slopes away from house. Drainspouts (D) discharge water far from foundation. Trees are planted where their growth and root systems do not interfere with moisture balance. Membrane at *M* prevents excess moisture from collecting under the house.

Figure 13-25 Piping in sediments of ancient Lake Bonneville, southern Idaho. (Courtesy Earl Olson.)

Figure 13-26 Subdivision at Bountiful City, Utah, built on fine-grained sediments of ancient Lake Bonneville. Note the extension of downspouts from homes in order to prevent foundation problems from the expansive soils. (Courtesy Earl Olson.)

3   Flood the area prior to construction.

4   Decrease soil density by compaction.

5   Change the soil properties by chemical treatment and injection. The basic principle is to increase the ionic concentration in the free water and base exchange ions. Lime replaces the cations with calcium which is less reactive and also prevents the flocculation of dispersed units. Fly ash, phosphates, and other organic compounds are also used.

6   Use physical barriers to prevent moisture transport. Such dams may be on the surface, perhaps paving over the adjacent terrain or vertical structures. These barriers are in-ground installations and may be any type of concrete or impervious membrane that prevents moisture migration. Even gravel aids to dissipate capillary buildup in soils.

7   Make a site design. A homeowner can do many things to minimize expansive soil damage, such as assuring that all gutters and downspouts divert water several meters from the foundation (Fig. 13-24), that the lot is graded to place the dwelling on a mound, that yard watering and planting of trees is not done near the house, that frame construction includes expansion sections at all the openings (doors and windows).

PIPING

Piping (not to be confused with pipelines) is the development of subsurface drainage conduits in nonlithified earth materials to depths above the nearby baselevel. It is both a natural phenomenon (Fig. 13-25) and a process that is accelerated and even initiated by human activities and structures (Figs. 13-26 through 13-28). Piping is a type of mechanical soil movement that appears to have been first proposed as an erosional process by Rubey in 1928. He described the tunneling effect whereby fine subsurface sediment moves in underground channelways through permeable materials. A free face is required for the material to exit from the interior passages. Rubey also described the type of pipe formation that develops in loose soils with a low water table. The soils become deeply cracked by repeated dry periods (Fig. 13-29). During rain, percolating water carries finer silt and clay particles and forms passageways. The small tunnels form just below the temporary underground water level of the rainy season and once initiated become cumulative with time. As the conduits expand, subsurface erosion becomes more vigorous with increased transport of sediment from the site. Roof collapse and subsidence features may form which intensify concentration of additional waters and

(a)

(b)

Figure 13-27 (a) and (b) Two views of pipes in fine-grained glaciolacustrine sediments on hillslopes undercut by highway construction near Chenango Forks, New York. Such activities have caused landsliding in the area. (Coates, 1977.)

may ultimately yield a new erosional cycle (Fig. 13-29). Headward erosion will then occur at the vertical cut or free face which then aids the development of its own drainage network.

There are three different types of piping based on their mode of origin.

1 **Desiccation-stress cracks.** This is a dominant type in the American Southwest and constitutes the evolution of features just described. These features are common near gulley outlets, sides of arroyos, and other types of embankments, natural or man-made. They may also occur in areas undergoing subsidence and even from animal burrows or rotted root tubes.

2 **Entrainment mode.** This type of piping chiefly occurs where construction activities have dewatered building foundations, or there is a rise of impounded water behind or under dams and levees (Fig. 13-30). The newly created hydraulic head causes subsurface channeling with entrainment of the water and saturated materials to outflow points at down-gradient sites. This entrainment type of piping rarely produces open subsurface conduits but may transport sufficient volumes of material to cause collapse of overlying sediment and superjacent structures. Sand boils and mud volcanoes are varieties of this process (Fig. 13-31).

3 **Variable permeability subsidence mode.** Piping produced by this process results when a sufficient hydraulic head already exists to move water through a stratum with sufficient velocity to transport fine-grained particles to the face of a gully, embankment, or steep slope. These passageways also intensify with time and grow headward from the outlet vicinity.

Piping has often been thought to be mostly associated with drylands, yet it can occur in wide climatic ranges. For example, piping in England is present at locations where there is a horizon of relatively low permeability with low aggregate stability. In order of most widespread occurrence, however, piping can be ranked according to the following causes: (1) susceptibility to cracking in dry regions with materials con-

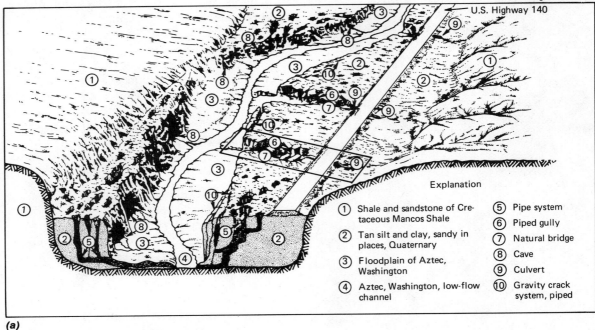

(a)

Explanation

① Shale and sandstone of Cretaceous Mancos Shale
② Tan silt and clay, sandy in places, Quaternary
③ Floodplain of Aztec, Washington
④ Aztec, Washington, low-flow channel

⑤ Pipe system
⑥ Piped gully
⑦ Natural bridge
⑧ Cave
⑨ Culvert
⑩ Gravity crack system, piped

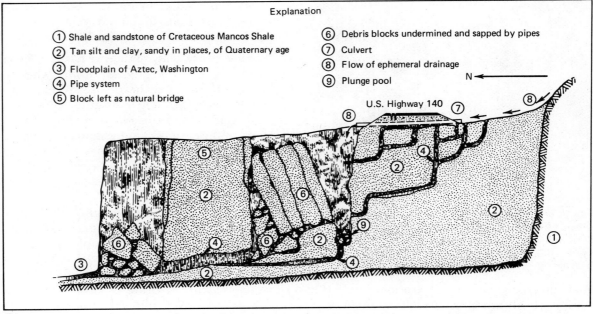

(b)

Explanation

① Shale and sandstone of Cretaceous Mancos Shale
② Tan silt and clay, sandy in places, of Quaternary age
③ Floodplain of Aztec, Washington
④ Pipe system
⑤ Block left as natural bridge

⑥ Debris blocks undermined and sapped by pipes
⑦ Culvert
⑧ Flow of ephemeral drainage
⑨ Plunge pool

N ←

Figure 13-28 Idealized diagrams of Aztec, Washington, along highway U.S. 140, southwestern Colorado. (Parker and Jenne, 1967.) (a) Dissected and extensively piped valley fill, old bedrock surface, and channel with drainage system along highway. (b) North-south cross section under highway. Note incipient piping system beneath roadway.

Figure 13-30 Entrainment-type piping from downspout, Bountiful City, Utah. Materials composed of silty fine sand from ancient Lake Bonneville. (Courtesy Earl Olson.)

Figure 13-29 Two views of piping, subsidence, and roof collapse near roadway areas of Fish Lake National Forest, Utah. (Courtesy Jerome DeGraff.)

taining high clay-silt content with a high percentage of swelling clays; (2) periods of high-intensity rainfall coupled with devegetation; (3) erodible layers above a base of high exchangeable sodium and high base exchange capacity of soluble salts (alkaline-type soils); (4) steep hydraulic gradient due to short-term downcutting or excavation.

We have caused piping during and after many of our construction enterprises. Parker and Jenne (1967) have discussed damages and the imperiled structures that result from piping in the American Southwest. Here bridge abutments, wingwalls, piers, culverts, and highways and their appurtenances are the most frequent victims—often caused by improper project engineering. To minimize losses thorough knowledge of the surficial geology and hydrology is needed. Where possible construction should avoid those areas where piping can easily develop. The structures and adjacent lands should

Figure 13-31 Sand boils created by the hydrostatic pressure system of the Mississippi River and the adjacent man-made levee. (Courtesy Charles Kolb.)

be designed and graded to prevent the concentration of runoff and to provide adequate drainage facilities to carry or divert runoff from entering critical sites.

In its penultimate development piping is capable of producing badland topography in which a type of pseudokarst terrane is formed with sinkholes, natural bridges, blind valleys, haystack hills, and a full array of box canyons and gullies. Aghassy (1973) described two ways that such badland topography was formed in the Negev region. The bedouins used camel-drawn plows that incised deep furrows in loess. This led to subsurface infiltration with the production of vertical pipes that then facilitated development of underground horizontal pipes. During a single cycle of erosion it was not unusual for the entrenchment and collapse of surface materials to reach 2 to 3 m depths. Additional badland topography formed as a result of construction of the Ottoman Railroad prior to World War I. Along this route incision and side-drainage channels created oversteepened slopes. Percolating waters developed steeper gradients with the creating of piping structures and the subsequent collapse of overburden into gullies, rills, and channelways. All such features become rejuvenated from time to time, and some have gone through several cycles of erosion in the past 80 years so that total entrenchment in some areas is measured in terms of 20 m and more. Construction of the Khodza-Kola Canal in the Soviet Union has caused piping along with subsidence to develop within every kilometer of the 20 km long excavation. Prior to 1963, 8 percent of all Australian and New Zealand dams exhibited piping failures. More recently the Teton Dam in the United States (see page 375) failed because of piping.

GEOTECHNOLOGY

Geotechnology is that discipline wherein the skills of the geologist, the engineer, and the manufacturer unite to produce a product and a design that will stabilize earth forces and processes. This specialty is concerned with choosing the best products that will perform at high standards to prevent landscape deterioration and enhance the structural integrity of developed sites. Such a field embraces a wide spectrum of facilities and materials, and we have time to discuss only certain aspects of three types of projects: those that prevent vertical water loss, those that inhibit channelway erosion, and those that facilitate stabilization of plantings.

Prevention of Seepage and Evaporation

Chapter 8 presents some ideas about water containment, but we will now add some other case histories of how this is accomplished.

1 Syracuse, New York The Westcott Reservoir had normal leakage of about 25 gpm or gallons per minute (95 liters/min) during its early life, but when it rose to 1000 gpm (3785 liters/min) in the 1960s something had to be done. The structure had been built in the 1930s with an oval, gunite and granite block-lined floor 327 m long, 165 m wide, and 10 m high. The 9 cm thick gunite lining contained an underlying clay blanket specifically designed to prevent leakage. To restore the reservoir the 110 million gal were drained, and all cracks sealed with a special sealant. Ten Hypalon Burke liners, each 93 m² in size, were also installed and overlapped up to 25 cm and sealed with adhesive to prevent separation. This synthetic rubber with 5-ply reinforced fabric is now preventing further water loss.

2 Salt Lake City, Utah The Baskin Reservoir is in the heart of one of the city's fine residential areas. Here a Roofloat Burke Hypalon cover was installed capable of withstanding severe temperature extremes, with superior resistance to sunlight, ozone, and water treatment chemicals. Because the cover floats on the water, the resource there is in no danger of contamination from wildlife or other airborne pollutants. Such covers reduce evaporation and provide for chlorine retention in addition to inhibiting algae production.

3 Monmouth, Illinois The first large anerobic treatment pond in the United States was successfully installed and covered with a flexible floating cover at this site. The 7.25 million gal treatment lagoon had to be covered because odorous gases were emanating during the processing of waste and were unable to be properly vented to meet acceptable levels of air pollution. The normal operation produces about 175,000 cu ft (4950 m³) of gas per day, so the cover allows its easy collection for safe venting and burning. The Roofloat that was installed requires no columns, cables, or other internal structure supports and is not affected by the methane gas. Such 5-ply heavy-duty liners have proved resistant to deterioration by the sun, the weather, and chemicals.

4 Charleston, South Carolina Prior to the installation of covers for the reservoir system, during prolonged periods of hot weather, there would be extensive evaporation of the chlorine. This caused an absence of residual chlorine in the water distribution system to users more than 5 km from the reservoir, posing a health problem where chlorine was absent and could not kill pollutants in the water. A floating cover was installed so that residual chlorine occurs as far as 24 km from the reservoir. Chlorine gas no longer escapes, and maintenance costs are reduced because contaminants no longer enter the reservoir.

5 Foley, Florida The Buckeye Cellulose Corporation completed an extensive waste treatment plan for its paper pulp mill in January 1971, but then disaster struck. Five days after the start of the final 120 acre secondary-treatment aeration lagoon, sink holes developed through the lagoon bottom. The lagoon was immediately taken out of service to prevent underground water contamination. After a testing program the incipient sinks were located and filled with sand, and fluidized blasting techniques helped move sand into the crevices. The lagoon was then compartmentalized into nine different cells so that each could operate separately. Hypalon-type linings were installed and the facility placed into operation in 1974. Since that time the massive 500 million gal capacity liquid waste lagoons have exhibited no seepage problems.

Figure 13-32 Benching in loess along road near Vicksburg, Mississippi. (Courtesy Dermot Brown.)

Embankment Stabilization

A wide variety of engineering techniques are employed to prevent erosion of slopes that humans have altered; some of these were discussed in Chapters 11 and 12. Although such additional cuts as in highway benching or terracing (Fig. 13-32) have wide application, in this section we primarily will discuss the emplacement of materials that conform to the newly established hillside contours. Such structures are called **revetments** and may consist of materials with widely contrasting shape and composition.

Gabions Gabions consist of some type of mesh or wire enclosures that are filled with rocks or rubble. This type of revetment dates back to the ancient Pharaohs who used wicker networks laced together and then filled with debris. Wire gabions have been extensively used in Europe for the past 100 years, but received little use in the United States until the last 15 years. Although they have been successfully used at many sites, they have a few drawbacks. (1) Heavy channel debris can tear the wire. (2) Water with high acid or salt content with debris can corrode wire and aid in deteriorating any coating over the wire. (3) Vandals can cut the wire and remove the rocks. (4) Long-range durability is unknown although in Europe some gabions are still effective without much damage after 75 years.

The channelization of the Humacao River in Puerto Rico is enabling the reclamation of many hectares of land for use in commercial, industrial, and residential development in the southeast corner of the island. For centuries the rich lowlands could only be used marginally for agricultural use because of frequent flooding. Runoff from the Central Mountain Range into this delta area became more serious yearly. The new 7.7 km channel from the sea inland past the town of Humacao will bring the area into production and provide a predictable environment for the rapidly expanding country. The construction began in 1969, and gabions were specified to stabilize the embankment because they permit differential settlement without fracture, which was a paramount consideration due to the unstable soils. The permeable gabion structure also combines free drainage with earth retention. When completed in the fall of 1977 it was the largest gabion structure in the Americas. It used 50,000 galvanized and PVC (polyvinyl chloride) coated steel wire gabions, 4 m by 1 m by 0.3 m. The project was the first large-scale use of the PVC thermobonded coating which provides extra protection from corrosion. Another benefit of porous gabions is they permit vegetation through the structure, which provides increased ecological and aesthetic qualities (Fig. 13-33).

Gabions were also used in certain reaches of the Eel River, California, where frequent floods washed out logging roads important to the economy of the region. The construction was done in 1973 and was the largest such design in the western United States and the third largest in the country. With the steep mountain slopes (Fig. 13-34) such structures require the minimum maintenance and allow hydrostatic pressure buildup behind the wall.

Fabriforms Another revetment technique is to cast structural shapes by injecting high-strength mortar into porous nylon fabric forms. Nylon is resistant to attack by acids, alkalies, microorganisms, and organic solvents. The fabric can be placed on nearly any type of terrain and, when injected with mortar, it automatically conforms to surface contours. Such revetment mattings have been used effectively for banks, canals, reservoirs, and other types of slope protection.

The 1.6 million kW nuclear powerplant of Virginia Electric and Power at the Surry Power Station has fabriform construction lining the banks and bottom of the discharge canal which returns cooling water to the James River. It was evident from the earliest plans that bank protection was necessary to prevent erosion of the loose saturated silt. Although riprap was employed in the outer breakwater extending out into the James River, it was not suitable for the banks. It would not only be more costly but riprap could not develop a sufficiently tight envelope to deter silt movement (Fig. 13-35).

Fabriform revetments were first introduced in 1965, but the first major use was in 1968 at the Kinzua Dam and Allegheny Reservoir in western Pennsylvania. In preparing the reservoir site, engineering specifications called for the installation of riprap while the impoundment area was dry and easily accessible to heavy equipment. However, as the water level rose rapidly during the spring of 1968, and wave action from prevailing winds started to create a dangerous erosion condition, it became apparent that quick and drastic measures were needed to protect a 1 km section of the reservoir on the southeastern side. Prompt action was required because the slope erosion would threaten a state highway along the crest of the embankment. The problem was twofold: the riprap was not readily available, and the necessary heavy equipment for emplacement had been demobilized. Even if these factors could be overcome, effective placement of massive stone would have been inefficient and not effective due to such rapid rising water conditions. To solve the problem the Corps of Engineers specified fabriform

Figure 13-33 Construction of the Humacao River Channelization Project, Puerto Rico, and installation of gabions for bank stabilization. (Courtesy Bekaert Steel Wire Corp.) (a) Part of the 7.7 km Humacao River channel as seen looking toward the town. Stone-filled gabions in the final stages of completion of the longest gabion structure in the United States and territories. (b) Close-up of filled gabion with top yet to be wired shut. Project used 50,000 gabions, each 4 m × 1 m × 0.3 m. (c) Crane lifts four-compartment bucket to waiting empty gabions. (d) Dragline excavators were used in wet delta areas, and motor scraper was used where subsoil permitted sufficient foundation digging and emplacement of gabions. (e) Workmen wire stone-filled gabions shut. The revetment in the background is completed.

Figure 13-34 Gabion structure, Eel River, California, needed for bank stabilization of logging road. (Courtesy Bekaert Steel Wire Corp.)

(a)

(b)

Figure 13-35 Surry Nuclear Plant of Virginia Electric and Power Corp., Virginia. The artificial channel that connects with the James River has been lined with a man-made revetment of fabriform concrete materials. (a) View of plant and upper part of channel. (b) View of channel, showing fabriform revetment for stabilization of the embankment, with the James River in the background. (Courtesy Prepakt Concrete Co.)

revetments for the job. The panels were installed and tied to an anchor trench about 2.1 m above the final high water line and then extended 16 m down the slope into the water (Fig. 13-36). As work progressed half of each panel was placed below water level. A total of 12,100 m² were installed and filled with mortar. The latest official inspection of November 1976 showed the structure was still doing the job and had not deteriorated.

Soil-protecting fabrics There is a large variety of other structural designs and materials used in bank stabilization. The effectiveness of such revetments can be enhanced by preparation of the soil and cover fabric to protect it from disturbance, and it is now becoming an approved technique to install some type of cover fabric on top of the soil prior to emplacement of revetment structures and materials. Several different types are on the market that have proved very effective. Furthermore, these monofilter or polyfilter fabrics can be used in other environmental settings where support of the soils to prevent erosion is vital (Fig. 13-37). These fabrics are designed to control loss of supporting solids surrounding such diverse structures as those armored with rock, with interlocking concrete blocks, concrete and sheeting type cribs, and gabions. These filter cloths are woven with monofilament polypro-

pylene yarns that have high strength and are resistant to acids and alkalies. They are also rot proof and mildew proof, with special ingredients that have proven effective against degradation by ultraviolet light.

A major concern in building the Honolulu International Airport's offshore runway was erosion of the fill from ocean wave attack. To

Figure 13-36 Fabriform revetment at Kinzua Dam, New York–Pennsylvania. (Courtesy Prepakt Concrete Co.)

protect the coral fill, 14.5 million m³ of rock, the entire oceanside structure was covered with poly-filter X fabric before the rock covering was placed on top. Another application of polypropylene filter sheets was the checkerboard block revetment at Jupiter Island, Florida (see Fig. 15-9). Here the fabric was placed over the sand, then covered by a crushed-rock course, and finally the fitted checkerboard concrete blocks. The fabric is sufficiently porous to allow water to pass through without providing much head pressure. It allows for the filtering out of fine particles. The revetment was constructed two weeks before Hurricane Donna struck and suffered no damage. Again in March 1962, high storm waves caused severe erosion to unprotected beaches and sea walls, but did not harm the revetment. The sea wall meets the full force of waves, and 60 percent of the force is deflected downward, where sand scour occurs. Instead, the sloping walls of the revetment pushes the waves up, the water withdraws with reduced energy, and sand is deposited. Thus such filters aid in eliminating hydrostatic pressure buildup while preserving the sandy base of the structure.

Another type of fabric, often termed "hold-and-grow control fabric," is used to aid the quick and stabilized establishment of vegetation for erosion control. These synthetic yarn nettings are interwoven with strips of paper (Fig. 13-38) and are designed to eventually degrade and become part of the soil mulch. Thus it has a two-stage action, providing maximum slope protection on freshly graded and seeded surfaces when sheetwash can severely degrade the soil. The fabric also breaks the force of rainfall while the slits between let moisture seep into the soil. Then as vegetation commences, the netting provides support for the new root structure. The same paper mulches the soil and holds seeds in place as well as protecting the available moisture, which can last throughout the entire growing season. Depending on the specific requirements of the site, the polypropylene-paper fabric can be designed to deteriorate within different time periods, thus allowing maximum protection for vegetal growth to be established. The hold-and-grow systems have been used successfully to stabilize slopes by using vegetation at the Dallas-Fort Worth International Airport (largest in the world); at the Georgia Power Dam at Milledgeville, Georgia; at Interstate 80 in Cedar Rapids, Iowa, to stabilize median landscaping; and at Gary, Indiana, to stabilize dunes at Miller Village Apartments.

Such erosion control methods must be engineered into the total strategy at construction sites for the prevention of erosion and its equally damaging production of sediment. Thus all design procedures should incorporate appropriate studies and methods that contain the following elements.

1   Identify both on-site and off-site areas that are especially vulnerable to erosion and make plans to mitigate them.

2   Divert runoff water originating upslope from the construction site so it will not flow over bare earth.

3   Limit the size of the area being graded at

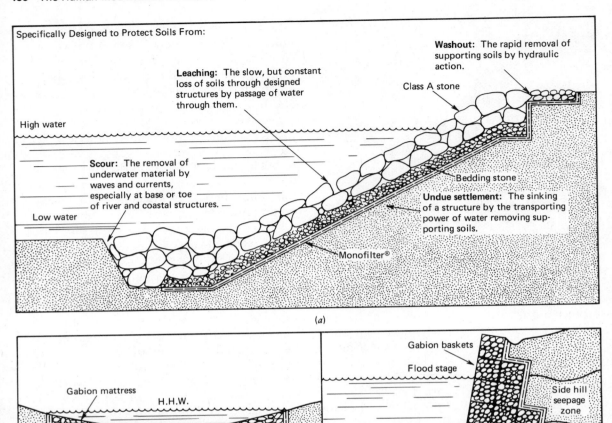

Specifically Designed to Protect Soils From:

Washout: The rapid removal of supporting soils by hydraulic action.

Leaching: The slow, but constant loss of soils through designed structures by passage of water through them.

Class A stone

High water

Scour: The removal of underwater material by waves and currents, especially at base or toe of river and coastal structures.

Low water

Bedding stone

Undue settlement: The sinking of a structure by the transporting power of water removing supporting soils.

Monofilter®

(a)

Gabion mattress

H.H.W.

Monofilter®

Gabion baskets

Flood stage

Side hill seepage zone

Channel bed
L.W.

Monofilter®

(b)

Figure 13-37 Use of monofilter fabric for earth stabilization. (Courtesy United States Filter Corporation.) (a) Protection of soils under a rock revetment from the erosion of hydraulic forces. (b) River and stream control structures serve to protect the existing shore against erosion and washout. Such structures must allow passage of water and yet retain supporting soils to remain stable and to prevent hydrostatic pressure build-up.

any single time so that areas of bare soil are not exposed.

4 Limit the time any area contains bare ground.

5 Decrease the velocity of runoff waters to retard erosion and enable sediment to settle on-site.

6 Prepare temporary or permanent drainage methods to handle heavy precipitation.

7 Trap sediment-laden runoff with temporary or permanent basins or filters.

8 Maintain erosion control inspections after each rainfall period to assure that all structures are in working order.

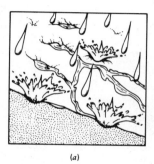

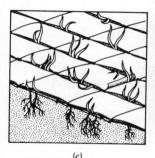
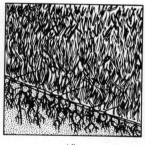

(a) (b) (c) (d)

Figure 13-38 Use of Hold/Grow fabrics for erosion prevention. (Courtesy Hold/Grow Erosion Control Systems.) (a) Raindrops can act as tiny bombs and displace soil. (b) The permeable fabric lets seepage moisture through, but prevents erosion by rain splash. (c) The fabric-type paper acts as mulch to hold moisture and degrades as new growth begins. (d) After paper degrades, the plastic net acts as a temporary root structure continuing to deter erosion.

Perspectives

The position I have adopted in this chapter views much of engineering as vital to society. However, by their very nature, construction activities do change the landscape—the form of the terrain and the processes that then operate within it. We now possess sufficient knowledge so that the damaging impacts created can in many cases be predicted and, with proper planning, their magnitude mitigated. Perhaps the matter can be summed up by saying those who are in charge of producing deleterious impacts should be held accountable for their actions. Hammurabi's Code recognized this more than 4000 years ago. Although none of us would champion the extreme punishment that the code mandates, perhaps those who tamper with nature or construct inadequate developments could design more environmentally sound projects if their accountability was heightened. (We return to this topic in Chapter 21).

Hammurabi's Code

If a builder build a house for a man and do not make its construction firm and the house which he has built collapse and cause the death of the owner of the house—that builder shall be put to death.

If it cause the death of the son of the owner of the house—they shall put to death a son of that builder.

If it cause the death of a slave of the owner of the house—he shall give to the owner of the house a slave of equal value.

If it destroy property, he shall restore whatever it destroyed, and because he did not make the house which he built firm and it collapsed, he shall rebuild the house which collapsed at his own expense.

If a builder build a house for a man and do not make its construction meet the requirements and a wall fall in, that builder shall strengthen the wall at his own expense.

Accountability for human action is as old as history. Although the engineering profession has been mandated to provide goods and services to mankind, along with this charge goes the responsibility to assure safety of structures that accomplish these objectives. Therefore, environmental affairs must become the province of all, if all are to benefit from the fruits of these human endeavors.

READINGS

Aghassy, J., 1973, Man-induced badlands topography: in Coates, D. R., ed., Environmental geomorphology and landscape conservation, Vol. II, non-urban regions, Stroudsburg, Dowden, Hutchinson & Ross, p. 124–136.

Arthur D. Little, Inc., 1972, Channel modifications; an environmental, economic, and financial assessment: Report to the Council on Environmental Quality.

Barnard, R. S., 1977, Morphology and morphometry of a channelized stream: the case history of Big Pine Creek Ditch, Benton County, Indiana: Studies in Fluvial Geomorphology, No. 4, Purdue Univ., 86 p.

Chen, F. H., 1975, Foundations on expansive soils: Amsterdam, Elsevier, 280 p.

Coates, D. R., ed., 1976, Geomorphology and Engineering: Stroudsburg, Dowden, Hutchinson & Ross, 360 p.

———, 1976, Geomorphology in legal affairs of the Binghamton, New York metropolitan area: in Coates, D. R., ed., Urban Geomorphology, Geol. Soc, Amer., Special Paper 174, p. 111–148.

Cockfield, W. E., and Buckham, A. F., 1946, Sink-hole erosion in the white silts at Kamloops: Roy. Soc. Canada Trans, 3rd ser., v. 40, p. 1–10.

Daniels, R. B., 1960, Entrenchment of the Willow drainage ditch, Harrison County, Iowa: Amer. Jour. Sc., v. 225, p. 161–176.

Emerson, J. W., 1971, Channelization: a case study: Science v. 173, p. 325–326.

Font, R. G., 1977, Engineering geology of the slope in stability of two overconsolidated north-central Texas shales: in Coates, D. R., ed., Landslides, Geol. Soc. Amer. Reviews in Engineering Geology, Vol. III, p. 205–212.

Gillette, R., 1972, Stream channelization: conflict between ditchers and conservationists: Science, v. 176, p. 890–894.

Gillott, J. E., 1968, Clay in engineering: Amsterdam, Elsevier, 296 p.

Isaacs, R. M., and Code, J. A., 1972, Problems in engineering geology related to pipeline construction: Natl. Research Council Canada Div. Bldg. Research Tech. Memo., 104, p. 147–178.

Johnson, A. W., 1961, Highway erosion control. Amer. Soc. Agr. Eng. Trans., v. 4, n. 1, p. 144–152.

Keller, E. A., 1976, Channelization: environmental, geomorphic, and engineering aspects: in Coates, D. R., ed., Geomorphology and engineering, Stroudsburg, Dowden, Hutchinson & Ross, p. 115–140.

Lagerwerff, J. V., and Specht, A. W., 1970, Contamination of roadside soil and vegetation with cadmium, lead and zinc: Env. Sci. and Tech., v. 4, n. 7, p. 583–586.

Lambe, T. W., 1960, The character and identification of expansive soils: U. S. Federal Housing Admin. Tech. Studies Rept. FHA-701, 51 p.

Legget, R. L., 1962, Geology and engineering, 2nd ed.: New York, McGraw-Hill, 884 p.

Lung, R., and Proctor, R., eds., 1969, Engineering geology in southern California: Arcadia, Calif., Assoc. Eng. Geol., 389 p.

Madison, J., 1972, Plague on all your rivers: Audubon, v. 74, p. 29–41.

Mathewson, C. C., Castleberry, J. P., and Lutton, R. L., 1975, Analysis and modeling of the performance of home foundations on expansive soils in central Texas: Assoc. Eng. Geol. Bull. v. 17, n. 4, p. 275–302.

Mathewson, C. C., and Clary, J. H., 1977, Engineering geology of multiple landsliding along I-45 road cut near Centerville, Texas: in Coates, D. R., ed., Landslides, Geol. Soc. Amer. Reviews in Engineering Geology, Vol. III, p. 213–223.

Miller, J. N., 1970, Rape on the Oklawaha: Reader's Digest, v. 96, June, p. 54–60.

Nakata, J. K., Wilshire, H. G., and Barnes, G. G., 1976, Origin of Mojave Desert dust plumes photographed from space: Geology, v. 4, p. 644–648.

Newton, J. G., Copeland, C. W., and Scarbrough, W. L., 1973, Sinkhole problem along proposed route of Interstate Highway 459 near Greenwood, Alabama: Geol. Survey Alabama Circ. 83, p. 19–37.

Paige, S., Chairman, 1950, Application of geology to engineering practice: Geol. Soc. Amer. Berkey Volume, 327 p.

Parizek, R. R., 1971, Impact of highways on the hydrogeologic environment: in Coates, D. R., ed., Environmental geomorphology, Publ. in Geomorphology, State Univ. of New York at Binghamton, p. 151–199.

Parker, G. G., and Jenne, E. A., 1967, Structural failure of western U.S. highways caused by piping: 46th Ann. Mtg. Highway Research Board, Washington, D.C., U.S. Geol. Survey Water Resources Division, 27 p.

Parkes, J. G. M., and Day, J. C., 1975, The hazard of sensitive clays—a case study of the Ottawa-Hill area: Geog. Rev., April, p. 198–213.

Rubey, W. W., 1928, Gullies in the Great Plains formed by sinking of the ground: Amer. Jour. Sci., 5th ser., v. 15, p. 417–422.

Ruhe, R. V., 1971, Stream regimen and man's manipulation: in Coates, D. R., Environmental geomorphology, Publ. in Geomorphology, State Univ. of New York at Binghamton, p. 9–23.

Scheidt, M. E., 1967, Environmental effects of highways: Jour. Sanitary Eng. Div. Proc. Amer. Soc. Civil Eng., v. 93, n. SA5, p. 17–25.

Schultz, J. R., and Cleaves, A. B., 1955, Geology in engineering: New York, John Wiley & Sons, Inc., 592 p.

Sheridan, D., 1978, Dirt motorbikes and dune buggies threaten deserts: Smithsonian, v. 9, n. 6, p. 67–75

———, 1979, Off-road vehicles on public land: Washington, D.C. Council on Environmental Quality, U.S. Govt. Printing Office, 84 p.

Terzaghi, K., and Peck, R. B., 1967, Soil mechanics in engineering practice, 2nd ed.: New York, John Wiley & Sons, 729 p.

Wilshire, H. G., and Nakata, J. K., 1976, Off-road vehicle effects on California's Mojave Desert: California Geology, v. 29, n. 6, p. 123–132.

Wilshire, H. G., Nakata, J. K., Shipley, S., and Prestegaard, K., 1978, Impacts of vehicles on natural terrain at seven sites in the San Francisco Bay Area: Env. Geol., v. 2, n. 5., p. 295–319.

Zaruba, Q., and Mencl, V., 1976, Engineering geology: Amsterdam, Elsevier, 504 p.

Chapter Fourteen
Engineering Impacts on Water Supply

Hoover Dam on the Colorado River with Lake Mead in the background. This multipurpose dam provides a water supply, electric power generation, wildlife enhancement, and a recreational area for the region. (Courtesy U.S. Bureau of Reclamation.)

INTRODUCTION

Because the topic of water is such a pervasive issue in environmental affairs, it has been discussed in many chapters throughout this book. In this chapter, however, we primarily address those effects that are produced when humans design engineering projects for the use of water as a **resource.**

As already discussed, water is the essential ingredient in modern civilization. Thus it is necessary to obtain, store, and distribute this precious commodity to society. In diverting water for human use, it is inevitable that the natural hydrologic regime and character will be changed. It should come as no surprise that such alteration of the normal water budget produces a range of feedback processes that lead to disruptions of terrain, which may yield economic and other losses to mankind. Such damages can occur when water is stored or diverted on the land or when it is mined from subsurface areas. These impacts range from increased erosion and sedimentation in stream channels and reservoirs to land subsidence and faulting.

Since civilization has had a long history of creating changes in water regimes, there is now a large body of information that is available for the prediction of impacts that will result from such modifications. Therefore, those who extract and use water resources should be able to anticipate ahead of time the likely consequences that will ensue whenever water is taken out of its natural setting. So although water is a necessary ingredient in human enterprises, there is no excuse for not being forewarned as to the price that must be paid whenever water is stored or used. The role of the geologist should be to alert water users to these inevitable changes and to aid in the planning that will mitigate damaging effects whenever possible.

DAMS

Although construction of dams dates back many millennia, the twentieth century has witnessed an explosion in the numbers, types, and sizes of dams. Perhaps the earliest masonry dam was built by Menes during the first Egyptian dynasty before 4000 B.C. It was 19 km south of the ancient city of Memphis and its purpose was to divert Nile waters for irrigation use. The 450 m long and 15 m high dam was maintained 4500 years and then abandoned. A 3.2 km long, 40 m high, and 150 m wide dam was constructed about 1700 B.C. in Yemen but was destroyed by a flood in about A.D. 300. As pointed out in Chapter 8, dams are now built for a variety of purposes that include water supply, hydroelectric power generation, flood control, and recreation. Controversies on the necessity of dams, especially when the main purpose is flood control, have become especially prevalent during the past 30 years. Principal debates have raged on whether upstream or downstream dams are more effective, whether the need for the dam is warranted, and whether the benefits outweigh the costs. Such considerations are discussed elsewhere in the book, and in this section we will explore the nature of dams and their impacts.

Types of Dams

The dam type, the location, and the geologic foundation and conditions determine its safety

Figure 14-1 Jackson Lake Dam and reservoir, with Grand Teton Mountains towering in the background. This is one of principal dams that comprise the Minidoka Project. (Courtesy U.S. Bureau of Reclamation.)

and some of the impacts that may result. Obviously a thorough geologic investigation should precede dam design and construction. Bore holes, testing programs, and rock mechanics analyses are vital to determine the rock composition, structure, and strength. Such factors as permeability of rock units and pore pressures are significant components in selecting the appropriate engineering design and construction methods. Such studies should also include an investigation of slopes and materials that will be innundated by the reservoir pool. Knowledge of these features can help prevent such disasters as the Vaiont tragedy and lessen damage by landslides, as in the case of the Grand Coulee Dam and its reservoir.

Gravity dam Such dams are composed of solid masonry that resists the forces made against them primarily by their weight (Fig. 14-1). Forces of water and sediment are stabilized by the vertical component of the dam's weight on the downstream face. The underlying rock must be sufficiently strong to resist stresses and still be well below the elastic limit along all places of the contact planes. Thus, the bearing power must be sufficiently large to carry the total load

Figure 14-2 Aerial view looking upstream at Trinity Dam Powerplant and Clair Engle Lake on the Trinity River near Lewiston, California. These features regulate a drainage area of 1600 km². The lake stores over 2.4 million acre-feet of water. The dam is a zoned earth-filled structure 162 m high, with a crest length of 735 m. The powerplant has two generators with a total capacity of 105,556 kW (Courtesy U.S. Bureau of Reclamation).

without rock movement of detrimental magnitude.

Arch dam A characteristic of this type dam is the transmission of large forces to the rock abutments by the arching action of the dam geometry. These dams are relatively thinner than gravity dams. The arch dam is keyed to the abutment, and its structural strength derives from the masonry and arcuate shape rather than sheer weight of the dam. There can be combinations of dams such as Hoover Dam which is an "arch gravity dam," wherein the loads near the top are supported by arches and those near the bottom by gravity.

Embankment dam These dams are constructed of excavated materials placed without addition of binding materials other than those inherent in the natural material. These materials are usually obtained near the site. An **earthfill dam** is an embankment dam primarily composed of compacted earth, but many also contain an impermeable core that in some cases

may consist of a masonry wall (Fig. 14-2). Most embankment dams require an ample spillway to assure water does not overtop the dam. Also, the height of the dam must be sufficiently great so that waves do not overtop the structure. The upstream and downstream dam slopes must be stable to prevent gravity movement or sheet-wash of materials. The distal slopes are usually protected by vegetation cover, and the proximal slope often contains riprap. **Rockfill dams** are embankment dams composed of rock fragments which provide mass to the structure along with an inner core of impervious materials, blanket, or membrane. These dams require the same safety features as earth dams.

There are also other dam types such as the **buttress dam.** This dam may be the preferred type of construction in mountain valleys with difficult access. These dams can be constructed with less material than that required in the gravity dam. The International Commission for Large Dams (ICOLD) reported that, of the 925 dams constructed during the 1965–68 period, 700 were mostly earth dams and 225 were masonry dams. The increased knowledge of soil mechanics has made possible the construction of earth dams of great heights that are both safe and economical.

Effects of Dams

Dams can produce many desirable benefits and are absolutely essential in many instances. The concept of multipurpose dams has become increasingly prevalent within the past 20 years. Unfortunately, some usages are contrary to others. For example, it is advantageous for recreation and power generation to have maximum water heights, whereas for flood control the reservoir pool should be as low as possible. Storing water for use during dry times with high river flow is a desirable conservation-type measure. Irrigation helps increase food supply, and even the silt trapped by the dam assures improved water quality downstream. Of course, water is basic for personal consumption, and

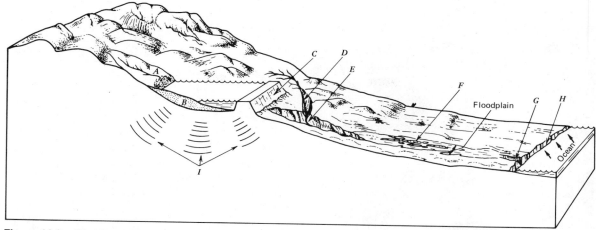

Figure 14-3 Diagrammatic sequence of impacts created by dams. (*a*) Upstream siltation that buries town. (*b*) Sedimentation in the dam. (*c*) Dam. (*d*) New cycle of erosion in tributaries. (*e*) Channel erosion immediately downstream from dam. (*f*) Channel filling and flooding from deposits eroded at (*E*). (*g*) Erosion in channel. (*h*) Erosion of shoreline deprived of sand nourishment because of the entrapment of sediment by the dam. (*i*) Earthquakes caused by pore water pressure changes and stresses from the weight of the dam, water, and sediments.

industry as well. Hydroelectric energy is pollution-free and in a way can be considered to be a renewable energy resource. Flooding mitigation can be a bulwark in the protection of life and property. Also, the recreational needs of workers with more and more leisure time may be fulfilled by those activities associated with artificial lakes. Unfortunately, there is another side to the story, and the list of impacts that may result from water impoundments is rather staggering (Fig. 14-3).

Reservoir and upstream sedimentation The first effect of a dam is to reduce flow velocity as the river enters standing water and, when coupled by the impoundment barrier, the entrained sediment is deposited on the floor of the reservoir (Fig. 14-4). In time there can be upstream growth and extension of the man-made delta whose sediments fill other structures and tributaries. The life span of reservoirs throughout the world is being greatly shortened by the influx of these silt and sediment loads. The Tarbela Dam in Pakistan was completed in 1975 as the world's largest earth and rockfill dam, but will be rendered useless by siltation in

less than 50 years. The Anchicaya Dam in Colombia was completed in 1955, but 21 months later was already 25 percent filled with sediment. In 7 years the reservoir had lost capacity to store water and the multimillion dollar hydroelectric plant to harness the energy now runs on river flow alone. Extensive dredg-

Figure 14-4 Destruction of forested areas immediately upstream of Garrison Dam, North Dakota. Damage caused by upstream migration of flooding and the sediment plume, created by the drowning action of the reservoir. (Courtesy John Conners.)

Figure 14-5 Lake Ballinger Dam, Texas, first built in 1920 and raised to the present height of 10 m in 1933. It was the city's reservoir until sedimentation forced its abandonment. The 1200 km² watershed produced about 50 cm of sediment after each major flood period. The original depth of 10.5 m had declined to 1.2 m when abandoned. (*a*) Lake Ballinger. (*b*) Ballinger Dam and the now-silted reservoir, September 1954. (Courtesy Soil Conservation Service.)

ing operations are necessary to prevent silt from ruining the turbines. The Shihmen Reservoir in Taiwan was predicted to last 71 years but during the 1963–68 period lost more than 45 percent of its capacity to siltation. A 1941 study in the United States showed that 39 percent of the nation's reservoirs had less than a 50-year life (Fig. 14-5). Lake Nasser of the Aswan Dam, Egypt, traps the 13 million m³ of silt that formerly was deposited in the floodplain and delta of the Nile. The results of this have greatly affected the downstream region.

Downstream erosion The hydrologic and geomorphic balance of the channel system is greatly altered due to the character of the water released from the dam. It is devoid of sediment load, and its chemistry and temperature are also changed. When measurements were made of

the Texacoma Reservoir behind the Denison Dam in the 1960s, there were 386 million tons of sediment, of which 20 percent (77 million tons) were sand. The amount of sand deposited is about the same as the amount eroded from the North Canadian River below the dam—67 million tons. Deprived of sediment, the clear water eroded the streambed 1.5 to 2.1 m in the first 16 km and caused channel erosion of 35,000 acre-feet (5.4×10^6 m³) in the 160 km downstream reach from Canton, Oklahoma.

Hoover Dam on the Colorado River caused a lowering of the river channel by 3 m, because stream load is only $1/15$ the original value (mostly from downstream tributaries). However, with the incorporation of the new materials channel siltation occurs further downstream when the eroded materials are deposited. This has led to the choking of the channel and to severe flooding problems from the changed flow regime, such as occurred at Needles, California. Similar events occur from impoundment of the Rio Grande River at Elephant Butte Dam, New Mexico. Here during a 30-year period the siltation plume advanced upstream from the reservoir site, depositing 3 m of material and burying much of the village of San Marcial. Downstream erosion has also led to sediment problems at Albuquerque, 160 km downstream, and to flooding problems at El Paso, Texas. Glen Canyon Dam, also on the Colorado River, built in 1963 has caused downstream erosion of as much as 6 m, has changed the character of many of the rapids, and has instituted a new geomorphic erosion cycle in some of the tributaries because of the change in baselevel conditions. The median size of suspended sediment is only $1/200$ the size of pre-dam material. The lowered flood peaks and number of floods have also produced a variety of changes—both physical and ecological. Indeed, if John Wesley Powell were to visit the region of his exploits after his conquest 110 years ago, he would probably not recognize many of the locations because of their severe alteration. For example, downstream tributaries deposit cobbles and boulders

Figure 14-6 Satellite view of Nile delta region taken by the *Gemini* 4 spacecraft in June 1965. Mediterranean Sea to the left; Suez Canal, Red Sea, and Sinai Desert in center. Most of the dark area throughout the delta is cultivated land. (National Aeronautics and Space Administration.)

into the main Colorado River channel during flood stage, but the reduced flow of the Colorado no longer can flush away these barriers. The result is an increased number of rapids in some spots and ponding and slack water at others. Downstream erosion even occurs from the damming of eastern U.S. rivers. For example, during a 20-year period more than a meter of Catskill beds were eroded from the channel of Ouleout Creek, which had been dammed by the East Sidney Dam, New York.

A similar story of downstream erosion can be told for other dams throughout the world. Although erosion of about 3 m is predicted for the region immediately downstream of the Aswan Dam, Egypt, other more devastating impacts are occurring because of this dam.

Downstream impacts from diminution of sediment supply In addition to erosion in some places and siltation in others, a variety of downstream changes occur due to sediment starvation in the natural systems. For more than 5000 years, Egyptian farmers depended on the an-

nual flooding of the Nile to provide nutrients and silt to their fields and renew the soil for cultivation. The flooding was a direct consequence of the summer monsoon rainfall over the catchment basin of the Blue Nile and the Atbara Highlands of Ethiopia. Now the 13 million m³ of upstream sediment is deposited in Lake Nasser, and such silt loss has produced an entire series of deleterious impacts. The silt had formerly supplied about 22 percent of fertilizing materials needed as plant nutrients which must now be compensated by more than 2 million tons annually of chemical fertilizers (both costly and wasteful of energy). Loss of silt has also caused terrible losses in the brickmaking industry. Because of the decrease in sediment, the Nile delta is undergoing erosion. In addition, the lower flow is allowing the saltwater wedge to move farther inland, affecting groundwater conditions and permitting soil salinization (Fig. 14-6).

Somewhat similar conditions occur in many other coastal environments. The Lake Volta Dam in Africa has also decreased downstream sediment which has led to a 48 km upstream advance of saline waters and is creating drastic changes in the estuarine environment. The dam has also largely destroyed the salt industry downstream from the dam. Formerly floodwaters trapped water in shallow basins along the floodplain and produced 10,000 tons of salt per year when water evaporated and salt was precipitated. Below the Clatworthy Reservoir of the River Tone (Somerset, England), there has been a reduction of channel capacity for an 11 km distance. At this point the catchment area contributing to the river is four times that draining to the reservoir. Thus channel capacity below the reservoir was reduced to 54 percent of original capacity. The changing pattern of sedimentation also occurs on the River Meary below the Burrator Reservoir, Dartmoor, England, and the River Derwent below its impoundment at Derbyshire, England. In each case the channel cross section has been reduced by 40 percent the pre-dam geometry.

Coastal erosion along shorelines that formerly depended on sand nourishment from inland rivers is also described in Chapter 15. The case of the Brazos River, Texas, is also informative. Here a series of impoundments causes a 71 percent decrease in the suspended sediment load of the river and has lowered river discharge five to nine times less the original. The storage capacity of Lake Waco (one of the man-made reservoirs) was reduced from 39,378 acre-feet in 1919 to 13,026 acre-feet by 1958. This led to coastal recession rates of as much as 4 m per year.

Water table changes Groundwater conditions are changed both upstream and downstream of dams. New water table gradients are established with the raising and lowering of water surfaces in the reservoir and the downstream channel. Within the reservoir area such changes, which produce drastic pore-water pressure alterations and bank storage shifts, can become sites for landsliding (page 347) and a threat to the safety of structures and society—as at Vaiont and in the Columbia River. In arid areas, fluctuations in the water table can accel-erate direct evaporation of soil water and lead to increases in soil salinity. This in turn effects the water quality and impairs its use for irrigation and other purposes.

Ecological changes Dams may destroy the habitat of special flora and fauna, by smothering sites or in the changed character of downstream releases. Water from impoundments is physically and chemically altered. Furthermore, the pH is often more acidic due to the organic decay of materials formerly washed out of the area. Dissolved solids are more concentrated because of evaporation, and the water temperature may be significantly different. Typical chemical changes include greater amounts of ferrous, manganous, and sulfide ions, with lower oxygen amounts. The TVA began work on the Tellico Dam in 1963, a project that would dam the free-flowing Little Tennessee River, flood 6400 ha of farmland, and jeopardize the snail darter, a small member of the perch family. A total of $119 million was allocated for the construction and, when the project was 90 percent complete, a court injunction halted construction under the Endangered Species Act of 1973. The Supreme Court ruled in June 1978 that construction could recommence, but Congress amended the Endangered Species Act later in 1978, giving a cabinet-level panel the authority to rule in such matters. On January 23, 1979, the panel voted unanimously to deny the project an exemption under the act, so construction once again was halted. However, congressional action later in 1979 once again permitted final construction and completion of the dam. Of particular interest was the removal of 710 snail darters in 1975–76—they had multiplied to a population of 2500 by 1979 in the adjacent Hiwasee River.

In another case, an unforeseen ecological change has occurred due to the different environmental conditions brought about by Lake Nasser. Here there has been an astonishing increase in diseases, including schistosomiasis, malaria, leischmaniasis, ancylostomiasis, filariasis, and arboviruses. For example, the snail population has exploded and it can carry the worm larvae of the schistosomiasis, which bore into the skin when in contact with humans. More than 15 million people are now afflicted with the disease. These small microorganisms multiply more than 100,000 times in the host snail.

Earthquakes   In Chapter 11 we took a close look at earthquakes. For a review of the effects of dams on the stability of underlying bedrock, see pages 299–300.

Failure   There is no guarantee of dam safety during or after construction. Some of the safety-related topics of dams are discussed in Chapters 8 and 12. The siting of a dam at a safe location is as important as the construction and materials that go into the dam. The St. Francis Dam failed March 13, 1928, with great loss in property and the loss of about 500 lives because a geologic investigation had not preceded construction (Fig. 14-7). The foundation was two-thirds on mica schist and one-third on fresh-water conglomerate that contained gypsum veinlets. When dry the rock was strong but disintegrated when wet. The Elwha River dam in the Olympic Peninsula, Washington, was footed on gravel and, with the reservoir pressure, the material was scoured, the lake emptied, and the dam left hanging above the channeled portion. The Hales Bar Dam on the Tennessee River, near Chattanooga, was built on limestone and eventually cost much more than planned because of the extensive grouting program necessary to prevent seepage losses. The original Bloomington, Indiana, reservoir was also built on a cavernous limestone site and never held water. It was shifted to a second site where the same thing occurred; finally, at the third site a location was selected with proper geologic investigation, and the reservoir proved successful. The Hondo Reservoir, New Mexico, was underlain by shale and gypsum beds which contained sinkholes, and the dam had to be abandoned. A dam at Lee, Massachusetts, failed in 1968 because it was constructed on materials that were easily piped, causing the loss of two lives

and $10 million in property damages. And as earlier pointed out, the Teton Dam also failed because of piping in the volcanic foundation.

In spite of failures that have occurred, most modern dams have many built-in safeguards and often can withstand even severe earthquakes. For example, in Japan an earthquake in May 1968 (7.8 M) damaged 93 earth dams, of which 85 were less than 10 m high. Most had been constructed before 1926 in accordance with standards of the time. However, the earthquake produced no failure in the large dams that had been constructed using modern design methods. Since 1900 earthquakes have affected only eight large dams and have destroyed none (although associated activity was instrumental in the collapse of the Baldwin Hills Dam).

Some argue that dams give downstream property owners a false sense of security, so people build on floodplains and develop lands accordingly. However, when the total number of dams is considered and the relatively small number of serious failures computed there are many more hazardous locations than those downstream from dams.

Other impacts   Dams produce many other environmental and societal effects. Dams and their reservoirs may simply trade upstream flooding for cessation of downstream floods. In some instances, the people and lands displaced are as important as localities downstream. The Aswan Dam displaced 100,000 as did the Lake Volta Dam, and 80,000 refugees were made by completion of the Kossov Dam, Ivory Coast. In dry areas a prodigious amount of water is lost by evaporation. In the American Southwest this may amount to 3 m of water from reservoir levels. The evaporation losses from Lake Nasser are calculated to be about 15 billion m³ of water. However, the effects of large man-made lakes also produce a microclimate nearby that may produce beneficial impacts to the local populace and crops. Even the weight of waters in the reservoir may produce changes. It has been reported that the former inflow of about 3 billion m³ of groundwater into the Lake Nasser

Figure 14-7A & B St. Francis Dam, California. (a) Prior to failure on March 12, 1928. (b) View downstream after failure, showing concrete blocks from dam. Note large landslide at left abutment. (Courtesy Los Angeles Dept. of Water and Power.)

Figure 14-8 California aqueduct, looking south toward Kettleman City, Kings County. This is the principal canal that exports water from northern California to the semiarid southern part of the state. (Courtesy California Dept. of Water Resources.)

area has been changed and that about an equal amount of water now percolates into subsurface aquifers.

And finally we should be reminded once again that dams produce vitally necessary benefits. For example, the Aswan Dam prevented great economic losses in Egypt during the drought years of 1972 and 1975 by providing the necessary water for irrigation and industry. It also prevented flood damages from exceptional rainfall in late 1975. The electricity generated has provided important revenues, and Lake Nasser has spawned a new fishing industry. Although the absence of downstream silt and nutrients has caused an increased use and need for artificial fertilizers, the cost for these man-made chemicals is only about one-third the former cost attributed to pre-dam dredging and removal of silt from canals.

CANALS

Canals are distributory arteries constructed to divert water into new locations. They are used for a variety of purposes that include navigation, water supply, wetland drainage, and flood prevention (Figs. 14-8 and 14-9). Extensive canal systems were constructed by ancient peoples in the Fertile Crescent of Mesopotamia to bring water from the Tigris and Euphrates rivers into the irrigated lands between the rivers. With time many of these canals became silted, and historians report that as much as 50 percent of the labor force was necessarily employed in cleaning and maintaining the canals. The nearly ubiquitous piles of earth materials throughout this region—sediment removed from the canals—attest to the severe problems that resulted from their siltation.

The design of canals requires very careful engineering, because if such structures are overdesigned (that is, if gradients are too steep) erosion will occur, whereas if underdesigned (with too shallow gradients) excessive siltation will result. Realizing such problems, the British engineers who designed the canals in India during the 1800s developed the principal water-flow equations for movement of water in constricted channels. The following brief case histories illustrate the types of impacts and considerations that are involved in canal construction projects. Of course, there are also some similar effects as those which occur during stream channelization (see pages 403–409).

Jonglei Canal

One of Africa's largest engineering projects began in April 1978 with the digging of the Jonglei Canal through the Sudan, at the edge

(a)

(b)

Figure 14-9 Canals of the U.S. Bureau of Reclamation. (a) View showing the 1966 completed portion of the San Luis Canal, Central Valley Project, California. The automobile and driver (bottom) illustrate the canal's vastness—11 m deep and 78 m wide at the top. (b) Aerial view of Greenfields Canal, part of the Sun River Project, Montana. Water for the project is supplied by the Gibson Dam on the Sun River. Note the meandering character of the canal that was specifically designed to minimize erosion or deposition by canal waters. (Courtesy Water and Power Resources Service.)

of the papyrus swamp known as "The Sudd." The 350 km long waterway, twice the length of the Suez Canal, will transport 20 million m³ of water a day or about one-fourth the annual flow of the White Nile River which flows through The Sudd. The Canal will eliminate the bend through The Sudd. This will save 4.7 billion m³ of water and increase the flow of the Nile—which presently contributes 29 percent of the flow below Khartoum but loses one-half by flowing into The Sudd and suffering evaporation losses there. The second phase of the project includes plans for storage dams in the Equatorial Lakes region with a second Jonglei

Canal, parallel to the first, carrying another 23 million m³ of water. The Sudan and Egypt consider the projects crucial to their development with the hopes of turning the region into the "breadbasket of the Arab world." In addition, the Jonglei Canal will greatly improve communications by linking north and south Sudan and eliminating the 300 km steamboat trip up the Nile River, because a road will be built along the canal. However, a series of deleterious impacts may also result.

The area most immediately affected by the canal will be the vast floodplain surrounding The Sudd, an area of more than 100,000 km²—about the size of Belgium, the Netherlands, and Switzerland combined. At present, this floodplain area is a gigantic fish farm when the river floods; at this stage the floodplain absorbs nutrients from the sediment, and fish breed rapidly. Thus, the canal will reduce fisheries and their breeding grounds. The decreased silt and nutrients into The Sudd, along with the lowering of water levels, will trigger other ecological imbalances, causing a breakdown of plant cover. This can reduce the absorptive capacity of the swamps and at the same time increase evaporation of residual waters. Downstream waters will also be affected because they will not have had the filtering action of the wetlands. Therefore, increased sediment load and salts will endure in canal and river waters. Other effects that may occur involve the lowering of water tables throughout the region and changed weather conditions, including a lowering of annual rainfall. Such conditions could add to the desertification process of the Sahara and destroy nomadic life-styles throughout the area.

With completion of the Jonglei Canal, increased flow will enter Lake Nasser to the north. Such waters can add an extra burden to the already high levels of the lake. Lake Nasser has been gradually expanding its 5200 km² size because more water is entering than is needed downstream. The top of Aswan Dam is 196 m above sea level, and the lake level is currently at an elevation of 182 m. If water levels get much higher they will jeopardize the safety of the rockfill dam because such a dam should not be overtopped. Furthermore, if excess water is released through the spillway downstream, detrimental effects will occur. The additional flow of sediment-free water will excessively erode riverbanks, destroy bridge foundations, and produce a hazardous condition to smaller dams downstream. The planned solution to the problem is the design of the Toshka Canal, which will act as a safety valve.

Toshka is a village 240 km upstream from Aswan. Construction is now underway to link Lake Nasser with the Toshka Depression, which is a valley 40 km west. When the water level of Lake Nasser reaches 178 m it will flow into the canal and be diverted to the desert. However, even this will produce problems because the present design does not possess any types of gates for monitored flow. Without such controls water flowing through the canal will become contaminated along the way by salts leached from the rocky bed and canal sides. Such saline water will then filter through the Toshka Depression and into the Nubian sandstone aquifer below. If this scenario is accurate it will preclude reclamation of the Toshka area and will further contaminate the groundwater supply of the New Valley to the north, which is one of Egypt's major reclamation projects. However, there is a raging debate as to when such contamination will reach a critical state.

The Everglades

Before human molestation the Everglades in south Florida covered double their present size of about 10,000 km². This "river of grass," named for the serrate-edged saw grass that is its predominant vegetation, is a unique water-based ecosystem that has been sadly exploited during the past seven decades. It has been dying, along with the alligators, egrets, and panthers.

In 1906 the state of Florida started an extensive drainage program to dewater the sloughs and mucklands using canals that diverted the water to Lake Okeechobee—the liquid heart of Florida and one of the largest freshwater lakes in the continental United States. This resulted in a range of impacts, including the drying out of the muck with concomitant land subsidence and a great increase in fires, as well as a lowered water table with failure of wells and intrusion of salt water. Although the Everglades National Park was established in 1947 (Fig. 14-10), a greatly amplified program for water diversion was initiated by Congress in 1948 with authorization of the Central and Southern Florida Flood Control Project. This was justified in the name of flood control but grew into a massive canal building, channelization, and diking enterprise. The Corps of Engineers, managers of the program,

Figure 14-10   Everglades National Park, Florida.

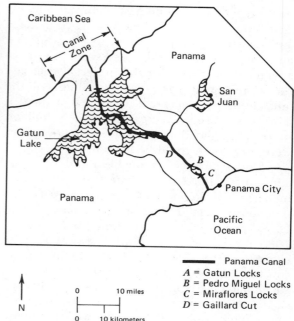

	Panama Canal
A = Gatun Locks	
B = Pedro Miguel Locks	
C = Miraflores Locks	
D = Gaillard Cut	

Figure 14-11   Map of the Panama Canal region.

have shrewdly named the fruits of their labors "Everglades Conservation Areas." Areas I, II, and III have become huge evaporation ponds that hold water for the large farm region to the north, called "The Everglades Agricultural Area." Water from these agricultural areas is backpumped or siphoned into Lake Okeechobee. This drainage reversal has introduced heavy volumes of nitrogen- and phosphorus-laden waters whose nutrients have now produced a virtual hypereutrophication of the lake. A 1976 U.S. Geological Survey report documents that 35 percent of the natural habitat of south Florida has been destroyed by agriculture, urbanization, and associated engineering projects of drainage. The wetlands that formerly occupied about 75 percent of the region now have water tables more than 2 m below pre-1900 levels, and wildlife and fisheries in marine and estuarine environments have been severely depleted. Numerous hammock sites have been destroyed, and the rich mucklands were oxided by direct exposure to the sun. Without their protective mantle, the organic-rich soils which require water for sustenance have disappeared through large areas. Indeed, water canals and distributary systems can have a devastating effect on fragile land-water ecosystems.

Panama Canal

The Panama Canal ranks with the greatest engineering projects of all time. This canal links the Atlantic and Pacific oceans at the Isthmus of Panama in Central America and has shortened the New York to San Francisco ship route from 21,000 km around the tip of South America to a distance of only 8360 km. The canal was finally constructed successfully by the United States during the eight-year period of 1906–1914. At the time it cost $380 million. The most expensive parts were the Gatun, Pedro Miguel, and Miraflores Locks and particularly the stabilization of the Gaillard Cut (especially the Culebra Cut section). In all, 161 million m³ of earth was removed, and 140 million m³ of this was from the Culebra Cut area (Fig. 14-11).

Fresh from his conquest and success with the Suez Canal, the Frenchman Ferdinand de Lesseps was looking for new fields to conquer. Such a venture did not materialize until 1882 when a French company he headed started construction of a canal at an Isthmus of Panama site. Unfortunately, a series of problems caused failure of the project, and the company went bankrupt in 1889. Dishonest politicians embezzled funds from the company, tropical diseases and epidemics took a fearful toll on workers, and the construction engineers did not have adequate machinery for the project and lacked the professional experience in dealing with the landsliding problems that developed at the Culebra Cut. Although they had dug 58 million m^3 from the site, the continual earth movements were more than could be handled.

Under rather questionable and devious circumstances, the United States inaugurated an international agreement that led to a treaty with the newly formed Republic of Panama. This Hay-Bunau-Varilla Treaty, signed in November 1903, authorized the United States to take charge of the Canal Zone and to undertake construction of a canal. The disease obstacle, which had killed one-fifth of all French workers, was largely overcome by the pioneering work of Colonel William C. Gorgas, who systematically wiped out yellow fever, bubonic plague, and greatly reduced malaria. Ultimate success was assured when the engineer in charge of digging the Culebra Cut, D. D. Gaillard, and his chief consulting geologist, C. P. Berkey of Columbia University, devised a strategy for dealing with the massive earth movements and stabilized the area. Although landsliding still occurs, in some years as much as 3.5 million m^3, the dredging equipment can easily handle such loads.

Other Canals

Numerous other examples of canals, their impacts, and geology could be cited. The Erie Canal was the first important waterway built in the United States. It was completed in 1825 and extended across New York from Buffalo on Lake Erie to Albany on the Hudson River. It helped New York City develop into the greatest financial center in the country. The original canal was 584 km long and averaged 10 m wide and 1.3 m deep. The Suez Canal is about 160 km long and provides a water link with the Mediterranean and Red seas. The canal is about 14 m deep, with a surface width of about 118 m. Dredges must constantly operate to remove blown sand and bank cave-ins.

The Netherlands is laced with canals, as is the city of Venice. Canals provide important waterway routes in the Inland Waterways System of the United States, the St. Lawrence Seaway, and the lower Mississippi River region. The largest canal for irrigation purposes in the United States is the All-American Canal, which connects the Colorado River from a site about 480 km above Hoover dam to the Imperial Valley of California. This 130 km long canal ranks as one of the largest irrigation canals in the world.

Perhaps the most awesome of all possible canal schemes is the one under current study in the Soviet Union, which would divert some of the water from north-flowing rivers to southern Russia. The first phase would involve excavation of 270 million m^3, a mass that would build a 1 m wide and 7.5 m high wall around the world. Construction would be in three phases: the early 1980s, the 1990s, and about 2030. Total excavation could be as much as 13 billion m^3, 75 times more than the Panama Canal, and could produce 300 km^3 of water per year. Even the first phase would produce 25 km^3 of water, sufficient to irrigate an area larger than California. The impacts from such a project could be stupendous. The salinity of the Arctic Ocean would be raised, which would reduce the formation of ice. However, the ocean would also receive less heat from the diminished river flow, which would increase ice formation. Whether such effects would be self-canceling is unknown. However, most analysts conclude weather conditions would change, with different patterns of rainfall and temperature highly probable. Proponents of the scheme argue that the additional lands that could be placed into irrigation in the Caspian region would offset any climatic deleterious effects.

Figure 14-12 Gullying that has intersected a part of a qanat system south of Teheran, Iran. The qanats are underground-dug passages for the obtaining and distribution of groundwater. (Courtesy Joseph Van Riper.)

GROUNDWATER

The remainder of this chapter will be largely devoted to groundwater considerations—impacts that result from excessive withdrawal of groundwater and remedies that have been undertaken to mitigate the damaging effects of such usage. Although groundwater has been used throughout historic times, severe problems have for the most part only emerged during the twentieth century. Jacob's well dates from biblical times, and the Qanats (Fig. 14-12) and chain-well systems throughout the Middle East have nearly an equally long history. The advent of powerful pumps, electricity, and petroleum have made possible high-speed extraction of groundwater where it has become a significant ingredient to life and irrigation in the drylands. However, when the water is withdrawn at faster rates than replenishment from recharge, an entire series of adverse impacts ensues. Such "groundwater mining" leads to lowered water tables, land subsidence, poorer quality of water, possible soil salinization and waterlogging, salt-water intrusion in coastal areas, shortened life of equipment, clogging of wells, and higher energy costs in lifting water greater distances. Clearly, this resource should be very closely

monitored and conservation practices used whenever possible.

Land Subsidence

When there is enormous groundwater pumping from unconsolidated and poorly consolidated aquifers, there is a loss in strength in the grain-to-grain surfaces, and the diminution of the buoyant force of water can cause readjustment in the packing arrangement of the earth materials. The weight of the overburden can add additional pressure, so that the final effect is both a reduced porosity of the aquifer materials and the sinking of the land surface to accommodate for the compacted and smaller volume (Fig. 14-13). The San Joaquin Valley, California, is the world's largest subsidence area, where 13,500 km² have been affected. The average land lowering exceeds 1 m and ranges up to 10 m. One 112 km long area has subsided more than 3 m, and the total volume in the valley of subsided materials is 186 km³. Subsidence has been correlated with the extensive irrigation development that began in the 1920s and constantly increased until the mid-1950s, at which time the annual subsidence rate in the west side of the valley reached 0.55 m per year. Thereafter the rate decreased, and by 1963 it was 0.33 m per year. In 1973, with the importation of surface water from northern California, the subsidence had stopped, and today there has been a recovery of the water table up to 60 m.

In the Santa Clara Valley, California, groundwater mining started as early as 1916 and increased from 4.9×10^6 m³ to 18.9×10^6 m³ per year by the 1960s. During this time, the artesian pressure head fell more than 75 m, the total volume of subsided materials was 3.5×10^8 m³, and average subsidence between 1934 and 1967 amounted to 2.4 m. As in all major subsidence areas, millions of dollars were spent realigning canals and ditches, building up roads, and renovating buildings. In addition, millions of dollars were required to reinforce

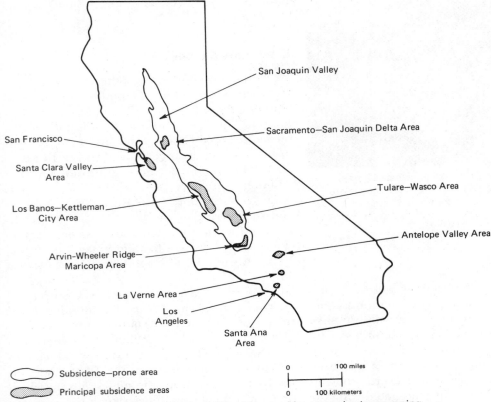

Figure 14-13 Areas of subsidence in California caused by groundwater pumping.

levees along the San Francisco Bay to prevent flooding in the subsided areas.

Texas has also had severe subsidence from groundwater overdrafts in the Houston-Galveston region. Here there was maximum subsidence of 2.7 m by 1973, and pumping had reached a rate of 1.9 million m³ per day. Extensive damages still result from coastal flooding, well casings protrude above ground surface (Fig. 14-14), and property losses are in excess of $30 million annually in a 2450 km² area.

The United States is not the only country afflicted with subsidence from groundwater pumping. In addition to Mexico City (see page 620) and Venice, other cities in England and Japan have been affected. The artesian pressure decline in London, England, started as early as 1820. It had declined 7.5 m by 1843 and by 1936 had been reduced to 100 m. Maximum subsidence is about 0.2 m. Land subsidence below sea level occurs in Tokyo under an area inhabited by 2 million people, and similarly in Osaka below 600,000 residents. Another type of subsidence occurs in the Netherlands, where about one-half the population lives below sea level. Here the water table must be maintained below land surface by pumps that discharge groundwater via the canals. This results in subsidence and the continuing need to implant dredge material to restore ground elevations. About 75 million m³ of dredged soil is used annually for such purposes and in other construction projects.

The dewatering in mines can also cause

Figure 14-14 Abandoned water well in Baytown, Houston-Galveston area, Texas. The top of the concrete platform was originally at ground level, showing subsidence due to groundwater withdrawal of 1.5 m. (Courtesy Charles Kreitler.)

Figure 14-15 Subsidence scarps from hydroconsolidation test ponds of the California Department of Water Resources in the southern San Joaquin Valley. (Courtesy Metropolitan Water District.)

ground surface changes. In order to extend mining in limestone quarries to lower depths, the rock was dewatered in Hershey Valley, Pennsylvania. Such pumping produced lowered water levels of more than 50 m, but also initiated the formation and extension of sinkholes. They ranged in size from 0.3 to 6 m in diameter, with depths of 0.6 to 3 m. In some drainages, such as Spring Creek, the total number of sinkholes ranged up to 100. Unfortunately, tragedy can even accompany these activities. A major dewatering program was initiated in 1960 for the Far West Rand Mining District near Johannesburg, South Africa, in order to extend gold mining to greater depths. This action produced the largest known man-induced sinkholes, up to 125 m in diameter and 50 m in depth. Collapse of materials during sinkhole formation killed 29 persons in 1962 and 5 in 1964.

Water application   A variation on the subsidence theme can occur by the extensive application of irrigation waters on loose, dry, low-den-sity soils. This phenomenon has been known only during the past 30 years. Large areas in North America, Europe, and Asia contain these soils. Such subsidence is becoming increasingly common in western United States where ground surface is lowered 1 to 2 m in many places and can reach depths of 5 m. The initiating cause for such land surface change is associated with the process of hydrocompaction (Fig. 14-15) and is especially prevalent in: (1) loose, moisture-deficient alluvial deposits that range from clayey water-laden sand and silt to mudflow materials (such as are common in alluvial fans); (2) loess and related eolian deposits; (3) materials that are reasonably fine-grained with a moisture deficiency and where seasonal rainfall rarely penetrates below the root zone. All these materials under natural

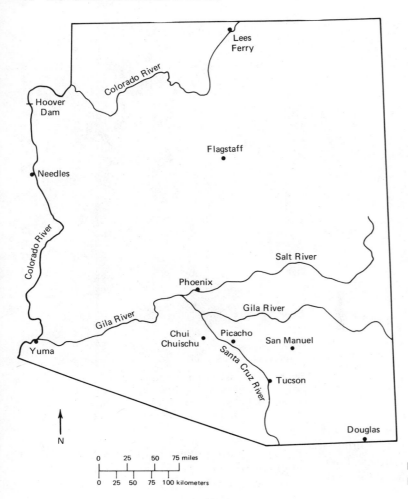

Figure 14-16 Location map showing places in Arizona.

conditions have sufficient high dry strength due to clay bonding, cohesion, and stacking to support overburden of a few hundred meters. However, when the dry strength is disrupted by wetting, the materials are forced to adjust to the new pressure system. This results in a different packing arrangement with subsequent subsidence. Damages from such topographic changes runs in the tens of millions of dollars yearly and affects irrigation ditches and canals, well casings, roads, pipelines, and houses. The largest single affected area in the United States is the San Joaquin Valley, California, where more than 500 km² are undergoing hydrocompaction subsidence. Other areas are in the Heart Mountains and Riverton areas of Wyoming, near Billings, Montana, various sites in the Missouri Basin, and near Pasco, Washington.

Arizona changes Groundwater usage in Arizona started to be important in the early 1900s and, by the 1920s, ground and surface water use in the Phoenix area had become so heavy that irrigated fields had become waterlogged (Fig. 14-16). However, 20 years later the water table trend had greatly reversed. With rapid development in the state in post-World War II times, and with the introduction of long-staple Egyptian cotton, acreage devoted to this crop vastly expanded and ushered in a new era of

(a)

(b)

Figure 14-17   View of earthcracks formed as a result of mining of groundwater, on Phoenix-Tucson Highway, September 1949. (*a*) Extension of crack across highway that needed frequent repair. (*b*) Close-up of crack. The barely visible geologic hammer shows scale of 0.6 m deep and 0.8 m wide at the top.

groundwater mining. Water levels in all irrigated areas throughout the state started their steep decline and, by the 1960s, water tables were 100 m or more lower in many areas.

Ground failure along faults associated with land subsidence caused by groundwater withdrawal was first recognized in central Arizona in 1949 (Fig. 14-17). Here pumping rates have been relatively constant since 1949 in the lower Santa Cruz Basin and the Salt River Valley Basin, averaging about 1.3 and 2.35 × 10⁹ m³ annually. Total pumpage through 1973 from the lower Santa Cruz Basin was 4.4 × 10¹⁰ m³ of which 69 percent was water removed from storage and lost to the atmosphere by evapotranspiration. During the same period, 8.42 × 10¹⁰ m³ was pumped from the Salt River Valley Basin of which 59 percent was lost. Thus, a mass of 8.02 × 10¹³ kg from these basins is indicated (Holzer, 1979). Depletion of these groundwater quantities produced water table declines of 48 m and 42 m during the 1915 to 1973 period. Local declines in excess of 100 m were common in both area (Fig. 14-18).

An interesting sidelight has also occurred in some areas from water withdrawal. First-order leveling surveys of the two regions established that during a 20-year period, 1948 to 1967, there was a *6.3 cm uplift* in the lower Santa Cruz Basin and a *7.5 cm uplift* in the Salt River Valley Basin! This uplift occurred throughout a 8070 km² area where 4.35 × 10¹³ kg of groundwater had been removed. The major areas of uplift coincide with the areas where crystalline bedrock is closest to the surface. Cause of the uplift is attributed to elastic rebound of the lithosphere when groundwater load was greatly reduced. An interesting by-product of such flexuring has been the creation of faults and other fissures. Such structures

Figure 14-18 Massive fissure-gully in south-central Arizona that formed as a result of declining water level and erosion. This site comprises part of the 310 km² area that has subsided more than 2.1 m since 1952. Several features, such as this one, are 3 m deep, 3 m wide, and more than 300 m long. (Courtesy U.S. Bureau of Reclamation.)

Figure 14-19 Picacho fault, Arizona. This is part of a 16 km long and 0.3 m high scarp in south-central Arizona. The recent movement was probably triggered by heavier groundwater pumping on one side of the fault than on the other. As a result, the ground has been moving slowly since 1961 and has caused highway and railroad maintenance problems. Since 1954 land surfaces in parts of this area have subsided 3.6 m due to the withdrawal of groundwater. (Courtesy U.S. Geol. Survey.)

seem to be related to the tensile failure produced by horizontal contraction in the dewatered zone.

The Picacho fault (Fig. 14-19) is the best documented one created by groundwater extraction. It is 15.8 km long with a vertical offset of 0.2 to 0.6 m. The scarp has been steadily increasing in height since first formed in 1961. It occurs along the east margin of the Eloy-Picacho subsidence bowl where more than 2.9 m of subsidence has occurred since 1934. On the other hand, fissures can occur suddenly and are best observed after heavy rains (Fig. 14-18). Many central Arizona fissures are first developed below ground level and later propagate and expand upward to the surface. Within a year or two after forming they become inactive, partially filled, and others are formed. The new ones may be along the same tread, although not necessarily continuous, and occasionally have an *en echelon* pattern. One fissure north of West Silver Bell Mountains continued through a mesquite tree which it split and opened a horizontal distance of 6.4 cm.

It should be pointed out that faulting associated with land subsidence has now been documented in several other states, including Raft River Valley, Idaho; Las Vegas Valley, Nevada; San Jacinto Valley, California; San Joaquin Valley, California; and along the Texas Gulf Coast. Lengths of these structures may exceed several kilometers, but depths are generally unknown due to the absence of external drainage. However, near Pixley, California, confirmed depths extend beyond 16 m, but displacements are usually small. More than 50 distinct active faults with an aggregate length of 220 km occur in the Houston-Galveston subsidence bowl, but this area has also been subjected to petroleum pumping.

Finally, we must not forget that subsidence can also be associated with dams. For example, a depression of about 18 cm has been associated with the impounding of 37.6×10^{12} kg of water in Lake Mead of Hoover Dam.

Salinization

Groundwater use for irrigation is only one of the ways that salt buildup can occur in soil horizons. When surface or groundwaters are

Figure 14-20   Salt scalding in western Australia. This is caused by the increase in the height of the water table due to deforestation. (Courtesy Karl H. Wyrwoll.)

used that contain high concentrations of dissolved minerals, they may remain within the soils after the pure parts of the water have been transpired and evaporated during plant growth and from loss by solar energy. In drylands even pure irrigation water may mobilize saline-type ions from the soil which in time become concentrated in amounts that cause losses in soil productivity. Deforestation is another process that produces changes in the water balance and can lead to a phenomenon known as **salt scalding,** as seen in Australia (Fig. 14-20). Elsewhere in this book other aspects of the salinization process are discussed (see pages 250–251). Groundwater quality deteriorates with water table decline and use of deeper aquifers because such water usually has been in residence for longer times, allowing for greater amounts of dissolved solids to be incorporated into the groundwater. Salinization problems are worldwide in scope and have afflicted civilization throughout history.

Groundwater Augmentation

Various engineering and conservational strategies are used in efforts to increase groundwater supply and to prevent its contamination. These man-induced methods for saving groundwater are responses to the increasing consumption of this precious resource as well as the solution to specific problems that threaten the abundance or purity of the water.

Water spreading   Water spreading is one of the oldest engineering techniques used to enhance groundwater production. One of its purposes, similar to dams, is to trap water on-site before it leaves the area and perhaps flows, unused, to the ocean. It is especially effective in dry environments because unlike surface water reservoirs the objective is to allow percolation into the substrate where the water will be free from evaporation loss, allowing greater volumes to be salvaged. This form of artificial recharge is designed to increase groundwater storage for eventual use by down-gradient users.

The idea of water spreading is to allow streamflow, often fanning out from mountain fronts, to pass over permeable and scarified sediments, or to be diverted by dikes into settling basins where the water can infiltrate below ground surface. In 1889 Denver began a modest-sized artificial recharge project to maintain domestic water supplies. By 1955 there were 120 public supply recharge projects to maintain groundwater supplies in 15 states, with 87 in California and 13 in Massachusetts.

Water spreading in California dates back to the early 1900s. The southern part of the state provides the ideal conditions because high mountains are close to the ocean, and high-intensity rainfall was mostly being lost to the ocean without beneficial use. Early attempts at spreading involved methods to slow stream velocity such as diverting flow to lower gradient ditches. The most common method today utilizes diked basins, into which streamflow is channeled (Fig. 14-21). In some areas regulated flood flow is released directly to the streambed. The spreading operation is highly desirable because it not only raises water level, but increases groundwater storage and supply (functioning as an underground reservoir) and also aids to prevent saltwater intrusion into coastal aquifers.

At the Saticoy spreading fields in the Santa Clara Water Conservation District, the water table was 24 m below ground surface at the start of the 1934–35 spreading season. The spreading of water on 97 acres—13,000 acre-

Figure 14-21 Hansen Dam Spreading Grounds in the San Fernando Valley, California. San Gabriel Mountains in the background. Water from the dam is diverted into this spreading basin for recharge downgradient of groundwater supplies. (Courtesy Los Angeles Dept. of Water and Power.)

feet—raised the water table to a position of only 3 m below the surface. The spreading fields on the south side of the Santa Clara River consisted of separate basins that received water via un-lined canals from the river. During the 1958–59 water year about 170,000 acre-feet were delivered to the Orange County Water District and the Los Angeles County Flood Control District

North

South

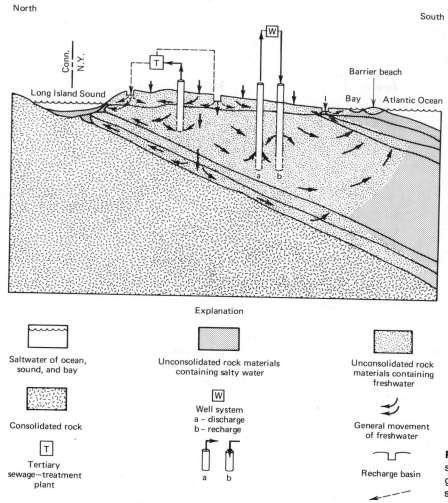

Explanation

Saltwater of ocean, sound, and bay

Unconsolidated rock materials containing salty water

Unconsolidated rock materials containing freshwater

Consolidated rock

W
Well system
a – discharge
b – recharge

a   b

General movement of freshwater

T
Tertiary sewage–treatment plant

Recharge basin

Movement of treated wastewater

Figure 14-22 Schematic cross section showing recharge of groundwater by tertiary sewage injection wells and recharge pits. Long Island, New York. (U.S. Geol. Survey.)

for spreading in the Santa Ana and San Gabriel river basins. Since 1949, and up to 1959, the Orange County Water District spread nearly 500,000 acre-feet of water which raised the water table more than 3 m.

Recharge pits Another method to prevent the loss of water and to allow its infiltration into shallow groundwater aquifers is to use "recharge pits" (Fig. 14-22). Although these are now in use in a number of cities, their most extensive development is on Long Island, New York. Instead of diversion into the small basins

from streamflow, these manmade depressions are designed to absorb the water flow from storm sewers. The use of such pits began here in 1935 with a twofold purpose: to raise the water table that was being drastically lowered by excessive pumping, and to help arrest saltwater intrusion. Such basins are also cost-effective because they aid in eliminating the construction of long storm sewers to discharge into streams or the ocean. The use of recharge basins had a very slow growth, and only 14 had been developed by 1950 in Nassau and Suffolk counties. However, by 1960, the number had increased to more than 700 and, in 1969, there were more than 2100 in the two counties. These recharge basins are open pits of different dimensions cut into highly permeable sands and gravels. The basins dispose of runoff from residential, industrial, commercial, and highway areas, and recently some have been used to infiltrate treated sewage. The basins range in area from less than a hectare to more than 13 ha, with an average size of 1 ha. Depths range up to 13 m, but those that are 3 to 5 m are more common. The major restriction placed on their construction is that the water table must be sufficiently deep so that infiltration will continuously occur and the basin does not become a standing water body.

Basins are divided into two categories: those with and those without overflow structures. The great majority are built with overflow piping to allow escape of water when filled into other basins or streams. A major problem is the retention of water within the basin. Also, some basins have very slow percolation rates. These problems arise because (1) the water table is too near the surface, which means the basin was sited in a poor location; (2) materials that line or occur under the basin are too fine-grained to permit ready percolation. These operating difficulties can be solved by (1) excavating the basin flow and scarifing if necessary; (2) installing diffusion wells below basin level; and (3) linking several basins into a coordinate drainage network so there is easy exchange between basins. Normal daily recharge on Long Island is 443,000 m³ from natural sources, but recharge through the artificial pits amounts to 560,000 m³. Thus man-induced flow now exceeds that from natural means.

Injection wells Excessive groundwater pumping from freshwater aquifers in many coastal cities throughout the United States has caused the intrusion of salty seawater into the aquifers and, in many cases, has caused abandonment of the well fields. This has been the case in many Florida cities such as Miami, Ft. Myers, Pensa-

cola, Daytona Beach, and others. Although saltwater intrusion also occurs on Long Island and south of Los Angeles, these two areas have undertaken massive injection-well programs designed to stop and control further saline encroachment into freshwater aquifers. Similar conditions occur along the European coastline where saltwater invasion became a problem as long as 70 years ago.

The first successfully operated seawater barrier formed by well-injection methods was constructed in California along the margin of Santa Monica Bay for the protection of inland groundwater basins. Although the first notice of saline instrusion was at Redondo Beach, California, in 1912, and although by the 1950s saltwater had advanced inland about 3.2 km in Southern California, the first experiments to control conditions were not undertaken until 1952. These experiments were successful so that in 1962 a full-scale project was authorized. The original pilot program was authorized by the California legislature which appropriated $750,000 for the project. A line of injection wells, nine in all, were installed by the Los Angeles Flood Control District to create a water barrier for a distance of about 17 km northward from the Palos Verde Hills. This barrier consists of a curved line of injection wells, with pumping wells that are somewhat seaward of the line. Such injection increases the artesian pressure head, setting a high-pressure ridge of freshwater that dams further advance of the saline plume.

Long Island Long Island, New York, has had a series of water and pollution problems for the past several decades. Originally European settlers dug wells into the shallow glacial aquifers to supply their water needs. The construction of cesspools recharged wastewater back into the aquifer. As population density increased, so did water use, creating a demand for large public water supplies. For a time cesspools and septic tanks continued to add waste effluent to the aquifers. While the net water balance remained about the same, the degradation of water quality reached such proportions that the glacial aquifer became unfit for human consumption. Shallow wells were replaced with deeper public supply wells that tapped the Jameco and Magothy bedrock aquifers. At the same time, domestic and industrial sewage was still being dumped into the glacial aquifer at accelerated rates. The increase of heavy-metal pollution and the migration of pollutants to nearby streams and wetlands caused additional health problems. For example, a plating plant near a recharge basin discharged chromium and cadmium into it, and contamination moved more than 1.6 km and surfaced in Massapequa Creek in Nassau County. Two leachate plumes associated with landfills in western Suffolk County migrated down the hydraulic gradient 1500 and 3000 m, respectively, into the base of a glacial aquifer and contained the heavy metals zinc, manganese, and iron. To control such degradation, additional sanitary sewering projects have been installed throughout many parts of Long Island. Although such pollution is now prevented, the cleansed waters are discharged into streams or the ocean with the loss of recharge water into groundwater aquifers. In addition, parking lots, urbanization, and some storm sewers have reduced recharge. Such depletions have caused hydrologic imbalance and a decline in the water table of more than 3 m in many areas. Heavy groundwater pumping formerly was also a contributor to water table decline.

In 1903 Kings County and Queens County, in the westernmost part of Long Island, were supplied with water from about 12 private water companies, from numerous ponds and wells owned by the City of New York, and from domestic and industrial wells. Most of the wastewater was exported from the area via public sewers. By 1933 nearly all the groundwater used for public supply, and a considerable amount of self-supplied industrial pumpage, was wasted to sewers, and an extensive deep cone of depression in the water table had developed (Fig. 14-23). Some of the public supply wells in Brooklyn became contaminated with seawater, and saltwater encroachment occurred in three saltwater wedges from the southeast. In some areas the deep and intermediate wedges advanced 1.6 km inland. The deepest salty wedge is the largest, thickest, and most directly influential on public water supplies. Its thickness ranges from zero at the leading edge to more than 120 m along the south shore of the island. The intermediate wedge contains up to 12,900 ppm chloride.

When it was discovered in 1933 that an area larger than 100 km^2 had water tables below sea level and that potability of the freshwater was being threatened with saltwater, the New York State legislature took action. It authorized establishment of a program that would require commercial-industrial users with wells of 70 gal/min (0.26 m^3) who used water for cooling and air conditioning to drill injection wells that would return such waters to the source aquifer. Special drilling permits had to be obtained from the state for any well capable of pumping 100,000 gal/day (378 m^3/day). From 1903 to 1936 more than 185 million m^3 of freshwater had been removed from groundwater aquifers, producing a maximum water table decline to 10.5 m below sea level. However, by the 1940s the trend had been reversed; importation of water from outside the region and more than 200 recharge wells had turned the tide. For example, by 1944 injection wells were returning 227,000 m^3 (60 million gal) daily back into the aquifer.

Phreatophytes Phreatophytes are plants with the ability to send roots below the water table for transpiration. Mohammed Karaji knew in the eleventh century that certain plants extended roots into the water table and cited the evidence of a man digging a well who found roots of bushes 45 m below ground surface. Along rivers, many plants (known as riparian vegetation) compete for available water and soil space. When a river dries up, hydrophytes

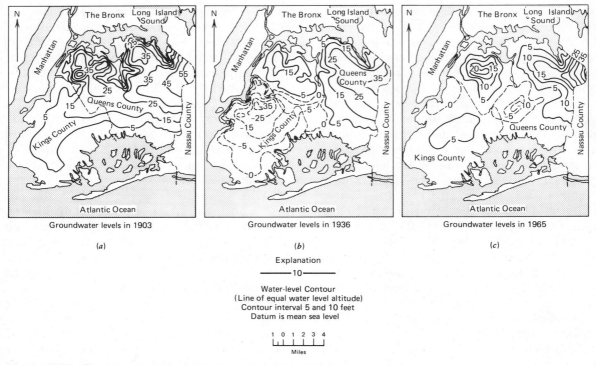

Figure 14-23 Changing groundwater levels on Long Island, New York. (*a*) 1903. (*b*) 1936. (*c*) 1965. (U.S. Geol. Survey.)

(water lovers) die first, followed by the meso-phytes. Phreatophytes, however, with deep-root plumbing systems, have the capacity to survive when other vegetation is killed (Fig. 14-24).

There are many species of phreatophytes but the most common in western United States are the salt cedar (tamarisk) and the mesquite. These trees and other phreatophytic vegetation cover more than 4 million ha in the southwestern states and annually consume more than 25 million acre-feet of water, mostly on floodplains. In six states—Arizona, California, Colorado, Nevada, New Mexico, and Utah—phreatophytes cover nearly 3 million ha. With such consumptive use of groundwater it is no wonder that some environmentalists are searching for methods that would eradicate the plants so as to save the transpired water.

The eradication view has been expressed by a Select Committee of the U.S. Senate, which stated that water consumed by phreatophytes serves a less economic value than if it were used for irrigation; thus they proposed removal as a way to salvage water. However, the issue is not so simple, and a large controversy has emerged among opposing groups of conservationists. Hydrologists favor eradication as a means to conserve water for human use, whereas zoologists wish to save the plants because of the shelter, protection, and food they provide wildlife. Although the issues are sharply drawn, many questions remain and all problems are not resolved. For example, which destruction method is most effective and yet environmentally safe? Will vegetation removal promote increased evaporation from wet soil because of the water level rise or rapid growth of xerophytic (using surface or soil water) species? Will replacement vegetation have as much of an erosion-inhibiting character as the phreatophytes that are destroyed? Which governmental agency should be charged with the program,

Figure 14-24 View of vegetation in and near the channel of the Santa Cruz River, Arizona. The tallest trees are cottonwoods, which have died because of the excessive lowering of the water table by human intervention. However, other vegetation still flourishes, such as the mesquite trees, because they are phreatophytes and can send deeper roots to tap the groundwater.

and how will it be financed? What is the true cost of removal? On this latter point the Corps of Engineers has estimated a cost of $50 per acre, whereas the U.S. wildlife agencies report a cost estimate of up to $300 per acre.

Advocates for phreatophyte removal argue that the savings in groundwater for reuse by humans in other economic pursuits more than counterbalances any adverse impacts from their elimination. A U.S. Geological Survey in Arizona, known as the Gila River Phreatophyte Project, showed that after removal of the plants, 2 acre-feet of water per acre were saved. Much of this savings occurs during the dry season when groundwater needs are crucial. It was also pointed out in the survey that xerophytic types of grasses can replace phreatophyte vegetation and still offer wildlife protection, have as much a stabilizing influence on soil erosion, use little water, and provide forage for cattle.

Those who urge retention of phreatophytes say the costs for removal would be excessive, that defoliates would be ecologically harmful, and that physical removal would ruin the fragile soil equilibrium. The establishment of other vegetation types may take 20 years or more, and excessive erosion and siltation will be the

product. Several studies have shown there is a direct relationship between the number of trees and the bird population—a partially cleared area had 580 pairs of birds per 100 acres, a less cleared area had 939 pairs, and an uncleared area had 1322 pairs. Dove, quail, gray hawk, and black-bellied tree duck were the most affected species. Other studies showed a decline in the fish population for cleared banks when compared with uncleared reaches. Bank rounding commonly occurred in the cleared reaches that reduced the undercut banks used for protection and spawning, and these changed banks also produced greater amounts of silt which damaged downstream fish habitats. An interesting by-product of mesquite trees is the abundant perennial grasses that grow under them, far more than occurs in open lands. A study in 1973 of this difference in grass density and vigor showed up to 3 times higher nitrogen content in soils under the mesquite, and grasses contained 15 times as much nitrogen. Sulfur and phosphorus were also made more available for grass growth near the mesquite. Thus, the biochemical growth pattern of the phreatophyte produces chemicals that are more freely abundant for use by other plants.

Other considerations There are other factors involved with pumpage of groundwater and the wells which provide the vehicle for its removal. Even the drilling of wells is important in determining their efficiency and the degree of impacts that may result. For example, it is important to test the quality and amount of water at different horizons to assure which units are of superior quality and quantity. Substandard zones with high ratios of dissolved solids should be cased off or cemented off so as not to contaminate purer supplies. Pumps must be carefully designed to fit the particular set of conditions as dictated by the specific yield ability of the aquifer. Pumps that are too powerful and that are geared to abnormal high-extraction rates should not be emplaced into aquifers that cannot accept such high-withdrawal rates. A too powerful pump can result in abnormal subsur-

Figure 14-25 Diesel well at Chiuschu on the Papago Indian Reservation, Arizona. Note that the principal canal is lined with macadam to prevent seepage losses.

face flow velocities that can lead to aquifer disruption and damage to the equipment.

To grow crops, such as irrigated cotton, pumps are forced to work day and night for week after week (Fig. 14-25). This steady use, season after season, causes great wear and tear on the equipment and can also produce mineral and other clogging particles in the screens or slots of the well casing. Eventually the pumping efficiency is so reduced that the well is no longer productive. If the principal trouble is in the incrustation of the openings in the casing, some-times remedial action can bring the well back into production (Fig. 14-26). In such circumstances a variety of techniques have been employed that include: (1) using a Calgon (Fig. 14-27), which can break down the mineral hardness of the clogging matter; (2) acidifying, which can dissolve the mineral incrustation; (3) dry icing, in which the turbulent action of gas under high pressure may be sufficient to unclog the opening; (4) surging, which is the rhythmic raising and lowering of the water column inside the well casing, flushing and plunging water to jar loose the contaminants; and (5) blasting, which may be used under unusual conditions when everything else has failed.

An additional factor to consider when developing well fields is the spacing of wells so that interference in cones of depression is minimized. Knowledge of safe annual yields, permeability, and other hydrologic conditions are necessary for the appropriate engineering of water projects. Pumping tests are also vital to determine accurately the specification for design of groundwater programs. Such tests provide the data base for development of the most efficient and environmentally sound hydrologic decisions.

Perspectives

A nearly endless series of impacts occurs when water is removed from its natural setting and stored, transported, and used—there is an increased rate of erosion and siltation, changing of water levels, subsidence, accelerated solution and precipitation of mineral matter, fissuring and faulting, and ecological changes, to name a few. Because water is vital to human endeavors, and because it is the engineer who is charged with providing this commodity, it should also be the engineer who seeks the services of a wide spectrum of scientists in order to minimize environmental effects from the changes introduced. Geologists can be especially helpful in providing information as to the range of physical impacts that will result, as well as assisting in the siting of dams, canals, and wells at locations that will be safe and as environmentally secure as possible. Thus, more than technical knowledge and construction design are required whenever nature is to be manipulated. The old credo of "Nature to be commanded must be obeyed" is still a sound policy. As populations continue to expand and as more and more of the best sites are used for a full range of activities that are required by society,

(a)

(c)

(b)

Figure 14-26 Dry icing of well at Chiuschu, Papago Indian Reservation, Arizona. This well had declined in pumping rate from 900 gpm (34 m³/min) to less than 300 gpm (11 m³/min) during a 4-year period due to encrustation of slotted openings in the well casing. In an attempt to restore a higher yield, the well was dry iced in August 1950. The process consisted of injecting about 100 kg of broken dry ice into the well bore. (a) The pump housing was reinserted over the well to prevent escape of vapor. (b) After sufficient pressure had developed from vaporized CO_2, the pump housing was quickly removed and the man-made geyser action started (c) and culminated with a blowout 12 m high (d). The flushing action of this process, plus surging of the well, loosened and removed a moderate amount of the materials that clogged the well; this resulted in an increased pumping rate to 600 gpm (23 m³/min).

(d)

the problems will become ever more acute: where and how are structures to be built, and when can we afford to alter the land-water ecosystem. Such decisions need professional judgments from many disciplines, and it is vital that geologists be part of this policy group. Their advice may well prevent a future Vaiont from happening.

READINGS

Bull, W. B., 1974, Geologic factors affecting compaction of deposits in a land-subsidence area: Geol. Soc. Amer. Bull, v. 84, p. 3783–3802.

Cohen, P., Franke, O. L., and Foxworthy, B. L., 1970, Water for the future of Long Island, New York: N.Y. Water Resources Bull. 62A, 36 p.

Culler, R. C., 1970, Water conservation by removal of phreatophytes: Amer. Geophys. Union Trans., v. 51, n. 10, p. 684–689.

Eckholm, E. P., 1975, Salting the earth: Environment, v. 17, n. 7, p. 9–15.

Foose, R. M., 1953, Ground-water bebavior in the Hershey Valley, Pennsylvania: Geol. Soc. Amer. Bull., v. 64, p. 623–646.

———, 1967, Sinkhole formation by groundwater withdrawal: Far West Rand, South Africa: Science, v. 157, v. 157, p. 1045–1048.

Gabrysch, R. K., 1969, Land-surface subsidence in the Houston-Galveston region, Texas: Pub. Inst. Assoc. Sci. Hydr., v. 88, p. 43–54.

Gottschalk, L. C., 1964, Reservoir sedimentation: in Cow, V. T., ed., Applied Hydrology, New York, McGraw-Hill, p. 17.1–17.34.

Graf, W. L., 1978, Fluvial adjustments to the spread of tamarask in the Colorado Plateau region: Geol. Soc. Amer. Bull., v. 89, p. 1491–1501.

Gupta, H. K., 1976, Dams and earthquakes: Amsterdam, Elsevier, 229 p.

Hamilton, D. H., and Meehan, R. L., 1972, Ground rupture in the Baldwin Hills: Science, v. 172, p. 333–344.

Harte, J., and Socolow, R. H., 1971, The Everglades: wilderness versus rampant land development in south Florida: in Harte, J., and Socolow, R. H., eds., Patient earth, New York, Holt, Rinehart and Winston, p. 181–202.

Figure 14-27 This well in Douglas Basin, Arizona, was subjected to Calgoning—the insertion of Calgon in an attempt to increase its yield. The flow of 250 gpm, as seen in this photo, was finally increased about 10 percent. Notice smokestacks from the Douglas copper smelter operation in the background belching pollutants (1949) without restriction. Such contamination killed or reduced crop yields in downwind fields for several kilometers.

Holzer, T. L., 1979, Elastic expansion of the lithosphere caused by groundwater depletion: Jour. Geophys. Research, v. 84, p. 4689–4698.

Holzer, T. L., Davis, S. N., and Lofgren, B. E., 1979, Faulting caused by groundwater extraction in southcentral Arizona: Jour. Geophys. Research, v. 84, p. 603–612.

Kiersch, G. A., 1964, Vaiont Reservoir disaster: Civil Eng., v. 34, n. 3, p. 32–39.

Leopold, L. B., and Maddock, T. Jr., 1954, The flood control controversy: New York, Ronald Press Co., 278 p.

Lofgren, B. E., 1969, Land subsidence due to the application of water: in Varnes, D. J., and Kiersch, G., eds., Geol. Soc. Amer. Reviews in Engineering geology v. II, p. 271–303.

Milne, W. G., and Berry, M. J., 1976, Induced seismicity in Canada: Eng. Geol., v. 10, p. 219–226.

Morgan, A. E., 1971, Dams and other disasters: Boston, Porter Sargent, 422 p.

Outland, C., 1963, Man-made disaster, the story of the St. Francis Dam: California, Arthur H. Clark Co., 249 p.

Parizek, R. A., and Myers, E. A., 1968, Recharge of ground water from renovated sewage effluent by spray irrigation: Proc Fourth Amer. Water Resources Conf., p. 425–443.

Peterson, E. T., 1954, Big dam foolishness: New York, Devin-Adair, 224 p.

Poland, J. F., and Davis, G. H., 1969, Land subsidence due to withdrawal of fluids: in Varnes, D. J., and Kiersch, G., eds., Geol. Soc. Amer. Reviews in Engineering geology v. II, p. 187–269.

Prokopovich, N. P., 1972, Land subsidence and population growth: 24th Intern. Geol. Cong. Proc., v. 13, p. 44–54.

Scudder, T., 1969, Kariba Dam: the ecological hazards of making a lake: Natural History, Feb., p. 68–72.

Seaburn, G. E., and Aronson, D. A., 1973, Catalog of recharge basins on Long Island, New York: 1969 N.Y. State Dept. Cons. Bull. 70, 80 p.

Stephens, J. C., 1958, Subsidence of organic soils in the Florida Everglades: Soil Sci. Soc. Amer. Proc., v. 20, p. 77–80.

Thomas, H. A., 1951, The conservation of ground water: New York, McGraw-Hill, 327 p.

UNESCO, 1969, Land subsidence: Publications of the Institute of Scientific Hydrology, 88, in 2 vols.

Wulff, H. E., 1968, The qanats of Iran: Sc. Amer., v. 218, n. 4, p. 94–105.

Chapter Fifteen
Coastal Environments

View of Outer Banks and Cape Hatteras, North Carolina from *Apollo* 9 space satellite. (Courtesy National Aeronautics and Space Administration.)

INTRODUCTION

For millennia, the oceans have inspired awe and reverence in mankind; however, this fascination has also introduced both practical and aesthetic results. The land-water interface became the locus of ports and commerce. Trade that flowed into settlements came from other regions that used the sea as waterways, and from the bounty of the nearby waters, there also came fishing and other marine resources. The ancient harbors were invariably nearly landlocked or somewhat inland of estuaries. Early man also respected the power of the sea and felt the nearby land, whether defended by rocks or sandy, was a bastion of safety and a refuge from the fury of the powerful coastal elements. But as populations increased, society began to challenge coastal forces and started building structures that were deliberately designed to modify natural processes of the coastal environment. Some recent developments have resurrected King Canute's arrogance. More often than not, the ignorance of natural forces yields unhappy results.

Our invasion of the coastal corridor has accelerated rapidly since World War II. This intrusion has taken many forms and includes greatly expanded industrial development, urban sprawl, and the nearly ubiquitous spread of second-home communities. Further pressures for coastal real estate take the form of recreational and sporting sites for leisure time activities. The popularity of shorelines is shown by demographic statistics. Although continental shelves and near-shore water comprise only about 5 percent of the world, about two-thirds of the world's population lives near the coast. In the United States, 90 percent of the population growth in the last decade has been in the 30 states that border the shores of the oceans and the Great Lakes. The 30 states contain nearly 75 percent of the total population and 12 of the 13 largest cities, and 50 percent of the people reside in coastal counties.

Coastlines may seem impregnable, but they actually constitute some of the most fragile and changing lands on the planet. When humans enter the picture, we invariably accelerate the degradation of this environment. To protect our investment and provide for our safety, we have altered natural processes to meet our needs. We have constructed breakwaters, jetties, groins, and seawalls that change the natural character of coastal processes. We have also changed the land configuration by building sand dunes or by dredging and filling wetlands. The final indignity is perpetrated by the discharge of massive amounts of municipal and industrial wastes. Although the quality of coastal waters is undergoing deterioration, the dynamic processes continue unabated and, with the relentless rise in sea level, it will continue to provide us with costly and damaging problems.

With the exception of the Netherlands, problems with the shoreline have generally been handled on only a local basis. In the United States, however, a large change has occurred in the last 10 years in which new concerns are being addressed on both the federal and state levels. These new elements in coastal management strategies are having a profound effect on our attitudes and our stewardship of this vital resource.

DYNAMICS OF THE SEAS

The shoreline is affected by a variety of coastal forces. These forces together constitute a geomorphic process which sculptures the seascape by eroding, transporting, and depositing materials. The sun and the moon provide the energy that drives the wind and water to do work on the land. Winds are caused by temperature and pressure changes in the atmosphere. Unequal heating of the earth's surface initiates temperature and pressure gradients and causes differential stresses in ocean waters and the subjacent air masses. Waves are the product of these stresses and are the ever-present features of all seas.

The gravitational attraction between the earth and the moon, and to a lesser extent the sun, creates the tides. These water bulges produce a breathing-like mechanism to the seas in which a rhythmic rise and fall of the sea level at the land interface regularly occurs every 12 hours and 25 minutes for a complete tidal cycle. The size of the tidal range depends on such factors as latitude, time of the year, relative position of the earth-moon-sun system, weather conditions, and land configuration. The tidal range for most coasts is generally only meters or so, but is in excess of 15 m in the Bay of Fundy between Nova Scotia and New Brunswick. At this site the landward rush of the tides, or **tidal bore,** is 2 m. Tides cause many coastal changes and even in the Hudson River the effects occur as far north as Troy, a distance of 240 km from the ocean where the tidal range is 1 m.

Oceans also contain currents—their own brand of rivers. The largest systems are **ocean currents,** which are caused by a combination of the differential heating of the water and the rotational momentum of the earth. These currents may consist of either warmer water than the water through which they move, as in the Gulf Stream, or colder water, as in the Humboldt Current. The principal importance of these currents is the influence they produce on climate and weather patterns. Other currents in the ocean are produced at and near the shore; these are called **longshore currents** and **rip currents.**

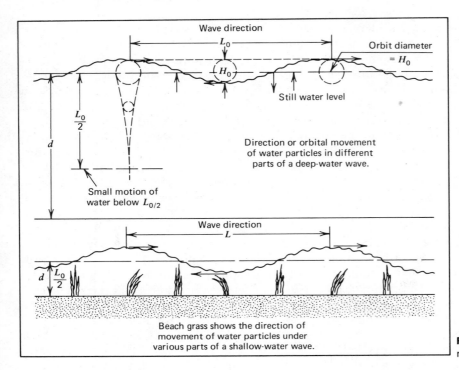

Figure 15-1 Water particle movement under wave action.

Waves

Waves (also referred to as wind waves, free waves, and oscillation waves) are created by air moving over the water surface and exerting a frictional stress. Although water movement within the wave form is orbital, forward motion is accomplished by the faster movement of water on the crests than in the troughs, resulting in speeds up to 3.2 km/h (Fig. 15–1). Wave height is a function of wind velocity and duration, distance of travel (**fetch**), and existing state of the sea. Winds of 16 km/h generate 0.6 m waves, whereas 80 km/h gales can theoretically create 18 m (60 ft) waves. Under unusual circumstances wave heights may become amplified and reach heights in excess of 23 m, as in the North Sea. Such unusual waves, called **rogue waves** or **nightmare waves,** have been responsible for sinking and destroying many ships.

The surface forms of water constitute a hierarchy that consists of **ripples, chop, waves, and swells,** in order of increasing wave length.

The wave length of wind waves is generally of the order of 7 to 20 times the wave height. However, when sufficient force is provided, as in large storms, waves may be propagated far beyond the location of the initiating storm. In traveling great distances the waves are transformed into swells, or ground swells. These may travel several thousand kilometers with low rounded heights of a few meters and wave lengths that range from 30 to more than 500 times the wave heights.

The Beach

A beach consists of sediments that extend landward from low sea level (Fig. 15–2). Waves form beaches by depositing materials and reworking them by a variety of coastal processes. The majority of beaches are composed of sand-size particles, but some low-energy beaches are muddy, and high-energy beaches may contain gravel-size particles called **shingle.** Beach steepness is mostly a function of energy and sediment

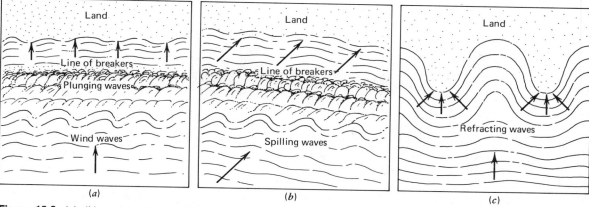

Figure 15-2 (a), (b), and (c) Types of waves.

size. High-energy (such as winter waves) flattens beaches, whereas large particles make beaches steeper. Beach erosion is most likely to occur from the short steep waves that occur from storms near the coasts, whereas beach deposition results from the long swells of distant storms. A usual pattern for coastlines in the mid-latitudes—for example, the Altantic Coast of the United States—is for beaches to erode during the winter months when more storms are near the shore and to accrete in summer.

As waves approach the shore, the front of the wave undergoes changes that are dependent on the configuration of the subaqueous topography and shoreline irregularities. When water depth is less than one-half the wave length the waves "feel bottom" and are refracted in the vertical plane in an attempt to conform to the bottom profile. This disruption causes asymmetry in the orbital, with inclination toward the land. When the depth is about 1.3 times the wave height, the velocity of water particle motion in the crest exceeds the velocity of the lower wave form. At this point the stability of the wave is broken, turbulence is initiated, and there is a landward thrust of water called **surf.** When this line of breakers is parallel with the land, plunging waves are formed; but if waves approach at an angle to the shore, spilling waves are created (Fig. 15–3). Any promontory projecting oceanward will receive frontal attack by

the waves, which also deform around the projection in refraction arcs. When the shore is somewhat straight and the incoming waves are oblique, the pressure of the water forces the already disrupted water to move as a current parallel with the beach. This river, in which one bank is the land and the other is the incoming ocean-pressure wave, is termed a **longshore (or littoral) current.** In special situations when the land is slightly irregular, or the ocean bottom is uneven because of bars, the continuity of the longshore current may be interrupted by **rip currents**, which are channelized flows of water that originated at the shore. Such water movements are hazardous to bathers who may become trapped in the strong oceanward-moving rip currents.

Representative beach landforms are depicted in Fig. 15–2. A submerged bar is formed at the position of breakers. The surf creates a thrust or runup of water on the foreshore (Figs. 15–4 and 15–5), which is termed **swash;** the seaward return of the water is the **backwash.** Displacement of beach materials in this zone is **beach drift** and, if the waves strike the beach at an angle, the materials have a vector of motion that moves the sediments in a "down-beach direction." When sediment becomes entrained in the water moving parallel to the beach, it is termed **longshore drift**. If the longshore currents predominantly move in one direction

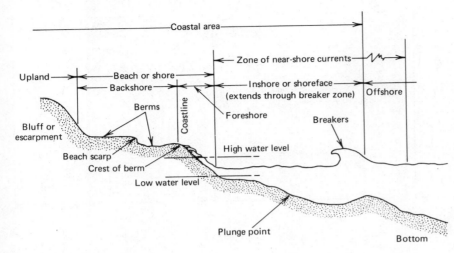

Figure 15-3 Terminology of beach features.

throughout the year, the volume of displaced sand that is transported as longshore drift may be as much as 460,000 m³/yr, which it is on the south shore of Long Island, or even greater.

Higher on the beach and just beyond the normal range of swash, a **beach scarp** may be developed that contains a treadlike upper surface. This combined feature is a **berm** and forms the division between the foreshore and the backshore. If the beach has strongly contrasting seasonal energy conditions, a winter berm will form at a higher position, whereas the summer berm will develop at a lower level during progradation. Further inland dunes may form at sites where wave action is generally minimal and wind is the dominant transporting process.

Unusual Storms and Tropical Cyclones

Our armoring of shorelines may minimize damage by normal coastal processes, but it is often insufficient to prevent destruction from storm waves and surge. The most spectacular and destructive forces are those associated with tropical cyclones (termed **hurricanes** in the Atlantic Ocean, **typhoons** in the Pacific Ocean, and **cyclones** in the Indian Ocean). These storms originate in the tropics and travel in an arc-shaped pattern with winds greater than 74 mph (118 km/h). Wave heights of 6 to 9 m are not unusual and produce water pileup and high water levels and surges on coastal lands. Such wind-driven storm surges killed 6000 persons in Galveston in 1900, and 1836 people were drowned in Florida in 1928 when waters of Lake Okeechobee were blown from their shallow basin onto adjacent lands. Tropical cyclones may also have an inland component where rains contribute to the damages. Hurricane Camille in 1969 was responsible for 248 deaths, 8000 injuries, and property losses of $1.4 billion. About one-half the deaths and 10 percent of the damage occurred inland from the intense rain that measured 787 mm and fell in five hours on already saturated soils.

Thus, the combination of high water, storm surge, high winds, and rain conspire to devastate coastal communities and human structures (Figs. 15–6 and 15–7). The most feared area for loss of life from tropical cyclones is Bangladesh where the Indian Ocean storm surges sweep out of the Bay of Bengal onto the low-lying coastal area with water depths of 9 m. In 1960 two storms killed 5149; 11,468 were killed in 1963; and 19,279 were killed in 1965. The November 12–13, 1970, storm was the worst ever, killing 300,000 and destroying 65 percent

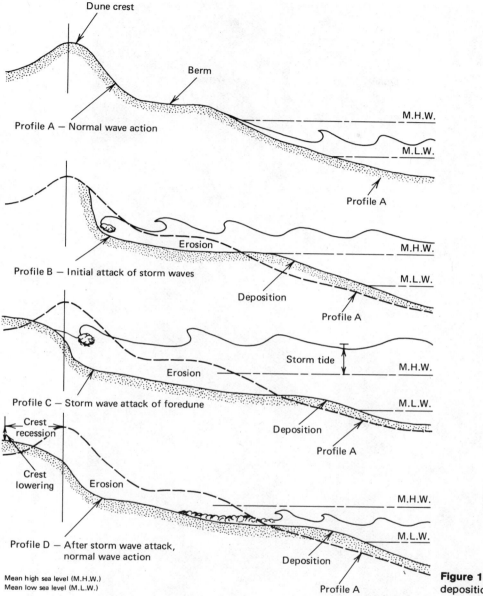

Dune crest

Berm

M.H.W.

M.L.W.

Profile A — Normal wave action

Profile A

Erosion

M.H.W.

Profile B — Initial attack of storm waves

M.L.W.

Deposition

Profile A

Storm tide

Erosion

M.H.W.

Profile C — Storm wave attack of foredune

M.L.W.

Deposition

Profile A

Crest recession

Crest lowering

Erosion

M.H.W.

M.L.W.

Profile D — After storm wave attack, normal wave action

Deposition

Profile A

Mean high sea level (M.H.W.)
Mean low sea level (M.L.W.)

Figure 15-4 Erosion and deposition of the beach zone.

of the total fishing capacity of the coastal region, thus seriously affecting the protein supply of the area. In addition, 280,000 cattled drowned, 400,000 houses were damaged or destroyed, and 99,000 fishing boats were lost. Japan is another country that sustains high losses from

tropical storms. During the 1945–61 period, nine damaging typhoons hit Japan, causing 15,861 deaths and destroying nearly 500,000 houses.

Significant damage to human installations and to beach landforms also can occur from

Figure 15-5 Directions of transport of sediment along a coast, shown by arrows. Finer particles are in suspension and, on occasion, coarser particles as bedload are transported parallel to the shore by longshore current. Dotted line shows path of a typical sand grain in the swash zone. The uprush of waves washes grains diagonally up the sloping beach; backwash carries them back at right angles to the shoreline. (Photo by A. Devaney, and courtesy Longwell, Flint, and Sanders, 1969.)

(a)

(b)

Figure 15-6 Barrier beaches of Long Island, New York. (Courtesy Fairchild photographs.) In September 1938, the first hurricane to hit New York and New England in decades did extensive damage to coastal communities and the barrier beaches. (*a*) Devastation in a settled community on the outer barrier beach. (*b*) One of the many new inlets cut through Fire Island barrier island on September 21, 1938.

other types of storms. In March 1962 a series of five high tides coincided with unusually high winds to cause more than $100 million damages to coastal communities in northeastern United States. The 1977–78 winter storms did extensive damages on both the east and west coasts

of the United States. In California 5 m waves coincided with midwinter tides that rose 3 m above mean sea level, causing extensive losses in numerous coastal communities. On Long Island, New York, and in Rhode Island, Connecticut, and Massachusetts, a "northeaster" combined with a blizzard to cause more than $500 million damages of which one-fifth were losses to shoreline installations. The tsunami, another ocean-bred catastrophe (Fig. 15–8), is discussed in Chapter 10.

EROSION

Emphasis in this section will be placed on coastal erosion, because we are generally not concerned with coastal deposition (except where it impedes navigation) and because sand is usually viewed as a resource and a benefit. Erosion is a natural phenomenon, and the power of the ocean is always working to create shoreline changes. It then becomes a question of what is the tolerable range of erosion, and what should be done when erosion is excessive and out of balance with the stabilizing depositional processes. Because sea level has been rising for several thousand years in most areas, causing new lands to be inundated, loss of land and erosion at new heights will be a continuing threat to mankind in the foreseeable future. Perhaps the best we can do is minimize erosion at critical sites and amortize the benefits of our investments before the sea can cause irreparable losses. However, coastal erosion is not the exclusive realm of natural processes; we can accelerate it when we build engineered structures or change by other means the coastal process system.

In 1971 the U.S. Corps of Engineers completed an evaluation of the 150,000 km coastline of the United States. Significant erosion is occurring along 33,000 km of coasts, and 4300 km of these are critical. It would cost $1.8 billion (1971 dollars) to provide some protection for the critical areas, with an annual $73 million expenditure to maintain them. Normal erosion

(a)

(b)

Figure 15-7 Results of a destructive northeaster that hit Cape Cod, Massachusetts, February 5, 1978. The aerial views (a) and (b) show summer homes knocked askew by waves that washed over barrier beaches on the east and Atlantic side of Cape Cod. Under normal conditions, there is a rather continuous barrier beach at this site, along which the houses were situated in a row. (Courtesy Dick Kelsey Airviews.)

rates in the eastern United States shorelines are 0.3 to 0.6 m/yr.

Louisiana is losing coastal land at a rate of 42.7 km²/yr. In the past 100 years erosion at Cape Hatteras has been 900 m, and of the 4800 km of shoreline on the Virginia part of Chesapeake Bay erosion removed 85 km² of land with 260 million m³ of material during the period of 1850–1950. Lands adjacent to the Great Lakes are not immune to erosion. In the 2.4 km long Lake Bluff area near Chicago, cliff recession was 80 m during the 1872–1975 period, of which 9.3 m occurred in the last 11 years of that period.

We can contribute to accelerated beach erosion in many different ways. On the West Coast

Figure 15-8   Demolition of the community at Lebak, Mindanao, Philippines, caused by a tsunami which resulted from the earthquake of August 16, 1976. (Courtesy Robert E. Wallace, U.S. Geol. Survey.)

time it reaches Miami—the remainder having been flushed out farther to sea at the dredged sites which allow magnified oceanward transport. A pass was dredged through Bolivar Peninsula in Texas. The pass changed the water circulation pattern so much so that in Galveston Bay and East Bay when Hurricane Carla hit in 1961 the new shore configuration allowed storm waves to erode more than 270 m from the peninsula which otherwise would have been naturally protected. The mining of 500,000 tons of offshore gravel near the English village of Hallsands in 1894 led to excessive damage to the village and beach from waves whose energy was no longer dampened by the former shallow underwater topography. There are additional ways in which we influence coastal erosion—for example, when we install groins, jetties, and breakwaters in the water or when we create artificial dunes and use vehicles on the beaches.

of the United States many rivers that formerly flowed freely to the ocean now contain flood control or water supply dams. Such structures have trapped more than 50 percent of the sediment that would normally be transported to the ocean and become available for replenishing the beaches. Rapid urbanization and paving in many cities also contributed to beach erosion by reducing the amount of sediment. Even if no new dams are constructed, these two sources— dams and urbanization—will continue to reduce another 30 percent of the sediment during the next 25 years. Silver Strand Beach in Southern California has no inland source for new sediment since Rodriques Dam was completed in 1937. During the 1941 to 1967 period the beach was artificially maintained by emplacement of 22 million m³ of sand.

Extraction of sediments for dredging or mining can also produce erosional side effects. There are 17 dredged inlets along the east coast of Florida, and they upset the normal coastal system. Littoral drift at Jacksonville is about 382,000 m³ but only 7600 m³ survive by the

GEOMORPHIC ENGINEERING OF SHORELINES

There are two results of coastal engineering projects—either material accumulates in an undesired area, or erosion occurs at a site where protection is needed. These unwanted conditions can be natural or man-induced. When changes must be made for some perceived benefit, the geomorphic engineer should take every precaution to assure that (1) the structures that are built are necessary and will accomplish their intended purposes; (2) construction is located at the optimum site that will cause minimum environmental disturbance; and (3) planning and management have accounted for environmental feedback on contiguous lands and waters. There are at least four categories of problems that must be faced by the geomorphic engineer whose purpose it is to protect human installations by altering natural processes. These include protection of the shoreline, the backshore, the harbor, and stabilization of inlets. Such a division is needed because of differences in the environmental setting and character of the coastal processes and the type of structural solutions required by each category.

Eroding beaches endanger near-shore structures and decrease the recreational value of the locality. The shoreline stabilization strategy is to maintain a constant beach width by erosion prevention, entrapment of littoral material, or by artificial nourishment of new materials. The choice of technique is determined by the sediment supply, local hydraulic conditions, and societal pressures.

Most structures in the coastal zone are located in the backshore area. The goal of backshore protection is to shield the site from destruction by abnormal storm surge. The engineer may choose to armor the valued property by structures, to increase the runup width of the beach by artificial nourishment, or to erect sand dunes.

Many coasts are characterized by offshore barrier islands that are separated from the mainland by lagoons and bays. Waterways that pass through or between islands are called **inlets** and are necessary as navigational channels and to allow dual circulation between the bay and ocean. Under natural conditions inlets constantly change, so to prevent such change we must attempt to stabilize the passage by constructing jetties and breakwaters, or by dredging and sand bypassing. Harbor protection is not only aimed at the prevention of shoaling, but also includes the maintenance of calm water conditions for shelter vessels and docking facilities.

Seawalls, Revetments, and Bulkheads

These structures are protective devices that armor the shore in an attempt to prevent direct wave attack on the beach or installations. They are erected when there is little or no beach as a last defense against destruction from ocean forces. No attempt is made to modify coastal processes and, at best, such structures can be considered only temporary expedients and should be combined with other types of protection whenever possible—**they are not permanent solutions**.

Seawalls, revetments, and *bulkheads* are structures emplaced at the land-sea interface, but they differ in function and shape. **Seawalls** are generally used to attenuate wave energy and therefore are larger and more massive than revetments and bulkheads. They have three basic shapes: vertical, concave, and sloping. Vertical walls permit use as a docking facility but resist wave attack poorly. Sloping walls dissipate wave energy, allow easy access to the beach, but can be easily overtopped by wave action. Concave structures are used when high-energy waves are more common and structural strength is required. A major defect in all these devices—seawalls, revetments, and bulkheads—is the increased erosion that generally develops in front of them. Reflected wave energy is concentrated at the foot and downdrift from the structure where sediment becomes more easily removed. This then requires the extension of the structures to the new erosion site. The toe erosion results in a steepening of the seabed profile and may lead to undermining of the structure in time.

Bulkheads do not require as much structural bulk as seawalls because they primarily serve as shoreline retainers of fill. They are vertical structures that are composed of steel, timber, or concrete pilings but are highly susceptible to erosion and can easily be undermined by waters that overtop the structure. Revetments conform to the contour of the shore and generally consist of interlocking stone or concrete. This armored slope dissipates wave energy, with less damaging effect on the beach than waves striking vertical walls. They can be effective in retarding the erosion produced by small waves and currents. Revetments installed in 1960 on Jupiter Island, Florida, proved to be particularly protective. Construction was completed only two weeks before Hurricane Donna hit the coast and the structures were undamaged. Again in March 1962, heavy seas caused severe erosion to unprotected beached and seawalls in the area (Fig. 15–9), but the revetments escaped without damage. In one 36 m section there was sand accretion of 1900 m³. The sloping walls had allowed wave runup, but subsiding water deposited the sand. The underlying filters eliminated hydrostatic pressure buildup and served to retain some surficial sand.

Jetties

Jetties are long, narrow damlike structures used to prevent (1) shoaling of inlets or harbors, (2) inlet migration, and (3) formation of tidal deltas. Many factors must be considered to design the type, size, and jetty placement for maximum benefit. Jetties may be constructed singly or in pairs, are usually perpendicular to the shore, and consist of steel, timber, concrete, or boulders. The angle the waves and littoral currents make with the shore and the quantity of drift determine jetty length and angle. Tidal flow must be sufficiently rapid to flush sediment from the inlet, but the flow direction should be diverted from areas where erosion would damage valuable property.

Environmental problems invariably result from jetty construction because of impounded sediment on the updrift side of longshore currents. The inlets which jetties are supposed to maintain provide an additional locale for sediment loss. Thus, the balance of the littoral drift process has been disrupted, and downdrift areas may experience severe erosion because the longshore currents are sediment-deficient and can now apply stress to these areas.

South Cape May [New Jersey] has virtually disappeared during the past 50 years due to the jetties to the northeast and to the southerly current. Assateague Island has been eroded at least 1500 feet because of the Ocean City jetties [Shepard and Wanless, 1971, p. 548].

McCormick (1973) has attributed increased ero-

(a)

(b)

(c)

Figure 15-9 Jupiter Island, Florida. (All photos courtesy of Carthage Mills.) These pictures catalog the destruction caused by Hurricane Donna which hit the area in 1960. (*a*) Erosion of cliffs at homesite caused by coastal processes prior to 1960. As a result, revetments were installed on this part of the beach. (*b*) This revetment, emplaced prior to the hurricane, was undamaged (photo taken after the storm). The checkerboard structure is composed of heavy precast interlocking concrete blocks that are laid on crushed rock with a polyvinylidene-chloride-resin monofilament cloth of woven yarn. The fabric is designed to prevent particles from passing through but is sufficiently porous to allow water penetration. (*c*) This view should be compared with (*b*). The seawall was greatly damaged by the hurricane waves and undermined in many spots, as shown here.

sion rates for part of the Long Island barrier beaches to artificial inlet stabilization by means of jetties. Erosion rate of the barrier beach east of Fire Island was 0.5 m/yr prior to inlet stabilization; thereafter the recession rate of the beach jumped to 2 m/yr. Most of the material lost 107,000 m³/yr, accumulated in tidal deltas associated with the inlets. One technique that can be used to help remedy this problem is sand bypassing.

Groins

Groins are generally smaller than jetties, but they both generally extend from the foreshore into the breaker zone for the purpose of changing the character of the coastal process. They are built in attempts to: (1) stabilize the beach, (2) reduce the rate of littoral drift, (3) widen the beach, (4) prevent loss of material from the beach, and (5) prevent accretion in downdrift areas.

Groins can be constructed of different materials, but their size specifications and orientation should be designed for the particular environmental setting. A groin length of 0.4 the distance from the shoreline to the breaking point of plunging breakers will usually produce minimal scour with maximum deposition in the downdrift direction. If groins are too long, sand may move around them to an offshore site and be lost to the littoral system. The height of groins should not exceed 1 m above mean high tide. Most groins are impermeable but, when designed to be permeable, the downdrift effects are reduced because some sediment can filter through the structure. Although most groins are installed perpendicular to the shoreline, under conditions of exceptionally strong littoral currents groins emplaced at angles of 110 to 120° in the direction of transport have proved less harmful to downdrift areas.

Because groins often cause erosion in a downdrift direction, this can lead to the development of additional groins. Such a series of groins is called a **groin field** (Fig. 15–10). Groins

Figure 15-10 Extensive groin and unique attached breakwater system along Sochi coast, Black Sea, U.S.S.R. Longshore drift is from the northwest (top right), with accumulation on updrift side of groins. Attached to the ends of these groins offshore in 5 m of water are submerged breakwaters to dampen wave energy. (Courtesy John J. Fisher.)

are commonly employed along coasts where there is not a high sediment supply. The groin locality may be artificially filled with sand so that any new material in the littoral drift is not trapped but instead continues movement along the shore causing no downdrift erosion. Thus, groins may aid to stabilize the artificial fill that may have eroded without structural support.

In Chesapeake Bay in Northhampton County, Virginia, five small groins were installed when the beach was 7.5 m wide. Within nine months the downdrift shoreline eroded 6 m. At a nearby site a property owner installed a series of plastic bags (filled with aggregate) in June 1974. Within three weeks the area between the sills, which had been formerly eroding, increased from 1.5 to 20 m with the addition of 600 m³, and the back part of the beach heightened 1.2 m. A severe storm on March 16, 1976, with water levels 1.2 m above mean high water and winds in excess of 100 km/h hit the area. The section with groins experienced erosion of 20 m and unprotected cliffs in the area eroded 3.6 m, but water did not reach closer than 6 m to the shoreline protected by the plastic bag sill.

Under certain conditions groins may provide short-term protection at a specific site, but they invariably produce deleterious feedback to adjacent areas. They have proved ineffective in

(a)

(a)

(b)

Figure 15-11 Beach conditions at Ocean Beach, Fire Island, New York. (a) One of two groins installed to protect the water tower; view is south. Note sediment buildup on the left, or updrift side of longshore currents, and lack of sediment on the right, or downdrift side. (b) View of groin looking east. The wood seawall was built to protect property downdrift from the groins. However, accelerated erosion by the sediment-starved current has partly destroyed the wall.

stopping beach erosion or trapping sand at Miami Beach, have tripled erosion rates in the downdrift of Westhampton Beach and Ocean Beach on Long Island, and failed to impede erosion at Lake Bluff, Illinois (Figs. 15–11 and 15–12).

(b)

(c)

Figure 15-12 Views of Miami Beach, Florida. (a) Photograph taken in November, 1974 showing the inadequacy of the groin field to prevent sand transport and the complete lack of sediment on the beach. (b) and (c) Artificial nourishment of the beach south of "hotel row" looking northward. (Courtesy Miami Beach Tourist Development Authority.)

Breakwaters

Shore features can be protected by an offshore breakwater if waves are dampened by an offshore barrier. However, massive breakwaters are more costly than onshore structures, so they are generally constructed for navigation and harbor purposes. Breakwaters have both beneficial and detrimental effects on the shore. They

Figure 15-13 Plastic bag filled with concrete breakwater on Lake Erie. (Courtesy Prepakt Concrete Co.)

Figure 15-14 Breakwater of floating used tires for marina in Florida. (Courtesy Florida Sea Grant Program.)

can reduce or eliminate erosion, but by interrupting the free movement of sand they can starve downdrift beaches and cause shoaling in undesirable places. A breakwater built at Santa Barbara, California, in 1929 to protect the harbor contributed to erosion of downdrift beaches for a distance of 16 km, amounting to 73 m in some areas.

Most breakwaters are constructed of concrete or the riprap of giant rocks, forming a permanently emplaced installation. However, when more modest costs are involved, greater flexibility needed, and less protection required, other types of breakwaters are possible. One method involves the use of flexible plastic bags (Fig. 15–13) that can be emplaced at only a fraction of the cost of rigid breakwaters. A variety of floating breakwaters are now in use and, although still in the experimental stage, have proved to be less costly. Concrete pontoons strapped together in modules of 7 m by 18 m have been used as a floating breakwater to protect fishing vessels in the Alaskan villages of Tenakee and Sitka. Floating tires as breakwaters have been used successfully in Rhode Island, Florida, Massachusetts, and on Lake Erie (Fig. 15–14). Such structures typically consist of several segmented units, about 5 m wide and 150 m long, that are tied on shore and pushed into the water and moored. Tires can be positioned vertically with only the top above water, which holds air and thus eliminates the need for expensive flotation structures. This design has

been successful in dissipating more than 60 percent of wave energy and decreasing heights to 1.5 m. Since the United States discards 200 million scrap tires each year, there should be a plentiful supply for construction of numerous and modest-sized breakwaters. An advantage of floating breakwaters is the ability to move them and change their characteristics to meet differing needs of the seasons. However, they can never replace permanent-type installations that are required for protection of major facilities and harbors. The large breakwaters are often designed as total systems and combined with jetties and dredging facilities for removal of shoals whose sediments are bypassed to nourish downdrift beaches.

Sand Dunes

Sand dunes are formed naturally by eolian deposition of sand blown off the berm by onshore winds. The sloping beach and berm are the outer defense areas that absorb most of the wave energy, whereas dunes are the last line of defense for inland property against storm surge. Although dunes may be periodically breached or destroyed by exceptional waves, if given sufficient time they will be reestablished by natural processes. However, humans are impatient and cannot wait for time to heal the wounds; therefore we embark on programs to

Figure 15-16 Cape Hatteras Lighthouse, North Carolina, the highest in the country. Severe erosion occurs at this outmost point of the land. Three groins have been installed as protection and the duneline has been artificially heightened and vegetated.

Figure 15-15 Slatted snow fences, Fire Island, New York, for the creation of man-induced dunes. The zig-zag pattern provides greater stability and a wider dune. When sand has largely covered the fencing additions, fences will be emplaced at higher levels to increase dune height.

create artificial dunes (Figs. 15–15 and 15–16). Furthermore, artificially constructed dunes, especially when properly vegetated, can offer more aesthetic vistas than seawalls or bulkheads, for the protection of backshore property.

In the 1936–40 period the National Park Service erected nearly 1000 km of sand fencing to create a nearly continuous barrier dune system along the Outer Banks of North Carolina, especially on Bodie and Hatteras islands (Fig. 15–17). Most construction was in a zone comprising the original low beach dunes and along a strip about 90 m wide behind the foredune. The sand that collected around the fencing was further stabilized with 2.5 million trees and shrubs and grass that covered 3254 acres (1316

ha) of dunes and sand flats (Dolan, 1973). Additional work in the 1950s increased the size and scope of the project so that a nearly continuous vegetative and dune mat extends from Nags Head to the south tip of Ocracoke. Islands south of Ocracoke were left in their natural state. Dolan (1973) and Godfrey and Godfrey (1973) have reported the effects of this human stabilization program. Whereas the unaltered beaches, as at Core Banks, range in width from 120 to 200 m, those on Hatteras and Bodie islands with artificial dunes have narrowed to about 30 m (Fig. 15–18). On Ocracoke Island, which was not altered until the 1950s, intermediate beach widths have occurred averaging about 70 m. Figures 15–19 and 15–20 compare altered beaches with natural beaches. This beach-narrowing process combined with permanent barrier dunes has created conditions of high wave energy that are concentrated in an increasingly restricted runup distance. This results in a steeper beach profile, increased water turbulence, and accelerated sediment attrition. The net effect produces further beach narrowing to a point whereby the wave uprush directly attacks the dune face, undercutting it and endangering man-made structures. Hurricane Ginger, with 5 m waves and a storm surge 2.4 m above mean sea level, hit the Outer Banks on September 30, 1971. Although the seas inundated much of Core Banks, negligible beach

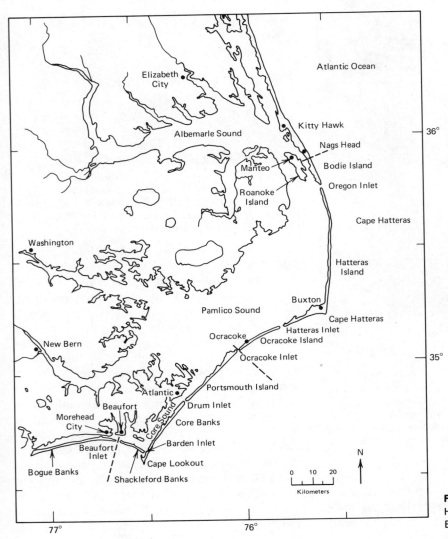

Figure 15-17   Map of Cape Hatteras region and Outer Banks, North Carolina.

recession occurred with the storm's passing. At Cape Hatteras where the dunes absorbed the full force of the storm surge, the duneline was eroded 3 to 4.5 m. Under certain environmental conditions it is possible that construction of artificial dunes may not increase the natural erosion rate. Studies by the Corps of Engineers on Padre Island, Texas, show man-made dunes at this locality have not caused accelerated erosion.

Beach Nourishment

Under the proper set of conditions beach structures are important and necessary in protecting human property. However, the most environmentally effective measures are those which simulate natural processes and landforms. The artificial nourishment of beaches provides a nonstructural alternative to coastal protection whereby we can work with, rather than against,

(a)

(b)

Figure 15-18 Aerial views of Outer Banks, North Carolina, showing comparison of natural beaches with those altered by man. (a) Central Core Banks south of Drum Inlet. This is a typical unstabilized and natural barrier island. Features are a wide berm; low, open dune zone; grasslands of varying density on dunes and overwash terrace; scattered shrubs and thickets; salt marshes that formed on tidal deltas and overwash deposits. (b) Eastern part of Ocracoke Island, looking south toward Ocracoke Village. This is a typical humanly altered, "stabilized" barrier island. Main features are the narrow beach; the single, man-made continuous dune; grassland vegetation directly behind the dune (in dense bands where fertilized), with advancing shrub lands; and the road. Old overwash terraces still visible behind the road. Beyond are salt marshes. The large pit in the salt marsh is a borrow area for sand, now being used to build a secondary dike along the road. (Godfrey and Godfrey, 1973.)

nature (Fig. 15–12b and 15–12c). The theory is that wide, high beaches offer the greatest deterrent to shoreline erosion. An important advantage of this method is that it seeks to cure the underlying cause of erosion—insufficient sediment supply. The downdrift coast also ben-

efits from the nourishment, unlike structural methods that usually produce deleterious erosional feedback in these areas. A number of other benefits can occur from a beach nourishment program: (1) the beaches may become suitable for recreational purposes; (2) the cost is invariably much less than structural methods; (3) the method does not involve a long-term management commitment; and (4) it is aesthetically pleasing.

Because beach nourishment instantly creates a new shoreline topography, it becomes subjected to coastal processes that must be evaluated in geomorphic and sedimentologic terms if the fill is to remain stable and perform its assigned task. It is vital that the new material which is mechanically or hydraulically introduced to the new beach be of similar or even slightly larger median grain size than the original beach sediments. Knowledge of wave conditions is necessary to calculate the width and height characteristics for a new and stable artificial berm. The berm width should generally be sufficiently wide to allow reacreational space and to provide sufficient distance for the runup of large waves to protect near-shore property. According to W. H. Vesper, three beach nourishment programs undertaken by the Corps of Engineers at Seaside Park, Prospect Beach, and Sherwood Island State Park (all in Connecticut) were all highly successful. Each of these beaches was nourished with about 380,000 m³ of sand. The beaches averaged 2250 m in length and cost about $24 per linear meter (1967 dollars). Annual sediment losses average about 11,400 m³, so that yearly maintenance of about $2 per linear meter is required.

Beach sand is a rapidly diminishing resource. Much of the beach sand in the coastal zone is an inheritance of lower water stands of the ocean during the Pleistocene glaciation. It is now being lost with the rising sea level. For example, in the New York Bight area, longshore currents carry material westward on the New York side and northward on the New Jersey side where it eventually is completely lost and

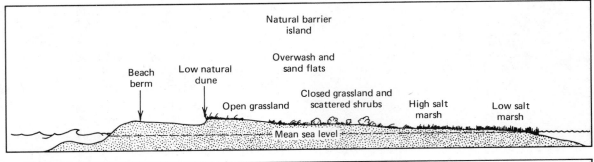

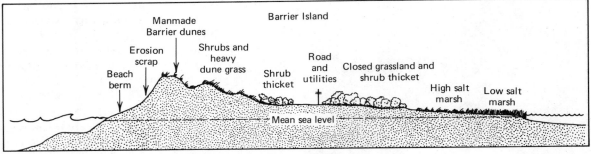

Figure 15-19 Cross sections of the two systems of barrier islands, natural and stabilized (or man-altered). Differences occur in the width of the beach berm and in the beach slopes which produce increased erosion of the stabilized island. (Dolan, 1973.)

deposited in the Hudson River Canyon. Thus another method for the preservation of beaches, as well as sand, is by means of sand **bypassing** or **forepassing.** When sediment supply is seriously depleted downdrift from a littoral obstruction such as an inlet, harbor, or breakwater, sand can by hydraulically removed from the sediment trap and reinjected into the system by placing it on the deprived beach. This maintains a constant source of material for action by coastal processes. Sand bypassing is the movement of this material in a downdrift direction, whereas sand forepassing is the man-induced transportation of the material in an updrift direction. Sand forepassing has the advantage of recycling the sediment and keeping it within the same system, thus causing the natural processes to rework the same materials. Such a conveyor-belt method can aid in prevention of

permanent loss of the material. Because the transfer operation must largely be a permanent installation built at substantial cost, sand bypassing is generally only used at major inlet obstructions. It has proved successful at Moriches and Fire Island inlets on Long Island and South Lake Worth inlet in Florida.

A variety of other strategies have been used to structurally protect and stabilize coastlines. In 1971 the Singapore Housing Board began reclaiming land along a beach, 8 km long and 180 to 660 m wide. To protect the reclaimed land, a revetment had first been planned. After it was partially built, wave action severely undermined it. An alternate scheme was designed with the construction of artificial headlands placed at about 240 m intervals and extended seaward 60 m. To protect the headlands two types of rockfill structures were used, gabions

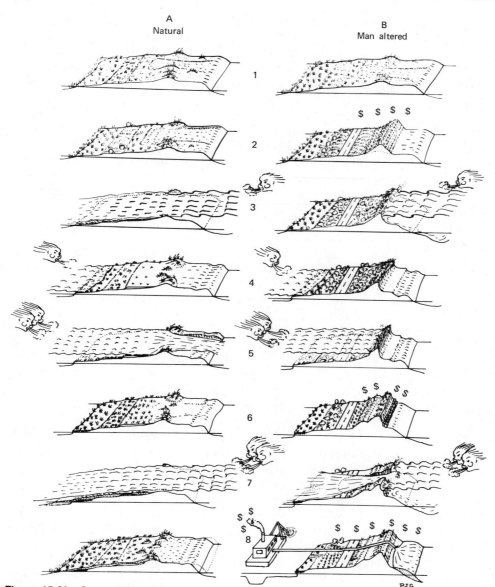

A
Natural

B
Man altered

Figure 15-20 Comparison of natural and man-altered barrier systems. 1A and 1B are alike; 2A—no change in the natural system; 2B—Construction of artificial dune to protect road; 3A—Storm overwash; 3B—Erosion of dune and foreshore; 4A—Overwash terrace recolonized by vegetation; 4B—Severely eroded dunes; 5A—Waves from sound cause overwash; 5B—Sound storm tide erodes inland side of dune; 6A—Natural barrier undamaged by storm; 6B—Human interference is renewed; 7A—Violent storm overwash but no permanent damage; 7B—Same storm produces a disaster; 8A—New stability is established; 8B—At great cost artificial beach nourishment attempts to repair narrowed beach.

and riprap. The gabions were 2 m³, extended 0.3 m above mean high sea level, and were 30 m long. The gabions produced a savings of more than 50 percent what the revetment would have cost and was 25 percent cheaper than riprap. Results of this plan have proved successful with sand filling the areas behind the headlands so that the beach now presents a scalloped stabilized coastline.

COASTAL RECLAMATION

Since early history man has created new lands from the oceans by various reclamation strategies. During Caesar's rule of Britian the drainage of coastal marshes was initiated and, by 1885, nearly 20,000 ha had been reclaimed in the Romney Marsh area of southeast England. The largest coastal engineering projects ever undertaken have been accomplished by the Dutch, where in the Netherlands nearly one-third of the 39,000 km² of the country is the hard-won prize of coastal and estuarine waters. Such works can be traced back to the first Dutch farms about 4000 B.C. By the year 1860 the Dutch had accomplished the following: (1) movement of 76 million m³ into small hills, (2) movement of 152 million m³ to build 2800 km of working dikes and another 38 million m³ for dikes later abandoned, (3) excavation of 611 million m³ for draining of the land, (4) excavation of 152 million m³ for canals, and (5) removal of 7 million m³ of peat. Such figures are staggering when compared to the largest pyramid built by the measuring Egyptians, 2.7 million m³, or the excavation of the Suez Canal, requiring removal of 76 million m³. The conversion of the Zuider Zee into the Ijsselmeer when completed in the 1980s will add another 2200 km² of new land. A still larger plan, The Delta Scheme, was initiated after the disastrous coastal flood of 1953 broke many existing dikes, flooded 152,000 ha, killed 1800 people, and crippled agriculture by spreading silt and salt over the flooded lands. The serious reclamation of the Netherlands started on a massive scale about A.D. 900. With much of the land below sea level, as much as 2.5 m, the method used was to cut off small areas by dikes and then pump out the water using windmills. These diked parcels, or **polders**, contain an elaborate network of drainage canals. By 1920 the country had 1600 km of sea dikes, and the entire change in the environmental setting was the largest engineering feat accomplished in the world.

Wetlands

Coastal wetlands are finally being recognized as valuable resources and environments that should remain untouched by man. They provide breeding, nesting, and feeding grounds for many forms of wildlife, waterfowl, and shore birds as well as cover to escape predators. They protect inland areas from coastal storm damage and provide countless recreational opportunities, for open space and aesthetic appreciation. The direct economic benefits from the Great South Bay wetland area of Long Island are more than $150 million annually, and the indirect economic values would double that figure. This reversal in public attitude is none too soon because coastal wetlands had usually been considered as unsightly, unhealthy, and worthless. The consensus was to get rid of them as quickly as possible—drain them, fill them, and put them to immediate use. Toward this end developers, real estate agents, and commercial enterprises joined in the conspiracy to demolish all such areas or to alter them for human use and economic gain. In the period 1933–36, New Jersey, Delaware, and Maryland drained more than 200,000 ha of tidal marshes for mosquito control (Fig. 15–21). In the 1954–68 period, New York and Connecticut drained 25 percent of their remaining wetlands. Between 1950–69, 260,000 ha were lost by dredging and filling, and one-fourth of the 500,000 ha designated as shellfish areas were polluted.

Figure 15-21 Drainage ditching for mosquito control and land reclamation, Pamlico Sound, North Carolina. This activity results in a loss of valuable estuarine habitat. (Courtesy of The Conservation Foundation.)

Figure 15-23 Collington Harbor, North Carolina. Typical situation where there is loss of a valuable estuarine habitat to provide solid ground for development and deep water for navigation. (Courtesy The Conservation Foundation, M. Fahay, photographer.)

Of course, the conversion of wetlands into highly developed areas, and even cities (Figs. 15–22 and 15–23), has produced economic benefits in many cases and has been necessary for the growth of many communities. However, it is time that limits are placed on unwarranted and heedless expansion. Rather belatedly, the Corps of Engineers has started to take a harder

Figure 15-22 Miami, Florida, with near view of the new Miamarina and the port facilities. This illustrates the metropolitan development of what formerly was a wetland environment. Extensive dredging, filling, landscaping, and construction have fashioned the area into one of high economic production. (Miami-Metro Dept. of Publicity & Tourism.)

look at its regulatory powers and is moving into a position as protector of wetlands. In a similar manner, many states now have laws that are highly protective of these areas, such as Connecticut, Massachusetts, and New York. Although the Corps of Engineers has had the power to control the dredging of coastal wetlands since 1899 under the Rivers and Harbors Act, it was not used for denial of developmental purposes until the classic Zabel-Russell case of 1967 (which denial was upheld by the U.S. Fifth District Court in 1970). In another important precedent the Corps denied two dredge and fill permits requested by the Deltona Corporation at its Marco Island property on the southwest coast of Florida. Deltona had previously sold 4000 coastal lots at average prices of $13,400 each. Such development would have destroyed the natural wetlands along with 840 ha of mangroves and hundreds of acres of grass-covered bay bottom which would have been used as fill. Under their current regulatory powers the Corps of Engineers has jurisdiction over wetlands construction that would involve dredging or filling below the mean high water line as well as those wetlands above that line that are periodically inundated by saline or brackish water and are characterized by salt-tolerant vegetation.

In New York, public ownership of wetlands dates back nearly 300 years under orders issued by Governor Dongan in the 1680s. He authorized that the "Freeholders and Commonality" be provided the unappropriated lands that were to be held " . . . for the use, benefit and behoof" of the beneficiaries of the trust who are residents, taxpayers, and heirs. The New York Tidal Wetlands Act of 1973 is the modern equivalent of the common law and is typical of similar acts passed by other states. All areas that fall within the classification of wetlands, which are usually defined in terms of water line elevations and vegetation index, are mapped and then protected under the law. Anyone wanting to alter such an area must file a permit with appropriate justifications; this permit is then reviewed by the New York State Department of Environmental Conservation. Most states mandate fines, as in Connecticut for $1000 a day until rectified, for those who unlawfully change wetlands and require the perpetrator to restore the site.

OTHER HUMAN CHANGES IN THE COASTAL CORRIDOR

We introduce change into coastal systems in a variety of other ways. The construction of buildings and roads alters the ecosystem of the area and generally reduces flora and fauna. To gain access to the beach, cuts are made in the duneline which later may become the locus for overwash during storm surge. In many localities it has been common practice for homeowners to cut away the tops of dunes so they will have an unobstructed view of the ocean. Recreationalists trample the dune and beach vegetation which allows for easier sediment removal by wind and water. One topic of concern that has only recently started to receive attention is the character of changes to the beach-dune system that are produced by the use of vehicles.

Vehicle Impacts

Unrestricted use of vehicles can cause deterioration of the beach-dune landform system in certain coastal environments. On Fire Island, New York, it has been demonstrated that vehicular traffic damages this environment in several ways.

1 Vehicle pressure indents the sand surface, and the displaced material invariably moves oceanward (Figs. 15–24 and 15–25). The displaced material or sand splay has a larger oceanward component for most turning ruts as well as for the rutted tracks that are parallel to the beach.

2 The ruts provide avenues that accelerate the movement of water. The swash that is

Figure 15-26 Fire Island, New York. Vehicle tracks resulting from travel at the duneline during high water. Such rutting aids in undercutting the dunes and causes accelerated, man-induced erosion.

Figure 15-24 Tracks produced by a turning vehicle, Fire Island, New York. Note the increased amount of sand splay on the ocean side. Atlantic Ocean in background.

Figure 15-27 Effects of vehicle use at Fire Island, New York. At this site the duneline is retreating three times faster than in areas of little vehicle use, or where dunes are protected from vehicle intrusion.

Figure 15-25 Tracks produced by vehicles traveling parallel to the waterline, Fire Island, New York. Note increased amount of sand splay on the ocean side and the movement of the splay ridge toward the ocean as erasure and filling of the track occur.

moving upslope develops greater turbulence when passing over the tracks, and the backwash uses them to channelize flow and to cause entrainment of sediment. Even erasure of the tracks has an oceanward movement component (Fig. 15–25).

3 In the Fire Island situation, vehicles travel high on the beach during high water and undercut the stability of the dunes (Figs. 15–26 and 15–27). In places this accelerated erosion is producing retreat of the duneline several times faster than natural erosion in untraveled areas. To protect the duneline from this invasion some property owners in the 20 villages of Fire Island

have erected a variety of structures to prevent traffic (Fig. 15–28).

4 Vehicles have also wantonly cut through the duneline (Fig. 15–29). This lowering of the dunes and the crossovers maintained for normal beach travel become sites for attack by storm surge and may produce a washover. Of the 95 washovers that broke through the primary duneline on Fire Island during the March 1962 Ash Wednesday storm, 55 occurred at positions where humans had interfered with the dunes.

Vehicles produced other changes, as when the splay billows into the air, and silt-size particles can be blown away with winds that exceed 9 mph (14 km/h; a figure that is commonly

Figure 15-28 Wooden piles emplaced at base of duneline to prevent vehicles from using the area, Fire Island, New York.

Figure 15-29 Primary duneline, Fire Island, New York. Vehicles have deliberately cut through the duneline in the background, causing its lowering. Such localities become washover sites during times of storm surge.

reached in the coastal environment). Vehicle effects also occur inland from the duneline when operators travel on sandy roads, become mired in sand, and take alternate routes around the soft sand. This destroys the vegetative mat and leads to the creation of blowouts.

For a typical Fire Island beach, our studies have shown that a vehicle traveling 48 km/h causes a seaward displacement of 1.3 m³ of sand for each kilometer traveled. The amount of sand moved is a function of the type of vehicle, vehicle speed, slope of the beach, and number of travelers on the same track. However, all indications to date show that vehicles have changed beach profiles and have caused deleterious changes on Fire Island beaches.

Pollution

The vastness of the oceans has given civilization a false sense of security. Mankind seems to think that because the oceans are so large that they can sustain a never-ending quantity of pollutants from industrialized society. Most coastal communities, and even inland populace, are to blame for the ever-increasing amount of pollution of all varieties that pours or is placed into the coastal waters.

Sewage effluent amounts to 2 percent of the discharge of all the world's rivers that finds its way to the oceans. Of the total solids in the river, 7 percent is man-made waste products. Solid waste now covers about 210 km² of New York Harbor. Such cities as Miami, which discharges 50 million tons of untreated sewage daily into the Atlantic, and Venice, which also releases untreated waste into the Mediterranean Sea, jeopardize the health of inhabitants and reduce tourist trade. The spatial distribution of DDT in zooplankton is exhibited as hot spots along the California coast, from north of San Francisco to 600 km south along the coast of Baja California. Australia also has pollution from agricultural sources. Oil spillage averages 2.4 million tons a year, and plastic is the ocean's most common flotsam, with 3600 pieces per square kilometer. Even the South Sea Islands are not free of the world's contaminants (see Fig. 20–6). Furthermore, sewage and industrial contamination have caused death and health impairment to humans and life in the sea. Mercury poisoning has killed people in Japan, and the incidence of hepatitis in coastal communities is linked to the eating of contaminated mollusks. Nitrates and phosphates contribute to the red tides, and the entire animal food chain can be seriously infected by the waste products of mankind.

Another type of change in coastal waters is produced by electric generating plants. Fossil fuel plants require the mechanical equivalent of 1.4 W of waste heat (coolant) for each watt of power generated, and nuclear plants need 2.1

W. For example, the power needed by California in 1980 will be 37,000 MW. The necessary coolant will be used at a rate of 1.2 calories/sec, which is the equivalent of heating a flow of seawater 15,000 m³/sec by 1°C. Such a flow would equal one-half the average flow of water over the California continental shelf or 3 percent the total flow of the California current. Such heat imbalances will produce ecological changes in the life of the sea.

COASTAL ZONE MANAGEMENT

Throughout most of the world, and certainly in the United States, it has been only recently that nations and states have started to take concerted action for the establishment of long-range objectives in the management of the coastal corridor. Such attitudes as "man against the sea" and "every man for himself" have been the rule not the exception. The famous decision rendered by Lord Tentreden in an English court in 1820 epitomizes this view. A landowner brought suit because a town-constructed groin caused erosion on his coastal property. The court ruled that, "Each landowner may erect such defenses for the land under their care as the necessity of the case requires, leaving it to others, in like manner, to protect themselves against the common enemy." This philosphy that nature is the enemy to be subdued and conquered is a recurring theme throughout the engineering literature and is witnessed in the construction projects that have been built or proposed. Instead of adopting the combative role unilaterally, it is being increasingly recognized that whenever possible it is preferable to work with the natural processes.

It is now important to draw attention to the changing attitudes in the United States, from those of local and piecemeal projects to the concerted efforts by regional, state, and federal agencies. Prior to 1930 the federal government only performed coastal works for the protection of federal property or for the improvement of navigation. In 1930 the Beach Erosion Board was authorized by Congress to undertake studies of beach erosion problems when so requested by other governmental levels. By the time the Coastal Engineering Research Center (CERC, U.S. Army) was established in 1963, the Corps of Engineers had been given much broader mandates by Congress which included study and protection of public beaches (in 1946), and of private beaches if instrumental to the safety of public shores or for public benefits (in 1956). In the years 1946–64, the Corps undertook more than 100 projects for the prevention of beach erosion at a cost of $180 million. In 1968 a study was initiated for the evaluation of erosion problems on all U.S. Coasts and resulted in a series of publications called the National Shoreline Study (including *Shore Management Guidelines, Regional Inventory Reports,* and *Shore Protection Guidelines,* 1971). Recently many regional areas have initiated their own studies. For example on Long Island, New York, the Nassau-Suffolk Regional Planning Board instituted a Regional Marine Resources Council that has evaluated six major problems of coastlines, including wetland management, shore erosion and stabilization, water use, and resource development.

Two federal programs are proving especially productive in terms of both immediate payoffs and in long-term benefits. The National Sea Grant Program was established by an act of Congress in 1966 whereby states that border the oceans or Great Lakes could apply for federal funds to undertake coastal work and research. The program is run by the U.S. Department of Commerce (through NOAA) and is patterned after the Land Grant College idea. Thus, a university can achieve status as a Sea Grant College when, after a period of funding for three years, the institution has demonstrated competency in its coastal program and has been responsive to the public welfare in providing services to the people. The multiple aims of research, education, and commercial interests are mandated and coordinated by the program guidelines.

In 1972 Congress passed the Coastal Zone Mangement Act (CZMA) to provide incentives and assistance to worthy programs carried on by the coastal states and territories. The preamble to the act states:

The Congress finds that the Coastal Zone is rich in a variety of natural, commercial, recreational, industrial, and aesthetic resources of immediate and potential value to the present and future well-being of the nation.

The act exhibits the dual recognition that the coasts are a key part of the national interests but that the states and local units of government should have the primary responsibility for management of coastal resources. The principal incentive for cooperation is federal funding that is provided on a cost-sharing basis for planning and implementation of a federally certified coastal management program. For example, New York was awarded $550,000 in 1974, the first year of funding (and increasing amounts in subsequent years) for the purpose of developing an integrated Coastal Managment Program for the state. The CZMA also requires the states to identify boundaries of the coastal zone, define permissible land uses within that zone, inventory designated areas of particular concern, develop criteria for controlling land and water uses, and consider priorities of use. After these tasks have been fulfilled, the state must

then demonstrate it has the legal statutes that provide the framework to implement the agreed on goals. Once approved the state not only obtains the commitment of federal agencies to conduct its activities in a manner consistent with the program, but can also receive additional federal funding to manage that program. In 1976 the CZMA was amended wherein the new bill authorized $1.6 billion to assist the coastal states and territories for developing and implementing coastal management programs and also for coping with impacts of offshore oil and gas production. The amendment also established a 10-year Coastal Energy Impact Fund which it funded for $1.2 billion. There is also a clause that states that the Federal government will manage the coastal area if the state does not. So although these acts can be considered environmental, they try to balance various sectors of society and explicitly encourage competing uses of the coast, as revealed by the policy statement:

. . . to preserve, protect, develop, and, where possible, restore or enhance, the resources of the nation's coastal zone for this and succeeding generations.

Thus, coastal zone management encompasses the full spectrum of human activities, and the coastal area cannot be looked on as the prerogative of a single and narrowly defined interest group.

Perspectives

The coastal zone has become the most popular locale for developmental projects. Many of the world's great cities are on coasts and in positions where shoreline terrane make possible good harbor facilities. Industry is also attracted to coastal environments because transportation costs can be minimized and waste products

easily disposed. Although beaches have provided recreational sites for many years, their utilization since World War II has vastly accelerated. The construction of entire new communities and the rush for second homes at the seashore have put a premium on coastal real estate values. Thus, we are deeply involved in all aspects that affect this land-water interface of the coastal corridor.

Shorelines constitute one of the most rapidly changing landforms on the earth. Coastal processes differ from other subaerial processes because the near-shore water movements are more complex. They are oscillatory, not unidirectional, as in fluvial and gravity systems, and they act at variable levels because of tidal and other sea level changes. Feedback operates in all geomorphological systems, but the relationship between the processes and responses in the coastal zone are particularly complex and include both negative, or self-regulating, and positive, or self-generating, loops. When a specific process creates a particular landform, that form in turn modifies the process.

The movement of ocean waters comprises a compounded array of interactions powered by energy derived from the sun and moon. The tidal regime generated by the gravitational force permits a wide zone of the beach to become influenced by changing sea levels. The relentless rise in sea level that is superimposed on this system offers a tremendous long-range challenge to coastal managers. Waves are oscillatory water movements that are initiated by wind-generated frictional drag on the upper water surface. When the waves strike the beach at an angle, longshore currents are created. This combination of tides, waves, and currents provides a continuing dynamo of energy that is constantly working to alter shorelines. Extraordinary storms, whether tropical cyclones or tsunamis, produce exceptional storm surges that are destructive to vulnerable human developments, and life itself. Worldwide average yearly damages amount to billions of dollars.

The youthfulness of the present coastlines accounts for the wide variety of coastal landforms. These can be related to the variations in energy levels, structural stability, wave climate, and geomorphic history. Such youthfulness means the coasts are not yet adjusted to ocean forces; so prediction of future behavior is vital whenever we need to modify the environment. A great paradox often emerges when we attempt to manipulate coastal areas in the name of conservation and preservation. Our very acts to conserve this environment frequently result in greater destruction. This can be the case when we engineer structures and change coastal processes by such structures as groins, jetties, seawalls, revetments, and bulkheads. Even the construction of artificial dunes or vehicular traffic on the sand can produce deleterious effects in this fragile environment.

The lesson is clear: whenever possible, we must work with the natural processes that affect shorelines instead of trying to fight, alter, and modify natural forces. Thus, artificial nourishment of beaches and sand bypassing are preferred solutions to many erosion and sediment problems. Another axiom is that coastal planning and construction should not be attempted in piecemeal fashion, but should require integrated and coordinated strategems that operate on a regional basis. In the United States such programs as the National Sea Grant Act, the coastal zone management acts now in effect on both the national and state levels, the authority of the Corps of Engineers to limit wetland modification, and the national seashore system of parks show increased public awareness of the importance for maintaining coastal resources. When these interdisciplinary management procedures are carried out, priorities assessed and implemented, and hazard perception placed in proper perspective, then mankind can more truly benefit and enjoy the blessings of the glorious coastal environment.

READINGS

Burton, I., and Kates, R. W., 1964, The flood plain and the seashore: Geog. Rev., v. 54, n. 3, p. 366–385.

Burton, I., Kates, R. W., and Snead, R. E.,

1969, The human ecology of coastal flood hazard in megalopolis: Dept. Geog. Res. Paper No. 115, Univ. of Chicago, 196 p.

Dolan, R., 1973, Barrier islands: natural and controlled: in Coates, D. R., Coastal geomorphology, Publ. in Geomorphology, State Univ. of New York at Binghamton, p. 263–278.

Godfrey, P. J., and Godfrey, M. M., 1973, Comparison of ecological and geomorphic interactions between altered and unaltered barrier island systems in North Carolina: in Coates, D. R., Coastal geomorphology, Publ. in Geomorphology, State Univ. of New York at Binghamton, p. 239–258.

Inman, D. L., and Brush, B. M., 1973, The coastal challenge: Science, v. 181, p. 20–32.

King, C. A. M., 1972, Beaches and coasts, 2nd ed.: London, Arnold, 570 p.

Komar, P. D., 1976, Beach processes and sedimentation: New Jersey, Prentice-Hall, 429 p.

Lizarraga-Arciniega, J. R., and Komar, P. D., 1975, Shoreline changes due to jetty construction on the Oregon coast: Sea Grant Program Pub. No. ORESU-T-75-004, Oregon State Univ., 85 p.

Morton, R. A., 1976, Effects of Hurricane Eloise on beach and coastal structures, Florida Panhandle: Geology, v. 4, p. 277–280.

New England River Basins Commission, 1976, The ocean's beach-identifying coastal flood hazard areas and associated risk zones: Digest of a Workshop, Boston, New England River Basins Comm., 91 p.

Niering, W. A., 1970, The dilemma of the coastal wetlands: conflict of local, national, and world priorities: in Helfrich, H. W., The Environmental crises, New Haven, Yale Univ. Press, p. 142–156.

Shepard, F. P., and Wanless, H. R., 1971, Our changing shorelines: New York, McGraw-Hill, 579 p.

Spits, A., 1970, Holland's struggle against the water: The Hague, The Society for Making Holland Better Known Abroad, 55 p.

U.S. Corps of Engineers, 1964, Land against the sea: U.S. Army Coastal Eng. Center Misc. Paper No. 4–64, 43 p.

———, 1971, Shore management guidelines, National Shoreline Study: Dept. of Army, U.S. Govt. Printing Office, 56 p.

———, 1971, Shore protection guidelines, National Shoreline Study: Dept. of Army, U.S. Govt. Printing Office, 59 p.

Van der Burgt, J. H., and Bendegom, L. V., 1949, The use of vegetation to stabilize sand-dunes: London, Int. Civil Eng. Conf. Proc., p. 158–170.

Veen, J. V., 1962, Dredge drain reclaim, the art of a nation: The Hague, Martinius Nijhoff, 5th ed., 200 p.

Vesper, W. H., 1967, Behavior of beach fill and borrow area at Sherwood Island State Park, Westport, Connecticut: U.S. Army Coastal Eng. Res. Center Tech. Mem. No. 20, 25 p.

Woodhouse, W. W., Jr., and Hanes, R. E., 1967, Dune stabilization with vegetation on the outer banks of North Carolina: U.S. Army Coastal Eng. Res. Center Tech. Mem. No. 22, 45 p.

As this book goes to press, Hurricane Allen is battering Caribbean and Gulf of Mexico coastal areas with the third worst storm of the century in this region.

Chapter Sixteen

Human Impacts on Soil

Severe erosion and badland topography caused by overgrazing and fire, Baja California, Mexico. (Courtesy U.S. Forest Service.)

INTRODUCTION

Because soil is nearly ubiquitous on land, all of our activities on the surface of the planet affect it in some way. This chapter will focus on those soil changes that are caused when we use soil to develop the living resources: crops, livestock, and timber. Soil modifications that result from mining, construction, and urbanization are discussed elsewhere.

Soil is the building block of civilization and, along with water, form the two basic ingredients for human survival. Thus knowledge of its location and characteristics is vital to the welfare of society. Soils differ widely in their fertility and ability to withstand erosion and misuse. Soil surveys and mapping should precede any anticipated use because certain soils may be too fragile or unsuited for an intended purpose. All soils have a **tolerance level,** or **threshold,** that when exceeded can cause deleterious degradation of the material. The lessons that can be learned from the past despoilation of soils show the importance of a comprehensive understanding and utilization of soil which is dependent on a multidiscipline approach—including agronomy, geology, geomorphology, forestry (silviculture), animal husbandry, pedology, and engineering. The cooperation of scientists in these specializations is crucial to optimize planning, use, management, and conservation of this essential resource.

No nation can afford to become complacent about its soil. The historical record paints a dim picture of man's ruination of many soils and, unfortunately, many of these depredations are continuing today. The nearly universal lack of land stewardship is a problem most communities and all governments must solve. To paraphrase a saying by Will Rogers, *it is important to save the soil because they aren't making any more of it.*

HISTORICAL BACKGROUND

The problem of soil erosion and siltation is not a new or twentieth-century phenomenon—it is as old as time. We are largely concerned in this chapter with the manner in which man accelerates this double-edged sword, being aware that a certain amount of geomorphic erosion (sometimes termed normal or natural) is inevitable. Neolithic hunters deliberately set forest fires in order to obtain game more easily. Much erosion resulted and the large thickness of deposits are recorded in river and delta silts in Europe which the anthropologists have termed **human sedimentation.** Since the advent of the agricultural revolution, we have constantly been

Figure 16-1 (a) Slash-and-burn agriculture in the Americas. Slash-and-burn deforestation cropping near San Jose de Buja in the Orinoco basin, northeastern Venezuela. Laboratory tests show that these soils are now infertile because the ecosystem was drastically changed and can no longer grow crops after a 2-to-4-year period. (Photo by Gerald W. Olson.)

Figure 16-1 (b) This area was cultivated after being deforested by the Mayan civilization about 10 centuries ago. In time, a new forest regenerated, which was again burned and cleared for cropping. This site in Valle de Naco, Honduras, is now nearly sterile, is suffering from erosion and rapid runoff, and has a lowered water table. (Photo by Gerald W. Olson.)

at work altering the soil ecosystem (see also Chapter 3). This is accomplished not only by erosion and siltation, but by waterlogging and salinization. Indeed, some of these changes are so massive that they may aid or cause desertification (Chapter 17). Few climates or terrains are immune from these ravages, and the continuing deforestation process has increased with renewed enthusiasm as the population explodes (Fig. 16-1). Soil erosion always affects the local cultivators and, when it reaches large proportions, it may influence the economy of an entire nation and contribute to the decline or fall of a civilization, as is conjectured in the cases of the Sumerian, Roman, Mayan, and Singhalese civilizations.

During most of human history while soil devastation was occurring, there were few prophets to announce such ruination. Plato seems to have been one of the earliest observers, but he described the spoilage after it had happened at Attica in these terms:

All the rich, soft soil has moulted away, leaving a country of skin and bones. . . . The annual supply of rainfall was not lost, as it is at present, through being allowed to flow over the denuded surface into the sea, but was received by the country, in all its abundance. . . .

George Perkins Marsh wrote the first book on the subject of soil in 1864 with the entire theme being the human destruction of our own habitat.

. . . man . . . changed millions of square miles, in the fairest and most fertile regions of the Old World, into the barrenest deserts.

The ravages committed by man subvert the relations and destroy the balance which nature had established between her organized and her inorganic creations. . . .

There are parts of Asia Minor, of Northern Africa, of Greece, and even of Alpine Europe, where the operation of causes set in motion by man has brought the face of the earth to a desolation almost as complete as that of the moon.

Deforestation and its aftermath were the principal causes for the collapse of the Singhalese civilization on Sri Lanka. Denudation of low-elevation forests and salinization of agricultural fields forced the Inca empire in Peru to move into higher valleys. The heroic measures necessary to continue food production contributed to a significant sapping of energy and strength of the population. The Mayan empire contained more than 1 million people in Guatemala, El Salvador, Honduras, and Mexico and

was especially powerful from 600 B.C. to A.D. 900. Although causes for the decline of this complex civilization are disputed, most researchers agree that the slash-and-burn type of farming and exorbitant sedimentation were important factors. For example, the maize yields declined to less than 25 percent after a site had been planted four times.

The early Indians that occupied such areas as Mesa Verde, Colorado—the Anasazi (the ancient ones)—numbered many thousands and thrived from about A.D. 1 to A.D. 1300. However, their destruction of the native forest and vegetation promoted erosion and siltation which ultimately depleted their agricultural base and caused the abandonment of the region. During the increase in population the construction of dams, terraces, and other water management systems could not sufficiently reduce soil loss to maintain fertility and productivity. Some of the siltation areas became the sites for small fields, but nutrients were depleted by cropping and erosion and the failure to apply fertilization. Even near Rome the natural soil erosion rates of 2 to 3 cm per 1000 years were increased an entire order of magnitude because of the intensive use of land during Roman times. Such losses contributed to the agricultural decline of the empire and may have been an important cause in the final demise of Rome (see Chapter 3).

As early as 1620 American farmers were recognizing the damaging effects of water erosion on cultivated lands and were attempting to develop practical ways to protect their soil. The worst cases of soil ruination stemmed from tobacco cultivation. In colonial times tobacco yielded profits six times higher than other crops. Lands were deforested in Virginia, Maryland, and the Carolinas and intensively cultivated 1 to 3 years. After this time the soils were usually exhausted and either abandoned to second-growth woods or occasionally replanted with corn or wheat. New lands were then cleared and the same cycle repeated. Such practices not only impoverished the soils but led to massive erosion and siltation. For example, dramatic effects were felt throughout the Piedmont rivers and estuaries draining the farmed uplands. Coastal ports such as Joppatowne and Elk Ridge in Maryland became choked with sediment and unfit for ocean ships. The loss in trade led to the abandonment of these river-port towns and harbors. During this period of about 70 years the average depth of soil loss on farms from Virginia to Alabama was 17.8 cm from the topsoil profile layers. More than 60 percent of this eroded sediment still remains in the watersheds as alluvium in floodplains and as colluvial sheetwash on hillslopes. Following the American Revolution some of the government leaders expressed concern over soil and erosion problems. Patrick Henry remarked "he is the greatest patriot, who stops the most gullies." George Washington was very concerned about soil losses on his property and in 1773 wrote:

. . . at present [the soil] is very much gullied, and if uncommon attention is not paid to it in the working . . . it will be unfit hereafter for grass even. . . .

Still on his mind four days before his death, he wrote:

The washed and gullied parts . . . ought to be levelled and smoothed . . . covered with litter, straw, weeds, corn stalks, or any other kind of vegetable rubbish, to bind together, and to prevent the earth from gullying.

AGRICULTURE AND GRAZING LANDS

Deforestation of lands and the subsequent use of land for farming and grazing have severely damaged soils throughout the world and produced as a by-product sedimentation of enormous proportions. Abusive land practices during several millennia in the Tigris-Euphrates river system have aided in adding deltaic deposits 320 km oceanward of the site of Ur, which formerly was a coastal city. More recently a 1974 ERTS (Earth Resources Technology Satellite) photograph revealed for the first time an immense new island that was forming in the Bay of Bengal, 103,000 km^2. Its sediments were being derived from erosion, much of it man-induced, in the Brahmaputra and Ganges rivers. For example, 75 percent of India's forest has been cut down in the past 20 years. In Nepal large tracts of land now contain less than 3 cm of topsoil. In this region annual erosion amounts to about 100 million tons, and in some cities there is insufficient timber to cremate the dead. Only eight of the 55 hill districts have sufficient food, and two-thirds of the deficient areas began eroding in the 1971–74 period.

Tropical soils are surprisingly fragile and vulnerable to ruination by improper farming methods. The verdant forests and foliage of the Amazon basin disguise an ecosystem deficient of nutrients; only 4 percent of the soils can be classed as medium to high fertility. Nutrients in tropical soils occur mostly in the plants, and the soil only offers mechanical support for growth. When the canopy protection is demolished, soil temperature is raised to destructive proportions which hasten biological and chemical deterioration of the remaining organic matter. Heavy rains can finish the job of leaching away remaining nutrients and removing by surface erosion the productive upper soil horizon. The slash-and-burn and shifting agricultural practices are proving especially devastating today in such areas as Oceania, Latin America, and Africa. Most soils after being denuded lose a great deal of their fertility within a 4-year period and must be abandoned for new sites. Regeneration, when permitted and possible, takes more than 20 years. Haiti (which means "green island") now has less than 9 percent forest, and the soil erosion and siltation of the other lands are leading causes of the extreme poverty in the country. El Salvador provides another case where the population explosion has led to ex-

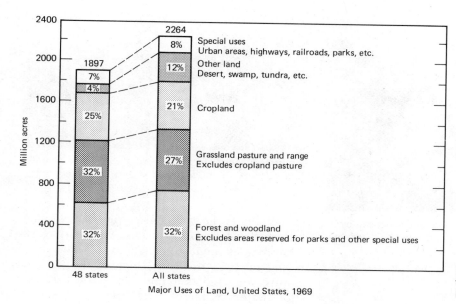

Figure 16-2. Land uses in the United States, 1969. (U.S. Dept. of Agriculture.)

Major Uses of Land, United States, 1969

ploitation of the limited soil resources, so that 77 percent of the land is suffering from man-induced accelerated erosion.

United States

Figure 16–2 shows the distribution of land uses in the United States. At present, about 19 percent of the total 2.27 billion acres (920 million ha) are in cropland. However, during the last 200 years, the food-producing ability of 200 million acres (81 million ha) has been lost by soil abuse, and one-third of the topsoil has been eroded from remaining cropland, causing a yield (potential) loss of about 15 percent. To offset soil erosion loss requires an extra 50 million barrels of oil for energy requirements each year. Since 1900, 125 million acres (50 million ha) of cropland were added, primarily by irrigation and drainage. Of this amount, 48.7 million acres (19.7 million ha) were converted since 1967, but 79.2 million acres (32 million ha) were lost between 1967–75, leaving a total cropland of about 400 million acres (162 million ha). Pastureland and rangeland increased 64 million acres (26 million ha) to 570.9 million

acres (285 million ha) during 1967–75, but forestland declined by 70 million acres (28 million ha) to 375 million acres (152 million ha).

At present, the United States is losing about 4 billion tons of soil annually, compared to 3 billion tons in 1934 (Figs. 16–3, 16–4, and 16–5). It would take a train of freight cars

Figure 16-3. Poor cropping practices and inadequate conservation measures produced this excessive gullying in Stewart County, Georgia, February 1937. (Library of Congress photograph.)

Figure 16-4 Severe erosion in loess soils caused by abusive farming without proper conservation measures in Lancaster County, Nebraska. (Photo by Soil Conservation Service, November 23, 1954.)

Figure 16-6 Dust storm on pasturelands, Tripp County, South Dakota. (Soil Conservation Service photograph on May 5, 1936.)

Figure 16-5 Sheet erosion on summer fallow that had been planted with winter wheat. Loss of soil at this site during the winter of 1947 was 125 t/a. (Photo by Soil Conservation Service, Lincoln County, Washington.)

633,000 mi (1 million km) long a distance of 24 times around the earth to transport the soil material lost annually. About 25 percent is wind erosion (Fig. 16–6) and the rest is the result of water and gravity movements. Much of the material does not reach the ocean but lays trapped on the land and accumulates in ditches, roadways, fields, streams, reservoirs, lakes, and harbors. Of the 4 billion tons lost annually, about 75 percent is from cropland, and the

remainder from rangelands, timberlands, and human construction activities. Various governmental assessments of such losses have been made. In 1965 the U.S. Department of Agriculture reported that immediate soil losses by "floodwater sediment, and related watershed damages" amounted to $1.3 billion. In 1975 CAST (Council for Agricultural Science Technology) estimated nutrient losses that should be replaced by fertilizers to be $1.2 billion annually. In addition, siltation damages occur in streams and can impair fish and wildlife habitats, reduce reservoir storage capacity, lower water quality and value for recreational and consumptive use, contain pollutants and other pesticides, and increase flood damages by altering river conveyance systems. A General Accounting Office study in 1977 of 283 representative farms in the Great Plains, Corn Belt, and Pacific Northwest showed that 84 percent of the lands were annually losing more than 5 t/a (tons per acre). Of these, 51 percent lost more than 10 t/a, and 13 percent lost more than 40 t/a. The report concluded that:

Not only has soil erosion been depleting the Nation's valuable topsoil, but it has also been creating pollution problems and increasing public and private expenditures for cleanup and repair. The productive soil of the Nation's agricultural land must be maintained and protected if the United States is to continue to meet its domestic food needs and help alleviate world food shortages.

In the United States not only were highly productive lands damaged by mismanagement, as were the original tobacco lands in such states as Maryland and Virginia, but many marginal lands were irreparably ruined, especially in the West. During the settlement period the erroneous adage "the rain follows the plow" provoked vast destruction of the terrain. Much virgin land was plowed, but commensurate rainfall failed to materialize to nourish the crops. Thousands of square kilometers of prairie lands were ruined. Finally, it was the work of John Wesley Powell that convinced many homesteaders of the fallacy of such plowing efforts.

Although there is sufficient literature on the topic, and although professional consultation by government agencies has been available regarding abatement of soil erosion for many years, there continues to be vast and ever-increasing soil erosion in the United States. A Soil Conservation Service (U.S. Department of Agriculture) report in 1967 showed that more farmers in the Corn Belt had used conservation measures in 1958 than were using them in 1967. Soil erosion was 22 percent greater in Iowa during the 1970s than it was in the 1960s. Some of this erosion is attributable to the storms of 1974 that severely eroded 1.8 million ha of the state's 8.4 million ha of corn and soybeans. However, another cause has been the rush to use more farmlands for soybean crops. Erosion of unprotected corn fields averages 5.2 t/ha, which represents two bushels of soil lost for each bushel of corn harvested! The soil-loss tolerance level for most Midwest soils is 1 to 2 t/ha, and these figures are being exceeded by the majority of farms. More than 70 percent of Wisconsin soils exceed this threshold by a factor of 3. In the Obion-Forked Deer River basin of northwest Tennessee, the average soil erosion for 500,000 ha is 3.7 t/ha; it is 7.6 t/ha on the upland soils.

Soil erosion by wind deflation continues to plague the Great Plains states 40 years after the Dust Bowl days (see also page 597). Wind damaged 500,000 ha during the dust storms of November–December 1973, causing three times the damage inflicted in 1972. In 1973 and 1974, an additional 650,000 ha of former pastures and woodlands was put into crops, in order to reap the benefit of higher prices for agricultural produce. However, by the end of 1974, the lands had suffered an average soil loss of 4.4 t/ha (three times the tolerable limit). In one part of the southern Great Plains where 20,000 ha had been newly planted, soil losses ranged from 51 to 65 t/ha. Again in 1977 wind damaged 560,000 ha, up by 25 percent from the previous year. Such losses were attributed to inadequate residue cover, insufficient moisture, and use of marginal lands unsuited to row crops.

The U.S. Department of Agriculture has pointed out an interesting paradox: although 67 percent of all cropland in the northcentral United States is in need of conservation treatment and about 85 percent of all lands have erosion that significantly exceeds tolerance levels, crop yields are higher than they were in the early 1900s. This increase in production is attributed to (1) abandonment of nonproductive land, (2) cultivation of highly producing crop varieties, and (3) much increased use of fertilizers, often as much as 16 times more than earlier years.

Grazing Lands

Overgrazing has been a continuing problem throughout the history of livestock domestication. Sheep and goats have been the primary ruination of many soils throughout the Mediterranean and Mideast regions (Fig. 16–7), whereas cattle and sheep have been responsible for land degradation in the United States and Australia (Figs. 16–8, 16–9, and 16–10). Goats have often been the preferred animal for the poor because of their unique ability to utilize food—even tree tobacco, prickly pear cactus, and poisonous locoweed. They are surefooted, can browse where other animals cannot, and can even climb some trees. Other appealing ownership aspects are: (1) their fairly small size

Figure 16-7. Erosion by sheetwash and soil slope failure, north Lebanon. The goats have caused this accelerated erosion by destroying nearly all the vegetation. The only remaining flora is *Calycotome,* a thorny, worthless shrub. (Photo by Joseph Van Riper.)

Figure 16-8. Overgrazing by cattle and sheep, western Australia. Notice the terracette trails and earth slumps, both caused by animal traffic and ruination of vegetative cover. (Photo by Charles Finkl.)

Figure 16-9. Aerial photograph of coastal area, western Australia. The light area in lower left was originally well-grassed rangeland that was severely overgrazed. All that remains now is a system of sand dunes. (Photograph contributed by William McArthur and CSIRO.)

and value, so that investment is minimal; (2) their ability to survive on forage of poor quality in harsh environments; and (3) their high reproductive rate (New Zealand studies have shown that if numbers are reduced 80 percent the population can recover to 90 percent the original population within four years). However, paradoxically, the same criteria that make goats attractive to many people also make them undesirable because they despoil the land (Fig. 16–11).

A problem that has caused much controversy in the past, and is still unresolved to the satisfaction of all, concerns the origin of events that initiated the nineteenth-century arroyo cycles in the American Southwest. John Rich (1911) was the first geologist to state unequivocally that the stream-trenching was man-induced and resulted from overgrazing practices. His thesis was that man, by introducing the open range along with millions of livestock— New Mexico had 4 million sheep by 1880, caused the ruination of the vegetative cover and the pulverization of the topsoil which prevented

Figure 16-10. Arroyo of the Rio Puerco near the village of La Ventana, Sante Fe National Forest, New Mexico. This arroyo is typical of those in Arizona and New Mexico whose development was accelerated by improper range management. (Photo by U.S. Forest Service, July 19, 1945.)

Figure 16-11. These limestone slopes near Beirut, Lebanon, were once covered with thick stands of cedars but are now worthless because of deforestation, goat browsing, and soil erosion. (Photo by Joseph Van Riper.)

plant regeneration. Such circumstances would lead to accelerated runoff with a consequent lowering of the water table which would also destroy the permanency of some streamflow. A contrasting view is presented by Luna Leopold (1951) who believes that a change in storm patterns with different frequency and intensity of rainfall was the principal cause for initiating a new arroyo cycle.

The Taylor Grazing Act was passed in 1934 to aid in the management of public rangelands, and in 1946 the BLM (Bureau of Land Management) was formed by uniting the General Land Office and the Division of Grazing. In the nation one-fifth of all land is federally owned, with 62 percent (470 million acres) under the jurisdiction of BLM. Public lands have always had low or no grazing fees, and the small fees have continued under BLM governance. Such practices have led to overgrazing and deterioration of rangelands. A BLM 1975 study reports that 33 percent of all rangelands are in poor or worse condition now than they were in 1934. Only 17 percent are in satisfactory condition, and it was predicted that the productive capacity of all lands would decrease 25 percent by the year 2000. Conservation is minimal on such lands because this work is financed by the grazing fees, and these insufficient monies prevent

adequate protective measures on the land.

When brushy and woody areas are converted to grasslands, erosion is increased. After conversion of brush areas to grasslands at the San Dimas Experimental Forest, the storms of 1969 caused soil losses of 470 t/a on the grass areas, but only 166 t/a for brush areas. A study of two burned-over watersheds in the San Gabriel Mountains, California, showed that Monroe basin which was kept in grass lost eight times more soil (about 2,300 m³) than Volfe basin which was allowed to revert to the native chaparral.

FORESTS

Forestlands are being radically reduced throughout much of the world and their regeneration is not keeping pace with losses. Furthermore, this denuded land is invariably abused by poor cropping management that produces excess erosion and siltation. Conservation methods for soil stabilization are rarely practiced, and even in the United States there is great conflict concerning harvesting methods and the value of fires. When properly utilized forests act as climatic stabilizers, inhibit flooding and sedimentation, and can produce fuel and wood products necessary in all societies.

Throughout history we have cleared the lands by firing the forests for settlements, driving game, and planting crop fields. Tierra del Fuego, or "land of fires," was named by Magellan in 1520 because of the extensive land burning near the tip of South America by the natives. The Plains Indians in the West expanded the prairies by burning forests for buffalo hunting. When von Humboldt visited Mexico in the early 1800s he recorded that the country had merchantable timber on 50 percent of the land, which has now been reduced to less than 10 percent. Forests are also denuded for the wood, which is used for construction purposes and fuel. Eric Eckholm (1976) has documented many case histories, for example, the Mediterranean forests, which were denuded during early Grecian and Roman times. The cedars of Lebanon have been decimated during 5000 years of timbering, and only a few remain (Fig. 16–11). At one time the Greeks had as many as 8000 men and 1000 pairs of draft animals cutting the cedars for export to Greece, whose forests had already been largely cut. Much of China was originally woodlands, but only 8 percent remain today. Even the moors of England and Scotland used to be heavily wooded and emerged as moors only after deforestation, which began during the Bronze Age and had been largely completed by the end of the Middle Ages.

The elimination of forests is continuing with unabated fury and is even increasing in many areas, as in the LDCs (Less Developed Countries). The FAO (Food and Agricultural Organization of the United Nations) reports that more than 8 million ha of forests are annually eliminated in Asia and an equal or greater amount in Latin America. The majority of this deforestation is caused by shifting cultivators who use the slash-and-burn technique for temporary farming. Others use the wood for fuel—indeed, one-half the world's population depend on wood for their primary source of energy. Even in the United States, some states are rapidly losing forests, such as Missouri which lost 800,000 ha in the 1960–70 period. However, some states, such as New York, have more forests now than they did in 1900.

Impact on Land-Water Ecosystem

Forests offer many benefits to society. The cooling capacity of one tree equals that of five average air conditioners operating 20 hours a day. Noise is reduced 8 decibels for 93 m^2 of woodlands, and 1 ha of forest can absorb all the CO_2 fumes from 20 cars. Tree roots provide extra shear strength to the soils to aid in resisting gravity movement and surface erosion. This influence may last 3 to 5 years after cutting. The forest canopy acts to dampen raindrop impact, and the tree's ground litter further reduces erosion potential. The decaying vegetal matter enriches the nutrients in the topsoil horizons and increases permeability to allow water infiltration. Thus direct runoff is minimized and sheet flooding prevented. Even the exceptionally heavy rains from Hurricane Diane in 1955 produced overland flow only on those slopes greater than 50 percent in New England and the Poconos of Pennsylvania. Sediment yield from forests is generally only a fraction of what it is for other land uses. Studies of the 900 km^2 Gunpowder Falls Basin, Maryland, showed the following sediment yields: (1) forests, 0.031 t/ha; (2) urban-suburban, 0.031 to 0.063 t/ha; (3) farmland, 0.78 to 3.1 t/ha; and (4) land undergoing construction activities, 16.5 to 31 t/ha. Wooded areas in the Potomac Basin with 20 percent woods yield 0.252 t/ha, but if 80 percent wooded yield only 0.028 t/ha of sediment. In North Carolina forested slopes of 10 percent produce an average of 0.001 t/ha, and in New Hampshire forests with 20 percent slope produce 0.004 t/ha sediment.

Deforestation or poor harvesting methods eliminate or reduce the number of benefits that forests produce. Deleterious influences that we create with wanton timber removal include: (1)

Figure 16-13. Shasta County, California. These chamise-manzanita brush slopes were burned over by fires in September 1950. Although the larger gullies were formed earlier, they have been reactivated in less than two months, and new rills and gullies formed as a result of loss in protection by the vegetation. (U.S. Forest Service photo taken November 22, 1950.)

Figure 16-12. Little Grass Valley Creek, California, June 24, 1952. Careless logging practices have ruined this trout stream by blocking water flow with excessive slash and sedimentation debris. (Photo by U.S. Forest Service.)

increased erosion and soil loss; (2) increased downslope sedimentation with all its accompanying damages such as more frequent flooding and silting of reservoirs; (3) upset in hydrologic regimes with lowering of water tables; (4) loss of soil nutrients; (5) destruction of wildlife habitats (Fig. 16–12); and (6) reduction of aesthetic qualities. Sediment rates in irrigation canals and dams is now four to five times greater than originally calculated by the designers. The forests in the foothills of the Indus and Ganges rivers that formerly inhibited heavy runoff are now mostly gone. Heavy rains in August and September 1973 that formerly would have produced only minimal influences instead inundated millions of acres of valuable cropland in six

Indian states and southern Pakistan and caused damages in the tens of millions of dollars. Extensive deforestation of 775 km² in the state of Espirito Santos, Brazil (north of Rio de Janeiro), in the last 20 years has led to the destruction of 450 varieties of plants and 204 species of birds. Papaya can no longer be grown because of the hordes of insects. This area was settled by light-skinned German immigrants, and 90 percent of the children examined in medical clinics have cancer lesions from the sun because their pigmentation does not contain sufficient filtering protection.

Of course, forests can be damaged and burned by "natural" causes (Fig. 16–13). A six-year battle was waged by the U.S. Forest Service in Targhee National Forest in an attempt to control the pine beetle that was ruining the trees. After spending $9 million the chemical spraying, which had poor success, was stopped. In a forest that once had 3 billion board-feet (1 board-foot = 144 in.³), less than half remains— a loss of $30 million. Such experiences have prompted heated debates among foresters concerning the desirability of instituting a program of controlled burning and the adoption of a policy (now in effect in some regions) of allowing natural fires, those produced by lightning, to burn. It is argued that under the appropriate conditions such burns are beneficial, for example, they eliminate pests and diseased trees. They can also kill immature trees that otherwise consume precious water, allowing the larger trees to mature more quickly. Whereas some of the conifers, sequoia, and Douglas fir seem to prosper with such a regime, the broad-leafed trees and deciduous hardwood forests of the East would be ruined by such practices of fire prescription.

U.S. Forests and Policies

The relatively strong health of the forest industry is attributable to a variety of factors: (1) good climate and fertile soils; (2) administrative vision and leadership of such men as Gifford Pinchot and Aldo Leopold; (3) federal passage of protective legislation; (4) a general enlightened attitude by commercial operators and programs for forest regeneration; and (5) a strong national economy and abundant other resources so that wood is only an incidental fuel.

A strong governmental position regarding forests was taken in 1891 with passage of the Forest Reserve Act. The first three presidents after passage of the act used these powers to preserve 27 million acres, and Theodore Roosevelt added another 148 million acres. The Organic Act of 1897 acknowledged that forests can perform important conservation functions:

No national forest shall be established except to improve and protect the forest within the boundaries, or for the purpose of securing favorable conditions of water flows, and to furnish a continuous supply of timber for the use necessities of citizens of the United States. . . .

The Weeks Forest Purchase Act of 1911 went even further to protect forestland. This law represented one of the earliest legislated mandates that proved that proper land-management practices can aid in retardation of runoff and reduction of erosion. It authorized the Secretary of Agriculture to take certain steps for "regulating the flow of navigable streams. . . ." A far-reaching law passed by the Congress in 1960, known as The Multiple-Use-Sustained Yield Act, has had profound influence on management of forestland (see p. 682).

Figure 16–2 shows the relation of forestlands to other terrain usages in the United States. Demand for wood products doubled in the 1940–75 period, and increasing public pressure for recreation and other activities that consume timber have further reduced commercial forest acreage. Although the total of all forests, private and governmental, only declined 8.5 million acres from 1962–70, the loss rate has been accelerating. In 1945 U.S. Forest Service lands produced 2.4 billion board-feet, but by 1975 10.8 billion board-feet were being harvested. This is 25 percent of the total production of timber. Despite increased wood use and shrinking acreage, commercial forest growth rate was 14 percent higher in 1972 than 1960 because of greater protection against fire, disease, and insects. Also in 1972 the net growing stock exceeded removals for the nation by more than 20 percent. Even greater savings could be realized if paper product reuse could be expanded from 17 percent, which it is today, to 40 percent, as it is in Japan and western Europe.

Clearcutting and Conservation

A great controversy is now waging in the lumber industry and in government forestry management concerning **clearcutting**—the practice of complete tree removal during the same cutting season. Since 1960 such methods have been replacing the selective cutting procedure. By 1970, 30 percent of the national forests that were logged in the West were clearcut, and this timber was 60 percent of the total harvest on all

Figure 16-14. Overview of part of the forestlands at the Coweeta Hydrologic Laboratory, North Carolina. (Courtesy James Douglas, U.S. Forest Service.)

Figure 16-15. Accelerated soil loss and gullying on privately logged timberland near Leadville, Colorado. This area has now become part of San Isabel National Forest. (Photo by U.S. Forest Service, 1915.)

Figure 16-16. Private lands on Smith River, California. Heavy rains in October 1950 subjected this improperly timber-harvested area to severe erosion and siltation. (Photo by U.S. Forest Service, 1951.)

public and private lands. In the East, 40 percent of the national forests were clearcut, which was 50 percent of all timber for the region. Depending on one's definition of "conservation," we can agree or disagree with the advocates of clearcutting who contend that this is a conservation practice. They point out the ease of harvesting, which is energy-saving, and the immediate economic benefits in reduced time and labor. They further maintain that clearcutting does not necessarily harm the ecology, and that any damage is more than compensated for by increasing the water supply of downstream users, thus adding water to reservoirs and benefitting hydroelectric users and irrigationists. In the high-rainfall areas of the Cascade Mountain forests in Oregon, clearcutting increases the annual water yield 46 cm per unit area. In the Hubbard Brook Experimental Forest, New Hampshire, clearcutting increased runoff 25 to

35 cm, and at the Coweeta Hydrologic Laboratory, North Carolina (Fig. 16–14), 40 percent runoff increases occur in the clearcut areas.

Opponents of clearcutting, however, reverse the water argument and document the damages that occur from excess flow. The additional water not only increases erosion and sediment yield by sheetflow and channeled flow in rills and gullies (Figs. 16–15 and 16–16), but also provides for more soil lubrication and a higher incidence of mass movements and landslides (Figs. 16–17 and 16–18). The mean annual sediment loss in clearcut forests of Oregon is 51 t/a, which is more than 100 times the

Figure 16-18. Accelerated erosion of a clearcut slope in northern California. This site is undergoing gullying and mass movements. (Photo by Douglas Swanston.)

Figure 16-17. Landslides induced by clearcutting in southeast Alaska. (Photo by Douglas Swanston.)

amount lost from selectively logged areas; landsliding was increased five times the normal amount. The extra water generated by clearcutting also deposits silt and debris to clog channels, and this increases the number and the height of flood waters. Even the summer temperatures of streams flowing through and down from clearcut areas are increased 10 to 12°F (6°C), and such increases can be harmful to the local ecology.

Water quality and residual soil nutrients are also severely altered in clearcut areas. Large quantities of nutrients are released because of the greater than normal amounts of water, sunlight, and heat. These imbalances result in accelerated decomposition of organic matter on the ground and within the soil horizons. Such ingredients are then rapidly flushed from the watershed, causing sterilization of the soil and pollution in the streams. At Coweeta, North

Carolina, nitrates increased to 90 ppm in the runoff, whereas normal concentrations fluctuated between 0.1 and 0.9 ppm. Losses in other elements such as calcium, potassium, and sodium increased 3 to 20 times the normal amount. At such rates of nutrient loss, Curry (1971) points out that sterility can occur in as short a time as 17.5 years after clearcutting.

One of the important factors that determines the magnitude of the destructive erosion process in clearcut areas is the method of harvesting and the design of roads and skidroads (Fig. 16–19). Because time is considered one of the economic benefits of clearcutting, roads receive heavy traffic during the harvest season. The average loss of sediment from roads and skidroads is several inches per year. In experimental plots turbidity increased 56,000 ppm where no road plans or drainage provisions were made in clearcutting—90 percent of it from the roads. Thus soil conservation in areas to be clearcut should start with the careful planning of roads (Fig. 16–20) and methods of tree extraction. Such management can usually reduce the number of skidrows in half, and 4

(a)

(b)

Figure 16-19.   Results of clearcutting. (a) Timber removal of clearcut area, Coweeta Hydrologic Laboratory, North Carolina. During such timber removal, sediment loss is severe along roads and skid roads. (b) Three years later, regrowth of new stand obscures some of the logging debris and sediment areas. (Courtesy James Douglas and U.S. Forest Service.)

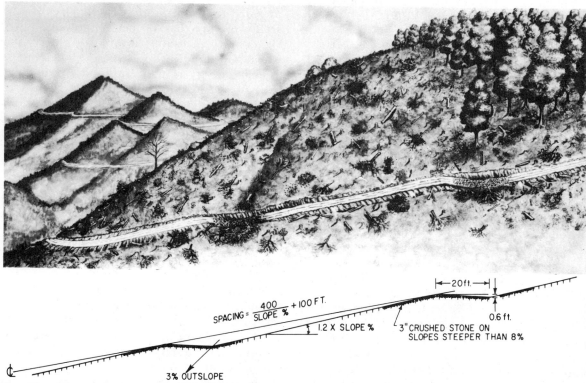

$$\text{SPACING} = \frac{400}{\text{SLOPE \%}} + 100\ \text{FT.}$$

20 ft.

0.6 ft.

1.2 X SLOPE %

3" CRUSHED STONE ON SLOPES STEEPER THAN 8%

3% OUTSLOPE

Figure 16-20.   Sketch of type of road construction that will produce minimum of soil displacement in clearcut timbering. The broad-base sags turn water out of forest roads and are designed to require little maintenance. (Courtesy James Douglas and U.S. Forest Service.)

percent of road grades can often be increased to 8 percent, thus requiring fewer drainage installations and less traveled miles. Drainage diversions must be designed to prevent overland flow from developing to a depth and velocity of water flow that can quickly incise rills and gullies. The drainlines should be spaced at intervals to intercept runoff and allow for percolation. Longer water retention lowers nutrient loss, dampens flood peaks, and maintains water table elevations.

The most obvious land conservation measure for soil is to regenerate forestlands whenever and wherever possible. In the 695 ha White Hollow basin, Tennessee, woodland planting of one-third the basin in 1935–36 reduced sediment yield from 2.9 t/ha to 0.12 t/ha in 1954–55—a 96 percent reduction; thereafter sediment yield has continued to drop about 15 percent each year. In the Pine Tree Branch basin, Tennessee, two-thirds of the basin was reforested during the 1942–45 period. By 1946–50, sediment yield was reduced from 9.7 t/ha to 3 t/ha, with further reduction to 0.9 t/ha in 1951–55, and 0.45 t/ha in 1956–60. In Mississippi, planting of Loblolly pine forests reduced sediment yields after 22 years to 0.008 t/ha. In adjacent fields that had been cut but abandoned, the sediment yield is 0.05 t/ha, whereas on nearby cultivated fields sediment yield is 14.7 t/ha. Even more startling is that the forest had been planted on lands where previous erosion had removed an average of 0.6 m of soil that also contained 1.5 m deep gullies. In a 9.2 ha experimental tract at Coweeta, North Carolina, a formerly forested area was cut, and 2.4 ha planted in corn, 2.8 ha put into pasture, and 4 ha acres allowed to regenerate with hardwood. The original forest had a sediment yield of 0.03 t/ha. During a 13-year period, the pasture-corn area annual losses averaged 0.4 t/ha. After these had been replanted with white pine and yellow poplar, soil losses were only 0.025 t/ha. Timber savings can be obtained with elimination of fires, insects, and disease, which annually damage more than $1 billion of commercial wood. How best to accomplish this and to what extent control of these forces should be exercised are highly controversial issues.

FOOD AND SOILS

Many of the world's problems in food, energy, and population rest on the character of our stewardship of the soil. These components are intricately interwoven, and it seems that unless drastic steps are taken, especially with population control, the entire fabric may come undone.

To feed a world population of 6 billion by 2000 will require a doubling of food production; however, to expand such production will use more than three times the current energy put into agricultural products. Already the LDCs use more than 60 percent of their available energy, including wood, for their food system, and on a worldwide basis 25 percent of all such energy is used in food production. Much of the world's arable land is already in use for crops and grazing. Outside of North America there are only modest amounts of forest suitable for continued cultivation, and such use would be at a high price because of the fertilizers and other controls that would be needed. Additional lands could also be irrigated, but again the technology is costly to keep them watered and productive. As recently as the mid-1930s western Europe was the only continent with a grain deficit, whereas by the mid-1970s only North America and Australia had significant surpluses and the rest of the nations had largely become grain importers. North America controls a larger share of the world's grain exports than the OPEC nations (or Middle East) control world oil exports. For example, in the 1965–75 decade North America doubled grain exports. The United States usually exports about 35 percent of crops such as wheat, soybeans, corn, and sorghum—more than 100 million tons, which is sufficient to feed

1 billion people one meal a day. In the past 20 years the United States has also given away $25 billion worth of food.

The three ingredients for crop and animal production are somewhat interchangeable—soil, energy, and labor. In the United States it takes 9 man-hours per acre to grow corn, but in Mexico 463 man-hours are required, the difference largely being in use of equipment which is fuel-dependent. When other factors are equal crop yields become highly dependent on use of fertilizers and pesticides. The poorer countries are especially seeing declines in yields because of inability to keep pace with rising petroleum and fertilizer costs. In the LDCs each ton of fertilizer would increase harvests by 10 tons, but even countries such as India can only afford about 50 percent of what is needed. The industrialized nations produce crops with 65 percent higher yields than the LDCs. However, because of soil erosion there are even limits in such areas as Illinois, where in 1968 five times as much fertilizer was necessary to produce the same crop yields as in 1949. Throughout the United States each ton of fertilizer increased corn production 15 to 20 tons, but by 1975 each ton of fertilizer yielded only 5 tons. Additional losses in the LDCs occur because of inadequate storage and distribution facilities. Rats, insects, fungus, and mildew destroy more than 25 percent of harvested grain.

The Green Revolution

The "green revolution" began modestly in the late 1940s when four scientists, headed by Dr. Norman Borlaug and under the auspices of the Rockefeller Foundation, gathered in the hills near Mexico City and began experiments on an unusually hardy, plump-grained, short-stemmed variety of wheat. In 1962, with an equally verdant rice strain, additional experiments were done at Los Banos, Philippines, with the cooperation of scientists representing the Rockefeller and Ford foundations. The preliminary verdict for such strains, called HYVs (high-yielding varieties), was that they were a near panacea for the starving nations of the world. In the 1960–70 decade, wheat yields doubled in the United States and tripled in Mexico. Using standard wheat varieties the highest harvest in India occurred in 1965 with 12 million tons; however, when farmers switched to HYVs production jumped to 17 million tons in 1967 and 1968, and further increased to 21 million tons in 1970. Similar increases were occurring in Pakistan, Iran, Afghanistan, Nepal, Tunisia, and Morocco. Turkey started an HYV program on 170,000 ha and the second year expanded it to 650,000 ha. Yields were three times higher than the native wheat. Similar increases occurred with HYV rice. Prior to 1965 the Philippines annually had to import 1 million tons of rice, but with HYV rice had become self-sufficient by 1970. West Pakistan reported a rice increase of 162 percent, and production was up 50 percent in India and 34 percent in Ceylon. Japan, for many years a rice importer, also became self-sufficient, and success also occurred in Malaysia, Burma, Indonesia, Kenya, Brazil, and Paraguay. However, by the 1970s disenchantment was setting in, and many critics were calling the green revolution "a hoax."

It had become apparent by 1970 that many shortcomings had surfaced to diminish the success of HYVs.

1 For best growth HYVs need more water than native grains—often as much as four times more. Thus, high yields are generally only obtained by extensive irrigation. This requires hydrologic structures and engineering know-how. The usual by-products of inadequate practices are occurring: salinization, water logging, siltation, and so on.

2 The HYVs, which in reality are hybrids, often lack the hardiness of regular varieties. Hence, there is a need to use pesticides which, in turn, increases costs and poses pollution and health hazards.

3 The HYVs require more fertilizer. This leads to skyrocketing production costs and use of important energy resources. Because extensive fertilizers must be added, this gradually leads to reduction of crop hardiness and yield, upsets the soil ecology, lowers grain protein content, and produces pollutants in the ground and surface waters.

4 The hybrids lack the nutritional content and flavor of natural grains. The increase in bulk content of HYVs is more one of carbohydrates not the badly needed protein. Where HYVs are grown they have commonly displaced more nutritive crops such as the legumes (beans, peas, lentils) which are higher in protein. Even the taste is disappointing to many. For example, in the Philippines HYV rice sells for 30 percent less than native rice because of its gluey consistency when cooked and its stickiness.

5 HYVs are also causing societal problems. It is an American technology package that is dependent on higher energy inputs—too formidable an obstacle for many LDCs to overcome. The green revolution is widening the gap between the rich and the poor, because only the large landowner with financial backing can afford the investment. The increased mechanization is displacing labor and creating unemployment, so that the jobless migrate to urban areas that are already bursting at the seams.

With so many problems it was inevitable that with time yields from HYVs would be reduced. For example, by 1973 per capita grain production in India was less than it was prior to 1967, when HYVs were first introduced. Yields have been dropping steadily; at one time they were four times higher than native varieties but are now only two times higher. These trends are of worldwide scope and will continue so long as the crop yields are dependent on, and a reflection of, the ability to apply huge energy investments in water, fertilizers, and pesticides.

The Food Bomb

A crucial problem with the population explosion is how to feed the millions of new mouths that arrive daily. This simmering time bomb is one that will become more critical in the next few decades. Of the 1 billion people that will be added to the earth in the next decade, 90 percent will be in the LDCs—those nations that can least afford to support themselves. Although 1977 was a good crop year in many countries, such production cannot be sustained without extraordinary efforts. In 1977 the United States produced (bushels) 1.8 billion soybeans, 2.04 billion wheat, and 6.1 billion corn. There was a 100 million ton surplus because of good harvests elsewhere and the inability of some countries to pay for needed food. Whereas U.S. exports are usually about 35 percent total production, it dropped to 15 percent in 1977. In addition, there was a 1.1 billion bushel carryover in wheat from 1976. However the Food and Agricultural Organization of the United Nations (FAO) has predicted there will be a serious imbalance among the non-Communist aligned LDCs by 1985 that will require five times as much food as they did in 1970. What will be the position of those countries capable of exporting grains? For example, such questions may arise:

1 Should food be produced at full capacity with little concern for long-term conservation of soil resources?

2 Should food exports be used as a political tool and provided to only countries that have the same ideologies as ours?

3 What should be the ratio and amount of food given away and when should we ask for payment?

4 What criteria should be used to determine which countries receive the needed food?

5 Should food power be used as a weapon in international politics?

6 If American surpluses are produced for export, what payment should be made to the farmers, and where do the monies come from?

7 Is it proper to have other nations depend on us for food, and is this then a type of nutritional imperialism?

8 What are the requisites for a sound food policy that can be internally beneficial and still yield dividends that will advance human welfare for the largest number of people?

The solution of the food problem needs multidisciplinary cooperation on an international basis. Of course, population control is vital, but other measures must also be taken. Human consumption of livestock is a very inefficient and costly manner to obtain protein. Substitutes—such as soybeans and associated high-yielding, protein crops—can replace at least part of this demand. Better conservation measures on current croplands and special care and attention must be devoted to new croplands. Some circumstances will continue to demand high water and fertilizer usage, so more efficient methods to optimize their distribution is required. There are 80,000 edible species of plants but only 50 are cultivated on a large scale. Continued experimentation and development of new strains can possibly produce important breakthroughs. For example, high-lysine corn (a hybrid whose soft kernel has 66 percent more protein than the regular hard kernel variety) and Tricale (a hybrid of wheat and rye largely used only for livestock) can be used. Better storage and transportation systems could save as much as 30 percent of the total crop. Better use and different types of pesticide control measures are also needed. There are more than 160 bacteria, 250 viruses, 800 fungi, and 8000 insect species that are injurious to crops. Insect and disease control require great care and expertise and can be costly. In the United States and Canada $1.5 billion is spent annually to control the wild oat weed. Other researchers are working on a variety of innovative projects that include: use of plankton as a food resource; farming of the oceans; construction of desert greenhouses on a massive scale of several kilometers; new induced methods for photosynthesis of edible ingredients; and hydroponics (farming on land using water and nutrients, but without soil).

SOIL CONSERVATION

The Romans Cato, Varro, and Columella laid the foundations for agronomy and soil conservation, and their writings formed the basis for early English husbandry books. Prior to A.D. first century most culturalists believed soils went through an aging process that was inevitable and that human intervention could not alter such decay. However, Columella showed that with conservation methods, such as manuring, soils could maintain their fertility.

By the eighteenth century many soil conservation measures were being employed in varying degrees throughout the world. As early as 1769 George Washington was conducting experiments at Mount Vernon to determine "whether the land was not preserved more by harrowing than by lying in furrows." He instituted a program to retire from cultivation those fields that had become washed and gullied. These were then smoothed with the harrow, covered with straw and other vegetal materials to prevent further erosion, and then left to resume the process of recovery. Washington constructed dams in gullies and filled them with various debris. By the early 1800s Virginia and Maryland farmers were using legumes and grasses for soil stabilization and were deep plowing, contour plowing, and hillside ditching—a prototype of the modern terrace. They believed sun rays could harm the soil and used shade crops whenever possible. In 1814 James Hall (a Virginian) had secured a patent for planting corn in an unplowed field, leaving unplowed the area between corn rows (the first conservation tillage). Thomas Jefferson's son-in-law was instrumental in introducing contour farming to America in 1793, and by 1810 Jefferson had become a staunch advocate. Jefferson was also especially fond of crop rotation. Both he and Washington adopted the practice of planting alternate rows of corn with vegetables and potatoes. By 1820 many Virginia farmers were using a rotation system in planting corn, then wheat, after which a 2 to 3 year rest was given with the fallow being plowed so fibers could aid in erosion prevention. Then corn was planted again in the fourth or fifth year.

The strategy for soil conservation is to counteract the manner in which the soil is ruined by water and wind. Therefore, the methods in use can be classed as engineering or biological in nature; their aim is to prevent unrestricted movement of water or air over barren surfaces. Soil loss depends on rain and wind intensity and duration; type and thickness of soil; type and character of the crop or vegetal cover; length of slope; slope gradient; and man-induced measures to abort shear forces. Of course, to be effective there must be the availability of technical and financial assistance; public awareness of the necessity of programs; and planning and implementation of conservation measures. No other single publication helped to stir up more public interest and governmental involvement than the spectacular publication *Soil Erosion—A National Menace* by Bennett and Chapline (1928). We, and other nations, have clearly benefitted from this article, which set the stage for the extensive soil conservation programs in the United States.

In 1947 a U.S. Department of Agriculture study showed that, for similar-sized farms, the average revenue for 1872 farms using conservation methods was $7332, whereas for 888 farms without conservation the revenue was $5950—a difference of $5 per acre. An ideal soil conservation program is one that in the long run increases farm receipts while reducing farm expenses. It is estimated that the money necessary to replace nutrient losses alone in a single year in the United States is $7 billion, and the loss of the nutrients reduce annual yields by a monetary value of $800 million. Since the 1930s the United States has spent $15 billion for soil conservation, and for the latest year of report (1977) the Soil Conservation Service was budgeted $214 million to help farmers develop conservation programs. By 1974 ten states had passed legislation setting guidelines for erosion and sediment control methodologies. However, much has yet to be done— 67 percent of the nation's croplands still need conservation treatment for soil erosion. Even in such states as Minnesota soil conservation practices are underway on only 1.3 million of the total 9 million acres of cropland.

Each soil, terrain, and climate has special qualities, so that the conservational practices that should be used must be specifically adapted to the particular environmental setting. So although there is now a set of generalized guidelines for the technology of soil conservation, they are not equally adaptable for every situation. What has proved to be a successful management ecosystem for one region may be intolerable in another region. Belgian agronomists learned this the hard way when they tried to export the European methods of agriculture into the Congo region. They cleared, burned, and uprooted the forest in experimental plots. The soil was plowed, sowed with a green manure crop, plowed again, and then sowed with a two-year rotation of rice, manioc (cassava), and peanuts, and finally sowed with another leguminous cover crop. The scientists were astounded at the abysmal results. The deep plowing had damaged the soil structure, and the row planting caused accelerated soil erosion and leaching of nutrients. Too much unprotected surface area aided solar destruction of important organics, causing drastic declines in crop yield. The microclimate of the soil was reduced to one of near sterility, and even after abandon-

Figure 16-21. This area in Marion County, Missouri, was formerly a severely eroded terrain with large gullies. The construction of terraces and ponds have stabilized erosion and the exposed areas have been reseeded to alfalfa and brome. (Soil Conservation Service photo, September 15, 1966.)

Figure 16-22. These peanuts are planted using contour-farming methods on 2 to 5 percent slopes. (Soil Conservation Service photo, June 1971, Terrell Co., Georgia.)

Figure 16-23. Terraces on mountainous terrain, Nepal. (Courtesy Brainerd Mears.)

ment for 10 years, allowing the site to be fallow, the soil could barely support scraggly vegetation.

The following techniques and procedures are those that can be effective in preventing erosion when properly applied. The best conservation results are obtained when there is a careful blending of one or more of these methods. Thus, there should be a well-orchestrated plan to utilize both engineering and biological procedures.

Check Dams and Ponding

With steep terrain or lands that have already been severely gullied, an important remedy is to build structures that will impede enlargement of the cuts and lower the volume and velocity of channelized flow (Fig. 16–21). The check dams may be constructed of a variety of materials, including earth, brush, wood, rocks, and cement. In some areas longer dams may be necessary, and the ponded waters can have many other benefits for recreation, wildlife, or recycled water use. A series of flood-retarding structures in the Washita Basin, Oklahoma, reduced sediment yields 48 to 61 percent.

Contour Cropping and Terracing

On steep terrain it is important to reduce the length of overland flow of sheetwash. The longer the flowline the greater the chance of causing incisement as a rill, gully, or stream channel. Thus, farming should conform to the natural hillside contours (Fig. 16–22), and when slopes are excessively steep they should be terraced (Fig. 16–23) so that the tread is flat or only moderately sloped. At Ithaca, New York, potatoes planted up and down the slope had soil erosion rates of 5.7 t/ha, whereas when planted on the contour had erosion of only 0.04 t/ha. However, contour farming resulted in a 5 to 7 percent increase in labor and fuel costs. In

Figure 16-24. Contour tillage and strip cropping, West Virginia, showing alternate strips of corn and alfalfa. (Soil Conservation Service photo, July 29, 1975.)

Figure 16-25. Contour tillage and strip cropping in the glaciated Appalachian Plateau, New York. (Soil Conservation Service photo, August 6, 1964.)

Georgia, cotton farming produced sediment loss of 35.6 t/ha when planted up and down, and the loss was reduced to 15.8 t/ha when planted on the contour at the same site. When the cotton was strip farmed with alternating strips of grass 72 m wide, the sediment yield was further reduced to 1.44 t/ha (Figs. 16–24 and 16–25). Cotton yields in Texas are 25 percent more on those farms that use contour farming, and in Illinois contour farming increased yields in corn by 12 percent, soybeans by 13 percent, and wheat by 17 percent. Table 16–1 illustrates typical reductions in sediment loss that might be expected when using contour, strip, and terrace farming methods.

Crop Rotation

Over a period of years, the cumulative yield of cash crops can be increased when they are not planted perennially. Instead, the crops are rotated with different types of cover crops and legumes to increase the soil-holding powers and to add nitrogen and other nutrients to the soil.

For example, interseeding a legume such as winter vetch can add as much as 25 kg nitrogen per hectare during a single season. On lands with 7 percent slope, yields of cotton grown in rotation were increased 30 percent while soil erosion was reduced from 9 to 5 t/ha. In Missouri annual erosion rates average 8 t/ha for continuous corn farming on 3.68 percent slope but, when rotated with wheat and clover, averaged only 1.1 t/ha. Runoff is also reduced with crop rotation; in Ohio 12 percent slopes with continuous corn produced 38 cm. of runoff with annual precipitation of 91 cm., but with the rotation of a corn-grass-grass-wheat-corn cycle the average runoff was only 15 cm.

Minimum Tillage

Although the practice of minimum tillage, also known as conservation tillage, has occasionally been used throughout the centuries, only within the last few decades has the technique grown in prominence. Revolutionary developments of chemical weed control that started about 1945 have made possible on large scale the use of minimum tillage cropping. Such reduced soil tillage inhibits erosion by leaving residues from the preceding crop on the surface as well as aiding retention of stubble between the planted rows. Moldboard-type plows, which tend to bury such residues, are replaced by equipment that diminish soil displacement during and after

Table 16-1 Reduction in soil loss from up-down slope tilling (in percent)

Slope	Contour tilling	Strip tilling	Terracing
1–12	50–60	25–30	10–12
12–18	80	40	16
18–24	90	45	18

Figure 16-26. Wind destruction of wheat fields grown on sandy soils, Victoria, Australia. (Courtesy Joseph Van Riper.)

preparation for row planting. Such residue breaks the impact of raindrops, reduces splash and the detachment of soil particles, and impedes water runoff. Minimum tillage is only effective when weeds, disease, and insects are controlled; otherwise they would thrive in the organic compost and stubble. In Iowa 2.2 million ha were planted in corn and soybeans by minimum tillage methods in 1973—a 100 percent increase from 1972. Whereas 5 years ago minimum tillage was negligible, it now constitutes 25 percent of total corn and soybean acreage. A 17-year study of the effectiveness of minimum tillage in Nebraska showed soil losses were reduced 50 percent during major storms and 67 percent for smaller storms. The U.S. Department of Agriculture estimates that minimum tillage of some sort is now practiced on about 10 percent of farmlands and states that if 80 percent of croplands used the method soil losses in the United States could be cut in half. Minimum tillage also produces important benefits, such as reducing field development time by 20 percent and cutting tractor bills by 50 percent.

Shelterbelts

The use of trees planted in rows can provide several beneficial results for contiguous soils and crops. Such plantings are referred to as **windbreaks, shelterbelts,** and occasionally **greenbelts.** They have a 100-year history of providing some protection from the high winds that are so

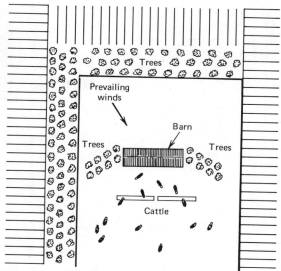

Figure 16-27. Sketch showing protection offered by windbreaks.

common throughout the Great Plains. The 1878 Timber Culture Act offered homesteaders 160 acres of land with the sole provision that they plant trees on 40 acres. Many plantings were made, but the law was repealed in 1891. In the 1930s the CCC (Civilian Conservation Corps) planted a total length of 32,000 km with more than 218 million trees in shelterbelts on 30,000 farms of the Great Plains. By 1965, 13 percent of the shelterbelts had been destroyed in the region, and in Oklahoma more than 20 percent had been removed by 1974. However, 9,250 ha were severely damaged by wind erosion during the storms of 1974–76 (Fig. 16–26).

Windbreaks in Nebraska create 2 to 4 percent higher humidity, reduce evaporation, increase soil moisture, trap more snow, and save contiguous farmhouses 28 percent in fuel that is used to heat and cool homes (Fig. 16–27). In Wuchenchao commune in the Gobi Desert of China, vegetation now flourishes from the shelterbelts of hardy trees that were planted to reduce wind speed and smooth sand dunes. Local plants have rebounded to cover 70 percent of the area. Saudi Arabia has recently planted 10 million tamarisk, acacia, and eucalyptus trees to aid in preventing dunes from

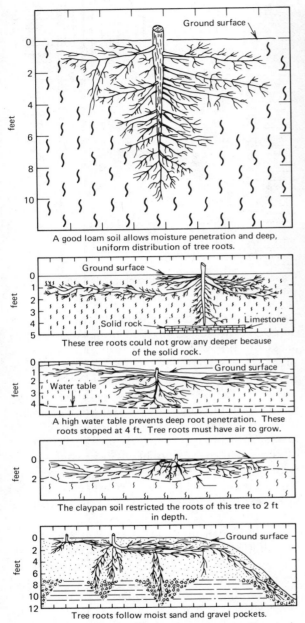

A good loam soil allows moisture penetration and deep, uniform distribution of tree roots.

These tree roots could not grow any deeper because of the solid rock.

A high water table prevents deep root penetration. These roots stopped at 4 ft. Tree roots must have air to grow.

The claypan soil restricted the roots of this tree to 2 ft in depth.

Tree roots follow moist sand and gravel pockets.

Figure 16-28. A series of cross sections of soil and plants, showing the importance of environmental conditions for healthy growth.

Figure 16-29. Before and after conditions, Presidio Co., Texas. (a) Drought and livestock overgrazing have combined to remove all grazeable forage. (b) During conservation and reclamation of the area, the site was disc plowed, and water spread through gated diversions on newly seeded ground. The locale now contains heavy grass growth and supports livestock again. (Photographs courtesy of Soil Conservation Service, Temple, Texas.)

overwhelming the Hofuf oasis. The usual effective range of shelterbelts extends to a distance of 2 to 5 times the height of the trees on the windward side and 20 to 30 times the height on the leeward side. Therefore, rather than forestation of a single large belt, more benefits are derived when they are grown by a corridor system.

Other Methods

There is a variety of other methods that may be classified as "conservational." For example, organic farming and manuring (and, of course, the use of other fertilizers) can be viewed as a type of soil conservation (Fig. 16–28). In Kansas soil losses were reduced from 2 t/ha to 0.2 t/ha by application of 12 t/ha of animal wastes. Manuring in Iowa with 6.4 t/ha for corn crops on 9 percent slopes reduced annual soil losses from 8.9 t/ha to 1.9 t/ha. In some terrains and climates dry farming techniques can be important. This method depends on the use of special equipment that will compact the subsoil so that water is retained as soil moisture only in the upper soil horizon, thus maintaining high water content when plants need it and preventing wind erosion. Other soils require deep plowing for certain crops. However, this can be a disaster if applied to the wrong types of soils, as in the Coastal Plain of Israel where bedouin plowing methods were instrumental in causing a new erosion cycle and badland topography (Aghassy, 1973). Water spreading and grass planting can also produce significant results when properly conducted in some climates and terrains (Fig. 16–29).

Perspectives

Soil problems can no longer be considered only a farmer's problem or a local problem. The accelerated loss of this vital resource transcends provincial boundaries and now affects policy on national and international levels. Unfortunately, we have not learned our history and have not benefitted from mistakes made by past civilizations. The lack of soil stewardship and appropriate management practices are condoned because of expediency—there are expanding population pressures and an escalation in farming costs that result from the high prices of equipment, fertilizers, and pesticides. Thus soil conservation measures, both in terms of engineering structures and biological systems, are receiving less attention now than they did 20 years ago. Much of the world is in need of grain, and most nations cannot feed their people. The only major grain exporting countries are the United States and Canada, and by 1985 pressure from many LDCs for adequate food will create international stress.

Soil, food, energy, and population all interact in a complex network. The problems caused by an exploding population influence the other parts of this network to cause severe problems. There are many opportunities for the geologist, especially in giving information on soils and energy. However, only a multidisciplinary and interdisciplinary approach can be successful in treating all the ramifications and, as in many environmental problems, societal and economic considerations become part of the political-governmental decisions. It is the hope of scientists that their advice can also contribute a voice in policy matters.

The soil is basic to civilization and our survival, and it is only a renewable resource when proper planning can be implemented by astute long-range conservation management. The following measures can contribute to maintaining soil fertility and decreasing loss of the products grown:

1   Use of conservation and management programs to reduce soil erosion and siltation.

2   Application of fertilizers and pesticides as required by specifics of the site.

3   Construction of adequate storage and distribution food systems.

4   Development of new grain strains that are harmonious with the particular environmental setting.

5   International cooperation that includes food banks, a tolerance and understanding of food habits of different people, and the realization that the quality of life depends on an adequate diet for all peoples.

A worldwide conscience is required to solve the intricate problems of food, soil, population, and energy. The world's cattle consume as much food as 8.7 billion people. The United States uses more fertilizer for ornamental purposes than India does for food production. To be self-sufficient, India requires one-half the total fertilizer produced in the world. Until we resolve these disparities, we will not have begun to understand this multifaceted issue.

READINGS

Aghassy, J., 1973, Man-induced badlands topography: in Coates, D. R., ed., Environmental geomorphology and landscape conservation, Vol. III nonurban regions, Stroudsburg, Dowden, Hutchinson & Ross, p. 124–136.

Ayres, Q. C., 1936, Soil erosion and its control: New York, McGraw-Hill, 365 p.

Bennett, H. H., 1939, Soil conservation: New York, McGraw-Hill, 993 p.

Borchert, J. R., 1971, The dust bowl in the 1970's: Ann. Assoc. Amer. Geog., v. 61, p. 1–22.

Bormann, F. H., et al., 1967, Nutrient loss accelerated by clear cutting a forest ecosystem: in Primary productivity and mineral cycling in natural ecosystems, Ecological Society of Amer. AAAS, Orono, Maine, Univ. Of Maine Press, p. 187–196.

Brown, C. B., 1950, Effects of soil conservation: in Trask, P. D., ed., Applied Sedimentation, New York, John Wiley & Sons, p. 380–406.

Brown, L. R., 1975, The world food prospect: Science, v. 190, p. 1053–1059.

———, 1978, The worldwide loss of cropland: Worldwatch Paper 24, Worldwatch Inst., Washington, D.C. 48 p.

Chancellor, W. J., and Goss, J. R., 1976, Balancing energy and food production, 1975–2000: Science, v. 192, p. 213–218.

Coppock, J. T., and Coleman, A. M., 1970, Land use and conservation: Geog. Jour., v. 136, p. 190–210.

Curry, R. R., 1971, Soil destruction associated with forest management and prospects for recovery in geologic time: Assoc. Southeastern Biologists Bull., v. 18, p. 117–128.

Eckholm, E., and Brown, L. R., 1977, Spreading deserts—the hand of man: Worldwatch Paper 13, Washington, D.C., Worldwatch Institute, 40 p.

General Account Office, 1977, To protect tomorrow's food supply, soil conservation needs priority attention: Rept. to Congress CED-77-30, Washington, D.C., Comptroller General of the United States, 59 p.

Glantz, M. H., ed., 1977, Desertification: Boulder, Colo., Westview Press, 346 p.

Glymph, L. M., and Storey, H. C., 1967, Sediment—its consequences and control: in Brady, N. C., ed., Agriculture and the quality of our environment, AAAS Sci. Publ. 85, p. 205–220.

Gray, D. H., 1969, Effects of forest clearcutting on the stability of natural slopes: Assoc. Eng. Geol. Bull., v. 7 (1 and 2), p. 45–66.

Harr, R. D., Harper, W. C., and Krygier, J. T., 1975, Changes in storm hydrographs after road building and clear-cutting in the Oregon Coast Range: Water Resources Research, v. 11, n. 3, p. 436–444.

Hewlett, J. D., and Helvey, J. D., 1970, Effects of forest clear-felling on the storm hydrograph: Water Resources Research, v. 6, n. 3, p. 768–782.

Jensen, N. F., 1978, Limits to growth in world food production: Science, v. 201, p. 317–320.

Le Houerou, H. N., 1977. The nature and causes of desertification: in Glantz, M. H., ed., Desertification, Boulder, Colo., Westview Press, p. 17–38.

Leopold, L. B., 1951, Rainfall frequency: an aspect of climatic variation: Trans. Amer. Geophys. Union, v. 32, p. 347–357.

Lull, H. W., and Reinhart, K. G., 1972, Forests and floods: USDA Forest Service Research Paper NE-226, 94 p.

Marsh, G. P., 1864, Man and nature: Lowenthal, D., ed., 1965 from the original The earth as modified by human action, originally published in 1864 by Scribner; Cambridge, Belknap Press of Harvard Univ. Press, 472 p.

O'Loughlin, C. L., 1974, The effects of timber removal on the stability of forest soils: Jour. Hydrology (NZ), v. 13, p. 121–134.

Packer, J. S., 1973, Slash and burn below the border: Smithsonian, v. 4, n. 1, p. 67–71.

Peterson, H. V., 1950, The problem of gullying in western valleys: in Trask, P., ed., Applied Sedimentation, New York, John Wiley & Sons, p. 407–434.

Pimentel, D., et al., 1976, Land degradation: effects on food and energy resources: Science, v. 194, p. 149–155.

Population Reference Bureau, 1968, The food-population dilemma: Population Bull. 24, n. 4, p. 81–99.

Reitz, L. P., 1970, New wheats and social progress: Science, v. 169, p. 952–955.

Rich, J. L., 1911, Recent stream trenching in the semi-arid portion of southwestern New Mexico, a result of removal of vegetative cover: Amer. Jour. Sci., v. 182, p. 237–245.

Sears, P. B., 1947, Deserts on the march: Univ. of Oklahoma Press, 178 p.

Smith, D. D., and Wischmeier, W. H., 1962, Rainfall erosion: Advance in Agronomy, v. 14, p. 109–148.

Stallings, J. H., 1975, Soil conservation: New Jersey, Prentice-Hall, 575 p.

Strahler, A. N., 1956, The nature of induced erosion and aggradation: in Thomas, W. L., ed., Man's role in changing the face of the earth, Chicago, Univ. of Chicago Press, p. 621–638.

Swanson, F. J., and Dyrness, C. T., 1975, Impact of clearcutting and road construction on soil erosion by landslides in the western Cascade Range, Oregon: Geology, v. 3, p. 393–396.

Swanston, D. N., and Swanson, F. L., 1976, Timber harvesting, mass erosion, and steepland forest geomorphology in the Pacific Northwest: in Coates, D. R., ed., Geomorphology and engineering, Stroudsburg, Dowden, Hutchinson & Ross, p. 199–221.

UNESCO, 1961, Salinity problems in the arid zones: Proc. of the Teheran Symposium, UNESCO, 395 p.

U.S. Department of Agriculture, 1969, Windbreaks for conservation: Soil Conservation Service. USDA Agri. Info. Bull. 339, 30 p.

U.S. Department of Agriculture, 1971, National inventory of soil and water conservation needs, 1967: USDA Statistical Bull. No. 461, 211 p.

U.S. Department of Agriculture, 1972, How to control wind erosion: Agri. Info. Bull. 354, 22 p.

U.S. Department of Agriculture, 1976, Erosion & sediment control guidelines, for developing areas in Texas: Soil Conservation Service, Temple, Texas, var. pages.

Wade, N., 1974, Green revolution (I): a just technology, often unjust in use: Science, v. 186, p. 1093–1096.

———, 1974, Green revolution (II): problems of adapting a western technology: Science, v. 186, p. 1186–1188.

Chapter Seventeen
Weather, Climate and Man

Devastation in Darwin, Australia, caused by Cyclone Tracy on December 25, 1974. Winds of 275 km/h destroyed more than 6000 homes and buildings. (Courtesy Australian Information Service.)

INTRODUCTION

The umbrella of gases that cloak the earth have profoundly affected our life-styles and the materials and processes that comprise our environment. However, this is a two-way street—climate is an important consideration in the management of many construction enterprises, and we are capable of changing some aspects of the weather. Indeed, even climate can be influenced by our behavior and our alteration of the natural hydrologic system. Climate, and the terrain it covers, will become even more important to the survival of mankind as we begin to outgrow our living space in the temperate and equable areas of the earth. Expanding populations will be forced to inhabit more foreboding areas in the future, including semiarid, tropical, and periglacial regions. By their very nature such environments are fragile and do not heal as readily as temperate climates. Thus we place in jeopardy those land-water ecosystems that can be easily damaged by our actions. Therefore, increased understanding of such areas—in terms of ability to sustain life, support human settlement patterns, and be conducive to buildings and to commercial and recreational activities—will take on new meaning in the decades that lie ahead.

Because weather and climate are so pervasive, various aspects of these atmospheric phe-

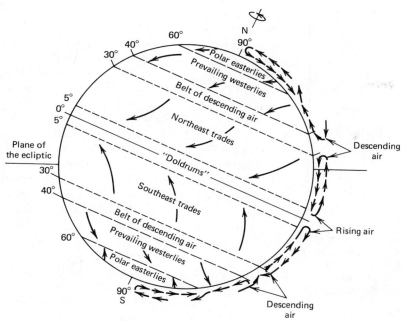

Figure 17-1. Idealized wind circulation on an earth without land masses. In reality, the wind belts shown prevail over oceans but are distorted over land areas. (Longwell, Flint, and Sanders, 1969.)

nomena are also discussed elsewhere in this book. For example, winds play an important role in coastal environments (see Chapter 15); air masses are a vital part of the hydrologic cycle (see page 63); and we may even be instrumental in aiding the desertification of some areas (see page 544).

In this chapter we explore to a large extent both deliberate and inadvertent weather changes produced by human activities. But we are also interested in citing as a case study one particular type of climate—the periglacial—that illustrates our dependency on the glacial legacy of the earth's past.

THE ATMOSPHERE

The spherical envelope of air that surrounds the solid earth is the **atmosphere.** The atmosphere is divided into several zones but the **troposphere,** the turbulent near-surface zone, is where weather occurs. **Weather** is the condition of the lower atmosphere at any given time as determined by the variables of temperature, precipitation, wind, humidity, air pressure, cloud cover, and visibility. The general atmospheric conditions at any given place over a long period of time (usually two or three decades) is **climate** (Fig. 17-1). Both weather and climate are subject to constant fluctuations.

The earth intercepts about one half-billionth of the sun's radiated energy. One-third of this incoming energy is directly reflected back into space. The remainder is absorbed by the atmosphere, oceans, and land, where it powers atmospheric and oceanic circulation and is used by plants in photosynthesis.

The atmosphere is similar to a large heat engine. The incoming solar energy warms the air, mainly by heat absorbed from land or water surfaces irradiated by the sun. Energy is lost to space by long-wave (infrared) radiation from the surface of the atmosphere (Fig. 17-2). Much of the energy transfer occurs via water, which absorbs large amounts of heat during evaporation and releases this heat during condensation and precipitation (Fig. 17-3). Wind circulation is also important in transferring heat to space.

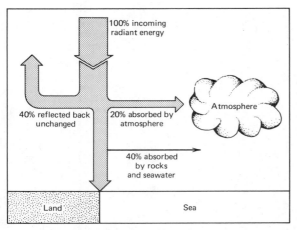

Figure 17-2 Paths followed by the earth's incoming radiant energy from the sun. The energy used in heating the atmosphere causes winds. Most of the energy is absorbed by land and sea and is used up in evaporation of water to form clouds and produce precipitation. (Flint and Skinner, 1977.)

Heat gain is at a maximum in the equatorial zones and heat losses are maximal in the polar regions. The cold, heavy air of the polar areas sinks and moves toward the equator as the warm tropical air rises and then spreads out toward the poles. This basically simple pattern is made highly complex by the deflecting forces of the earth's rotation, the irregularities of the land surface, and the distribution of land relative to water bodies.

The amount of heat absorbed by the earth must be balanced by that lost, otherwise global temperatures will change. This delicate **heat balance** is the crucial factor in climatic change. We are modifying the earth and its atmosphere in many ways that could conceivably alter climate, perhaps without our being aware of it until the scales have been irreversibly tipped one way or the other. Our understanding of weather and climate is still very rudimentary, and the entire field of meteorology and clima-

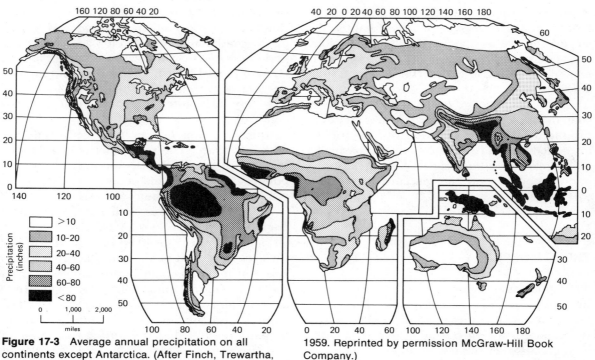

Figure 17-3 Average annual precipitation on all continents except Antarctica. (After Finch, Trewartha, and Shearers, 1959, Physical geography. Copyright © 1959. Reprinted by permission McGraw-Hill Book Company.)

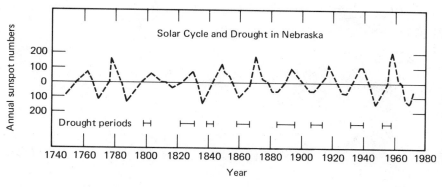

tology is fraught with uncertainties. It is generally conceded that climatic changes are under way and that human influence could help trigger major climatic alterations of global significance. Any change in global climate would have massive repercussions for mankind.

WEATHER MODIFICATION

The influence of weather on human activities is well known to all: the farmer whose crops suffer from the lack of rain, or whose crops cannot be harvested because of too much rain; the traveler stranded in a sudden snowstorm; the picnic ruined by rain; the perfect day for hiking; the beauty and inspiration of a magnificent sunset. We cannot review here the many important relationships between our existence and the daily weather, but we will concentrate on human attempts to modify the weather, either to avert damage or to provide direct benefits.

Precipitation Augmentation

Many parts of the earth are subject to cycles of wet and dry periods. In the U.S. Great Plains, for example, a 22-year drought cycle has been documented (Fig. 17-4). Although the exact mechanisms are not known, the drought appears to be related to magnetic polarity reversals and sunspot cycles. Such weather cycles follow alternate periods of minimal sunspot activity

when the sun's field orientation is in the same direction. These solar cycles influence the shape of earth's magnetic field, and this in turn may affect high cloud formations or global circulation patterns, such as the undulations in the jet stream (Fig. 17-5). It is highly questionable whether we can knowingly and deliberately influence a drought cycle such as this one. However, we can help minimize the harmful effects of it and of other lesser droughts by stimulating precipitation.

Rainmaking is an ancient tradition among many peoples, but only recently has it emerged as a large and serious business enterprise. The first cloud-seeding experiment was performed in 1946 using dry ice. The most successful and widely used technique is to "seed" clouds from aircraft, usually with silver iodide or dry ice (Fig. 17-6). Cloud seeding does not increase the moisture in the clouds—it merely provides additional nuclei on which ice will sublimate. When the ice crystal becomes sufficiently heavy, it will fall to earth, changing to rain if temperatures nearer the surface are warm (Fig. 17-7). Results are highly variable depending on such factors as cloud depth, cloud lifetime, available moisture, weather conditions, and seeding techniques (too many nuclei may form many small ice crystals, none of them heavy enough to fall to earth).

When properly administered under favorable conditions, cloud seeding can produce up to 20 or 30 percent more rain from a given cloud (Table 17-1). This amount could easily

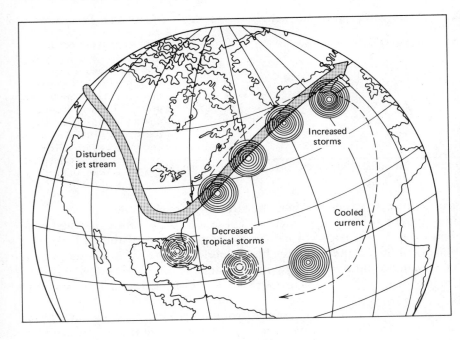

Figure 17-5 A disturbed jet stream increases the cooling effect of storms on the ocean, resulting in decreased tropical storms. With fewer tropical storms, causing less moisture perturbations, the jet stream reverts toward an undisturbed configuration, thus beginning another cycle. (Ferrar, 1976, The urban costs of climate modification. John Wiley & Sons.)

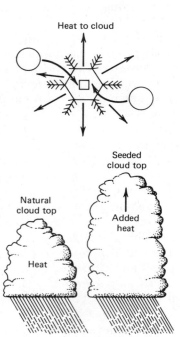

Figure 17-6 The effect of added heat from cloud seeding. The additional ice crystal formation makes the cloud more buoyant and permits the cloud to grow to greater heights than normal. (U.S. Dept. of Commerce.)

spell the difference between crop failure and success for farmers suffering drought (Table 17-2). Although most cloud seeding is done to benefit agriculture, winter seeding to induce snowfall for skiing areas or to increase snowpack for spring and summer water supply is becoming more common. Winter seeding of clouds with a temperature of 11 to 12°C has produced 10 to 30 percent more precipitation in Colorado. In 1977, clouds were seeded over 7 percent of the United States, and 73 other countries were experimenting with cloud-seeding techniques (Fig. 17-8).

The data available on the environmental effects of precipitation augmentation are very sparse. The most serious problems have to do with undesirable side effects. It has not been proven whether seeding can precipitate severe storms but some think it can, and has. Cloud-seeding operations in the Black Hills of South Dakota preceded the "cloudburst" that resulted in the June 9, 1972, Rapid City flash flood. Some survivors of that flood sued the federal government, charging that cloud-seeding experiments were responsible for the excessive

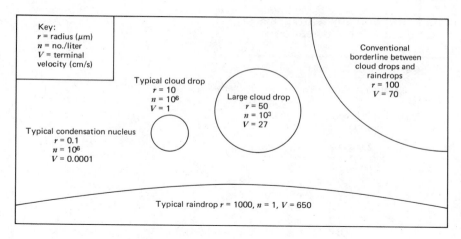

Key:
r = radius (μm)
n = no./liter
V = terminal velocity (cm/s)

Typical cloud drop
r = 10
n = 10^6
V = 1

Large cloud drop
r = 50
n = 10^3
V = 27

Conventional borderline between cloud drops and raindrops
r = 100
V = 70

Typical condensation nucleus
r = 0.1
n = 10^6
V = 0.0001

Typical raindrop r = 1000, n = 1, V = 650

Figure 17-7 Relative sizes of typical cloud condensation nuclei, cloud drops, and raindrops. (U.S. Dept. of Commerce.)

precipitation that produced the flood.

Most disputes about precipitation augmentation stem from the redistribution of rainfall that results (Fig. 17-9). Because no moisture is added to the air, it is apparent that rain made to fall in one place is not available to fall elsewhere. Increasingly, legal skirmishes are occurring over cloud-seeding operations. In the severe drought of early 1977 in the Pacific Northwest, Idaho accused Washington of "cloud rustling." Farmers in eastern Colorado worry about the effects of seeding in the mountains to the west. North Dakotans are worried about loosing needed rainfall to cloud-seeding operations in Montana, while Minnesota farmers wonder the same thing about North Dakota seeding. Such problems could escalate into serious conflicts, especially between poor countries where rainfall is vital to crops needed for subsistence.

Hail Suppression

Hail causes hundreds of millions of dollars of damage each year in North America alone (Fig. 17-10). In less developed countries where many people lack shelter, such as India, hundreds have been known to perish during a single hailstorm. Only one death in the United States has been attributed to hail.

Table 17-1 Comparison of average rainfalls from seeded and control clouds

	Rainfall from		Difference (acre-feet)	Difference (%)
	Seeded clouds (acre-feet)	Control clouds (acre-feet)		
Water calculation (total)				
Without RCC[a]	237	110	127	115
With RCC	237	97	140	144
Water calculation (relative to standard)				
Without RCC	167	40	127	318
With RCC	167	49	118	241

Source: W. L. Woodley, 1970, Rainfall enhancement by dynamic cloud modification: *Science,* v. 170, p. 128. Copyright ©, 1970 American Association for the Advancement of Science.

[a]RCC = radar control clouds.

Table 17-2   Annual estimated corn and soybean losses due to various weather conditions in the Midwest corn belt area

	Average annual loss (bushels/acre)					
	Western corn belt (Nebraska, western Iowa)		Central corn belt (eastern Iowa, Illinois)		Eastern corn belt (Indiana, Ohio)	
	Corn	Beans	Corn	Beans	Corn	Beans
Hail	3.5	2.4	1.9	1.4	1.2	0.9
Wind	3.7	1.0	3.9	0.9	3.8	1.2
Drought	7.3	2.7	5.4	2.7	8.7	3.7
Excessive moisture	2.7	1.6	4.9	2.6	6.9	2.8
Excessive heat	3.5	0.8	2.3	1.1	3.0	1.7
Excessive coolness	0.3	.3	1.5	0.4	1.5	0.7
Freeze or frost	1.1	.6	0.9	.4	1.4	.4
Total loss	22.1	9.4	20.8	9.5	26.5	11.4
Total as percent of total yield	38.0	32.0	28.0	31.0	36.0	38.0

Source: Weather Modification Advisory Board, 1978, The management of weather resources: U.S. Department of Commerce.

[a]From Changnon (1972).

The same techniques and materials are used in hail suppression as are used in rainmaking in the United States. Cloud seeding causes more numerous nuclei to form. These cannot grow to the large size that fewer nuclei would, and they usually will melt before striking the ground (Fig. 17-7).

The Soviet Union reports a benefit-cost ratio of 10:1 for suppression techniques that have reduced hail damage 80 to 90 percent. Their method involves radar detection of hail-spawning cloud areas and seeding them from the ground by use of cannon grenades. Although some feel these figures may be exaggerated, hail suppression has been successful in many cases.

Lightning Suppression

At any given moment of the day or night, 1800 to 2000 thunderstorms are in progress on the earth. The sudden discharge of static electricity buildup by convection currents in cumulonimbus clouds produces lightning. Lightning produces nitrogen oxides and ammonia compounds that are flushed to earth, enriching the soil. Of greater concern, however, are the damages inflicted by lightning.

Lightning kills over 150 people, injures more than 1000 people, and causes over 100 million dollars in property losses in the United States each year. In addition, more than 1 million ha are burned in some 10,000 lightning-caused forest fires in the United States each year. The cost of fighting these fires is well over $100 million annually.

Overseeding of thunderstorms with silver iodide appears capable of reducing the dangerous ground strikes of lightning by up to 50 percent while increasing intracloud strikes. Another technique introduces chaff (small metallic needlelike particles) into charge centers of a storm, causing a reduction of built-up electrical charges by corona discharge, thus reducing the frequency of lightning.

Fog Dispersal

Fog consists of minute water particles and is essentially a cloud that occurs at the ground or

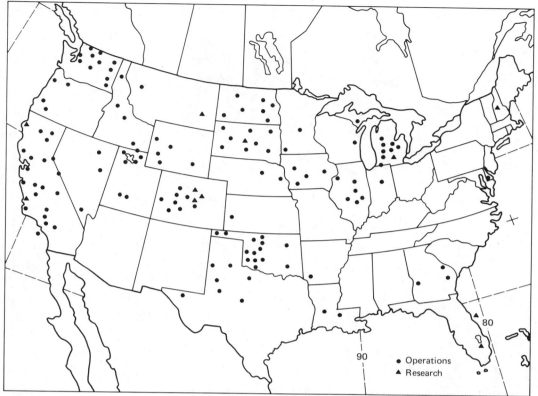

Figure 17-8. Locations of research and operational weather modification projects in the United States, 1973–1977. Two other operational projects were in south-central Alaska; no projects were in Hawaii. (Courtesy of NOAA, Weather Modification Reporting Program.)

water surface. It forms when air is cooled to its saturation point. The occurrence of heavy fog (0.3 km or less visibility) is highly variable and occurs on an average of more than 100 days per year in some areas and less than 5 days per year in others (or zero in deserts).

Restricted visibility caused by fog is a hazard to travelers. The greatest problem ensues when an airport is located in an area of high fog frequency. Tens of millions of dollars are spent annually by airports in attempts to dissipate fogs. Dispersal of supercooled fogs can be accomplished quite effectively with the injection of dry ice or silver iodide.

Unfortunately, 95 percent of the fogs affecting aiports are warm fogs. Unlike supercooled fog, warm fog is in a stable state and no

satisfactory dispersal method is known. Among the approaches that are sometimes attempted, with limited success, are pumping of hot and/or dry air into the fog using helicopters, ground fans, or jet exhausts and seeding with hygroscopic materials.

Modification of Severe Storms

The two most destructive types of severe storms are the hurricane and the tornado. **Hurricanes** are intense tropical cyclones with winds of 119 km/h or more. They average about 320 km in diameter, and winds often exceed 160 km/h. An average of two hurricanes a year strike the U.S. mainland (see page 478).

Tornadoes consist of violent funnel-shaped

Figure 17-9. Who owns the clouds? Cloud seeders can earn the wrath of many people.

clouds extending to the ground from the base of cumulonimbus clouds. With rotating winds which may reach 480 km/h, these small, erratic vortexes are the most dangerous and destructive of storms. From 1953 to 1971, the mean annual number of reported tornadoes in the conterminous United States was 655. They caused an annual average of 118 deaths, 2000 injuries, and $200 million in property damage. A record number of 1109 tornadoes were reported in the United States in 1973, and the number has remained exceptionally high (over 800 a year) ever since. Wichita Falls, Texas, was devastated in April 1979, with losses of hundreds of millions of dollars.

The most intelligent approach to reducing damages from many of the effects of severe storms is improved land use regulation to min-

imize the people and property in vulnerable locations, such as the floodplains and unstable, exposed areas near shorelines. Where settlement already exists and permanent evacuation is unfeasible, a combination of protective measures, warning systems, and emergency procedures can greatly reduce damages and loss of life.

The dynamics of severe storms are not well understood, and storm modification techniques are still in their infancy. The sudden and sporadic nature of tornadoes makes any kind of control extremely difficult. Hurricanes are carefully watched and, when they threaten populated areas, overseeding of the "wall clouds" near the eye may help to diminish the intensity of the storm (Fig. 17-11). It is believed that the heat released by condensation may alter the

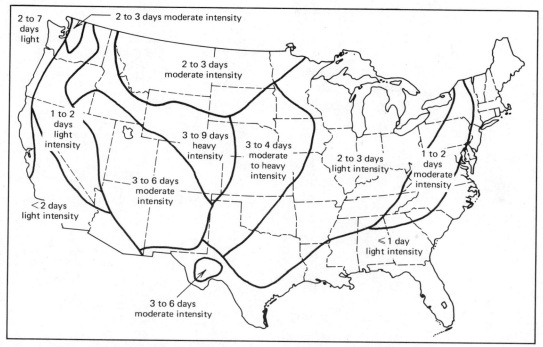

Figure 17-10. Hail regions in the United States showing average annual number of days with hail at any point in a region and the average intensity of hail.

(Changnon, et al., 1978, Hail suppression and society; Science, v. 200, p. 388. Copyright © 1978 American Association Advancement of Science.)

pressure gradient and diminish wind velocity. Even a small reduction in wind velocity is significant, because damage is proportional to the square of the wind speed.

It is always risky to tamper with natural phenomena, especially when there are so many unanswered questions about the weather. Thus far, however, negative impacts have, with only a few exceptions, been minimal. Carefully planned efforts to reduce the severity of harmful weather phenomena or to redistribute precipitation so as to be more beneficial have met with sufficient success to encourage greatly increased efforts in this direction. The United States now spends $16 to $19 million annually on weather modification. The Weather Modification Advisory Board recommends doubling that expenditure in 1981. The board feels that we are "within sight" of consistently achieving a 10 to 30 percent increase in mountain snow-pack by the early 1980s, a 10 to 30 percent increase in rainfall in the Midwest and High Plains by the late 1980s, a 10 to 20 percent reduction in hurricane winds by the 1990s, and a 50 percent hail reduction by the 1990s.

CLIMATIC INFLUENCES ON SOCIETY

Climate profoundly influences the human food supply, energy and material needs, and lifestyles, and any change can have most serious consequences in today's crowded world with its numerous volatile political and social problems. Better comprehension of natural climatic change and our possible influences on it are therefore major priorities among scientists today.

An understanding of the weather and cli-

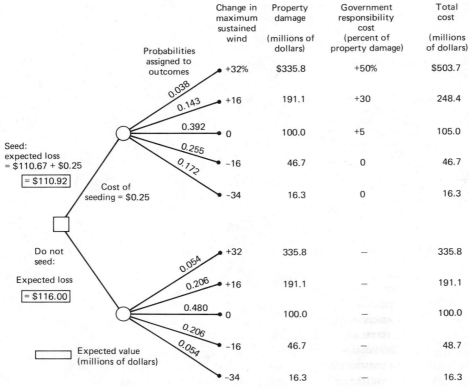

	Change in maximum sustained wind	Property damage (millions of dollars)	Government responsibility cost (percent of property damage)	Total cost (millions of dollars)
	+32%	$335.8	+50%	$503.7
	+16	191.1	+30	248.4
	0	100.0	+5	105.0
	−16	46.7	0	46.7
	−34	16.3	0	16.3
	+32	335.8	—	335.8
	+16	191.1	—	191.1
	0	100.0	—	100.0
	−16	46.7	—	48.7
	−34	16.3	—	16.3

Probabilities assigned to outcomes

0.038
0.143
0.392
0.255
0.172

Seed:
expected loss
= $110.67 + $0.25
= $110.92

Cost of seeding = $0.25

Do not seed:

Expected loss
= $116.00

0.054
0.206
0.480
0.206
0.054

Expected value (millions of dollars)

Figure 17-11 Points considered in the decision to seed the nominal hurricane, including government responsibility cost. (Howard, et al., 1972, The decision to seed hurricanes: Science, v. 176, p. 1197. Copyright © 1972 American Association Advancement of Science.)

matic variations of the past is vital to a working knowledge of present and future climates. In recent years great advances have been made in the methods of determining past climatic changes.

More and more, climatologists are investigating anthropogenic evidence related to climate in the records of past civilizations as deciphered from artifacts, structures, writings, and other remains. Many of the migrations and downfalls of past cultures, which were previously attributed largely to the invasions of enemies or other cultural, social, or political factors, may in fact have instead been caused by changing climates. In the American Southwest, where the arid climate is especially sensitive to climatic fluctuations, it is possible to trace the migrations of the Indian peoples from higher to lower elevations, or vice versa, in response to changing climates. A wide variety of natural evidence helps us determine the nature of those changing climates.

Many living organisms, more often plants, can survive only within a relatively narrow range of climatic conditions. Analysis of pollen and other remains found in deposits, especially in wetlands, will reveal the types of vegetation that existed in the past, giving us a very good idea of the climate. Improvements in the dating of materials, usually with radiometric techniques using carbon-14, are providing a more detailed picture of earth's paleoclimates. **Dendrochronology** (the study of tree rings), study of buried soils formed during past climatic episodes, past erosional and depositional cycles, and paleomagnetic evidence are also contribut-

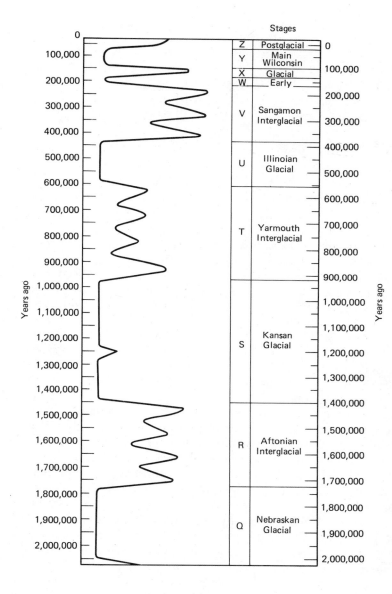

Stages

Z	Postglacial
Y	Main Wisconsin
X	Glacial
W	Early
V	Sangamon Interglacial
U	Illinoian Glacial
T	Yarmouth Interglacial
S	Kansan Glacial
R	Aftonian Interglacial
Q	Nebraskan Glacial

Figure 17-12 Pleistocene time scale and generalized climate curve based on study of deep-sea sediment cores. (Ericson and Wollin, 1968, in Science v. 162, p. 1233. Copyright © 1968 American Association Advancement of Science.)

ing significantly to our understanding. Because the proportion of oxygen-18 to oxygen-16 in the seas and atmosphere varies in response to temperature, analysis of these oxygen isotope ratios in glacial ice or ocean sediments can supply information on the temperature at the time those materials were formed. By combining these and other techniques with the geomorphic evidence left from past glaciations and

our knowledge of existing glaciers, a detailed scenario of the earth's changing climates is emerging.

There can be no question that climates during the past 3 million years or so (the Quaternary period) have fluctuated widely (Fig. 17-12). Weather conditions during the Quaternary have been abnormally cold and erratic compared to most of geologic time, and the climatic

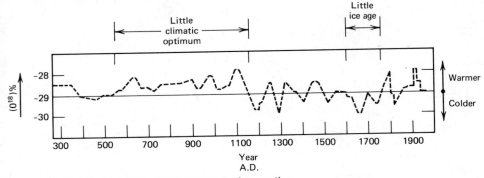

Figure 17-13. Oxygen-18 isotope concentration on the ice core at Camp Century, Greenland, and approximate climatic stages. (U.S. Army, CRREL.)

fluctuations have had profound effects on all living things and natural surface processes.

At the height of the most recent major period of glaciation, about 18,000 years ago, nearly 30 percent of the world's land surface was covered by ice and sea level was 150 m lower than at present. Glaciers thousands of meters thick covered most of North America north of the Ohio and Missouri rivers. Long Island consists of material deposited at the margin of these mammoth ice sheets.

For the past 11,000 years we have experienced an **interglacial stage**—a period in which the great ice sheets have shrunk and milder climates prevail. However, we are still very much in the "ice ages"—10 percent of the world's land area is presently covered by ice, and over three-quarters of the world's freshwater is in the form of ice.

Recent evidence suggests that planetary movements have a strong influence on the climatic fluctuations that have characterized the Quaternary. On this basis, the long-term trend will be back toward cooler climates and major glacial advances. The time required for the change from an interglacial to a glacial age is a matter of great conjecture. Most past theories assumed that thousands of years were involved, but recent evidence indicates that such climatic shifts may have occurred in several hundreds of years. One somewhat radical theory, the "snowblitz" theory, even suggests that a major

glaciation could develop in as little as 7 to 10 years by buildup of thin but very extensive perennial snow and ice cover produced by a series of snowy winters followed by cool summers. Snow (with high albedo due to whiteness) reflects 85 percent of the sunlight; thus it tends to create its own cold-climate environment that inhibits melting. Once a thickness of about a meter is attained over large areas of Canada and northern Eurasia, a major ice age could be initiated.

However, it would not require a climatic change of the magnitude of a major glacial age to seriously disrupt human activities. Within the present interglacial stage, climatic fluctuations that have had serious consequences for mankind have been frequent (Fig. 17-13).

For example, in A.D. 981, the Vikings established a colony in Greenland, which they named for the lush grasses growing there at that time. By about 1271, the climate was becoming colder, and glaciers advanced filling the northern seas with icebergs. Some consider this the beginning of the "Little Ice Age." Viking exploration (which had probably extended to North America) then ceased, and Greenland was eventually overrun with ice. The fourteenth century was a disastrous one for Europe, with many freezing winters and wet summers that depleted crops and led to major famines and disease epidemics. In 1348 the bubonic plague (Black Death) appeared—and over the next 300 years

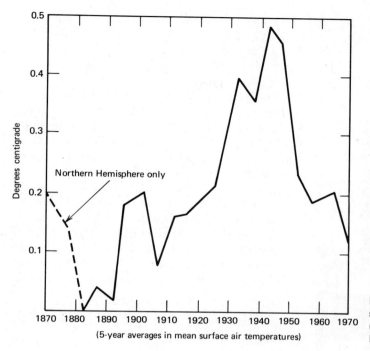

Degrees centigrade

Northern Hemisphere only

1870 1880 1890 1900 1910 1920 1930 1940 1950 1960 1970
(5-year averages in mean surface air temperatures)

Figure 17-14. Global temperature changes since 1870 in the Northern Hemisphere. (National Center for Atmospheric Research.)

this dread disease would take some 25 million lives in Europe. Erratic weather characterized by severe cold periods and droughts continued into the nineteenth century, triggering repeated crop failures and epidemics. The last in a series of blights that destroyed most of the Irish potato crop from 1845 to 1847 precipitated the great Irish migration to the United States. But that was the last of the severe spells of weather because, beginning about 1850, a warming trend started and the Little Ice Age came to an end. From 1880 to 1940, temperatures increased an average of about 0.6°C (1°F), and glaciers retreated resulting in the warmest period in the Northern Hemisphere in at least 800 years (Fig. 17-14). Note that the mean temperature change need not be very large to produce very significant climatic changes. In the last 7 million years, mean global temperatures may not have been as high as those in the first two-thirds of the twentieth century for more than about 5 percent of the time.

This benign, yet ominously anomalous, pe-

riod of warm climate has witnessed, and is in part responsible for, the unprecedented explosion of human population and resource exploitation. From 1940 to 1970, there was a 0.3°C (0.5°F) drop in mean global temperature, and the experts generally agree that we cannot expect the favorable weather of the recent past to continue.

The 1970s have already provided some unsettling evidence that climate may be changing for the worse. The world's snow and ice cover increased by 10 to 15 percent in the early 1970s. From 1970 to 1976, 3 million people died from drought in six countries of the African Sahel (the southern border region of the Sahara Desert). Millions of hectares turned to desert in that area and 30 to 70 percent of the livestock perished. In 1972, the Russian wheat crop failed as the worst drought in several centuries struck the Moscow region. Floods, droughts, abnormally cold winters, shorter growing seasons, and generally errant weather occurred with much greater frequency in the

Figure 17-15. Plumes of air pollution belching from smokestacks, Cleveland, Ohio. (Grant Hellman.)

1970s than during the previous 70 years. The probability is that more unpredictable and extreme weather is here to stay for many years.

What will happen with a permanent return to more "normal" weather conditions such as that of the Little Ice Age? Could we still manage to feed over 4 billion people? How will the many unstable and deeply divided nations of today react to the crisis that may be approaching? Such concerns led the U.S. Central Intelligence Agency to conclude that changing climate is "perhaps the greatest single challenge that America will face in coming years."

AIR POLLUTION

Smog alerts, visibility indices, and pollution levels have become an important part of weather reporting in virtually every major modern city (Fig. 17-15). The 5.8 thousand billion tons of air surrounding the solid earth is one of our three absolutely essential resources (the others being water and soil). The eventual effects of the industrialized world's habit of using the atmosphere as a handy and inexpensive dump for pollutants are still very poorly understood (Table 17-3). The more immediate results of uncontrolled air pollution—in terms of loss of life, health, property and aesthetic values—are often painfully obvious. The cost of air pollution cleanup will run into many billions of dollars in the United States alone, but what is not adequately emphasized is that the costs of *not* cleaning up are far greater.

Atmospheric contaminants are generated in many ways from many sources. In the United States as a whole, roughly 60 percent of the air pollution comes from motor vehicles, 17 percent from industry, 14 percent from electric generating plants, and 9 percent from space heating and incineration (Table 17-4). Not considered in these estimates are the greatly increased atmospheric burdens of such substances as soil particles, pesticides, and plant matter (many of them allergens, such as pollen, spores, and "smust"—smog and dust) that result from human earth-moving and agricultural activities.

The major goal of the 1970 Clean Air Act was healthful air over the entire country by 1977. But battles against massive pollution are not easily won and, in 1977, estimates were that 4000 deaths and 4 million sick days were caused annually by auto emissions alone. The auto makers were granted a fourth extension (this time until 1982) to meet emission standards. In 1977, Congress also weakened provisions intended to prevent relatively unpolluted areas, such as national parks and wilderness areas, from degradation of air quality. A detailed 1977 study by the National Academy of Science concluded that up to 21,000 people east of the Mississippi River die each year from power plant emissions. A major move to coal as a principal energy fuel could nearly double that death rate.

Table 17-3 Some questions and comments on atmospheric pollutants

	Question	Comments		Question	Comments
1.	What atmospheric process is the discharge likely to affect?	Controls should be based on known or postulated effects of the pollutants.	9.	What are the lifetimes and sinks of the pollutants?	Rapid sinks depress concentrations, but the throughput may remain high and the deposition concentrated.
2.	Is the discharge large or small relative to the concentration of the same or other agents in the atmosphere which act similarly?	A discharge must cause an appreciable change in the concentration of the same or similar agents in order to affect an atmospheric process.	10.	Are the pollutants water-soluble or hygroscopic?	If so, they will concentrate in clouds, will be potentially active in modifying warm clouds, and will experience relatively short residence times in the atmosphere with possibly large local deposition.
3.	Are all the constituents of the effluent known?	Trace substances might have more impact on the atmosphere than the primary constituents.	11.	Are the pollutants surface-active?	Surface-active molecules may coat cloud droplets and affect cloud and aerosol evaporation and coagulation.
4.	Will any of the chemicals in the discharge react with atmospheric constituents?	Beware of regenerating chemical chains, especially those known to involve OH and nitric oxide (NO). Small influences on these two species may have large secondary effects (for example, photochemical reactions leading to smog formation).	12.	Do the solid pollutants have crystal structures and dimensions similar to those of ice (that is, hexagonal) and are they insoluble?	If so, beware! They may nucleate ice in cold clouds and affect precipitation processes, even at very low concentrations.
5.	Can chemicals in the discharge interact in the atmosphere to amplify their effects?	If so, beware! [For example, the reaction between SO_2 and ammonia (NH_3) in the gaseous phase is slow, but in the presence of cloud droplets it is accelerated enormously to produce sulfates.]	13.	Do the chemicals interact with visible light? If so, (i) Do they photodissociate? (ii) Do they form reactive fragments such as free radicals and metastable atoms? (iii) Do they sensitize other chemicals?	If so, they may contribute to atmospheric heating. Affirmative answers to these questions imply activity in photochemical processes affecting smog.

Table 17-3 (Continued)

	Question	Comments		Question	Comments
6.	Is the discharge concentrated (a point source) or diffuse (an area source)?	The first may stimulate more complaints, but the latter more damage.	14.	Do the chemicals interact with infrared radiation?	If so, they will affect radiation transfer, especially if they absorb between the strong lines of H_2O and CO_2. Absorption at these wavelengths will cause warming near the earth's surface but cooling in the stratosphere.
7.	What concentrations are to be expected downwind from sources of pollutants?	Simple model calculations can be used to estimate these concentrations. (The results are not always reliable.)	15.	Most critical of all, do the pollutants affect aerosols which play a role in cloud processes?	If so, the structure and distribution of clouds may be affected by the pollutants, thereby causing changes in precipitation and optical scattering. Research priority should be given to this area.
8.	Is the weather stagnant or dispersive?	Controlled emissions are generally best vented in dispersive conditions. However, in some cases discharges in dispersive conditions might produce more total damage by being distributed over a wider area.			

Source: P. V. Hobbs et al., 1974, Atmospheric effects of pollutants: *Science*, v. 183, p. 914. Copyright © 1974 American Association for the Advancement of Science.

Table 17-4 Major industrial sources of carbon dioxide (ton $\times$ 10^6) emitting to the U.S. atmosphere

Year	Power plants			Other combustion			Total fossil fuel	Cement manu- facture	Total CO_2 emis- sions	From power plants (%)
	Coal	Oil	Gas	Coal	Oil	Gas				
1890	10	1	0	356	21	1	388	2	422	2
1900	12	1	0	510	34	9	555	4	559	2
1910	25	1	1	960	118	28	1163	7	1170	2
1920	77	3	5	1191	204	66	1546	13	1559	5
1930	96	4	8	1023	485	99	1715	16	1731	6
1940	176	8	11	970	614	154	1933	18	1951	10
1945	183	11	20	1209	792	201	2416	14	2430	9
1950	217	38	39	910	1089	307	2600	33	2633	11
1955	340	38	68	706	1223	486	2861	43	2904	15
1960	418	43	101	509	1400	589	3060	48	3108	18
1965	673	59	134	495	1510	669	3540	55	3595	24
1970	1031	92	150	448	1751	870	4342	69	4411	29
1975	1510	108	160	405	2028	998	5209	82	5291	34
1980	2060	203	160	366	2353	1131	6273	99	6372	38
1990	2403	277	155	300	3162	1392	7689	146	7835	36
2000	2197	254	150	245	4252	2227	9325	216	9541	35

Source: F. A. Rohrman et al., 1967, Industrial emission of carbon dioxide in the United States: *Science*, v. 156, p. 931. Copyright © 1967 American Association for the Advancement of Science.

However, the air in many of the most severely polluted U.S. urban areas has been notably improved, and overall emissions of particulates, carbon monoxide, hydrocarbons, and sulfur dioxide have been decreased. But recent studies reveal that the harmful effects of air pollution are more widespread than previously believed. Los Angeles smog can be detected (visually, and identified by chemical analysis) in Grand Teton National Park in Wyoming. Emissions from midwestern sources is more harmful to people living in the eastern United States than to local inhabitants.

It is becoming increasingly apparent that vegetation is highly sensitive to airborne pollution, with over $25 million worth of annual damages to crops in California alone. Grapes in New York State are suffering from polluted air carried by prevailing westerlies from northern Indiana and Ohio. Annual tree ring growth in ponderosa pines in the San Bernardino Mountains of California showed that 20 board-feet of merchantable wood were produced per tree between 1910 and 1940, but only 5 board-feet were produced in similarly aged trees during the more polluted years from 1944 to 1974.

We will now take a closer look at two types of air pollution with far-ranging long-term consequences. The influence of human-induced atmospheric alterations on climate will be discussed later in this chapter.

Acid Rain

One of the most widespread and potentially dangerous results of air pollution is acid rain (Fig. 17-16). As discussed in Chapter 6, a major impurity in coal is sulfur. When burned, much of the sulfur is released as sulfur dioxide, a gas which then combines with atmospheric water to produce sulfuric acid. The acidic water returns to earth as acid rain. Electric utilities, largely through coal-burning power plants, were responsible for 65 percent of all sulfur dioxide pollution in the United States in 1976. Under the proposed National Energy Plan schedules

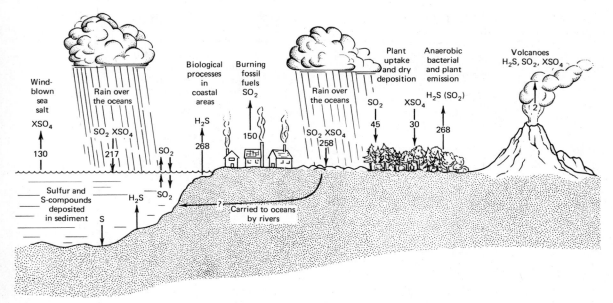

Figure 17-16. Sources and characteristics of acid rain.

that call for greatly increased dependency on coal-burning plants, 19 to 23 million tons of sulfur dioxide per year would be released into the atmosphere by U.S. plants by 1990.

Acid rain is a growing problem downwind from all industrialized and fossil fuel consuming areas. Normal rainwater has a pH of about 5.7, but much of the precipitation in Europe and the eastern United States has an average pH of 4, over 50 times more acidic than normal. Rainwater pH readings as low as 2.1 have been measured.

Acid rain raises the pH of soil, killing soil organisms, leaching nutrients from the soil, and decreasing the soil's ability to support crops. Farm crops have recently been shown to suffer up to 20 percent damage within an 800 km radius of the new coal-fired power plants at Colstrip, Montana. Acid rain can harm the plants directly in a variety of ways, including the solution of plant tissue and sulfur dioxide exposure which has been observed to decimate lichen populations important to soil stability, to reduce soil insect and microorganism populations, to alter metabolism of grasses, and to

decrease protein content and digestibility of range grasses. The costs of such damage to agricultural and forestry enterprises run into the hundreds of millions of dollars each year.

The surface runoff of acid rain lowers the pH of streams and lakes, threatening valuable ecosystems. If pH drops below 5.0, fish usually disappear completely and plant life is drastically altered. Because of acid snow melt, stream acidity is often highest in spring when eggs are being laid. Many fish eggs have been shown to be extremely sensitive to even slight drops in pH. More than 15,000 lakes in Scandinavia and Canada have been damaged by acid rain, and 90 percent of the lakes above a 2000 ft elevation in the Adirondack Mountains of New York have a pH below 5 and support few fish. In the 1930s, less than 4 percent of these same lakes were below a pH of 5 and trout were abundant.

Acid rain also induces the corrosion of solids (Fig. 17-17). Each year billions of dollars of damage is done to sculptures, cathedrals, and other stone or metal structures around the world. One common chemical reaction combines the sulfuric acid in acid rain with lime-

Figure 17-17. Acid rain not only affects natural systems but is also costly to society.

stone or marble ($CaCO_3$):

$$CaCO_3 + H_2SO_4 + H_2O \rightarrow CaSO_4 \cdot 2H_2O + CO_2 \uparrow$$

The corrosion of the limestone by the acid produces gypsum ($CaSO_4 \cdot 2H_2O$). Tiny amounts of gypsum crystallize in small interstices of the stone as it drys out. This accelerates the natural weathering process of **exudation** (salt weathering), wherein the pressure exerted by the tiny growing crystals wedges apart minute cracks in the rock, causing granular disintegration or spalling of the rock, which exposes more of the stone to attack by acidic waters. Because of such accelerated chemical and physical weathering, important buildings such as the Lincoln Memorial in Washington and the Greek Parthenon have deteriorated more in the past 20 years than in all previous years since construction.

Sulfur dioxide also reduces visibility, even at low pollution levels. This is a source of growing concern in the western coal regions (northern Great Plains and Colorado Plateau) where new and planned power plants threaten some of the clearest air in the country.

The most acute effects of sulfur dioxide have been observed within 40 km downwind of emission sources. It has been known to cause women's stockings to dissolve while walking along rainy city streets. New, taller smoke stacks, while reducing the local air pollution level, are spreading the problem over ever-increasing areas. A 360 m high stack can cause acid rain 1000 km away. There is also evidence that tall stacks cause the sulfur dioxide to remain airborne longer, permitting a greater percentage of it to be oxidized into sulfuric acid.

If the sulfur in coal is in discrete minerals such as pyrite or marcasite, it can be mechanically separated from the coal before combustion (Fig. 17-18). Unfortunately, sulfur is often bound to the organic molecules of the coal, and effective means of "cleaning" this coal have yet to be developed.

Much sulfur that is burned must be captured by air-emission control techniques. The material thus collected must also be used or disposed of. Smokestack scrubbers are effective in removing much gaseous pollution, but they are also expensive, energy-intensive, and produce 0.3 m³ of sludge per ton of coal burned. A 1000 MW plant using high-sulfur coal would need to dispose of 2400 m³ of sludge each day.

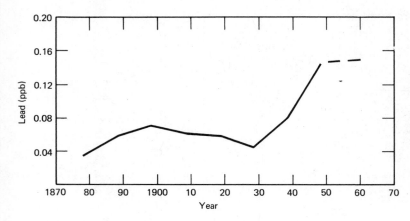

Figure 17-18 Variation of anthropogene lead dust fall at Camp Century, Greenland, from about 1880 to 1960, showing the rapid increase in recent times of manmade particulates in the atmosphere. Lead is measured in parts per billion (ppb). (After Bryson, 1974. A perspective on climatic change: Science, v. 184, p. 758. Copyright © 1974 American Association Advancement of Science.)

Regenerative scrubbers that would reuse scrubbing agents and produce usable sulfuric acid and elemental sulfur in the process are now being developed. Other methods using adsorptive materials such as activated carbon also look promising.

Recently proposed (1979) regulations in the United States would require .all new coal-fired plants to control 85 percent of their sulfur dioxide emissions. Industry will strongly resist such requirements, and it is highly questionable whether this goal can be achieved. It is worth noting that, in Japan, 93 percent sulfur dioxide removal is routinely achieved in a cost-effective manner. In early 1978, less than two-thirds of existing U.S. coal-fired plants were meeting the then-current air emissions standards, and only 120 of the nation's 970 fossil-fuel-powered plants had installed scrubbers. Even at the 85 percent cleanup level, sulfur dioxide emissions in 1985 are expected to be much greater than 1978 levels because of increased reliance on coal. Of course, other particulates are also dispersed by combustion processes caused by human activity (Fig. 17-18).

Stratospheric Pollution

Very little mixing occurs between the upper and middle stratosphere and the troposphere. Therefore, pollutants introduced into the stratosphere will remain there for a very long time.

The major source of pollutants emitted in the stratosphere come from high-flying aircraft such as the supersonic transport (SST). The exhause of jet planes includes water, carbon dioxide, sulfur dioxide, nitrogen oxides, and particulates. The ultimate effect of these materials is conjectural—an increase in water vapor content of the stratosphere favors temperature increases but the particulates provide nuclei for ice crystal formation, which increases reflectivity and favors cooling. In any event, pollution levels are still low and no significant effects are foreseen for the immediate future as a result of aircraft pollution.

A more serious threat involves the use of certain gaseous chemicals at the earth's surface that can infiltrate upward to the stratosphere. The greatest concern to date has centered around the chlorofluoromethanes (CFM), or freons, that are commonly used as refrigerants and propellants in aerosol cans. It has been shown that these chemicals can destroy the ozone that forms a protective layer in the stratosphere.

Stratospheric ozone forms when solar ultraviolet (UV) radiation reacts with oxygen. The ozone layer then absorbs additional UV radiation, protecting the biosphere from this severe form of radiation. Exact effects of increased levels of UV radiation are uncertain, but all biological processes would presumably be influenced to some extent. It is possible that no land

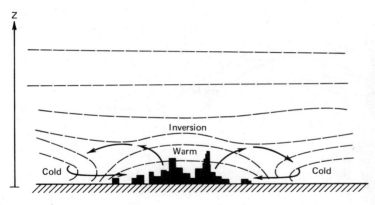

Figure 17-19 Idealized scheme of nocturnal atmospheric circulation above a city in clear, calm weather. The diagram shows the urban heat island and the radiative ground inversions in the rural areas, a situation that causes a "country breeze" with an upper return current. Dashed lines are isotherms; arrows are wind; Z is the vertical coordinate. (After Landsberg, 1970, Man-made climatic changes: Science, v. 170, p. 1270. Copyright © American Association Advancement of Science.)

life was able to evolve on earth until an ozone layer was built up nearly 400 million years ago. The UV rays have also been linked to skin cancer in humans. Malignant melanoma (a severe form of skin cancer) is increasing at rates of 3 to 9 percent per year in all countries, and death rates from melanoma have doubled in the last 15 years.

According to the World Meteorological Organization, average global ozone depletion, based on 1977 rates of clorofluoromethane use, will reach 5 percent in 20 years and then continue to increase until a steady state is reached at 15 percent depletion. Other studies predict 20 percent depletion over the next 50 years. A 15 percent reduction in ozone means a 30 percent increase in UV radiation at the earth's surface and a 10°C rise in temperature of the upper stratosphere. The climatic effects of stratospheric warming are not know, but a 30 percent rise in UV radiation could have deadly consequences for many living things. Concern over ozone depletion led to major cutbacks in CFM usage in the United States in the mid-1970s (1976–78). However, its use is still widespread around the world, and the chemicals will continue to accumulate in the stratosphere for many years after all use is ended.

Stratospheric nitrous oxide, which may originate from aircraft exhausts, nuclear explosions, and fertilizer use, can also destroy ozone but the available evidence does not show this to be a significant source of depletion at the present time.

THE HUMAN INFLUENCE ON LOCAL CLIMATE

The results of intentional weather modification and of air pollution, if continued for any length, will undoubtedly influence local climates. Localized changes in climate can also be an inadvertent result of a wide variety of modifications that we have imposed on the earth. Changes produced in any of the following are likely to produce a response in local weather and climate: surface color (albedo or reflectivity); terrain roughness and orientation; the nature of surface or near-surface materials (permeability, thermal characteristics, etc); local sources of heat and/or moisture (water bodies, thermal areas, cities, power plant complexes, etc); vegetation; dust sources.

Perhaps the "heat island" effect of cities has been the most widely recognized climatic influence (Figs. 17-19 and 17-20). Cities are often up to 10°C (18°F) warmer at night and in calm winter weather and 1 to 2°C (1.8 to 3.6°F) warmer at other times than the surrounding countryside (Table 17-5). There are two major reasons for this: (1) the rate of energy generation and conversions with their inevitable pro-

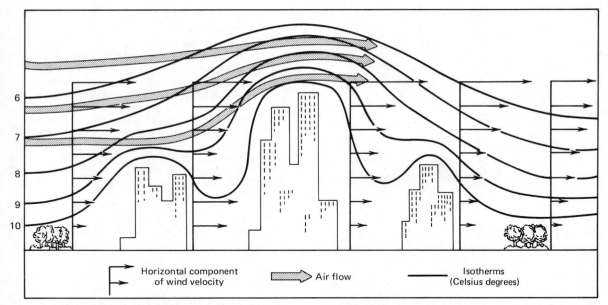

Figure 17-20. Urbanization affects airflow and temperature. Notice how buildings obstruct air patterns. Porosity of building may produce low level peaks of the downwind profiles, and heat retained in concrete-stone buildings is released with other heat, causing increased air temperature. This heating causes isotherms to rise from the countryside and bulge over building clusters to form the "heat island effect," making air less stable above and more stable within the heat island. (Ferrar, 1976, The urban costs of climate modification. John Wiley & Sons.)

duction of waste heat are greater in cities, and (2) asphalt, concrete, buildings, and so on, are heat traps. A plume of heat and material pollutants extends downwind from cities, produc-ing, among other things, cloudier conditions in those areas (Fig. 17-15). Cities themselves re-ceive about 10 percent less solar radiation than would be received normally (with no city). Such

Table 17-5   Maximum urban-rural differences in summer rainfall and severe weather events, expressed as percent of rural values and based on historical data

City	Rainfall	Thunderstorms	Hailstorms
St. Louis	+15	+25	+276
Chicago	+17	+42	+246
Cleveland	+27	+38	+90
Indianapolis	0	0	0
Washington, D.C.	+9	+36	+67
Houston	+9	+10	+430
New Orleans	+10	+27	+350
Tulsa	0	0	0
Detroit	+25	No data	No data

Source: Weather Modification Board, 1978, The management of weather resources: U.S. Department of Commerce.

[a]From Changnon (1973).

Table 17-6 Weather changes (in percent) resulting
from major urbanization in the Northern Hemisphere

Weather phenomenon	Annual	Average changes compared to rural conditions	
		Cold season	Warm season
Contaminant volume	+1000	+2000	+500
Solar radiation	−22	−34	−20
Temperature (°C)	+2	+3	+1
Relative humidity	−6	−2	−8
Visibility frequency	−26	−34	−17
Fog frequency	+60	+100	+30
Wind speed	−25	−20	−30
Cloudiness frequency	+8	+5	+10
Rainfall	+14	+13	+15
Snowfall	±10	±10	—
Thunderstorm frequency	+15	+5	+30

Source: Weather Modification Board, 1978, The management of
weather resources: U.S. Department of Commerce.

[a]From Changnon (1976).

figures vary widely from city to city, of course,
depending on pollution controls and sources.
London, England, is an excellent example of
what can be accomplished with controls—aided
by stable population and slow industrial growth,
pollution cleanup has resulted in 80 percent
more sunshine today than London was receiving in 1955.

The air in cities is slightly less humid, but
due to rising air currents and the pollutants
that serve as condensation nuclei, precipitation,
thunderstorms, hail, and fog tend to occur
somewhat more frequently in cities and also
downwind than in the remainder of the surrounding countryside (Table 17-6). In addition,
wind speed and direction is altered and the
movements of major weather fronts can be
retarded by urban areas. Abundant trees and
plants within the urban area can improve air
quality and lower urban temperatures as well as
contributing to aesthetic appeal.

Water vapor from power plants or industrial cooling towers and ponds may form plumes
extending for many kilometers under certain
conditions. Local fog and icing conditions near
these sources may become much more severe

than is normal. This has generated concern
over the local climatic effects of proposed giant
energy complexes.

Large artificial reservoirs can modify local
temperature extremes; increase local precipitation, humidity, and fogs; influence local winds;
and even extend the average growing season
along the shore. Such effects are usually limited
to within a few kilometers of the shore and to
less than 1 km into the air. Schemes to haul
mammoth icebergs from Antarctica to Southern
California or the Arabian nations could produce
similar effects—both in the atmosphere and in
the oceans.

Where flights of jet planes are frequent,
vapor trails may generate enough cirrus cloud
cover to diminish significantly the sunlight
reaching the surface.

Deforestation has been shown to lower minimum low-level temperatures from 1 to 7°F in
all seasons and to increase maximum temperatures 1 to 8°F in warm seasons and 1 to 3°F
during cold seasons. Humidity decreases of 2 to
25 percent and rain reductions of 1 to 10
percent have also been documented in deforested areas.

Agriculture totally alters native vegetation and modifies terrain characteristics over vast areas. Farming tends to reduce water retention in the soil, increase heat conduction to the atmosphere, and reduce evapotranspiration, resulting in decreased humidity and precipitation. The albedo from cropland is much greater, especially in winter, than it would be with native vegetation, and this leads to colder temperatures and more ground freezing. The burning of crop residue in many areas fills the air over thousands of square kilometers with dense smoke, diminishing sunlight and leaving fields exposed to accelerated erosion.

Irrigation, on the other hand, can increase humidity and precipitation, as documented in such places as Phoenix, Arizona, where high local humidities have proven a great disappointment to people who moved to those areas mainly to enjoy an arid climate. In the Great Plains, less severe but more widespread changes have been attributed to increased irrigation.

Overgrazing, perhaps the most widespread and damaging of all land use practices, destroys vegetation with effects similar to dryland overfarming. These practices reduce available ground moisture, humidity, and temperatures and add dust to the atmosphere. These results tend to stabilize the local atmosphere and lead to sinking and cooling of the air which inhibits precipitation, tending to create a semipermanent condition of aridity.

Approximately one-fifth of the earth's land surface is grassland. This is forage land for animals, which provides most of the world's meat, dairy products, wool, leather, and other products, in addition to supporting the draft animals that still till one-third of the world's cropland. The carrying capacity of this land is being exceeded on an enormous scale all around the world (see Chapter 16). With continued overgrazing or natural drought, this deteriorated land, much of it in particularly vulnerable semiarid or high relief areas, is likely to be stripped of vegetation and soil and turned into nonproductive, barren wasteland, a process called **desertification** (Fig. 17-21). Overfarming, depletion of water supplies, and deterioration brought about by a wide variety of misuses—from firewood and dung gathering in poor countries to off-road vehicle use in affluent countries—can also create desertlike wastelands.

The arid climate that takes hold as a result of such massive alteration of the land surface may be mollified by natural regeneration if overgrazing and other misuses are halted. Other possible remedies for desertified lands require massive, often hopelessly costly, inputs of materials and energy to reclaim them.

Unfortunately, with 200,000 people being added to the world's population every day, mostly in the LDCs, the need for such lands continues to increase. Deserts are expanding on every continent and the rate of desertification is accelerating. Some 78 million people still live on lands that have been rendered almost useless by devegetation, dune formation, erosion, and salt encrustation. The Sahara Desert is expanding in all directions. Most publicity has centered on its southward advance but, along its northern tier in Morocco, Algeria, Tunisia, and Libya, some 100,000 ha of rangeland and cropland is being desertified each year, and studies in the Sudan indicate a southeastward advance of 90 to 100 km from 1958 to 1975.

OUR INFLUENCE ON GLOBAL CLIMATE

As noted earlier in this chapter, small changes in climatic parameters, especially temperature, can be of enormous consequence to humans. Mean global temperature today is only about 4°C warmer than it was during periods of maximum glaciation. A decline of just 1°C in the mean annual temperature of the Northern Hemisphere will cut an average of two weeks off the growing season and cause a northward shift of the monsoon belt, threatening the sur-

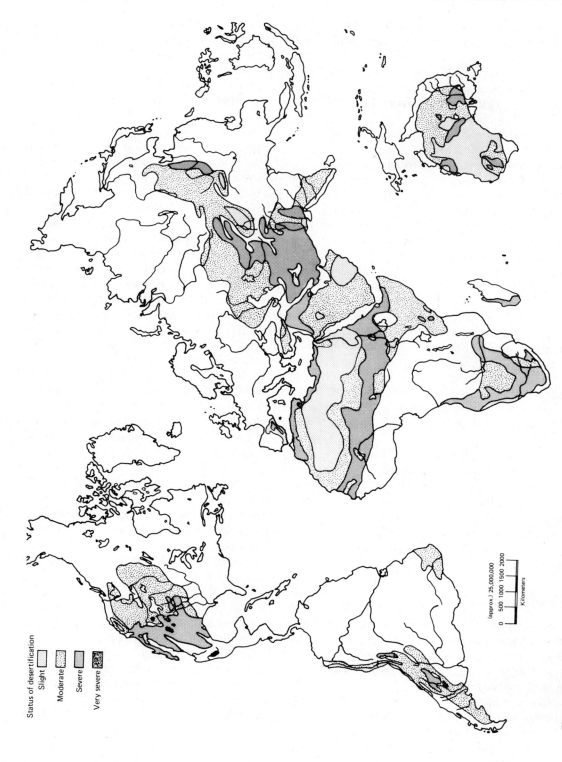

Figure 17-21. Desertification map of the world. (UNESCO-FAO, 1977.)

Status of desertification
Slight
Moderate
Severe
Very severe

(approx.) 25,000,000

0   500 1000 1500 2000
Kilometers

vival of millions of people. If the earth's albedo were to be increased by one percentage point, a 1.7°C (3.1°F) drop in average temperature would occur. Present cloud cover of the earth is 31 percent—an increase to 36 percent would depress the average global temperature by 7.2°F to the level of the glacial ages. A 2 percent decrease in the sun's energy output could lead to glaciation of many land areas.

Climate is equally sensitive to increases in temperature. During the Altithermal period (postglacial thermal optimum), 8000 to 4000 years ago, average temperatures were only about 1.8 to 3.8°C warmer than at present; yet many presently productive croplands, such as the American corn belt, were arid plains and other areas now too cold or dry for farming supported abundant vegetation. A temperature increase of 4°C (7.2°F) over the present temperature would melt the Arctic ice areas, raise sea levels, and cause a major northernward shift of earth's major climatic zones, bringing drought to much of the world's most heavily populated and productive temperate zones. Even small rises in sea level resulting from the melting of Arctic or Antarctic ice would jeopardize millions of people in densely populated coastal regions around the world.

It is therefore understandable why there exists great concern over our influence of an already highly uncertain global climate. Some have warned of a possible man-induced "hot earth" disaster, while others suggest that we may trigger a "runaway glaciation." Let us examine the major ways in which we might alter the global climate.

Greenhouse Effect

A major consideration in any climatic management scheme is the realization of the complexity and feedback mechanisms that operate in the global system (Fig. 17-22). Certain gases, notably water and carbon dioxide (CO_2), cause the atmosphere to act like a one-way mirror, in that incoming solar radiation is allowed to reach the earth while outgoing radiation to space is inhibited (Fig. 17-23 and Table 17-7). This is because CO_2 and water molecules absorb the outgoing infrared radiation (or "heat rays"), thus warming the atmosphere much like a greenhouse—hence the term "greenhouse effect" (Fig. 17-24).

The major concern is over the enormous quantity of carbon dioxide that we are disgorging into the atmosphere—mostly by combustion of fossil fuels (see Chapter 6). Over 600 million tons per year are released from U.S. coal-fired power plants alone. Although CO_2 constitutes only 0.03 percent of the atmosphere, it is estimated that near-surface temperatures are about 10°C warmer because of it. On the average, about 53 percent of the CO_2 we put into the air will remain there increasing the natural content. Since the Industrial Revolution began, atmospheric CO_2 has been increasing at an exponential rate roughly proportional to the consumption of fossil fuel (about 3 to 4 percent each year). Carbon dioxide content in 1860 was about 285 ppm; in 1977 it was 330 ppm and it is expected to reach 400 ppm by the year 2000. By 2150, it could be four to eight times the preindustrial level; this would produce an increase in mean global temperatures of at least 6°C and perhaps three times that in polar areas. A doubling of the level of the mid-1970s is possible by 2050, producing an anticipated temperature increase of 3°C.

Large-scale deforestation may amplify CO_2 (carbon dioxide) buildup because it not only removes trees, which are effective absorbers of CO_2, but the decaying wood and exposed soil release additional CO_2. By the year 2000, the estimates for the continued buildup of CO_2 caused by the burning of fossil fuels range from 10 to 160 percent higher than present-day levels. Factors governing the absorption of atmospheric CO_2, the production of other pollutants with either cooling or heating effects, natural climatic cycles, and other circumstances may intervene causing unforseen results.

It is generally acknowledged that CO_2

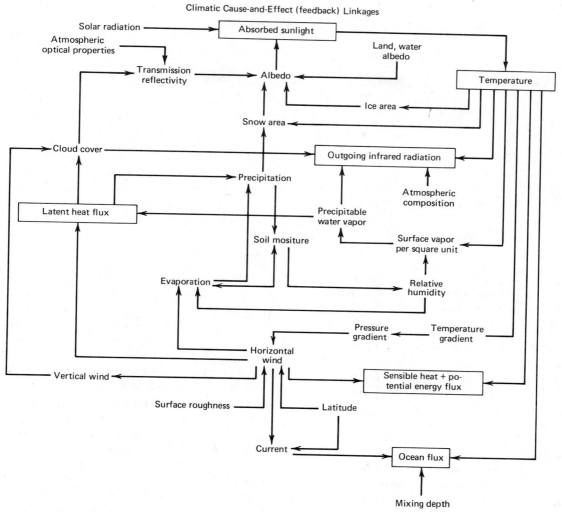

Climatic Cause-and-Effect (feedback) Linkages

Figure 17-22. Schematic presentation of climatic feedback linkages. A difficulty with such models is that the feedback processes have influences that are comparable in magnitude but opposite in direction.

(After Kellogg and Schneider, 1974, Climate stabilization for better or for worse: Science, v. 186, p. 1164. Copyright © 1974 American Association Advancement of Science.)

buildup will tend to increase temperatures, regardless of complicating factors. For the first 40 years of the twentieth century, temperatures did rise but, since about 1940, they have declined in spite of the ever-increasing carbon dioxide output. Some attribute this to a natural return to colder climates; others to the increase in atmospheric particulate matter.

Particulates

There is general agreement that large quantities of small particles in the atmosphere can profoundly affect climate. This has been amply demonstrated by volcanic eruptions. The April 1815 eruption of the Indonesian volcano Tambora blasted some 52 km of pulverized rock

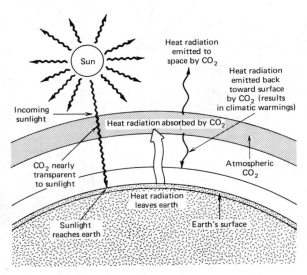

Figure 17-23 Schematic view of how the "greenhouse effect" is produced. Solar radiation enters the atmosphere and hits the earth, but outgoing radiation is absorbed by CO_2, which causes climatic warming.

into the sky. The clouds of ash were so dense that complete darkness prevailed for three days on the island of Madura, 800 km away. Finer particles carried by winds completely surrounded the Northern Hemisphere, reducing solar insolation and producing erratic weather conditions. In London, England, mean summer temperature in 1816 was 2 to 3°C below normal. Crop failures were widespread and, in the northeastern United States, there were widespread snows in June of 1816—"the year without a summer."

Much disagreement exists as to the effects of man-produced particulates in the atmosphere. Particle size, shape, color, density, and composition all influence the effect of incoming and outgoing radiation. Some low-level concentrations of particulates can lead to increased temperatures. However, most who have studied particulate pollution believe that the dominant effect is to scatter incoming sunlight and favor increased cloud cover, resulting in an increased albedo and lower surface temperatures.

Estimates of our influence on particulate increases vary from 5 percent to over 100 per-

cent, with some of the most recent and wide-ranging studies giving figures of 20 to 30 percent. We produce some 300 million tons of airborne dust (often called aerosols because their tiny mass allows them to remain suspended in air for lengthy periods of time) each year, and the total atmospheric particulate content is increasing 4 percent per year. One estimate indicates that atmospheric dust may increase 400 percent within 100 years, resulting in a mean global temperature decrease of 4°C! Deflation from the world's arid regions adds about 2 million metric tons of dust to the atmosphere each year. Virtually all of human activities, such as construction, deforestataion, overgrazing, warfare, and especially agricultural malpractice, add to this burden.

Particulate emissions from fossil fuel combustion (especially coal) is also a growing problem, even though mechanical devices such as electrostatic precipitators can remove 99 percent of these materials by weight. The billions of minute submicron particles cannot be completely removed, and these constitute by far the most serious health hazards and haze producers. The Four Corners coal-fired plant in northwestern New Mexico, even with controls, releases more than 46 tons of particulates each day, and serious degradation of air quality and visibility from this and other plants is apparent in the Colorado Plateau—an area reknowned for its remarkable scenery and, at one time, for its clear air.

A brownish haze due at least in part to human practices now surrounds the entire earth. The haze over the Amazon Basin may be responsible for precipitation increases up to 40 percent in parts of that region. Cloud cover over the north Atlantic Ocean and parts of Europe and North America have increased 50 percent according to some studies. Ten percent less sunlight reached the Soviet Union in 1967 than in 1940.

Whether such observed effects as these and other recent climatic variations are caused by human actions, and what their long-term significance may be, cannot be determined with

Table 17-7 Reconstruction and prediction of atmospheric CO_2 contents based on fuel consumption data

Year	Chemical fuel CO_2 ($\times 10^{16}$ g)	Excess atmospheric CO_2[a] ($\times 10^{16}$ g)	Excess atmospheric CO_2 (%)	Excess atmospheric CO_2 (ppm)	CO_2 content of the atmosphere[b] (ppm)	Global temperature increase[c] (°C)
1900	3.8	1.9	0.9	2	295	0.02
1910	6.3	3.1	1.4	4	297	0.04
1920	9.7	4.8	2.2	6	299	0.07
1930	13.6	6.8	3.1	9	302	0.09
1940	17.9	8.9	4.1	12	305	0.11
1950	23.3	11.6	5.3	16	309	0.15
1960	31.2	15.6	7.2	21	314[d]	0.21
1970	44.0	22.0	10.2	29	322[d]	0.29
1980[e]	63.0	31.0	14.0	42	335	0.42
1990[e]	88.0	44.0	20.0	58	351	0.58
2000[e]	121.0	60.0	28.0	80	373	0.80
2010[e]	167.0	83.0	38.0	110	403	1.10

Source: W. S. Broecker, 1975, Climatic change: are we on the brink of a pronounced global warming: *Science,* v. 189, p. 461. Copyright © 1975 American Association for the Advancement of Science.

[a]On the assumption that 50 percent of the CO_2 produced by the burning of fuel remains in the atmosphere.
[b]The preindustrial atmospheric partial pressure of CO_2 is assumed to be 293 ppm.
[c]Assumes a 0.3°C global temperature increase for each 10 percent rise in the atmospheric CO_2 content.
[d]Value observed on Hawaii for 1960, 314 ppm; value for 1970, 322 ppm (8).
[e]Post-1972 growth rate taken to be 3 percent per year.

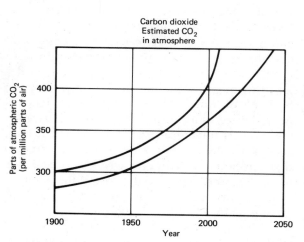

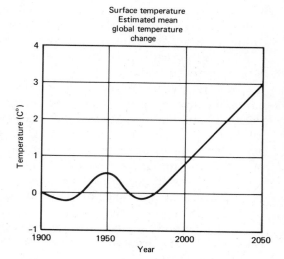

Figure 17-24 Comparison of CO_2 content in the atmosphere, showing possible global temperature changes.

precise accuracy. However, cause for concern clearly exists, especially in view of the high sensitivity of climate to small temperature changes and triggering effects.

The majority of climatologists view CO_2 buildup and the growing emphasis on coal-fired power plants with alarm. Their consensus is that present and recent cooling may be related to natural cycles, but that perhaps the greenhouse effect will soon reverse this. Another school of thought, led by Reid Bryson of the University of Wisconsin, feels that particulate pollution will more than offset the CO_2 buildup and augment global cooling. In either case, results on civilization could be catastrophic.

While CO_2 and particulates are regarded as the major man-made threats to global climatic stability, several other possibilities merit brief comment. Any large-scale change in sea ice cover in the Arctic could produce far-reaching, though far from agreed on, results. Some schemes have suggested damming the Bering Strait or blackening Arctic ice to induce melting of the ice pack, in the belief that this could result in moister, warmer polar air masses and milder winters, less drought, and improved agriculture for the Northern Hemisphere. Others believe it would increase mid-latitude droughts and even trigger a major ice age, because the additional moisture from an open Arctic Ocean could lead to greatly increased snow accumulation in northern regions.

Massive hydroengineering schemes for damming and diverting major Arctic-feeding rivers, notably the Ob, Yenisei, and Mackenzie, would increase the salinity of the Arctic Ocean. The effects again are very uncertain, but the potential for upsetting climatic balances with catastrophic results are considered high.

Clearing large tracts of tropical forests could also have global climatic effects, Decreasing precipitation in some South American countries has already been related to deforestation in the Amazon Basin.

Direct heat from human energy conversions is still small compared to solar input.

However, continued growth of human heat production could trigger global changes. The intricacies of climatic cause-and-effect relationships are shown in Fig. 17-22.

CLIMATE EXTREMES

As our numbers increase, there will be more pressure and a greater need for populations to inhabit more marginal lands and to live in regions with climatic extremes—the cold, the hot, and the dry areas of the earth. In this section we will discuss some aspects of cold climates and use the example of frozen-ground terrane as a case study to illustrate the relationship between climate and society.

Cold Climates

During their greatest influence, the Pleistocene ice sheets covered 30 percent of the earth's land areas. Today the glaciers have shrunk so that only 10 percent is now covered. Not only did glaciation greatly affect landforms and the character of surficial materials covered by the ice, but also regions beyond the ice margin. The lands immediately adjacent to the ice experienced abnormal changes caused by a periglacial climate. Even areas far beyond the margin were also changed. For example, the great loess deposits throughout the midwestern states are the heritage of winds deflating silt from deposits near the ice margin. The giant outwash sands and gravels for more than half of Long Island, New York, were washed by meltwater streams that traveled away from the ice. Thus one needs to understand the glacial process in order to evaluate properly the origin and location of terrain features and their materials.

There are also regions on earth that are still permanently cloaked with a mantle of ice that exists under the ground surface. These areas have what is called **permafrost,** and the

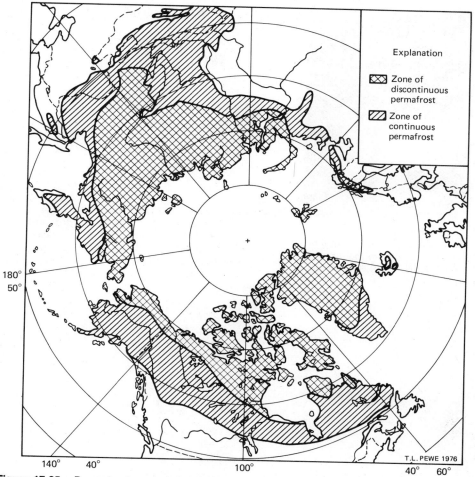

Figure 17-25. Permafrost map of the Northern Hemisphere. (Courtesy Troy Péwé.)

lands are referred to as **frozen-ground terrane.** Approximately 20 percent of the world's land area falls within the permafrost zone (Fig. 17-25). Nearly 80 percent of Alaska and 50 percent of Canada and the Soviet Union is underlain by permafrost. Such areas are experiencing new population growth, and their development takes the form of mineral extraction and new settlement patterns.

Frozen-ground terrane is exceptionally intolerant of even slight disruptions. This is largely a consequence of the unique physical behavior of water, especially near its freezing point.

When water freezes, it expands 9 percent. Numerous freeze-thaw cycles therefore repeatedly agitate the earth's surface in this harsh environment, inducing very active downslope movements even on gentle slopes, which includes gelifluction (soil flowage under permafrost conditions) and soil and boulder creep (Figs. 17-26 to 17-33). Frost heaving and shattering are also very frequent. These processes can cause great damage to roads, buildings, and other structures.

If active glaciers are nearby, these can threaten lives and human works by catastrophic

Figure 17-26.   Foliated ground ice mass in late Quaternary silt at Mamontova Gora. This is the left limit of the Aldana River, 310 km upstream from the junction with the Lena River, Central Yakutia, U.S.S.R. (Photo by Troy Péwé, July 22, 1973.)

Figure 17-27.   Abandoned Alaskan tractor trail used during the summertime about 10 years prior to this photo's date, in May 1973. Disruption of vegetation and surface mat has induced thaw in underlying permafrost with the resultant subsidence of more than 1 m. Note trees leaning toward subsiding trail. (Courtesy William McMullen.)

Figure 17-28.   Mass movement and solifluction in the hillside of a man-made cut into permafrost, Alaska. (Courtesy William McMullen.)

Figure 17-29. Entrenched trail and polygonal microrelief developed since scraping in mid-1950s near Umiat, Alaska. Polygons are 20 to 30 m in the well-developed area. (Courtesy Richard Haugen.)

Figure 17-30. Frost-heaved piling of bridge spanning outlet of Clearwater Lake, 13 km southeast of Big Delta, Alaska. (Photo by Troy Péwé, August 15, 1951.)

Figure 17-31. Fracturing of 10 cm thick steel-reinforced foundation of an apartment building in Fairbanks, Alaska, due to frost heaving in silty ground. Incomplete foundations were left unfinished during the winter. (Photo by Troy Péwé, May 8, 1954.)

Figure 17-32. View of schoolhouse at Glennallen, Alaska. Air vents which are open allow cold air to enter crawlway in window to counteract heat from the building. The vents are closed during the summer. Jacks are used to counteract any differential settlement caused by the permafrost. (Photo by Troy Péwé, May 4, 1954.)

Figure 17-33. Construction on permafrost terrane in the U.S.S.R. Brick apartment building on concrete piles frozen into the permafrost, Yakoutsk, Siberia. (Photo by Troy Péwé, July 28, 1969.)

outburst floods from unstable glacier-dammed lakes or by rapid advances of the glaciers themselves (glacial surges). Avalanches and ice falls are serious hazards in steep alpine regions.

During the spring thaw, rivers in cold climates frequently become congested with ice blocks forming ice jams. These jams can cause local flooding and dangerous flash floods of frigid ice-laden water which result if ice jams collapse suddenly.

By far the greatest problems in cold environments are attributed to the permafrost. When more freezing of water in the ground occurs during cold seasons than can melt during the warm seasons, a permanent zone of frozen ground develops. It may be relatively "dry" permafrost, with only small quantities of frozen interstitial water, or it may contain a continuous "cement" of ice, often including lenses and wedges of pure ice (Fig. 17-26). The permafrost may be continuous or discontinuous (patchy) and may reach depths of several thousand feet.

Overlying the permafrost is an **active layer** that thaws during the summer. A dense mat of vegetation protects the underlying permafrost from melting. A very delicately balanced thermal regime exists between the permafrost and the surface environment. Once the fragile surface is disturbed and the insulating vegetation destroyed, the exposed permafrost will begin to melt, forming a mushy quagmire that typically enlarges well beyond the limits of the initial disturbance (Fig. 17-27). Any activity that heats, adds pressure to, or disturbs the overlying surface materials will tend to upset the precarious equilibrium and induce melting of the permafrost (Fig. 17-28). All such phenomena can raise havoc with attempts to use the land for buildings, farming, and other endeavors.

Much care, usually at great expense, must be taken in cold regions to avoid the problems of permafrost. Buildings, roads, bridge foundations, and other structures need to be constructed to preserve the permafrost (Figs. 17-31, 17-32, and 17-33). The most common methods are to place a broad insulating platform of permeable material, usually sand or gravel, beneath structures such as roads or to elevate heated buildings on pilings driven deep into solid permafrost. Water lines and sewage pipes must be insulated and heated to prevent freezing. Artificial refrigeration is also needed frequently to prevent ground thawing. In still other cases, the best procedure may be to intentionally thaw the ground prior to any construction, especially if permeable gravels underlie the site. Permafrost problems can be avoided by building whenever possible on solid bedrock. When this is impossible, sand and gravel provides the best foundation material because it is less susceptible to frost heaving and will allow the drainage of excess water. Techniques used on the Alaska Pipeline provide good examples of some approaches to construction in permafrost. The methods included refrigeration in underground sections, placement on large gravel banks or steel pilings, and encasement in concrete and burial beneath rivers.

The best source of water in cold regions is usually groundwater from alluvial deposits. If natural unfrozen groundwater does not occur, the alluvium can be artificially thawed to provide water, but problems of freezing wells and pipes and the formation of thaw pockets will still need to be controlled. If construction or mining activity disrupts groundwater flow, springs frequently will develop at the surface where the water freezes, forming large **icings** that can be destructive and very difficult to control.

Perspectives

The Mark Twain adage that "Everyone talks about the weather but no one does anything about it" is now an anachronism. Instead the problem today is, everyone does something about the weather, either inadvertently or deliberately. We now influence the weather in numerous ways, and our ability to control certain aspects of the weather is well proven. Whether this is always a good thing and what the by-products will be are still debatable questions.

Weather and climate will become even more important in future decades because of population pressure on the rapidly dwindling areas of temperate climates. Since about 40 percent of the planet is extra dry, 20 percent contains permafrost, and 15 percent is tropical, there will be a greater need to inhabit these areas of climatic extremes. Thus we will need to know more about climate in order to cope with such environments. One problem with these regions is their vulnerability to rapid degradation, because they are more fragile than terrain in temperate zones. So although we are very adaptable, a heavy price must be paid for developing areas with extremes of moisture, heat, and cold.

Society is only belatedly discovering the immensity of its impact on weather and climate. Our power to control and change the gaseous envelope that enshrouds us must be very carefully monitored, because the arrogance of indiscriminately subduing atmospheric conditions could produce disastrous feedback results. The political and legal questions concerning cloud seeding will vex society for years. So we must keep in mind that weather is mobile, and transitory—very different from other parts of the human environment. Changing the air or weather in one locality can possibly produce unforeseen effects on another locality which experiences them at some subsequent time. This is especially crucial when matters of health are involved. Thus pollutants that are man-induced are an infringement on the rights of others to a heritage that should permit clean air.

Weather and climate characteristics are a concern to many earth scientists—the hydrologist, the glacialist, the geomorphologist, and the pedologist—because water, its availability and temperature, affect water budgets, ice balance, erosion and sedimentation style and amount, and soil properties. The effects on soil are especially vital to food production and animal vitality.

When we consider the management of the atmosphere, we should remember a few historical facts: (1) climate is not fixed; in fact, it can change rather rapidly; (2) cultural changes often accompany climate changes; (3) what may seem like a normal climate today, may not have been the average climate for much of past history; (4) cool periods are generally accompanied by greater weather and climate instability than warm periods.

READINGS

Broecker, W. S., 1975, Climatic change: are we on the brink of a pronounced global warming?: Science, v. 189, p. 460–463.

Brown, J. E., 1972, Rainmaking is now almost a science but its legal status is still murky: Smithsonian, v. 3, n. 7, p. 50–57.

Bryson, R. A., 1967, Is man changing the climate of the earth?: Sat. Review, Apr. 1., p. 52–55.

———,1974, A perspective on climatic change: Science, v. 184, p. 753–760.

Changnon, S. A., Jr., Farhar, B. C., and Swanson, E. R., 1978, Hail suppression and Society: Science, v. 200, p. 387–394.

Ferrar, T. A., ed., 1976, The urban costs of climate modification: New York, John Wiley & Sons, 284 p.

Ferrians, O. J., Jr., Kachadoorian, R., and Greene, G. W., 1969, Permafrost and related engineering problems in Alaska: U.S. Geol. Survey Prof. Paper 678, 37 p.

Haugen, R. M., and Brown, J., 1971, Natural and man-induced disturbances of permafrost terrane: in Coates, D.R., ed., Environmental geomorphology, Publ. in Geomorphology, State Univ. of New York at Binghamton, p. 139–149.

Hobbs, P. V., Harrison, H., and Robinson, E.,

1974, Atmospheric effects of pollutants: Science, v. 183, p. 909–915.

Hodge, C., and Duisberg, P., eds., 1963, Aridity and man: Amer. Assoc. Adv. Sci., Publ. No. 74, 584 p.

Honeyborne, D. B., and Harris, P. B., 1958, The structure of porous building stone and its relation to weathering behaviour: The Colston Papers, v. 10, p. 343–365.

Howard, R. A., Matheson, J. E., and North, D. W., 1972, The decision to seed hurricanes: Science, v. 176, p. 1191–1201.

Huntington, E., 1917, Climatic change and agricultural exhaustion as elements in the fall of Rome: Quart. Jour. Econ., v. 31, p. 173–208.

Kellogg, W. W., and Schneider, S. H., 1974, Climate stabilization: for better or for worse?: Science, v. 186, p. 1163–1172.

Landsberg, H. E., 1970, Man-made climatic changes: Science, v. 170, p. 1265–1274.

McMullen, W. B., et al, 1975, Design & construction of roads on muskeg in arctic and subarctic regions: 16th Ann. Muskeg Conf. Natl. Research Council of Canada, 51 p.

McPhail, J. F., McMullen, W. B., and Murfitt, A. W., 1976, Yukon River to Prudhoe Bay Highway—lessons in arctic design and construction: Civil Eng., v. 46, n. 2 p. 76–82.

National Academy of Sciences, 1973, Weather & climate modification: Washington, D.C., 258 p.

Norwine, J., 1977, Hot or cold: Environment, v. 19, n. 8., p. 7–27.

Pack, D. H., 1964, Meteorology of air pollution: Science, v. 146, p. 1119–1128.

Sears, P. B., 1953, Climate and civilization: in Shapley, H., ed., Climatic change, Cambridge, Mass., Harvard Univ. Press, p. 35–50.

Sewell, W. R. D., et al., 1973, Modifying the weather: Western Geographical Series, v. 9, University of Victoria, 349 p.

Swinzow, G. K., 1969, Certain aspects of engineering geology in permafrost: Eng. Geol., v. 3, p. 177–215.

Taubenfeld, H. J., ed., 1970, Controlling the weather: New York, Dunellen Company, Inc., 275 p.

UNESCO-FAO, 1977, United Nations Conference on desertification, 29 August–9 September 1977: United Nations.

Weather Modification Advisory Board, 1978, The management of weather resources, Vol. I: Rept. to the Secretary of Commerce, Washington D.C. Department of Commerce, 229 p.

Winkler, E. M., 1973, Stone: properties, durability in man's environment: New York, Springer-Verlag, 230 p.

Woodley, W. L., 1970, Rainfall enhancement by dynamic cloud modification: Science, v. 170, p. 127–132.

PART FIVE

Humans vastly change the terrain into an artificial "cityscape" by prodigious environmental management feats. The New York City skyline with the typical smog from pollutants. (Ellis Herwig/Stock, Boston.)

Environmental Management

I have chosen to allow some biases to shine through in this part of the book—call it "author's license." Although the theme of environmental management is present throughout the book, I have selected four chapters that I feel emphasize those features that are endemic to the topic. Thus this section opens with Chapter 18 which sets the stage for policy, planning, and decision making. The next two chapters provide extended case histories of themes that reflect problems and solutions in environmental affairs. Chapter 19 is close to my heart, because I live in a metropolitan area of 200,000 and have become very concerned with the management of urban areas. Chapter 20 discusses those environmental problems that must quickly be resolved if we are to live on this planet with clean land, water, and air and not become deluged by our own debris. The final chapter, Chapter 21, clearly shows what I consider the culmination of our journey through environmental geology. It is possible for a geologist to do magnificent technical and scientific work as part of an enterprise, but, if the entire projects fails to receive societal approval or becomes flawed by legal procedures, such work has been an effort in futility. It is not only important to do a superb professional job, but it is equally important that scientific involvement in the policy-making arena be sufficiently staunch and articulate to see the successful conclusion of the venture. Having participated in more than 100 legal actions, I have become increasingly aware that geologists must be an indispensable part of legal actions and legislative policies that concern the environment.

Environmental management is in the "eyes of the beholder." What may appear to be a managed environment to some, may to others be unmanaged, mismanaged, or a hopeless morass. Environmental management is interpreted to mean the deliberate planning and administration of policies that determine the status of the biosphere in which we live. It covers the broad spectrum of activities involved with the air, water, and land on the planet Earth. Management ranges from the local to the national and international level; from policies instituted by private persons or special-interest groups to those formulated by industry and governmental bodies. Our concerns in Part Five involve general policy that has developed from legal and mandated directives with the concurrence of societal objectives and aspirations. It will be assumed that decisions reached by responsible environmental managers will have been made after careful examination of a pertinent data base that was established for this purpose.

Before any type of environmental management is orchestrated, there must be the recognition that something is in need of being managed, that a problem already exists, or that one is likely to arise in the future. Thus an awareness and perception of how we are affecting our world, and how nature influences us, are prerequisites to programs and schemes for environmental management. Not only is a political and societal support base necessary, but the objectives must be consistent with what can realistically be accomplished with funds that will be available.

Objectives that are adopted in programs requiring environmental management are a reflection of many factors that include: (1) assignment of priorities, (2) decisions on constraints, and (3) character of accommodation to the physical systems. Priorities that are established provide the controlling guidelines of how, and for what purpose, the environment will be managed. Three competing viewpoints are often present—the utilitarian-developmental ethic, the conservation ethic, and the preservation ethic. Thus the management of environmental affairs can become dependent on the type of ethic that is adopted.

As in all man-made programs, there are constraints that operate which provide boundary conditions for the system being implemented. The size of the area to be managed is of utmost importance. Political subdivisions can constitute unnatural barriers that interfere with the complete integration of the system. For example, a watershed comprises a natural entity, but it may be impossible to obtain rights to administer it as a coordinated unit. Furthermore, feedback mechanisms that operate in the process systems provide unwanted effects when the entire region is not managed. Downstream channel areas may be subjected to erosion when sediment is entrapped behind dam barriers. Thus, management of such a land-water ecosystem must consider the side effects that occur at other locales, instead of being only concerned with geologic conditions at the dam site.

A second constraint for environmental management concerns the time frame of the system being designed. What recurrence interval of a natural event should be considered when planning floodplain developments, highway construction, flood levees, and so on? The rate of allowable resource depletion and the degree materials and processes should be preserved for future generations also place limits on utilization.

The degree of sophistication and understanding of environmental components place further restrictions on its management. Most environments are highly complex, interrelated, and possess thresholds that when exceeded may produce irreversible changes. Clearly, interdisciplinary efforts and skills are needed for comprehensive environmental management.

As in many human endeavors, decisions are generally made on the basis of some type of benefit-cost analysis. An environmental program will not be deemed feasible if the predicted benefits are judged as considerably less than the costs. Thus, it is the manipulation of strategies concerned with how to determine the true budgetary ledger that often decides whether a program will be approved, financed, and implemented. It should now be obvious that environmental management is a very broad field, but in Part Five we will emphasize those facets that are in the domain of the geologist.

Since the material in Chapters 18 through the end of the book are not intended to be material for objective examinations, the publisher has set these chapters in three columns. This serves the purpose of saving space without sacrificing readability.

READINGS

Cooke, R. U., and Doornkamp, J. C., 1974, Geomorphology in environmental management: Oxford, Clarendon Press, 413 p.

Geyer, A. R., and McGlade, W. G., 1972, Environmental geology for land-use planning: Env. Geol. Rept. 2, Penna. Dept. Env. Res. 44 p.

Meadows, D. H., et al. (eds.), 1972, The limits to growth: New York, Club of Rome, Universe Books, 207 p.

Nichols, D. R., and Campbell, C. C., eds., 1971, Environmental planning and geology: U.S. Dept. Housing and Urban Development and U.S. Dept. Interior, DCPD-32, 204 p.

Chapter Eighteen
Policies and Decision Making

Beautification and flood control project at Coeburn, Virginia. This site can now safely carry the same volume of water that had formerly flooded the area. (Courtesy Tennessee Valley Authority.)

INTRODUCTION

Decision making is a process in which the ultimate objective is to determine a course of action. It is the response that is necessary to solve a perceived problem. Policy comprises the operational mandate that is prescribed which guides the planning and management of human affairs. Policy is therefore a resultant vector that is set in motion by the selection of a particular framework for action. Thus decision making and policy are two sides of the same coin and are the necessary ingredients that constitute environmental management.

By its very nature, the establishment of environmental policy always breeds controversy. The principal problem is that the environment belongs to all people, so there are always multiple constituencies involved in matters dealing with the earth's land-water ecosystem. In the final analysis, environmental policy becomes the prerogative of governments and is institutionalized through legislation, jurisprudence, and the policing powers of governing bodies. In recent years many government agencies have taken on new powers through their perception and extension of

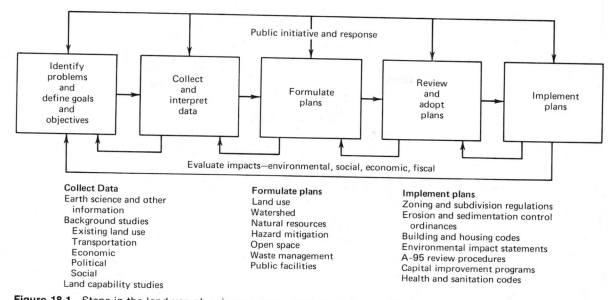

Collect Data
Earth science and other
  information
Background studies
  Existing land use
  Transportation
  Economic
  Political
  Social
Land capability studies

Formulate plans
Land use
Watershed
Natural resources
Hazard mitigation
Open space
Waste management
Public facilities

Implement plans
Zoning and subdivision regulations
Erosion and sedimentation control
  ordinances
Building and housing codes
Environmental impact statements
A-95 review procedures
Capital improvement programs
Health and sanitation codes

Figure 18-1 Steps in the land use planning process. (U.S. Geol. Survey Circ. 721, 1976.)

policies that seem to fall in their area of interest.

Policy generation stems from a variety of concerns and motives. It can result from a very necessary, but long delayed, action. For example, the St. Francis Dam disaster produced policy in California in which engineering geology analysis became prerequisite to all major construction in California. The floods and landslides that caused costly damages to Southern California in 1951–52 prompted the policy of requiring geologists to aid in the planning of all hillside developments as stated in the grading ordinances. The Mississippi River and tributary floods that caused extraordinary losses in 1927 and then in 1935–36 led to the passage of federal legislation in each period that revolutionized policy for dealing with flooding hazards.

Other policy matters are designed to set the tone of action and provide a suitable framework that will conserve and preserve vital resources. Thus decisions are made to determine the most fea-

sible manner in which to mine, transport, and use natural resources. This requires the cooperation and input of many diverse fields. Thus communication becomes a vital component when science, technology, politics, economics, and society are all involved in the policy solution that emerges. Such multidisciplinary and interdisciplinary liaisons can only be effective when there is a commitment to the necessity for coordination and a mutual empathy for the final results that must be achieved.

PLANNING

Environmental planning, decision making, policy, and management are intimately related. Various authors have compiled elaborate flow charts showing the connecting and overlapping linkages. Figure 18-1 shows one point of view on the matter and indicates some of the steps and procedures that

should be used in arriving at those judgments that will provide environmental governance.

The purpose for establishing environmental policy is to set a course of action in which it is hoped the number of conflicts will be minimized and the environmental impacts will be reduced—both to society and to natural systems. This is accomplished by providing a planning and assessment base wherein alternative actions have been considered and both short-term and long-term effects of the project determined. Such evaluation will aid to subdue adversary militancy and enhance the management phase by securing necessary funding and legal support. Policymaking may not be 100 percent scientific, neither will a given policy please all the people all the time. Vital decisions will always ultimately depend on the perception of the problem and on the values that are perceived and how they are expressed through the political-legal system. Also, decision making may be a continuing process, instead of a single act, and commitments and choices may change during the tenure of a particular program. After all, policy is instituted because of some perceived benefit that is predicted to accrue as a result of the action. Benefits may be in terms of financial rewards or viewed as necessary for health, safety, and welfare. Problems always emerge when value judgments must be made on intangibles and the benefit-cost ratios are subject to questions with indirect effects.

Definition of the problem is prerequisite to planning, regardless of whether the decision maker is an individual or a group. We can define a problem in terms of:

1 *Objectives:* Those goals that are to be achieved or those values that are to be maintained.

2 *Alternatives:* Those options that also need scrutiny to determine the most effective course of action.

3 *Environmental setting:* Those factors that will be involved because of the action. This includes not only the narrow confines of the specific action, but its relationship to the greater arena of social, economic, political, and biophysical systems that will be affected. No longer can proposed projects be considered in isolation from other dimensions of the human scene.

Without all three of these components, there is no problem to solve, no decision to be made, and consequently no plan to be formulated.

The character and effectiveness of environmental planning affects the quality of life. However, too often planners must function within jurisdictional boundary lines that are artificial in terms of natural systems. To cope with such difficult odds and other constraints that circumscribe solutions requires unusual analytical tools and multidisciplinary efforts with the decision makers.

Problems and Conflicts

Should phreatophytes be eradicated to save groundwater at the expense of valuable wildlife habitats? Should a community deliberately restrain its growth in order to possess more open space? Should a wild river be dammed to provide a water supply for new consumers? Should costly antipollution devices and abatement be required of an industry that may lose its competitive edge and thereby cause widespread unemployment? These and many questions like them must be solved by environmental planners. Such questions have one common theme—they pose problems of evaluation wherein comparisons must be made between alternative actions. How to judge the severity of the environmental impact and how to relate it to some type of benefit-cost analysis become exercises in the assignment of values, risk, and economics.

Not only can there be wide gaps between scientists and lay citizens in their comprehension of environmental processes, hazards, and resources, but the gap between the scientist and the planner can be equally broad. A typical planner may be so heavily burdened with day-to-day operational matters that there is insufficient time to keep up to date on environmental data. Planners, furthermore, must deal with a set of variables that may not be internally consistent. They must develop the project in terms that are technologically possible, economically feasible, socially and politically acceptable, and legally unimpeachable. Unfortunately, the environmental world is one of complex interactions, and decision makers may not have the time, patience, or training to understand the relationships that exist among the integral parts of an ecosystem—the resources, the energy, the waste, and the land use aspects of the natural-human biosphere. Figure 18-2 shows some of these characteristics and the kinships among the materials and processes that are vital to civilization.

Thus it is incumbent on the planning process that consideration be given to three basic flow systems. These are: transformations of materials and energy, the transport and storage of material goods, and the controls that determine flow rates and distribution of materials and energy. These transformations occupy three sectors: the geosphere, the human sector, and the natural sector. And these sectors contain built-in feedback mechanisms that affect each other. The utilization of the resource base depends on the full spectrum of society.

There is an old Hindu fable of six blind men describing an elephant. Each in turn feels a wall, a spear, a snake, a tree, a fan, a rope. Thus each was partly right, but all were wrong about the total system. Because society is pluralistic, we all don't see the same thing, and the reconciliation of these disparities is the job of decision makers. In viewing agricultural land near urban centers, some see it as a means of providing important food and fiber to the community and the nation, others see it as providing open space for recreation, a chance to improve air and water quality, an opportunity to develop and enhance the local economy and tax base. To attempt a solution concerning the disposition of land use problems requires definition of the time and space domains. Urban planners often think in

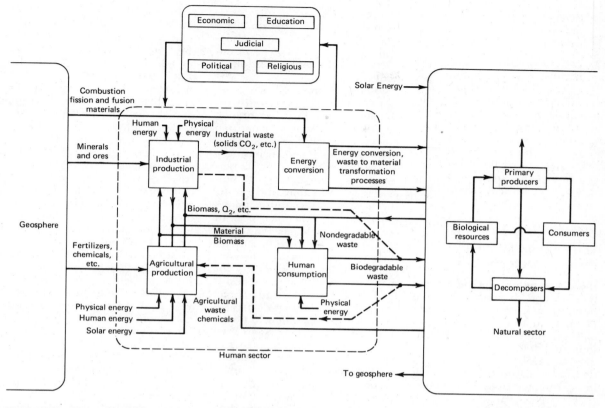

Figure 18-2 Flow diagram of macro features in the material transformation and transmission process and in the institutional control process. (National Science Foundation.)

terms of 25-year master plans or planning to the year 2000. Economists commonly develop econometric models that have 5 to 10 year ranges. Politicians, however, are generally tuned to thinking of 2, 4, or 6 year intervals, depending on the election cycle of their office. Geologists may think in terms of geologic time—decades and centuries. Obviously, the time frame needs careful delineation when dealing with the environment. Spatial dimensions may also provide stumbling blocks for developing a consensus among decision makers and their consultants. Natural systems consist of a mosaic of different rock types, soil, rivers, valleys, hillsides, and vegetational changes. The func-

tional role of each of these disparate parts of the ecosystem also contributes to different constraints in the environmental planning matrix. So although we may integrate human works and activities with our natural environment, there will always be trade-offs and judgments to be made. Three obstacles commonly emerge:

1   Planners are usually generalists, in both education and community outlook, but very few are schooled in geology.

2   Only recently have geologists started to regard the environmental planner as a consumer of their know-how. In turn,

the planner has been skeptical of the skills and expertise the geologist can bring into the "real" world. All too often resources were viewed as being inexhaustible, and hazards as acts of God that were inevitable. Under such circumstances, economic and political goals became the primary determinants of environmental policy. The planner is involved with, and subject to, society's mandates, whereas the geologist is evaluated by peers in accordance with scientific standards of truth. Thus the geologist is often uncomfortable when faced with political compromise and disheartened when solid geologic analysis is rejected by decision makers for political, social, or economic reasons.

3 Each profession has its own particular jargon, and the lack of precision in planning nomenclature and the highly specialized terms in geology cause mutual confusion on all sides. Most geologic reports are not tuned to the frequency of decision makers and, because of the language barrier, can often be neglected and bypassed (see page 612).

The Present Mood of Planning Determinants

Increasing public concern over alteration and degradation of the human environment is causing planners to focus new strategies and priorities on indiscriminant industrial development and waste generation. Some vocal extremists demand a halt to further growth in the commercial-business-industrial complex. Part of this revulsion stems from the Madison Avenue type of ethic that has been instrumental in fostering a consumption-oriented society. This has produced a proliferation of goods that were abundant and cheap, with a supporting technology that was highly successful in making and marketing the products. Even those goods that were not very abundant or inexpensive were made so artificially by governmental policies (as in the energy field) that either deferred part of the true cost or transferred expenditures to other sectors of the economy. Thus, many environmental economists now argue that all industrial activities must absorb both direct and indirect costs associated with production of goods as a means of cleaning up the environment and offsetting the unacceptable pol-

lution and waste by-products. Such concerns and demands are generating new laws and regulations that will have significant impacts on our national resource base.

The dilemma arises because Americans need but also distrust "big business." In a recent poll people were asked to name the biggest potential threat to the nation—10 percent said big business, 15 percent said big labor, 32 percent big government, and 32 percent said a combination of the three. Another paradox enters our discussion when we speak of *technology,* which has long been regarded as the force that drives the engine of progress. However, blame is given to technology for many present environmental difficulties but, at the same time, assuming technology will also solve all the problems. Long-range success can only be achieved by long-range planning. So this necessitates the uniting of all facets that operate in the human sector.

This diversity of opinion poses a constant challenge to the planner who must accommodate those priorities that are most pressing. Many writers champion the theory that wealth is the "environmental culprit." Lewis Moncrief finds that several factors, including the increase of individual wealth, have led to an aggressive attitude toward nature and have directly contributed to the environmental crises. Jean Mayer writes that it is the rich who wreck the environment because they "occupy much more space, consume more of each natural resource, disturb ecology, litter the landscape, and create more pollution." Leo Marx argues that environmental problems stem from the propensity of Americans to consume more material goods at increasing rates. And James Ridgeway scathingly decries the ulterior mo-

tives of government in claiming that (1) antipollution legislation is written to benefit Wall Street, (2) the Federal Power Commission is an aid to industrial manipulation of the energy market, and (3) the Neo-Malthusian doctrine is a manipulative scheme aimed at controlling the poor in the interests of the wealthy.

How many times have you heard there is no energy crises, or gas shortage, that it is all a contrived device to drive prices higher for increased profits of industry? A skeptical public ironically increased their gasoline usage after the oil embargo of 1973 was lifted. Even pollution programs must be carried out with a somewhat disenchanted public. A majority of people when asked if they would contribute $15 per year in direct taxes for pollution abatement said they would not do so voluntarily.

If Congress is any barometer of the specialties of people who are decision makers, it means that the sciences are highly under-represented in the policy arena. Nearly one-fourth of the first Continental Congress had members with science backgrounds, whereas today less than 1 percent in today's Congress are scientists. A study of more than 10,000 bills that have recently been introduced showed that 40 percent of them had significant scientific and technological content. One might well wonder with what degree of expertise these bills have been evaluated by the decision makers.

Since funding is always limited, the assignment of priorities becomes crucial in planning decisions that affect the environment. The prediction of future trends is vital in knowing how to cope with those problems that are most plaguing to society. For example, is the metropolitan growth pattern of urban areas reversible? Will energy considerations produce new living styles and spatial considerations? What types of incentives and rewards should be offered to those who need extra encouragement to prevent environmental degradation? What are the most effective stratagems in passing needed legislation? What are the most appropriate indices for measuring those factors that must be environmentally managed?

The remainder of this chapter will be devoted to an analysis of planning and policy that affects primarily the nonurban regions. Other chapters in Part Five will analyze the specific problems of waste disposal and pollution, urban areas, and the legal framework that operates to control environmental conflicts.

LAND USE POLICIES

Background

Thomas Jefferson once said "The face and character of our country are determined by what we do with America and its resources." **Land use** is a unit of the landscape that can be categorized by its state of existence or management. The way in which a part of the landscape serves a functional purpose—for example, for the mining of natural resources—defines its land use. **Landscape values** are determined by groups or individuals. Resource allocation is a nonissue if deemed valueless or unimportant by decision makers. **Landscape assessment** is the method by which landscape values are estimated. The concept is analogous to property tax assessment, wherein an assessor provides a specific tax appraisal based on information and criteria that are currently available to decide the relative worth of a parcel.

Historically land use planning and control were largely concerned with land as an economic and cultural entity. Natural environment processes received scant attention as means of differentiating one parcel from another. Policies were made in utilitarian terms, considering only the ability of the land to accommodate manmade structures. Until the 1950s, major landscape planning was the exception and was practiced in only a limited number of localities. However, within the past two decades, the new awareness and perception of land use in terms of landscape values and assessment has led to multiple approaches for determining policy. It is now recognized in the terms of Kenneth Hare that there are three types of environments: (1) the natural environment, which encompasses the physical-biological world that exists outside of society; (2) the social environment, which arises from the matrix of people and their culture—which must coexist within the community confines; and (3) the built environment, which recognizes that man-made structures must provide a place to live and work.

Most policies that now deal with land use and its landscape setting make some attempt to account for:

1 Air and water properties, which are fluid and flow through the entire ecosystem occupied by humans.

2 Processes that create threats to life and property.

3 Loss of soil and mineral resources.

4 Those landscape elements that need protection in their own right, yet which may be in jeopardy due to possible development.

Thus land is recognized as constituting both a commodity and a resource. Furthermore, land use policies have largely become governmental and public matters because of the extensive power that governments wield in its control. This power is manifested in: (1) eminent domain powers, (2) taxation power, (3) police power, as in zoning ordinances and restrictions, and (4) power of the public purse by which the landowner is influenced by the manner in which the government allocates funds, subsidizes, protects, provides services, and provides information on land use technology (Fig. 18-3).

Data and Methods

Since land use involves spatial arrangements and considerations of different parcels, some type of map and classification system become essential ingredients in planning procedures that lead to policy decisions. Other chapters have already addressed the problem of providing specialized maps, such as hazard maps for volcanic activity (see page 285) or maps for the study of flooding (see page 613). In this chapter we are primarily interested in generalizing the approach to land use policies and giving an overview of the decision-making process.

In 1976, the U.S. Geological Survey adopted a **land use and land cover classification system** that was to be used in compiling maps for all states (Table 18-1). This system divides the landscape into nine basic Level I categories, which are in turn separated into 37 more detailed Level II groups. The purpose of this program is to provide a land use inventory for the entire country by 1985. As of November 1978, completed maps had been compiled for Arkansas, Florida, Kansas, Louisiana, Maryland, Missouri, and the New York City metropolitan area (Fig. 18-

Table 18-1 Land use and land cover classification system used on maps prepared by the U.S. Geological Survey

Level I		Level II
1.	Urban or built-up land	11. Residential
		12. Commerial and services
		13. Industrial
		14. Transportation, communications, and utilities
		15. Industrial and commercial complexes
		16. Mixed urban or built-up land
		17. Other urban or built-up land
2.	Agricultural land	21. Cropland and pasture
		22. Orchards, groves, vineyards, nurseries, and ornamental horticultural areas
		23. Confined feeding operations
		24. Other agricultural land
3.	Rangeland	31. Herbaceous rangeland
		32. Shrub and brush rangeland
		33. Mixed rangeland
4.	Forestland	41. Deciduous forestland
		42. Evergreen forestland
		43. Mixed forestland
5.	Water	51. Streams and canals
		52. Lakes
		53. Reservoirs
		54. Bays and estuaries
6.	Wetland	61. Forested wetland
		62. Nonforested wetland
7.	Barren land	71. Dry salt flats
		72. Beaches
		73. Sandy areas other than beaches
		74. Bare exposed rock
		75. Strip mines, quarries, and gravel pits
		76. Transitional areas
		77. Mixed barren land
8.	Tundra	81. Shrub and brush tundra
		82. Herbaceous tundra
		83. Bare ground tundra
		84. Wet tundra
		85. Mixed tundra
9.	Perennial snow or ice	91. Perennial snowfields
		92. Glaciers

Source: U.S. Geological Survey.

4). The U.S. Geological Survey had also been involved with the Department of Housing and Urban Development with what to date (1970–75) was the most detailed mapping of a large region—

the nine county area comprising the San Francisco Bay Region. Here a five-year program mapped many environmental components of the landscape, including mineral deposits, soil types, soil stability, and susceptibility to landsliding, flooding hazards, and so on. Ecoregions of the United States, as another type of environment mapping, are shown in Fig. 18-5.

Geologic maps date back into antiquity. Probably the oldest is entitled "A geologic map of an old gold mine exploited in Pharoanic Time during the reign of King Set I, 19th Dynasty." The map which includes adits and contains geologic symbols, is dated about 1350–1205 B.C. The four western surveys of the United States in the 1870s (Hayden, King, Powell, and Wheeler Surveys) contained environmental reports and maps that were influential in the future settlement and development of the region. These maps were used for water assessment, mining operations, and road and railroad design, and the newly instituted system of contour maps was justified on the basis of military necessity.

Prior to the national effort, numerous environmental maps had been prepared on state, county, and local levels. For example, New York State undertook in 1970 a Land Use and Natural Resource Inventory (LUNR) of the state (Table 18-2). Each of the 10 principal land categories were further divided into more specialized uses. Thus 10 main land categories are recognized, and 22 subsets are refined into an additional 136 classes. All mapping was done on a 1:24,000 scale.

Land Capability

It is not enough to simply classify lands, but they must also be cate-

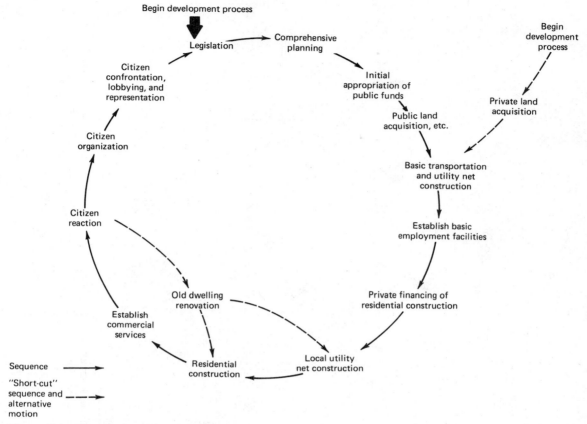

Figure 18-3 The local sequence flow of land use development. (National Science Foundation.)

gorized in terms of potential uses and capabilities. Table 18-3 shows the major land uses in the United States (Fig. 18-6). Usage is best decided by matching land characteristics to their most appropriate purpose. Although up until a few years ago the dictum of the appraiser, as reinforced in the courts, was "the highest and best use," this was generally considered only in terms of short-term financial benefits; now many other considerations enter the picture. The evaluation of physical features of an area in regard to different types of land use is termed a **land capability study.**

Such investigation is a means of determining the relative physical meits for specified land uses, and it is vital in establishing policies and management in the area. A study may be a descriptive narrative that provides a physical assessment of the area, but in its more sophisticated forms it will also provide a quantification or weighting procedure that gives relative merits to the various land parcels. At least five steps are necessary in such a procedure:

1 Identify the type of land use for which land capability is to be determined.

2 Assess which natural factors have a significant effect on the capability of the land to accommodate each use.

3 Develop a scale of values for rating each natural factor in relation to its effect on land capability.

4 Assign a weight to each natural factor that indicates its importance in a hierarchy with other factors.

5 Establish land unit determinants that reflect the weighted aggregates for each land use factor.

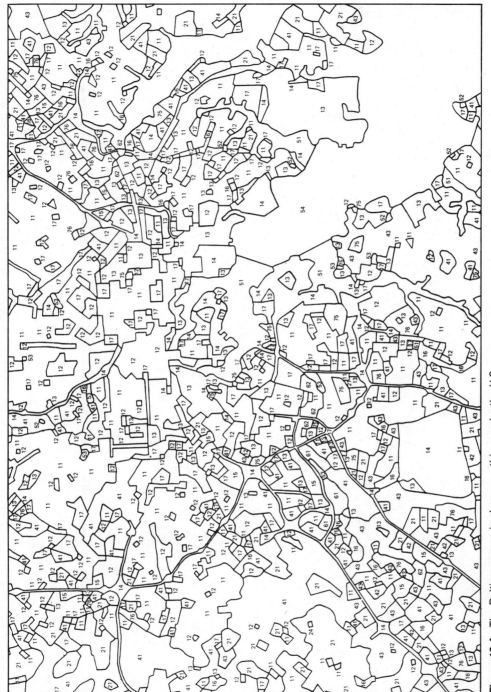

Figure 18-4 The Baltimore area is shown on this map by the U.S. Geological Survey illustrating land use and land cover for part of Maryland. Each land use or land cover category is identified by a two-digit number. This part of the map depicts 77.6 km². (U.S. Geol. Survey.)

Figure 18-5 Ecoregions of the United States. (Bailey, 1978.)

One example of this type of procedure was provided by Robert Bailey in his study of the Lake Tahoe region (Fig. 18-7). On the other hand, most of the newer U.S. Department of Agriculture county soil map series contain tables that provide "suitability" uses for various soil types. For example, the Broome County, New York, soils report categorizes all soils according to use as "topsoil, granular material, fill material, highway location, embankment foundation, foundations for low buildings, farm ponds, agricultural drainage, irrigation, diversions, and waterways."

As previously mentioned (see page 389), susceptibility to hazards and risk have repeatedly been used as criteria for land use decisions. Table 18-4 provides a listing of those processes that have a potential for causing costly losses, and Fig. 18-8 shows potential losses that can be incurred in California unless appropriate policy and management is undertaken.

Land Use Indices and Constraints

Just as economic indicators are needed to determine the financial health of a nation (or a business)—unemployment, cost of living, market basket processes, gross national product, inventory of goods—so are indices needed to determine the environmental status of the lands, the waters, and the resource potentials. Such factors are vital in order to evaluate the present and develop future policy and programs. Environmental indices should establish limits, trends, thresholds, and provide insight that gives some measure of the success of federal, state, and local programs in dealing with environmental problems. There are many obstacles in providing unequivocal measurement systems. Some of these are conceptual questions that must first be resolved. For example, what are the most important elements of land use to measure, and what system should be employed to yield the least ambiguous answers? Furthermore, the development of good

Table 18-2 Classification system used by the Land Use and Natural Resource Inventory (LUNR) of New York State

I. Agriculture
   1. Active areas
   2. Inactive areas
   3. Agricultural enterprises
II. Forestland
III. Water resources
   1. Lakes and ponds
   2. Streams and rivers
   3. Wetlands
   4. Marine lakes, rivers, and seas
IV. Residential land use
   1. Residential areas
   2. Cottages and vacation homes
   3. Apartment buildings
   4. Mobile homes
   5. Rural nonfarm residences
V. Commercial and industrial land uses
   1. Commercial areas
   2. Industrial areas
VI. Outdoor recreation land use
VII. Extractive industry land use
   1. Open mining
   2. Underground mining
VIII. Public and semi-public land uses
IX. Transportation land uses
   1. Highways
   2. Railways
   3. Airport
   4. Barge canal
   5. Marine shipping
   6. Communications and utilities
X. Nonproductive land

Source: Office of Planning Coordination, 1971, Albany, New York.

indices depends on good data, and representative sampling is a prerequisite. Specifically, how much park space does an urban dweller want, or indeed need? Should a new causeway be constructed to an offshore island when the increased use of the fragile beach may result in its deterioration?

Benefit-cost analysis Since the early 1900s the Corps of Engineers has used a type of benefit-cost analysis to determine the merits for construction of river and harbor improvements. This administrative device was not derived from traditional economic theory and, until the 1930s, such assessments were applied only to tangible benefits and costs. The great depression that followed in the wake of the 1929 stock market crash prompted governmental programs aimed at reviving the national economy. Massive public works projects were instituted, but to justify them it was found that the old benefit-cost formulas were too restrictive. Therefore, the Water Resources Commission of the National Resources Board recommended the extension of benefit-cost analysis to intangible factors (Fig. 18-9). They were now to include benefits to whomever they might accrue; for example, they would now consider as a possible benefit the employment of a labor force that might otherwise be idle. These ideas became the foundation for new principles in governmental projects in which the decision to undertake engineering structures was weighed against construction costs and benefits to be derived. In referring to benefits, the Flood Control Act of 1936 stated, "if the benefits to whomsoever they may accrue are in excess of the estimated costs," a project was deemed justifiable. This triggered attention to **intangibles** throughout the field of governmental construction projects. However, the controversy is continuing. What constitutes an intangible? How far can secondary benefits and costs be pushed? How much is one life worth? Elsewhere in the book some of the abuses of these methods have been mentioned (see page 245).

Regulation and Control Systems Costs

One consideration in policies that pertain to environmental safeguards and controls relates to a type of cost-accounting involved in such systems. The antipollution laws have raised domestic steel prices up to $8 a ton and this has placed U.S. steelmakers at a big competitive disadvantage when compared with foreign imports. The entire range of automobile safety and emission pollution devices add an average of $600 to the price of a new car. However, a Bureau of Labor Statistics study concluded that for each $1 billion spent on pollution control, including capital investments and operating and maintenance costs,

Table 18-3   Major uses of land in the United States, 1969[a]

	Acres (millions)	Hectares (millions)	Percent
Cropland	472	191	20.9
Cropland used for crops	333	135	—
Cropland harvested	286	116	—
Crop failure	6	2	—
Cultivated summer fallow	41	17	—
Soil improvement crops and Idle cropland	51	21	—
Cropland used only for pasture	88	35	—
Grassland pasture and range[b]	604	245	26.7
Forestland	723	293	31.9
Grazed	198	80	—
Not grazed	525	213	—
Special uses[c]	178	72	7.9
Urban areas	35	14	—
Transportation areas	26	11	—
Rural parks	49	19	—
Wildlife refuges	32	13	—
National defense, flood control, and industrial areas	26	11	—
State-owned institutions and miscellaneous other uses	2	1	—
Farmsteads, farm roads, and lanes	8	3	—
Miscellaneous land[d]	287	116	12.6

Source: J. R. Anderson et al., 1976, U.S. Geological Survey Professional Paper 964.

[a]H. T. Frey, 1973. Does not include area covered by water in streams more than one-eighth of a mile in width and lakes, reservoirs, and so forth, of more than 40 acres in size.

[b]Includes pasture that is to be included with cropland in the U.S.G.S. classification system.

[c]Except for urban and built-up areas and transportation uses, these special uses will be classified by dominant cover under the U.S.G.S. classification system.

[d]Tundra, glaciers, and icefields, marshes, open swamps, bare rock areas, deserts, beaches, and other miscellaneous land.

66,900 jobs were added to the labor market. During the 1970–75 period an EPA (Environmental Protection Agency) study showed that only 75 plants closed, with a loss of 15,710 workers because of unbearable pollution control requirements that could not be met. The government alone spent $15.7 billion on pollution control systems in 1975.

The visible spending of regulatory environmental agencies has increased from $745 million in 1970 to $4.8 billion during the 1978–79 fiscal year. The cost for complying with the mandated regulations forced industry to spend $7.8 billion during 1976, while at the same time EPA was spending $416 million on its program. There will be high costs attached to compliance with the far-ranging regulations—for instance, the 1975 Hazardous Materials Transportation Act and the 1976 Toxic Substances Act. Current projects indicate that the costs of regulation may exceed the $100 billion figure (spread over several years), an amount that approaches the annual amount spent on national defense!

Taxation in land use   The selection of land use policy instruments has generally been a choice between incentives and controls. Each has merits, and which to use or the character of the blend depends on the specific situation. Differential property tax assessment has been a favorite tool in agricultural land preservation policies, but it is becoming increasingly apparent that it can be even more effective if coupled with land use control measures. Simply zoning for agricultural use may be insufficient unless there are incentives for farming and the land is truly economically productive.

There are three types of differential assessment policies:

1   **Preferential assessment,** where land is valued according to its current use with no penalty if later converted to other use.

2   **Deferred taxation,** where land is taxed on current value but a penalty is exacted against the land when that use changes.

3   **Restrictive agreement,** where the landowner and local government agree to restrict land use in return for differential assessment.

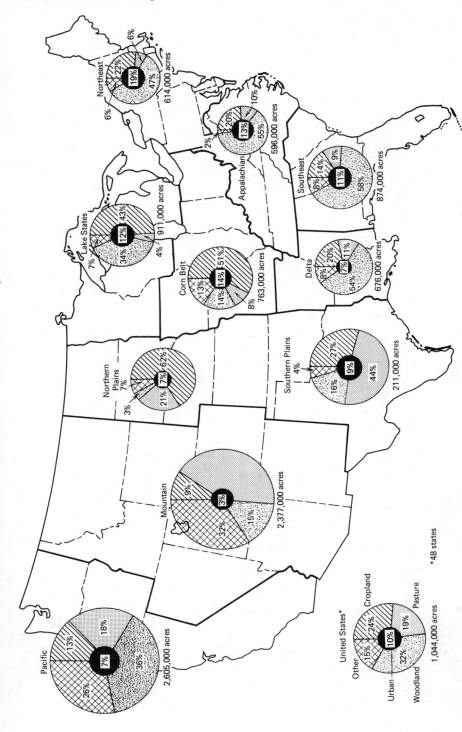

Figure 18-6 Distribution of land uses, by region, in the United States for 1970. (U.S. Dept. of Agriculture.)

Legend

Capability levels	Tolerance for use	Slope percent	Relative erosion potential	Runoff potential	Disturbance hazards
7	Most	0–5	Slight	Low to moderately low	Low hazard lands
6		0–16	Slight	Low to moderately low	
5		0–16	Slight	Moderately high to high	
4		9–30	Moderate	Low to moderately low	Moderate hazard lands
3		9–30	Moderate	Moderately high to high	
2		30–50	High	Low to moderately low	
1a	Least	30 +	High	Moderately high to high	High hazard lands
1b		Poor natural drainage			
1c		Fragile flora and fauna			

Figure 18-7 Land capability legend that depicts manner for mapping Lake Tahoe Basin. (Bailey, 1974.)

In 1973, 9 states had preferential assessment laws, 18 had deferred taxation, and 10 used restrictive agreement.

The California Land Conservation Act of 1965 authorizes local governments to create agricultural preserves and to contract with volunteering landowners the forfeiture of their development rights for a 10-year period (renewable each year on a 1-year basis). In fiscal 1973–74, 5.1 million ha, or 35 percent of the farmland was in the program.

The New York Agricultural District Law of 1971 encourages farmers to preserve valuable agricultural land in organized districts using a five-point program:

1 Exempts from taxation any land value considered strictly a farm.

2 Limits ordinances that affect farming land.

3 Instructs state agencies to encourage farming.

4 Makes eminent domain proceedings more restrictive.

5 Restricts public funds to nonfarm development that jeopardizes the farming activity.

By 1974 there were 117 districts encompassing about 400,000 ha, or 10 percent, of the farmland of the state.

Land Use Planning Examples

The federal government has no comprehensive land use policy or law, although several have been proposed and defeated in Congress. However, land use programs are in affect from state levels on down. The following examples provide a range of these approaches.

1 What some have described as the "quiet revolution in land use control" started with the Hawaiian Legislature's passage of the Land Use Law in 1961. It gave state agencies a degree of control over the use of the land resources that was far in excess of any pre-

Table 18-4 Natural processes important to land use planning

Process	Description of hazard
Flooding	Overtopping of river and stream banks by water produced by sudden cloudbursts, prolonged rains, tropical storms or seasonal thaws; breakage or overtopping of dams; ponding or backing up of water because of inadequate drainage
Erosion and sedimentation	Removal of soil and rock materials by surface water and depositing of these materials on floodplains and deltas.
Landsliding	Perceptible downslope movement of earth masses.
Faulting	Relative displacement of adjacent rock masses along a major fracture in the earth's crust.
Ground motion	Shaking of the ground caused by an earthquake.
Subsidence	Sinking of the ground surface caused by compression or collapse of earth materials; common in areas with poorly compacted, organic, or collapsible soils and commonly caused by withdrawal of groundwater, oil, or gas; or collapse over underground openings, such as mine workings or natural caverns.
Expansive soils	Soils that swell when they absorb water and shrink when they dry out.
High water table	Upper level of underground water close to ground surface causing submergence of underground structures, such as septic tank systems, foundations, utility lines, and storage tanks.
Seacliff retreat	Recession of seacliffs by erosion and landsliding.
Beach destruction	Loss of beaches owing to erosion and (or) loss of sand supply.
Migration of sand dunes	Wind-induced inland movement of sand accelerated by the disturbance of vegetative cover.
Saltwater intrusion	Subsurface migration of seawater inland into areas from which freshwater has been withdrawn, contaminating freshwater supplies.

vious state legislation in the nation. It created a state Land Use Commission and directed it to divide the state into four districts for: conservation, agriculture, rural, and urban uses. The law authorized that land in the urban district be used for whatever purpose is permitted under local zoning regulations. Land in the agricultural and rural districts were to be used only in compliance with regulations of the Land Use Commission. Conservation districts must comply with rules of the state Department of Land and Natural Resources.

2 Starting in the late 1950s the peaceful state of Maine started to come under attack on two fronts—from the sea by oil interests, and on the land by tourists and new developers. The state viewed these invaders with mixed emotions. The new revenue and possible jobs would boost the economy and diminish unemployment. However, by so doing the development would create serious environmental threats. With the influx of the tourist industry, the drive to build numerous second homes and other real estate development, and the push to use the deep ports for the mammoth new oil tankers, the state finally awakened to the fact that there were no land use policies or laws to govern such activites. The legislature responded with a package of legislation (1969–71), including a new Site Location Law that requires approval by a state agency for certain types of new development. For example, commercial and industrial projects covering more than 20 acres (8 ha) must obtain permits from the state Environmental Improvement Commission. Such policies dramatically illustrate governmental response to the public's concern over the vulnerability of coasts and forests to unplanned or minimally regulated development. There was

Table 18-4 *Continued*

Process	Description of hazard
Liquefaction	Temporary change of certain soils to a fluid state, commonly from earthquake-induced ground motion causing the ground to flow or lose its strength.

Source: U.S. Geological Survey Circ. 721, 1975.

also public resentment over the exploitation of the state's riches by outsiders for temporary advantage.

3 On the county level, Bucks County, Pennsylvania, evaluated the entire county in 1971 with the Natural Resources Plan. Phase I in 1971 prepared an inventory of natural resources, including prime agricultural soils, forests, wetlands, steep slopes, lakes and ponds, floodplains, extractive resources, scenic areas, and aquifers. Each natural feature was defined, weighted, and computer mapped on a grid. Phase II in 1972 set forth specific policies for the protection of each natural resource in terms of the percentage of each grid cell containing a particular natural resource that would remain in open space use. The land use consequences of fulfilling the policies are depicted on a Natural Resource Protection Map, which shows prime agricultural lands, parks, scenic areas, and the open space requirements for resource protection.

4 Ian McHarg has had a tremendous influence on landscape architects, and his planning assignment of the Baltimore, Maryland, area is typical of his style for "designing with nature." Here the 180 km² area on the Baltimore outskirts was first divided into five major systems: valley floors, wooded valley walls, unwooded walls, wooded plateaus, and unwooded plateaus. Each unit was described in terms of development, and protection recommendations were provided. For example, in connection with the unwooded plateaus it was proposed that "forest and woodland sites on the plateaus should not be developed at densities in excess of one house per acre." Such recommendations were then substantiated using general ecological principles.

5 In Orange County, New York, an area larger than Manhattan Island has been the site for another type of land use strategy. Sterling Forest is 78 km² of primeval woodlands, with clean mountain lakes, an area once called "Land of the Rainbows" by the Iroquois Indians. This oasis is only 56 km from New York City, but is the home for not only deer,

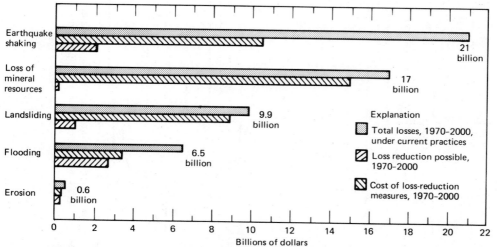

Figure 18-8 Estimated dollar losses from environmental problems that could be alleviated by the use of earth-science information and technology. The amount of losses that could be reduced and the estimated cost of loss-reduction measures are also shown. (Adapted from U.S. Geol. Survey Circ. 721, 1976.)

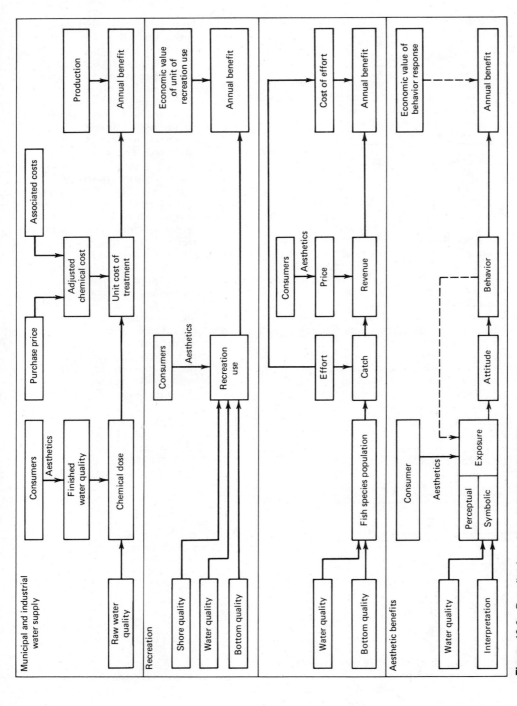

Figure 18-9 Benefits from improved water quality. (Redding, 1973, Aesthetics in planning: Environmental Protection Agency.)

ducks, and trout, but some of the nation's most prestigious research-oriented firms. These include Union Carbide, International Paper, International Nickel, Reichhold Chemical, International Business Machines, and New York University research laboratories. This type of environmental-ecological mix was started in the 1950s and is now a complete and planned community. Three objectives were the guiding lights for development of the pristine area: (1) to fashion a totally planned community for education, research, light industry, corporate office centers, commercial businesses, and residences; (2) to service the community with cultural and recreational facilities for human satisfaction and creative expression; (3) to preserve and enhance the natural beauty and resources of the land, as trustees of the forest, and provide a living and working environment for the community.

Refinements continue to be made in the mapping and quantification of land use properties. The advent of remote sensing by satellite technology has expedited assessment of very large areas. Also, computer methodologies are becoming common in many planning operations (Fabos and Caswell, 1977).

NATURAL RESOURCES

A consistent policy in natural resource planning is fundamental to the development of an effective administration and management of those ingredients vital to society. Such policies must address the basic question of whether a particular resource at a locality should be developed or utilized. An environmentally acceptable plan must involve:

1   Conservation considerations of natural resources that provide sustained yield.

2   Preservation of those areas that contain unique ecological, historical, scenic, and scientific value.

3   Utilization of resources necessary to society.

Natural resource management involves the rate of resource extraction and utilization, the impacts of resource development, and programs that define conservation, preservation, and reclamation measures. Policy recommendations are developed to resolve conflicts between conservation and utilitarian ethics.

Many, indeed probably most, decisions to develop natural resources are based on economic judgments, which in turn affect the degree and sequence of development. Nearby resources are usually developed first. For example, if the initial resource proves inadequate, such as water, construction considerations will influence the alternative sources, whether to impound, import, recycle, or enforce strict conservation. Judgments about the economic factors should only be made after a strong geologic data base has been obtained. This study should also include adverse environmental impacts that would accrue by utilization of the various choices. A harmonious blending of human activities and natural materials and processes can only be achieved with consummate skill being exercised by all who occupy decision-making roles.

Water Management

The Colorado River basin was selected by the Committee on Water of the National Academy of Science for the institution of a pro-gram that would account for a full range of water management alternatives (White, 1968). The Colorado River system is the largest river in the United States that flows mainly through lands that have chronic deficiencies of water for crops. Thus irrigation is necessary for farm production in these areas. The river is 2316 km long and drains a 632,000 km² basin, which covers parts of seven states. Despite its small flow, the Colorado water is exported to other regions more than any other river in the country. Because of the growing economy of the region and adjacent areas, the river water is oversubscribed, especially when considering the need to allow flowage into Mexico. When maximum use of water does occur, and all parties start taking their allotted amounts, there are only three alternatives available for policymakers in the future management of the river: (1) to allow regional development to adjust to available water supplies; (2) to seek new sources of water by importation from outside the basin, such as desalting of brackish water and rainfall augmentation; and (3) to reduce total diversion requirements by reducing consumptive uses and waste and by increasing the reuse of water.

In approaching the problem, the committee analyzed the various alternatives and choices in order to alert the decision makers of the possible options that were available. The objectives that were identified as significant in planning the management of Colorado River water included: economic efficiency, income redistribution, political equity, natural environment control, preservation, and aesthetics. For example, to many people in the basin, and even those in other parts of the nation who may pay part of the cost of any development program, economic efficiency is a significant objective. As reflected in the Central Arizona Project, income redistribution is a goal of a substantial group, which considers it desirable to aid the farmers who are confronted with declining water supplies as well as to maintain the economic base of those who live by providing services to farmers. Political equity is the basis for the interstate compact and other arrangements for dividing the water of the basin among the states and Mexico. Environmentalists continue to argue their case for natural environmental amenities throughout the region to ensure the conservation and preservation of these features.

Another aspect of the Colorado River problem is the recent controversy surrounding the role of the National Park Service (NPS) in management of the Grand Canyon National Park. It permits 21 private companies, or concessioners, to offer package-type tours to the public and limits the total number of user-days in the river. Concessioners took up 92 percent of the days, and 80 percent of the river trips were by motorized craft. Fearing the continuing despoilation of the canyon, the NPS in 1978 instituted a three-year phaseout of motors because they were "a detriment to normal, relaxed conversation and [have] a great negative effect on the ability of the guide to interpret the resource." Along with the motor prohibition, NPS also proposed doubling the user-days. But the types of problems that emerge from such policies are numerous. The multimillion dollar concessioner business is under attack from private citizens who want to use the river but have been unable to obtain permits. The citizens point out that the split between concessioners and private users is highly unbalanced. One group, the Wilderness Public Rights Fund, has sued NPS saying citizens should have priority because the

river is public land. Another lawsuit brought by four Arizona rivermen charges that any allocation between private and commercial runners is illegal and violates constituitional rights by classifying citizens and treating them differently. The Sierra Club has also filed suit, on behalf of the Grand Canyon National Park, alleging NSP violated the public trust doctrine by not protecting its lands for future generations. Because of the politics of the heated controversy it will involve the highest officials in the Department of Interior. It is a classic case that not only will force us to decide what constitutes wilderness, but it will also illustrate government's obligation to private business.

Of course, there are many other types of governmental rules, regulations, and laws that in themselves constitute environmental management policy. A good example is the 1954 Watershed Protection and Flood Prevention Act (P.L. 566). This champions a system of land use planning and control that involves soil conservation, flood prevention, agricultural water management, municipal and industrial water supply, recreation, fish and wildlife enhancement, rural area development, and water quality amelioration.

LANDSCAPE RESOURCES

In this section we will discuss those aspects of the landscape whose indirect benefits are necessary to society and human activities—for the most part policies that deal with nonurban parks, wilderness, and wild rivers. Such undespoiled and natural areas are important in minimizing degradation of the land-water ecosystem and are necessary as locales for human spiritual and recreational enjoyment.

Parks

George Catlin, celebrated painter, after a trip up the Missouri River in 1832 was the first to mention the need for a national park system. He recommended that the Great Plains "might in future be seen preserved in their pristine beauty and wildness, in a magnificent park.... A nation's park containing man and beast, in all the wild and freshness of their nature's beauty." In 1832 Hot Springs, Arkansas, was made a federal reservation and public health resort, and Yosemite Valley was deeded by the federal government to California in 1864 "upon the express conditions that the premises be held for public use, resort, and recreation." However, it was not until 1872 that the first national park was established, Yellowstone, in the northwest corner of Wyoming. Here more than 2 million acres (800,000 ha) were designated "a public park or pleasuring-ground for the benefit and enjoyment of the people." In the first years after its establishment, Yellowstone National Park was beset with stormy battles and its retention was threatened from many quarters. Commercial interests wanted to develop the area. Prospectors wanted to exploit the region for mineral claims, and a railroad right-of-way was proposed through the park. By 1886 the Department of Interior had won out, and the new philosophy of the park was established by the Secretary, Lucius Q. C. Lamar, for "the preservation of wilderness of forests, geysers, mountains . . . with a view to holding for the benefit of those who shall come after us . . . affording the student of nature and the pleasure tourist a restful contrast to . . . busy and progressive scenes." Although many national parks were added prior to its official beginning in 1916, the National Park Service now has jurisdiction over a huge domain of national parks, historic monuments, and other landscape features necessary to our physical well-being and our cultural heritage.

Wilderness

The National Park Service defines wilderness as "an area whose predominant character is the result of the interplay of natural processes, and large enough and so situated as to be unaffected, except in minor ways, but what takes place in the non-wilderness around it." Furthermore, wilderness criteria are said to include "A scene or vista of unusual natural interest or beauty. . . . An area secluded or removed from the sight, sounds, and odors of mechanization." Thus wilderness is both an ecological condition and a state of mind.

The first recognition of these features into public policy occurred with the 1885 establishment of a land preserve in the Adirondack Mountains of New York. In the 1880s mining and lumbering interests were stripping and defacing the area, but the real trigger that aroused the outside political and commercial interests was the decline of water levels in the Erie Canal and Hudson River. Control of the watershed areas was argued as being necessary to conserve downstream commercial investments. Thus on May 15, 1885, a New York bill was enacted to establish a "Forest Preserve" of 715,000 acres (290,000 ha) that was to permanently remain "as wild forest lands." In 1892 the area was expanded to include more than 3 million acres and was relabeled the "Adirondack State Park." In 1894 a constitutional convention of the state guaranteed permanent preservation of the Adirondack wilderness.

The concept of wilderness on federal lands, and in forest preserves, became the famous battleground (Figs. 18-10 and 18-11) between the preservation ethic of John Muir and followers and the conservation-utilitarian ethic of Gifford Pinchot (see also page 23). However, it was Aldo Leopold, along with Arthur H. Carhart, who were responsible for the incorporation of the first federal wilderness area. Their work and publications culminated with the establishment on June 3, 1924, of the Gila Wilderness Area in New Mexico. Here 574,000 acres of the Gila National Forest were set aside for preservation of its natural state. This breakthrough in national policy led to a series of other wilderness sites, so that by 1940 the Forest Service had designated 31 such areas. In addition, different management procedures were undertaken, depending on whether the area was designated as "wild," "wilderness," or "primitive." Passage of the 1964 Wilderness Act ushered in a new era of governmental land use management of nonurban regions.

The Wilderness Act originally envisioned an area 40 to 50 million acres, but the law formally designated 54 areas out of the National Forests, totaling 9.1 million acres, as components of the National Wilderness Preservation System. The act also directed the secretaries of the Department of Interior and the Department of Agriculture to review within 10 years certain additional areas from the National Forests and National Park systems as well as National Wildlife Refuges to determine their suitability for preservation as wilderness. The existing primitive areas protected in National Forests were to remain unchanged until further study and recommendation. In addition, all roadless areas of 5000

contiguous acres or more in the federal jurisdictions were to be evaluated for wilderness potential.

In January 1975, the Eastern Wilderness Areas Act (P.L. 93-622) acknowledged the need for wilderness areas in the eastern United States and designated 16 areas of 207,000 total acres for preservation as wilderness; another 125,000 acres in 16 other sites were to be considered as future wilderness possibilities.

By March 1977, the National Wilderness Preservation System included 163 wilderness areas, totaling more than 14 million acres (Table 18-5). Another 71 units pending at the time would add 24 million additional acres, with the recommendation for future study of 299 different sites, totaling 39.5 million acres.

The Bureau of Land Management of the Department of Interior administers about two-thirds of all federal lands, or about 450 million acres (180 million ha) of public domain. They have designated 11 primitive areas with 234,000 acres (95,000 ha). There are about 46 million acres (18.6 million ha) of roadless areas outside of Alaska; a roadless area is defined as containing 5000 acres (2000 ha) or more. The Alaska Lands Bill illustrates a classic confrontation between those policies which seek further development and the opening of frontier lands and the preservation ethic which is attempting to classify public lands into components of the national parks and wilderness systems.

Wild Rivers

The United States has more than 5 million km of rivers that pour their waters into the sea. Many have already been harnessed for flood control, irrigation, hydroe-

"We need more wilderness."

Some people tramp the woods, are refreshed and want more lands preserved. Others see needs for woodland resources escalating and want more commercial latitude. Who's right?

Many dream of wilderness, to visit occasionally, for recreational experience. To escape the rush and pressure of today's world. To be relaxed by recollections of crystal waters, misty alpine meadows. To save for the future, often for unexpressed reasons. Wild. Beautiful. "Let's expand this wonderful heritage," they say, "put more lands away in wilderness holdings."

Others see impounded wilderness lands as entombed. No roads, shelter, communications, conveniences. No mineral exploration or woodland harvests. Today, wilderness covers nearly 19 million acres. Another 105 million Alaskan acres have been set aside by executive order. And 56 million other national forest acres are under study. Those acres supply over a quarter of our softwood timber. Woodsmen in particular say "lock up that much potential and lumber costs will climb. Jobs will be lost. Let's have expanded commercial acreage, not less."

What's the answer? We desire forest recreation, wildlife sanctuaries. We want wood products too. But these seeming opposites needn't be incompatible. Though woods are upset in cutting, the 25 to 50 years between harvests offer beauty, serenity, wildlife cover. Not that all our woodlands should be put into lumber. Some areas—unique ecosystems, remote valleys — should be preserved. But those recommendations should come from professional forest managers. With legislators making the final decision on land use to balance aesthetic needs with sustained outflow of forest products.

"We need more wood."

Figure 18-10 Wilderness conflicts. (Courtesy Caterpillar Tractor Co.)

Figure 18-11 Land use managers constantly need to resolve differences between developers and environmentalists

lectric power, municipal and industrial water supply, and navigation. Cities, factories, and homes have been built on their floodplains. Their banks and waters have been the dumping grounds of waste materials and their waters the recipient of industrial and municipal wastes. In many ways, the beauty and purity of these streams have been destroyed in the name of progress. An *affluent* society has become an *effluent* society. It has degraded drinking water and destroyed the values of fish, wildlife, and scenic and recreation resources.

In October 1968, after some six years of discussion and debate, the Wild and Scenic Rivers Act (P.L. 90-542) became law. The act established the basic principle that certain selected rivers of the nation that, with their immediate environments, possess outstanding scenic, recreational, geologic, fish and wildlife, historic, cultural, or other similar values, are to be preserved in a freeflowing condition and protected for the benefit

and enjoyment of present and future generations.

The clear intent of Congress was to establish a system of areas distinct from the traditional concept of a National Park. Rather than acquiring a massive land area, the federal agency managing a component of the National System is to acquire only a narrow strip of land sufficient to ensure protection of the river environment. Fee acquisition of lands is limited to an average of no more than 100 acres per river mile, and the power of eminent domain is suspended when fee title to 50 percent of the authorized area is in public ownership. In addition, scenic easements may be acquired in sufficient amount to make the total acquisition an average of 320 acres per mile, equivalent to a river corridor one-half mile in width. The effect is to permit continued agricultural and residential use near rivers in the system, but to preclude heavy development that would impair the character of the river.

The act further encourages the inclusion of state rivers in the system, providing that, on request of the governor of that state, rivers will be designated by the state legislature in the same manner and meet the same standards as criteria set forth by Congress.

Since 1968, there have been six amendments to the act which have resulted in the addition of seven rivers to the National System under federal administration and the designation of 31 other rivers for study as potential components of the system. Originally there were 8 rivers covered in the act, but the number had increased to 19 rivers or river segments by January 1977, totaling 2663 km (1655 mi). As can be seen in Table 18-6, 15 rivers are administered by federal agencies and 4 by states. The act also established three classes of rivers. The primary difference among them is the degree to which there is evidence of human presence in the river environment. **Wild rivers** essentially provide primitive experiences, with little or no evidence of human presence. They are generally inaccessible except by trail. **Scenic rivers,** while largely primitive and relatively undeveloped, are accessible in some places by road. **Recreational rivers** are readily accessible by road or railroad and have some development along their shorelines.

In keeping with recent federal environmental legislation, states are encouraged to pass similar measures and manage their resources in styles consistent with the federal regulations. The hope is that states will even extend these legislative mandates. In 1972 the New York State Legislature passed the NYS Wild, Scenic, and Recreational Rivers System Act. This act designated 264 km of 15 Adirondack rivers for immediate inclusion into the system. In addition, about 1600 river kilometers

Table 18-5 The Natural Wilderness Preservation System as of March 1, 1977

Components of the system		Units	Acres
National Forest Wildernesses		94	12,605,405[a]
National Park Wildernesses		17	1,120,213
National Wildlife Refuge Wildernesses		52	718,087
	Total	163	14,443,705[a]
Wilderness proposals before Congress		Units	Acres
National Forests		17	3,163,887
National Parks		22	14,018,285
National Wildlife Refuges		34	7,520,142
	Total	73	24,702,314
Potential wilderness remaining to be studied		Units	Acres
National Forest		276	12.5 million[a]
National Park		19	13.4 million
National Wildlife Refuge (these are areas on which studies have been delayed by the Alaska Native Claims Settlement Act.)		4	13.6 million
	Total	299	39.5 million[b]

Source: U.S. Department of Interior.

[a]Some acreages estimated pending map compilation; also, totals in all charts do not add because of rounding acreage in some instances. Totals from U.S. Forest Service, National Park Service, and U.S. Fish and Wildlife Service.

[b]Bureau of Land Management areas of public domain still under consideration as required by P.L. 94-579 are not included in this total.

were incorporated for study as possible inclusion into the system. The Department of Environmental Conservation manages the system. To qualify a river must be relatively free of pollution and be within the following guidelines:

Wild river. Must be free-flowing, accessible only by water, foot or horse trail, have no structures but foot bridges, and be at least 5 mi long. Less than 10 percent of all study rivers qualify as wild rivers, and most of these are on state-owned lands.

Scenic river. Must be free of impoundments, except log dams, and be mostly primitive and undeveloped. Limited road access is permissible, but no minimum length is required.

Recreational river. Must be relatively pure and natural, but may be readily accessible by road or railroad, may have undergone some impoundment or diversion in the past, and may have development in the river area. No minimum length is required.

Aesthetic values have become increasingly used as criteria in land use planning. (This topic is more fully discussed in Chapter 21, page 688.) There are many different systems that attempt to develop both qualitative and quantitative indices for the evaluation of aesthetic factors. Typical analyses include the consideration of landscape unit, spatial arrangements, relations of lines, masses, light, and color, vista aspect, veg-etation, spaciousness, serenity, naturalness, uniqueness, human interference, pollution, and so on (Fig. 18-12). The most comprehensive matrix was developed by Luna Leopold, who uses a 46-factor checklist to determine the relative uniqueness of a locality or region. He was especially interested in showing that Hells Canyon of the Snake River had such highly unusual qualities that it deserved preservation and protection from developers.

NATURAL HAZARDS

The natural features of the earth continually undergo change. The scale and pace of change may vary from tiny inobstrusive modifications to spectacular alterations on massive amounts in short time intervals. Some of these processes are important hazard-creating agents, but we also can produce changes that ultimately reach hazard-producing status.

Society has generally depended on government as its protector from all enemies, both foreign and domestic. The geologic hazards associated with earthquakes, volcanic eruptions, hurricanes, floods, and landslides are viewed as inescapable phenomena that the government must somehow prevent or control. To obtain the public trust, environmental managers must provide effective policies and plans that mitigate risks and costs to citizens.

The development of public policy in dealing with hazards requires answers to such questions as, how safe is safe, and what standards should be used to judge the extent of the hazard? Thus, several assessments are necessary before appropriate decisions can be reached. The presence of the hazard must be first recognized. It is then important to evaluate

Table 18-6 National Wild and Scenic Rivers System, as of October 1976

River (present units in the national system)	Administering agency	Miles by classification			
		Wild	Scenic	Recrea-tional	Total miles
1. Middle Fork Clearwater, Idaho (P.L. 90-542—10/2/68)	USFS	54	—	131	185
2. Eleven Point, Missouri (P.L. 90-542—10/2/68)	USFS	—	44.4	—	44.4
3. Feather, California (P.L. 90-542—10/2/68)	USFS	32.9	9.7	65.4	108
4. Rio Grande, New Mexico (Rio Grande Mgt. by Agency) (P.L. 90-542—10/2/68)	BLM/USFS (BLM) (USFS)	51.75 (43.90) (7.85)	— — —	1 (0.25) (0.75)	52.75 (44.15) (8.60)
5. Rogue, Oregon (Rogue Mgt. by Agency) (P.L. 90-542—10/2/68)	BLM/USFS (BLM) (USFS)	33 (20) (13)	7.5 — (7.5)	44 (27) (17)	84.5 (47) (37.5)
6. St. Croix, Minnesota and Wisconsin (P.L. 90-542—10/2/68)	NPS	—	181	19	200
7. Middle Fork Salmon, Idaho (P.L. 90-542—10/2/68)	USFS	103	—	1	104
8. Wolf, Wisconsin (P.L. 90-542—10/2/68)	NPS	—	25	—	25
9. Allgash Wilderness Waterway, Maine (Secretarial designation—7/19/70)	State of Maine	95	—	—	95
10. Lower St. Croix, Minnesota and Wisconsin (P.L. 92-560—10/25/72—27 mi) (Secretarial designation—6/17/76—25 mi)	NPS States of Minnesota and Wisconsin	— —	12 —	15 25	27 25
11. Chattooga, N.C., S.C., and Georgia (P.L. 93-279—5/10/74)	USFS	39.8	2.5	14.6	56.9
12. Little Miami, Ohio (Secretarial designation—8/20/73)	State of Ohio	—	18	48	66
13. Little Beaver, Ohio (Secretarial designation—10/23/75)	State of Ohio	—	33	—	33
14. Snake, Idaho and Oregon (P.L. 94-199—12/31/75)	USFS	32.5	34.4	—	66.9
15. Rapid, Idaho (P.L. 94-199—12/31/75)	USFS	31	—	—	31
16. New, North Carolina (Secretarial designation—4/13/76)	State of North Carolina	—	26.5	—	26.5
17. Missouri, Montana (P.L. 94.486—10/12/76)	BLM/FWS	72	28	59	159
18. Flathead, Montana (P.L. 94-486—10/12/76)	FS/NPS	97.9	40.7	80.4	219
19. Obed, Tennessee (P.L. 94-486—10/12/76)	NPS/State of Tennessee	46.2	—	—	46.2
Total		689.05	462.7	503.4	1,655.15

Source: U.S. Department of Interior.

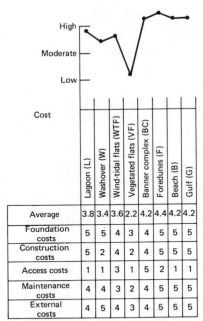

	Lagoon (L)	Washover (W)	Wind-tidal flats (WTF)	Vegetated flats (VF)	Banner complex (BC)	Foredunes (F)	Beach (B)	Gulf (G)
Average	3.8	3.4	3.6	2.2	4.2	4.4	4.2	4.2
Foundation costs	5	5	4	3	4	5	5	5
Construction costs	5	2	4	2	4	5	5	5
Access costs	1	1	3	1	5	2	1	1
Maintenance costs	4	4	3	2	4	5	5	5
External costs	4	5	4	3	4	5	5	5

Figure 18-13 Cost matrix for Padre Island. (Mathewson and Piper, 1975.)

Figure 18-12 Erosion along footpath in Kokanee Glacier Provincial Park, British Columbia, Canada. Mankind can cause landscape deterioration even in remote wilderness terrain. (Courtesy John Conners.)

the likely severity and frequency of occurrence. Finally, the degree of risk must be determined. This should account for what can be done to reduce the hazard in terms of public costs and benefits. Four basic methods can be used to reduce the risks associated with natural hazards.

1 Land use may be restricted or regulated.

2 Prevention or control may be used against the hazard-producing process.

3 Structural efforts that reduce property loss can be implemented.

4 Monitoring and warning systems that permit evacuation time can be developed.

All of these methods have environmental, economic, political, and social effects that must be measured in some type of benefit-cost equation. The decision maker can then make policy based on the acceptable level of risks that can be endured.

One type of cost and risk matrix system has been developed by Mathewson and Piper (Figs. 18-13, 18-14, and 18-15). They developed a map for the Padre Island, Texas, site of potential development and assigned the var-ious land parcels an index based on cost, risk, and geo-economic values. Land investment decisions in such a coastal corridor are commonly determined by the economics of the site, which are a function of location, demand, aesthetics, and developmental costs. Because builders and real estate brokers are generally not tuned to geologic nomenclature, a system that uses numerical ratings may have more meaning in arriving at value judgments. At Padre Island, eight different environments were identified and evaluated in a matrix, with numbers ranging from 1 (low) to 5 (high). Each area can then be evaluated in terms of the type of development that is best suited for that particular environment. Such an approach was justified on the basis that insurance companies always assign risk in their policy rate determinations. Thus environmental consumers should also become

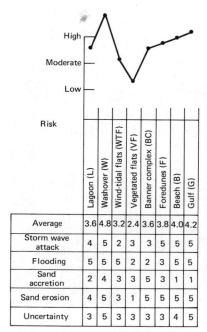

	Lagoon (L)	Washover (W)	Wind-tidal flats (WTF)	Vegetated flats (VF)	Banner complex (BC)	Foredunes (F)	Beach (B)	Gulf (G)
Average	3.6	4.8	3.2	2.4	3.6	3.8	4.0	4.2
Storm wave attack	4	5	2	3	3	5	5	5
Flooding	5	5	5	2	2	3	5	5
Sand accretion	2	4	3	3	5	3	1	1
Sand erosion	4	5	3	1	5	5	5	5
Uncertainty	3	5	3	3	3	3	4	5

Figure 18-14 Risk matrix for Padre Island. (Mathewson and Piper, 1975.)

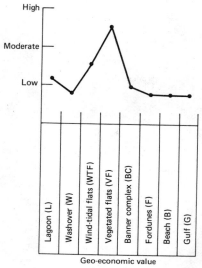

Figure 18-15 Geological-economic value. Land-value graph of Padre Island. (Mathewson and Piper, 1975.)

accustomed to a numbering system that makes decisions as objective as possible.

RECLAMATION

The policies set by environmental managers should not only cover the development of natural resources and human construction efforts, but should also include the planning and rehabilitation that may be necessary to minimize damages that may result from such actions. Other chapters in this part discuss such planning enterprises in the urban sector and in waste disposal systems.

The term **reclamation** as currently used embraces all of our endeavors to bring into more useful production lands that had previously been nonproductive or were damaged by human or nat-

ural processes. Some authors prefer to use the term **rehabilitation** in describing those lands made derelict by human activity that are then renovated for beneficial purposes. Early examples of land reclamation include those in Holland (the Netherlands) and the "fens" on the east coast of England (page 494). The Bureau of Reclamation (page 237) is the major U.S. government agency charged with the task of reclaiming many western lands which would otherwise be idle except for the advent of irrigation.

There are many faces to reclamation. After the Dust Bowl years in the 1930s, in the Great Plains states massive reclamation efforts were undertaken to restore to productivity an area that had been devastated by several continuous years of degrading storms and erosion cycles (Fig. 18-16). The moors of the Scottish Highlands are also being re-

claimed (Fig. 18-17). Although there is peat production in some areas, the wetlands of others are being drained and reforested for what is hoped will become cash crops. Other wetland environments throughout the world are also being reclaimed for agricultural purposes. Probably the largest region now being affected is in the Amazon Basin where millions of hectares of tropical forests are being cut down along with the installation of extensive road networks. Such large-scale reclamation efforts will undoubtedly produce many deleterious environmental impacts on the ecology of the area and may even produce atmospheric changes that can have regional, even global, significance. Projects of this kind need more critical evaluation than they seem to have received. Other areas in South America (Fig. 18-18) are also being changed by reclamation programs.

(a)

(b)

Figure 18-16 Before and after effects of conservation reclamation, Beadle Co., South Dakota. (a) View of barns with drifted sand caused by Dust Bowl winds, September 17, 1935. The sand is 2 m deep in places. (b) Emergency cover-crop planting of cane and sudan grass restored control of the area. These crops were seeded in 1937 and, by August 1937 when photo was taken, these measures had proved effective in protecting the site from further losses. (Courtesy Soil Conservation Service.)

(a)

(b)

Figure 18-17 Reclamation of Scotland moors. (a) Drainage ditches provide avenues for movement of wetland waters. (b) Reforestation after water table has been lowered.

Figure 18-18 Dredging swampy areas in the Maracaibo basin near El Vigia, northwestern Venezuela. Dams, levees, channelization, drainage works, and other engineering structures help to control water and sediment and to open area to more productive land uses. (Olson, 1976.)

The prime candidates for reclamation are those areas that were destructively abused during surface mining operations. Although several states, especially the eastern coal mining states, had some laws and regulations about strip mining, it took several decades of bitter debate before Congress finally passed a federal law, Public Law 95-87, on August 3, 1977, that sets standards for strip mining and the accompanying controls that are mandated during the reclamation process.

The state of New York has set policy and regulations that govern mining activities and these are summed up in the New York Mined Land Reclamation Act of 1975. The three main purposes of the act are to:

1 Foster and encourage the development of the mining in-

dustry and the mineral resources of the state.

2 Prevent pollution associated with mining activity.

3 Assure the reclamation of mined lands in such manner as to render such land suitable for further productive use.

This legislation, typical of most other eastern mining states, requires that all mined lands have their owners submit a report to the Department of Environmental Conservation that describes the current type of mining activities; the report must also provide a reclamation plan of how the land will be restored after mining ceases. A bond must be filed and its size depends on the amount of land, and its status, that is under development. Such monies will serve as a revenue base and guarantee that the land will be properly reclaimed. Typical costs for land reclamation range from $1000 to $5000 per acre for quarries and $3000 to $8000 per acre for coal strip mines.

Coal Mine Reclamation

Perhaps the rehabilitation of surface coal mining offers the most dramatic testimony of what proper policy and planning can accomplish for an area that would otherwise be one of complete destruction and desolation. Such renovation techniques provide the epitomy of prudent multipurpose and sequential land use management of natural resources (Fig. 18-19).

It is imperative that a complete reclamation plan and operation be in force during all phases of mining activity. Only by such long-range policies can minimum environmental damage be accomplished. Figures 18-20, 18-21, and 18-22 show some of the results that have been achieved by proper

environmental management procedures. The following methods describe the nature of the geomorphic engineering that should be instituted for the reclamation of surface coal mines:

1 Spoil piles should be covered, protected, and planted as soon as possible to reduce their erosion and the resulting siltation on downgrade lands and watercourses. Such cover will also prevent oxidation of the sulfur minerals which are the main villains in the production of acid water drainage.

2 The mine face and working area should be as restricted as possible, to minimize exposure of new rock to weathering processes. The longer the rock is exposed to atmospheric weathering, the larger the amount of pollutants that will result.

3 Slopes should be graded so as not to exceed the angle of stability. More gentle grades will reduce landsliding and also minimize sheetwash and channeling.

4 Catchment basins should be located to trap all sediment that might flow from the mine area.

5 Access roads should be designed to encompass a minimum area. They should also contain roadbed material that produces minimal fugitive dust to prevent drainage from degrading adjacent streams.

6 Revegetation of the area should always occur throughout the entire mining operation and, on completion of mining activities, these efforts must be coordinated with the entire grading and recontouring of the disturbed terrain.

TENNESSEE VALLEY AUTHORITY

The Tennessee Valley reclamation project is the most comprehensive and wide-ranging environmental program in the world (Fig. 18-23). In 1933 President Franklin Roosevelt signed the Tennessee Valley Authority Act for the environmental management of the 106,000 km² Tennessee Valley region. The stage had been set earlier for such a massive approach to the planning and operation of a vast area by President Theodore Roosevelt when he established the Inland Waterways Commission of 1907. This action embraced the principle of unified river basin management, regardless of state boundaries, by combining flood control, power development, irrigation, resource control, drainage, and pollution programs. The Tennessee Valley area contains many important resources—coal, iron, manganese, copper, marble, limestone, zinc, sand, and gravel. It also was nearly annually subjected to severe flooding that caused devastating damage to farmlands and cities (Figs. 18-24 and 18-25).

Wilson Dam at Muscle Shoals, Alabama, had been built under the National Defense Act of 1916, with one purpose—that of supplying nitrate for powder in World War I. The war ended before it became operational, but the dam acted as a catalyst for the further development of the region. With the inauguration of the Tennessee Valley Authority (TVA) Act, a unified program was set in motion to contruct dams to control flooding and create hydroelectric power, to deepen rivers for shipping, to plant new forests along with instituting other land conservation measures (Fig. 18-25), to control timbering operations, and to mine and manage mineral resources. Since 1933,

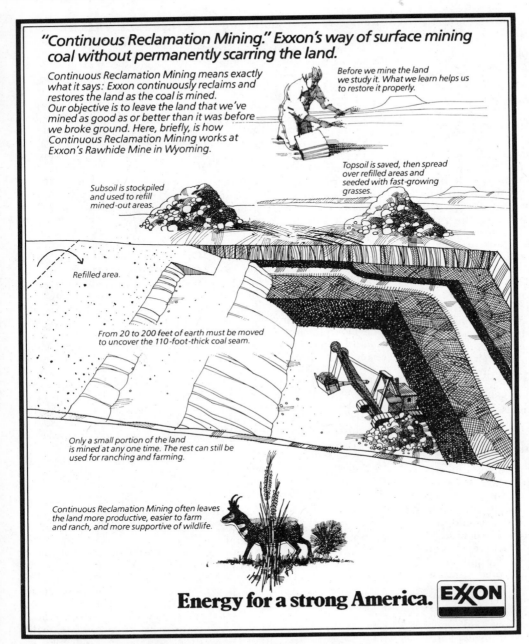

Figure 18-19 Continuous reclamation mining. (Courtesy EXXON Corp.)

the TVA has built 20 dams which provide one of the most highly regulated river systems in the world. Fontana Dam (Fig. 18-26) is the highest—146 m; Kentucky Dam is the largest—63 m high, 0.8 km long, impounding a 295 km artificial lake. In addition, the reservoirs provide 18 billion m³ of water storage at the beginning of the flood season. The Tennesse River now provides a 1000 km

Figure 18-20 A helicopter spreads grass and tree seed to help reclaim the strip mined land. The seed is discharged from a motor-driven spreader on a hopper hung below the helicopter. Such techniques allow rapid large-scale seeding of rugged terrain. (Courtesy West Virginia Surface Mining Assoc.)

Figure 18-22 An Illinois wheat field that formerly was a mined-out site. (Courtesy National Coal Association, 1971.)

Figure 18-21 Lands reclaimed by the Peabody Coal Co., Montana. The shaped and blended spoil banks bear little resemblance to the former mined-out area. This part of Big Sky Mine is partially seeded with a mixture of native grasses. Oats and alfalfa will act as a cover crop. (Courtesy U.S. Bureau of Reclamation, photo taken April 4, 1974.)

route for boats with 3 m drafts and is connected to the inland waterway system. By 1955, the TVA had reforested 100,000 ha, and the power plants were generating 53 billion kWh/yr. Furthermore, the program has given impetus to the increased community planning, with cooperative arrangements in financing worked out with local governments.

The TVA electric network affects a 207,000 km² area and is distributed to more than 150 municipal and cooperative electric systems that serve more than 1.5 million consumers. The phosphate fertilizers that are produced are sold nationwide, and coal mining has become another export industry. However, with the rapid rise in energy use throughout the

entire region, hydroelectric power no longer is sufficient. Coal-burning plants now produce additional power and the plants have become one of the major coal users in the country.

NATIONAL ENVIRONMENTAL POLICY ACT OF 1970

The National Environmental Policy Act (NEPA) states in the preamble that its purpose is "To declare a national policy which will encourage productive harmony between man and his environment." It provided for the inception of an Environmental Protection Agency (EPA) and a Council of Environmental Quality (CEQ). However, one feature which at first received little notice, but which has increasingly proved to be a major aspect of the policy, was the institution and requirement for providing "environmental impact statements" (EIS). Such statements must now be filed by government agencies and by other groups when major construction activites are proposed. The EIS must contain specific information as required by the five clauses:

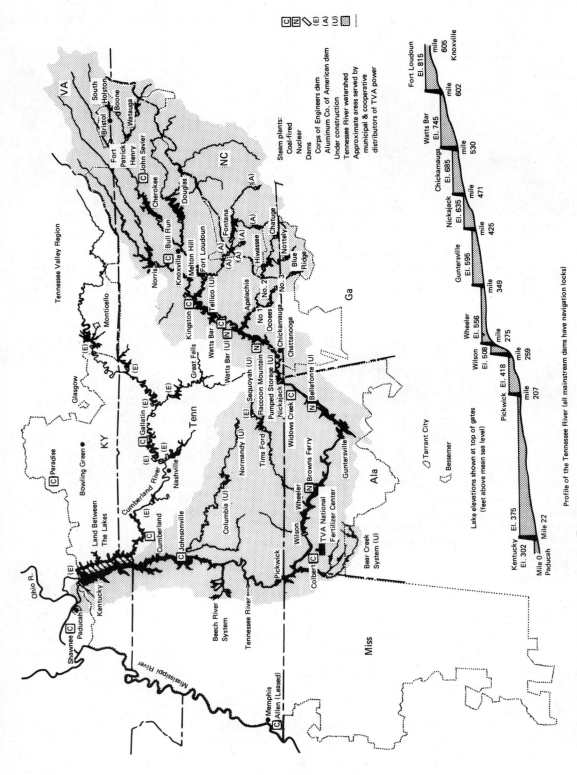

Figure 18-23 Map of TVA region showing Tennessee River profile and dams. (Courtesy TVA.)

Figure 18-24 Typical badly eroded terrain on lands of the Tennessee Valley region before the reclamation and conservation programs sponsored by the TVA. (Courtesy TVA.)

(a)

(b)

Figure 18-25 Before and after channelization of the Tennessee River at Sevierville, Tennessee. (Courtesy TVA.) (a) Nearly annual floods overwhelmed much of the floodplain prior to channelization. (b) Manmade levees now serve to contain most floodwaters. Note that the meandering character of the river has been maintained, aiding to maintain stability.

1 The environmental impact of the proposed action.

2 Any adverse environmental effects which cannot be avoided should the proposal be implemented.

3 Alternatives to the proposed action.

4 The relationship between local short-term uses of the environment and the maintenance and enhancement of long-term productivity.

5 Any irreversible and irretrievable commitments of resources which would be involved in the proposed action should it be implemented.

During the early years of NEPA, misunderstandings developed concerning its scope and intent, and debates are now raging on the degree of its effectiveness. Some extremists have charged that NEPA was invented to silence the "ecofreaks." Others decry the mountains of paperwork that must be filed to be in compliance with the law, and some argue that the act has actually mitigated the environmental movement. However, Russell W. Peterson, Chair-

man of the Council on Environmental Quality in a 1976 statement pointed to many positive results. The Alaska Pipeline project was redesigned to avoid adverse impacts; placement and permanent storage of nuclear wastes was postponed until more thorough studies had been performed; dams and other water resource projects that would have destroyed valuable natural systems have been modified or canceled; and scores of major highways and airport projects have been redesigned or eliminated.

Perhaps the principal criticism of NEPA is the view that an

EIS is invariably filed "after the fact," that it is some type of rationalization that is provided after a decision has been reached to undertake a project. Instead of this approach, the first question that a construction agency should ask is: Do we need and can the environment tolerate the project? In a July 31, 1976, audit by the General Accounting Office (GAO), four agencies which had prepared 59 percent of the 5144 final EIS were evaluated. Of these, half were written by the Federal Highway Administration, and others by the Corps of Engineers, EPA, and General Services Administra-

Figure 18-26 Fontana Dam is TVA's highest at 165 m. It is the highest concrete dam east of the Rocky Mountains and produces 238,500 kW. (Courtesy TVA.)

tion. One conclusion was that the filing of an EIS generally delayed the construction process, and in the case of highways the average delay was about one year. By the end of 1976, a total of 7500 EIS were on file. Of these, about 700 had been challenged in the courts as being inadequate. Temporary injunctions halted 123 projects with delays of a few weeks to the time it took to prepare a satisfactory EIS. Only in six cases were permanent injunctions given, and even in these projects the agency was not precluded from undertaking the project if additional study and work produced a favorable EIS.

The principal weaknesses of NEPA include:

1 Absence of a format or guides to determine the significance of an impact. There are no minimal analytical standards and no indication of the appropriate scope or level of detail of the report.

2 Emphasis is on a disclosure type of statement rather than on quantitative indices of impacts.

3 The government reviewing agency has no authority to cancel projects of other government agencies, but can only refer unsatisfactory statements to CEQ for arbitration.

4 There is no regulatory machinery to supervise construction conduct or police-resulting actions.

5 Incentive is lacking for preparation of top-caliber reports because there is little reward or recognition to the authors.

6 There is nothing to prevent excessive adversary action that unduly postpones needed projects.

In spite of many drawbacks, many NEPA supporters cite it as heralding a major Legislative policy breakthrough that has reformed the administrative decision-making process. The requirements that necessitate consideration of alternatives, prediction of impacts, and involvement of the public are all pluses in planning and management procedures. Furthermore, as a federal policy it imposes a responsibility on geologists to engage in a critical analysis of their habitat. NEPA has resulted in project funding that assures, in most cases, that scientific input is vital to the completion of an appropriate EIS. Thus, it provides geologists the opportunity to participate in the policy process and to communicate with the decision makers. Such blessings can only have beneficial results for professionals engaging in geological studies and produce an increased public awareness of political realities.

Perspectives

The magnitude of the environmental dilemmas that face the world, and the United States, must be impressed on those who make decisions and determine policy. In the next 20 years, the United States must grow, mine, transport, build, manufacture, and distribute as much in the way of material goods as it has done in all previous American history. We have shown great arrogance in thinking nature was invented solely for our benefit and in assuming that the human species always knows best how to use the earth. We have the ability to greatly modify our habitat, but this is often done at a price that becomes economically unjustifiable, physically disastrous, and aesthetically detrimental. Population growth and its supporting technology have been instrumental in creating a myriad of environmental problems. The major question to policymakers is how to control human activities so as to optimize society and individual well-being while reducing the harmful effects on health and the environment. Two kinds of efforts are required as a prudent basis of decisions. One is the estimation of the actual risks or effects associated with the technological options. This is difficult because of

the diversity of potentially harmful pollutants, the large number of organisms and ecosystems that must be considered, and the complexity of the pathways of the destructive forces. The second effort is equally difficult, because it involves value judgments—what level of risk is acceptable? A typical problem, still without solution, concerns the biological effects of ionizing radiation. After 20 years of research and more than $2 billion in funding, acceptabe exposure levels are still controversial.

Of paramount importance in policy decisions is the need to increase the public's understanding of what in our system of political institutions and economic incentives it has cost us in terms of severe environmental degradation and resource depletion. Is one solution to charge fees to

those who produce effluent and indirectly despoil the environment? If so, there are numerous culprits—industries, municipalities, mining, agriculture, and so on. Would another solution be to institute some type of "science-technology court," in the same way there is a civil court for citizen actions? Could a science and technology court arrive at conclusions that would be beneficial and help direct policy, rather than having to react after the fact as a postmortem to a system already in great jeopardy? Perhaps the only agreement among all groups is that the public should be well informed, that there is no substitute for a strong educational program, and that at best the purpose of policy is to provide for the health, safety, and welfare of most of the people most of the time.

READINGS

Anderson, J. R., et al., 1976, A land use and land cover classification system for use with remote sensor data: U.S. Geol. Survey Prof. Paper 964, 28 p.

Bailey, R. G., 1974, Land-capability classification of the Lake Tahoe Basin, California-Nevada: U.S. Dept. Agriculture-Forest Service, 32 p.

————, 1978, Description of ecoregions of the United States: U.S. Dept. Agriculture-Forest Service 79 p.

Bartelli, L. J., et al., eds., 1966, Soil surveys and land use planning: Soil Sci. Soc. Amer., 196 p.

Bosselman, F., and Callies, D., 1971, The quiet revolution in land use control: Washington D.C., Council on Environmental Control, 327 p.

Fabos, J. G., and Caswell, S. J., 1977, Composite landscape assessment: Mass. Agri. Ex. Sta. Research Bull. 637, 323 p.

Hardin, G., 1968, The tragedy of the Commons: Science, v. 162, p. 1243–1248.

Leopold, L. B., 1969, Landscape esthetics: Natural History, Oct., p. 35–46.

————, 1969, Quantitative comparison of some aesthetic factors among rivers: U.S. Geol. Survey Circ. 620, 16 p.

McAllister, D. M., ed., 1973, Environment: a new focus for land-use planning: Washington, D.C., National Science Foundation, 328 p.

Ministry of Housing & Local Government, 1963, New life for dead lands: London, H. M. Stationery Office, 30 p.

Moss, E., ed., 1975, Land use controls in New York State:

New York, The Dial Press, 368 p.

Nash, R., 1967, Wilderness and the American mind: New Haven, Yale Univ. Press, 256 p.

Nichols, D. R., and Buchanan-Banks, J. M., 1974, Seismic hazards and land use planning: U.S. Geol. Survey Circ. 690, 33 p.

Office of the Chief of Engineers, 1976, A perspective on flood plain regulations for flood plain management: Department of the Army, Washington, D.C., EP 1165-2-304, 156 p.

Office of Planning Coordination, 1971, LUNR: Land use and natural resource inventory of New York State: Albany, Office of Planning Coordination, State of New York, 21 p.

Palmer, L., 1976, Application of land-use constraints in

Oregon: in Coates, D. R., ed., Urban geomorphology, Geol. Soc. Amer. Spec. Paper 174, p. 61–84.

Paone, J., Morning, J. L., and Giorgetti, L., 1974, Land utilization and reclamation in the mining industry, 1930–71: U.S. Bureau of Mines Info. Circ. 8642, 61 p.

Redding, M. J., Proj. Officer, 1973, Aesthetics in environmental planning: Washington, D.C., U.S. Environmental Protection Agency, 187 p.

U.S. Geological Survey and U.S. Department of Housing and Urban Development, 1971, Program design for San Francisco Bay Region Environmental and Resources Planning Study: Menlo Park, Calif., 123 p.

U.S. Water Resources Council, 1976, A unified national program for flood plain management: Washington, D.C., U.S. Water Resources Council, var. pages.

William Spangle and Associates, et al., 1975, Earth science information in land-use planning—guidelines for earth scientists and planners: U.S. Geol. Survey Circ. 721, 28 p.

Chapter Nineteen
Urban Areas

Cars on a Los Angeles freeway—the ubiquitous machine in the urban landscape. (Courtesy U.S. Dept. of Housing and Urban Development.)

INTRODUCTION

Throughout recorded history, cities have been closely linked to the rise and fall of empires. Although the concept of cities is millennia old, worldwide massive urbanization is largely a twentieth-century phenomenon. It is in the city where humans can exert their most domineering presence over the natural world. Here we completely alter the original terrain to suit our own needs and pleasures. We cut and fill, rearranging the landscape by emplacing our own brand of bricks, mortar, stone, asphalt, wood, and steel into gigantic monoliths and impervious surfaces that express our utilitarian presence—occasionally reflecting our artistic and aesthetic senses—into a metamorphosed terrane, the **cityscape**. Fortunately in some countries, such as the United States, this race to completely change the natural world has been counterbalanced in the twentieth century by the concepts of wilderness and preservation.

Mankind has a host of reasons for concentrating in congested areas, and these crosscut all facets of life—social and cultural, economic and commercial, political and military. However, high population densities become so

Figure 19-1 Aerial view of Sydney, Australia, looking north along the Northern Expressway over the Harbour Bridge. The firm ground and excellent harbor provide good building sites, as well as a highway network and an unexcelled transport system. (Courtesy Australian Information Service.)

interdependent that they cannot sustain themselves without supporting systems from rural regions. So although cities may be the heart of a civilization, they are useless without a proper import-export system that permits new sustenance and raw materials from the outside and exterior trade and transport for the removal of waste products.

The location of a successful city is dependent on a variety of factors, and many of these relate to geological, hydrological, and topographical features of the site (Fig. 19–1). Furthermore, just as commerce has usually been the reflection of the urbanizing process, its vitality is governed by the availability of resources—materials, agricultural land, water—and by the ease of transportation, which is a function of available energy, topography, and navigable waterways.

As cities continue to grow, and more and more land is used in development, the involvement of geologists in the planning and management of urban areas is be-

coming increasingly important. Policymakers are becoming more aware of this as evidenced in the employment of geologists on planning staffs and the increased financial support for government surveys and maps on all levels— local, county, state, and federal. As prime land becomes more scarce, as resources become more difficult to find and mine, as the dangers from geologic hazards increase, and as water becomes more precious, the need for earth scientists will become more vital. Cities also produce a prodigious number of impacts on the environment. Thus environmental assessment prior to new development is being required in most cities, and continual monitoring and surveillance is necessary to determine violations to many laws and ordinances that are already in force. The manner in which geologists fulfill their mission and articulate the requirements for a safe and sustained geologic environment may well determine the fate of many people and their quality of life.

HISTORICAL BACKGROUND

Although the urbanization revolution has greatly accelerated in the twentieth century, the rise of cities started more than 7000 years ago. In ancient times cities became nearly synonymous with empires, and their rise and fall coincided with that of the particular associated civilization. True cities were flourishing by 3000 B.C., and since that time the study of history is closely linked with the study of urban graveyards. Cities emerged when several conditions became coincident: the growing of surplus foodstuffs, development of writing, water technology, and the necessity for a common defense. A sociopolitical structure was required to manage such a large human concentration, which led to a ruling and governmental class. This new organization manifested itself as "empire" and became jealous and covetous of neighboring areas, as well as possessive of its own territory. Urban centers were therefore crucial to the power of empires as rallying locations where military supremacy could be maintained. The cities were also administrative focal points so that resources could be monitored and disseminated; they were also places where commerce could enrich the establishment. Thus cities evolved because they were the most efficient systems for propagating and protecting the individual and his possessions.

The location and growth of cities are most commonly related to topographic-hydrographic features and to what can be labeled as the **civilization drive**—the social, economic, and political pressures that control society and its destiny. Factors that influence city sites include: (1) amenable climate, (2) trade routes, (3) easily defendable position, (4) access to water, (5) nearby presence of natural resources, and (6) suitable building sites. The ancient cities in the Middle East and Far East, Egypt, India, and China were first located in rich agricultural floodplain lands with nearby water sources. As commerce increased, such areas became well-traveled trade corridors, as were the coastal sites when man conquered the water with ocean-going vessels. Many cities in Europe developed from the fortified villages of the Middle Ages, as did some of the New World cities, such as Pittsburgh, Detroit, Quebec, and Montreal. However, the early large cities of the United States usually formed where there were good, safe harbors—for example, Boston, New York City, Philadelphia, Baltimore, and Charleston. With the "wonders" of modern civilization, we can now build cities in areas that would be considered marginally inhabitable 70 years ago. Air conditioners make Albuquerque, Los Angeles, Phoenix, and Tucson livable and desirable for health and therapeutic

reasons. Siberian cities have developed for political and economic reasons in a harsh environment, and Brasilia, Brazil, was selected to be the showplace of the nation and exhibit the technological skills as administered by the government. The profit incentive is behind some of the rather spectacular new towns of the United States, such as Reston, Virginia, and Columbia, Maryland. Thus the location of cities is dependent on a very wide range of pressures, and their geography has changed through the ages.

Although we discovered that cities had several advantages over rural areas more than 7000 years ago, the growth in population in urban centers was very slow for several millennia. For example, the walls of Babylon enclosed only 8.3 km² and Ur with its canals, harbor, and temples was only about 100 ha. The major cities in the Indian subcontinent were Mohenho-Daro, which covered about 2.5 km², and Harappa in the Punjab with a walled perimeter of 4 km. These populations could not have exceeded 15,000. Such small-size cities can be attributed to several factors: (1) labor in agriculture was inefficient, requiring 50 to 100 farmers to support a single city dweller, and (2) transportation and storage facilities were inadequate and food distribution often inequitable. The dramatic increase in city size in the second millennium B.C. coincided with the Iron Age and the development of sturdier and more labor-saving equipment. So that by 1600 B.C., Thebes, the capitol of Egypt, had a circumference of 22.5 km and a population of about 200,000. During the Greco-Roman period, 600 B.C. to A.D. 400, the major cities were in coastal areas because of the ability to import necessary items cheaply via water routes. Thus Athens, Syracuse, and Carthage all ex-

ceeded 150,000 in population during their zenith, and Rome and Constantinople were larger than 300,000. However, such cities had a basic flaw: they depended on an agricultural economy, either local or imported, that was commanded by military strength rather than obtained commercially. The growth of other European cities, although the sites may have had a different original purpose, during the next 1000 years was stimulated by economic factors and the clever use of a full range of handicrafts and manufacturing skills of the time. Thus through trade they could afford the purchase of raw materials and resources that were not of local origin. Antwerp, London, and Nuremberg are typical of this second rise in urban development. The gradual rise in mechanization started during the Renaissance in the fifteenth century, but took a quantum jump with the extensive use of coal and new methods for mass production inaugurated with the Industrial Revolution. The necessity for concentrating coal, iron, and factories ushered in a new era of urbanization by the 1700s, so that by 1900 England was the first nation where more than half the population lived in the urban environment.

URBANIZATION AND THE MODERN GROWTH OF CITIES

The concept of **urbanization** is quite different from the physical reality involved with the growth of a city. Urbanization is a process that involves an inward migration of people from dispersed areas to a centralized core area. It is the "de-population" of the rural environment and the concomitant expansion of metropolitan regions. Historically urbanization occurs when a nation develops from an agrarian to an industrial society. Thus, the growth of cities was more a function of emigration from the countryside than a result of population increases. In the twentieth century, however, a new element entered the equation that is being acutely accentuated in nations of the "third world." Not only is there an overwhelming exodus of people from rural communities, but a population explosion is vastly accelerating the growth of cities. This coupling of two forces is producing runaway expansion and blight in and adjacent to the metropolitan areas.

Early Urbanization and the Future

In 1800 about 1 percent of the world's population resided in cities of 100,000 or more people. By 1930 the figure was 11 percent, and it increased to 20 percent by 1960. If present trends continue, the population of the planet will be 6.3 billion in 2000 (it is 4 billion now), and 60 percent of the people will live in urban areas. In 1790 the United States population was 3.9 million, and 5 percent lived in towns of 2500 or more. By the year 2000, 80 percent of the populace will live in urban areas.

The size of American cities has grown dramatically during its history. In 1800 there were no cities with a population of 100,000, but by 1900 there were 37 such cities. During this time New York City grew from 50,000 to 5 million! In 1900, 28 large cities were in the northeast metropolitan belt, whereas only 9 were elsewhere in the country. The first two urbanized areas to reach 100,000 by 1820 were New York City and Philadelphia. Boston and Baltimore followed by 1840, Chi-

cago by 1860, and by 1880 there were also Pittsburgh, Cleveland, Cincinnati, and Buffalo. Between 1900 and 1920 the number of urban areas increased from 37 to 70, and the number more than doubled between 1920 and 1960 to 160. By 2000 there will be 223 major urbanized areas. Clearly, the twentieth century is the urbanization period for the United States. By 1967, 66 percent of the population lived in cities of 50,000 or more. However, the growth rate in central cities had slowed, and two-thirds the increase occurred in suburban communities.

In Europe, cities with more than 100,000 contained only 1.6 percent of the total population in 1600. The figure was 1.9 percent in 1700 and, in 1800, 2.2 percent. Thus by the eve of the Industrial Revolution Europe was overwhelmingly still an agrarian society. By 1801 one-tenth the people in England and Wales lived in cities having a population over 100,000, and this doubled by 1840 and again in another 60 years so that by 1900 much of Britain was urban. Latin America presents the most startling figures for urbanization, with a growth of cities averaging about 4 percent per year and a doubling rate for urban populations every 17 years! By the year 2000, the urban centers will have 500 million with 19 metropolitan areas containing one-half the city population.

Deurbanization

In the United States, and several other countries, a reversal in migration patterns started to emerge by 1970. The urban core was no longer attracting immigrants; instead, there has been what some have described as a flight, or stampede, from the central city to the outlying areas. The U.S. Bu-

reau of the Census describes **urbanized areas** in terms of the Standard Metropolitan Statistical Area (SMSA), in which there is at least one city of 50,000 inhabitants. During the 1960s only one of the 25 largest SMSAs decreased in population but, by 1974, 10 of the largest declined and, of the 265 SMSAs in the country, 15 percent are losing population. Stated another way, one out of three Americans live in a metropolitan area that is declining in population. This is in marked contrast to the 1965–70 period when for every 94 people that moved out of a city, 100 moved in. By 1975 for every 100 that moved in, 131 moved out. However, it is still too soon to determine what **urban homesteading** (the encouragement and monetary incentives for the reoccupation of derelict city dwellings) and the energy crises will do to population mobility.

Overview of Urbanization Effects

Although profound changes occur in society and in the life-style of the urban-suburban dweller in contrast to the rural family, here we are concerned with how cities affect natural systems. It is obvious that the cityscape completely changes the normal land-water ecosystem and disrupts and violates natural processes. Many of these effects are of direct interest to geologists and fall within their area of expertise. For example, with proper designs and locations for buildings and services, minimum damage to the environment will occur, which in terms of costs may provide significant savings, both in terms of human safety and construction safety and in long-range economic benefits.

Urbanization is intricately associated with the availability of energy and the production of food surpluses. With the increased density of inhabitants in urban areas, this *should* free more lands for other purposes. However, as we will discover, these factors produce problems and paradoxes. The growth of cities in advanced countries enabled agricultural areas to be consolidated, allowed for more effective capitalization, and in general increased efficiency on the farm. To accomplish these results, prodigious amounts of energy are required to satisfy food production, housing needs, transportation corridors, and amenable leisure activities for shortened work weeks. To be competitive, each farmer must increase production, which could only be done through increased mechanical equipment; this in turn becomes energy-dependent. In the United States many brag of this "success story"—indeed 5 percent of society is so efficient that it feeds the remaining 95 percent, with sufficient leftovers that the surplus is sent abroad. Such exports have steadily increased, and it is significant that this period coincides with the urban growth in the nondeveloped countries. United States grain exports increased from 5 million tons in 1936 to 23 million tons in 1950; this figure jumped to 56 million tons in 1970 and to 94 million tons in 1976. What is hidden in such a story is the enormous energy expenditure; only 25 percent of this energy went into actual food production—75 percent was used to transport, process, and distribute the food once it left the field. With the decline in cheap energy, and with the dependence of other nations on food imports, a new politics of food will undoubtedly emerge before the twenty-first century.

The urban area not only affects the immediate terrain, water, and air of the developed sites, but also has a profound impact on the surrounding countryside as well as distant places (see page 547). The city can be viewed as a resource debtor and as parasitical in deriving its strength and vitality from the nonurban regions. Not only must all agricultural produce be imported, but the raw materials to fuel industry and commerce and often the water necessary to sustain and cleanse society must also be imported. Thus the influence of cities on the nation is vast and, although many geologists may not realize it, they are in many instances performing studies and research that are directly related with urban affairs—whether they do these investigations in, near, or at a great distance from the consuming urban environment. Indeed, many geologists still associate environmental geology with a related theme—urban geology. The highways, on which goods are transported, and the large reservoirs, which may be 100 km or more from the city, bear witness to the changes in the rural areas that were necessary to make city living possible.

THE ROLE OF GEOLOGY

For the remainder of this chapter, we will be devoted to those matters that specifically involve geologists or, in some cases, the more general subject of earth science. In terms of the siting of cities or their continued growth, there are many geologic constraints that need consideration for the safe and efficient operation of a congested area. Here a constraint involves both geologic hazards and other dangers from geologic processes, as well as those features that affect the natural world and the quality of life (Fig. 19–2). Yet the role of the geologist in urban affairs is not only a necessary one, but an ironic one as well. For example, without the geologist's remarkable success in discovering oil the automobile would not have been able to change the entire mobility index of city dwellers. Such rapid transportation, however, has allowed a massive reordering of communities and a resulting sprawl that has greatly added to the desecration of the environment. There also would not have been the need to lace the countryside with a myriad of roads that chop up the normal terrain.

Geologists can play an important role in city planning. The urban area should be free from most natural dangers and geologic hazards. However, the surface and groundwaters are changed in quantity and quality by human use and refuse, and their hydrology is altered by installation of impervious surface coverings. The natural resources are either mined out or masked by construction, and the surface topography is changed by cut-and-fill operations. The production of waste products intrude into landfills, the air, and the waters, and the resulting pollutants produce onerous feedback to the community and adjacent areas. Indeed, if the city is our ultimate achievement and the mark of civilization, it can also be a pathetic reminder of our perception of "good living" and a reflection of our quality of life.

Whenever possible, geologists should become involved in those governmental forums that make policy regarding urban areas. It is vital that they communicate with the decision makers and provide the type of documentation that can prove that there is a need to include geologic considerations in the matrix of land use management. Through the compilation of appropriate data and the presentation of re-

(a)

(b)

Figure 19-2 Photographs of Chattanooga, Tennessee, from Lookout Mountain taken 100 years apart. (a) Only the high ground in the central area at Missionary Ridge escaped this March 1867 record flood. (b) This 1967 view was taken after completion of most major flood control projects in the region. (Courtesy TVA.)

Figure 19-3 Ogden City, Utah. This suburban community is nestled along the Wasatch Mountains, but it is a precarious existence—the Wasatch fault trace is immediately adjacent. (Courtesy Earl Olson.)

ports that are sufficiently clear and understandable, geologists can serve an important function. Given reliable facts, city planners cannot afford to disregard the danger signals, as they did in the Anchorage area, where such denial contributed to the unnecessary loss of life and property during the 1964 Alaska earthquake (Fig. 19–3).

More and more geology reports are becoming available for urban areas. Some of these are specific in terms of a single hazard or process, such as the West Virginia landslide reports (see page 337) and the many flood reports—for instance, "Water in Urban Planning, Salt Creek Basin, Illinois" and "A Community Decision—Managing the Binghamton Area Flood Plain" (Figs. 19–4 and 19–5). Many cities have geology reports that describe and map all important geological aspects, such as Saskatoon (Christiansen, 1970) and Washington, D.C. In addition, entire regions are now being mapped and reports are being written that approach the subject from the point of view of urban geology; for example, a report of the San Francisco Bay region was a combined U.S. Geological Survey and Department of Urban Housing venture that took five years; another similar report, "Urban Geology of Madison County, Indiana," was completed in 1975 (Wayne, 1975; Fig. 19–6). The urban geologist is not only interested in resources and foundations, but in the input-output of water systems, waste disposal, hazards, surface disruptions of erosion-sedimentation, subsidence, and open space corridors. This entire spectrum takes on new meaning when the pressures of congested areas are placed into the matrix of the total environment. A $55 billion price tag (in terms of 1973 dollars) has been placed on the costs California can expect—mostly in urban commu-

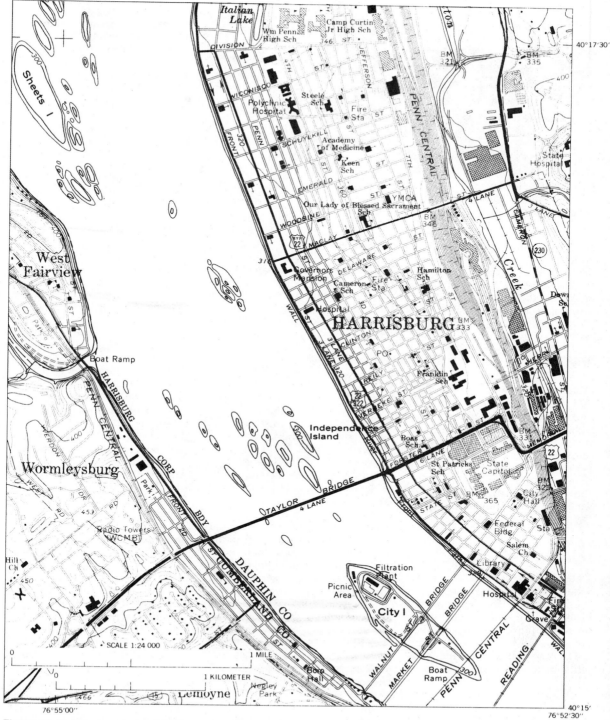

Figure 19-4 Flood-prone area of Harrisburg, Pennsylvania (U.S. Geol. Survey Circ. 601-J, 1974.)

Figure 19-5 Flooding of Harrisburg, Pennsylvania during Hurricane Agnes, 1972. (Courtesy Arthur Socolow.)

nities—during the 1970–2000 period from improper planning related to geologic phenomena (Alfors and Gay, 1973). Who says geology in the urban environment is unimportant?

WATER

Water is basic to urban dwellers and to the services that keep them and the wheels of commerce and industry humming (Fig. 19–7). Its treatment as a resource and its natural processes affect more people and cause more problems than any other physical aspect of the urban environment. Thus water has been aptly described as "a necessity and a nuisance."

Water as a Resource

Urban areas require extensive water resources. Most large cities cannot obtain sufficient quantities locally, so they must become water debtors and import from other areas—sometimes, as in the case of Los Angeles, from areas hundreds of kilometers away. Of the 35 largest cities in the United States, five have supplies from the Great Lakes, ten from major rivers, three from groundwater, and the remainder obtain water from distant reservoirs. In fact, one out of eight Americans uses water that is transported more than 120 km. The California Water Plan has been called "the most expensive faucet in history," and total expenditures will amount to about $10 billion. In order to supply the heavily populated metropolitan areas in Southern California, entire drainages in the north were diverted. This necessitated the construction of 21 dams and reservoirs, 22 pumping stations, 1100 km of tunnels, pipelines, and canals, and a 6400 ha artificial lake behind the highest dam in North America which has a 210 m deep reservoir. Industry is the largest water user in the urban environment, requiring 800 million m^3 per day. Adjacent irrigated lands that supply foodstuffs to the large urban cities of the Southwest also use prodigious quantities of water, both from reservoirs and from groundwater sources.

When water is used as a resource, the part that is nonconsumptive commonly is polluted. Its disposal is costly when treated and, if untreated, deterioration of adjacent waters damages the environment. Table 19–1 provides a cataloging of various hydrologic impacts that occur during urbanization. For example, the excessive pumping from groundwater sources can produce deleterious effects in the form of ground subsidence in some localities as well as the intrusion of saltwater into formerly freshwater aquifers in coastal cities.

Hydrology

We can alter the water budget of a city in many ways. We create a microclimate in which wind, temperature, and precipitation patterns are changed by buildings and other stone, brick, and paved structures. For example, one effect is the "heat island." It is not unusual for cities to have 3 to 6°C higher temperatures than adjacent land. Studies in several European cities also indicate that cities receive 10 percent more precipitation. United States studies show cities have 10 percent more cloud cover than adjacent areas and 24 percent more rain during the week than equivalent time periods on weekends. Such changes in the atmosphere when coupled with the burning of hydrocarbons and other industrial contaminants greatly accelerates the weathering rate of construction materials on buildings and monuments (Fig. 19–8). Fumes in Athens, Greece, are severely pitting the 2400-year-old Parthenon and rusting the iron bars used in attempted restoration. In Agra, India, the Taj Mahal is now an endangered shrine because of the chemical air pollution that is degrading its exterior. The original outdoor statues in Venice are being replaced with replicas because of deterioration from the pollution of the adjacent industrial city of Mestre. The most often cited example is Cleopatra's Needle in New York City. This obelisk of granite originally had deeply notched hieroglyphics which remained essentially unchanged during the 3500 years it was in Egypt. However, within 50 years after

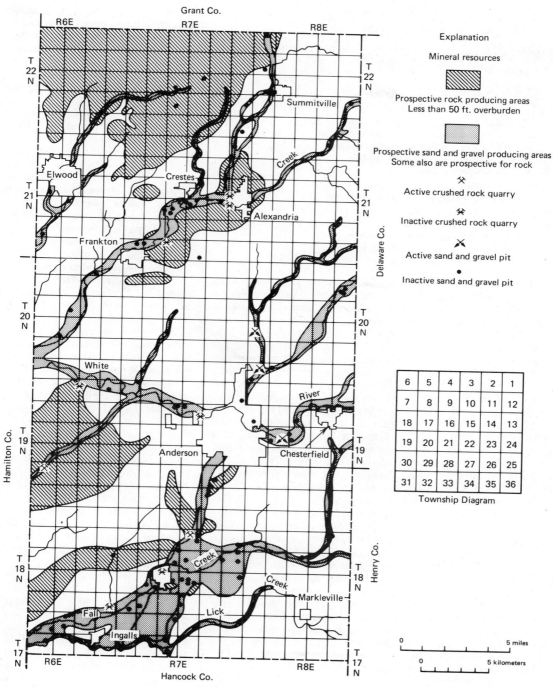

Figure 19-6 Prospective rock, sand, and gravel producing areas in Madison County, Indiana. (Wayne, 1975.)

Figure 19-7. San Manuel, Arizona. This city was completely built and designed for a community of 6000 whose workers were to operate the newly discovered Magma Mine, the largest copper discovery in the twentieth century. Before the new town was opened for settlement (this 1951 photo shows the completed community prior to being inhabited), a sufficient water supply had to be discovered and verified by the U.S. Geological Survey. After a bountiful water supply was assured, the mine and community received the largest Reconstruction Finance Corporation loan up to that time—$117 million.

being brought to Central Park in New York City, the writing had become illegible. Such destruction is usually attributed to atmospheric pollutants, but the most devastating changes are produced during inversions of air above cities. When the low air is trapped by warmer air above and the pollutants cannot dissipate, severe health hazards can occur. This happened at Donora, Pennsylvania, on October 27, 1948. The stagnant air persisted four days and the soot and sulfur dioxide fumes killed 20 and caused illness to 14,000. The worst disaster of this type killed 4000 in London in 1952 and, in 1930, 60 died in the highly industrialized Meuse Valley of Belgium.

Surface water Surface water in urban areas is completely altered in the urban environment, and stream hydrographs (Fig. 19-9) have different geometries than those in natural areas. Thus the normal streamflow regime is disrupted in urban areas. The slope of the recession curve on the hydrograph is much steeper because there is less bank storage and groundwater flow into the

stream channel. The **baseflow** of the steam thereby suffers, and less water in the channel is available during dry periods without rainfall. Because of the reduced base flow any pollutants reaching the stream will have an amplified contamination effect on the water. The streamflow regime is governed by the percentage of area made impervious and the delivery rate of water being transmitted across the land. The ground that is made impermeable by buildings, streets, parking lots, and so forth, cannot absorb precipitation by normal percolation (Fig. 19–10). Instead, such water is quickly channeled into man-made conduits of gutters and storm sewers that reach adjacent streams in highly shortened time periods. This causes very high and quick flood peaks in rivers, which are reflected in hydrographs of the flow character. Furthermore, the inability of precipitation to penetrate through the surface and into the water table produces a deficit in groundwater recharge. This also shows on the hydrograph by the diminution of groundwater discharge into stream channels. Studies show that this combined

constraint of paving and sewering produces more flood water, higher flood peaks, and larger number of floods (Fig. 19–11). For example, for an area that is 50 percent paved and 50 percent sewered, there would be four times as many floods for a similar area in the natural state. In Houston, Texas, the increase in impervious surfaces from 1 to 35 percent increased the magnitude of the 2-year peak flood by a factor of 9 and of the 50-year flood by a factor of 5. Urban streams have higher temperatures than rural streams, such as the populated central part of Nassau County, New York, where streams are 5°C warmer than those in adjacent, less densely populated terrain.

Although flooding is discussed in Chapter 12, it should be mentioned here that urban areas are in part responsible for some of their own flooding, as well as increasing the hazard in downstream localities. Because of level ground, floodplains have often been a favorite site for urban construction and even homesites. A U.S. Geological Survey study of 26 urbanized areas in the United States showed the percentage of developed lands in the floodplain had a median value of 10.5 percent with a weighted average of 16.2 percent—ranging from 2.4 percent in Spokane, Washington, and 81 percent in Monroe, Louisiana. When the business district and factory sites were only considered, the median was 57 percent and the weighted average 52.8 percent. No wonder flooding is such a hazard to cities! In New York there are 330 communities with a population greater than 2500, and 260, or 79 percent, have problems of local flooding and drainage. Although the pace of encroachment on floodplains has greatly slowed, both because of increased awareness and new laws and ordinances (see Chapter

Table 19-1 Hydrologic effects during a selected sequence of changes in land and water use associated with urbanization

Change in land or water use	Possible hydrologic effect
Transition from pre-urban to early-urban stage:	
Removal of trees or vegetation, construction of scattered city-type houses and limited water and sewage facilities	Decrease in transpiration and increase in storm flow; increased sedimentation of streams
Drilling of wells	Some lowering of water table
Construction of septic tanks and sanitary drains	Some increase in soil moisture and perhaps a rise in water table; perhaps some waterlogging of land and contamination of nearby wells or streams from overloaded sanitary drain system
Transition from early-urban to middle-urban stage:	
Bulldozing of land for mass housing; some topsoil removal; farm ponds filled in	Accelerated land erosion and stream sedimentation and aggradation; increased flood flows; elimination of smallest streams
Mass construction of houses; paving of streets; building of culverts	Decreased infiltration, resulting in increased flood flows and lowered groundwater levels; occasional flooding at channel constrictions (culverts) on remaining small streams; occasional over-topping or undermining of banks of artificial channels on small streams
Discontinued use and abandonment of some shallow wells	Rise in water table
Diversion of nearby streams for public water supply	Decrease in runoff between points of diversion and disposal
Untreated or inadequately treated sewage discharged into streams or disposal wells	Pollution of streams or wells; death of fish and other aquatic life; inferior quality of water available for supply and recreation at downstream populated areas
Transition from middle- to late-urban stage:	
Urbanization of area completed by addition of more houses and streets, and of public, commercial, and industrial buildings	Reduced infiltration and lowered water table; streets and gutters act as storm drains, creating higher flood peaks and lower base flow of local streams
Larger quantities of untreated waste discharged into local streams	Increased pollution of streams and concurrent increased loss of aquatic life; additional degradation of water available to downstream users
Abandonment of remaining shallow wells because of pollution	Rise in water table
Increase in population requires establishment of new water-supply and distribution systems, construction of distant reservoirs diverting water from upstream sources within or outside basin	Increase in local streamflow if supply is from outside basin
Channels of streams restricted at least in part to artificial channels and tunnels	Increased flood damage (higher stage for a given flow); changes in channel geometry and sediment load; aggradation

Table 19-1 (*cont.*)

Change in land or water use	Possible hydrologic effect
Construction of sanitary drainage system and treatment plant for sewage	Removal of additional water from area, further reducing infiltration recharge of aquifer
Improvement of storm drainage system	
Drilling of deeper, large-capacity industrial wells	Lowered water-pressure surface of artesian aquifer; perhaps some local overdrafts and land subsidence; overdraft of aquifer may result in saltwater encroachment in coastal areas and in pollution or contamination by inferior or brackish waters
Increased use of water for air conditioning	Overloading of sewers and other drainage facilities; possibly some recharge to water table, owing to leakage of disposal lines
Drilling of recharge wells	Raising of water-pressure surface
Wastewater reclamation and utilization	Recharge to groundwater aquifers; more efficient use of water resources

Source: U.S. Geological Survey.

Figure 19-8 Air pollution in New York City. This midday view on July 19, 1970 shows a thick blanket of smog enshrouding the urban area; it was a clear day in the region outside the metropolitan area. (Wide World Photos.)

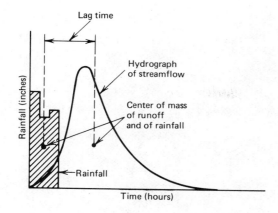

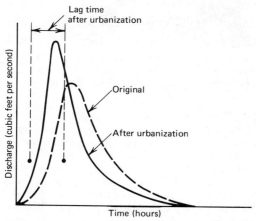

Figure 19-9 Comparison of stream hydrographs for urban and nonurban conditions. (Leopold, U.S. Geol. Survey Circ. 554, 1968.)

Figure 19-10 Detroit, Michigan, showing complete impervious cover of original ground by buildings, streets, and parking lots. (Courtesy U.S. Dept. of Housing and Urban Development.)

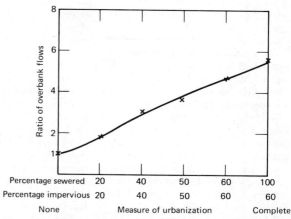

Figure 19-11 Urbanization produces an increase in the number of large floods and flows in an area. The increase in number of flows per year equal to or exceeding original channel capacity for a 1 mi² drainage area is graphed as a ratio to the number of overbank flows before urbanization as reflected by different degrees of development. (Leopold, 1968.)

21), most cities will have to continue to live with their historical sins (Fig. 19–12).

The changes in rivers already mentioned are sufficient to give environmentalists continuing nightmares, but the devaluation in the attractiveness of some is an equally appalling indictment of city ills. The natural beauty of the rivers is reduced because the channel is often gradually enlarged (to say nothing of its common channelization), owing to increased floods. This produces unstable banks, devegetated slopes, scoured or muddy channel beds, and accumulated debris. The beer cans, tires, lumber, and assorted flotsam of civilization, along with increased production of algae, plankton, turbidity, and odors, bear testimony to the unsavory side of human progress.

Figure 19-12 Aerial view of Memphis, Tennessee, April 4, 1973. This photo was taken during the disastrous Mississippi River flood, but Memphis was largely spared because of its massive levee structures. (Courtesy Corps of Engineers.)

Figure 19-13 The crack in this house was caused by ground subsidence from heavy groundwater pumping, Las Vegas, Nevada. (Courtesy U.S. Geol. Survey.)

Groundwater The reduction in percolation of surface waters causes a lowering of the water table. This depletes groundwater storage for potential use in wells and also reduces the amount of groundwater flow that can recharge streams. Thus streamflow is diminished during nonstorm periods. Only 3 of the largest 35 cities—Miami, San Antonio, and Memphis—obtain their water resources entirely from groundwater. Houston had relied on wells until the population exceeded 500,000; now it must supplement supplies from Lake Houston.

Other effects from the city use of groundwater produce some of the subsurface changes discussed in Chapter 14. Intrusion of formerly freshwater aquifers by saline waters has occurred in numerous coastal cities, and land subsidence has afflicted the United States as well as many countries—Venice, Italy, Savannah, Georgia, and Mexico City (Fig. 19–13). For example, Mexico City rests on the bed of a drained lake—Texcoco. The Aztecs had settled on an island in the lake in 1325. By a system of drainage, dikes, and

causeways, they expanded the settlement to about one-half million, the capital of an empire that extended over most of central and southern Mexico—one of the greatest metropolitan centers of its time. With the building of modern Mexico City, additional drainage was engineered and the permeable unconsolidated materials contributed water for the growing city. However, the earth materials do not have great strength when dewatered, and some parts of the city have subsided as much as 7 m since 1900 (Fig. 19–14). Between 1948 and 1950, the subsiding rate reached 75 cm per year in the most beleaguered areas. Sewer lines were tilted and debris flowed in the opposite direction. Thus unusual engineering is necessary to stabilize large structures. For example, the 43-story Latin American Tower has 360 steel and concrete piles 32 m deep that support half the weight. The rest of the structure floats in the muddy subsoil with a boxlike

foundation 13 m below the surface. Water content in the soil is kept constant by automatically injecting more water whenever the pressure drops. The tower has not sunk at all and, in 1957, it escaped unscathed the severest earthquake in Mexican history which greatly damaged 975 large buildings in the city.

Unusual results can occur when policies are changed regarding the pumpage of groundwater. Brooklyn prior to the 1930s used prodigious quantities of groundwater and greatly lowered the water table (up to 15 m). With declining use of groundwater, because of importation from upstate New York and installation of recharge wells, the water table recovered as much as 12 m. Structures built when the water table was low did not always possess waterproofing materials, so with the rise of groundwater many buildings became flooded and seepage in several subway stations increased from less than 20 gpm

Figure 19-14 Differential settlement of the Guadalupe Shrine, Mexico City. The dewatering of the underlying lake beds had led to a loss of pore pressure, and different thicknesses of the strata produce different amounts of compaction. The photo was taken so that the border at the bottom is horizontal. (Courtesy Donald Doehring.)

Figure 19-15 New Orleans, Louisiana. A "typical" urban area with large buildings, highway networks, bridges, and parking lots, situated for convenience near a major river, the Mississippi, for trade and commerce. (Courtesy Corps of Engineers.)

(gallons per minute) in 1947 to as much as 1000 gpm in 1961.

CONSTRUCTION AND BUILDINGS

When one thinks of an urban area, the picture that comes most readily to mind is a congested place with numerous, and often large, structures and buildings (Fig. 19–15). The lifeline of a city also consists of the numerous services necessary to provide the goods for consumption or export: highways, pipelines, tunnels, subways, bridges, water and sewer lines, electrical installations, and so on. And, of course, dwelling places for citizens. All of these rest on or in the terrain, and their construction produces changes in the geomorphic setting. Geologic aspects of some of these features have also been discussed in other chapters (for example, pages 401–439).

Clearly, the designers and builders of cities must face many potential geological problems, including durable foundations, safe locations, and strong structural frameworks. Furthermore, site planning and preparation also must include many geological considerations (Fig. 19–16).

Foundations

The perfect example of an unfavorable site is the oft-repeated joke about the Leaning Tower of Pisa, in which a person representing the budget director is telling the architect, "And we can save 700 lira by not taking soil tests." The tilt is caused by differential settlement of a clay stratum 8.5 m below the ground. However, the other joke is that the city planned it that way in order to attract the tourist trade . . . which has proved a boon to the 100,000 city population. However, even borings must be carefully scrutinized and are not infallible in some cases

(see page 402). Unlike the Tower of Pisa, the skyscrapers of Manhattan Island, New York City, rest on firm bedrock with only a shallow soil veneer, which has proved the ideal foundation for the massive towers of that city. Foundations in different climates can be subjected to unusual forces, such as those in permafrost of cold regions (see Figs. 17–31 and 17–32). The expansive clays and quick clays can also be the undoing of otherwise ruggedly designed structures. The expansive clays cause more than $2 billion damages to structures in the United States, and the quick clays have produced billions of dollars of destruction during earthquakes and landslides (see Fig. 10–21).

Locations

City developments can be undercut by a double-edged sword (the syndrome of *double jeopardy*). When construction occurs on flat lands, such sites can often be flooded; when built on hillslopes, the structures can be in "landslide territory." The safety of buildings is a

Figure 19-17 Uncontrolled erosion in a housing development caused by the removal of all natural vegetation and the lack of preventive measures, at South River, New Jersey. (Courtesy Soil Conservation Service.)

Figure 19-16 The Skid Row area of San Francisco, California. This section was built with muds from San Francisco Bay. Settlement in many areas is 6 m. These buildings have subsided so much that now entry is made on the second floor. (Courtesy Donald Doehring.)

responsibility of both government agencies, which carry out ordinances that prohibit development at hazardous sites, and of contractors who should inform themselves of probable restrictions and ensuing penalties for violations.

Structural Framework

Building codes must be sufficiently strict so that structures can maintain their integrity when severely stressed. Thus buildings that are likely to be flooded must possess special flood-proofing materials; if they may be subjected to earthquakes, the design must incorporate the latest safety-tested materials and construction methods. It is especially vital that all public places have these safeguards. When designing the building even the surrounding grounds should be considered, because even trees can cause grievous structural damage. For example, trees can dewater the soil and thereby create settlement of the foundation as much as several centimeters, as in Ottawa, Canada.

Site Development

Once the geologic foundation has been evaluated, the location decided, and the type of construc-tion determined, the actual site can be excavated for emplacement of its new host. Again, how this is done is crucial to the environment because it will govern the amount of degradation of the terrain and associated physical systems. Erosion and sedimentation—the twin hazards of undesired terrain change—affect in some degree all development projects (Figs. 19–17, 19–18, and 19–19). Some of the deleterious impacts that result from increased sedimentation in urban areas include: (1) aggradation and increased levels of flooding, (2) siltation of recreation sites, (3) clogging of drains, (4) turbid water unsuited for municipal or industrial use, (5) damage to pumping equipment, and (6) change in channel geometry. At specific sites, some studies indicate that sediment production may be 20,000 to 40,000 times higher in construction of highways than on rural adjacent lands. The urbanizing of Little Falls Branch, Maryland, produced 3200 t/mi^2 of sedimentation compared with nearby rural Watts Branch and Rockville, Maryland, with only 516 t/mi^2. In 1977 a newly urbanized watershed in central New Jersey also produced five times more sedi-

Figure 19-18 Severe gullying at a commercial shopping center, Newport Beach, California. The exposed areas comprise part of the Lower Pliocene Imperial Formation, a friable sandstone. The Bolo knife is 31 cm. (Courtesy Allen Hatheway.)

Figure 19-19 Erosion and sedimentation caused by improperly engineered sidewalk construction, Bountiful City, Utah. (Courtesy Earl Olson.)

(a)

(b)

Figure 19-20 The effective use of straw bales at storm-drain inlets which allow the deposition of sediment from surface runoff. (a) A six-bale filter with deposition of approximately 1 tonne of sediment. (b) A four-bale filter showing the ponding effect of surface runoff. (Guy, 1976.)

ment than that of rural lands of the area.

Guy (1976) provides a series of guidelines for the reduction of sediment loads: (1) reduce site construction to the smallest area possible, (2) quickly establish permanent vegetation, (3) use temporary vegetation and mulches on exposed soil (Fig. 19–20), (4) limit the length of steep slopes, (5) reduce the volume and velocity of water crossing disturbed areas by engineering works, (6) use hardened or established channels for transporting increased runoff, and (7) create sediment detention basins.

The thousands of settling basins and recharge pits excavated on Long Island serve the dual purpose of minimizing sediment entrainment into streams and recharging the groundwater. A study of 78 small watersheds in the Philadelphia region showed that channel enlargement, which ranged from 0.7 to 3.8 times, was the common result of urbanization. In the Patuxent River, Maryland, urban channels hold about 15 times as much sediment as rural channels, and their size is 3 times greater. Such changes amplify flooding damage in downstream reaches. Thus site development and the character of runoff delivery to urban streams, and their response, are intricately interwoven.

RESOURCES

Natural resources are the fundamental building blocks for an industrial and urban society. Cities could not be built without construction materials; many urban jobs would not exist without minerals; factories could not function without fuel; disease and human welfare depend on water purity; and people could not be fed without rich and productive agricultural land. These resources may not be abundant in the urban or adjacent area, so it is of utmost importance that they be identified and judiciously used if the urban community is to retain its vitality.

The process of producing mined resources can often be in conflict with perceived goals of maintaining the quality of the environment. Open pits are ugly; underground operations can cause subsidence; and the dust, noise, pollution, and truck traffic further desecrate the beauty of the countryside. If reasonable re-source costs are to be obtained and the raw materials supplied to a growth-city economy, then accommodations must be made between the producer and the consumer. This can take the form of ordinances and laws, which permit sequential-zone and multiple-phased land use, or education of the general public on the necessity of keeping the resource pipelines open.

Perhaps the gravest problems arise with the high-bulk, low-value resources that are "transportation-sensitive." Sand and gravel and stone aggregates cannot be shipped great distances to the market because of high transport costs. Such materials cost about $3 per ton at the quarry and 30 cents per mile for delivery. Thus the "zero demand distance" is about 10 miles—that is, the distance whereby transportation costs are greater than the raw commodity. Urbanizing communities must realize this during their growth stages. For example, the Los Angeles metropolitan region requires 15 million tons of aggregate for highways alone. Austin, Texas, did not have this in mind when real estate developers were permitted to build over the last remaining sand and gravel deposits in the nearby area. Thus present building costs are higher because aggregate must be trucked into the city from great distances. Unless appropriate resource management policies are adopted, California stands to lose $17 billion in natural resources during the 1970–2000 period. This lesson has been learned by Pima County, Arizona, which has adopted an ordinance that prohibits surface development over potentially mineable copper deposits. The oil pumps in commercial, industrial, and even residential areas of Los Angeles and Oklahoma City also bear testimony to priorities that have been placed on the extraction of resources in some urban areas.

It is possible to harvest natural resources in such a manner that the land surface can later be used for other purposes. Such planning and development involves: (1) recognition of resource areas, (2) determination of geologic and hydrologic constraints, (3) evaluation of environmental consequences of development, (4) rehabilitation programs for land restitution, (5) adoption of guidelines for sequential land uses, and (6) a blend of imaginative management and enlightened public response. A showcase example of multiple use has occurred in Denver where a land parcel was originally used as agricultural land and was then converted into a sand and gravel operation. When the material was mined out, the hole was used for a sanitary landfill. When the hole was filled, the ground was leveled and became the site for the Denver Coliseum, its parking lot, and overhead ramps of Interstate-70. Multiple use of underground mining has also proved feasible in a number of localities.

Underground Space

The first human inhabitants of underground space (subsurface space) were troglodytes, or cave dwellers. Although modern man largely abandoned the underground openings for greater geographic mobility, a resurgence in the utilization of subsurface caves, mined-out corridors, and newly constructed space is now occurring. Because many world cities will double in size by the year 2000, and surface lands are becoming increasingly used and abused, a great potential exists for development of underground space in many urban areas. The geoengineering aspects of underground tunneling and structures

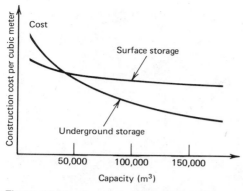

Figure 19-21 Comparison of construction costs for underground and surface oil storage plants in Sweden. (Courtesy Sentab, Sweden.)

Figure 19-22 A Swedish air force base carved out of crystalline Precambrian rocks, entirely underground. (Courtesy Goran Wettlegren.)

were examined in Chapter 13; here we are interested in the advantages that the use of subsurface areas have for city dwellers and their environment. Some of the benefits that are possible from subsurface construction include: (1) thermal dampening of daily and seasonal temperature fluctuations. (2) reduction in energy consumption of as much as 75 percent; (3) lower building costs (Fig. 19–21), especially when occupation is made of mined-out lands; (4) protection from tornadoes, hurricanes, fire, and so forth; (5) self-supporting roofs can offer both above-ground and below-ground uses; (6) conservation of resources because few building materials are needed; (7) camouflage, concealment, and shelter from possible attack. Underground space can be used for a great variety of human activities, services, and necessities: transportation routes in subways and highway tunnels, storage facilities, offices, stores and shopping districts, water reservoirs, industries, barracks and military operations, shelter, parking lots, among other uses.

The rock of Gibraltar is ho-neycombed with tunnels and excavated spaces in the limestone. The first extensive excavations started in the eighteenth century, and during World War II more than 750,000 m³ were excavated to provide shelter, services, and a large underground reservoir. The London headquarters of General Eisenhower during the war was also located underground in a refurbished subway. Underground space is in wide use in both American and Russian military forces, to house nuclear armaments and also command posts; for example, at Colorado Springs granite was excavated for the North American Air Defense Command (Fig. 19–22).

The multiple use of underground space is an environmentally appealing and conservation-oriented strategem. Perhaps one of the best examples of multiple use is Kansas City where the Bethany Falls Limestone, ranging up to 7.5 m in thickness, was mined for more than a century as aggregate. In the past quarter century, this vast mined-out area has been converted to a large network of storage space, offices, salesrooms, and manufacturing concerns. Frozen-food energy savings alone is 33

percent lower than similar storage above ground. Nearly 1 million m² of space has been utilized by the various enterprises at 24 different sites. One-seventh of Kansas City's warehouse space is now located underground, and 50 ha is newly mined each year to keep pace with increasing demands for use. Multiple use can also be made of the materials removed from deep excavations. The spoil from digging of foundations in Toronto was used as a new breakwater, 5 km long, at a fraction of the cost of deliberate mining. Such uses should be part of city planning, unlike in Los Angeles when between the years 1960–1966 private construction moved 101 million m³ in the Los Angeles area without a concerted program for its disposal.

In terms of population size Sweden has more programs that utilize underground space than any other country. The underground storage of oil (Fig. 19–23) has proved to have many advantages over above ground storage—costs are 20 to 40 percent less for construction and 66 percent less for maintenance (see Fig. 14–21). In addition, there is increased

(a)

(b)

Figure 19-23 Oil storage caverns, Stockholm, Sweden. (a) Roof of cavern is 20 m below the water table. Height of cavern is 30 m and width is 20 m. Bedrock is granite. (b) Unlined pumproom for dry pumps, 5 m below the bottom of the oil storage cavern. Roof is 55 m below water table. (Courtesy Sentab, Sweden.)

Figure 19-24 The underground Uddevalla sewage plant. This is a good example of localizing a dirty process below the surface in an environment where people live, work, and have their leisure activities. It is situated in the city center close to the pleasure boat harbor. The bedrock is Precambrian gneiss and is 1.75 billion years old. (Courtesy Svenska Vägaktiebolage.)

Figure 19-25 Underground storage of herring at Kungshamm, Sweden. This cavern was built as a bomb shelter during World War II. (Courtesy Goran Wettlegren.)

protection against fire, explosion, or sabotage, and potential pollution is greatly reduced. Important Swedish airbases are also underground. The Uddevalla sewage treatment plant is all underground and handles 41,000 m³ per day (Fig. 19–24). A bomb-proof underground shelter at

Kungshamm, a village on the Swedish west coast, is now used as a herring storage for the fishing industry (Fig. 19–25).

LANDSCAPE ABUSE

In addition to the changes created in hydrologic regimes (Fig. 19–26) and stream channels, urbanization imposes other forces that work to desecrate the environment—the physical pollution of urban sprawl and the biochemical pollution of the water systems.

Urban Sprawl

This term was introduced by William Whyte to signify the rapacious wastage of lands adjacent to cities by uncontrolled and unplanned growth (Fig. 19–27). The automobile plus the development of high-speed highways are the chief culprits in the rapid expansion of fringe and bedroom communities (Fig. 19–28). The U.S. cities of the 1960s used three times as much land as those of the 1920s. Sixty percent of Los Angeles is devoted to automobile-selling servicing, roads, and parking lots, and most of the $62 million flood damages in 1969 occurred in sprawl areas without zoning codes to delineate hazards. Even from 1950 to 1970, the population density of urban areas declined from 2110 per km² to 1316 per km². Many people have been taking the advice of Will Rogers who said, "Buy land. They ain't making any more of it."

    Commonly, urban sprawl consumes seven times the area of land as compared to planned communities. This is highly wasteful of energy, natural resources, and land. In the 1967–75 period, urbanizing areas with their support facilities converted 6.9 mil-

Figure 19-26 Chicago's Central Water Filtration Plant on Lake Michigan. It contains the world's largest filtration system capable of chemically treating and filtering 1 million gal/min. Built in 1964 at a cost of $1 billion, it serves a population of 5 million in a 1100 km² area, including Chicago and over 70 suburbs. (Courtesy Chicago Water Department.)

Figure 19-27 Row houses and urban sprawl, Belair Housing Development, Bowie, Maryland. (Courtesy U.S. Dept. of Housing and Urban Development.)

lion ha to such uses, and 3.2 million ha of this was prime agricultural land. If this rate continues unabated, all remaining prime agricultural lands that are not currently under the plow will disappear in the next 40 years. Several communities, counties, and states have taken important steps to arrest

this creeping paralysis of farmlands. Towns such as Petaluma, California, and Boca Raton, Florida, have declared moratoriums on continued expansion. Suffolk County, New York, has earmarked $55 million for the purchase of development rights to 3700 ha of rich agricultural lands adjacent to

Figure 19-28 The United States' "love affair" with the automobile. (a) cars at a shopping center, Anytown, U.S.A. (b) Complex geometric highway networks allow the automobile to invade all areas of the landscape, as here in California. (Courtesy U.S. Dept. of Agriculture.)

(b)

the metropolitan communities. Several states have passed legislation whereby agricultural property is to be taxed only as farming lands and not as "highest and best use," which to some governmental appraisers always meant home development sites. The other ills of urban sprawl are legion: it increases floods, erosion, and sedimentation; it initiates higher costs for public services, sewage, water systems, and electricity; and it frequently destroys any possibility for future planned constructions and efficient and economic mass-transit systems. Typical cost increases are 40 percent on streets and 63 percent on utilities; 11 percent more energy is used; and pollution levels are 30 percent higher. Fortunately, many states are now passing state Environmental Quality Acts (as followup to NEPA), such as New York and Florida, which both require developers of large projects to file detailed impact statements that specify the effects on natural resources, the economy, public facilities, and adjacent regions. Of course, this action was taken by other states back in the 1970s

(Hawaii, Vermont, and Maine, for example).

Pollution

Pollution can take many forms (also see Chapter 20). For example, excess sedimentation is one form of pollution and derelict lands another (those areas created by abandoned and nonrehabilitated mining operations or by deterioration and out-migration of destroyed dwellings). At this time we are principally interested in calling attention to urban waste as part of the problems associated with cities. Although the United States contains less than 6 percent of the world's population, it consumes about one-third the world's production of energy and natural resources and produces about half the world's pollution. Urban areas alone produce 500,000 tons of sewage daily. The price tag in 1974 to renovate the sewage systems of the cities exceeded $50 billion, and this price would be much higher now.

Solid waste is known by a variety of terms—garbage, refuse,

junk, rubbish, trash, and litter. Regardless of its name, these by-products of civilization represent the metabolism and refuse of the city. A major difference exists between modern civilization and earlier ones. Ancient cities commonly built on their own debris, and in archeological digs it is not unusual to find as many as a dozen earlier cities built on top of each other. For example, London sits on 3 m of debris from past ages. In contrast, modern cities usually deposit their materials outside the city, which raises the topographic surface of the perimeter, or exports the waste via waterways to some distant burial ground. The nearly $5 billion annual cost for handling solid wastes in American cities represents one of the costliest acts of government, exceeded only by expenditures for schools and roads. Nearly 75 percent of the nation's total solid wastes are discarded in open dumps. Of the 14,000 land disposal sites, only 13 percent are operated within recognized sanitary landfill standards, which require earth coverage of all deposits each day and prohibit burning.

The concept of "sanitary" landfill did not emerge until about 1940, but by 1945 almost 100 cities were using the practice, and by 1960 more than 1400 cities had adopted the policy.

OPEN AREA MANAGEMENT

One answer to the haphazard developments that stem from urban sprawl (sometimes called *sluburbia* or *Californication*) is the planned community idea associated with open space, greenbelts, new towns, and garden cities. Although these ideas are not new, they have received more study and have been practiced on a larger scale in the twentieth century. The redesign of Paris in the mid-1850s became the forerunner of what was to become a twentieth-century revolution in planning. The seeds for "the commons" were planted centuries ago in England, but it took additional legislation in 1947 to enforce the principle in that country.

Open Space and Greenbelts

A 1959 California statute (Sec. 1, Chap. 12, Div. 7. Title 1) defines **open space** as:

. . . any space or area characterized by (1) great natural scenic beauty or (2) whose existing openness, natural condition, or present state of use, if retained, would enhance the present or potential value of abutting or surrounding urban development, or would maintain or enhance the conservation of natural or scenic resources.

The principle of open space dates back at least to the thirteenth century B.C. when the Levitical cities were surrounded by pasturelands for use by city dwellers. The city of Gezer contained 59 ha, with an open landbelt 15 times larger. The 1515–1516 writings of Sir Thomas More provide greenbelt ideas, where the imagined cities of *Utopia* were surrounded by agricultural lands; when cities were filled, new cities were built beyond the agricultural belts. In England the open space concept was rooted in the royal forest and hunting preserves, which evolved into public places—"the commons"—for gathering of firewood, grazing of animals, and recreation. When faced with increasing pressures from landlords and business to partition the urban commons, Parliament passed the Enclosure Act of 1845 which placed restrictions on commons-type land. With the help of the Green Belt Act of 1938 and the Town and Country Act of 1947, the original 47 London commons, each averaging 160 acres (64 ha), had expanded to 840 mi² (2175 km²) in 1959 with concentric rings of agricultural and recreational land mixtures. The Moscow greenbelt constitutes 30 percent of the city's total area, and the Ottawa greenbelt will contain 14,800 ha for farming, forestry, recreation, and water supply protection. Scottsdale, Arizona, has recently designed a greenbelt for the control of erosion and sedimentation for an 11 km reach through the city. It will also provide 500 ha of environmentally compatible open space for a variety of activities. Planned Unit Development (PUD) is a special concept that retains many elements of open space but still provides utilization of an area for homes and shopping purposes (Figs. 19–29 and 19–30).

In summation, open space can serve a diversity of functions: it can (1) provide resource production for forestry, agriculture, minerals, and water supply; (2) preserve natural and human resources; (3) aid the health, welfare, and well-being of inhabitants by providing water and air quality, recreation, and visual amenities; (4) enhance public safety regarding floods, unstable hillslopes, airplane flight paths, and fire zones; (5) provide corridors for power transmission and transportation systems; and (6) offer a hedge for unseen necessities in the urban expansion of commerce, industry, housing, and public service. Studies in San Francisco show that one healthy acre of mature white pine trees spaced 14 ft apart (about 200 trees) will produce enough oxygen in a single growing season to keep 200 mature persons alive for a year. In addition to reducing carbon dioxide content, vegetation filters dust, pollen, ash, and other airborne particles. Green spaces in Leipzig, Germany, were found to reduce dust from 210 to 50 particles per cm³. The air in areas surrounding Central Park has 100 percent more sulphur dioxide than air in the park. Vegetation also produces benefits in the microclimate of cities and helps compensate for the "heat island" effect by cooling the air through evapotranspiration. Ancient Babylon used trees to block the dry, hot desert winds, and trees are used in Moscow to aid in the regulation of summer heat. Thus, the advantages of open space and greenbelts are myriad and worthy of fighting for and keeping.

One final word about the geometry of open space and greenbelts—their configuration should not be artificially and rigidly circumscribed to follow a mathematical geometry. Instead, they should be so planned as to be conforma-

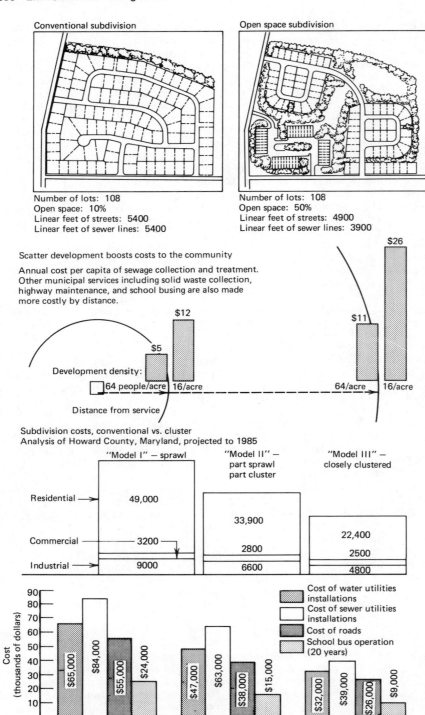

Conventional subdivision

Number of lots: 108
Open space: 10%
Linear feet of streets: 5400
Linear feet of sewer lines: 5400

Open space subdivision

Number of lots: 108
Open space: 50%
Linear feet of streets: 4900
Linear feet of sewer lines: 3900

Scatter development boosts costs to the community

Annual cost per capita of sewage collection and treatment.
Other municipal services including solid waste collection,
highway maintenance, and school busing are also made
more costly by distance.

$26

$12

$11

$5

Development density:

64 people/acre   16/acre   64/acre   16/acre

Distance from service

Subdivision costs, conventional vs. cluster
Analysis of Howard County, Maryland, projected to 1985

"Model I" — sprawl

"Model II" —
part sprawl
part cluster

"Model III" —
closely clustered

Residential →   49,000   33,900   22,400

Commercial —   3200   2800   2500

Industrial →   9000   6600   4800

Cost of water utilities
installations
Cost of sewer utilities
installations
Cost of roads
School bus operation
(20 years)

Cost
(thousands of dollars)

90
80
70
60
50
40
30
20
10

$65,000 $84,000 $55,000 $24,000 $47,000 $63,000 $38,000 $15,000 $32,000 $39,000 $26,000 $9,000

Figure 19-29 Comparative
maps and costs for
conventional and open-space
subdivision development.
(Courtesy New York State
Dept. of Environmental
Conservation.)

Figure 19-30. A Planned Unit Development (PUD) at the Operation Breakthrough site, Macon, Georgia. Compare this community with that in Figure 19-27. The PUD grouping saves land and permits construction of community-wide recreation and park facilities. (Courtesy U.S. Dept. of Housing and Urban Development.)

Figure 19-31 Reston, Virginia, a "new town." (Courtesy U.S. Dept. of Housing and Urban Development.)

ble with important topographic features and to take advantage of the terrain and watershed forms whenever possible. This entails an accommodation with nature and can only be accomplished through careful obeyance to topographic and geologic maps. This also means that geologic hazards have been accounted for, and such zones are the perfect reason for the requirement of nondevelopment ordinances.

New Towns

The term "new town" has been used to describe those emerging communities where environmental concerns play an important role in their design and management. They were first developed on an important scale in Britain and were an outgrowth of Ebenezer Howard's classic book *Garden Cities of Tomorrow* (1902; first published in 1898 as *Tomorrow: A Peaceful Path to Reform*). He sought to relieve the congestion of London by long-range planning and balanced growth. He also hoped to capture and utilize the facilities of modern technology without sacrificing the social advantages of a city. Garden cities would have

their own industries and achieve an internal financial balance. They would reduce commuting problems and would be protected from chaotic sprawl. Thus they are not merely suburbs or bedroom-community developments. The first Garden City Ltd. was the new city of Letchworth, England, started in 1903, 60 km north of London. A later one, also designed by Howard in 1919, became known as Welwyn Garden City, and by the late 1930s both towns had become prosperous. Since 1947, 21 new towns have been designated in Great Britain with about 650,000 inhabitants. The garden city concept, as well as the greenbelt idea, has guided much of urban-area planning in Britain and the interfusing of growth with open space and agricultural lands.

The United States has examples of most types of "new towns," and they exhibit a wide range of environmental awareness and success in preserving natural areas.

1   *Company towns*   Growth of this type of town peaked before 1900, but would include such towns as Gary, Indiana, Kohler, Wisconsin, Pullman, Illi-

nois, Kingsport, Tennessee, and Lowell, Massachusetts.

2   *Garden cities*   Typical of some of the early ones is Radburn, New Jersey, founded in 1928. Columbia, Maryland, and Reston, Virginia (Fig. 19-31) can in part be classed within this category.

3   *Greenbelt community development*   The 1935 Executive Order created the Resettlement Administration, which included provision for greenbelt towns. Three areas were selected for development: Greenbelt, Maryland (on the outskirts of Washington, D.C.), Greenhills in Cincinnati, Ohio, and Greendale in Milwaukee, Wisconsin. Only the Maryland town was a partial failure, but it did provide helpful guidelines for planning future communities.

4   *Power, resource, and reclamation projects*   Boulder City, Nevada, was established in 1930 by the Bureau of Reclamation and, in 1937, it emerged as a permanent town with a master plan. Norris, Tennessee, was built by the Tennessee Valley Authority between 1933 and

1935, to house workers employed on the Norris Dam project.

5 *Atomic energy towns* Examples include communities developed by the Atomic Energy Commission, such as Oak Ridge, Tennessee, Hanford, Washington, and Los Alamos, New Mexico.

Just as there has been a variety of stimuli for new town developments, there are also diverse approaches to accomplishing urban renewal and, with it, concomitant beautification programs in the satellite communities. Many plans have been conceived largely with the automobile in mind, and their designs are dominated by freeway construction, parking facilities, and urban auto-oriented centers for shopping and business. Other recent approaches hold that the car is inimical to healthy cities and seek high-speed public transportation combined with pedestrian-oriented centers of living and commerce. Reston (Fig. 19–31) and Columbia have emphasized clustered development of buildings around recreation lakes surrounded by green areas of open space; such towns are essentially an urban village within an urban area within a major metropolitan region. The purpose of the smaller, unified communities is to provide the individual with greater opportunity and scope for activities in the affairs of society and a greater feeling of personal identity in governmental structures and attitudes. Columbia, Maryland, was built on lands that formerly constituted 165 farms, 6250 ha, and cost the developer $23 million to purchase. The town was designed for a population of 100,000, with 30,000 people in its plants, offices, stores, and institutions. There are five lakes, stream valleys, forests, 42 km of riding trails, as well as parks, recreation areas, and other open space designs.

The siting of new towns and their growth can contain many obstacles. In 1971, HUD (Department of Housing and Urban Development) began financing 13 new towns, but the operation was such a flop that by 1978 HUD had abandoned financial control of seven, and several others were in severe jeopardy. Most of the towns were poorly conceived, executed, and built at the wrong time and place. In Park Forest South, Illinois, the land costs were 89 percent higher than planned, and sales were 58 percent fewer than expected. Developers of seven of the HUD-financed new towns defaulted on interest payments, leaving the agency to pay the bondholders $149 million and take title of the bankrupt areas. Newfields, near Dayton, Ohio, and Riverton and Gananda, near Rochester, New York, were begun when the nearby metropolitan areas were losing jobs. Others like Flower Mound, Texas, were located outside the path of growth, and after six years had only attracted 420 residents out of a proposed population of 61,000. Indeed, the road is not all roses in new town developments!

Of course, there is also extensive new town construction throughout the world. In the 1917–1965 period, the Soviet Union built 900 new towns, during the time urban population was expanding from 26.3 million to 124.8 million. In Brazil the government built the entire new showplace city of Brasilia, with artificial lakes that arc around the city on three sides. New towns have also been completed in Sweden, Finland, France, and Germany. It is a worldwide phenomenon.

Perspectives

Cities are becoming increasingly indentured for their survival to conditions and resources that are distant from their borders. Because most of the political power in many countries stems from the more populous urban centers, they now dictate not only policies for the city dweller, but for the rural population as well. Thus the problems of cities influence the entire country, in both political and environmental ways. We can only hope that geologists are not part of the problem, but part of the solution.

One scenario that is highly possible, unless greater efforts are placed in management, population growth, and resource use and allocation, might be the following: A diminution in food production will create food shortages in cities. The high cost of energy will impede economic development and increase the level of unemployment. The concomitant inflation will increase the nutrition gap, and there will be a growing dependence on external food and energy imports. This will produce an added indebtedness that will cause social unrest and political instability. Such a condition may reach the third-world nations before the older industrialized societies.

Many cities throughout the world will double in size by the

twenty-first century, and prime agricultural lands and the supporting resource base are becoming increasingly jeopardized. Thus city planning is different from rural planning—there is no time or place for mistakes and experiments. When a development project is made, it "locks into the environment" and is not subject to inexpensive change if the design is faulty. Therefore, it is vital that such new urbanization sites receive thorough prior scrutiny and planning for potential geologic impacts on the land-water ecosystem and resource base of the community.

There are a number of strategems that geologists working in concert with city planners can institute to enhance the quality of life and minimize environmental deterioration. These range from a compilation of appropriate reports and maps which are presented in clearly understandable language, to the appearance of geologists on planning boards and in public office. Resource depletion, erosion-sedimentation changes, water budget and flow regime modifications, and foundation structure potentials should be foremost in city planning and design. Benefit-cost analysis can be important in convincing environmental managers of the need for attention to the earth processes in the city. The liaison of the geologist and the lawmaker can be seen in many efforts—for example, in parking ordinances that require cars to periodically discontinue parking along curbs. Since as much as 75 percent of dirt and sediment in cities occurs within 15 cm of the curb, street sweeping could alleviate much of this man-made debris.

The pulse of a city may throb to the heartbeat of commerce, but its lifeline is determined by the indispensable flow of natural and agricultural resources which are the province of the geologist.

READINGS

Alfors, J. T., and Gay, T. E., Jr., 1973, Urban geology master plan for California: Calif. Div. Mines and Geology Bull, 198, 112 p.

Bryant, W. R., 1975, The effects of urban expansion on farming in Wayne County, New York: A.E. Res. 75-28, New York State College of Agriculture and Life Sciences, 59 p.

Chapin, F. S., Jr., 1965, Urban land use planning, 2nd ed.: Urbana, Univ. of Illinois Press, 498 p.

Clawson, M., 1962, Urban sprawl and speculation in suburban land: Land Econ., May, p. 99–111.

Coates, D. R., ed., Environmental geomorphology and landscape conservation, Vol. II, urban areas: Stroudsburg, Dowden, Hutchinson & Ross, Inc., 454 p.

———, ed., 1976, Urban geomorphology: Geol. Soc. Amer. Spec. Paper 174, 166 p.

Davis, F. F., 1972, Urban ore—a new resource opportunity: California Geology, Calif. Div. Mines and Geol., v. 25, n. 5, p. 99–112.

Detwyler, T. R., and Marcus, M. G., eds., 1972, Urbanization and environment: North Scituate, Mass., Duxbury Press, 287 p.

Dobrovolny, E., and Schmoll, H. R., 1968, Geology as applied to urban planning: an example from the Greater Anchorage Area Borough, Alaska: 23rd Internt. Geologic Congress Proc., Section 12, p. 39–56.

Fox, H. L., 1976, The urbanizing river: a case study in the Maryland Piedmont: in Coates, D. R., ed., Geomorphology and engineering, Stroudsburg, Dowden, Hutchinson & Ross, p. 245–271.

Gulick, L., 1958, The city's challenge in resource use: in Jarrett, H., ed., Perspectives in conservation, Johns Hopkins Univ. Press, p. 115–137.

Guy, H. P., 1970, Sediment problems in urban areas: U.S. Geol. Survey Circ. 601-E, 8 p.

Guy, H. P., 1976, Sediment-control methods in urban development: some examples and implications: in Coates, D. R., ed., Urban Geomorphology, Geol. Soc. Amer. Spec. Paper 174, p. 21–36.

Hammer, T. R., 1972, Stream channel enlargement due to urbanization: Water Resources Research, v. 8, p. 1530–1540.

Hansen, W. R., 1973, Geomorphic constraints on land development in the Front Range urban corridor, Colorado: in Coates, D. R., ed., Urban geomorphology, Geol. Soc. Amer. Spec. Paper 174, p. 85–109.

Howard, E., 1902, Garden cities of tomorrow: London, S. Sonnenschein & Co., 167 p.

Johnson, S. L., and Sayre, D. M., 1973, Effects of urbanization on floods in the Houston, Texas metropolitan areas:

U.S. Geol. Survey Water-Resources Investigation 3-73, 50 p.

Koppelman, L. E., 1964, A plan for open-space in Suffolk County: Suffolk County Planning Committee, Hauppauge, N.Y., 121 p.

Legget, R. F., 1973, Cities and geology: New York, McGraw-Hill, 624 p.

Leighton, F. B., 1976, Urban landslides: targets for land-use planning in California: in Coates, D. R., ed., Urban Geomorphology, Geol. Soc. Amer. Spec. Paper 174, p. 37–60.

Leopold, L. B., 1968, Hydrology for urban land planning: U.S. Geol. Survey Circ. 667, 32 p.

———, 1968, Hydrology for urban land planning—a guidebook on the hydrologic effects of urban land use: U.S. Geol. Survey. Circ. 554, 18 p.

Metropolitan Washington Council of Governments, 1968, Natural features of the Washington metropolitan area: Washington, D.C., The Council, 49 p.

Rantz, S. E., 1970, Urban sprawl and flooding in southern California: U.S. Geol. Survey Circ. 601-B, 11 p.

Savini, J., and Kammerer, J. C., 1961, Urban growth and the water regimen: U.S. Geol. Survey Water-Supply Paper 1591-A, 43 p.

Schneider, W. J., and Goddard, J. E., 1974, Extent and development of urban flood plains: U.S. Geol. Survey Circ. 601-J, 14 p.

Schneider, W. J., Rickert, D. A., and Spieker, A. M., 1973, Role of water in urban planning and management: U.S. Geol. Survey Circ. 601-H, 10 p.

Slosson, J. E., 1969, The role of engineering geology in urban planning: State of Colorado, Geol. Survey Spec. Pub. 1, p. 8–15.

Strong, A. L., 1971, Planned urban environments: Baltimore, Johns Hopkins Press, 406 p.

U.S. Department of Agriculture, 1970, Controlling erosion on construction sites: Soil Cons. Service Agri. Inform. Bull. 347, 32 p.

Wallace, D. A., ed., 1970, Metropolitan open space and natural process: Philadelphia, Univ. of Pennsylvania Press, 199 p.

Wayne, W. J., 1975, Urban geology of Madison County, Indiana: State of Indiana, Geol. Survey Spec. Report 10, 24 p.

Williams, E. A., 1969, Open space: the choices before California: San Francisco, Diablo Press, 187 p.

Wolman, M. G., and Schick, A. P., 1967, Effects of construction on fluvial sediment, urban and suburban areas of Maryland: Water Resources Research, v. 3, p. 451–464.

Chapter Twenty
Solid Waste and Pollution

A car graveyard in Illinois. We can only hope that the metal is being recycled for further use. (Courtesy U.S. Dept. of Housing and Urban Development.)

INTRODUCTION

The earlier environmental movements in the United States were largely concerned with matters associated with various aspects of conservation of natural materials and soils, but the environmental crusades that started in the 1960s covered a much broader area and have increasingly become devoted to the cleaning up of the environment—specifically those aspects of degradation connected with the management of pollution and solid wastes. Industrialized societies consume enormous amounts of materials. The manufacturing, mining, and ultimate disposal of the by-products of civilization impose a tremendous alteration of natural systems. Such refuse is not only harmful to the normal living processes of plants and animals, but disruptive to human society and, in some cases, lethal to our very existence.

The sheer bulk and amount of waste materials of all descriptions is staggering and is increasing at the rate of 4 percent per year. Environmental contamination is present in the air, on the land, and in the waters. Its abatement and control is costly and has been the fighting ground of government agencies, industrial and municipal polluters, and environmental watchdogs.

ENVIRONMENTAL LEGISLATION

The U.S. Congress has passed an unusually large number of bills that specifically relate to pollution and wastes produced by society. In many cases the states, and even more local governing bodies, have also passed laws that seek to curb even more of this type of environmental degradation. The following list provides a sample of the Federal legislation:

- Rivers and Harbors Act (Refuse Act) 1899
- Water Pollution Control Act of 1948, 1961, 1972
- Clean Air Act 1963, 1966, 1967, 1977
- Solid Waste Disposal Act 1965
- Clean Water Restoration Act 1966
- Air Quality Act 1967
- National Environmental Policy Act 1970
- Water Quality Improvement Act 1970
- Clean Water Act 1972
- Environmental Pesticide Control Act 1972
- Resource Conservation Recovery Act 1976
- Toxic Substances Control Act 1976
- Surface Mining Control and Reclamation Act 1977

A striking feature of this list is the large number of acts passed after 1960. Many acts are further amended on nearly a yearly basis, and invariably the amendments provide additional environmental safeguards. Many of the acts have had unusual histories. Although the Refuse Act forbids the dumping of refuse into navigable waters without permission from the Corps of Engineers, it was not enforced until the 1960s at which time it was resurrected and further interpreted that the term "refuse" included pollutants. The original passage of NEPA (National Environmental Policy Act) took place rather quietly and only gradually was it discovered that the Environmental Impact Statement (EIS) clause would become such a powerful force in the management of environmental affairs. For example, the Environmental Protection Agency (EPA) which it fostered has 1000 workers on the pesticide staff alone. Some of the laws seemingly speak to a particular aspect of the environment, such as the Resources Conservation and Recovery Act, yet have clauses that cover a much broader mandate. For example, this 1976 law has a section that gives authority to the EPA in the management of hazardous wastes. Some of the laws were passed easily, without great controversy, and within a year after their original proposal. Others, such as the Surface Mining Control and Reclamation Act, took more than 20 years for passage, and only after many bitter debates, postponements, and failures.

Although there are many environmental laws, statutes are barren without proper enforcement or the knowledge of what constitutes infractions. For example, it is estimated that it will take up to 15 years to determine which of the 30,000 different pesticide varieties are harmful to society. The costs of completely obeying all legal mandates pose severe hardships for some. From 1973–78,

$74.3 billion was spent on water pollution controls to meet government standards, but the General Accounting Office estimates another $174 billion must be spent by 1984 if the nation is to achieve goals mandated by the Clean Water Act. One favorite argument used by many is that environmental controls for pollution and waste abatement are too costly and help speed inflation. However, studies have shown that during the 1975–78 period the total costs for environmental protection materials and equipment increased the rate of inflation only 0.5 percent. In such areas as air pollution the expenditure is about $90 per person for a year, but this pollution causes damages in terms of health and property of $240 per person.

WATER POLLUTION

The necessity of water to mankind and all living things has already been discussed in other chapters. But water can be more a curse than a blessing when its quality is degraded. From earliest history surface water has provided us with a cheap and convenient medium for the disposal of wastes. It was not discovered until the nineteenth century that bacteria in human and other wastes caused disease. We will never know how many millions of lives have been lost by usage of contaminated water supplies. However, underground seepage from privy vaults (septic tanks) resulted in 616 deaths from a single communal well in London during a 40-day period in 1854. Many notables throughout history have died from waterborne diseases, including King Louis VIII of France, Charles X of Sweden, Prince Albert of England, his son Edward VII, and even his grandson

George V. Ironically Louis Pasteur's two daughters were reported to have died from typhoid fever.

Although the slow filtration of groundwater through the earth and the dilution of wastes by oxygen-rich surface waters are both capable of absorbing and degrading various quantities and types of wastes, these self-purification capacities are easily overburdened. The oxygen in water consumed by microorganisms feeding on organic pollution is referred to as the biochemical oxygen demand (BOD). When the oxygen supply is exhausted by excessive BOD, organisms begin to die and putrification commences. With unprecedented growth in human populations and the concentration of people in cities, it became obvious that the ability of natural water to assimilate impurities was being exceeded on a massive scale. Thus we were forced to find methods that would cleanse contaminated waters. The early treatment methods simulated nature's own processes by the use of settling ponds, beds of sand and gravel for a filtering medium, and exposure to sunlight.

In recent decades the water pollution problem has continued to escalate, and ever more costly and sophisticated techniques are necessary to rid waters of harmful substances. The quantity of wastewater requiring treatment generally amounts to 75 to 200 gal per day per capita (0.7 m³/day) in U.S. cities today. Domestic wastewater contains body wastes, kitchen wastes, and a large array of household chemicals (many of which are now being shown to be highly toxic). Solid matter usually constitutes less than 0.1 percent of domestic wastewater.

Although there is virtually no likelihood that the U.S. government's goal of stopping all direct industrial and municipal pollution by 1985 will be met, some notable progress has been made in the last decade. Many of the most polluted lakes and streams have shown significant improvement in recent years. Organic wastes and bacteria levels have dropped in 25 percent of the nation's waterways and, in 1977, 3400 of the United States' 4000 major industries were keeping pace with federal cleanup guidelines. This reflects tough new regulations and billions of dollars in recent expenditures. In spite of this we are barely holding the line against water pollution. From 1969–79 about $20 billion was allocated to local governments for water treatment by the EPA, but by 1979 two of every three cities were still releasing inadequately treated sewage into rivers. So although it is costing $100 per person for pollution abatement each year, many treated waters still contain harmful substances (Fig. 20–1). For example, phosphates and nitrates are not removed by treatment methods, and agricultural runoff contributes large quantities of these substances. The phosphates and nitrates can overfertilize water, stimulating excessive algae growth and other plants. This overburdens the water with dead and dying organic matter, resulting in smelly, murky water, largely without life except anaerobic bacteria and other nuisance and disease-carrying organisms. Shallow bodies of standing water are especially susceptible to this degrading process, called **eutrophication.**

More than 70,000 different chemicals, many of them toxic, are commercially produced in the United States, and their number is growing about 1000 per year—of which about 100 are significant polluters. A total of nearly 50,000 industrial plants add wastes containing such chemicals to municipal sewer systems. These hazardous chemicals cannot be effectively controlled by the standard water treatment systems that are designed primarily for treating biological wastes.

In recent years it has been discovered that many hazardous substances, some of them **carcinogenic** (cancer causing), are finding their way into the nation's drinking water supplies. The sources of these chemicals include treatment plants not equipped to handle special chemical wastes, leakage from holding tanks and waste storage facilities, accidental spills, and direct dumping into rivers and lakes. The last of these may be illegal, accidental, or in accordance with the EPA's controversial "permit-to-pollute" program aimed at giving industrial polluters sufficient time to comply with new regulations. An average of 7000 chemicals are tested each year and, in 1977, of those tested 1500 were found to be carcinogenic. New regulations requiring special treatment to remove toxic chemicals from drinking water supplies in cities with more than 75,000 people were issued by EPA in 1978. Unfortunately such treatment is very costly, more than $1 billion, and the results are not guaranteed.

A major source of water pollution continues to be **nonpoint sources,** over which most programs have little control (Figs. 20–2, 20–3, and 20–4). More than half the hazardous substances entering U.S. waterways are from widespread indistinct nonpoint sources—cropland, streets and highways, construction sites, farms and homes, mines, airborne pollutants, vehicles, and so forth. In Montana, one of the least polluted states, 6400 km of streams are closed to fishing and swimming because of nonpoint pollution. Acid rain is becoming an increasing and insidious threat to all water systems.

"This water is still a mess."

"I shudder at the cost to clean it up."

One dreams of crystal waters. Another talks cost. Which is most important?

Generations have used our waterways for recreation, barging, garbage disposal. Many rivers, lakes and streams so fouled by sewage and wastes they literally stink from abuse.

Yet, water is essential for health, manufacturing, for outdoor fun. No wonder enthusiasts demand "clean up the despoiled waterways regardless of cost."

But the dollars are frightening. Modernizing a plant to treat waste water for 100,000 people can run $40 Million. And there's the continuing job of maintaining and running a facility which only controls "man-made" waste. When you talk erosion runoff, muddying our waters or total clean up, the whole job may be beyond economic feasibility.

Agreed, we can't use our waterways as open sewers. We must protect them from damaging levels of impurity. It can be done as proven by Oregon's Willamette, the Houston Ship Channel and elsewhere. But we must recognize a natural capacity of our waterways to carry on the purification work begun in modern treatment facilities. We must set, and work to, realistic guidelines for future, attainable clean water goals. Support your community in its efforts to keep our waterways clean.

Figure 20-1 Factors in environmental decision-making processes. (Courtesy Caterpillar Tractor Co.)

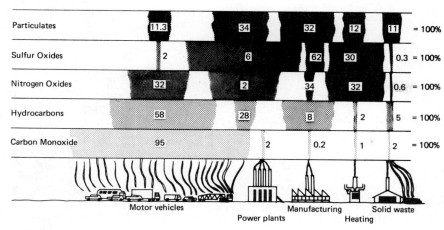

Particulates	11.3	34	32	12	11	= 100%
Sulfur Oxides	2	6	62	30	0.3	= 100%
Nitrogen Oxides	32	2	34	32	0.6	= 100%
Hydrocarbons	58	28	8	2	5	= 100%
Carbon Monoxide	95	2	0.2	1	2	= 100%

Motor vehicles   Power plants   Manufacturing   Heating   Solid waste

Figure 20-2 Sources of air pollution. (New York State Dept. of Environmental Conservation.)

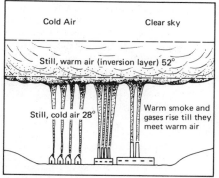

Cold Air   Clear sky

Still, warm air (inversion layer) 52°

Still, cold air 28°   Warm smoke and gases rise till they meet warm air

Inversion layer results from stable high-pressure air mass remaining stationary for several days.

Because of cloudless skies, nighttime radiation heat losses from earth leave cold air close to the ground, held down by a "lid" of warm air.

This inversion of the normal situation, where air is warmer close to the ground, results in air stagnation in the absence of vertical convection current, which normally keep air moving.

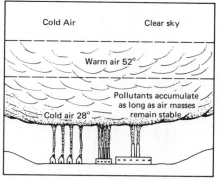

Cold Air   Clear sky

Warm air 52°

Cold air 28°   Pollutants accumulate as long as air masses remain stable

Pollution danger is highest in valleys, where the least horizontal air movement takes place.

But danger also occurs over large level areas when the stable air mass is extensive and winds do not rise.

Figure 20-3 How a temperature inversion traps industrial pollution. (New York State Dept. of Environmental Conservation.)

Coastal Areas

Water pollution takes on special significance in coastal communities throughout the world. Many towns and cities still dump raw or partially treated sewage directly into coastal waters. Each day more than 500 million gal (2×10^5 m³) of untreated waste is disposed at sea by New York City, resulting in a large area of fetid, garbage-strewn debris in offshore waters of the New York Bight (Fig. 20–5). Polluted coastal waters contain dangerous organisms and hazardous chemicals that have caused severe economic costs, disease, and even deaths. Recreational and tourist facilities may be closed, and fish and shellfish often become contaminated. In the United States more than 400,000 ha of prime shellfish beds have been closed due to coastal pollution.

An even greater threat looms behind these immediate and obvious effects—the oceans are the ultimate sump into which most of civilization's wastes eventually accumulate (Fig. 20–6). Contamination from oil spillage amounts to 10 million tons each year. The final long-term result of all these pollutants on the ocean ecosystem

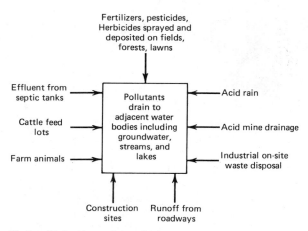

Figure 20-4 Nonpoint pollution sources.

Figure 20-5 Solid waste disposal in the Atlantic Ocean, New York Blight Area. (Courtesy U.S. Environmental Protection Agency.)

Figure 20-6 Fiji Islands. The dark band of material near the water edge is flotsam debris, most of human origin, that has accumulated from a large ocean area. (Courtesy Joseph Van Riper.)

is one of today's most vital unanswered questions. The fate of all mankind may depend on the answer.

WATER TREATMENT

There is a wide variety of mechanical, chemical, and biological methods that can be used to reduce water pollution. The techniques depend on the quality of the incoming water (**influent**) and the intended use or destination of the outgoing water (**effluent**). Some sewage treatment plants remove only the coarsest fraction of the pollution and discharge the remainder into a river, creating a major source of pollution. Others may yield an effluent that is of drinkable (potable) quality. We will now review some of the standard water purification processes (Fig. 20–7).

Sedimentation

Suspended silt, clay, and other materials will settle to the bottom of standing or very slow-moving water. Settling, or sedimentation, tanks and basins are usually less than 6 m deep and may detain water from 1 hour to several days. The settled material may then be scraped, suctioned, or drained from the bottom of the settling tank either continuously or at appropriate intervals.

Coagulation

Coagulation is the combination of flocculation of fine particles and colloidal materials in water to form large particles which then settle out. Iron or aluminum salts are usually added to the water in a settling tank to produce insoluble hydroxide precipitates. Lime,

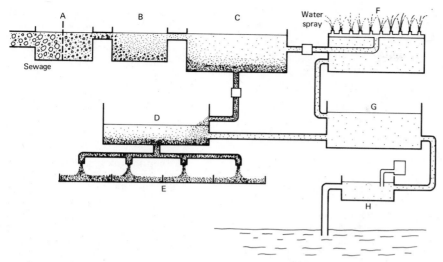

A. Sewage first passes through a bar screen that catches large objects.
B. Next, the liquid flows slowly through a grit chamber, allowing sand and gravel to settle to the bottom.
C. In the settling tank, solids settle to bottom and are called "sludge."
D. Sludge is pumped to a digestion tank.
E. Digested sludge is dried. It may then be burned, buried, or used as fertilizer.

F. Liquid from the first settling tank is sprayed on coarse rockbed (trickling filter), where other impurities are removed by action of bacteria.
G. To remove any remaining solids, liquid is piped to a second settling tank.
H. Liquid is chlorinated to kill any remaining harmful bacteria before it is discharged into the receiving stream.

Figure 20-7 Components of a typical municipal sewage disposal system.

caustic soda, clay, or other materials may also be added to aid the coagulation process. The light, feathery "floc" that results must then be collected from filters or from the bottom of the tanks and treated, reclaimed, and/or disposed of. Sedimentation and coagulation are the most important processes in most water and sewage treatment plants.

Filtration

Although meshed screens may be used to remove coarse debris prior to sedimentation, more sophisticated filtration procedures usually follow sedimentation or coagulation. The purpose of filtration is to remove the full range of colloidal material, suspended solids, bacteria, and other organisms.

Slow sand filters consist of beds of fine (0.2 to 0.5 mm in diameter) sand that are usually 50 to 122 cm thick. They range from less than 0.5 ha to several hectares in area and allow water to pass through at rates of about 2.5 to 10 million gal per acre (1 to 4 million gal/ha) per day. The uppermost layer of sand must periodically be removed as it becomes clogged with filtered-out particles.

Rapid sand filters use a uniform, coarser-grained (0.4 to 1.0 mm in diameter) sand that allows more rapid infiltration—about 125 to 250 million per acre (50 to 100 million gal/ha) per day. The filter bed area ranges from 150 to 1500 ft² (14 to 140 m²). Such filter systems are widely used in public water supply plants. The filters are cleansed by backwashing—a process that forces water up through the filter thereby washing away the trapped particles. Prior

to 1953, these waste products were usually released directly into adjacent rivers. However, by 1965, only five states had failed to pass legislation prohibiting such a procedure.

Both slow and rapid sand filters are underlain by perforated pipes, beds of gravel, tile, or other devices to drain off the filtered water. In addition, sand filters can be complemented with a diatomaceous earth filter, used for small water systems and in swimming pools, and with trickling filters, in which sewage nutrients are deposited as thin films on specially coated materials.

Aeration

Aeration exposes water to air in order to add oxygen to the water and to remove carbon dioxide, hydrogen sulfide, and other waste-

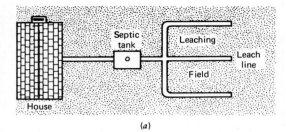

(a)

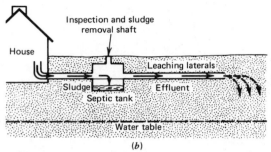

(b)

Figure 20-8 Elements of a septic tank system for disposal of household wastes. (a) Plan view showing septic tank, leaching field, and lateral drains. (b) Cross section showing sludge effluent and travel pathways.

and odor-producing gases. Aeration is achieved by breaking water into small droplets by using sprays or artificial waterfalls and rapids, by forcing air through water, or by a combination of these.

Water Softening

Water "hardness" is caused by calcium and magnesium salts which make washing difficult, produce harmful scale in pipes and boilers, and create scums and stains during household use. Municipal treatment plants usually incorporate water softening where natural hardness exceeds 150 ppm. To remove hardness, water may be passed through a permeable cation exchanger that substitutes sodium ions for the calcium and magnesium ions, or lime and soda ash may be added to precipitate the calcium and magnesium as

carbonate and hydroxide, as in the coagulation process.

Disinfection

The most common method for killing disease organisms, mainly pathogenic bacteria, is **chlorination.** Chlorine is generally added as chlorine gas in large systems and as a salt in small treatment operations. The chlorine dosage for drinking water exceeds the actual amount that can be assimilated by the impurities, so that there is excess chlorine available to prevent contamination along the distribution route and in storage reservoirs.

Chlorine can chemically react with other impurities in water, such as algae and chemicals, to produce objectionable tastes and odors, and even carcinogens. However, it is generally argued

that the benefits of its use far outweigh the dangers. Ammonia or sulfur dioxide may be used to reduce chlorine content. Activated charcoal is effective in reducing the disagreeable effects of chlorine, but it is considerably more costly than simple chlorination.

Septic Tanks

The septic tank system is the principal method for disposing of wastes in households and other small facilities that are not connected to municipal sewage lines. Millions are in use throughout the United States and, when properly designed, they can be an effective means for the containment of pollution. A septic tank system consists of a central underground tank that usually has laterals extending from it that comprise a leaching field (Fig. 20–8). The tank serves as a settling basin for solids and sludge, whereas the laterals transport away the excess fluids. The sewage decomposes because of anaerobic bacteria, with the undigested sludge settling to the tank bottom. Such sludge must be periodically removed and transported by truck to a treatment plant or buried in an isolated area. Most states require single residences to have septic tanks with at least 500 gal capacity. This size tank will usually become filled by more than 50 percent within a five-year period with normal use by a family of four. In most cases the tank is not pumped until signs of clogging or overflow occur, but the efficiency would be enhanced with more frequent pumping.

The effluent from septic tank systems is highly contaminated and foul smelling. Less than 60 percent of the suspended solids are usually removed by simple septic tanks. Their performance can be improved by constructing

Figure 20-9 Two-stage 30-acre secondary effluent treatment lagoon of the Great Northern Paper Co., Maine. The lagoon is divided into two cells and holds 150 million gal of wastes. The first 25 million gal cell holds effluent one day to provide additional settling after the liquid is processed through a primary operation. The second 125 million gal pond holds the liquid 5 days and is used for aeration. After the liquid is thoroughly treated, it is returned to the west branch of the Penobscot River. More than 1.3 million ft³ of Hypalon synthetic rubber liner was installed by the Burke Rubber Co. to prevent waste infiltration into the ground. (Courtesy Burke Rubber Co.)

additional tanks, compartmentalized tank designs, use of filter mediums, and frequent evacuation by pumping to other disposal systems. Even at best, septic tanks can be a significant source of surface and groundwater pollution. Many shallow aquifers on Long Island have become severely contaminated, and the toxins pose a health threat in several communities.

Sewage Treatment Plants

When sewage first enters a sewage facility it undergoes **primary treatment** in which mechanical operations remove the coarser particles. Screens capture large objects that are then ground up and disposed of or returned to the treatment cycle. Following the screening, the suspended particles are removed by sedimentation in settling tanks.

Secondary treatment is primarily designed to remove the fine colloidal organic matter along with dissolved chemicals. Biodegradable matter is effectively consumed by a complex biological process initiated by bacteria as the water slowly moves over a trickling filter (90 to 95 percent removal of the BOD) or through a sand filter (more than 95 percent BOD removal). Industry also occasionally maintains its own special waste disposal systems (Figs. 20–9, 20–10, and 20–11).

The activated **sludge process** is another popular method of secondary treatment that typically removes 95 to 98 percent of the BOD. The sewage is seeded with bacteria and contained four to eight hours in an aeration tank where biodegradation occurs. The bacteria-laden organic-rich solids (activated sludge) are removed from a settling tank and then treated for disposal, or some parts may be used for bacterial seeding of additional incoming sewage.

Tertiary treatment consists of any number of additional processes aimed at removing specific contaminants that remain after primary and secondary treatment. These persistent pollutants may include phosphates, nitrates, nonbiodegradable synthetic organic compounds, toxins, and many other chemicals. Chemical proc-

Figure 20-10 Liquid waste control for a lumbering operation in Georgia. This pulp mill operation requires a 120-acre waste treatment lagoon that holds 500 million gal divided into 11 cells. Percolation into the shallow water table is prevented by a 2-ply nylon-reinforced Hypalon lining. (Courtesy Burke Rubber Co.)

Figure 20-11 This is the first anaerobic treatment pond in the United States to be successfully covered with a flexible floating cover. This 7.5 million gal tertiary lagoon is capable of handling effluent from a 150,000 population, but at present treats sewage waste from the Wilson Packing Co., Monmouth, Illinois. The 5-ply heavy-duty Hypalon synthetic rubber cover floats on the lagoon and does not require columns, cables, or other internal supports. Being gas-tight, it traps the methane gas generated in the pit so it can be safely removed for disposal or utilization. (Courtesy Burke Rubber Co.)

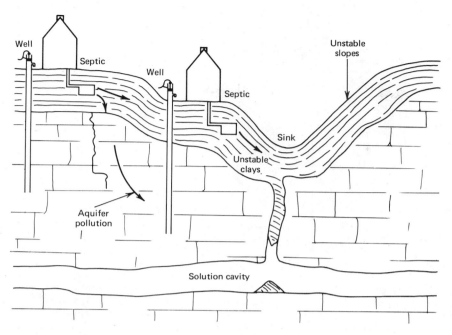

Figure 20-12 Geologic limitations of water supply and waste disposal in karst topography. (Source, DuMontelle, Jacobs and Bergstrom in Coates, 1971.)

esses that are used include chlorination, flocculation, adsorption by activated carbon, and reverse osmosis.

Because of the recent growth of small towns and the increasing numbers of rural homes and housing developments, many large regional sewage treatment plants have been proposed throughout the country. Local developers and other growth advocates often strongly push for these facilities because convenient sewer systems encourage new growth and the federal government pays for 75 percent of the cost. This often results in costly, oversized treatment plants; it accelerates planned "sprawl" with its attendant problems; and it incites widespread ill feelings between the pro- and anti-growth constituencies. Geologic conditions further impose constraints on development (Fig. 20–12).

The recent upswing in sewage plant construction has had both good and bad effects. All too

often conventional high-technology plants are overpriced, inefficient, and do not represent the best way to solve water pollution problems. Proof of these recent shortcomings is seen in the new EPA policy of providing 25 percent funding for experimental alternatives to conventional treatment. Among the alternatives are collective septic tanks, compost toilets, land treated with wastes, and separation of drinking water and "graywater"—water suitable for washing cars and lawn watering, for example.

LAND TREATMENT METHODS

Various methods for disposal of sewage wastes on land have been practiced for thousands of years. For example, vegetables, fruits, forests, grains, and other products have been successfully grown

on land treated with sewage effluent. The methods include placement of sewage in dry or liquid form; hand, bucket, and wheelbarrow emplacement; and new techniques of "spray irrigation." The sewage should be free from dangerous chemicals which could be absorbed by plants or animals. Well-designed land treatment operations have been shown capable of achieving higher quality effluent than conventional plants. Furthermore, costs are generally lower, and the by-products earn money by enriching the soil and producing crops.

Sewage farming dates back earlier than the Chinese "night soil" collectors. Sewage was also an important ingredient in farming methods of the Roman Empire and in the agricultural husbandry techniques championed by Columella nearly 2000 years ago. During the 1850–1900 period, such cities as Berlin, Edinburgh, Leipzig, London, Manchester, Nottingham, and Paris

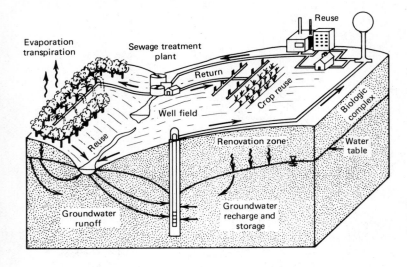

Figure 20-13 Experimental spray irrigation operation at Pennsylvania State University. (Courtesy Richard Parizek.)

operated sewage farms. Today about 1000 U.S. communities fertilize crops with sewage effluent, and each year larger numbers of food-processing and industrial concerns are using the method. Current methods in use include rapid infiltration, slow infiltration, and overland flow spreading. **Slow infiltration** provides the maximum cleansing system for effluent waters. The success of experiments by scientists at Pennsylvania State University using **spray irrigation** methods (Fig. 20–13) was largely responsible for the development of the largest spray irrigation system in the world, at Muskegon, Michigan (Figs. 20–14, 20–15 and 20–16). Whereas the Pennsylvania usage was on a small experimental scale that started in 1962, the Muskegon plant is a major facility that went into full-scale usage in 1974. It is expected to generate about a $400,000 a year savings from sale of grass and grain crops and, by 1990, will be collecting 43 million gpd (160,000 m³/day) of raw sewage and industrial wastes.

Spray irrigation cannot be used in all terrains or on all soils. Leachate from the Broome Coun-

ty, New York, sanitary landfill was sprayed on adjacent forestlands. Because of a combination of steep terrain, dense soils, and excessive spraying, the leachate killed the forest, contaminated the adjacent streams, and led to a lawsuit by the property owner of the adjacent land that had been polluted. The Deltown Chemurgic Corporation's creamery plant was a major polluter of the West Branch Delaware River, the main source of waters for the Cannonsville Reservoir, which is a major water source for New York City. When the extent of pollution became known, the Delaware River Commission issued an order to the corporation to abate the pollution or suffer an injunction and fine. A consulting engineering firm designed and constructed a system they believed would solve the problem (Fig. 20–17). A holding pond was built to collect sewage effluent from the plant, and the fluids were pumped onto adjacent lands as a spray irrigation project. Unfortunately, the river continued to receive pollutants much to the chagrin of the company and the engineers. As a consultant I was asked to solve the riddle. The

answer was simple: the coarse gravels of the floodplain terrane which received the sprayed fluids were so permeable that the polluted water traveled through them so quickly that they were not appreciably cleansed. In addition, the large quantity of sprayed water raised the water table, so that the increased gradient accelerated water flow velocity. The company was finally required to install a regular sewage treatment plant.

SOLID WASTE DISPOSAL

Virtually every facet of modern civilization—industrial and commercial activities, mining and energy production, food and agriculture enterprises—generates solid waste (see also page 126). Each year the United States accumulates a staggering 400 million tonnes of mining waste, and 1.8 billion tonnes of agricultural waste (Table 20–1).

Municipal solid waste in the United States increased from 1.2 kg per person per day in 1920 to

Water resource

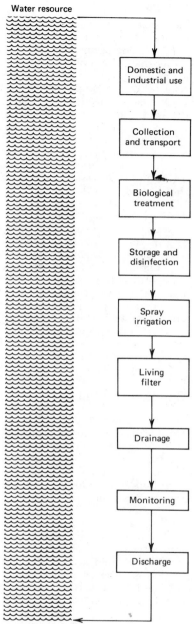

Figure 20-14 Components of the Muskegon, Michigan, Wastewater Management System.

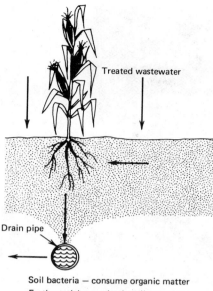

Treated wastewater

Drain pipe

Soil bacteria — consume organic matter

Earth particles — adsorb pollutants and viruses, mechanically strain suspended solids

Root system — takes up nutrients

Figure 20-15 The Wastewater Management System, Muskegon, Michigan. Diagram of the soil-plant living filter. The system provides tertiary treatment for the sprayed wastewater. Organic matter is decomposed by bacteria. Nutrients are used by plants or are removed within the soil. Suspended matter and color are removed as the water percolates through the soil. Heavy metals are adsorbed by organic and clay particles.

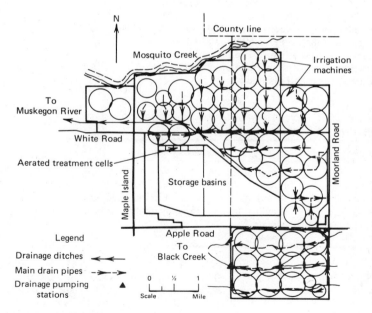

Muskegon, Mona Lake subsystem

Figure 20-16 Map of the entire Muskegon waste disposal system.

(a)

(b)

Figure 20-17 Deltown Chemurgic Corp. dairy plant, Delhi, New York. (a) Holding pond for effluent materials. Note turbid character of water and thick effluent on far side of pond. (b) Spray irrigation site for effluent shown in (a). Note the very coarse character of the floodplain deposits. They are composed of glaciofluvial gravels. Photo was taken only 15 minutes after area received a layer of sprayed effluent. The fluid has already percolated below the surface, attesting to the permeability of the materials.

Table 20-1 Waste produced by households in the United States

Percent of total	Type of waste
50	Paper products
12	Food
10	Wood, garden residue, bricks, rocks, dirt
10	Plastics, leather, rags, synthetics, rubber
9	Metals
9	Glass

2 kg in 1965, and 3 kg in 1975. By weight, municipal solid waste is roughly composed of 50 percent paper, 12 percent food wastes, 9 percent metal, 9 percent glass, 7 percent yard residues, and 13 percent plastics, wood, cloth, rubber, and other materials.

The increasing volumes of municipal solid waste reflect the dramatic population growth, from 76 million in 1900 to 225 million today, and its shift from rural areas to the urban sector. The problem is exacerbated by the trend toward processed foods,

wasteful packaging, and an economy that encourages quick disposal and replacement of materials, including planned obsolescence rather than conservation, durability, and repair of used products. As a result, landscapes across the United States, and indeed throughout the world, are littered with solid wastes. Mountains of trash, containing millions of dollars worth of usable materials, are scattered and piled into unsightly heaps across the countryside, buried in the ground, and dumped at sea. Billions of dollars each year are spent in the United States for solid waste cleanup, transportation, and disposal. Because of the unwillingness or inability of some communities to fund efficient and environmentally sound waste management programs, the haphazard and poorly planned methods generate friction and ill will among various groups of the community.

Sanitary Landfill

Within the last 15 years "sanitary landfill" has become the preferred method for the elimination of solid urban wastes. In a sanitary landfill, the refuse is deposited (one hopes) at a carefully chosen site, then spread and compacted into layers usually 2 to 3 m thick. The waste is then covered each day with a layer of earth 15 cm thick (Fig. 20–18). No open burning is permitted at the site, and fencing is generally placed around the site to hide it from view and trap blowing paper and other refuse.

Advantages of sanitary landfill

1 Most economic method of solid waste disposal where land is available.

(a)   *(b)*

Figure 20-18 Contrast in solid waste disposal methods. (a) Unsightly open dump. (b) Sanitary landfill operation.

2 Initial investment is low compared with other methods.

3 The operation can begin within a short time period.

4 All types of wastes can be deposited, eliminating separate collections.

5 Completed sites may be used for other purposes.

Disadvantages of sanitary landfill

1 In densely populated areas, suitable land may not be available at economic hauling distances.

2 Daily maintenance is required and constant surveillance.

3 Methane and other gases may be produced and create a nuisance and hazard.

4 Unless properly engineered, leachate may be a continuing problem for years.

Planning and site development

The sanitary landfill site must be carefully selected keeping in mind hauling distance, transportation routes, possible impact on nearby developments, and geologic conditions. For example, it is important that the base of the fill should rest on impermeable material, that there be sufficient nearby cover material, and that the topography be conducive to containment of all contaminants (Fig. 20–19).

When water comes into contact with buried refuse, leachate—a foul liquid containing hazardous dissolved chemicals and perhaps pathogenic bacteria—is produced. It is therefore necessary that the refuse be as high above the water table as possible, preferably more than 15 m. Furthermore, the cover material should be an impermeable layer that prevents rapid infiltration of precipitation into the refuse. A completed landfill tends to be more permeable than the surrounding area, which can accelerate movement of leachate away from the site by downward or laterally moving water. This problem can be minimized by using techniques such as lining or enclosing the refuse with impermeable materials (for example, clay) or by using man-made plastic liners and the placement of perforated pipes to drain leachate away. Because leachate poses a threat to all water supplies—streams, aquifers, and local wells—its containment, monitoring, and disposal are vital concerns for the successful operation of a sanitary landfill.

To minimize excess erosion and infiltration problems, the topography of the site should not permit surface water to run into or through the landfill. Level upland areas, the heads of ravines, and dry depressions (provided they are not in highly permeable materials, such as gravel pits and sinkholes) often provide excellent landfill sites.

Methods Three methods of sanitary landfill are commonly employed—area, trench, and ramp methods. On fairly flat ground or in broad depressions the **area method** is preferred (Fig. 20–20). Wastes are spread and compacted on the ground surface and covered with earth to form a "cell," which usually represents one day's waste. The same procedure is followed the next day to form an adjacent waste compartment. Additional waste may be deposited on top of finished cells. On completion, the uppermost waste layers are capped with a thick layer of cover 0.6 m and often planted with vegetation.

In a **trench sanitary landfill,** a broad trench is excavated and filled with compacted refuse as described above. Cover material is also supplied by the excavated trench material. This method is well suited to level terrain where

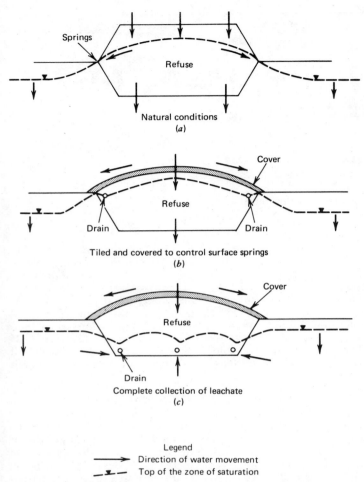

Springs

Refuse

Natural conditions
(a)

Cover

Refuse

Drain   Drain

Tiled and covered to control surface springs
(b)

Cover

Refuse

Drain

Complete collection of leachate
(c)

Legend
———⟶ Direction of water movement
— ▼ — Top of the zone of saturation

Figure 20-19(A) Subsurface hydrology of a sanitary landfill, with various schemes to control spread of leachate. (After Bergstrom, Illinois Geological Survey.) *(A)* Earth materials have low permeability, and the site is in a groundwater recharge area. Landfill intersects water table. *(B)* Earth materials have high permeability, and the site is in a groundwater recharge area. Landfill intersects water table. *(C)* Earth materials have high permeability, and the site is near a groundwater discharge area. Landfill intersects water table. *(D)* Earth materials have high permeability, and the site is in a groundwater recharge area. Landfill does not intersect water table.

technique can be adapted to almost any area and is especially economical for small operations.

Completed landfill sites can be used for a variety of other purposes, including parking or storage areas, gardens, parks, and playgrounds. In some cases, solid waste sites have later been shaped into hills, ramps, and other forms for use as ski slopes, raceways, and other amusement facilities. Landfills are generally poor sites for buildings because settlement may occur for many years, and leachate and escaping gases may be a continuing problem. However, some sites (see Denver, page 624) have successfully been used for highway structures and even a sports coliseum.

Sanitary landfills represent a vast improvement over open dumps, but they also have problems, not least of which is space. On the average, a good sanitary landfill operation requires 1 ha of land each year for every 25,000 people. As the mountains of waste continue to grow, places to put the waste are becoming ever more difficult to locate. Land for disposal near large population centers is especially difficult to obtain. Landfill sites are becoming increasingly expensive to operate, and more and more legal suits are being brought against them by nearby homeowners. We all want our garbage disposed of, but nobody wants it disposed of in their own backyard!

Incineration

Controlled combustion of wastes reduces the weight and volume of waste and is especially desirable in cities where land disposal sites are not available within economic hauling distances. The unburned solid residue remaining after incineration is usually inert and makes a satisfactory fill material.

the water table is quite deep (Fig. 20–21). The **ramp or slope method** is a simple technique that can be used on sloping terrain or in conjunction with area and trench methods. Refuse is spread over the hillslope, compacted, and covered with earth, which is typically excavated from a cut at the base of the ramp (Fig. 20–22). This

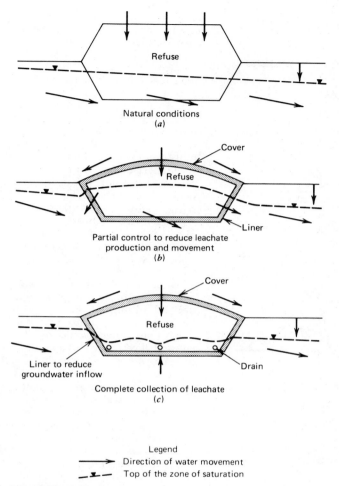

Natural conditions
(a)

Partial control to reduce leachate
production and movement
(b)

Complete collection of leachate
(c)

Legend
⟶ Direction of water movement
--⊻-- Top of the zone of saturation

Figure 20-19 (B.)

A major impediment to incineration operation is the incorporation in municipal waste of substances that release gases or particulates containing cadmium, vinyl chloride, mercury, and other hazardous and difficult-to-control emissions during combustion. Costly pollution controls are therefore required for modern incinerators. Many older incinerators have been closed down because they have not met air pollution standards. The capital investment and operating costs of incinerators are generally considerably higher than for landfill operations without incineration.

Despite such problems, modern, well-designed, and strategically located incinerators with proper pollution controls are an effective and efficient means of reducing the solid waste burden while simultaneously generating usable energy. Fluidized-bed incinerators burn pellets of solid waste at high pressure, producing hot gases which then power a turbine and generate electricity. Municipal solid waste has a respectable heating value of 2268 BTU/kg, approximately half that of a good grade coal (see also Chapter 7).

Seventeen energy-generating plants using wastes were in operation in the United States in late 1978, with 12 more scheduled to begin operation soon. The electricity can be sold, helping to offset the huge costs of solid waste management, and at least three-quarters of the heat in the gas turbine cycle can be put to use for heating buildings, steam generation, drying sewage sludge, aquaculture, and saline water conversion.

Recycling

As noted previously (in Chapters 5 and 7), the classification and separation of wastes allow us to recycle valuable materials, isolate particularly hazardous components for special treatment and disposal, and use nonrecyclable combustible wastes in energy recovery plants. When properly planned and managed, this is the most efficient and economical approach to conserving precious raw materials, obtaining maximal use from our resources, and at the same time reducing the solid waste burden (Fig. 20–23). Although recycling and conservation are increasing, stronger incentives are necessary to significantly reduce the profligate waste that continues to characterize modern business and life-styles for millions of affluent people. In 1975, 69 percent of U.S. municipal trash was "disposed of" in open dumps or in oceans; 22 percent was deposited in landfills; 8 percent was incinerated; and only 1 percent was recycled. The importance of recycled materials in our economy was emphasized in 1978 by the report of the National Association

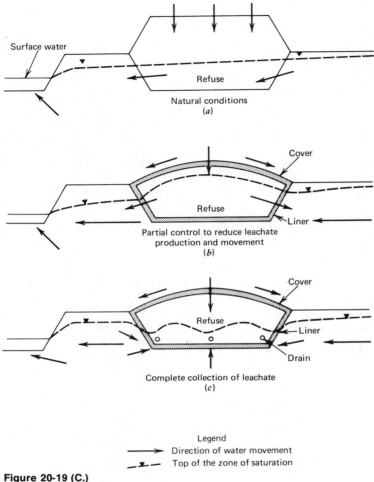

Natural conditions
(a)

Partial control to reduce leachate
production and movement
(b)

Complete collection of leachate
(c)

Legend
→ Direction of water movement
─▼─ Top of the zone of saturation

Figure 20-19 (C.)

of Recycling Industries, which found that 40 percent of copper, 30 percent of stainless steel, 25 percent of aluminum, 20 percent of paper, and 14 percent of zinc were recycled in the country (Fig. 20–24).

The ideal solid waste recycling program will maximize salvage and utilization of materials whenever possible. The diverse nature of solid wastes requires a variety of strategies and multiple techniques. One method, that is used at Franklin, Ohio, feeds a refuse mix to a large tank of water. Heavy materials sink to the bottom where they are removed by rotating blades and passed under an electromagnet that separates ferrous metals. Additional processes use centrifuges, screens, and other devices to extract cellulose fiber, glass, and a mixture high in aluminum and other nonferrous metals. Markets for such products already exist in many areas, and new uses are being devised. For example, the cellulose waste can be used in paper and biofuels production; it can also be treated and converted into high-protein supplements for animal feeds and perhaps even for human consumption.

Raw organic components of solid waste may also be composted under controlled conditions of temperature, ventilation, and moisture to yield sanitary, humuslike material valuable as a soil conditioner. Similar processes can generate gaseous or liquid fuels from organic wastes (page 202).

The waste delivered to complex, multifunction waste-repository facilities can be efficiently reduced by sorting and reusing "waste" materials at their source. This may take the form of home composting or biogas-generating devices, subsidiary operations at industrial sites to utilize materials that would otherwise be discarded, and direct reuse of manufactured products such as bottles, machine components, and the bodies and frames of refrigerators, stoves, and other items. At the start of 1979, seven states had passed bans on throwaway beverage containers, and a long-continuing debate over a national bottle bill is still undecided.

In addition to reducing solid waste and its attendant problems, recycling conserves material and energy resources. Reprocessing aluminum saves 96 percent of the energy needed to manufacture new aluminum—in similar fashion, energy savings for reprocessed copper is 87 percent, 70 percent for paper, and 63 percent for zinc and lead. However, in order to be even more effective, laws and rate structures will need changing. For example, depletion allowances favor production of virgin ores over recycled metals, and rail transportation costs for mined ore is one-third cheaper

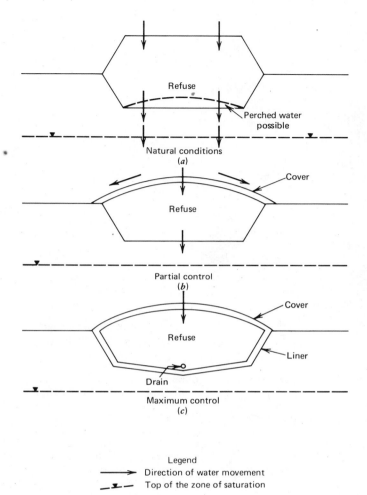

Refuse

Perched water
possible

Natural conditions
(a)

Cover

Refuse

Partial control
(b)

Cover

Refuse

Liner

Drain

Maximum control
(c)

Legend
Direction of water movement
Top of the zone of saturation

Figure 20-19 (D.)

than transportation of scrap iron.

Open Dumps

Open dumping—often in a gully, natural depression, abandoned gravel pit or rock quarry, or at sea—is the dominant means for solid waste disposal in the United States. The dumping is commonly accompanied by open burning of combustibles, resulting in gross air pollution. Decay of putresci-bles produces a breeding ground for disease organisms. Hazardous chemicals and pathogenic micro-organisms may escape the dump site as air pollutants, in surface runoff, or via seepage into groundwater. Dumps are ugly, malodorous, attract rats, flies, and other vermin, and should be regarded as a severe breach of environmental ethics. The Resource Conservation and Recovery Act of 1976 champions a phaseout of open dumping by 1983. However, because of the widespread use of open dumps, and the lack of policing measures for numerous "sanitary landfills," it is doubtful if the 1983 goal will be achieved.

There are other types of waste that need disposal (see Fig. 20–25). However, whether to use animal waste as fertilizer, for biogas, or for burial requires careful thought and planning.

HAZARDOUS CHEMICAL WASTES

A national awakening to the severity and magnitude of hazardous chemical wastes is finally evident. Only within the last decade have some of the problems surfaced, and only within the last five years has the data base indicated the seriousness of the threat to society. Chemical (and biochemical) hazardous wastes include those that are ignitable, corrosive (pH < 3 or > 12), infectious, and toxic. The sources of such wastes vary, ranging from domestic and municipal disposal, to industrial, mining, and farming effluent and debris. Hazardous waste at landfill sites is increasing at a 10 percent rate each year—30 million tons in 1974, 45 million tons in 1978, and a projected accumulation of 70 million tons by 1983. In 1975, the United States dumped 100 million tons of poisonous sludge, industrial wastes, and dredge spoils into the ocean (Table 20–2).

One of the earliest reports linking man-induced contaminants with health was a paper read in 1775 by Percival Pott at the London meeting for the Association of Chimney Sweeps. Pott argued that soot was the cause for the increasing incidence of cancer. However, the more recent alarm

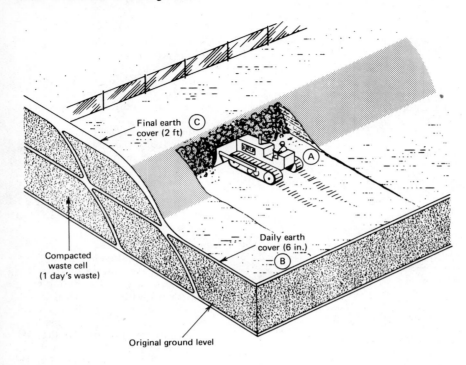

Final earth cover (2 ft) Ⓒ

Ⓐ

Daily earth cover (6 in.) Ⓑ

Compacted waste cell (1 day's waste)

Original ground level

Figure 20-20 Area method for a sanitary landfill.

was not sounded until 1968 when five died and 1000 people were stricken in Japan after eating food containing cooking oil contaminated with PCB (polychlorinated diphenyl). Since then PCB has been increasingly identified as a highly hazardous chemical and one that is carcinogenic. Prior to 1966, analytical tests had not distinguished it from DDT, but it is now known to occur in a wide range of products, such as toilet soap, primer paints, degreasers, ironing board covers, electric insulaters, and plastics. Unfortunately, PCBs are very persistent and the chemical is not destroyed until temperatures of 1400°C are reached. During a 25-year period, two General Electric capacitor plants discharged more than 270,000 kg of PCBs into the Hudson River. In an out-of-court settlement they paid a $7 million fine, and cleaning operations to remove a major

part of the contaminants will cost in excess of $30 million. Waters of Waukegan Harbor, Illinois, in 1978, contained PCB concentrations of 246,000 ppm whereas the safe tolerance is considered to be 10 ppm.

The number of other devastating poisonous chemical wastes is staggering. The EPA estimates that there are about 25,000 hazardous waste sites in the country and that decontamination costs would run $16 billion annually for several years to bring them into conformance with current regulations, as stated in the Resource Recovery and Conservation Act of 1976. However, there still is no comprehensive strategy for the safe and efficient disposal of these deadly materials. Seventy-five percent of major disposal sites lack adequate safeguards, and in one EPA study 40 out of a sample of 50 industrial waste la-

goons and landfills were leaking hazardous substances into the groundwater.

Although all health authorities do not agree on what percentage of deaths and severe physical impairments can be attributed to environmental causes, there is a consensus that hazardous wastes are causing increased fatalities and injury. Cancer, heart, and lung diseases accounted for 17 percent of U.S. deaths in 1900, but now cause 60 percent of deaths. Obviously, there are additional reasons for this increase other than environmental factors. However, reputable studies have shown that thousands of deaths each year are attributable to power plant pollutants, and that automobile emissions are responsible for 4000 deaths and 4 million sick-days in the labor force.

Two recent cases have aided to rivet health agencies to the in-

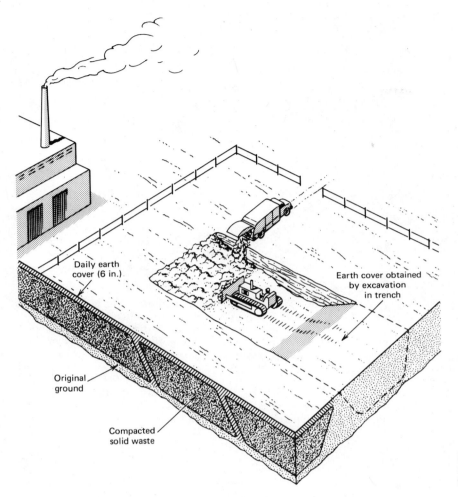

Figure 20-21 Trench method for a sanitary landfill.

creasing threat of chemical hazardous wastes—increasing each year at the rate of 4 percent. At Hopewell, Virginia, Kepone (hexachlorocyclopentadiene), a pesticide, was discharged into a municipal sewer that flowed into the James River and then into Chesapeake Bay. The materials were ingested by fish and shellfish, and 70 people eating the seafood became stricken with severe neurological disorders. Even workers who handled the material at the

industrial source suffered grave illness. The other case involves the Love Canal, Niagara Falls, New York. As early as 1942, the Hooker Chemical and Plastics Corporation had started to use the abandoned canal as a burying ground for chemical wastes. Most of the materials were buried in drums during the 1947–52 period. During this time at least 200 tons of trichlorophenol (one of the most toxic synthetic chemicals of that time containing a dioxin)

and 21,800 tons of other chemical wastes were placed into metal drums. In 1953, the site was backfilled with a 1.2 m cover of clay. (It should also be mentioned that the same company deposited a total of 400,000 tons of other chemical wastes throughout the Niagara Falls and North Tonawanda areas.) After the filling in and grading processes at the Love Canal, developers built homesites, and many were within 30 m of the canal. During home construc-

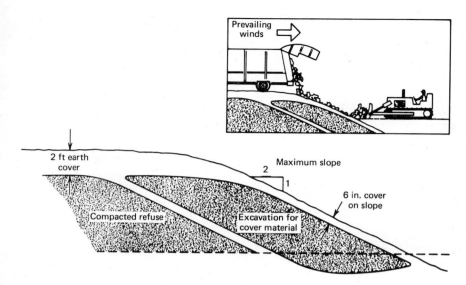

Figure 20-22 Ramp method (or slope method) for a sanitary landfill.

tion some of the clay cap material was used as fill for the housing tract. After heavy rains in 1976, it was discovered that chemical wastes were seeping into home sump systems and even formed surface pools of waste in some lawns. Heavy rains again in 1977 continued to aggravate the problem, and health authorities undertook an investigation. They discovered more than 80 different toxic organic chemicals in the wastes. Furthermore, anomalously high rates of birth deformities, miscarriages, mental retardation, and blood abnormalities occurred in families near the canal. In August 1978, the New York State Health Department declared an emergency and began the evacuation and relocation of 240 families nearest to the canal. By 1980 more than 700 families were being relocated.

Hazardous wastes become entrained in the environment by air pollution that discharges cadmium, mercury, antimony, and other derivatives during combustion of fossil fuels. They also enter the land-water ecosystem from organized disposal sites and from nonpoint sources. The groundwater on Long Island has recently been shown to be dangerous in many localities. A study of shallow wells showed that 80 percent of the groundwater had traces of trichoroethylene (a chemical used in cleaning solvents, which causes cancer in mice). The National Solid Wastes Management Association is championing proposals that would establish a program to provide a "perpetual care" fund to monitor all dumping facilities and compensate victims and heirs of deadly wastes. However, most present emphasis is on the development of standards and regulations, and the funding of research and personnel involved in hazardous waste programs are woefully inadequate. Hazardous wastes will be one of the severest environmental problems of the 1980s.

DELIBERATE CONTAMINATION

Whereas most of the cases previously discussed involved inadvertent pollution, we in our attempt to control environmental factors at times deliberately spread poison on the land and on vegetation in the form of herbicides and pesticides. The publication of Rachel Carson's *Silent Spring* in 1962 did more than any other single event to halt the proliferation of wanton and indiscriminant poisoning of the environment. For example, the chlorinated hydrocarbon insecticides, along with the herbicides and fungicides, are now some of the most widely distributed synthetic chemicals in the biosphere. For many years DDT (trichlorophenylethane) was considered the wonder chemical of the century and a product that would wipe out all deleterious insects. However, DDT is very insoluble in water, but quite soluble when in contact with organic solvents. These properties provide an environmental curse. Because of water insolubility, DDT can move freely through aqueous systems without serious attenuation. But because of its organic solubility, DDT becomes absorbed by living organisms through the food chain. By the process of **biological**

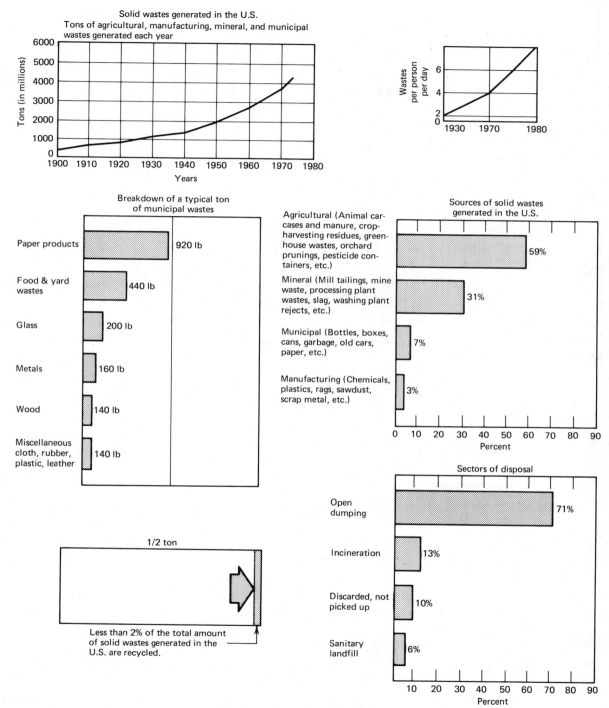

Figure 20-23 The magnitude of the U.S. solid waste problem and its sources. (After data from U.S. Environmental Protection Agency.)

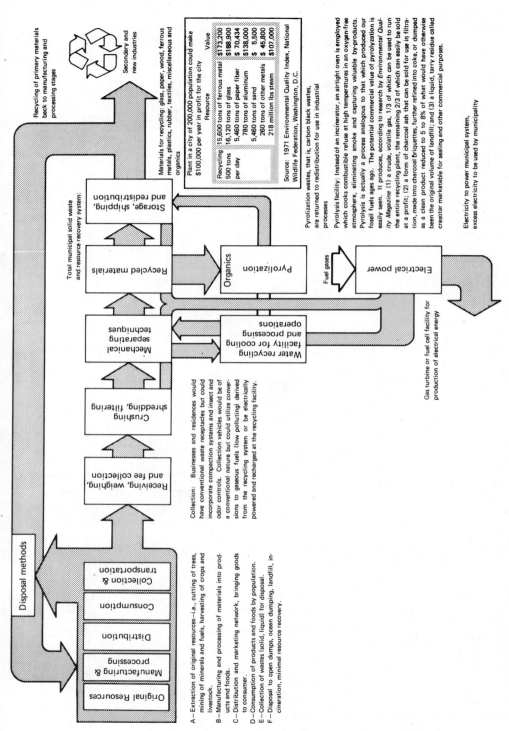

Figure 20-24 A proposal for a recycling program. (Environmental Alert Group.)

Figure 20-25 Cattle feedlot companies rely on soil surveys to find proper locations for the disposal of wastes, in this case at Lockney, Texas. (Courtesy Soil Conservation Service, Temple, Texas.)

Table 20-2 Potentially hazardous wastes, by type, in New York State industry

Type of industry	Estimated total number of plants	Estimated number included in DEC[a] inventory
Batteries	12	10
Inorganic chemicals	70	70
Organic chemicals, pesticides and explosives	106	85
Electroplating	322	200
Paint and allied products	137	100
Petroleum refining	2	2
Pharmaceuticals	150	80
Primary metals	200	125
Textile dyeing and finishing	215	50
Leather tanning	97	40
Electronic components	200	100
Rubber and plastics	100	100
Printers and publishers	40	40
Photographic chemicals	30	20
Paper manufacturing and finishing	100	20
Metals mining	3	3
Machinery manufacturing	420	200
Aircraft and motor vehicle manufacturing	20	10
Miscellaneous manufacturing industries	250	40
Totals	2474	1295

Source: Conservationist, v. 32, 1978.
[a]New York State Department of Environmental Conservation.

Table 20-3 Steps in the management of radioactive wastes

1. Collection of the various waste forms
   (i) spent fuel (if designated as waste)
   (ii) mine tailings, fabrication plant wastes
   (iii) decommissioned plants and other irradiated equipment
   (iv) low-level wastes from reactors
2. Initial handling prior to reprocessing
   (i) onsite storage and packaging
3. Reprocessing of spent fuel (with variations for military and commercially produced wastes)[a]
4. Interim storage
5. Solidification
6. Long-term to ultimate disposal

Source: T. R. La Porte, 1978, Nuclear waste: increasing scale and sociopolitical impacts: *Science,* v. 201, p. 25. Copyright © 1978 American Association for the Advancement of Science.

[a]This phase should include partitioning of actinides if space disposal option is analyzed.

magnification each member of the food chain passes along the DDT, which becomes increasingly concentrated in higher animal forms. Eventually such accumulations reach toxic proportions and become lethal to the host. Carnivores are the most susceptible animals to DDT poisoning because they depend on other animals for their diet.

There are at present more than 30,000 different known pesticides and numerous other herbicides and fungicides. It would take decades to test each to determine toxicity levels and degrees of environmental safety. The use of these chemicals is often for short-range environmental benefits, such as the immediate killing of pests or the defoliation of unwanted flora. However, their long residence time in the soils and waters guarantees problems for generations in the future. Their thoughtless use can no longer be tolerated by the world community.

RADIOACTIVE WASTE DISPOSAL

The origin and dangers of radioactive wastes were mentioned in Chapter 7. The unique and insidious character of these extremely lethal wastes has created one of the most pressing and difficult of all waste disposal problems. It is imperative to keep these hazardous materials out of the living environment. One estimate of the toxicity of the more than 60 million gal (230,000 m³) of intensely radioactive wastes stored at Hanford, Washington, states that only 3 gal (3.78 liters) if disseminated among the world's population would bring every person to the danger level for radiation exposure. Some radioactive wastes will remain deadly for many thousands of years, so their ultimate interment and disposal must be permanent (Table 20–3).

Radioactive wastes are divided into three categories depending on radioactivity content: low-, medium-, and high-level wastes. **Low-level wastes** may be diluted and released to the environment after removing long-lived radioisotopes, solidified and buried, or placed in barrels and dumped at sea. In 1975, the Nuclear Energy Agency supervised the disposal of 4500 tons of low-level wastes from Europe at one site in the Atlantic Ocean 1300 km west of France. There had previously been 35,000 tons dumped at the site. A 1976 EPA report estimated that low-level waste in the United States would be at 28 million m³ by the year 2000.

Intermediate-level wastes are often placed in concrete trenches (cribs) in arid areas where the radioisotopes, including plutonium-239, cesium-137, and strontium-90, are allowed to percolate into the substrate and where they will presumably be retained in dry rock and soil long before reaching the deep water table. In many cases, low- and intermediate-level waste disposal sites have been selected with no prior geologic or hydrologic investigation.

At present, special attention has been focused on **high-level wastes,** which are now temporarily stored in cooled, stainless steel and concrete underground tanks such as those at the Hanford nuclear complex in south-central Washington state. The wastes are extremely corrosive, and tanks must be continuously monitored for possible leaks. In the 1958–74 period, 18 leaks occurred with the loss of 450,000 gal (1700 m³) of high-level wastes into the ground. The largest single leak went 48 days before its discovery in June 1973. By that time there was a loss of 115,000 gal (440 m³), and the hazardous plume spread 14 m below the tanks. It has been assumed (hoped) that the wastes will remain in the nearby earth materials and will not reach the water table or the nearby Columbia River. There has also been the loss of several hundred gallons of waste because of container leakage at the Savannah River nuclear facility.

Another danger exists wherever fissionable radioisotopes such as plutonium are present. As plutonium seeps from containers, it may accumulate in local zones within the earth. When the amount of fissionable material reaches a certain "critical" mass, it is possible for a chain reaction to happen. This could result in a low-grade explosion or "nuclear fizzle," but lethal quantities of radioisotopes could be released by such an event. Plutonium-contaminated soils beneath trenches at Hanford have been excavated, processed, and the wastes stored separately to prevent this type of "accident."

It is believed that such a critical accident occurred in the southern Ural Mountains of the Soviet Union during 1958. In the lake district between the cities of Chelyabinsk and Sverdlovsk, a large military facility produced plutonium that was used in the fission reactors and chemically separated for use in the weapons program. Although the incident has never been officially reported, other sources state that a region larger than 1500 km² suffered a major catastrophe. The site remains closed to this day, and tens of thousands of people were evacuated. Deaths may have run into the hundreds, even thousands. It is believed that the radioactive wastes had been stored in trenches, but that a critical threshold was reached when a mud volcanolike nuclear eruption occurred, pouring radioisotopes into the atmosphere and the surrounding countryside.

Radioactive wastes are generated by military nuclear weapons development, nuclear power plants, uranium mining and milling, and in lesser quantities from other sources. For example in the United States, weapons work produces 7.5 million gal/yr (28,500 m³/yr) of highly radioactive waste, with 72 percent of storage in salt cages at Hanford, 25 percent at Savannah, and the remainder in dry form at Idaho Falls.

The mining wastes are not in as concentrated a form as the lethal brews that result from nuclear power plant and weapons programs. However, the low-level radioactivity associated with mining activities represents a long-term hazard to nearby populations. A particularly dangerous site is one that is 30 city blocks from the state capitol in Salt Lake City. Here on 128 acres (51 ha) 1.8 million tons of "radwaste" rest in earth materials in the midst of one-half million people. At Grand Junction, Colorado, more than 300,000 tons of contaminated radwaste tailings were used as fill material. After there was recognition of "elevated" levels of radioactivity in many of the thousands of buildings constructed on the tailings, a program was undertaken to remove the tailings. By 1976, $11 million had been spent on this removal, with the cost of removing a few bushels of tailings from a single dwelling running more than $25,000. To date, the U.S. government has identified more than 25 sites, from a list of 1000 in western regions, where radwaste contamination poses sufficient risks to warrant expensive cleanup. The government will pay 90 percent of the costs.

Another source of radwaste generation is the fuel-reprocessing plant, where spent fuel from nuclear power plants is treated to separate reusable radioisotopes from wastes. A series of severe problems concerning the technology of operations has caused long delays and prevented the reprocessing program from being effective. The only functional commercial reprocessing plant was operated from 1963 to 1972 at West Valley in western New York State (Fig. 20–26). It ceased operations and was closed, however, due to financial difficulties and the inability to meet health and environmental standards. To date, no decision has been reached concerning the ultimate fate of the approximately 600,000 gal (2270 m³) of high-level waste and 56,600 m³ of low-level waste in temporary storage at the site. Cost of solidification of the waste has been estimated at $500 million.

At nuclear plants and facilities throughout the United States, there is a growing backlog of radwaste accumulating for which there is not yet a disposal solution. In 1970, the government spent $28 million on disposal technology, but with the growing seriousness of the problem $450 million was spent in 1979. At present, there is 5200 tons of spent fuel in temporary storage from the 72 nuclear reactors, and by 1990 this waste will amount to eight times the present amount. It is estimated that uranium tailings contain 140 million tons of radwaste materials, and these spoilage sites grow at a rate of about 15 million tons/yr. A typical reactor produces 45,000 ft³ (1270 m³) of contaminated wastes per year.

Most long-term nuclear waste disposal strategies require the solidification of wastes (Table 20–4). The tremendous heat that is generated by the decay of short-life radioisotopes requires a delay of about five years before solidification is attempted. Another five years of cooling down is usually needed before wastes are ready for long-term storage. During the

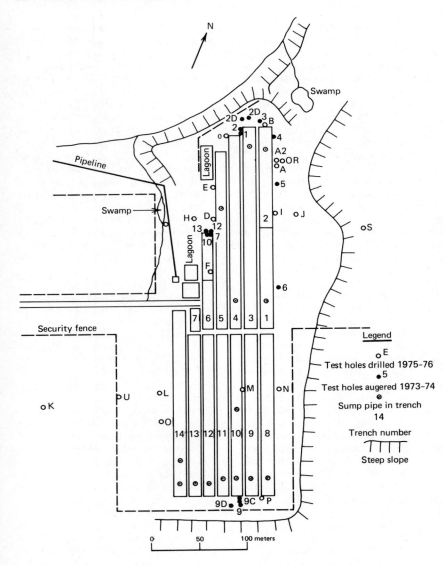

N

Swamp

Pipeline

Swamp

Security fence

2D · · 2D
2 · 3 B
0 · 1 · 4
A2
OR
A
· 5

Lagoon

E o

H o   D o
13 · 12
10   7
F o

2 o I   o J

o S

· 6

7 6 5 4 3 1

Legend

o E
Test holes drilled 1975–76
· 5
Test holes augered 1973–74
o
Sump pipe in trench
14
Trench number

Steep slope

o K   o U   o L   o M   o N

o O

14 13 12 11 10 9 8

9D   9C   o P
9

0   50   100 meters

Figure 20-26 Map of West Valley, New York, radioactive waste disposal site. (Prudic and Randall, U.S. Geol. Survey, 1977.)

solidification process, wastes are mixed with about twice their volume of inert material, such as a durable ceramic or glass. In 1978, Australian scientists reported the development of a new method of freezing wastes into the crystalline structure of synthetic minerals. Because this type of waste is locked inside the crystal lattice,

and not just part of a solid mixture, leaching of wastes can be minimized. Such a synthetic rock is stable and nearly insoluble in water. However, this product would still be highly radioactive and would need further encasement in some type of alloy cylinder, such as nickel-iron, prior to final burial.

The annual high-level waste production from one 1000 MW reactor would occupy about 1000 ft³ of space after final solidification. Present plans in the United States call for solidified wastes to be placed in steel canisters 3 m long and 0.3 m in diameter. The U.S. nuclear industry could fill about 80,000 of these high-level

Table 20-4 Summary of the technical feasibility of alternative nuclear waste management systems

Storage method	General characteristics relative to feasibility	
	Favorable	Unfavorable
Storage in geologic formations		
Solid waste emplaced in mined cavity; no fluid cooling or melting	Ion exchange as backup	
Solid waste emplaced in mined cavity; initial water cooling; melting[a]	Ion exchange of rocks as backup	Irreversible high temperature in rock
Solid waste emplaced in man-made structure in mined cavity; initial air cooling; no melting	Ion exchange of rocks as backup; provides ready interim retrievability	Requires interim operation by man
Solid waste emplaced in man-made structure in mined cavity; initial water cooling; no melting	Ion exchange of rocks as backup; provides ready interim retrievability	Requires interim operation by man
Solid waste emplaced in matrix of drill holes; no fluid cooling or melting	Ion exchange of rocks as backup	Very poor retrievability and monitorability; many penetrations to surface
Solid waste emplaced in deep holes; no fluid cooling; melting or nonmelting[b]	Ion exchange of rocks as backup; large distance from man's environment	Very poor retrievability and monitorability; deep geology unknowns
Storage in ice sheets		
Self-melt through ice[c]	Great distance from man	Extended transport; poor retrievability
Anchored storage or disposal[c]	Low temperature for cooling	Extended transport
Ice surface storage or disposal[c]	Possible international solution	Many technical unknowns
Antarctica (subsurface burial in ice-free areas)	Great distance from man; low temperature for cooling; possible international solution	Retrievability and monitorability good
Storage in the seabed		
Subduction zones and other deep-sea trenches[b,c]	Great distance from man; water for dilution	Extended sea transport; mobility of seawater
Stable deep-sea areas	Ion exchange of sediments as backup	Concentration by ecology
Rapid sedimentation areas[c]	Possible international solution	Very poor retrievability and monitorability

Source: E. E. Angino, 1977, High-level and long-lived radioactive waste disposal: *Science*, v. 198, p. 887. Copyright ©1977 American Association for the Advancement of Science.

Note: Favorable characteristics include: fair distance from the human environment; safety from storms and most human activities. Unfavorable characteristics include: some potential for penetration by humans in future; poor retrievability and monitoring; possible groundwater transport. Differences from these general points are indicated to the right of each method under the appropriate column.

[a]This method can also involve in-place melting and conversion to a rock-waste matrix.
[b]Cannot be implemented with today's technology.
[c]These have an uncertain potential for providing adequate safety.

waste storage canisters during the next 25 years.

The location for a final disposal site for radwastes is a very volatile, emotional, and scientifically debatable problem (Table 20–5). As of mid-1978, seven states had passed laws banning radioactive waste repositories. Dozens of areas across the United States are currently under investigation as radwaste sites. The Energy Re-

Table 20-5 Technical irreversibility of selected methods for the management of wastes with high levels of activity

Disposal method	Irreversibility		
	Physical	Social	Technical
Retrievable surface storage	Very low for water cooling to moderately low for above ground convection cooling	Very low	Very low
Sealed mausoleums	Low to moderate	Very low	Low to very low
Mined caverns in salt	Moderately low	Low	Low
Drilled- or solution-mined cavities in salt	Depends on groundwater, low to moderate	Low for domes to moderate for bedded salt	Low to moderate
Seabed, emplacement in bottom sediments	Moderate, depends on nature of sediments	Moderate to moderately high	Moderate to moderately high
Mined rock cavity, partial melting	Moderately high	Moderate to moderately high (away from other minerals)	Moderate to moderately high
Mined rock cavity, complete melting	Moderately high	Moderately high (away from other minerals)	Moderately high
Seabed emplacement in basement rock, no melting	Moderately high to high, depends on geologic activity	High	Moderately high to high
Deep rock melt, drilled hole	High, depends on geologic activity	High (located away from other minerals)	High
Deep rock melt, self-descending capsule	High to very high, depends on geologic activity and sinking depth	High to very high	High to very high
Space disposal, outer space mission Transmutation	Very high to complete Complete	Very high to complete Complete	Very high to complete Complete

Source: G. I. Rochlin, 1977. Nuclear waste disposal: two social criteria: *Science*, v. 195, p. 27. Copyright © 1977 American Association for the Advancement of Science.

search Development Authority of the government has launched a massive study on siting and its problems. An experimental operation is underway in the salt beds near Carlsbad, New Mexico, even though the area may have potential for potash and natural gas production. Originally, the local residents were amenable to the plan, but now they have become increasingly restive over the idea.

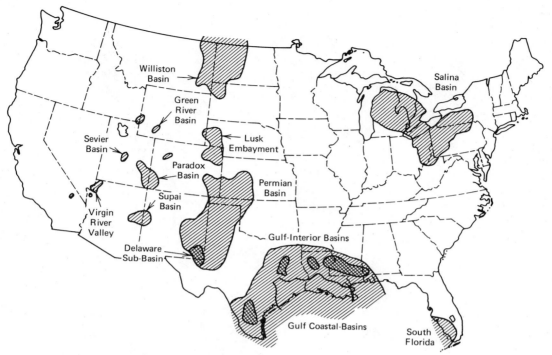

Figure 20-27 Salt formations map of the United States. (U.S. Geol. Survey.)

In the following sections, we will review some of the ideas that have been suggested for storage of radwaste materials.

Underground Rock Storage

Permanent storage in deep underground rock formations is currently regarded as the most promising site for disposal. Proposals of this type generally seek to locate thick impermeable rock formations more than 500 m below the surface. The location must be geologically stable with a minimum risk of seismic disturbance. If located in northern states, another criterion is that burial be sufficiently deep so that erosion from a new glacial age would not uncover the radwaste. Deep structural basins are considered as possible candidates. The preferred

bedrock for the excavation of burial chambers is still being argued. Many consider salt beds the ideal host, although others champion different rock types such as granite, shale, and even basalt (Figs. 20–27, 20–28, and 20–29). Desirable qualities of salt are its behavioral characteristics under subsurface pressures. It behaves like a very viscous fluid (technically a rheid) and in time would slowly entomb the wastes, perhaps effectively sealing them from the environment for thousands or even millions of years. Others argue, however, that salt is highly soluble and in time groundwaters would entrain the contaminants and transport the products to locales hazardous to society. An important aspect to all programs is the retrievability of the radwaste. This would be necessary if some new technological breakthrough would

occur and the materials would need to be recovered for additional treatment. Some plans call for keeping all chambers open for easy monitoring. Also, the materials could be electrically refrigerated to reduce corrosion and prevent melting (Figs. 20–30, 20–31, and 20–32).

Unfortunately, the subsurface of most developed countries, such as the United States, has been extensively probed, and it is difficult to locate potential sites that are not riddled with old drill holes and natural fractures. This was the prime reason for abandonment of the proposed nuclear waste depository in the salt beds near Lyons, Kansas. It was believed possible that the heat of decaying wastes could induce abnormal flowage, and that the stresses in the salt, or even other rock, could trigger additional

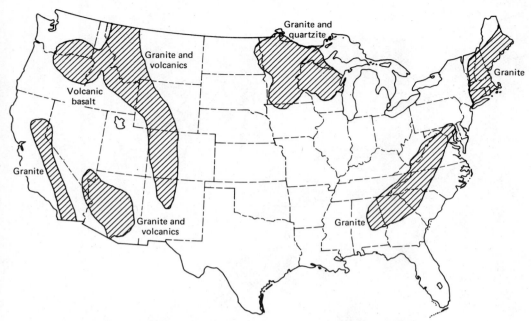

Figure 20-28 Crystalline rock map of the United States. (Office of Waste Isolation.)

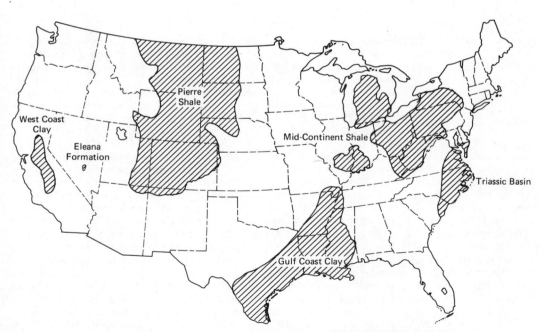

Figure 20-29 Map of argillaceous rocks in the United States. (Office of Waste Isolation.)

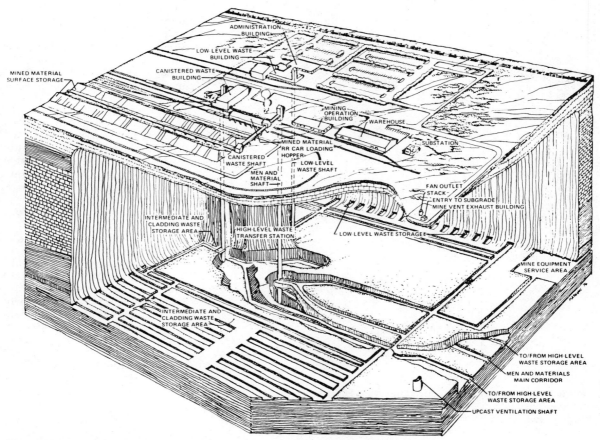

Figure 20-30 Schematic presentation of a possible nuclear waste disposal facility. (Office of Waste Isolation.)

Figure 20-31 Experimental and testing facility equipment for a nuclear waste disposal operation. Radioactive wastes are lowered through a shaft onto this vehicle which then transports the cannisters to the burial underground site. (Office of Waste Isolation.)

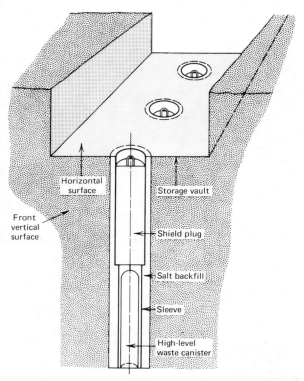

Retrievable high-level waste salt emplacement.

Figure 20-32 The cannisters moved in Figure 20-31 are then emplaced into drill holes in a specially designed storage vault, which is constructed so that the materials can be retrievable. (Office of Waste Isolation.)

fracturing in the surrounding beds.

The direct injection of liquid radioactive wastes into deep geological formations is also under consideration. Some suggest the excavation of chambers thousands of meters below the surface, and others believe injection can be accomplished without prior excavation. Deep-well injection of hazardous wastes produced earthquake activity at the Rocky Mountain Arsenal (see page 299), so the site would have to be especially carefully located in areas that would continue to be aseismic. If wastes get out of control beneath the surface, their collection and entrapment for further treatment become a near impossibility.

Another variation on subsurface disposal is to mix wastes with a cement slurry and inject them into thick, dense shale or claystone formations. Burial in basalt is under study in Oregon and Washington. Silty-argillaceous interbeds in the basal could help to retard leaching of wastes in areas of low precipitation. However most basalts are highly fractured and permeable, and seismic risks are moderate to high in such volcanic terranes.

Oceans

Another set of proposals would seek to dispose of radwastes in the oceans. Unregulated dumping of nuclear wastes at sea is now prohibited, but there is still much supervised dumping of low- and intermediate-level wastes into the ocean. Disposal of high-level wastes in areas of rapid sedimentation (where wastes would quickly be covered over with a blanket of deposits), in deep-ocean trenches (where subduction processes could drag the waste deeper into the earth), and on tectonically stable parts of the ocean floor are among the schemes that have been seriously considered. However, all of these must be viewed with serious reservations because of incomplete knowledge on all facets of the problem. The dangers of intensely radioactive wastes escaping into the 'global seawaters needs no further elaboration! Such processes as ocean sedimentation and subduction take thousands and millions of years to be effective, whereas radwaste container leakage and breakage might be accomplished in only years or decades.

Glaciers

Some proposals suggest that burial of wastes in the Greenland or Antarctic ice sheets might work. Waste containers placed on the surface of a 3000 m thick glacier would melt its way down to bedrock; the hole would refreeze and isolate the wastes beneath the ice for thousands of years. However, glaciers are not static features. They are dynamic, moving masses of plastically deforming ice that can respond dramatically to even minor disturbances and perturbations in their regime. Drilling into glaciers has sometimes caused pockets of melted waters under pressure to gush to the surface when penetrated. A semipermanent heat source with the ability to leak radioactive wastes into enclosing pockets of water, or to the ice directly, and to basal meltwaters is too great a risk to take.

Transmutation

It is possible to produce a controlled nuclear reaction, as in a high-neutron flux process, that will change deadly radioactive isotopes into less dangerous, or even nonradioactive, isotopes. Such transmutation processes are extremely expensive and treat only small quantities of waste. Thus,

they are generally not regarded as feasible solutions to the radwaste disposal problem. It might be possible, however, to transmutate some of the most hazardous elements, such as americium.

Other Schemes

A wide variety of other schemes has been proposed for the elimination of radwaste. These proposals include some bizarre ideas,

such as rocketing the material to outer space, construction of pyramidlike tombs in out-of-the-way places such as wilderness areas, and transportation to other desolate parts of the world and the incarceration of the waste. These ideas are not taken seriously by most responsible authorities. Apparently, it is a problem we are forced to live with, and the storage and confinement of these wastes will most likely be on or under our own soil.

Perspectives

Pollution and solid wastes are the dues civilization must pay for inhabiting the earth with so many people (Fig. 20–33). Environmental contamination is the by-product of a consumer society that places little effort on conservation and the maintenance of natural systems. The general public still does not take seriously the massive expenditures that will be needed to mount effective cleanup and remedial programs to arrest the harmful effects of inadequate waste disposal facilities. Part of the trouble stems from the inadequacy of educational efforts to alert the populace to the insidious damage that occurs from improper waste control methods. Another factor is the low-key character of deleterious effects. Sometimes it takes years before the cumulation of pollutants takes its toll. Contaminants also act quietly and seem intangible—not like flood hazards or earthquakes that are physically demonstrable and visually apparent to all who experience such disasters.

Elimination and treatment of wastes at first glance seem to be largely the province of biologists, chemists, and engineers. However, because the unwanted residue of civilization needs disposal, its final resting place is the soil,

rock formations, and the waters of the earth, and this is the domain of geologists. Thus, the geological sciences must enter the picture of waste management and control and provide expertise concerning how all waste disposal

systems should be planned, monitored, and designed. Only by such interdisciplinary efforts can mankind achieve a cleaner earth, and one that will sustain a higher quality of life.

Figure 20-33 Waste disposal in the open canals, Venice, Italy.

READINGS

Albert, J. G., Alter, H., Bernheisel, J. F., 1974, The economics of resource recovery from municipal solid waste: Science, v. 183, p. 1052–1058.

Angino, E. E., 1977, High-level and long-lived radioactive waste disposal: Science, v. 198, p. 885–890.

Cain, J. M., and Beatty, M. T., 1973, Disposal of septic tank effluent in soils: Soil and Water Cons. Jour., v. 20, p. 101–105.

Carson, R., 1962, Silent Spring: Boston, Houghton Mifflin Co., 368 p.

Claus, G., and Halasi-Kin, G. J., 1972, Environmental pollution: in Fairbridge, R. W., ed., Encyclopedia of geochemistry and environmental sciences, New York, Van Nostrand Reinhold, p. 309–337.

de Marsily, G., Ledoux, E., Barbreau, A., and Margat, J., 1977, Nuclear waste disposal: can the geologist guarantee isolation?: Science, v. 197, p. 519–527.

Deutch, J. M., Chairman, 1978, Nuclear waste management: Rept. to the President by the Interagency Review Group— Draft, Washington, D.C., U.S. Dept. Commerce, 95 p.

Donaldson, E. C., Thomas, R. D., and Johnston, K. H., 1974, Subsurface waste injection in the United States: fifteen case histories: U.S. Bureau of Mines Info. Circ. 8636.

Foose, R. M., and Hess, P. W.,

1976, Scientific and engineering parameters in planning and development of a landfill site in Pennsylvania: in Coates, D. R., ed., Geomorphology and engineering, Stroudsburg, Dowden, Hutchinson & Ross, p. 289–312.

Hughes, G. M., 1972, Hydrologic considerations in the siting and design of landfills: Env. Geol. Notes, n. 51, Illinois State Geol. Survey, 22 p.

Hughes, G. M., Landon, R. A., and Farvolden, R. N., 1971, Hydrogeology of solid waste disposal sites in northeastern Illinois: U.S. Environmental Protection Agency Rept. SW-12d, U.S. Govt. Printing Office, 154 p.

La Porte, T. R., 1978, Nuclear waste: increasing scale and sociopolitical impacts: Science, v. 201, p. 22–28.

McCarty, P. L., and King, P. H., 1966, Movement of pesticides in soils: 21st Ind. Waste Conf. Proc., Part I, Purdue Univ., p. 156–171.

Micklin, P. P., 1974, Environmental hazards of nuclear wastes: Science and Public Affairs, v. 30, p. 36–42.

Muskegon County, 1973, Muskegon County, Michigan wastewater management system no. 1: Bauer Eng. Inc., 16 p.

Parizek, R. R., et al., 1967, Waste water renovation and conservation: Penn. State Univ. Studies No. 23, 71 p.

Prudic, D. E., and Randall, A.

D., 1977, Ground-water hydrology and subsurface migration of radioisotopes at a low-level solid radioactive-waste disposal site, West Valley, New York: U.S. Geol. Survey, Open-file Report 77-566, 28 p.

Ridker, R. G., 1972, Population and pollution in the United States: Science, v. 176, p. 1085–1090.

Rochlin, G. I., 1977, Nuclear waste disposal: two social criteria; Science, v. 195, p. 23–31.

Schneider, W. J., 1970, Hydrologic implications of solid-waste disposal: U.S. Geol. Survey Circ. 601F, 10 p.

Sheets, T. J., 1967, The extent and seriousness of pesticide build-up in soils: in Brady, N. C., ed., Agriculture and the quality of our environment, AAAS Publ. 85, p. 311–330.

Sorg, T. J., and Hickman, H. L., Jr., 1968, Sanitary landfill facts: Bureau of Solid Waste Management, U.S. Dept. Health, Education and Welfare, Rept. SW-4ts.

U.S. Dept. Agriculture, 1971, Soils and septic tanks: Agriculture Info. Bull. 349, U.S. Govt. Printing Office.

Waldrip, D. B., and Ruhe, R. V., 1974, Solid waste disposal by land burial in southern Indiana: Water Resources Research Center, Bloomington, Indiana Univ., 110 p.

Chapter Twenty One
Environmental Law

Yellowstone National Park, the first national park in the United States. Old Faithful epitomizes the beauty and majesty of nature that can be preserved by environmental legislation.

INTRODUCTION

Environmental law as a subdiscipline of the legal profession became popular starting in the late 1960s. Although prior to this geologists and lawyers were involved with legal matters that were concerned with the environment, there were no formal university courses or law societies that emphasized the subject. Law consists of societal rules established by the governing authorities. The regulations as formulated have the force of the general public behind them, and noncompliance and disobeyance can result in penalties. Such punishment can take a variety of forms, including fine, imprisonment, or even death. Laws may also be written or unwritten, such as the English Common Law. Common law consists of a wide range of legislative statutes, ordinances, agency regulations, and codified directives of recognized public officials. The establishment of the common law has developed through custom and tradition, and its guidelines and limits of power have been formulated through action in the courts and the judicial decisions pertaining to these actions. There are few absolutes when dealing with laws and legal affairs, and it

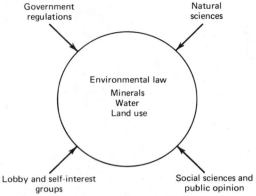

Government
regulations

Natural
sciences

Environmental law
Minerals
Water
Land use

Lobby and self-interest
groups

Social sciences and
public opinion

Figure 21-1 Constituencies that influence the character of environmental law.

has been said that the best laws are those that are in tune with the times. In this aspect, laws are a reflection of what is customary and the accepted practice and behavior of the general public. However, many laws are somewhat out-of-phase with the times; in fact, it is not uncommon for there to be a few years of legislative lag on the part of lawmakers. Occasionally, laws are made that are ahead of their times or that require societal adjustments because they impose change. Environmental law is now being interpreted to cover an extremely wide range of human behavior and actions. This chapter will emphasize the geologic aspects of the discipline, with special attention devoted to the physical effects of the relationship.

HISTORICAL ROOTS

The basis of an ordered society, indeed civilization, is law. Environmental law recognizes that society's best interests are served when humans and their habitats

are protected from wanton degradation. To accomplish this, laws have been enacted that use both the carrot and the stick philosophy. If an environmental conscience can be engendered through friendly persuasion and incentives, then this method is more effective and popular. However, if this approach fails, machinery must be available to punish the offender, and this response becomes part of the criminal justice code. The modern clean water, clean air, and land use laws exhibit the continuing growth and sophistication of human behavioral patterns and reflect a new attitude toward nature (Fig. 21-1). Other more ancient laws, hundreds of years even millennia ago, however, have helped pave the way for this new day.

Water was such an important part of the lives of earlier civilizations that laws and rules for its governance were common. The Code of Hammurabi (ca. 2300 B.C.) stated:

If a man have opened his trenches for irrigation in such a careless way as to overflow his neighbor's field he shall pay his

neighbor in grain . . . .

Plato, reflecting the Greek law of his day (ca. 300 B.C.), commented:

He who likes may draw water from the fountainhead of the common stream on to his own land, if he do not cut off the spring which clearly belongs to some other owner.

If some one living on the higher ground recklessly lets off the water on his lower neighbor, and they cannot come to terms with one another, let him call in a warden and obtain a decision determining what each of them is to do.

In A.D. 534, Justinian described the common Roman law, which was to become the forerunner of riparian rights,

By natural law, these things are common to all: air, running water, the sea, and as a consequence the shores of the sea.

All rivers and harbours are public; consequently the right of fishing in a harbour and in rivers is common to everyone.

All use of river banks is public . . . like the use of the river itself; and so everyone is free to put in at the bank . . . just as everyone is free to navigate the stream. But the ownership of the banks . . . is vested in the riparian proprietors.

Pollution of the air is not unique to the twentieth century. By A.D. 1306 air pollution had become such a problem in English cities that a royal proclamation was issued that curtailed the use of coal in London. Violation was punishable by death. By 1536, the foundation for the nuisance theory of environmental law had been formulated in England with passage of the Public Nuisance Law. Spain was one of the first countries to uphold the importance of conservation in edicts and, by 1550, the law that a new tree must be planted for each tree cut down was enforced not only in Spain, but throughout all her colonies.

The Napoleonic Code of 1804 formalized the rights of property owners and became the basis for mining law. It stipulated that ownership of land entitles the owner to possess materials that are above and below the land surface. This has also been interpreted to include groundwater, except when restricted to flow properties that impinge on riparian rights.

English common law has had a long history of development. This is law that is unwritten, but has the force of society behind it and whenever challenged has received specificity through court decisions. For example, the public trust doctrine was developed early in Roman times and spread to Spain, France, and England during Roman occupation of those lands. The royalty of England recognized that the public has rights and, although the lands were officially held in the name of the crown, the public was entitled to certain "common" usage.

What seems to be an unusual ruling was handed down by the English courts in a decision by Lord Tentreden in the 1820s regarding shorelines. When a landowner brought suit because a town-constructed groin caused erosion on his property, it was judged that:

Each landowner may erect such defenses for the land under their care as the necessity of the case requires, leaving it to others, in like manner, to protect themselves against the common enemy.

Not only did this establish a freedom to act on their own behalf, without regard for others, or littoral property owners, but the philosophy of "man against nature" was enunciated in a manner that would be repeated by countless environmental managers and architects.

PRESENT STATUS

The types of laws that can be considered to be environmental cover a very wide range of legislative statutes and more localized ordinances and regulations. English common law, legal theories for bringing court grievances, and the Constitution and understandings that stem from the Bill of Rights provide much of the basis of environmental law in the United States. The traditional environmental lawsuit until the 1960s was often based on principles that involved nuisance, trespass, and negligence. In the past 20 years, more and more litigation is stemming from a land use type of legislation and from property owner's rights in eminent domain and inverse condemnation, and in other ramifications of what is called the "trust doctrine."

The basis for many pollution lawsuits involves some form or combination of nuisance, trespass, negligence, and the Bill of Rights. Although **nuisance** is one of the most widely used theories for bringing suit, it is also one of the most confusing and conflicting terms. The law differentiates between a public and a private nuisance. It is generally held that to bring a successful lawsuit the grieved party must have suffered uniquely from the pollution; when all people suffer from the same problem, only a governmental body can intervene. The type of judgment rendered by courts in nuisance cases involves considerations of the degree of injury to the plaintiff, the level of society's interest for the defendant, and the reasonable methods available that could prevent damages. In the lawsuit *Costas* v. *City of Fon Du Lac, Wisconsin* (1964), the plaintiff contended that the city's sewage disposal plant created excessively offensive odors that damaged his motel business. The defendant argued that the odors had not been sufficiently harmful because the motel made a profit. However, the plaintiff argued that he would have been able to make a larger profit and that business had declined. The court ruled that damages had occurred and issued an abatement order with stipulation that the city must modernize its plant.

The original interpretation of the **trespass** theory stated that unlawful and deliberate intrusion onto one party's property was punishable to the extent that damages had occurred. Physical trespass constituted the standard lawsuit, but the concept has now been extended to include air space and damages inflicted by noise and vibrations. For example, it is standard practice for many road contractors to include a no-liability clause in their bidding, regarding wells of property owners. In highway construction it is usual to pay damages to well owners who claim their water supply was damaged by blasting during construction operations. These losses and those damages to walls and windows, which may be cracked by blasting or by airplane sonic booms, fall within the category that may also be referred to as **inverse condemnation** or **indirect trespass.** The more usual type of trespass suit is illustrated by the *Renken* v. *Harvey Aluminum, Inc.* (1963) Oregon case. The aluminum plant was constantly releasing airborne fluorides that fell on the plaintiff's property and damaged his orchards. It was shown that similar plants throughout the country had installed cell hoods to prevent escape of the harmful materials, and that the plant was derelict in not using modern technology. The court mandated that similar antipollution devices be installed within one year or an injunction would be served to close the plant. Such a decision had elements of trespass, negligence, and even nuisance.

When the theory of **negligence** is used in environmental lawsuits the plaintiff must prove that there was willful, deliberate, and careless behavior on the part of the defendant. Thus the burden of proof is often difficult because a *judgment* of the conduct of the offender is being rendered. The case of *Boomer* v. *Atlantic Cement Company* (1970; New York) provides an interesting example of how the court attempted to balance equities. The Boomer property was located one-half mile downwind from a limestone quarry. For 2½ years the mining operation had produced many air pollutants, lime dust, coal dust, and other chemicals that landed on the plaintiff's property. They damaged the land and the buildings. The decision, by majority, of the court awarded $185,000 for permanent damages and property depreciation, but did not call for an injunction for the plant to change their method of operation because it was performing a needed public service. However, the court did state that the decision "does not foreclose public health or other public agencies from seeking proper relief in a proper court."

The Constitution

The theories of nuisance and trespass are interrelated with the trust doctrine and many stated and implied articles and amendments in the U.S. Constitution. For example, it has become ingrained in the American people that they have the rights (and these cannot be deprived) of life, liberty, property, and the pursuit of happiness. Thus in *Reynolds Metals Company* v. *Martin* (1964) the court defined trespass as "an actionable invasion of a possessor's interest in the exclusive possession of land," and defined nuisance as "an actionable invasion of a possessor's interest in the use and enjoyment of his land." The Fifth and Fourteenth Amendments provide that no person can be deprived of life, liberty, or property without due process of law and that just compensation must be made when property is taken for public use. Such statutes provide the basis for the doctrine of **eminent domain**—the right of governments to acquire private land in the interests of the public good. The Ninth Amendment is also being increasingly used by environmental groups when particular issues arise that do not seem to have been specifically mentioned in the Constitution. The force of this amendment is that people may have additional rights to those enumerated in the Constitution, and just because they were not specified does not mean such rights should be disparaged or denied. Thus there is an extensive arsenal of weapons that are available to individuals and groups to help serve them to protect their health and their habitat. The new specific environmental laws that started to be passed in the 1960s were mindful of these principles, such as the clean water and clean air acts. Earlier legislation that provided guidelines for new governmental agencies, such as the Soil Conservation Act of 1935, expanded the role of older agencies, such as the Flood Control Acts of 1928 and 1936, which extended the public trust theory to the environmental protection of society.

The new land use legislation also builds on the ethic of public trust—that the government is responsible to hold certain lands and waters in the public domain for the use and enjoyment of everyone. Such principles were first enunciated with the establishment of national parks, Yellowstone Park in 1872, and embodied later in the Wilderness Act of 1964. The various states have also used the same principles in formulating their policies of state lands, and similar ideas are used on the local levels as well. Of course, the National Environmental Policy Act (NEPA) of 1970 represents the culmination and synthesis of the government's position:

To declare a national policy which will encourage productive and enjoyable harmony between man and his environment; to promote efforts which will prevent or eliminate damage to the environment and biosphere and stimulate the health and welfare of man; to enrich the understanding of the ecological systems and natural resources important to the Nation; and to establish a Council on Environmental Quality.

Litigation Processes

When an environmental grievance occurs a potential plaintiff must have "standing" in the eyes of the court. The issue of **standing** is most critical when lawsuits are brought against the government. The courts have repeatedly ruled that "only those with a genuine and legitimate interest can participate in a proceedings." Thus the person or group bringing a suit must be harmed or damaged in some way. In a group or **class action** lawsuit the plaintiffs must show their members will suffer personally. In 1970, Michigan became the first state to recognize that individual citizens had standing rights to sue the government on environmental grounds, whereas prior to this only class action suits had been permitted. Also prior to 1970, environmental grounds for bringing lawsuits to the U.S. Supreme Court had to be framed largely with damages that were in terms of money or health impairment. In *Association of Data Processing Organizations* v. *Camp* (1970), the the court extended the scope of environmental lawsuits to also cover "aesthetic, conservational, and recreational . . . values."

Environmental lawsuits involve not only lawyers on the opposing sides but also expert witnesses. Thus geologists who by their education, training, and experience of specific and relevant subjects become expert witnesses will find themselves on both sides of the issue at hand in the courtroom. Lawyers have always pointed out that there are at least two sides to any question, and so it is not uncommon for geologists to disagree on the interpretation of "facts" and what constitutes "truth." It is difficult to generalize on the geologist's role in environmental litigation. This is due to the wide range of cases and the relative importance of the geologist, whether peripheral or incidental to the case or central and dominant in the proceedings. This involvement may be merely a short discussion with the appropriate lawyer and a few minutes on the witness stand or may include periods of lengthy involvement. For example, in *Archie Campbell* v. *City of New York* (1955) I discussed geology principles with the plaintiff's lawyer 11 hours a day for six days and was on the witness stand 12 hours during a two-day period.

Role of Expert Witness

Trial law in environmental cases rarely has the excitement and drama of a Perry Mason performance. Many of the proceedings are boring, dull, and filled with motions and countermotions that are framed in legalese. The courtroom is no place for the timid or the ulcer-prone geologist. As an expert witness there will be times of frustration, tedium, and anxiety. Since the job of opposing lawyers is to discredit testimony that does not support their case, the expert witness can expect innuendos and carefully worded questions that are designed to further the cause of the questioning attorney. Only the most careful preparation and presentation of the data can serve as a bulwark for the defense of sound geological opinions.

Expert witnesses generally contribute to the preparation of the lawsuit. Their work is mission-oriented in effect, because the battlelines are already drawn. They are used in the courtroom only when it is believed their testimony will measurably aid a particular viewpoint. Geologists that appear at the trial invariably meet with the lawyer, show and tell him the results of investigation, and suggest questions for the lawyer to ask on direct examination that will most clearly allow the expert to present the supporting views. In addition, the expert may be called on to aid in the preparation of materials and questions that can be used to cross-examine the opposing expert(s). It is also not uncommon for the expert to listen to the testimony of the opposing side and to suggest how the best rebuttal can be made. Extreme care should be taken, however, lest it appear that the geologist is overzealous and is "playing the advocate."

There are several basic guidelines that the expert witness should observe when providing court testimony:

1 Don't be arrogant. It is bad practice to try to give the appearance of a "know-it-all." Neither should one be a Casper Milquetoast. Instead, one should strive to achieve an air of quiet confidence and competence without being overbearing.

2 Don't be caught unprepared. The best expert witnesses do their homework and have anticipated possible questions so that appropriate answers will provide the most cogent and effective response. An opposing shrewd lawyer is constantly probing for weaknesses and laying traps. Such ploys should have been resolved in the mind of the expert before taking the stand.

3 Don't be pedantic or preachy. Such tactics will give the appearance of condescension and obfuscation. Furthermore, they will bore the court. The testimony given should be simple, as brief as possible, but still showing necessary ramifications and relevant data.

4 Don't play the advocate. It is important to maintain objectivity, fairness, and complete honesty and integrity. Although it is natural to want to provide maximum help for the side that has engaged the expert, the expert must uphold the precepts of science. The courtroom is not the place for crusading, because it can be interpreted that too great

a concern may have biased some judgments.

5 Don't volunteer information beyond what your lawyer has mentioned. It is easy for experts to get so involved in testimony that they venture far beyond the scope that was intended. The witness may not be aware of other ramifications of the case, and this lack of conversance could upset the case and the strategy of the lawyer.

6 Do maintain your credibility. This is the paramount rule. It is not easy. One of the most difficult tests arises when the opposing lawyer demands that a question be answered "yes" or "no." Because geology can be interpretative, it is often very difficult to provide such a simple answer without indicating the qualifications that may apply and the complexities of the question. In such a case, the expert hopes his lawyer will aid in allowing a complete answer and also hopes for a sympathetic judge.

There is one very important difference in providing court testimony when compared with presentation of ideas at a professional meeting—namely, the data presented in court must be your own or that collected by your close and working associate. Ideas and data of others, as in published materials, are often not admissible in court because the authors cannot be cross-examined by the opposing lawyer. Thus nearly complete reliance must be placed on the opinions, information, and data that the expert witness has made and discovered. In the courtroom, it is both what you say, and how you say it, that are important to the case.

Geology and Legal Affairs

Geologists are being increasingly used in all aspects of legal affairs, in litigation proceedings, with general environmental laws, and with specific laws that mandate the involvement of geologists. Geologists are needed in court cases on a diverse range of subjects, such as damage to wells; calculation of quantity and quality of natural resources in condemnation cases; causes of erosion, flooding, landsliding, and other processes that may have been man-induced; prediction of what impact a certain developmental action may precipitate; and many others. Geologists are also involved in all of the environmental legislation, such as the soil conservation acts, flood control acts, mining acts, water acts, and land acts governing parks, open space, and wilderness. Then there are laws and ordinances that specifically state geologists are to be used in the data gathering and/or decision-making process. These include the following statutes and regulations:

1 The grading ordinances in Southern California—for example, the City of Los Angeles Grading Ordinance of 1952, amended in 1963 and 1969.

2 The Raymond Basin Adjudication of 1949.

3 The Pennsylvania Solid Waste Management Act of 1969.

4 New York State Department of Transportation 1972 Requirements for Geologic Source Reports.

Of course, geologists are in the direct employment by numerous government agencies whose duties are mandated by law. In performance of their work they obtain raw data, prepare guide-lines of how to use information, and provide conclusions based on their expertise of environmental matters. In action agencies such as the Corps of Engineers, Bureau of Reclamation, and Soil Conservation Service, geologists are called on to aid in the design work and route location for engineering projects. Thus the scope of geologic work in environmental legal affairs is as broad as the discipline itself. It should also be remembered that much of the environmental legislation, and the involvement of geologists in policy matters, originated from geologic events, often disasters. The California Field Act was passed as a result of the 1933 Long Beach earthquake; it required earthquake-resistant construction in school buildings. The Federal Flood Control Act of 1936 was enacted as a consequence of severe flooding disasters in 1935 and 1936. The failure of the St. Francis Dam in 1928 and the loss of more than 500 lives prompted California to require that all foundations for dams in that state receive the approval of qualified geologists.

MINING LAW

In the United States there is not uniformity in laws pertaining to the mining of mineral resources. The states and their mining legislation can be grouped into three categories.

1 The original 13 states, along with Kentucky, Maine, Tennessee, Texas, Vermont, and West Virginia. In these states U.S. laws regarding mining have no application because such laws apply only to states where the federal government

originally owned the land. Litigation that arises is decided by common law or by state statutes.

2 Arkansas, Illinois, Indiana, Iowa, Michigan, Missouri, and Wisconsin. In these states the land originally belonged to the federal government, and suspected mineral lands were ordered to be reserved from sale except those containing lead and copper which were sold. In addition, Congress by special legislation then exempted from federal laws the states of Alabama, Kansas, Michigan, Minnesota, Missouri, and Oklahoma.

3 In all other states the federal government is the proprietor of mineral lands, so that the governing laws are derived from federal legislation; however, common law, state laws, and even local customs can also provide important directions in legal affairs.

The principle of English common law, as extended from the Napoleonic Code, holds that landowners possess everything on and below their property. The exception was gold and silver, which was owned by the crown as a regalian right. Such a right was practiced by some of the states and, in 1866, Texas was the last state to renounce it.

In early mining considerations, the term "mineral" was often used to imply a metalliferous deposit. However, in 1836, in the *Gibson* v. *Tyson* case (Pennsylvania) the court ruled that the chromate of iron was a mineral, and in *Hartwell* v. *Camman* (1854) a New Jersey court ruled that a mineral need not contain metal. In *Armstrong* v. *Granite Company* (147 New York, 495) the court extended minerals to include granite, and in *Brady* v. *Smith* (181 New York, 178) the New York Court of Appeals ruled limestone could be a mineral and stated:

The adoption of arbitrary definitions in reference to mineral substances buried in the earth is not permissible. The word "mineral" standing by itself might, under a broad, general, popular definition embrace the soil and all that is to be found beneath its surface; under a strict definition it might be limited to metallic substances, and under a definition coupling it with mines, it covers all substances taken out of the bowels of the earth by the process of mining.

Mining that resulted from the California Gold Rush of 1849 led to the establishment of new mining laws. The miners established rules that (1) discovery and mining gave the right to the mineral, and (2) the discoverer had the right to continue mining a vein even when it extended under the surface claim of another. Such common laws were incorporated into federal statutes in 1866. These acts provided the method to obtain lands and how to patent a claim for lands in the public domain. They also provided the background legislation for the Mining Act of 1872, which stated:

All valuable mineral deposits in lands belonging to the United States, both surveyed and unsurveyed, are hereby declared to be free and open to exploration and purchase, and the lands in which they are found to occupation and purchase, by citizens of the United States and those who have declared their *intention to become such, under regulations prescribed by law, and according to the local customs or rules of miners in the several mining districts, so far as the same are applicable and not inconsistent with the laws of the United States.*

The states and lands that come under this law are Alaska, Arizona, Arkansas (partly), California, Colorado, Florida, Hawaii, Idaho, Louisiana, Mississippi, Montana, Nevada, New Mexico, North Dakota, Oregon, South Dakota, Utah, Washington, Wyoming, and forestlands. Exceptions include Indian lands, parks, and military reservations.

Mineral laws govern those deposits in place, not debris, wash, or float material. After discovery, a claim must be filed and recorded with the accurate location and traceable boundaries. A working shaft, usually at least 10 ft (3m) deep, must be sunk within a specified time, often 60 to 90 days. In order to continue a valid claim, yearly developmental work worth $100 must be accomplished to classify it as "assessment work." This work can either be done by laborers in digging or, under Public Law 876 of September 2, 1958, it can be charged to geological, geophysical, or geochemical surveys conducted by qualified experts.

The Federal Mineral Leasing Acts of 1917 and 1920 removed coal, petroleum, oil shale, potash, phosphate, and sodium salts from mining controls and placed them into a different set of regulations whose administration and enforcement falls under the jurisdiction of the U.S. Geological Survey and the Bureau of Mines. The Multiple Mineral Development Law of 1954 attempted to resolve the Mineral Leasing Laws and the general Mining Laws to prevent the damage of leasing minerals by other mining operations. The Multiple Use Act of 1955 stipulated that underground mining operations were subordinate to the extraction of land containing proved and marketable sand, stone, gravel, pumice, pumicite, cinders, and clay, except in unusual cases where particularly valuable products are involved. Furthermore, the United States reserved the right to remove and dispose of the timber contained on the property. The revised statutes of the United States governing mines and minerals have been codified, and the principal statutes are contained in Volume 30 of U.S. Code Annotated. These laws provide the basic provisions under which all rights to mineral deposits on public lands may be acquired.

Oil and Natural Gas

Because of the mobility of oil and natural gas, the laws that govern them are different from fixed mineral laws. These natural resources have been held by the courts to be minerals, but their fluid character prevents them from being the absolute property of the landowner until they are extracted as a possession. In a 1866 case (*Funk* v. *Holdeman*, 53 Pa. St., 229) oil was defined as a mineral, and this was reaffirmed by the Pennsylvania Supreme Court (*Gill* v. *Westan*, 110 Pa. St., 313):

It is a mineral substance obtained from the earth by a process of mining, and lands from which it is obtained may with property be called mining lands.

The same conclusion was reached by a West Virginia Court in *Williamson* v. *Jones* (19 S.E., W. Va., 436):

. . . the authorities now very

generally—universally so far as I have examined them—hold petroleum to be a mineral, and as much a part of the realty as timber, coal, or iron ore.

In spite of its qualification as a mineral, however, petroleum was not considered property, even with possession, until the classic Supreme Court decision in *Ohio* v. *Indiana* (177 U.S. 190):

*True it is that oil and gas, like other minerals, are situated beneath the surface of the earth, but except for this one point of similarity, in many other respects they greatly differ. They have no fixed **situs** under a particular portion of the earth's surface within the area where they obtain. They have the power, as it were, of self-transmission. No one owner of the surface of the earth, within the area beneath which the gas and oil move, can exercise his right to extract from the common reservoir, in which the supply is held, without, to an extent, diminishing the source of supply as to which all other owners of the surface must exercise their rights. The waste by one owner, caused by a reckless enjoyment of his right of striking the reservoir, at once, therefore operates upon the other surface owners.*

Although in virtue of his proprietorship the owner of the surface may bore wells for the purpose of extracting natural gas and oil, until these substances are actually reduced by him to possession, he has no title whatever to them as owner. That is, he has the exclusive right on

his own land to seek to acquire them, but they do not become his property until the effort has resulted in dominion and control by actual possession.

WATER LAW

Modern water law is a collage of statutes and decisions from many different countries. Important contributions have come from Roman law, Moslem law, Spanish law, the Code of Napoleon, and English common law. The courts have invariably ruled that water, like the soil, is not mined in the same sense as a mineral deposit. However, in certain local situations, as in the "mining of groundwater," the courts have agreed to a similarity. There are some differences in laws that govern use of surface water when compared with those for groundwater use. Water law is still being refined by the courts, and many cases are decided on a case by case basis.

Surface Water Law

The two principal theories governing watercourses are the riparian doctrine and the prior appropriation doctrine. In the western United States, pueblo rights and the principle of prescriptive rights are used under specific circumstances.

Under the riparian doctrine, the right to use the water of a stream accompanies ownership of land adjacent to the stream. The right to use water exists even if the water is not physically used. However, the owner can only use water on riparian land and cannot sell or export water for use elsewhere. Each riparian owner has equal right to use the water and cannot deprive another of equal opportunity for its use. The owners are subject to two allocation theories—the natural flow theory and the reasonable use theory.

Under the **natural flow theory,** each owner is given the responsibility of maintaining the flow of the water in as much its natural state as possible, without appreciable diminution of its quantity or impairment of its quality. Obviously, any use of the water causes changes, and so the question that courts must resolve is the amount and degree of change that will be tolerated. Many of the more recent pollution and environmental laws governing waters are based on the natural flow theory. Two early Massachusetts cases helped establish the natural flow theory. In *Colburn* v. *Richards* (1816) a downstream mill owner was deprived of water because an upstream farmer had erected a dam and used the water for irrigation. The court ruled that a valuable business was damaged because the upstream owner consumed too much water without allowing return flow to the stream. *Cook* v. *Hull* (1820) was a similar case where an upstream owner diverted water for irrigation so that during certain periods the downstream mill owner could not operate his machinery. The judgment of the court enjoined the upstream owner from use of water for irrigation.

The **reasonable use theory,** although well established in court decisions, is subject to a wider latitude of interpretations than other principles involved with surface water use. The *Merritt* v. *Brinkerhoff* (New York) case documented this principle in 1820. Here a downstream mill owner had to shut down his operation at various times because an upstream owner used water to heat iron. When the water was released, it overflowed the mill dam in such large torrents that much was wasted and not usable. The court ruled:

. . . had the defendants allowed the plaintiffs that reasonable participation in the use of the water, which by law they had a right to require, the damage sustained would have been materially less, if not inconsiderable.

In 1854 a Massachusetts court defined what reasonable use entailed in *Thurber* v. *Martin*:

In determining what is such reasonable use, a just regard must be had to the force and magnitude of the current, its height and velocity, the state of improvement in the country in regard to mills and machinery, and the use of water as a propelling power . . . and all other circumstances bearing upon the fitness and propriety in the use of the water in the particular case.

Thus riparian owners have a fundamental right to use adjacent waters, and there is to be no advantage to ownership by virtue of being upstream or downstream from a watercourse. What use is made of the waters must be both reasonable and beneficial. Even the relative economic merits and investments of property owners should not be a consideration in determining rights, as stated in *Strobel* v. *Kerr Salt Co.* (New York, 1900):

While the courts will not overlook the needs of important manufacturing interest, nor hamper them for trifling causes, they will not permit substantial injury to neighboring property, with a small but long-established business, for the purpose of enabling a new and great industry to flourish. They will not change the law relating to ownership and use of property in order to accommodate business enterprise.

The riparian system of water law is especially adaptable in the eastern United States where water is abundant, but it proved a failure in the more arid west where the urgent need for water was in mining and irrigation. Thus in the arid parts of the west a different system of water rights evolved, called the **prior appropriation theory,** or doctrine. This doctrine makes land ownership irrelevant to water rights. Instead, such rights are acquired by first taking water and using it beneficially. Unlike riparian doctrine, a water right can be lost when use of the water ceases. Furthermore, there are no restrictions on where the water can be used. Thus, with persons competing for water from the same source, rights are not determined on the "reasonableness" basis, but on which party had prior use of the water. Priority in time is the principal determinant of rights when the supply cannot satisfy all who wish to become users. Whereas riparian rights coexist with private ownership, under prior appropriation doctrine water rights are initially public and become private after use. This doctrine was not part of English common law. Instead, it evolved locally from diverse backgrounds, such as the Spanish settlements in the southwest, gold miners in California, Mormons in Utah, and homesteaders throughout the West.

When the West started to be settled, the best lands were generally in valleys and basins where rainfall was insufficient for land use activities. Thus water had to be imported from the distant mountain sources. This usage was condoned and placed into legal rights by the California Supreme Court in *Tartar* v. *Spring Creek Water and Mining Company* (1855):

> . . . *a prior appropriation of either [wood or water] to steady individual purpose establishes a quasi-private proprietorship, which entitles the holder to be protected in [his] quiet enjoyment against all the world but the true owner.*

The "true owner" was interpreted to be the first user of the source. Congressional legislation in 1866 and 1870 approved the appropriation doctrine that was developed by water users of public lands. The Desert Land Act of 1877 reaffirmed the rights of states to adopt for all nonnavigable waters of the public domain their own governing laws.

Colorado was the first state to repudiate the riparian doctrine in 1882. Other states that followed suit were Nevada in 1885, Arizona in 1887, Idaho in 1888, New Mexico and Utah in 1891, Wyoming in 1896, and Montana in 1921. The two doctrines—riparian rights and prior appropriation—exist concurrently in a number of states that have vastly differing climates, such as California, Oregon, and Washington. The Great Plains states of North Dakota, South Dakota, Nebraska, Kansas, Oklahoma, and Texas also have joint doctrines, as do Mississippi and Florida.

Groundwater Law

Legal affairs dealing with surface waters are in less dispute than those relating to groundwater. Because the amount and movement of groundwater are hidden from view, there is more mystery and ambiguity concerning its ownership. An 1861 Ohio decision handed down by the court in *Frazier* v. *Brown* typifies the point:

> . . . *the law recognizes no correlative rights in respect to underground waters percolating, oozing, or filtrating through the earth: and this mainly from considerations of public policy. 1. Because the existence, origin, movement and course of such waters, and the causes which govern and direct their movement, are so secret, occult and concealed that any attempt to administer any set of legal rules in respect to them would be, therefore, practically impossible.*

The French civil code, through the Napoleonic Code, became part of English common law as enunciated in the courts, as in *Acton* v. *Blundell* in 1843. This case held that owners had rights to the land and minerals above and below the ground, including the water. This was construed to mean only percolating waters and not water in "underground streams," which was governed by riparian doctrine. An 1899 court decision in Illinois (*Edwards* v. *Halger*) is representative of such legal rulings:

> *Water which is the result of natural and ordinary percolation through the soil is part of the land itself, and belongs absolutely to the owner of the land, and, in the absence of any grant, he may intercept or impede such underground percolations, though the result be to interfere with the source of supply of springs or wells on adjoining premises.*

However, groundwater usage, as in surface water, was judged in terms of whether it was reasonable and beneficial.

In arid parts of the West, the prior appropriation doctrine has also been used in conjunction with applicable parts of the Desert Land Act of 1877. This legislation gave the states the power to use water on public lands according to their respective laws and courts. The New Mexico statutes of 1927, 1931, and 1953 illustrate such ownership rights:

> *All underground waters of the State of New Mexico are hereby declared to be public waters and to belong to the public of the State of New Mexico and to be subject to appropriation for beneficial use within the State of New Mexico.*

Limitations in groundwater use In the twentieth century it is becoming increasingly apparent that groundwater withdrawals are placing in jeopardy many environmental concerns of the people, ranging from declining water tables and loss of water to salinization of freshwater aquifers. Several measures have been taken to meet this threat. The decline of the water table in the heavily populated and industrialized western part of Long Island, New York, was being caused by increasing rates of groundwater withdrawal. To alleviate this problem, the State Legislature passed a conservation law in 1933 that required approval of the State Water Power and Control Commission before a well with a capacity of 100,000 gal/day (380 m³) could be constructed. To prevent continued overdevelopment, the commission required that water pumped from new wells for cooling and similar purposes be pumped back into the ground by means of a recharge well.

A different legal approach to water management was taken in the case of *Pasadena* v. *Alhambra,* also known as the Raymond Basin Adjudication. Heavy overdrafts in the Raymond Basin of California threatened groundwater supplies for all users and would stifle additional development in the region. The City of Pasadena filed a complaint against the City of Alhambra and 24 other defendants. The Division of Water Resources of the State was assigned the job of determining the groundwater facts. The investigation showed that pumpage exceeded the safe annual yield, which led to a judgment in 1944 and a confirmation in 1949. The court decision was to formulate a new doctrine of mutual prescription whereby the original water users were given correlative rights. The allowed amount of pumping was thereby allocated to the various users in accordance with a water usage formula calculated by the Watermaster as designated by the court.

LAND USE LAW

The previous two sections, on mining law and water law, discussed the legal aspects of resources when they are withdrawn from the earth. This section deals with laws that affect the management of the land. These mandated policies govern land utilization by society and the implementation of policies that ensure health, safety, welfare, and sustenance for our survival. There is a great variety of laws and legal actions that affect geologists in this area of land use and environmental management. Such laws—whether federal, state, or local—are necessary to delineate policy, set guidelines, provide funds and machinery for implementation, and establish agencies for continuing planning and administration.

Water and Soils

In the early years of the United States, flood control was deemed to be only a local problem, but the Swamp Land Acts of 1849 and 1850 granted unsold swamp and overflow lands to Louisiana, Arkansas, and other states containing similar lands. These acts made such lands subject to disposal by state legislatures and required the resulting monies to be applied to drainage, reclamation, and flood control projects. In 1874, Congress provided for the appointment of a commission of engineers to investigate and devise a permanent plan for the reclamation of that part of the Mississippi River basin that was subject to inundation. The resulting report in 1875 discussed methods of flood control, including cutoffs, diversion of tributaries, reservoirs, outlets, and levees. This led to the establishment of the Mississippi River Commission in 1879 (the same year the U.S. Geological Survey was founded) with the duty to consider plans that would improve navigation and "prevent destructive floods." A Flood Control Act in 1923 authorized expenditure of $60 million for flood control work on the lower Mississippi, but it took the disastrous floods of 1935 and 1936 for Congress to enact the much more inclusive Flood Control Act of 1936. Although the first concepts for retardation of runoff and curbing of erosion were formulated as early as 1911 in the Weeks Forest Purchase Act, such legislation had not been implemented. Thus the 1936 federal law was the first nationwide recognition of the need to use soil conservation measures on watersheds to prevent downstream floods. This statute when coupled with the Soil Conservation Act of 1935 put the federal government in the business of being large-scale land managers. The Department of Agriculture (through the Soil Conservation Service) was to assume the responsibility for upstream flood control measures, and the U.S. Army Corps of Engineers was authorized to undertake large projects on navigable rivers throughout the country. In 1954, the role of the Soil Conservation Service was expanded by passage in 1954 of the Watershed Protection and Flood Prevention Act. It provided a new approach to soil and water resource development, in which projects were locally undertaken with federal help. A size limitation was placed on construction costs to $250,000, and the volume of water impounded behind dams could not exceed 4000 acre-feet of total capacity. Within a 10-year period, 100 projects had been completed, and those completed or authorized for construction involved more than 30 million acres (12 million ha).

The Desert Land Act of 1877 authorized sale of 640 acres at $1.25 per acre to any person who would irrigate within three years, and thus began governmental involvement to help settle and irrigate the West. In 1890, Congress limited acreage to 320 for a single claimant, and all programs were limited to providing certain water uses on public lands for specified purposes. However, with passage of the 1902 Reclamation Act, Congress established irrigation in the West as a national policy. Several lawsuits were brought against the government challenging such authority, but the 1902 act was sustained as a proper exercise of governmental proprietor power, and the use of such funds was deemed beneficial for the common welfare. The Supreme Court has also affirmed the principle of governmental promotion for the general welfare through "large-scale projects of reclamation, irrigation, or other internal improvement" (*United States* v. *Gerlach Livestock Co.*, 339 U.S., 723, 738, 1950). During the early years of the Reclamation Service (which became the Bureau of Reclamation in the Department of Interior in 1923), the principal focus was on development of irrigation projects because the objective was to change drylands into productive farmlands by use of imported water. Today only about one-third of the bureau's projects are directed toward irrigation; the remainder have purposes to establish benefits in navigation, flood control, fish and wildlife enhancement, power production, recreation, and water supplies for industries and municipalities. A further inducement for development occurred in 1916 with passage of the Smith Act which authorized the inclusion of public lands into irrigation districts, subject to specified conditions. The usual irrigation district included only those lands that received irrigation benefits, and so assessment and charges were made on the irrigable lands. This resulted in a flourishing of urban settlements that prospered because of reflected benefits for which they bore none of the cost burden. This has caused some states to enact provisions whereby adjacent areas that are benefited are included within the improvement district.

A different, and unique approach, to land use, water control, and soil conservation was undertaken in 1933 with enactment of the Tennessee Valley Authority (TVA) Act. This marked an unusual legislative mandate for federal involvement in the development of a single geographic area. The management of the region by TVA was in terms of multiple uses, including navigation, flood control, reforestation, power generation, and provisions to aid in proper utilization of marginal lands, as well as growth of agricultural and industrial development.

Rangelands and Forestlands

The first federal environmental legislation governing forested lands occurred in 1891 with the Forest Reserve Act. It was passed by Congress as the result of heated public debates about the desirability of withdrawing forested public lands from development for the purpose of restraining floods and for the protection of water supplies and streamflow. The law enabled the President to "set apart and reserve . . . in any part of the public lands wholly or in part covered with timber or undergrowth, whether of commercial value or not, as public reservations." However, the Forest Management Act of 1897 made it clear that such areas were not to be wilderness, since one purpose was to ensure a continuous timber supply, as well as "for the purpose of securing favorable conditions of water flow." These acts were generally relevant only in the West, where most of the public lands were located. The Weeks Act of 1911 laid the groundwork for federal programs throughout the United States. This act recommended the federal acquisition of lands in watersheds of navigable rivers for management in much the same manner as lands in national forests. Priority was to be given to forested as well as denuded and cut-over lands within the watersheds of navigable streams for their regulation and for the production of timber.

The Clarke-McNary Act of 1924 authorized the Secretary of Agriculture to recommend forest fire prevention and suppression systems "with a view to the protection of forest and water resources." The purpose was to assure a continuous supply of timber, protect navigable streams, and secure water supply for domestic and irrigation use. The McSweeney-McNary Act of 1928 stated that the management of forest resources should include the purpose "of maintaining favorable conditions of water flow and the prevention of erosion."

Starting in 1944, a much broader-based policy was inaugurated for forested lands with passage of the Sustained Yield Forest Management Act. It authorized the Department of Agriculture and the Department of Interior to cooperate with programs under their respective jurisdictions, to secure the benefits of forests "in the maintenance of water supply, regulation of streamflow, prevention of soil erosion, amelioration of climate, and preservation of wildlife." Thus the federal government went on record, as with TVA, for multipurpose planning strategies. This culminated with the Multiple Use-Sustained Yield Act of 1960, which specified two economic goals and three services that the U.S. Forest Service was to accomplish in its management of forests. The act stated that the National Forests must be maintained in perpetuity for the purpose of assuring: (1) timber from trees, (2) forage from grazing lands, (3) recreation, (4) wildlife management, and (5) protection of rivers and consideration for their constant flow. Whereas grazing was an important component and benefit to be derived from these lands, other lands had already received governmental attention for such purposes through the Taylor Grazing Act of 1934. This had been the first federal legislation that directed attention to a public policy for the establishment of grazing districts and the raising of forage crops. Under this act the Secretary of Interior was empowered "to continue the study of erosion and flood control and to perform such work as may be necessary . . . to protect and rehabilitate the areas." The management of the grazing districts sought to establish a balance in grazing use, including game animals and conservation programs. Thus it crystallized into policy various disjointed types of land classification for grazing and conservation purposes.

Parks and Wilderness

Public policy for the establishment of parks in habitated areas dates back several millennia. More modern parks in urban areas were designed as part of city renovations—for example, in Paris and Central Park in New York City, which was established in 1857. The establishment of parklike areas stems from the idea of public trust in Roman law and from the idea of "the commons," derived from English tradition. The first U.S. government legislation for the development of a specific park was passage of the Yellowstone Act by Congress in 1872, which made Yellowstone National Park the first of its kind in the world.

Although several other national parks were established prior to 1916, it wasn't until that date that the program for national parks was organized into the newly formed National Park Service. Thus governance for the preservation of recreational nonurban lands was placed under the Department of Interior's National Park Service and the Department of Agriculture's Forest Service.

The idea of the outdoors as a resource has prompted several other governmental agencies to become involved and has seen the enactment of much additional legislation. For example, the Outdoor Recreation Act of 1963 declared:

Congress finds . . . it to be desirable that all American people of present and future generations be assured adequate outdoor recreation resources.

This was the first act that treated recreation as a human resource

Figure 21-2 Failure in this cribbing accelerated erosion and gullying and resulted in a court injunction that stopped construction and maintenance on this part of Interstate 88. The judge decided that a proper Environmental Impact Statement had not been presented. (For additional information see Coates, 1976, Urban Geomorphology.)

that needs planning, classification, and evaluation just as any other resource. The Land and Water Conservation Fund Act of 1965 authorized the Secretary of Interior to grant matching funds to the states for the planning, acquisition, and development of outdoor recreation lands and facilities.

LEGAL RIGHTS FOR NATURAL FEATURES

The previous section emphasized the value gained by society in conserving and preserving land and its resources for our use and benefit. However, it has become increasingly apparent that to have total environmental balance the natural features and objects must also be given certain rights. Since ecosystems cannot speak for themselves in a legislature or a court, advocates for nature must speak as surrogates, such as the Sierra Club, Nature Conservancy, and Friends of the Earth. This viewpoint holds that some natural features and objects must have their integrity preserved, and that despoilation is not only harmful to the environmental balance, but is also ruinous to humans in the long run. Thus it becomes an important consideration to provide some type of "standing" for environmental systems in the courts and in law-making procedural bodies. To pursue this objective, the NEPA of 1970 was passed, and it may become one of the farthest reaching and most encompassing legislative mandates in history (Fig. 21-2). The states, and the federal government, have been responding through a number of statutes in an attempt to preserve the integrity of natural features (see page 593).

Another part of the landscape singled out for special attention in legislation, in addition to lands and rivers, has been coastlines. The Delaware Coastal Zone Act of 1971, the first of its kind, barred heavy industry along a 2 mile zone of the state's 185 km coastline. The stated purpose was to "protect the natural environment of its bay and coastal areas." In a 1972 referendum California voted for the protection of the 1725 km shoreline from uncontrolled commercial development. Many coastal states also have special wetland laws for the preservation of these fragile environments. For example, Connecticut law imposes a $1000 a day fine for violators who dredge or fill wetlands without a permit and further requires payment of costs to restore lands to their former natural state. Local governments such as the Town of Islip on Long Island, New York, have resorted to other tactics to reduce degradation of the coastal environment. A Vehicle Ordinance passed in 1964 severely limits the number of vehicles that are permitted registration stickers to drive on Fire Island beach (see also page 497).

CASE HISTORIES

During the past 20 years, a number of circumstances have led to the large increase in environmental-type lawsuits. Contributing factors for this increase include: (1) a new environmental awareness on the part of the general public, (2) passage of new laws, and reinterpretation of old statutes, that facilitates "standing," the ability to bring suit against an environmental polluter or degrader, (3) the rise of new advocate organizations with increased funding to initiate litigation, (4) more lawyers and advances made in environmental law, and (5) the willingness of more scientists to become involved in legal matters. As we will find in the following cases, the participation of geologists and environmental scientists in litigation proceedings are called for in an extremely broad spectrum of topics, covering the full breadth of ecosystems. First, we will look at some classic lawsuits

and, then, we will examine examples of the interplay of science, environment, and law in representative cases in several different subject areas. This section differs from the rest of the chapter because we are now putting emphasis on the involvement of science in rather traditional legal affairs, instead of enumerating those cases that set court precedents and established new law.

Hetch Hetchy Case

This was one of the earliest and most notable court battles between wilderness adherents and utilization interests. The two chief antagonists were John Muir and Gifford Pinchot. The Hetch Hetchy Valley is a spectacular high-walled segment of the Tuolumne River on the western slopes of the Sierra Nevada Mountains in California. When Yosemite National Park was created in 1890, the Hetch Hetchy part and its surrounding area were designated as a wilderness preserve. In the 1900s, San Francisco started action that requested use of the valley as a reservoir site. Although the request was first denied by the Secretary of Interior, it was approved on May 11, 1908, apparently as an aftermath of the 1906 earthquake, when fire had devastated the city. Thereafter, a five-year struggle ensued that encompassed the administrations of Theodore Roosevelt and Woodrow Wilson. Muir, voicing a position in favor of the preservation of wilderness areas, stated:

. . . for everybody needs beauty as well as bread, places to play in and pray in where Nature may heal and cheer and give strength to body and soul alike.

These temple destroyers, devotees of ravaging commercialism, seem to have a perfect contempt of

Nature, and instead of lifting their eyes to the God of the Mountains, lift them to the Almighty Dollar [Nash, 1967].

Pinchot, on the opposite side, presented the utilitarian viewpoint:

. . . I am fully pursuaded that . . . the injury . . . by substituting a lake for the present swampy floor of the valley . . . is altogether unimportant compared with the benefits to be derived from its use as a reservoir.

. . . the fundamental principle of the whole conservation policy is that of use, to take every part of the land and its resources and put it to that use in which it will serve the most people [Nash, 1967].

It is interesting to note that Roosevelt, who usually supported Pinchot and originally had favored the reservoir, changed his view toward such areas and stated in his eighth annual message on December 8, 1908, that Yellowstone and Yosemite "should be kept as a great national playground. In both, all wild things should be protected and the scenery kept wholly unmarred."

Hundreds of newspapers throughout the country carried articles and editorials about the Hetch Hetchy controversy in support of the preservation view. Groups such as the Sierra Club and the Appalachian Mountain Club kept the issue in the public eye. However, in order to prosecute the case, the wilderness enthusiasts were obligated by law to combine their cases under a single organization, which was called the California Branch of the Society for the Preservation of National

Parks. Although the preservation arguments were that the dam and reservoir would desecrate the landscape, proponents of the project shrewdly pointed out that, instead of spoiling the beauty, the artificial lake would enhance the park and allow for walks and roads in the region, so that it could be used for public recreation in much the same way as those areas already developed in European mountain-lake resorts. In 1913, Congress approved the Hetch Hetchy Bill, and in signing it on December 19, 1913, President Wilson announced that "the bill was opposed by so many public-spirited men ... that I have naturally sought to scrutinize it very closely. I take the liberty of thinking that their fears and objections were not well founded" (U.S. Congressional Record, 1913).

Florida Jetport Case

The controversy surrounding the proposed jetport in south Florida is typical of court battles that are being waged on environmental issues in the last quarter of the twentieth century. In 1947, the Everglades National Park was established in which 2035 mi² (5270 km²) of the southern part of Forida was to be preserved in a near-wilderness type of condition. The Dade County Port Authority in 1968 purchased 39 mi² (100/km²) of land in the Big Cypress Swamp for an international jetport whose location would be only 11 km north of the park. Conservationists were convinced that construction of such a large facility would produce ecological disasters throughout the region. Groups such as the National Audubon Society provided adversary testimony at the public hearings held as information forums about the proposed project. This testimony resulted in dramatic and favorable coverage for the environmen-

tal position in the news media. As a consequence, investigations and reports were made by three different groups: the Department of Interior, the National Academy of Sciences, and a private group headed by former Interior Secretary Udall. All reports expressed alarm over the possible deleterious impact of the jetport on the National Park.

Development of the proposed jetport and its attendant facilities will lead to land drainage and development for agriculture, industry, housing, transportation, and services in the Big Cypress swamp that will destroy inexorably the south Florida ecosystem and thus the Everglades National Park. . . . [This] will take place through the medium of water control, through land drainage and changed rates of discharge. It will come about through decrease in quality of water both by eutrophication and by the introduction of pollutants, such as pesticides [Leopold, 1969].

The south Florida problem is merely one example of an issue which sooner or later must be faced by the Nation as a whole. How are the diffused general costs to society to be balanced against the local, more direct and usually monetary, benefits to a small portion of the society? Concurrently, the society must ask itself whether the primary measure of progress will indefinitely be the degree of expansion of development, such as housing, trade, and urbanization, even at the expense of a varied and, at least in part, a natural landscape.

. . . the training airport is intolerable . . . because the collateral effects of its use will lead inexorably to urbanization and drainage which would destroy the ecosystem. . . . Elimination of the training airport will inhibit land speculation and allow time for formulation of public awareness of environmental degradation which is the prerequisite for effective and practical action in the field of planning and land-use control [U. S. Department of Interior, 1969].

Such actions and reports were instrumental in restraining the Port Authority from continuing the development at this particular site.

Property Damage Cases

1 *Demoski* v. *State of New York* (1957) A series of flood control measures was undertaken by the Corps of Engineers starting in 1938 for the Binghamton, New York, region which had been devastated by floods in 1935 and 1936. As part of this continuing program, in 1955 the Corps initiated a channel improvement program in the upstream part of the Susquehanna River. In the spring of 1956, three floods hit the area and destroyed property and a retaining wall on the lands of Paul Demoski (Fig. 21-3). His home site was on the outside bend of a meander, a naturally vulnerable position. However, because his property had not previously been damaged from flood waters during the 11-year life of the retaining wall, he brought suit against the Corps and the State of New York claiming the construction had caused the damage. The court testimony was in disagreement about whether it was natural

Figure 21-3 View of the Demoski property near Binghamton, New York. Floodwaters destroyed retaining wall, removed embankment and soils, and partly undermined house. The government had to pay damages according to the court ruling which found that proper precautions were not taken to assure property integrity during channelization of the Susquehanna River.

erosion or man-induced accelerated erosion that produced the damage. In my testimony for the plaintiff, I pointed out that there had been 23 floods in the area during the prior 11-year period of the retaining wall's existence, and that the wall had remained intact and undamaged. By straightening the channel, removing all trees and brush along the channel and on islands, and bulldozing the banks, the construction had increased the velocity of the river. The court held that the government was responsible for the damages and stated:

We are satisfied that the flood control plan of the project was defective insofar as the premises of these claimants are concerned in that it exposed said premises

and the aforementioned retaining wall to excessive abnormal and erosive conditions reasonably foreseeable unaccompanied by precautions to guard against such conditions [*New York State Supreme Court, 1957*].

2 *Robert H. Barnes* v. *State of New York* (1966) This case involved the principles of a class action suit. For the construction of Interstate-81, north of Binghamton, the Chenango River was diverted 150 m east of its normal position. There were 15 property owners with houses along the former riverbank who claimed their wells went dry, or nearly dry, because of the river relocation. It was decided that all cases would be bound to the single case, or

class action suit, initiated by Barnes. The entire lawsuit rested on hydrogeologic considerations—namely, was the river-groundwater system of the influent type, as contended by the claimants, or of the effluent type, as I argued for the New York Attorney General. I based my testimony on several types of data. Chemical analysis of the river and of well water showed huge differences. They should have been similar if the water in wells was lost immediately after river diversion, as claimed by the defendants. A series of four test holes were drilled at right angles to the river, and water levels in the holes established that the water table sloped toward the river, and not away from the river as contended by the plaintiffs in their proposed influent system. I also gave testimony that the area had been in drought conditions for four years, and that other wells throughout the region had also had declines in water level (Coates, 1976). The following decision was given by the New York State Court of Claims (1966):

. . . there was no causal connection between the diversion of the river and the loss of the wells.
. . . the proof established that the area had been suffering from a severe drought. . . . The Court agrees that the loss of water to these claimants has caused great hardship to them and that it was a coincidence that the loss of water and the construction were momentarily the same, but we regard the evidence as insufficient to carry the burden of proof that rested upon the claimants to sustain the claim. The motion made by the Attorney General to dismiss the claims . . . is now granted. All

other motions upon which decision was reserved are now denied.

3 *Agins* v. *City of Tiburon, California* (No. 79-602, 1980) In this recent case the Supreme Court ruled unanimously that a zoning ordinance that limits development in the name of conservation does not necessarily violate the constitutional rights of the affected property owners. Bonnie and Donald Agins had purchased in 1968 five acres of land overlooking San Francisco Bay. In 1973, the city enacted an open-space zoning ordinance that restricted development of the claimants' property. Agins argued that the law constituted a "taking" of their lands without appropriate "just compensation," and that they were entitled to $2 million damages because of inverse condemnation. However, the decision of the court continued to back zoning curbs for environmental protection, even though broad social policies may occasionally infringe on private property rights.

Condemnation Cases

1 *Archie Campbell* v. *City of New York* (1955) In an expansion of their water supply network, the City of New York was granted eminent domain status to condemn property in the Catskill Mountains for reservoir sites. Campbell owned more than 20 acres of property on lands and hillslopes adjacent to the East Branch Delaware River near Downesville, New York. In the appropriation of his property for construction of the Pepacton Reservoir (Fig. 21-4), Campbell sued the City of New York for what he thought was a deflated evaluation of his lands. In the first trial, the court awarded him $16,000, which was $2500 more than the City had

Figure 21-4 View of Downesville Dam and the riprap lining of the embankment facing the Pepacton Reservoir. This rock was removed from the rock cut beyond the control house in the background. The court ruled the property owner must receive "just compensation" from the City of New York because the stone was used as an economic product in construction of the dam. Delaware River valley, New York.

offered. He appealed the decision, feeling the award was too low and that insufficient consideration was given to rock resources on his land. On the retrial I provided testimony concerning the rock materials. I was able to show that the City had prior knowledge through an extensive drilling program of the quality and quantity of the massive sandstone that underlay the property. By siting the dam at the Campbell property, the rock that was excavated for the overflow spillway of the dam was used as riprap to line the proximal part of the dam. These rocks found at the site saved hundreds of thousands of dollars that would have been spent in importing the rocks from distant sources. It was also demonstrated that other rocks in the vicinity were more shaly and thin bedded and thus could not be used as riprap. Even though Campbell had not used the rocks on his property for riprap, the fact that the City had had prior intent to use the rocks, and had never informed him of such usage, in the eyes of the court

entitled Campbell to a larger settlement. He was accordingly awarded $62,000 as a result of the retrial.

2 *Bojo and Vetrone* v. *State of New York* (1969) The claimants' property was located adjacent to the Delaware River near Hancock, New York, in the Catskill Mountains (Fig. 21-5). The total property consisted of about 69 acres, of which the State of New York appropriated 12 acres for the relocation of the four-lane Route 17. Although this was a complex claim, the major issue was the price per acre of the land as a reflection of the quality and quantity of the sand and gravel resources. In the first trial, the judge awarded an average value of $1400 per acre ($560 per ha) to the claimants for loss of sand and gravel land, which totaled $35,482.40. The case was appealed by the State, and in the retrial I provided testimony for the attorney general. In the new trial, the plaintiffs contended that the lost or impacted sand and gravel lands were worth $86,686, or an average of $3325 per acre ($1300 per ha). This determina-

Figure 21-5 View of Hawk Mountain roadcut near Hancock, N.Y. East Branch Delaware River in background and four lanes of Southern Tier Expressway below, and old Route 17 above. This terrain provided several condemnation lawsuits brought by property owners against the State of New York because of alleged low payments for appropriated lands. The Bojo-Vetrone property is in far background adjacent to the river, and the Sleepy Hollow Estates property is in foreground. For further information see Coates, D.R., 1971.

tion was made on the basis of what they claimed was a comparable property in Buffalo, New York, 230 miles (370 km) away! The plaintiffs' geologists also contended that the deposits averaged 60 ft (18 m) in thickness. In my testimony for the case I reported the following:

1 The sand and gravel occur in small and isolated kames with mixtures of materials that range from clay to boulders and till.

2 There are more than 310 million yd^3 of sand and gravel of similar or superior quality in a 10 mi (16 km) radius of the property.

3 The materials are of poor quality and were rejected by the State for use in hard-surface roads because of test results that showed excessive loss in the magnesium sulphate soundness test.

4 The average sand and gravel thickness is 22 ft (7 m) as shown by a series of seven bor-

ing logs drilled specifically to determine the character of subsurface materials.

The New York State Court of Claims reconsidered decision was given April 10, 1969. The claimants were awarded an average of $400 per acre for sand and gravel lands that had been affected by the construction, with a total damage figure of $9632—only about one-fourth the amount that would have been provided if the original decision had been confirmed (Coates, 1971).

Pollution Cases and Aesthetic Concerns

One of the most difficult, and perhaps subjective, aspects of environmental law concerns the status of aesthetics in damage cases. How to measure and quantify possible losses are only now starting to reach an acceptable "status of the art." Searles (1969) stated:

. . . the law of aesthetics is becoming more important today

than it has been in the past. The courts have come to the realization that the aesthetic factors previously considered outside the context of public use are equally important as other factors not only in connection with value but also with the public health and welfare.

In *Crance* v. *State of New York* (1955) the Court of Appeals awarded the claimant $10,000 because a road caused an impaired view of Seneca Lake along with a loss of easy access to the lake. The wording of the decision by the Appellate Division in New York, in *Keinz* v. *State of New York* (1957), helped establish a precedent for consideration of aesthetic values in making environmental policy:

Two properties might be physically identical yet their market value could be different because of their surroundings. . . . The "view" might be a mountainside or a valley as well as a lake. In either event, the view augments the value of the premises and if a portion thereof is taken and the view spoiled the market value of the premises remaining is reduced. It may be a matter of judgment but it is also a matter of dollars and cents and the constitutional policy requires that such reduction in value not be borne by the owner whose property is taken for a public purpose without his consent.

In a slightly different approach, the erection of billboards near a major highway was prevented by the court in *New York State Thruway Authority* v. *Ahley Motor Court* (1962):

. . . billboards can be as destructive of the beauties of the countryside as a plague of locusts. "Beauty may not be queen but she is not an outcast beyond the pale of protection or respect. She may at least shelter herself under the wing of safety, morality or decency."

Perhaps the classic case concerning change to a locality that might cause aesthetic pollution was *Scenic Hudson Preservation Conference* v. *Federal Power Commission* (1966). At a picturesque site of the Hudson River, the Consolidated Edison Company proposed a pumped storage reservoir power project. The issue was fought by environmentalists for seven years in the courts. The resulting furor aided in preventing construction of the project. In *Dennison* v. *State of New York* (1969) the decision by the Court of Claims was upheld by the Court of Appeals, which stated:

As a result of the appropriation, a new highway has been constructed. . . . In place of the beautiful view of forest and mountain [Lake George, N.Y.], which claimants could see from their westerly windows and living areas . . . [there] has been substituted the new highway supported by an embankment approximately 27 feet above grade . . . . All of the sylvan beauty afforded by the forest pre-existing the highway and the privacy and quiet it provided are gone. . . . In awarding damages . . . the Court of Claims took into consideration "the loss of privacy and seclusion, the loss of view, the traffic noise, lights and odors, all as factors causing consequential damage to the remaining property."

Thus, aesthetic considerations provide another important weapon for environmentalists and property owners in their search to maintain the status quo of natural objects and lands.

Perspectives

There are many facets to environmental law because not only are the laws significant, but their interpretation by the courts and the police powers of the government are also involved. Furthermore, during the past two decades the self-perceived mandates of governmental agencies have increased in scope and importance. To be effective, environmental laws and decisions must consider a wide latitude of objectives and interests which often include conflicting views of: (1) economics with loss or gain of revenue, (2) individual liberty, (3) societal and political goals and imperatives, (4) health, safety, and humanistic and aesthetic needs, and (5) science and engineering data.

The very nature of law is to provide a set of rules for governing. However, the legal system operates with a series of checks and balances, because built into it is the advocate role as well as the adversary position. Such views are often antagonistic. In the final analysis, the environmental posture established by legislation and the courts is one of balancing extreme elements, thereby leading to different types of compromise solutions.

The role of geology in legal affairs is becoming enlarged, and there are many opportunities for involvement by those scientists with an interest in the decision and policy arena. The allegiance of geologists to certain views as professionals may also come into conflict with their opinions as individuals. With the expansion of the environmental law field, it will be necessary for more and more geologists to become engaged with pursuits that will enhance the place of science in law and, we hope, also help science play a part in arriving at fairer and more equitable legislation and court rulings. The National Environmental Policy Act (NEPA) has provided an unparalleled forum in which the scientist can move and operate. Reserach and data can be the very heartstone of the environmental impact statements. One other value of NEPA is that it forces the construction agencies to present alternatives and to make all data available to the general public before a decision on an action is reached.

There are many problems that arise from legal aspects associated with environmental conflicts, such as the status, strength, and strategy of adversary groups. Some adversary groups have become so entrenched that they become involved in undeserving cases and needlessly protract hearings and litigation. In other situations, a more deserving case at times lacks sufficient adversary expertise and becomes overwhelmed by the series of experts that governmental or industrial sources are able to muster. Obviously, balance is needed, and the pendulum should not swing off its base.

READINGS

Anderson, F. R., 1973, NEPA in the courts: a legal analysis of the National Environmental Policy Act: Washington, D.C., Resources for the Future, 324 p.

Brew, D. A., 1974, Environmental impact analysis: the example of the proposed Trans-Alaska Pipeline: U.S. Geol. Survey Circ. 695, 16 p.

Burchell, R. W., and Listokin, D., 1975, The environmental impact handbook: Center for Urban Policy Reserach, Rutgers, The State University, 234 p.

Clements, W. C., et al., eds., 1971, The common law of water: Ann Arbor, Mich., Great Lakes Foundation, 36 p.

Coates, D. R., 1971, Legal and environmental case studies in applied geomorphology: in Coates, D. R., Environmental geomorphology, Publ. in Geomorphology, State Univ. of New York at Binghamton, p. 223–242.

———, 1976, Geomorphology in legal affairs of the Binghamton, New York metropolitan area: in Coates, D. R., ed., Urban geomorphology, Geol. Soc. Amer. Spec. Paper 174, p. 111–148.

Grad, F. P., Rathjens, G. W., and Rosenthal, A. J., 1971, Environmental control: priorities, policies and the law: Columbia Univ. Press, New York, 311 p.

Gray, O. S., 1973, Cases and materials on environmental law: Washington D.C., The Bureau of National Affairs, 1420 p.

Heffernan, P. H., and Corwin, R., eds., 1975,

Environmental impact assessment: San Francisco, Freeman, Cooper & Co., 277 p.

Hill, A., and McClosky, M., 1971, Mineral King: wilderness versus mass recreation in the Sierra: in Harte, J., and Socolow, R. H. eds., Patient earth, New York, Holt, Rinehart and Winston, p. 165–180.

Jahns, R. H., 1969, Seventeen years of response by the City of Los Angeles to geologic hazards: in Olson, R. A., and Wallace, M. M., Geologic hazards and public problems, Office of Emergency Preparedness, Washington, D.C., U.S. Govt. Printing Office, p. 283–295.

Kendall, J. H., 1967, Water law: streamflow rights in New England and New York State: Boston, Mass., New England Interstate Water Pollution Control Commission, 47 p.

Kiersch, G. A., and Cleaves, A. B., 1969 ed., Legal aspects of geology in engineering practice: Geol. Soc. Amer. engineering geology case histories, No. 7, 112 p.

Landau, N. J., and Rheingold, P. D., 1971, The environmental law handbook, New York, Ballantine Books, 496 p.

Lehr, J. H., et al., 1976, A manual of laws, regulations, and institutions for control of ground water pollution: Final Rept., Washington, D.C., U.S. Environmental Protection Agency, var. pages.

Leopold, L. B., 1969, Everglades jetport, summary: National Parks & Conservation Magazine, v. 43, n. 11, p. 11–13.

Libby, L. W., and Bugliari, J. B., 1968, The role of easements in New York's open space planning: A.E. Res. 272, Dept. of Agri. Econ. New York State College of Agriculture, 38 p.

Mann, J. F., Jr., 1969, Ground-water management in the Raymond Basin: Geol. Soc. Amer. Eng. Geol. case histories, No. 7, p. 62–74.

Mitchell, J. G., and Stallings, C. L., eds., 1970, Ecotactics; the Sierra Club handbook for environment activists: New York, Pocket Books, 288 p.

Murphy, E. F., 1971, Man and his environment: law: New York, Harper & Row, 168 p.

Nash, R., 1967, Wilderness and the American mind: New Haven, Yale University Press, 256 p.

Piper, A. M., 1960, Interpretation and current status of ground-water rights: U.S. Geol. Survey Circ. 432, 10 p.

Pollock, J. P., 1969, When is an engineering geologist liable?! in Lung, R., and Proctor, R., Engineering geology in southern California, Assoc. Eng. Geol., p. 219–224.

Reitze, A. W., Jr., 1972, Environmental law, 2nd ed.: Washington D.C., North American International Vol. 1, various pages.

Rose, J. G., ed., 1974, Legal foundations of environmental planning: cases and materials on environmental law: The Center for Urban Policy Research, Rutgers, The State Univ. of New Jersey, 319 p.

Sax, J. L., 1970, Defending the environment: New York, Vintage Books, 252 p.

Scullin, C. M., 1969, History, development, and

administration of excavation and grading codes: in Lung, R., and Proctor, R., Engineering geology in southern California, Assoc. Eng. Geol., p. 227–236.

Sherrod, H. F., 1970, Environmental law—a definition and survey: New York State Bar Jour., v. 42, n. 5, p. 387–400.

Searles, S. Z., 1969, Aesthetics in the law: New York State Bar Jour., v. 69, n. 3, p. 210–217.

Sive, D., 1970, Some thoughts of an environmental lawyer in the wilderness of

administrative law: Columbia Law Review, v. 70, n. 4, p. 612–651.

Thomas, H. E., 1970, Water Laws and concepts: U.S. Geol. Survey Circ. 629, 18 p.

Trelease, F. J., 1964, The concept of reasonable beneficial use in the law of surface streams: in Smith, S. C., and Castle, E. N., eds., Economics and public policy in water resource development, Iowa State Univ. Press, p. 272–292.

U.S. Department of Agriculture, 1972, Rural zoning in the United States: analysis of

enabling legislation: Econ. Research Service Mis. Publ. No. 1232, 170 p.

U.S. Department of Interior, 1969, Report on the Florida jetport: U. S. Dept Interior, Washington, D.C.

Water Resources Policy Commission, 1950, Water resources law, Vol. 3: Washington, D.C., U.S. Govt. Printing Office, 777 p.

Yannacone, V. J., Jr., Cohen, B. S., and Davison, S. G., 1972, Environmental rights and remedies: San Francisco, Bancroft-Whitney, 2 volumes.

PART SIX

Water is an essential ingredient for human life and society. The All American Canal carries water from the Colorado River through the desert and desolate regions for use in Southern California. (Courtesy U.S. Depart. Agriculture.)

Synthesis and Epilogue

In preparing this book, 120 instructors who teach a course in environmental geology were polled for their various attitudes on the subject. One question concerned the composition of chapters that should be included in such a text. Unfortunately, if all the ideas had been incorporated 100 percent, this book would take on the dimensions of an encyclopedia. Although many environmental subjects are mentioned and discussed, several have been omitted. Two of these are briefly used as case histories at the end of Part Six—medical and military geology—to illustrate what is nearly a never-ending spectrum of topics that can be considered to fall within the domain of environmental geology.

There are several reasons why I felt the need for a Part Six. One purpose is to provide a type of "wrap-up" that contains a summation and conclusion of those points that I have especially tried to stress throughout the text. Another objective can be viewed as an assessment of the "status of the art," along with a prognosis that attempts to capture the essence of the timeli-

ness of the subject. And, finally, there is a section on additional themes that might have been included in a more extensive treatment as well as examples of the vast reach of the topic into multidisciplinary and interdisciplinary such material may take on the appearance of a smorgasbord, but the intent is to provide the reader with a type of overview of the field of environmental geology and its relevancy and significance in today's world of never-ending problems and confrontations.

PRINCIPAL ENVIRONMENTAL THEMES

The following ten points are representative of those ideas that have been repeatedly stressed throughout the book. Such a list should not be viewed as comprehensive, but only suggestive of points that can provide us with some cohesive and perennial themes in such a varied field of endeavor.

1 History repeats itself. Many present-day environmental problems have been prevalent throughout mankind's tenure of the planet. A careful reading of history can aid in alerting planners so that past mistakes will not be repeated.

2 The primary goal of environmental geology should be the prediction of natural events and the knowledge of impacts that are created when we do decide to alter the environment. When forewarned, strategies can be

adopted that will minimize harmful effects.

3 The "environment" and mankind form a reciprocal relationship. Nature directly effects our habitat and activities through the geologic setting and through events such as floods, earthquakes, volcanic activity, landslides, and so forth. However, we also change the environment through our activities and construction enterprises.

4 Nature possesses a dual personality. When properly respected and understood, nature can act as a servant, but, when abused and degraded, nature can be a tyrant and adversely affect society. Francis Bacon wrote in 1620 that "Nature to be commanded must be obeyed."

5 The three most awesome environmental problems are closely interconnected—population, energy, and food.

The severity of many of our deleterious changes in the environment would be largely mitigated if population control was exercised. The rapid expansion in human numbers has created a nearly intolerable burden in many countries that will ultimately affect the entire world community.

6 The solution of environmental problems must rest on a very broad interdisciplinary and multidisciplinary base. None of the scientific, technical, or sociopolitical fields has a monopoly on methods to resolve environmental conflicts and abuses.

7 There has been more and more government involvement in environmental management through time because private citizens and industries have usually forfeited their stewardship role. Environmental debasement has resulted from self-interest, self-profit, and local expediency.

8 Policy matters must be concerned with the proper assessment of priorities and the search for reasonable alternatives. Such jurisdictional matters should involve as large a regional planning posture as possible and the denial of local or parochial solutions whenever possible.

9 Geologists must become involved in the decision-making process. The collection and dissemination of data and reports are inadequate if not followed by the implementation of a course of action, through planning, management, and the coordination of goals.

10 It is not the purpose of this book to present strident and advocate positions on all environmental questions. However, it is appropriate to make informed judgments when factual data and research have demonstrated a clear course of action.

A SPECIALIZED RÉSUMÉ

The number of people who need food, water, shelter, and material goods has been increasing at a rate of about 65 million per year, whereas the available resources to satisfy these requirements have been decreasing at an alarming rate. Throughout most of the earth, forests are being depleted; rangelands are being badly overgrazed and their carrying capacity reduced; farmlands are being overcropped; prime agricultural lands are being lost to urbanization; and energy and mineral resources are being used at rates that will soon exhaust many of them. The increasing rate of all phases of earth

exploitation is placing an accelerated stress on all ecological systems that are vital to a healthy, habitable planet. Virtually all prospects for the future indicate continued growth and demand in nearly all environmental areas.

For most of the twentieth century, growth rates associated with human development were exponential. In less than one generation of the twentieth century, we have been using more of our natural resources than we had used in all previous history. Obviously, this cannot be allowed to continue. The French have illustrated this point by considering a lily pond on which there is a single leaf. Each day the number of leaves

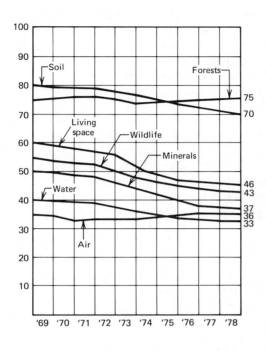

1969–1979
EQ Summary

Wildlife: down
Though widespread habitat loss continues,
important new laws and conservation
programs have been established.

Air: up
Prodded by tough federal standards, air
polluters have slowly cleaned up their act.
But in many areas, air is still not safe to
breathe.

Soil: down
In the face of rampant soil erosion and urban
sprawl, prime farmland has declined while
the nation's food needs have grown.

Minerals: down
As the U.S. plodded toward a national energy
and conservation policy, the days of cheap
fuel and abundant minerals vanished forever.

Forests: same
A battery of tough forestry laws ushered in
a new era of timber management, offering
hope that the U.S. can avoid wood shortages.

Living space: down
Without effective land use planning, vast
stretches of America's land have remained
vulnerable to topsy-turvy development.

Water: down
As cleanup efforts got underway, the
complexities of the country's water
pollution problems became more apparent.

Six-1 Summary of
environmental quality for
1969–79. (Source: National
Wildlife Federation.)

doubles until the pond is completely covered by leaves after 30 days have passed. On which day was the pond only half covered? The answer is, on the 29th day!

Complicating this already disturbing situation are the gross imbalances in the distribution of the world's wealth and resources. Geologic processes responsible for the varying amounts of raw materials are unevenly distributed and do not recognize political boundaries. Too often the assertion is made, or assumed, that technology and engineering along with human ingenuity will solve all problems: as shortages arise, substitutes will be found, materials will be recycled, social systems will evolve and adjust, and civilization will persevere and prosper. Others contend that the earth is already overpopulated and overexploited and that even at present demand levels the quality of life is not sustainable. In addition, modern society has the military capacity to annihilate virtually all higher life forms on earth. With the present inability to cope successfully with unstable political, economic, and social conditions, the probability for large-scale war is increased. It is therefore obvious that international cooperation on political, economic, social, and technocracy levels is essential if we are to become capable of meeting the enormous challenges posed by so many threats to the environment and human existence.

Regardless of personal opinions, there do appear to be certain grim realities which we must face, expecially those related to the limited capacities of the earth's life support systems to indefinitely absorb all the changes being introduced into the environment through overexploitation. Intelligent, humane solutions to the problems created by uncontrolled growth and disregard of nature must be sought now. The frontiers have vanished. There are no greener pastures to inhabit. We simply cannot afford to continue behaving in the fashion described by Adlai E. Stevenson when he remarked, "Man is a peculiar animal who can read the writing on the wall only when his back is up against it."

THE SCOPE OF ENVIRONMENTAL GEOLOGY

The field of environmental geology is so broad and comprehensive that it is impossible to cover all topics. As a final section of the book, however, we will look at two additional topics, which, if time had permitted, would have merited much fuller treatment. These topics—medical and military geology—share a common theme in having similar goals that bear on the health, safety, and welfare of mankind.

Medical Geology

Although not yet recognized as a special subdiscipline, the topic of geomedicine is taking on ever-increasing importance throughout the world. Environmental influences on society have received the attention of some researchers for many years; Hippocrates wrote of terrain and climate influences on the human physical constitution, and Huntington believed climate was important in determining the vitality of a nation. Elsewhere in this book mention is made of additional natural and man-made pollutants that affect human health—black lung disease in coal mining, acid rain effects on the fish population, increased death and disease from coal-burning plants, and accelerated disease incidence from Lake Nasser. However, one topic that is just now starting to emerge concerns the relation of trace metals to human health. The reason for such tardiness is that only since 1968 have the analytical procedures been perfected to a state of precision whereby very minute amounts of trace elements could be accurately measured.

There are about 60 chemical elements that occur in biological systems and, of these, at least 30 are necessary for life and health. Oxygen, carbon, hydrogen, nitrogen, calcium, and phosphorus constitute 98.5 percent of the human body. Sulfur, potassium, sodium, chlorine, and magnesium comprise another 0.8 percent. Those elements that make up less than 0.05 percent may be viewed as **trace elements**, and at least 14 have been identified as important in human health, these include chromium, cobalt, copper, fluorine, iodine, iron, manganese, molybdenum, nickel, selenium, silicon, tin, vanadium, and zinc. The problem lies with the exceedingly narrow range of tolerance for these elements, because either too much or too little can provide ill health or toxicity to the body. As in all nature, such materials are unevenly distributed in the lands and waters of the planet. Thus, geochemical studies are vital to determine the prevalence of such elements so that proper human adjustments can be made.

The study of diseases and the possible relationship of geology and environmental factors are very complex. In all instances, caution is necessary so that a spurious cause-effect relationship is not adopted. For example, in 1959 a New York medical report stated there was a relationship between the high rate of malformed children in the Adirondacks and granitic rocks that were assumed to have high levels of radioactivity. Subsequently, other factors were shown to negate this conclusion. Whooping cough was shown to be less common in the copper belt of Zambia, but instead of the copper

A. "Etiology"—for example,
specific agents:
   Physical
   Chemical
   Biological, including vectors and
      reservoirs

1. Hereditary factors—for example,
specific genetic defects as manifested
by:
   G6PD deficiency
   Hemachromatosis
   Diabetes mellitus
Or genetically determined characteristics:
   Hb structure
   Skin color
   Immunologic responsiveness
   Gastric secretion

B. Contributory causes—for example,
exposure to the primary agents (with
consideration of dose and portal of entry)

   Factors increasing susceptibility:
      Physical stress (heat, cold, dehydration,
      low or high barometric pressure, ex-
      haustion, trauma)

   Mental stress
   Other diseases or conditions
      (malnutrition, anemia, infection, cancer,
      drug therapy)

2. Environmental factors—for example,
   Geochemical
   Climatic
   Politico-economic
   Cultural

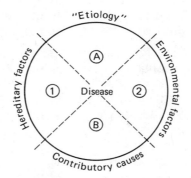

Six-2 Four major categories of disease causal factors. Although shown by the diagram in equal segments, they rarely contribute equally to the complete cause of the disease. Each of the four broad categories includes several subcategories. (Source: Hopps, H.C., 1971).

abundance being the cause of lower frequency, the immunization program at the mine hospitals was found to be the reason for the reduced rate of the disease. Even now the relationship of higher incidences of cardiovascular disease to "hard" drinking water is being furiously debated.

The following list represents our current knowledge of trace elements and the range of impacts that can result in the field of geomedicine.

Asbestos. This fibrous material is carcinogenic when inhaled in excessive amounts. Miners of this element have higher cancer rates, and asbestos products have been taken off the market in the United States; these products include ceiling tiles and hairdryers.

Cadmium. Cadmium and zinc are chemically similar and can compete within the biological system. However, cadmium has an adverse effect, whereas zinc is an essential nutrient. Cadmium is a factor in hypertension and can cause osteomalacia (deterioration

and softening of the bones) and also pulmonary emphysema. High cadmium-zinc ratios in eastern North Carolina may have contributed to the severity of atherosclerosis in that state. The presence of cadmium in soils and water results from mine wastes and smelters and is often associated with lead and mercury.

Calcium. This element is vital for bone development and for the transport of metabolic waste products through the cell walls. Along with magnesium, it can buffer soils and make many metallic elements less reactive and inhibit their toxicity. Calcium deficiency can produce Urov disease, which is incomplete calcification of bones.

Chromium. This element is related to glucose tolerance and may be a causal factor in diabetes. Bronchitis and emphysema have also been noted as being caused by a chromium imbalance in the human body.

Cobalt. Cobalt, nickel, molybdenum, tungsten, and vanadium are all important enzyme catalysts in the body. When in trace amounts, they are beneficial; but when they are too abundant, they act with toxic effects.

Copper. This element is required for cell respiration, and a deficiency can cause loss of muscular control and anemia. As early as 1931, it was discovered that cattle developed weakness, failed to fatten, and had rough hair, swollen joints, and anemia when restricted to forage sites with sandy soils or peat soil; it was found that these soils were deficient in copper and iron.

Fluorine. This element is especially known for its ability to inhibit caries (cavities and tooth decay) in young children. The optimum amount in water is about 1 ppm. When too little fluorine is present in drinking water, it leads to tooth decay; however, too much, in amounts of 1.7 to 4.0 ppm, causes brown spotting and enamel mottling. Highly deficient fluorine waters can lead to an increased incidence of osteoporosis. The volcanic eruption of Hekla, in Iceland, produced fluorine gases that settled on grass and killed more than 100,000 sheep from fluorine poisoning.

Iodine. A deficiency in this element can cause goiter and has been known to increase the incidence of breast cancer in some women.

Lead. It is not known whether this element is necessary in human growth, but when ingested it can become highly toxic and carcinogenic.

Lithium. A deficiency in this element can lead to mental disorders.

Manganese. In the body this element acts as an enzyme catalyst, but is toxic in high amounts.

Mercury. The effects of mercury on the central nervous system is well documented. In some Japanese fishing villages, mercury poisoning has caused severe health problems.

The element is cumulative, so that in the food chain the higher animals have concentrated amounts in their diets (as in DDT).

Molybdenum. This element can produce abnormality in animals and humans. Mining of clay in Missouri has caused soil and water changes in both the spoil piles and in downstream areas. The metabolism of cattle grazing on contaminated soils has been affected, producing molybdenosis—a disease that interferes with normal growth, nutrition equilibrium, and reproduction; the disease may also cause xanthine (growth of kidney stones).

Nitrate. Although nitrogen is basic to all life, nitrate (NO_3) can be poisonous to humans when levels exceed 45 ppm and cause "blue" babies. Furthermore, nitrates can reduce to nitrites, which react with amines to produce carcinogenic substances, deformities, and harmful mutations. Excessive nitrates result from untreated waste and sewage, fertilizers, and tail-end irrigation waters.

Radon. This gaseous element may be emitted from many radioactive minerals and is present in uranium mines and tailings. Being a gas, it enters the body through breathing and affects the lungs. Uranium miners have four times the possibility of lung cancer than the rest of the population. Workers at nuclear plant sites, such as at the Kittery, Maine shipyards, contract cancer at double the normal rate and leukemia at five times the normal rate.

Selenium. This element was known to be toxic to livestock as early as the 1920s when ranchers in Wyoming learned they could not graze sheep on certain Cretaceous-age geologic formations. Wet years were especially bad, and selenium poisoning caused sheep to lose muscular control and many died. Certain New Zealand soils had a similar history of sheep kill. Toxic amounts of selenium in humans cause motor ataxia, damage to the alimentary canal, and produce disease in blood-forming organs. Recent studies have indicated, however, that selenium may be an anticancer agent and a beneficial trace element in small amounts.

Silicon. This substance is now thought to play a role in the aging process. Certain levels are necessary in calcification processes and in bone and cartilage formation.

Zinc. This element is necessary for plant growth, but should be in amounts less than 0.07 ppm. Zinc promotes cell growth, and if not present will stunt biologic growth. It also aids in the regeneration of tissue and promotes healing, perhaps even inhibiting some cancers. However, too much zinc also produces toxicity.

The subdiscipline of geomedicine shows great promise and will become increasingly important in the future. For example, only within the past two years have the many hazardous substances in the environment been studied, and these materials contain numerous trace elements and toxic substances that are lethal to all animal species as well as humans.

Military Geology

Although the use of geologic-type endeavors in military affairs dates back into antiquity, its deliberate use as planned and implemented by geologic specialists is largely a twentieth-century phenomenon. All of the great military commanders of history had an understanding of suitable terrain and were able to maximize their advantage and position by the shrewd use of topographic and hydrologic features.

In many respects, the military and nonmilitary uses of geology are similar—a principal goal being to aid in the health, security, and well-being of the clients that are served. When in military service, the geologist is charged with selecting sites for roads and bridges; obtaining potable water supplies; providing for sanitation services; preparing maps; and finding the resources necessary for various types of operations. Thus, geologists are problem solvers and their work is strictly mission-oriented. However, as the following discussion will show, there are other considerations and goals in military geology that are not duplicated in civilian enterprises.

Perhaps the earliest large-scale incorporation of geologists into the military occurred in World War I, when the German high command involved geologists with planning operations. Their importance was also recognized by the requirement that at least one professional geologist be assigned to every infantry division. By contrast, the American Expeditionary Force had few geologists, with only 17 professionals in an army of more than 2 million. Such an oversight contributed to the loss of countless thousands as a result of nonbattle-associated disease. For example, the influenza epidemics and other ailments associated with inadequate working conditions were in many cases caused by the cold, wet, and unsanitary trenches where troops were forced to live. Many trenches were improperly designed and located and failed to be properly geologically engineered in terrain that would allow for adequate drainage.

One of the crucial battles of World War I was the struggle to control Verdun, France. Numerous tactical and geologic mistakes were made on both the German and French sides. The Meuse River occupies an entrenched valley with a river that meanders throughout its floodplain. In advancing on the fortress, the German troops had to cross the river many times. However, at the meander spurs, many troops were trapped when the French soldiers would sweep across the necks of the spurs, imprisoning the troops on a peninsula that was surrounded on three sides by an unfordable river. On another occasion, the German artillery was laying down a barrage and the French commander ordered his troops to dig trenches on the plateau overlooking the fort. Unfortunately, tough limestone capped the terrain with only a thin topsoil. Unable to dig deep enough for protection, thousands lost their lives during the ensuing bombardment.

World War II made use of geologists on both sides, in the Allied and in the Axis maneuvers. However, there were some notable catastrophes for the Americans before there was full involvement by geologists. In the battle for Tarawa Island, a small Pacific atoll, the invading U.S. marines became "sitting ducks" for the Japanese when their amphibious vehicles were stranded by an outer coral reef. The planners of the attack had failed to calculate properly the sea level position, so that at the hour of invasion the ocean surface was lower than predicted and didn't allow sufficient freeboard for vehicles to move over the reefs. The United States was also ill prepared for operations in Alaska. With the threat of a Japanese invasion, roads, airfields, and other installations had to be installed quickly. Unfortunately, many sites were located on permafrost, and there was insufficient geologic information on how to build in such terrane. Many costly mistakes were made, and it took much longer than planned to mount a program that would aid in stabilizing the construction projects. The U.S. North African campaign got off to a very costly beginning when American troops were severely defeated at the Battle of Kasserine Pass. Not only were the troops and commanders "green," but they lacked knowledge of the terrain. The battle maps contained insufficient topographic detail, so that the German General Erwin Rommel was able to capitalize on maneuver mistakes and inflict great damage. It is interesting to note that when Hitler gave command of the German Afrika Corps to Rommel he is reported to have asked Rommel if he had any special

requests. The answer was "600 geologists," and the request was granted.

The deepest involvement of American geologists in World War II was in the establishment of the Military Geology Unit under the administration of the U.S. Geological Survey. Most of its work was for the Military Intelligence Division, Office of Chief Engineers. The principal reports of the unit were published as map folios and were part of a series of Strategic Engineering Studies. These were entitled Terrain Intelligence and consisted of a series of maps and explanatory tables, each stressing a special terrain topic, including items such as terrain appreciation, rivers, water supply, problems of road and airfield construction and maintenance, trafficability for vehicles, engineering properties of soil and rocks, and distribution and availability of construction materials, fuels, and mineral resources. Such reports were prepared in Washington, D.C., and were primarily used at strategic planning levels by the War Department and by General Headquarters in the various theaters of operations.

To meet the problem of serving the needs of the tactical planners, teams of geologists were selected from the unit for the different theaters, where they were attached to the staff of the Theater Engineer. The Military Geology Unit was coordinated with other groups working on problems of terrain intelligence, such as the Hydrographic Office in the Navy Department and the Beach Erosion Board of the Office of Chief Engineers.

During the war the Military Geology Unit prepared 313 reports, of which there were 140 Terrain Intelligence Folios. The Unit was often rushed by military demands; for example, a report on eastern Sicily, covering 1600 km², was assembled in 10 days. During this time, 16 quadrangles on a scale of 1:100,000 were compiled, along with 57 other maps, 2 terrain diagrams, 36 large tables of text, and 39 photographs.

Although military geology may seem like

an unusual extension of environmental geology, in these days of troubled times the science of geology must fulfill a wide spectrum of demands, even those associated with such unpalatable activities as war.

READINGS

Cannon, H. L., and Davidson, D. F., eds., 1967, Relation of geology and trace elements to nutrition: Geol. Soc. Amer. Special Paper 90, 64 p.

Cannon, H.L., and Hopps, H.C., eds., 1971, Environmental geomchemistry in health and disease: Geol. Soc. Amer. Memoir 123, 230 p.

Ebens, R. J., et al., 1973, Geochemical anomalies of a claypit area, Callaway County, Missouri, and related metabolic imbalance in beef cattle: U.S. Geol. Survey Prof. Paper 807, 24 p.

Freedman, J., ed., 1975, Trace element geochemistry in health and disease: Geol. Soc. Amer. Special Paper 155, 118 p.

Nash, R. C., and Woolson, E. A., 1967, Persistence of chlorinated hydrocarbon insecticides in soils: Science, v. 157, p. 924–927.

National Wildlife Federation, 1979, Environmental quality index: 1969–1979, a decade of revolution: Washington, D.C., Natl. Wildlife Fed., p. 17–32.

Shacklette, H. T., Sauer, H. I., and Miesch, A. T., Geochemical environments and cardiovascular mortality rates in Georgia: U.S. Geol. Survey Prof. Paper 574-C, 39 p.

Woodwell, G. M., 1967, Toxic substances and ecological cycles: Sc. Amer., v. 216, n. 3, p. 24–31.

GLOSSARY

A

A-horizon The top part of the soil profile. May contain humus or organic-rich matter.

Aa A Hawaiian term denoting very rough and jagged lava.

Abrasion Erosion accomplished by direct impact of particles contained within the moving geologic medium.

Acid rain Precipitation from the atmosphere of water whose pH is acidic. Some would like to restrict the term to the acid increments caused by man-made, airborne pollutants such as sulfurous compounds.

Aerosols Minute particles, liquids, or gases dispersed within another gas.

Aesthetics The consideration and study of beauty as a value and a goal in itself.

Aggradation The process of building up a surface by means of a depositional process.

Aggregate Substances such as sand and gravel or crushed stone used for construction purposes.

Alluvium The general name applied to those sediments deposited by streams on land. The opposite of *colluvium*.

Anaerobic Of or describing an environment that does not contain oxygen.

Angle of repose The maximum slope or angle for stablity of loose rock or soil.

Anthracite The highest ranking coal, with a high percentage of fixed carbon and low amounts of volatiles, formed by the metamorphism of lower ranking coals.

Anthropogene A product, landscape, or feature primarily caused by human activities.

Anthropogeography That special subdiscipline of geography that evaluates the role of nature in influencing human behavior and activities, such as settlement patterns.

Anticline A fold that is convex upward. The rock strata dip away and down from the crest, and oldest rock are in the center of the structure.

Appropriation doctrine That type of water law in which the original water user has the favored status as the owner of water rights. Also known as *prior rights.*

Aquiclude Those rocks below the water table that do not transmit water readily and cannot be used as sources of water supply.

A quifer A body of permeable earth material through which groundwater moves readily, thus making possible its use as a water supply source.

Area (landfill) method That technique used to entomb solid waste when terrain is flat.

Area (strip) mining The type of strip mining operation that is used on level terrain.

Artesian system Groundwater flowage where water rises above the local water table into a well or spring. This is commonly caused by upslope hydrostatic pressure where the rocks form a valley-dipping envelope with an aquifer that is enclosed by impermeable rocks.

Artificial recharge Technique used to increase the amount of groundwater in storage; includes injection wells, infiltration pits, and manmade ponds.

Asbestos A mineral fiber used in many industries, but hazardous to health when inhaled.

Ash (volcanic) Rock particles less than 4 mm in diameter blown out of a volcano.

Asthenosphere A zone within the earth's mantle where plastic rock movement occurs. The zone roughly extends from 50 to 100 km to depths of 500 km and is regarded as the material over which the lithospheric plates move during global tectonics.

Atoll A coral-type island system with a central mass surrounded by an arcuate series of reefs and a lagoon.

Avalanche A large mass of material that is suddenly displaced from a superior elevation and rapidly moves and cascades down hillsides. The composition may include debris, rock, regolith, snow, and ice.

Azonal soil Youthful surficial materials, such as alluvium or colluvium, that have had insufficient time to develop a pronounced soil profile.

B

B-horizon The soil zone of accumulation below the A-horizon. Contains the leached materials derived from the soil above.

Barrier island A low offshore sandy

island. Sometimes called *offshore bar* or *barrier beach*, but a barrier island usually has multiple beach ridges and is roughly parallel to the mainland with an intervening salt wetland or lagoon.

Basalt An igneous extrusive rock formed by volcanic surface activity. It is fine-grained, dark, and composed of ferromagnisian minerals. The most common igneous rock.

Baseflow That part of streamflow derived from discharge into the stream by groundwater.

Bauxite A rock composed of one or more hydrous aluminum oxides. The most common ore of aluminum.

Beach A body of wave-washed sediment on the coast between ocean or lake and the mainland.

Bed A stratum of rock. Also the floor of a stream channel.

Bedding plane The top or base of a rock stratum. A rock surface that contains some continuity because of slightly different deposits of contiguous strata.

Benefit-cost ratio A number that expresses all benefits that would result from a construction project in relation to the costs that would be involved. A greater than 1.0 ratio is necessary before the project can be authorized.

Berm A nearly level part of the beach or backshore formed by the deposit of materials from the action of waves. These areas may resemble small terraces and, in some instances, there may be a winter berm and a summer berm.

Biodegradable A term used for those substances that can readily be decomposed by living organisms.

Biomass The total weight of all living organisms that exist in a particular environment.

Biological oxygen demand (BOD) The number is a quantification of the amount of oxygen required to decompose organic materials in a unit volume of water.

Biosphere The part of the earth and its atmosphere capable of supporting life.

Bituminous coal Medium rank coal. The hydrocarbons contain low volatiles and moderately high carbon content.

Breakwater A coastal structure em-placed some distance from the shore and designed to interrupt the power of waves. It can be built of different types of materials with a geometry dependent on the specific purpose and terrain setting.

Breeder reactor A nuclear reactor that is designed to produce more fissionable material than it consumes. This is accomplished by the atomic changes that occur with U-235, U-238, and thorium.

Brittle A solid that ruptures without undergoing plastic flow.

Buttress An emplaced hillside structure used to prevent downslope movement of material. Weight of the material, whether on a hill or in dam construction, acts as a stabilizing influence and increases the forces of resistence.

C

C-horizon The lowest soil zone that contains partially decomposed or disintegrated parent material.

Cadmium A metallic element that can cause serious health problems.

Calcite A carbonate mineral ($CaCO_3$) that is the principal constituent of limestone. It is dissolved by water, and mass solution creates sinkholes and caverns.

Canal An artificially constructed waterway, commonly used for navigation and for the movement of water for irrigation.

Capillary action The rise of water in small, hairlike openings within the soil. The amount of rise depends on soil type and solar energy.

Carbonate materials Those rocks and minerals that contain the radical (CO_3^{--}).

Carcinogen A substance capable of producing cancer in humans or animals.

Carrying capacity The maximum number of individuals of a species that is capable of being supported by a particular environment.

Cementation The binding together of particles into a cohesive mass.

Cesspool A cistern built into the ground to collect waste products from houses.

Channelization The deliberate modification of a stream, by straightening, deepening, widening and narrowing, and constricting the channel.

China syndrome Term applied to a catastrophic meltdown from a nuclear reactor accident. The "hot" nuclear materials penetrate through the containment vessel and into the underlying earth materials.

Circum-Pacific belt The arcuate-shaped sequence of lands and islands that girdle the Pacific Ocean. Also known as "ring of fire" because it coincides with both the major volcanic and earthquake zones of the earth.

Cirque A bowl-shaped landform near a mountaintop formed by the erosive power of an alpine-type glacier.

Clay Refers to both a specific mineral, such as montmorillonite, and particles whose diameter is less than 1/256 mm. Clays produce many environmental problems and do not allow easy water movement.

Clearcutting A technique of timbering whereby all trees are cut down and removed at the same time. Its environmental impacts are highly controversial.

Climate Average conditions of the weather at a place over a period of years.

Coal A sedimentary rock consisting chiefly of partly decomposed plant matter and containing less than 40 percent inorganic matter.

Collapse A gravity phenomenon that produces a sudden, mostly vertical, displacement of earth material without a free face.

Colluvium Loose earth materials moved downslope by the action of gravity and sheetwash processes. The sediment is not stratified but may possess a fabric of aligned material.

Columnar jointing Elongated joints with fracture patterns that produce columns that are at right angles to the cooling surface and contain polygonal-shaped cross sections. Most commonly formed in tabular lava flows.

Commons Original usage dates to "the commons," a place where English citizens had equal use of the land by royal decree.

Compaction The reduction of pore space within earth materials in response to the weight of overlying materials or to other forces.

Compression An inward directed stress on earth materials. Generally causes reduction in volume and distortion of the material.

Condemnation The governmental action of land acquisition by the power of eminent domain.

Cone of depression A conical-shaped depression in the water table caused by the withdrawal of water from a pumped well.

Conglomerate A clastic sedimentary rock containing rounded pebbles in a finer grained matrix.

Connate water Water trapped within enclosing rocks during their formation.

Conservation The ethic for management of earth, soil, crop, and timber resources that will provide for long-term utilization. The programs used to accomplish conservation may involve measures to reduce soil, water, or other losses to resources, for example, soil conservation practices.

Contact metamorphism The type of metamorphism produced in rocks adjacent to a cooling magma. Also called *thermal metamorphism* because the changes are largely caused by elevated temperatures.

Continental drift The theory that the continents are in lateral motion and mobile.

Contour farming The technique of planting crops in rows that conform to similar hillside elevations.

Contour (strip) mine The mining method that extracts earth materials along cuts that are made parallel to similar hillside elevations.

Convection Transmission of energy and matter by motion of the material. Excess heat causes material to rise to places of less gravitational pull or pressure.

Convergent plate boundary The locus of two lithospheric plates where one is forced under the other into a subduction zone.

Corrosion The deterioration of material by chemical means.

Creep The imperceptible downslope movement of surface materials by gravity processes.

Crust The outer layer of the planet Earth. Generally considered to be composed of rocks that have brittle characteristics. Also referred to as *lithosphere*, with sialic (silicon and aluminum) rocks under the continents and simatic (silicon iron and magnesium) rocks under the ocean basins.

Crystal A solid bounded by natural, regular plane surfaces that reflect an orderly interior atomic arrangement.

Cyclone 1. Atmospheric motion of an air mass in counterclockwise direction in Northern Hemisphere. 2. The term used to describe large tropical disturbances in the Indian Ocean; the equivalent of hurricanes in the Atlantic and typhoons in the Pacific. Air movement exceeds 74 mph. 3. Name applied to high-velocity wind vortices in North America which can also be highly destructive.

D

DDT A persistent chlorinated hydrocarbon pesticide; its full chemical name is dichlorodiphenyltrichloroethane.

Debris Any accumulation of unsorted materials. It can be deposited through either natural or human means.

Debris slides and flows Downslope movements of earth materials within the regolith. They represent types of landslides.

Decomposition The chemical alteration of rock material, such as produced by oxidation and hydration.

Deep-well disposal The surface elimination of waste products by pumping materials into the ground.

Deflation Erosion of surface materials by the action of the wind blowing loose sediments away.

Deforestation The cutting down and removal of all trees in an area. It generally implies an absence of conservational measures and the lack of replanting provisions.

Degradation 1. Lowering of the land surface by erosional processes. 2. A man-made action that causes deterioration of earth substances.

Delta A body of sediments deposited by a stream flowing into a standing water body.

Depletion allowance A percent tax reduction for companies involved with exploration, mining, and extraction of mineral and fuel resources. It is similar to a depreciation allowance for manufacturing businesses.

Deposition Synonymous with *sedimentation*; the process by which earth materials accumulate when dropped or precipitated from geologic phenomena, such as rivers, oceans, glaciers, and so on.

Desertification The process of changing terrain from nondesert conditions to one where plant growth is severely inhibited or impossible. The term is usually associated with land use or atmospheric changes caused by humans.

Desiccation The process of dehydration of earth materials.

Deterioration The process of lowering the utility of land and water.

Developed country A nation with relatively high economic base, high per capita income, and high gross national product. Generally refers to industrialized nations; also known as MDCs (more developed countries) as compared to the LDCs (less developed countries).

Developmental ethic The system of land, water, and resource management that favors rapid utilization of materials. It implies an absence of conservation or preservation measures.

Diabetes A disease in which the human body does not properly metabolize sugar. Insulin must be taken to make up for this deficiency.

Diastrophism The process of large-scale movement and deformation of solid rock in the earth's crust.

Dike A tabular body of intrusive igneous rocks that cuts through older rocks.

Dilatancy The inelastic increase in rock volume.

Dimension stone Bedrock that is fashioned into specific sizes. Marble and granite are typical rocks that are cut and sawed for use in buildings and other structures.

Dip The angle in degrees between a horizontal plane and an inclined plane. The measurement is taken perpendicular to the strike and is always the maximum angle of inclination.

Disaster A sudden and extraordinary calamity that results in above-average losses in terms of property damage or lives. Volcanoes, earthquakes, floods, and landslides produce geologic disasters.

Discharge The quantity of water that passes a reference point in a given time unit. Commonly measured in cubic feet per second (cfs) or cubic meters per second (m³/sec).

Disseminated mineral deposit Ore

minerals that are scattered throughout a rock body. The host rock is usually igneous, but its composition may have changed by surficial processes, such as the prophyry copper disseminated ores.

Divergent plate boundary The position between two lithospheric plates where mantle processes are creating new crustal material with concomitant movement of the plates away from the position, such as in the Mid-Atlantic Ridge region.

Drainage basin The area enclosed by a superior divide whereby all surface water flows toward the mouth of the master stream. Synonymous with *watershed.*

Drainage network The system of streams that occupies a drainage basin.

Drawdown The measured amount of lowering in the water table from a pumping well.

Drift Materials transported by glaciers or longshore currents. Also material deposited by glacial meltwater processes.

Drought A prolonged dryness caused by the absence of rain.

Ductile A material that undergoes large permanent deformation without fracturing.

E

Earthflow A type of landslide that involves movement of regolith materials when they contain a high moisture content. Often represents some combination of movement, such as slumping and mud and debris flowage.

Earthquake The sudden shaking of the earth in response to rock movement, as along a fault, or volcanic activity.

Ecology A subdiscipline of biology that studies the interrelationships among plants and animals within their environment.

Economic geology That field of geology involved with the discovery, evaluation, and mining of rocks and minerals.

Ecosystem A grouping of animals and plants that lives together and interacts in a particular environment.

Effluent The deliberate discharge of materials into adjacent water bodies. It often implies discharge of harmful waste products.

Effluent stream The hydrologic occurrence of a stream that gains water from the groundwater zone.

Elastic The type of rock behavior in which materials resume their former size and shape after the force has been removed.

Elastic rebound theory The concept that earthquakes result from the motion of rocks elastically springing back after a fault has occurred.

Element A unique combination of protons and electrons that cannot be broken down by ordinary chemical means.

Eminent domain A governmental power that allows land to be confiscated for use in the public interest. Such condemnation and appropriation requires just compensation to the original owner.

Emphysema A disease of the lungs that creates abnormal extension of the air cells and introduces oxygen into the blood.

Endogenic Of or relating to the forces operating within the earth's crust and mantle, such as diastrophism and vulcanism.

Energy The capacity for doing work. It takes many forms such as potential, kinetic, heat, chemical, electrical, mechanical, atomic, and so on. It is interchangeable with mass.

Energy crises A condition that threatens the normal use of energy in which the demand for energy is much greater than the supply.

Engineering Those endeavors in which the properties of matter and the forces of nature are made useful to society. This involves the reconstitution and change of earth materials, processes, and structures.

Engineering geology The use of geology in engineering projects for the purpose of assuring that geologic factors have been considered in design and construction.

Environment The total combination of the surroundings and circumstances that denote a particular set of conditions for a locality.

Environmental crises A condition that threatens the normal status, behavior, and relationship between society and nature, involving irreconcilable losses.

Environmental geology The practical application of geologic principles in the solution of environmental problems.

Environmental impact statement (EIS) A required report under the National Environmental Policy Act of 1970 that discusses effects of a construction project on the environment.

Enzyme A complex organic substance capable of changing other organic materials by catalytic processes.

Ephemeral Nonenduring features, such as a stream that flows only during certain times of the year. A temporary condition.

Epicenter The position on the earth's surface vertically above the earthquake focus.

Erosion A general term to describe the physical or chemical removal of material from a site. Usually accomplished by geomorphic surface processes such as streams, glaciers, wind, and so on. An active force is involved that necessitates transportation, as contrast to weathering, which is static and occurs at the site.

Erratic A transported rock dropped by ice on terrain with dissimilar rocks.

Esker A sinuous body of ice-contact stratified glacial drift formed by meltwater.

Estuary A drowned river mouth with tidal action that moves inland.

Ethic A system of values that governs behavior patterns.

Eutrophication The enrichment of a water body with nutrients that deteriorate the quality for human use and consumption.

Evaporite A sedimentary rock with constituents precipitated from water by evaporation, such as rock salt.

Exfoliation The separation by weathering of successive concentric shells from the main rock mass.

Exogenic Of or relating to forces operating on the earth's surface that include running water, gravity, glaciers, wind, ocean, and groundwater.

Expansive soil A general term for those earth materials that undergo swelling with added moisture. Other terms include expansive sediments, expansive clays, and so forth. Clay minerals such as montmorillonite cause maximum volume changes in the material.

Exponential growth A rate of increase, as in population, that is logarithmic not arithmetric.

Externalize The inclusion of all costs associated with a product into the final market price. Special examples would include the consideration of damage costs caused by pollution as part of the pricing system of the industry that has created the pollution.

Extrusive rocks Igneous rocks that form on the earth's surface by volcanic action.

F

Fault A fracture in earth materials along which the opposite sides have been relatively displaced.

Feedback A resultant process caused by actions that change another part of the interrelated environmental system.

Feldspar A family of minerals that constitute about 60 percent of the earth's crust. Feldspars can be grouped into two classes, the orthoclase and the plagioclase feldspars. These groupings depend on the percentages of potassium, sodium, and calcium, and all feldspars are silicates.

Ferromagnesian A term used to refer to minerals that contain abundant iron and magnesium, such as hornblende. Rocks with a high percentage of such elements may also be called *mafic rocks*.

Fission A nuclear reaction that involves the splitting of an atom and the release of energy.

Fissure eruption Extrusion of volcanic materials along an extensive fracture.

Flood High streamflow that overtops natural banks or man-made constraining structures and covers adjacent land.

Floodplain That part of the river valley capable of being flooded. The terrain is generally relatively level and contains alluvial deposits.

Flood proofing The structural changes made to a building to make it stronger and more flood resistant.

Floodway That part of the floodplain nearest the river that is subjected to most floods. For planning purposes this is often designated as the 100-year flood zone.

Fluid Material that offers little or no resistance to forces acting to change its shape: such as liquids and gases.

Fluorine A chemical element important in human nutrition and in the prevention of tooth decay.

Fluvial Of or relating to a process associated with the action of rivers.

Fly ash Particulate matter that results from the burning of fossil fuels, especially coal.

Focal depth Vertical distance from an earthquake focus to its epicenter.

Focus, earthquake The position within the earth where the original earthquake energy was first released.

Fold A bend or wave in earth materials. Usually associated with stratified material.

Foliation The feature of metamorphic rocks that creates parallel fabrics of platy or elongated minerals.

Footwall One side of a fault that occurs under the fault zone.

Force Stress on a body that tends to produce a motion or change in the material.

Formation A stratigraphic term used to describe a rock mass that has distinct properties that can be mapped.

Fossil fuel Fuels that have formed geologically by changes in animal and plant life. The resulting hydrocarbons formed in past geologic time and their development is measured in thousands to millions of years.

Fracture Any break or rupture in earth materials. Minerals have special fracture types, and rock fractures can be divided into faults, joints, and cleavage.

Friction The resistance to motion between two bodies in physical contact.

Fumarole A surface earth opening from which gases escape in volcanic and geyser regions.

Fusion The atomic combining of light elements to form heavier elements with the release of energy.

G

Gabbro A coarse-grained intrusive igneous rock with mafic silicate minerals and plagioclase.

Gabion A wire mesh basket filled with rocks and emplaced on hillsides and stream banks for stability.

Gasification The production of gas from the burning of coal or other solid hydrocarbons.

Geochemistry The use and application of chemical principles to the study of geology.

Geohydrology The science of subsurface water. Some prefer the term *hydrogeology*, but there is disagreement on which term is most appropriate, to emphasize the movement of water, or to emphasize the character of the geologic host rocks.

Geologic cycle The system of interrelated earth cycles that includes the hydrologic cycle, the rock cycle, and the tectonic cycle.

Geology The science of the earth. It includes the study of earth materials, processes, and time relations.

Geomedicine The study that relates mineral, rock, and water properties to human health.

Geomorphology The study of the earth's surface features and the exogenic processes that sculptured them.

Geophysics The use and application of physics to the study of geology.

Geophysical prospecting The application of physics-type equipment for the exploration of mineral and fuel deposits.

Geothermal energy The utilization of heat and gases produced by subsurface geologic processes.

Geyser An earth opening that periodically erupts steam and boiling water.

Glacier A mass of moving ice on land.

Global tectonics The theory that describes the entire range of lithospheric plate motions and the deformational features that result.

Gneiss A banded coarse-grained metamorphic rock.

Goiter An enlargement of the thyroid gland. One cause is iodine deficiency.

Gossan A yellow or yellow-red-brown mixture of hydrated iron oxides at the earth's surface. Such an occurrence may indicate the presence of deeper sulfide ore deposits.

Granite A coarse-grained intrusive igneous rock composed of quartz, orthoclase, and about 10 percent ferromagnesian minerals. Most common intrusive rock.

Gravel Unconsolidated sedimentary deposit with pebbles greater than 2 mm in diameter.

Gravimeter An instrument used to detect differences in subsurface gravity fields.

Greenbelt Open space that occurs in peripheral areas to a city and may encircle the urbanized core.

Greenhouse effect The process of warming the earth's atmosphere by absorption of solar infrared radiation by gases such as carbon dioxide and water vapor. The term is unfortunate because man-made greenhouses stay warm mostly by heat retention from the conservation of warm air.

Groin A shore protection structure built from the beach into the water to trap sediment in the littoral zone and prevent coastal erosion.

Gross national product (GNP) An economic index that measures the commercial productivity of a country.

Groundwater Water in the zone of saturation below the water table.

Grout A man-made mixture of cementing materials injected into rock openings and cracks to enhance their strength.

Gypsum An evaporite mineral with the chemical composition $CaSO_4, \cdot 2H_2O$.

H

Habitat The place or locality where an organism thrives best.

Half-life The time required to reduce the number of atoms of a radioactive isotope by one-half.

Hanging wall One side of a fault that occurs over the fault zone.

Hazard A process or locality that has the potential for producing a disaster.

Hazardous substances Those elements or chemical compounds that are injurious to human health.

Heat island effect The elevated temperatures produced within a city that cause it to be warmer than the surrounding region.

High-yield varieties (HYV) Crops developed during the "green revolution" with very high yields.

Hot spot The locality where a plume of magmatic heat is rising.

Humus The decomposed residue of plant and animal tissue that occurs in the top part of the A-horizon of the soil.

Hurricane A tropical storm with counterclockwise winds in the Northern Hemisphere that exceed speeds of 74 mph. These tropical cyclones are called hurricanes in the Atlantic, typhoons in the Pacific, and cyclones in the Indian Ocean region.

Hydration The process of adding water to earth materials.

Hydraulic mining The use of high-pressure hoses to destroy the compaction of unconsolidated materials to extract ore minerals.

Hydrocarbon An organic compound containing oxygen and carbon, such as coal and petroleum.

Hydroelectric power Energy created at dams and other localities where water falls and its force is sufficient to move turbine vanes and create electricity.

Hydrofracturing Injection of water by pumping into the ground and allowing high pressure buildup in order to create rock fractures and increase their permeability. A method used to increase oil or water production.

Hydrograph A graphic depiction of water changes over a period of time.

Hydrologic cycle The global system for movement of water by evaporation from water bodies, entrainment of water into the atmosphere, precipitation on land, movement on and through earth materials, as in streams and groundwater, and finally its return to the ocean.

Hydrology The science of water. Usually divided into surface water hydrology and geohydrology.

Hydrothermal ore deposits Mineral deposits that originated from hot water and gaseous materials by magmatic action.

I

Ice age A time in earth history when a significant part of the land is covered by glaciers. The most recent ice age was largely confined to the Pleistocene Epoch, which started more than 2 million years ago. Some argue we are still in the Ice Age because 10 percent of the land is still cloaked by glaciers.

Igneous rock Rock formed by the solidification of molten material. If crystallization occurs below the ground, such rocks are called "intrusive"; if formed on the surface, they are called "extrusive."

Impermeable Describing earth materials that prevent the easy flow of fluids. Also called *impervious*.

Infiltration The movement of surface water into subsurface materials.

Influent stream The hydrologic condition in which streamflow percolates into the groundwater zone because the water table is lower than the streambed.

Intensity earthquake The relative damaging power of an earthquake at a particular locality. Commonly measured by the Modified Mercalli Scale.

Irrigation The act of emplacing additional water on crops to augment natural rainfall. Irrigation water may have either groundwater or surface water sources.

Island arc A curved group of volcanic islands, commonly associated with deep-ocean trenches and occurring at a plate boundary subduction region.

Isotope Those elements that have different forms due to differences in the number of neutrons.

J

Jetty A man-made structure at coastal inlets designed to prohibit erosion and protect navigation routes. The structure generally extends from land into the water.

Joint A fracture that contains no appreciable movement parallel to the break.

K

Kame An ice-contact hill of glacial meltwater deposits that contains stratified sediments.

Karst topography An assemblage of topographic depressions such as sinkholes, caused by underground solution of carbonate rocks. Named for the Karst region of Yugoslavia.

Kimberlite An ultrabasic igneous conduit-type intrusion that is the source area for diamonds.

L

Lahar Avalanching landslide of pyroclastic materials on the flank of a volcano.

Landform A single terrain feature, such as a mountain or a valley.

Landscape The combination of individual landforms that comprise the total terrain system.

Landslide The rapid movement end of the mass movement spectrum.

Specifically, the motion and the landforms created as a result of rapid movement of rock and/or regolith on the surface. Such movement can occur by falling, sliding, or flowing.

Land use planning The entire scope and strategy involved with systems analysis related to issues involving utilization of the lands and waters of a particular site or region. It includes a range of environmental assessments with reports, maps, and documentation for implementation of preferred programs.

Land-water ecosystem The manner in which the terrain and hydrology of a site or region are interrelated and in balance. Man-made changes can affect the harmony of the system and produce deleterious effects.

Laterite A soil formed by chemical weathering in hot, humid-tropical regions that produces material rich in oxides of iron and aluminum.

Lava Molten silicate materials formed by volcanic action on the earth's surface.

Law of aphasy Describes the effects produced when the climate changes more rapidly than organic life can change in order to adapt to the new environment.

Law of equifinality The concept that similar appearing landforms can be created by different geologic processes.

Law of parsimony When faced with a choice, the most reasonable selection is the alternative that is the least complex.

Law of variables The development of earth features is complex and involves the interaction of many factors.

Leachate The contaminated wastewaters that are produced as a result of water reacting with materials in a landfill. Such waters contain many trace metals and hazardous substances.

Leaching The continual removal by water of soluble matter in the soil, regolith, or bedrock, generally by downward percolation.

Less developed countries (LDC) Those nations with very low living standards and a low gross national product (GNP). In general, countries with few industries. Also called nations of the *Third World*.

Leukemia Cancer of the blood.

Levee A broad low ridge of unconsolidated earth material along the side of a stream. Natural levees are produced by water spreading out of the channel during floods. Artificial levees are constructed to prevent damages from flooding.

Lignite A low rank coal with less than 8300 BTU.

Limestone A sedimentary rock comprised mostly of calcite.

Liquefaction The rather sudden release of water from a saturated sediment, resulting in deformation of the remaining sediments and migration of the water. Earthquake shocks can liquefy materials such as quick clays and sensitive sediments.

Lithification The conversion of unconsolidated materials into sedimentary rocks.

Lithosphere The outermost earth layer about 100 km thick that contains rocks of the oceanic and continental plates.

Littoral current The streamlike movement of ocean water parallel and adjacent to the shore. Synonymous with *longshore current*.

Littoral drift The sediments that are in transport by littoral currents.

Loess A deposit of wind-blown material, consisting mostly of silt-size particles.

Longshore current Same as *littoral current*. The streamlike movement of ocean water that is parallel and adjacent to the shore.

M

Mafic rock A rock containing more than 50 percent ferromagnesian minerals. Usually refers to igneous rocks.

Magma Molten silicate materials created by igneous activity.

Magnetite This mineral is the important iron ore, Fe_3O_4.

Magnitude, earthquake A measure of the total energy released by an earthquake. Usually expressed on a logarithmic scale, such as the Richter Scale, in which the magnitude is given by a series of numbers.

Management The total system of operations necessary to organize, govern, and control a project or enterprise.

Mantle The 2900 km thick zone between the lithosphere and crust of the earth.

Marble A nonfoliated carbonate metamorphic rock, usually formed from limestone.

Mass movement The motion of earth surface materials by gravity. Synonymous with *mass wasting*. Rapid movements are landslides, and slow movements are creep.

Meanders The sinuous geometry of streams and channels.

Mechanical weathering The breakdown of rock in place at or near the earth's surface by physical processes not associated with moving forces. Identical to physical weathering.

Mercalli scale A scale of earthquake intensity with divisions from I to XII that rate the relative motion at a particular site.

Metamorphic rock A rock formed in the solid state from a preexisting rock by changes in temperature, pressure, and chemistry. Such transformation occurs deeper within the earth than zones of weathering.

Metasomatism In ore deposits, the process that involves replacement of one mineral by new solutions that produce a new mineral.

Mica A common rock-forming silicate mineral with extremely thin sheets.

Microclimate A small-scale system of climatic factors that differs from the general climate on contiguous lands.

Microseism Feeble earth tremors, not necessarily generated by earthquakes.

Mineral A naturally occurring substance with characteristic physical properties and chemical composition with crystalline formation.

Mineral resources Those minerals that are useful and needed by society. In the general sense, both minerals and rocks are often included in this classification.

Minimum tillage The farming technique that keeps soils as intact as possible and does not remove other growing plants. Seeds for new crops are planted with minimum terrain disturbance.

Monocline A fold with strata having the same dip and direction on either side of the bend.

Montmorillonite A clay-mineral family of hydrous aluminum silicates. This mineral has maximum shrink-swell potential and can produce costly structural damages.

Moraine A general term used to denote a glacial deposit, usually consisting of hilly terrain.

Mudflow A mixture of water and unconsolidated materials that flows rapidly in stream channels and more rarely on hillsides.

N

National Environmental Policy Act (NEPA) An act that in 1970 established a Council on Environmental Quality and requires Environmental Impact Statements (EIS) from government agencies and other groups that plan to construct major projects.

Natural resources Original earth materials of use and importance to mankind.

Nature Inherent earth processes and materials not related to mankind.

Negligence A litigation in which a plaintiff sues a defendant who reportedly has failed to correct a knowingly damaging situation.

Noachian Deluge The biblical account of Noah and the flood.

Nonrenewable resources Those geologic resources that are not being sufficiently renewed by natural processes.

Normal fault A fracture in which the hanging wall has been displaced downward relative to the footwall.

Nuclear reactor In an atomic power plant, the chamber in which the nuclear fission reaction is controlled for energy production.

Nuée ardente An incandescent cloud consisting of superheated gases and hot, fine-grained tephra emitted during volcanic activity.

New town A village or city that has been planned from the beginning with certain objectives in mind. The beauty of the physical environment is retained when possible.

O

Ocean current Major oceanic river-like movements of massive bodies of water, such as the Gulf Stream and the Humboldt Current.

Ocean thermal energy conversion (OTEC) The engineering methodology and equipment that produce energy by the difference in temperature between the surface ocean water and the colder subsurface water.

Off-road vehicles (ORV) Motorbikes, dune buggies, and four-wheel drive vehicles that can be driven on unpaved areas.

Oil Synonymous with *petroleum*, except it does not include natural gas.

Oil shale A body of fine-grained sedimentary rocks rich in hydrocarbon derivatives. Such deposits are capable of yielding petroleum through rock cracking and distillation.

OPEC The name applied to the group of nations that constitute the Organization of Petroleum Exporting Countries. This is a cartel of 13 countries that fix the price of oil to be sold on the world market.

Open pit mining The surface mining of mineral resources. As distinct from strip mining, open pit mining usually extracts the ore from a single or a multiple series of pits.

Open space The planning concept that sets aside open areas and disallows their development for commercial purposes. Usually situated in or contiguous to urbanized areas.

Ore Metallic minerals in sufficient concentration that can be profitably mined.

Orogeny A general term for a period of mountain building.

Osteoporosis A disease that attacks the bone and causes its destruction.

Outcrop A surface exposure of earth materials.

Outwash A glacial deposit of materials transported by meltwaters beyond the ice margin containing stratification of sediments.

Overburden Earth materials that are on top of another deeper group of materials.

Overgrazing The process of land deterioration caused by an overabundance of grazing animals which exceeds the capacity of the vegetation to regenerate.

Oxidation The chemical process of adding oxygen to an earth material. It is also one of the weathering processes.

P

P wave The primary or fastest earthquake wave capable of transmission through both liquids and solids.

Pahoehoe A Hawaiian term for a type of lava that is ropy.

Particulates Small solids emitted by man-made combustion-producing

activities.

Pathogen Any disease-causing substance.

PCB The full chemical name for this substance is polychlorinatedbiphenyl. It is a hazardous chemical waste compound that has only recently been discovered to cause serious health problems to humans and animals.

Peat A lightweight mixture of partly decomposed plant tissues containing about 60 percent carbon and 30 percent oxygen.

Pedology The science of the soil.

Pegmatite An exceptionally coarse-grained igneous rock. Usually occurs in veins or dikes and may contain important ore minerals.

Perched, water A saturated zone of subsurface earth materials above the regional water table.

Percolation test The mandated method to determine if a particular soil will accept infiltrated water sufficiently fast and in volumes that will permit use for a septic tank system.

Permafrost Permanently frozen ground.

Permeability The ability of earth materials to transmit fluids.

Petroleum Gaseous, liquid, or solid substances occurring naturally and consisting of chemical compounds of carbon and hydrogen. Includes both oil and natural gas.

Petrology The study of rocks.

pH A numerical chemical scale that measures the degree of acidity or alkalinity of a solution; a pH of 7 is neutral; higher numbers are alkaline (or basic), and lower numbers are acidic.

Phosphate A chemical radical of the form PO_4. Also refers to those minerals that can be classified into the phosphate group.

Phreatic Pertaining to the groundwater zone.

Phreatophytes That type of vegetation capable of extending roots into the groundwater zone and using the water for growth.

Physiographic A term used in the description of terrain features for large regions.

Piezometric surface A geohydrologic term that relates to water level position when influenced by a pressure head. Generally coincident with the water table.

Piping The formation of small passages and conduits by percolating surface water into unconsolidated sediments above the water table.

Placer deposit An ore deposit formed by surface processes that have winnowed away less dense or softer materials and left behind a lag of heavy or resistant minerals, such as gold, diamonds, and tin-bearing minerals. Streams produce most placers.

Planning The operational preparation for a scheme of action.

Plastic deformation The strain that results when a body is stressed and is changed in shape without being fractured.

Plate tectonics Similar to global tectonics, but with emphasis being placed on movement of the lithospheric plates.

Playfair's law The concept that tributary streams join master streams at concordant positions of similar elevations.

Podzol A soil with B-horizon containing aluminum- and iron-leached materials that forms in cool temperate climates.

Point bar The geomorphic landform of alluvial sediments that occurs on the inside bend of a meander.

Pollution The deleterious man-made products that impair the quality of some part of the environment.

Pool, stream That part of a stream channel deeper than other sections. Such deeps occur at distances in equilibrium channels equal to about six times the channel width.

Population explosion The extraordinary rapid growth in the number of humans throughout the world that started during the twentieth century.

Pork barrel legislation The type of statutes passed by government bodies designed to channel funds into specific localities in recognition of special-interest groups or because of some favor and special treatment.

Porosity The proportion of openings in earth materials given in percent.

Potable water Drinking water that is safe and free from contaminants.

Pothole An erosional depression, such as road potholes or stream potholes.

ppm Parts per million.

Precursory event A warning signal that a hazardous process is about to occur, such as preliminary phenom-ena that herald the onslaught of an earthquake or a volcanic eruption.

Preservation The ethic that believes that mankind should not disturb natural features but retain them in as pure a form as possible, such as in wilderness areas.

Prior rights Synonymous with *appropriation law*. The first land owner beside a water body is entitled to first use of the water, regardless of uses requested by later property owners.

Pyrite An iron sulfide mineral. Important in surface mined areas and man-made terrain cuts because its weathering produces substances that cause acid buildup on adjacent lands and water.

Pyroclastic activity Volcanic eruptions that produce tephra during explosive phases.

Pyrolysis The process of decomposing material by heating in the absence of air. One form of solid waste elimination, as well as valuable residue, may result.

Q

Quartz Silicon dioxide composition. A very common rock-forming mineral.

Quartzite A quartz-rich nonfoliated metamorphic rock.

Quick clay A water-bearing clay capable of rapid liquefaction when disturbed.

R

Radioactive waste Unwanted nuclear products that result from nuclear power plants or other manufacturing industries that use radioisotopes.

Radioactivity The spontaneous decay of the atoms of certain isotopes by emission of alpha, beta, and gamma particles.

Radon A radioactive gaseous element.

Reasonable use The doctrine that requires beneficial and moderate use of water supplies in order to qualify for continued use of the resource.

Recharge The addition of water to the groundwater zone. When we deliberately engineer a system, such as one using injection wells and recharge pits, the process is called "artificial recharge."

Reclamation The act of restoring previously used lands for new productive purposes, or the change in natural systems to accommodate human utilization, such as the draining of wetlands.

Recurrence interval The amount of time between recurring events, such as floods or landslides.

Recycling The reuse of materials that are reprocessed from waste debris.

Regolith Unconsolidated rock fragments that cover the bedrock.

Relief, topography The height difference between high and low places in a given terrain.

Remote sensing The procedure in which instruments and equipment that are used to obtain data and measurements of earth materials have a recording device that is not in contact with the feature being calculated. Images from satellites are one example.

Renewable resources Those resources that can be naturally regenerated, such as timber and water.

Reserves Potential resources that have been determined to be useful and can be profitably mined under existing economic and technologic conditions.

Residence time The length of time a particular phenomenon occurs at a site, such as the number of days the floodwaters cover the land.

Resistivity The measure of retardation of electricity through earth materials. It is the opposite of *conductivity*.

Resources Earth materials useful and needed by society. In the broadest sense, they include minerals, rocks, soil, organic matter, water, and air.

Reverse fault A fault with upward movement of a hanging wall relative to the footwall block.

Revetment A sloping surface lined with stone or resistant man-made materials designed to prevent water erosion.

Riffle That part of a stream channel where the bed is higher and water flow is more rapid. Such features alternate with pools and occur at similar intervals.

Riparian rights The privileges of waterfront owners to use stream and lake water. Such use must be reasonable. Under · water law, the water must be returned to the stream, but the owner has rights to the water, which is supposed to be maintained in its normal quality and quantity.

Riprap Broken rocks or masonry used to cover an embankment for its protection against water erosion.

Rock In geology, an aggregate of minerals. In engineering, any earth material that must be blasted in order to excavate the site.

Rock cycle Those processes that form igneous, sedimentary, and metamorphic rocks.

Rock mechanics The study of physical properties of rock, especially those features related to breaking, crushing, and deformation of the material.

Rockfall Rockfall, rockflow, and rockslide are all landslide types. A rock avalanche is a special category of landslide in which dry broken rock attains extraordinarily rapid movement and the landslide occurs on a massive scale.

Rotational landslide Movement of earth material that is scooplike, with top materials moving less than basal materials, in a downslope direction.

S

S wave The secondary or shear wave generated by an earthquake that propagates only through solid material.

Safe yield The amount of water that can be pumped from a groundwater reservoir that equals recharge to the reservoir.

Salinity The measurement of total dissolved solids in water.

Salinization The process of increasing the salt content of earth materials and water.

Salt dome An up-arched deformation in sedimentary rocks created by upward movement of a salt plug. Such sites may contain oil and gas deposits.

Sand Sediment particles with diameters larger than 1/16 mm and smaller than 2 mm.

Sand dune A geomorphic feature consisting of a hill created by wind-blown sand.

Sandstone A clastic sedimentary rock consisting primarily of sand-size particles.

Sanitary landfill A solid waste disposal method in which rubbish, garbage, and waste products of all descriptions are dumped on the ground and covered daily with a layer of compacted fine-grained earth material.

Saturation The filling of all openings in earth materials by a fluid.

Scenic resources Visual amenities that have aesthetic value.

Schist A well-foliated metamorphic rock comprised of visible flaky minerals.

Schistosomiasis A crippling or lethal tropical disease caused by a worm that infests snails which then become the carriers.

Sea floor spreading The growth of ocean basins by the movement of lithospheric plates. The counterpart to the earlier concept of *continental drift.*

Sea wall A coastal embankment emplaced along the shore to prevent water erosion.

Secondary enrichment Weathering processes that increase the concentration of minerals so that they become sufficiently common to be mined as ore.

Secondary treatment That part of the operation in a waste disposal and sewage plant in which the processed fluids are biologically treated for their purification.

Sedimentary rock A rock formed by lithification of a sediment at or near the earth's surface.

Sedimentology The study of environments and processes involved in the formation of sediments.

Seismic Pertaining to shock waves of earthquakes.

Seismogram The record made by a seismograph.

Seismograph An instrument that records seismic waves.

Seismology The study of seismic waves and earthquakes.

Septic tank A buried container designed to hold fluid and solid waste products from a single residence or small facility. The holding tank permits breakdown by anaerobic bacterial action of waste materials, and fluids drain from the tank into lateral structures that comprise a leaching field.

Sewage sludge Solid waste that remains behind after treatment of sewage by a sewage disposal plant. Such material may occasionally be used as fertilizer or for other purposes when not contaminated.

Shale A sedimentary rock composed of fragments of microscopic size, less than 1/256 mm in diameter, that has not been cemented.

Shear Inward directed forces that are not along the same stress plane, creating an action similar to the movement of a deck of cards.

Sheeting A weathering process that creates large slabs of exfoliated rock. Often associated with the denudational unloading of overburden.

Shrink-swell A feature associated with sediment that changes size depending on moisture content. Occurs especially in montmorillonite-bearing materials.

Silicate minerals The most common earth minerals which contain silicon dioxide as an important constituent.

Silt Sediment with diameters between 1/16 and 1/256 mm.

Siltation The process of accumulation of silt and closely associated sediments.

Sinkhole A solution cavity open to the sky that is generally formed by subsurface erosion in carbonate rocks and the caving in of overlying rock.

Slate A fine-grained foliated metamorphic rock that splits readily along smooth thin planes.

Slump The downward and outward movement of a coherent body of earth material representing one type of landslide motion.

Soil In pedology, that part of the regolith that can support rooted plants. In engineering, any earth material that can be excavated without blasting.

Solar energy The utilization of the sun for production of energy.

Solid waste The full range of non-fluid waste produced by society.

Solifluction The imperceptably slow movement downhill of saturated regolith. Most common in cold climates, where thawed surface materials move over permafrost.

Special-interest group An organization designed to obtain particular benefits that are self-serving. Lobbyists usually perform such services.

Spheriodal weathering The exfoliation of thin rock shells on a small scale by weathering processes.

Spoils, mining The unused debris and tailings that result from mining operations.

Spring The emergence of underground water that flows from a ground surface opening.

Standard project flood That particular flood stage that is used for flood planning purposes. Often it is the 100-year flood.

Stewardship Unselfish and wise management of the land-water ecosystem that permits its long-range use and enjoyment.

Strain Any change in earth material produced by stress, such as change in shape or size.

Strata Layered sediments. Synonymous with *bedded rocks*.

Stream Any natural, clearly defined, and constrained body of flowing water on land. The size range includes creeks, brooks, rivers, and washes. The *stream channel* or *stream bed* describes the bottom over which the water moves.

Streamflow The water that moves in a stream.

Strength The resistance an earth material offers to a stress before final rupture. It can be measured in terms of short-term, long-term, and ultimate strength.

Stress Force per unit area. Typical stresses include tension, compression, and shear.

Striation Small-scale scratches on rock caused by rock abrasion. Striations may result from the action of glaciers, faults, landslides, and so forth.

Strike The compass direction of the horizontal line in an inclined plane.

Strip mining The method of surface mining that progressively removes new rock layers row after row.

Subduction zone The location of a lithospheric plate where it descends under another plate.

Subsidence A differential sinking of the earth's surface in respect to surrounding terrain.

Subsurface water All water that occurs below the ground surface. It includes soil water, perched water, and phreatic water.

Sulfuric acid A highly corrosive chemical compound, H_2SO_4.

Surcharge The loading or extra weight burden emplaced or infiltrated into a body of earth materials.

Surf Waves of translation between the line of breakers and the shore.

Suspended load Sediment entrained within a moving fluid.

Syncline A downbend in rock strata. This type of fold has the youngest rocks in the center with limbs that are inclined downward to a common linear trough.

Synergism A condition in which the total effect is greater than the sum of the parts.

System An assemblage of parts, materials, and processes that are united in a common manner or purpose.

T

Taconite Iron-bearing silicon dioxide.

Tailings The waste residue from mining operations.

Talus An apron of earth material sloping outward from a cliff that supplied the rock.

Tar sands A sand or sandstone whose pores contain hydrocarbons in the solid state.

Tectonic Pertaining to rocks that have been deformed by endogenic processes.

Tension Forces directed outward from a common locus; a pulling apart.

Tephra A collective term for all ejected volcanic particles, regardless of size, shape, or composition.

Terrace A relatively elongated flat surface bounded by steeper slopes on both sides of the tread. Such landforms can be created by many different processes.

Texture, drainage Refers to the closeness of spacing of streams in a basin.

Texture, rock The size, shape, and arrangement of rock particles or crystals.

Third World nations Those countries that are economically poor and industrially backward.

Threshold The condition in which a slight change will produce a much greater effect than the final force that triggered the impact. That situation of equilibrium that can easily be disturbed and result in a magnified change.

Tidal energy The use of tides to create electricity.

Till An unconsolidated glacial sediment that is directly deposited by ice without being reworked by other processes. Usually contains a lack of bedding and sorting.

Tolerable erosion The amount of earth material that can be removed from a soil without jeopardizing its productivity. The amount of soil loss is considered to be about equal to the rate of new soil formation.

Topography The status of terrain in terms of shape, length, and height of hillslopes and accompanying hydrographic features.

Toxic Substances that are injurious or lethal to humans and animals.

Translation landslide That family of landslides in which the mass of earth materials moves downslope on a clearly defined geologic horizon.

Trespass, law of A legal basis for environmental lawsuits when contaminants from a polluting source invade the property of a landowner.

Tropical cyclone The general name applied to extraordinarily large and strong ocean air mass storms with winds that exceed 74 mph. Hurricanes and typhoons are examples.

Tsunami Ocean waves generated by displacement of the sea floor. They can be caused by earthquakes, landslides, and volcanic action, and the extra long wave lengths of water produced pile up on shore where water levels reach great heights. Incorrectly called *tidal waves* in popular literature.

Tuff A fine-grained pyroclastic volcanic rock consisting of ash that has been lithified.

Typhoon The name applied to tropical cyclones in the Pacific Ocean region.

U

Ultimate strength The final breaking point of rock. The highest point on the stress-strain diagram.

Unconfined aquifer An aquifer whose water movement is not restricted.

Undeveloped nations Those countries with minimal industrialization and low living standards. Also called *LDC* (less developed countries).

Urban sprawl The housing and commercial condition that results when land is developed in areas adjacent to cities without planning and management of the land-water ecosystem.

Urbanization The process of inward migration to metropolitan sites. It involves the movement of people into the city from rural and other areas, causing a growth and expansion of the area consumed by the increased development.

V

Vein A crack or fissure in rock that is filled with minerals.

Viscosity The property of a substance to resist flow. Water has low viscosity and tar has higher viscosity.

Volcanic Pertaining to igneous rocks on the earth's surface.

Vulcanism Pertaining to all igneous rocks, regardless of location.

Vulcanology The study of volcanic rocks.

W

Water cycle See *Hydrologic cycle*.

Water logging That condition of saturation of surface materials associated with a rise of the water table which is caused by excessive irrigation.

Watershed See *Drainage divide*.

Water table The top surface of the groundwater zone. The demarcation between saturated earth materials below and nonsaturated materials above.

Wave A configuration of matter that transmits energy from one position to another.

Weather The condition of the atmosphere at a specific time and place, including its moisture, winds, temperature, and pressure. A short-range status compared to climate, which averages conditions for a long time period.

Weathering The disintegration and decomposition of rock materials during surface exposure to air, moisture, organisms, and chemicals. It involves both physical and chemical rock changes in place, as distinct from erosion, which causes movement of material away from the site.

Wetland A general term that applies to all water-saturated ground. The term replaces such previous words as bog and swamp. There are both saltwater and freshwater wetlands.

Z

Zero demand distance The distance beyond which the cost of transportation for minerals and rocks is greater than the value of the product at the mine.

Zinc An important trace element necessary for human life. A metal with many industrial uses.

Zonal soil A soil with a well-developed profile that is in balance with the climate.

Zone of aeration The region of earth materials above the water table where all pore spaces are not filled with water, but contain some air.

Zone of saturation The region of earth materials below the water table in which all openings are filled with water.

Appendix A Geological time scale, showing major events in geological history of the United States

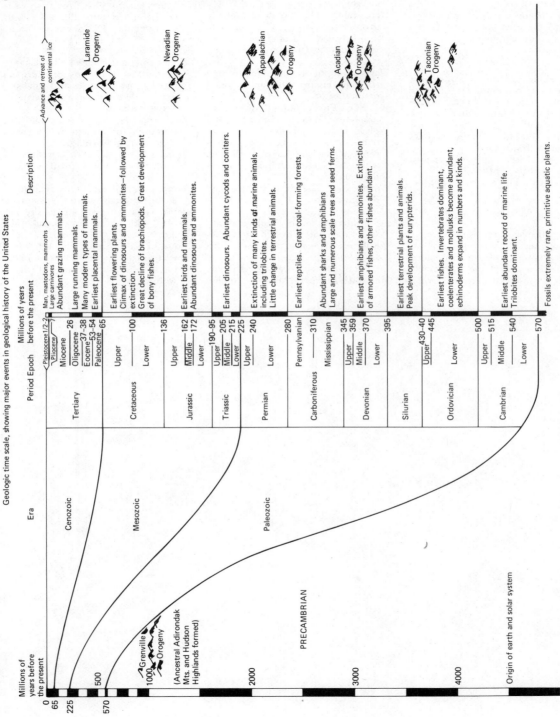

A1

Appendix B Classification of Igneous Rocks

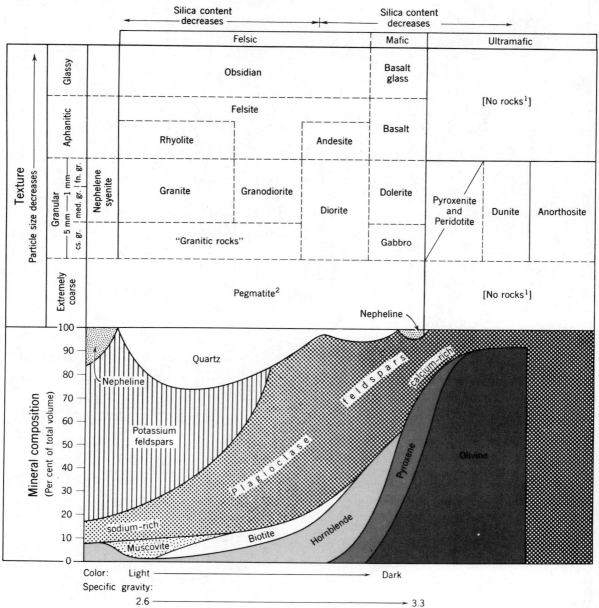

From C.R. Longwell, R. F. Flint, and J. E. Sanders, 1969, *Physical geology:* New York, John Wiley & Sons.

Note: Gradational boundaries are shown by dashed lines. Only *four* kinds of aphanitic rocks and *two* kinds of glassy rocks are shown, because it is not possible to identify others without a microscope or chemical analysis.

[1]No rocks having this composition and texture are known.

[2]Pegmatite designates extremely coarse-grained igneous rocks of any composition. When used alone, *pegmatite* implies granitic composition. Pegmatites having other compositions are named by including the appropriate rock term, as *diorite pegmatite* or *gabbro pegmatite*.

Appendix C Classification of Sedimentary Rocks

	Partly igneous, partly sedimentary	Rocks formed from sediments		Accumulations of plant debris and hydrocarbons
Major classes of sedimentary rocks	Partly igneous, partly sedimentary	Rocks formed from sediments		Accumulations of plant debris and hydrocarbons
Major classes of sediments	Pyroclastic	Clastic	Nonclastic — Precipitates	Nonclastic — Organic
Particles: Kinds of particles	Tephra (particles blown out of the Earth's crust)	Detritus (particles of, or mineral grains derived directly from or as a result of weathering of preexisting rocks or sediments)	Whole shells, crystals, and other particles (particles precipitated out of water at site of deposition by biological and/or physical-chemical processes)	Pieces of plants or material derived from plant debris
Particles: Composition of particles	Crystals; crystalline particles; rock particles; glass particles. Minerals are typical of igneous rocks	Particles of silicate rocks; quartz; feldspar; micas; heavy minerals; clay minerals / Particles of carbonate rocks; calcite; dolomite	Aragonite Calcite Dolomite / Broken shells and shell debris[1] / Halite Gypsum Anhydrite	Organic compounds
Texture		Clastic (Fragmental)	Crystalline	Organic
Method of transport of particles	Physically transported through atmosphere[2]	Physically transported; may be abraded or broken during transport	Material transported in solution; precipitated as particles or as coatings on particles at site of deposition / Particles not transported	Some accumulated in place

A3

Classification of sediments and rocks by particle size

Tephra (pyroclastic)

Rock name	Sediment name and limiting diameter of dominant particles	
Agglomerate	Bombs (particles molten when ejected) / Blocks (particles solid when ejected)	
		32mm
Lapilli tuff	Lapilli	
		4mm
Tuff	Ash	
		$\frac{1}{16}$mm
	Dust	

Tephra[4]

Particle size increases →

Clastic and chemical/biochemical sediments and rocks

Sediment name and limiting diameter of dominant particles		Rock name
Gravel		Conglomerate (particles rounded) / Sedimentary breccia (particles angular)
	2mm	
Sand		Sandstone, Arkose, Graywacke
	$\frac{1}{16}$mm	
Silt		Siltstone
	$\frac{1}{256}$mm	
Clay		Claystone, Shale (if fissile)

Limestone and Dolostone — [No rocks[3]]

Evaporites: Rock salt (mineral halite), Gypsum rock (mineral gypsum) etc.[4]

Reef rock[4] (mostly limestone and dolostone)

Peat, Coal[4]

From C.R. Longwell, R. F. Flint, and J. E. Sanders, 1969, *Physical geology:* New York, John Wiley & Sons.

[1] Some skeletal material is composed of silica.

[2] Tephra can be transported also by other carrier agents, such as running water.

[3] No rocks having this grain size, composition, and origin are known.

[4] Tephra, salt, gypsum, reef rock, peat, and coal are set vertically to emphasize the fact that they are unrelated to the particle diameters shown in the scale within the diagram.

Appendix D Classification of Metamorphic Rocks

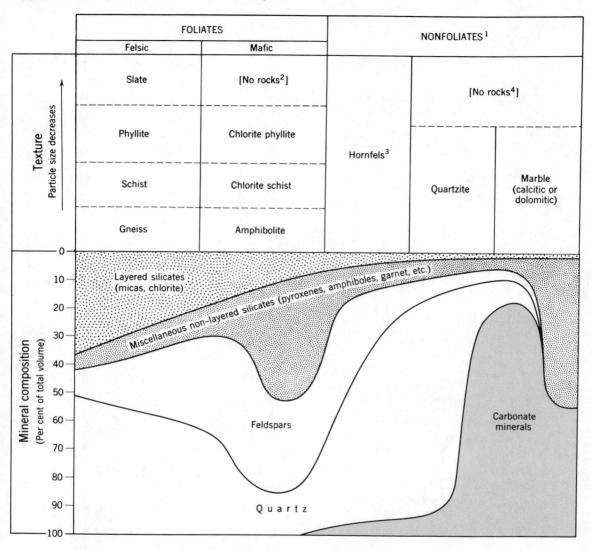

	FOLIATES		NONFOLIATES [1]		
	Felsic	Mafic			
	Slate	[No rocks [2]]		[No rocks [4]]	
	Phyllite	Chlorite phyllite	Hornfels [3]		Marble (calcitic or dolomitic)
	Schist	Chlorite schist		Quartzite	
	Gneiss	Amphibolite			

Texture — Particle size decreases

Mineral composition (Per cent of total volume)

Layered silicates (micas, chlorite)

Miscellaneous non-layered silicates (pyroxenes, amphiboles, garnet, etc.)

Feldspars

Quartz

Carbonate minerals

From C. R. Longwell, R. F. Flint, and J. E. Sanders, 1969, *Physical geology:* New York, John Wiley & Sons.

Note: Horizontal dashed lines show gradational particle-size boundaries.
[1]Most rocks that appear in hand specimens to be nonfoliates actually are foliated on a large scale when seen in large natural exposures.
[2]In a simple classification, no rocks of this kind need be considered.
[3]The name hornfels is generally applied only to contact-metamorphosed rocks. Typical hornfels is fine-grained, but coarse-grained varieties are known.
[4]Fine-grained quartzites and marbles are rare. Although fine-grained parent rocks exist, recrystallization enlarges their particles.

Appendix E Origin of Mineral Deposits and Proposed Classification System[a]

Process	Deposits	Examples
1. Magmatic concentration	I. Early magmatic:	
	A. Disseminated crystallization	Diamond pipes
	B. Segregation	Chromite deposits
	C. Injection	Kiruna magnetite(?)
	II. Late magmatic:	
	A. Residual liquid segregation	Taberg magnetite
	B. Residual liquid injection	Adirondack magnetite, pegmatites
	C. Immiscible liquid segregation	Insizwa sulphides
	D. Immiscible liquid injection	Vlackfontein, S. Africa
2. Sublimation	Sublimates	Sulphur
3. Contact metasomatism	Contact metasomatic: Iron, copper, gold, etc.	Cornwall magnetite, Morenci (old), etc.
4. Hydrothermal processes A. Cavity filling	Cavity filling (open space deposits):	
	A. Fissure veins	Pachuca, Mexico
	B. Shear-zone deposits	Otago, New Zealand
	C. Stockworks	Quartz Hill, Colo.
	D. Ladder veins	Morning Star, Australia
	E. Saddle-reefs	Bendigo, Australia
	F. Tension-crack fillings (pitches and flats)	Wisconsin Pb and Zn
	G. Breccia fillings:	
	a. Volcanic	Bassick pipe, Colo.
	b. Tectonic	Mascot, Tenn., Zn
	c. Collapse	Bisbee, Ariz.
	H. Solution-cavity fillings	
	a. Caves and channels	Wisconsin-Illinois Pb and Zn
	b. Gash veins	Upper Mississippi Valley Pb and Zn
	I. Pore-space fillings	"Red bed" copper
	J. Vesicular fillings	Lake Superior copper
B. Replacement	Replacement:	
	A. Massive	Bisbee copper
	B. Lode fissure	Kirkland Lake gold
	C. Disseminated	"Porphyry" coppers
5. Sedimentation (exclusive of evaporation)	Sedimentary: Iron, manganese, phosphate, etc.	Clinton iron ores
6. Evaporation	Evaporites:	
	A. Marine	Gypsum, salt, potash
	B. Lake	Sodium carbonate, borates
	C. Groundwater	Chile nitrates
7. Residual and mechanical concentration		

Process	Deposits	Examples
A. Residual concentration	Residual deposits: Iron, manganese, bauxite, etc.	Lake Superior iron ores, Gold Coast manganese, Arkansas bauxite
B. Mechanical concentration	Placers:	
	A. Stream	California placers
	B. Beach	Nome, Alaska, gold
	C. Eluvial	Dutch East Indies tin
	D. Eolian	Australian gold
8. Surficial oxidation and supergene enrichment	Oxidized, supergene sulphide	Chuquicamata, Chile. Ray, Ariz., copper
9. Metamorphism	*A.* Metamorphosed deposits	Rammelsberg, Germany.
	B. Metamorphic deposits	Graphite, asbestos, talc, soapstone, sillimanite group, garnet

[a]A. M. Bateman, 1950, Economic ore deposits, 2nd ed.: New York, John Wiley & Son's.

Appendix F HAZARD AND DISASTER EVENTS FROM 1975 TO 1979

1975

Date	Locality	Killed	Remarks
		Earthquakes	
Sept. 6	Lice, Turkey	2,400	12,000 homes damaged, 6.8 M
		Landslides	
Jan. 25–26	Philippines	11	Triggered by tropical storm
Aug. 17–24	Japan	94	Triggered by typhoon rains and floods
Apr. 6	Swiss Alps	10	Avalanche
		Flood	
Jan. 11	South Thailand	131	10,000 homeless, much property damage
Feb. 22	South of Cairo, Egypt	15	20 villages devastated
Mar. 13	Neuquen Province, Argentina	20	6,000 homeless
May 11	Burma	187	Caused by tropical storm
July 28	Punjab area, Pakistan	63	River overflowed
Aug. 26	Yemen	70	Thousands homeless
Sept. 6	Patna, India	67	From drowning and cholera
Sept. 16–20	West Indies and United States	86	Results from hurricane Eloise

1976

Date	Locality	Killed	Remarks
		Volcanic activity	
Sept. 2	Santiaquito, Guatemala	1	Volcanic ash dammed rivers, broke dams, damaged farmlands
Oct. 7–17	Taal volcano, Philippines		Five towns had to be evacuated
Oct. 30	La Soufriére, Guadeloupe		Residents of area were evacuated
		Earthquakes	
Feb. 4	Guatemala	23,000	74,000 injured, 1 million homeless, 7.9 M
Mar. 19	Afghanistan	50	
Apr. 19	Northern Ecuador	10	
May 6	Northeast Italy	900	Much destruction and deaths from landslides
June 26	West Irian, Indonesia	5,000	Many deaths from landslides
July 14	Bali, Indonesia	440	3,000 injured, numerous houses destroyed
July 27–28	Tangshan, China	750,000	City and industrial area destroyed, 7.6 M
Aug. 17	Mindanao, Philippines	3,000	Great destruction largely from tsunami, 8.0 M
Oct. 29	New Guinea	133	Much destruction in mountain terrain
Nov. 7	Northeastern Iran	15	Occurred in mountain area
Nov. 24–27	Van, Turkey	4,000	8,000 injured, 50,000 homeless, destroyed 100 villages, 7.4 M
		Landslides	
Feb. 11–13	Esmeraldas, Ecuador	60	Destruction from rockslides and mudslides
Feb. 12	Bitlis, Turkey	27	Avalanches
May 2	Fresno, Colombia	13	
June 25	Southwest Japan	24	Triggered by torrential rains
Aug. 25	Hong Kong	21	Caused by torrential rains
		Floods	
May 20–24	Luzon, Philippines	215	60,000 homeless, typhoon

Date	Locality	Killed	Remarks
June 5	Teton Dam, Idaho	11	Flooded 780 km², damaged property and farmlands when dam broke, about $1 billion
July 15–25	Central Mexico	20	
July 31	Big Thompson Canyon, Colorado	136	Flash flood, destroyed roads and homes
Aug. 7	Chongju, South Korea	25	Torrential rains
Aug. 20	Pakistan	311	Indus River overflowed
Sept. 1	Kyushu, Japan	57	Typhoon caused flooding
Sept. 10	Southwest Japan	12	
Oct. 1	La Paz, Mexico	500	14,000 injured, hurricane
Oct. 6	Pereira, Colombia	58	Dike burst from rivers swollen by rain
Nov. 13–14	Indonesia	136	
Nov. 29	Hawaii	1	$4 million damage from tsunami

1977

Date	Locality	Killed	Remarks
	Volcanic Activity		
Jan. 10	Gomo, Zaire	70	Overwhelmed by lava
Apr. 5	Great Comboro Island, Indian Ocean		Three villages destroyed, 4,000 evacuated
Aug. 7	Usu, Japan		25,000 evacuated
	Earthquakes		
Mar. 4	Bucharest, Romania	1,300	Damaged many parts of the city, 7.2 M
Mar. 22	Southeast Iran	167	
Mar. 25	Eastern Turkey	30	
Apr. 6	South central Iran	350	Many villages destroyed in 150 km² area
Aug. 19	Sumbawa Island, Indonesia	187	Most damage from tsunami, many injured
Nov. 13	San Juan, Argentina	50	Damaged several villages
Dec. 20	Central Iran	589	16 villages ruined, 4,000 homeless
	Landslides		
Feb. 27	Jujuy Province, Argentina	16	Avalanche
June 28	Southern Philippines	17	Landslides caused by heavy rains
Oct. 22	Sri Lanka	27	
	Floods		
Jan. 2–7	North central Bolivia	16	
Mar. 12	Transvaal, South Africa	12	3,000 homeless
Apr. 4	Central Appalachians, Kentucky, Virginia, West Virginia	10	Cumberland and other rivers flooded
July 19	Johnstown, Pennsylvania	73	2,000 homeless
July 25	South Taiwan	28	20,000 homes destroyed, typhoon
Aug. 15	Madurai, India	35	
Sept. 12–17	Kansas City, Missouri	26	1,200 homeless
Nov. 3	Athens, Greece	26	
Nov. 6	Toccoa, Georgia	39	Heavy rains caused dam collapse
Nov. 12–19	Southeast India	8,400	Cyclone from Indian Ocean

Date	Locality	Killed	Remarks
	Volcanic Activity		
May–June	Mt. Mayon, Philippines		23,000 forced to evacuate or flee
Oct. 24	Usu, Japan	3	
	Earthquakes		
Jan. 14	Izu Peninsula, Japan	25	132 injured, many from landslides, thousands of houses damaged
Apr. 15	Northern Sicily	5	400 houses destroyed
June 12	Honshu Island, Japan	27	1,100 injured
June 20	Salonika, Greece	51	City greatly damaged, thousands homeless
Aug. 13	Southern California		$15 million damages
Sept. 16	Tabas, Iran	25,000	Many cities ruined
Dec. 14–15	Southwest Iran	76	17 villages damages, 6.3 M
	Landslides		
Feb. 4–5	Alps in France and Austria	21	Avalanches
July 29	South central Guatemala	5	Several villages damaged
	Floods		
Mar. 20–25	Mozambique	45	Zambezi River flooded
July 9	Border of Afghanistan-Pakistan	122	
Aug. 3–6	Central Texas	52	
Aug. 18	Acajutla, El Salvador	100	Drowned in tsunami
Sept. 13	Arkansas–southeast Texas	11	
Sept. 22	South Korea	29	
Sept. 27	Southeast Asia	111	Flooding throughout several countries
Oct. 27	Manila, Philippines	20	
Nov. 25	Sri Lanka	500	70,000 homeless from typhoon Much damage from cyclone of Indian Ocean

1979

Date	Locality	Killed	Remarks
	Volcanic Activity		
Feb. 10	Ambrym, New Hebrides		Destroyed vegetation, contaminated water
Feb. 21	Indonesia	175	Also 1,000 injured by volcanic eruption on Java
Apr. 13	Carran, south Chile		Eruption damaged agricultural land with ash, 125 people evacuated
Apr. 13	Soufriére, St. Vincent		17,000 evacuated, crops severely damaged
July	Mt. Etna		Ash destroyed 400 ha of orchards
	Earthquakes		
Jan. 16	East Iran	1,000	1,000 injured, three villages destroyed
Feb. 16	South Peru	14	Arequipu area damaged
Mar. 14	Guerrero, Mexico	5	7,500 homeless, 8 villages destroyed
Apr. 15	Yugoslavia	235	1,400 injured, 100 km of Adriatic Sea developments very damaged

Date	Locality	Killed	Remarks
May 30	Bali Sea area, Indonesia	22	Lombok Island damaged
July 9	East China	41	2,000 injured in Liyang area
Sept. 12	West Irian, Indonesia	100	Many houses destroyed
Sept. 19	East central Italy	5	Umbria area damaged
Nov. 14	Northeastern Iran	500	A 6.7 M earthquake near Meshed
Nov. 23	Colombia	60	A 6.4 M earthquake that also ruined 1,000 buildings
Dec. 12	Colombia-Equador border region	400	A 7.8 M earthquake that also destroyed several villages
Dec. 18	Bali, Indonesia	25	A 6.1 M earthquake that also injured 200 inhabitants

Landslides

Date	Locality	Killed	Remarks
March 12	Northern India	230	Rock and snow avalanches in Himalayan mountain villages in Himachal Pradesh state
Apr. 30	Sumatra, Indonesia	80	Heavy rains caused landslides, which also damaged five villages

Floods

Date	Locality	Killed	Remarks
Apr. 12	Pearl River, Mississippi		Greatest flood damages in history of Jackson area, total loss $600 million
Apr. 18	Philippines	12	Typhoon caused $3.5 million in damages
May 14	Southeast India	600	Indian Ocean cyclone left 1 million homeless in flooded region
Aug. 13	West India	3,000	Heavy rains in Machu River basin caused collapse of dam and great water damage in city of Morvi
Aug. 15	Southern Bangladesh	50	Cyclone from Bay of Bengal
Aug. 25	South Korea	43	Two days of flooding also caused landslides and left 20,000 homeless
Aug. 29–Sept. 6	Caribbean and Eastern United States	1,300	Hurricane David devastated many areas and particularly damaged Dominican Republic
Sept. 10	Caribbean area and United States	1,300	Hurricane David devastated entire region
Sept. 12	U.S. Gulf Coast	13	Florida, Alabama, and Mississippi were hit by hurricane Frederick; at least 4,700 were injured; worst storm on record in Gulf States from insurance status with $752 million in claims; total damages $1.7 billion
Sept. 17	Gulf of Mexico area and United States	9	Damages exceeded $1.7 billion in Alabama–Florida region
Oct. 16	French Riviera	13	Tsunami, apparently induced by dredging in coastal waters

INDEX